Lecture Notes in Artificial Intelligence 2336

Subseries of Lecture Notes in Computer Science
Edited by J. G. Carbonell and J. Siekmann

Lecture Notes in Computer Science
Edited by G. Goos, J. Hartmanis, and J. van Leeuwen

Springer
Berlin
Heidelberg
New York
Barcelona
Hong Kong
London
Milan
Paris
Tokyo

Ming-Syan Chen Philip S. Yu
Bing Liu (Eds.)

Advances in Knowledge Discovery and Data Mining

6th Pacific-Asia Conference, PAKDD 2002
Taipei, Taiwan, May 6-8, 2002
Proceedings

Springer

Series Editors

Jaime G. Carbonell, Carnegie Mellon University, Pittsburgh, PA, USA
Jörg Siekmann, University of Saarland, Saarbrücken, Germany

Volume Editors

Ming-Syan Chen
National Taiwan University, EE Department
No. 1, Sec. 4, Roosevelt Road, Taipei, Taiwan, ROC
E-mail: mschen@cc.ee.ntu.edu.tw

Philip S. Yu
IBM Thomas J. Watson Research Center
30 Sawmill River Road, Hawthorne, NY 10532, USA
E-mail: psyu@us.ibm.com

Bing Liu
National University of Singapore, School of Computing
Lower Kent Ridge Road, Singapore 119260
E-mail: liub@comp.nus.edu.sg

Cataloging-in-Publication Data applied for

Die Deutsche Bibliothek - CIP-Einheitsaufnahme

Advances in knowledge discovery and data mining : 6th Pacific Asia conference ; proceedings / PAKDD 2002, Taipei, Taiwan, May 6 - 8, 2002. Ming-Syan Chen ... (ed.). - Berlin ; Heidelberg ; New York ; Barcelona ; Hong Kong ; London ; Milan ; Paris ; Tokyo : Springer, 2002
(Lecture notes in computer science ; Vol. 2336 : Lecture notes in artificial intelligence)
ISBN 3-540-43704-5

CR Subject Classification (1998): I.2, H.2.8, H.3, H.5.1, G.3, J.1, K.4

ISSN 0302-9743
ISBN 3-540-43704-5 Springer-Verlag Berlin Heidelberg New York

Springer-Verlag Berlin Heidelberg New York
a member of BertelsmannSpringer Science+Business Media GmbH

http://www.springer.de

Typesetting: Camera-ready by author, data conversion by Boller Mediendesign
Printed on acid-free paper SPIN: 10869781 06/3142 5 4 3 2 1 0

Preface

Knowledge discovery and data mining have become areas of growing significance because of the recent increasing demand for KDD techniques, including those used in machine learning, databases, statistics, knowledge acquisition, data visualization, and high performance computing. In view of this, and following the success of the five previous PAKDD conferences, the sixth Pacific-Asia Conference on Knowledge Discovery and Data Mining (PAKDD 2002) aimed to provide a forum for the sharing of original research results, innovative ideas, state-of-the-art developments, and implementation experiences in knowledge discovery and data mining among researchers in academic and industrial organizations.

Much work went into preparing a program of high quality. We received 128 submissions. Every paper was reviewed by 3 program committee members, and 32 were selected as regular papers and 20 were selected as short papers, representing a 25% acceptance rate for regular papers. The PAKDD 2002 program was further enhanced by two keynote speeches, delivered by Vipin Kumar from the Univ. of Minnesota and Rajeev Rastogi from AT&T. In addition, PAKDD 2002 was complemented by three tutorials, XML and data mining (by Kyuseok Shim and Surajit Chadhuri), mining customer data across various customer touchpoints at e-commerce sites (by Jaideep Srivastava), and data clustering analysis, from simple groupings to scalable clustering with constraints (by Osmar Zaiane and Andrew Foss). Moreover, PAKDD 2002 offered four international workshops on "Knowledge Discovery in Multimedia and Complex Data", "Mining Data across Multiple Customer Touchpoints for CRM", "Toward the Foundation of Data Mining", and "Text Mining". Articles from these workshops have been published separately.

All of this work would not have been possible without the dedication and professional work of many colleagues. We would like to express our sincere appreciation to all contributors to the conference for submitting papers, offering tutorials, and organizing workshops. Special thanks go to our honorary chairs, David C. L. Liu from the National Tsing Hua University and Benjamin Wah from the University of Illinois, Urbana-Champaign, for their leadership and advice on planning this conference. We are also deeply grateful to Huan Liu for serving as the Workshop Chair, Yao-Nan Lien for being the Tutorial Chair, Chia-Hui Chang and Vincent Tseng for serving as the Industrial Chairs, Show-Jane Yen and Yue-Shi Lee for being the Publication Chairs, and Chun-Nan Hsu for acting as Local Arrangement Chair. Last but not least, we are indebted to the Steering Committee Members (Chaired by Hongjun Lu) for their timely guidance and unwavering support. All of them deserve our very sincere gratitude.

Finally, we hope you all had a very pleasant stay at PAKDD 2002 in Taipei, and we wish you great success in your KDD endeavors.

May 2002

Arbee L. P. Chen
Jiawei Han
Ming-Syan Chen
Philip S. Yu
Bing Liu

PAKDD 2002 Conference Committee

Honorary Chairs:

David C. L. Liu	National Tsing Hua University, Taiwan
Ben Wah	University of Illinois, Urbana-Champaign, USA

Conference Co-chairs:

Arbee L.P. Chen	National Tsing Hua University, Taiwan
Jiawei Han	University of Illinois, Urbana-Champaign, USA

Program Chairs/Co-chairs:

Ming-Syan Chen	National Taiwan University, Taiwan (Chair)
Philip S. Yu	IBM T.J. Watson Research Center, USA (Co-chair)
Bing Liu	National University of Singapore, Singapore (Co-chair)

Workshop Chair:

Huan Liu	Arizona State University, USA

Tutorial Chair:

Yao-Nan Lien	National Chengchi University, Taiwan

Industrial Chairs:

Chia-Hui Chang	National Central University, Taiwan
Vincent S. M. Tseng	National Cheng Kung University, Taiwan

Publication Chairs:

Show-Jane Yen	Fu Jen Catholic University, Taiwan
Yue-Shi Lee	Ming Chuan University, Taiwan

Local Arrangement Chair:

Chun-Nan Hsu	Academia Sinica, Taiwan

PAKDD Steering Committee:

Hongjun Lu	Hong Kong University of Science & Technology, Hong Kong (Chair)
Hiroshi Motoda	Osaka University, Japan (Co-chair)
David W. Cheung	The University of Hong Kong, Hong Kong
Masaru Kitsuregawa	The University of Tokyo, Japan
Rao Kotagiri	University of Melbourne, Australia
Huan Liu	Arizona State University, USA
Takao Terano	University of Tsukuba, Japan
Graham illiams	CSIRO, Australia
Ning Zhong	Maebashi Institute of Technology, Japan
Lizhu Zhou	Tsinghua University, China

PAKDD 2002 Program Committee

Roberto Bayardo	IBM Almaden Research Center, USA
Michael Berthold	U.C. Berkeley, USA
Chia-Hui Chang	National Central University, Taiwan
Edward Chang	UCSB, USA
Meng-Chang Chen	Academia Sinica, Taiwan
David W. Cheung	The University of Hong Kong
Umeshwar Dayal	Hewlett-Packard Labs, USA
Guozhu Dong	Wright State University, USA
Ada Fu	Chinese University of Hong Kong, Hong Kong
Yike Guo	Imperial College, UK
Eui-Hong Han	University of Minnesota, USA
Jiawei Han	University of Illinois, Urbana-Champaign, USA
David Hand	Imperial College, UK
Jayant Haritsa	Indian Institute of Science, India
Robert Hilderman	University of Regina, Canada
Howard Ho	IBM Almaden Research Center, USA
Se-June Hong	IBM T.J. Watson Research Center, USA
Kien Hua	University of Central Florida, USA
Ben Kao	University of Hong Kong, Hong Kong
Masaru Kitsuregawa	University of Tokyo, Japan
Willi Klosgen	GMD, Germany
Kevin Korb	Monash University, Australia
Rao Kotagiri	University of Melbourne, Australia
Vipin Kumar	University of Minnesota, USA
Wai Lam	Chinese University of Hong Kong, Hong Kong
C. Lee Giles	Penn State University, USA
Chiang Lee	Cheng-Kung University, Taiwan
Suh-Yin Lee	Chia-Tung University, Taiwan
Chung-Sheng Li	IBM T.J. Watson Research Center, USA
Qing Li	City University of Hong Kong, Hong Kong
T. Y. Lin	San Jose State University, USA
Huan Liu	Arizona State University, USA
Ling Liu	Georgia Tech Institute, USA
Xiaohui Liu	University of Brunel, UK
Hongjun Lu	Hong Kong UST, Hong Kong
Yuchang Lu	Tsing Hua University, China
Hiroshi Motoda	Osaka University, Japan
Raymond Ng	UBC, Canada
Yen-Jeng Oyang	National Taiwan University, Taiwan
Zhongzhi Shi	China
Kyuseok Shim	KAIST, Korea
Arno Siebes	Holland
Ramakrishnan Srikant	IBM Almaden Research Center, USA
Jaideep Srivastava	University of Minnesota, USA
Einoshin Suzuki	Yokohama National University, Japan

Ah-Hwee Tan	Kent Ridge Digital Labs, Singapore
Changjie Tang	Sichuan University, China
Takao Terano	University of Tsukuba, Japan
Bhavani Thurasingham	MITRE, USA
Hannu T. T. Toivonen	Nokia Research, Finland
Shin-Mu Tseng	Cheng-Kung University, Taiwan
Anthony Tung	National University of Singapore, Singapore
Jason Wang	New Jersey IT, USA
Ke Wang	Simon Fraser University, Canada
Kyu-Young Whang	KAIST, Korea
Graham Williams	CSIRO, Australia
Xingdong Wu	Colorado School of Mines, USA
Jiong Yang	IBM T.J. Watson Research Center, USA
Yiyu Yao	University of Regina, Canada
Clement Yu	University of Illinois, USA
Osmar R. Zaiane	University of Alberta, Canada
Mohammed Zaki	Rensselaer Poly Institute, USA
Zijian Zheng	Blue Martini Software, USA
Ning Zhong	Maebashi Institute of Technology, Japan
Aoying Zhou	Fudan University, China
Lizhu Zhou	Tsinghua University, China

Table of Contents

Industrial Papers (Invited)

Survey Papers (Invited)

Association Rules (I)

Classification (I)

Interestingness

Sequence Mining

Clustering

Web Mining

Association Rules (II)

Semi-structure & Concept Mining

Data Warehouse and Data Cube

Bio-Data Mining

Classification (II)

Temporal Mining

Classification (III)

Outliers, Missing Data, and Causation

Network Data Mining and Analysis: The $\mathcal{NEMESIS}$ Project

Minos Garofalakis and Rajeev Rastogi

Bell Labs, Lucent Technologies

Abstract. Modern communication networks generate large amounts of operational data, including traffic and utilization statistics and alarm/fault data at various levels of detail. These massive collections of *network-management* data can grow in the order of several Terabytes per year, and typically hide "knowledge" that is crucial to some of the key tasks involved in effectively managing a communication network (e.g., capacity planning and traffic engineering). In this short paper, we provide an overview of some of our recent and ongoing work in the context of the $\mathcal{NEMESIS}$ project at Bell Laboratories that aims to develop novel data warehousing and mining technology for the effective storage, exploration, and analysis of massive network-management data sets. We first give some highlights of our work on *Model-Based Semantic Compression (MBSC)*, a novel data-compression framework that takes advantage of attribute semantics and data-mining models to perform lossy compression of massive network-data tables. We discuss the architecture and some of the key algorithms underlying $\mathcal{SPARTAN}$, a model-based semantic compression system that exploits predictive data correlations and prescribed error tolerances for individual attributes to construct concise and accurate *Classification and Regression Tree (CaRT)* models for entire columns of a table. We also summarize some of our ongoing work on warehousing and analyzing network-fault data and discuss our vision of how data-mining techniques can be employed to help automate and improve fault-management in modern communication networks. More specifically, we describe the two key components of modern fault-management architectures, namely the *event-correlation* and the *root-cause analysis* engines, and propose the use of mining ideas for the automated inference and maintenance of the models that lie at the core of these components based on warehoused network data.

1 Introduction

Besides providing easy access to people and data around the globe, modern communication networks also *generate* massive amounts of operational data throughout their lifespan. As an example, Internet Service Providers (ISPs) continuously collect traffic and utilization information over their network to enable key network-management applications. This information is typically collected through monitoring tools that gather switch- and router-level data, such as SNMP/RMON probes [13] and Cisco's NetFlow measurement tools [1]. Such tools typically collect traffic data for each network element at fine granularities (e.g., at the level of individual packets or packet flows between source-destination pairs), giving rise to massive volumes of network-management data over time [7]. Packet traces collected for traffic management in the Sprint IP backbone

M.-S. Chen, P.S. Yu, and B. Liu (Eds.): PAKDD 2002, LNAI 2336, pp. 1–12, 2002.

amount to 600 Gigabytes of data per day [7]! As another example, telecommunication providers typically generate and store records of information, termed "Call-Detail Records" (CDRs), for every phone call carried over their network. A typical CDR is a fixed-length record structure comprising several hundred bytes of data that capture information on various (categorical and numerical) attributes of each call; this includes network-level information (e.g., endpoint exchanges), time-stamp information (e.g., call start and end times), and billing information (e.g., applied tariffs), among others [4]. These CDRs are stored in tables that can grow to truly massive sizes, in the order of several Terabytes per year.

A key observation is that these massive collections of network-traffic and CDR data typically hide invaluable "knowledge" that enables several key network-management tasks, including application and user profiling, proactive and reactive resource management, traffic engineering, and capacity planning. Nevertheless, techniques for effectively managing these massive data sets and uncovering the knowledge that is so crucial to managing the underlying network are still in their infancy. Contemporary network-management tools do little more than elaborate report generation for all the data collected from the network, leaving most of the task of inferring useful knowledge and/or patterns to the human network administrator(s). As a result, effective network management is still viewed as more of an "art" known only to a few highly skilled (and highly sought-after) individuals. It is our thesis that, in the years to come, network management will provide an important application domain for very innovative, challenging and, at the same time, practically-relevant research in data mining and data warehousing.

This short abstract aims to provide an overview of our recent and ongoing research efforts in the context of $\mathcal{NEMESIS}$ (NEtwork-Management data warEhousing and analySIS) , a Bell Labs' research project that targets the development of novel data warehousing and mining technology for the effective storage, exploration, and analysis of massive network-management data sets. Our research agenda for $\mathcal{NEMESIS}$ encompasses several challenging research themes, including data reduction and approximate query processing [2,5,6], mining techniques for network-fault management, and managing and analyzing continuous data streams. In this paper, we first give some highlights of our recent work on *Model-Based Semantic Compression (MBSC)*, a novel data-compression framework that takes advantage of attribute semantics and data-mining models to perform lossy compression of massive network-data tables. We also describe the architecture and some of the key algorithms underlying $\mathcal{SPARTAN}$, a system built based on the MBSC paradigm, that exploits predictive data correlations and prescribed error tolerances for individual attributes to construct concise and accurate *Classification and Regression Tree (CaRT)* models for entire columns of a table [2]. We then turn to our ongoing work on warehousing and analyzing network-fault data and discuss our vision of how data-mining techniques can be employed to help automate and improve fault-management in modern communication networks. More specifically, we describe the two key components of modern fault-management architectures, namely the *event-correlation* and the *root-cause analysis* engines, and offer some (more speculative) proposals on how mining ideas can be exploited for the automated inference and maintenance of the models that lie at the core of these components based on warehoused network data.

2 Model-Based Semantic Compression for Network-Data Tables

Data compression issues arise naturally in applications dealing with massive data sets, and effective solutions are crucial for optimizing the usage of critical system resources like storage space and I/O bandwidth, as well as network bandwidth (for transferring the data) [4,7]. Several statistical and dictionary-based compression methods have been proposed for text corpora and multimedia data, some of which (e.g., Lempel-Ziv or Huffman) yield provably optimal asymptotic performance in terms of certain ergodic properties of the data source. These methods, however, fail to provide adequate solutions for compressing massive data tables, such as the ones that house the operational data collected from large ISP and telecom networks. The reason is that all these methods view a table as a large byte string and do not account for the complex dependency patterns in the table. Compared to conventional compression problems, effectively compressing massive tables presents a host of novel challenges due to several distinct characteristics of table data sets and their analysis.

• **Semantic Compression.** Existing compression techniques are "syntactic" in the sense that they operate at the level of consecutive bytes of data. Such syntactic methods typically fail to provide adequate solutions for table-data compression, since they essentially view the data as a large byte string and do not exploit the complex dependency patterns in the table. Effective table compression mandates techniques that are *semantic* in nature, in the sense that they account for and exploit both (1) existing data dependencies and correlations among attributes in the table; and, (2) the meanings and dynamic ranges of individual attributes (e.g., by taking advantage of the specified error tolerances).

• **Approximate (Lossy) Compression.** Due to the exploratory nature of many data-analysis applications, there are several scenarios in which an exact answer may not be required, and analysts may in fact prefer a fast, approximate answer, as long as the system can guarantee an *upper bound on the error of the approximation*. For example, during a drill-down query sequence in ad-hoc data mining, initial queries in the sequence frequently have the sole purpose of determining the truly interesting queries and regions of the data table. Thus, in contrast to traditional lossless data compression, the compression of massive tables can often afford to be *lossy*, as long as some (user- or application-defined) upper bounds on the compression error are guaranteed by the compression algorithm. This is obviously a crucial differentiation, as even small error tolerances can help us achieve much better compression ratios.

In our recent work [2], we have proposed *Model-Based Semantic Compression (MBSC)*, a novel data-compression framework that takes advantage of attribute semantics and data-mining models to perform guaranteed-error, lossy compression of massive data tables. Abstractly, MBSC is based on the novel idea of exploiting data correlations and user-specified "loss"/error tolerances for individual attributes to construct concise data mining models and derive the best possible compression scheme for the data based

protocol	duration	byte-count	packets
http	12	2,000	1
http	16	24,000	5
ftp	27	100,000	24
http	15	20,000	8
ftp	32	300,000	35
http	19	40,000	11
http	26	58,000	18
ftp	18	80,000	15

(a) Tuples in Table

(b) CaRT Models

Fig. 1. Model-Based Semantic Compression.

on the constructed models. To make our discussion more concrete, we focus on the architecture and some of the key algorithms underlying *SPARTAN*[1], a system that takes advantage of attribute correlations and error tolerances to build concise and accurate *Classification and Regression Tree (CaRT)* models [3] for entire columns of a table. More precisely, *SPARTAN* selects a certain subset of attributes (referred to as *predicted* attributes) for which no values are explicitly stored in the compressed table; instead, concise CaRTs that predict these values (within the prescribed error bounds) are maintained. Thus, for a predicted attribute X that is strongly correlated with other attributes in the table, *SPARTAN* is typically able to obtain a very succinct CaRT predictor for the values of X, which can then be used to completely eliminate the column for X in the compressed table. Clearly, storing a compact CaRT model in lieu of millions or billions of actual attribute values can result in substantial savings in storage.

Example 21 *Consider the table with 4 attributes and 8 tuples shown in Figure 1(a), where each tuple represents a* data flow *in an IP network. The categorical attribute* protocol *records the application-level protocol generating the flow; the numeric attribute* duration *is the time duration of the flow; and, the numeric attributes* byte-count *and* packets *capture the total number of bytes and packets transferred. Let the acceptable errors due to compression for the numeric attributes* duration, byte-count, *and* packets *be 3, 1,000, and 1, respectively. Also, assume that the* protocol *attribute has to be compressed without error (i.e., zero tolerance). Figure 1(b) depicts a regression tree for predicting the* duration *attribute (with* packets *as the predictor attribute) and a classi-*

[1] [From Webster] **Spartan:** */'spart-*n/* (1) of or relating to Sparta in ancient Greece, (2) a: marked by strict self-discipline and avoidance of comfort and luxury, b: sparing of words : TERSE : LACONIC.

fication tree for predicting the protocol *attribute (with* packets *and* byte-count *as the predictor attributes). Observe that in the regression tree, the predicted value of duration (label value at each leaf) is almost always within 3, the specified error tolerance, of the actual tuple value. For instance, the predicted value of* duration *for the tuple with* packets = *1 is 15 while the original value is 12. The only tuple for which the predicted value violates this error bound is the tuple with* packets = *11, which is an marked as an outlier value in the regression tree. There are no outliers in the classification tree. By explicitly storing, in the compressed version of the table, each outlier value along with the CaRT models and the projection of the table onto only the predictor attributes (*packets *and* byte-count*), we can ensure that the error due to compression does not exceed the user-specified bounds. Further, storing the CaRT models (plus outliers) for* duration *and* protocol *instead of the attribute values themselves results in a reduction from 8 to 4 values for* duration *(2 labels for leaves + 1 split value at internal node + 1 outlier) and a reduction from 8 to 5 values for* protocol *(3 labels for leaves + 2 split values at internal nodes).* ∎

To build an effective CaRT-based compression plan for the input data table, $\mathcal{SPARTAN}$ employs a number of sophisticated techniques from the areas of knowledge discovery and combinatorial optimization. Below, we list some of $\mathcal{SPARTAN}$'s salient features.

• **Use of Bayesian Network to Uncover Data Dependencies.** A Bayesian network is a directed acyclic graph (DAG) whose edges reflect strong predictive correlations between nodes of the graph [12]. $\mathcal{SPARTAN}$ uses a Bayesian network on the table's attributes to dramatically reduce the search space of potential CaRT models since, for any attribute, the most promising CaRT predictors are the ones that involve attributes in its "neighborhood" in the network.

• **Novel CaRT-selection Algorithms that Minimize Storage Cost.** $\mathcal{SPARTAN}$ exploits the inferred Bayesian network structure by using it to intelligently guide the selection of CaRT models that minimize the overall storage requirement, based on the prediction and materialization costs for each attribute. We demonstrate that this model-selection problem is a strict generalization of the *Weighted Maximum Independent Set (WMIS)* problem [8], which is known to be $\mathcal{NP}$-hard. However, by employing a novel algorithm that effectively exploits the discovered Bayesian structure in conjunction with efficient, near-optimal WMIS heuristics, $\mathcal{SPARTAN}$ is able to obtain a good set of CaRT models for compressing the table.

• **Improved CaRT Construction Algorithms that Exploit Error Tolerances.** Since CaRT construction is computationally-intensive, $\mathcal{SPARTAN}$ employs the following three optimizations to reduce CaRT-building times: (1) CaRTs are built using random samples instead of the entire data set; (2) leaves are not expanded if values of tuples in them can be predicted with acceptable accuracy; (3) pruning is integrated into the tree growing phase using novel algorithms that exploit the prescribed error tolerance for the predicted attribute. *SPARTAN* then uses the CaRTs built to compress the full data set (within the specified error bounds) *in one pass*.

An extensive experimental study of the *SPARTAN* system with several real-life data tables has verified the effectiveness of our approach compared to existing syntactic (`gzip`) and semantic (fascicle-based [10]) compression techniques [2].

2.1 Overview of Approach

Definitions and Notation. The input to the *SPARTAN* system consists of a n-attribute table T, and a (user- or application-specified) n-dimensional vector of *error tolerances* $\bar{e} = [e_1, \ldots, e_n]$ that defines the *per-attribute* acceptable degree of information loss when compressing T. Let $\mathcal{X} = \{X_1, \ldots, X_n\}$ denote the set of n attributes of T and $dom(X_i)$ represent the domain of attribute X_i. Intuitively, e_i, the ith entry of the tolerance vector $\bar{e}$ specifies an upper bound on the error by which any (approximate) value of X_i in the compressed table T_c can differ from its original value in T. For a numeric attribute X_i, the tolerance e_i defines an upper bound on the *absolute difference* between the actual value of X_i in T and the corresponding (approximate) value in T_c. That is, if x, x' denote the accurate and approximate value (respectively) of attribute X_i for *any* tuple of T, then our compressor guarantees that $x \in [x' - e_i, x' + e_i]$. For a categorical attribute X_i, the tolerance e_i defines an upper bound on the *probability* that the (approximate) value of X_i in T_c is different from the actual value in T. More formally, if x, x' denote the accurate and approximate value (respectively) of attribute X_i for *any* tuple of T, then our compressor guarantees that $P[x = x'] \geq 1 - e_i$. (Note that our error-tolerance semantics can also easily capture *lossless* compression as a special case, by setting $e_i = 0$ for all i.)

Model-Based Semantic Compression. Briefly, our proposed *model-based* methodology for semantic compression of data tables involves two steps: (1) exploiting data correlations and (user- or application-specified) error bounds on individual attributes to construct data mining models; and (2) deriving a good compression scheme using the constructed models. We define the model-based, compressed version of the input table T as a pair $T_c =< T', \{\mathcal{M}_1, \ldots, \mathcal{M}_p\} >$ such that T can be obtained from T_c within the specified error tolerance $\bar{e}$. Here, (1) T' is a small (possibly empty) projection of the data values in T that are retained *accurately* in T_c; and, (2) $\{\mathcal{M}_1, \ldots, \mathcal{M}_p\}$ is a set of data-mining models. A definition of our general model-based semantic compression problem can now be stated as follows.

[Model-Based Semantic Compression (MBSC)]: Given a multi-attribute table T and a vector of (per-attribute) error tolerances $\bar{e}$, find a set of models $\{\mathcal{M}_1, \ldots, \mathcal{M}_m\}$ and a compression scheme for T based on these models $T_c =< T', \{\mathcal{M}_1, \ldots, \mathcal{M}_p\} >$ such that the specified error bounds $\bar{e}$ are not exceeded and the storage requirements $|T_c|$ of the compressed table are minimized. ∎

Given the multitude of possible models that can be extracted from the data, the general MBSC problem definition covers a huge design space of possible alternatives for semantic compression. We now provide a more concrete statement of the problem addressed in our work on the $\mathcal{SPARTAN}$ system.

[$\mathcal{SPARTAN}$ **CaRT-Based Semantic Compression]:** Given a multi-attribute table T with a set of attributes $\mathcal{X}$, and a vector of (per-attribute) error tolerances $\bar{e}$, find a subset $\{X_1, \ldots, X_p\}$ of $\mathcal{X}$ and a set of corresponding CaRT models $\{\mathcal{M}_1, \ldots, \mathcal{M}_p\}$ such that: (1) model $\mathcal{M}_i$ is a predictor for the values of attribute X_i based solely on attributes in $\mathcal{X} - \{X_1, \ldots, X_p\}$, for each $i = 1, \ldots, p$; (2) the specified error bounds $\bar{e}$ are not exceeded; and, (3) the storage requirements $|T_c|$ of the compressed table $T_c =< T', \{\mathcal{M}_1, \ldots, \mathcal{M}_p\} >$ are minimized. ∎

Abstractly, $\mathcal{SPARTAN}$ seeks to partition the set of input attributes $\mathcal{X}$ into a set of *predicted attributes* $\{X_1, \ldots, X_p\}$ and a set of *predictor attributes* $\mathcal{X} - \{X_1, \ldots, X_p\}$ such that the values of each predicted attribute can be obtained within the specified error bounds based on (a subset of) the predictor attributes through a small classification or regression tree (except perhaps for a small set of outlier values). Note that we do not allow a predicted attribute X_i to also be a predictor for a different attribute. This restriction is important since predicted values of X_i can contain errors, and these errors can cascade further if the erroneous predicted values are used as predictors, ultimately causing error constraints to be violated. The final goal, of course, is to minimize the overall storage cost of the compressed table. This storage cost $|T_c|$ is the sum of two basic components:

1. *Materialization cost*, i.e., the cost of storing the values for all predictor attributes $\mathcal{X} - \{X_1, \ldots, X_p\}$. This cost is represented in the T' component of the compressed table, which is basically the projection of T onto the set of predictor attributes. The storage cost of materializing attribute X_i is denoted by $\texttt{MaterCost}(X_i)$.
2. *Prediction cost*, i.e., the cost of storing the CaRT models used for prediction plus (possibly) a small set of outlier values of the predicted attribute for each model. (We use the notation $\mathcal{X}_i \rightarrow X_i$ to denote a CaRT predictor for attribute X_i using the set of predictor attributes $\mathcal{X}_i \subseteq \mathcal{X} - \{X_1, \ldots, X_p\}$.) The storage cost of predicting attribute X_i using the CaRT predictor $\mathcal{X}_i \rightarrow X_i$ is denoted by $\texttt{PredCost}(\mathcal{X}_i \rightarrow X_i)$; this does *not* include the cost of materializing the predictor attributes in $\mathcal{X}_i$.

2.2 $\mathcal{SPARTAN}$ System Architecture

As depicted in Figure 2, the architecture of the $\mathcal{SPARTAN}$ system comprises of four major components: the DEPENDENCYFINDER, the CARTSELECTOR, the CART-BUILDER, and the ROWAGGREGATOR. In the following, we provide a brief overview of each $\mathcal{SPARTAN}$ component; for a more detailed description of each component and the relevant algorithms, the interested reader is referred to [2].

• <u>DEPENDENCYFINDER.</u> The purpose of the DEPENDENCYFINDER component is to produce an *interaction model* for the input table attributes, that is then used to guide the CaRT building algorithms of $\mathcal{SPARTAN}$. The main observation here is that, since there is an exponential number of possibilities for building CaRT-based attribute predictors, we need a concise model that identifies the strongest correlations and "predictive" relationships in the input data.

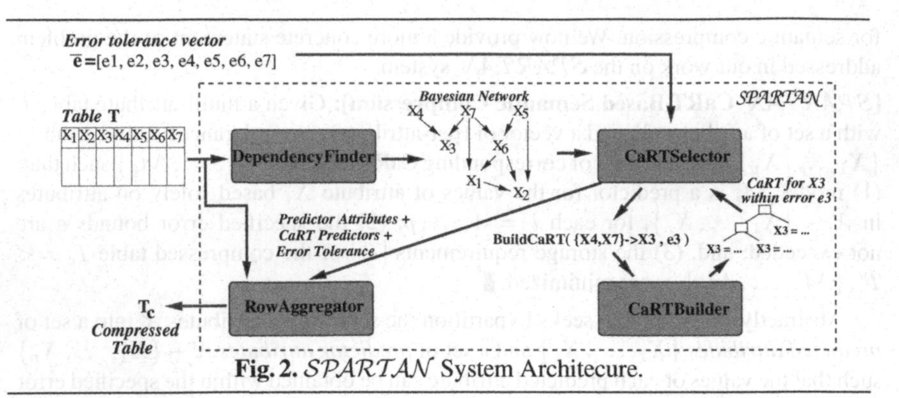

Fig. 2. $\mathcal{SPARTAN}$ System Architecure.

The approach used in the DEPENDENCYFINDER component of $\mathcal{SPARTAN}$ is to construct a *Bayesian network* [12] on the underlying set of attributes $\mathcal{X}$. Abstractly, a Bayesian network imposes a Directed Acyclic Graph (DAG) structure G on the set of nodes $\mathcal{X}$, such that directed edges capture direct statistical dependence between attributes. (The exact dependence semantics of G are defined shortly.) Thus, intuitively, a set of nodes in the "neighborhood" of X_i in G (e.g., X_i's parents) captures the attributes that are strongly correlated to X_i and, therefore, show promise as possible predictor attributes for X_i.

• CARTSELECTOR. The CARTSELECTOR component constitutes the core of $\mathcal{SPARTAN}$'s model-based semantic compression engine. Given the input table T and error tolerances e_i, as well as the Bayesian network on the attributes of T built by the DEPENDENCYFINDER, the CARTSELECTOR is responsible for selecting a collection of predicted attributes and the corresponding CaRT-based predictors such that the final overall storage cost is minimized (within the given error bounds). As discussed above, $\mathcal{SPARTAN}$'s CARTSELECTOR employs the Bayesian network G built on $\mathcal{X}$ to intelligently guide the search through the huge space of possible attribute prediction strategies. Clearly, this search involves repeated interactions with the CARTBUILDER component, which is responsible for actually building the CaRT-models for the predictors (Figure 2).

We demonstrate that even in the simple case where the set of nodes that is used to predict an attribute node in G is *fixed*, the problem of selecting a set of predictors that minimizes the combination of materialization and prediction cost naturally maps to the *Weighted Maximum Independent Set (WMIS)* problem, which is known to be $\mathcal{NP}$-hard and notoriously difficult to approximate [8]. Based on this observation, we propose a CaRT-model selection strategy that starts out with an initial solution obtained from a near-optimal heuristic for WMIS [9] and tries to incrementally improve it by small perturbations based on the unique characteristics of our problem.

• CARTBUILDER. Given a collection of predicted and (corresponding) predictor attributes $\mathcal{X}_i \rightarrow X_i$, the goal of the CARTBUILDER component is to efficiently construct CaRT-based models for each X_i in terms of $\mathcal{X}_i$ for the purposes of semantic compression. Induction of CaRT-based models is typically a computation-intensive process that requires multiple passes over the input data [3]. As we demonstrate, however, $\mathcal{SPARTAN}$'s CaRT construction algorithms can take advantage of the compression semantics and exploit the user-defined error-tolerances to effectively prune computation. In addition, by building CaRTs using data samples instead of the entire data set, $\mathcal{SPARTAN}$ is able to further speed up model construction.

• ROWAGGREGATOR. Once $\mathcal{SPARTAN}$'s CARTSELECTOR component has finalized a "good" solution to the CaRT-based semantic compression problem, it hands off its solution to the ROWAGGREGATOR component which tries to further improve the compression ratio through row-wise clustering. Briefly, the ROWAGGREGATOR uses a fascicle-based algorithm [10] to compress the predictor attributes, while ensuring (based on the CaRT models built) that errors in the predictor attribute values are not propagated through the CaRTs in a way that causes error tolerances (for predicted attributes) to be exceeded.

3 Data Mining Techniques for Network-Fault Management

Modern communication networks have evolved into highly complex systems, typically comprising large numbers of interconnected elements (e.g., routers, switches, bridges) that work together to provide end-users with various data and/or voice services. This increase in system scale and the number of elements obviously implies an increased probability of faults occurring somewhere in the network. Further, the complex interdependencies that exist among the various elements in the network cooperating to provide some service imply that a fault can propagate widely, causing *floods* of alarm signals from very different parts of the network. As an example, a switch failure in an IP network can cause the network to be partitioned resulting in alarms emanating from multiple elements in different network partitions and subnets, as they detect that some of their peers are no longer reachable. To deal with these situations, modern network-management platforms provide certain *fault-management* utilities that try to help administrators make sense of alarm floods, and allow them to quickly and effectively zero in on the root cause of the problem.

Typically, the architecture of a fault-management subsystem comprises two key components: the *Event Correlator (EC)* and the *Root-Cause Analyzer (RCA)*, as depicted in Figure 3. The goal of the Event Correlator is improve the information content of the observed events by filtering out uninteresting, "secondary" alarms from the alarm flood arriving at the network-management station [11,14]. (Secondary alarms or *symptoms* are observable events that are directly caused by other events observed in the network.) This filtering is implemented with the help of a set of *fault-propagation rules* that the Event Correlator uses to model the propagation of alarm signals in the underlying network. The output of the Event Correlator, i.e., the "primary" alarm signals in the observed set of alarms, are then fed into the Root-Cause Analyzer whose goal is to produce a set of possible root causes for the observed problem along with associated degrees of confidence for each "guess" (Figure 3).

The fault-propagation rules that model the propagation of alarms throughout the underlying network form the basic core of the Event Correlation engine. In general, these rules try to capture the *probabilistic causal relationships* that exist between the

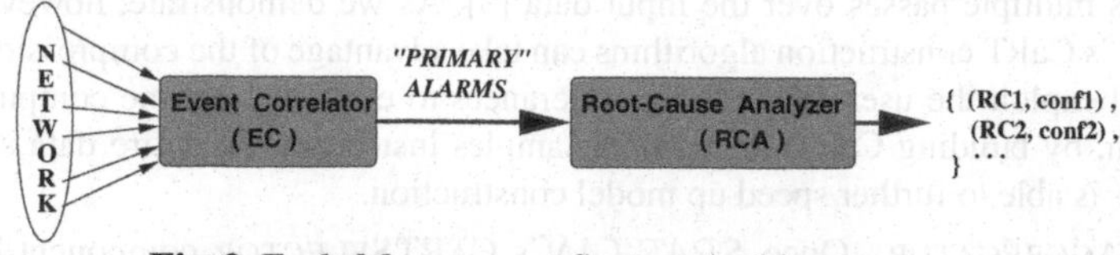

Fig. 3. Fault-Management System Architecture.

various alarm signals in the network. As an example, Figure 4 depicts a small subset of such fault-propagation rules; based on the figure, alarm signal A_1 causes the occurrence of alarm signal A_3 with probability p_{13} and that of alarm signal A_4 with probability p_{14}. Thus, the fault-propagation rules that lie at the heart of the Event Correlator are essentially equivalent to a *causal Bayesian model* [12] for network alarm signals. Given such a causal model for network alarms, the problem of filtering out secondary events in the Event Correlator can be formulated as an optimization problem in a variety of interesting ways. For example, a possible formulation would be as follows: Given a confidence threshold $\theta \in (0, 1)$ and the set of all observed alarm signals A, find a minimal subset P of A such that $P[A|P] > \theta$ (i.e., the probability that A was actually "caused" by P exceeds the desired confidence θ).

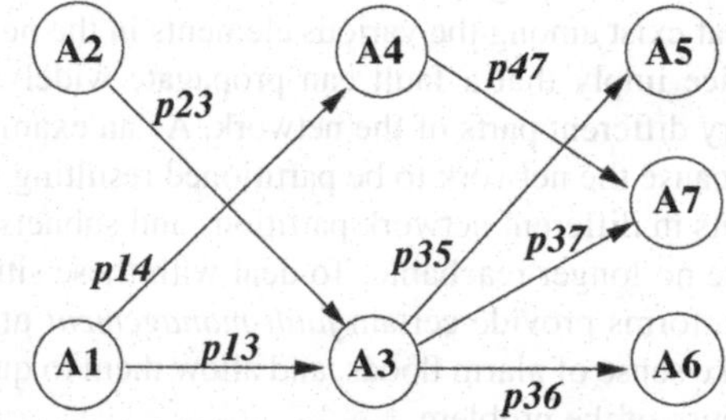

Fig. 4. Example Fault-Propagation Model for EC.

Current State-of-the-Art. There are several commercially-available products that offer event-correlation services for data-communication networks. Examples include SMARTS InCharge [14], the Event-Correlation Services (ECS) component of the HP OpenView network-management platform, CISCO's InfoCenter, GTE's Impact, and so on. A common characteristic of all these Event Correlators is that they essentially force the network administrator(s) to "hand-code" the fault-propagation rules for the underlying network using either a language-based or a graphics-based specification tool. This is clearly a very tedious and error-prone process for any large-scale IP network comprising hundreds or thousands of heterogeneous, multi-vendor elements. Furthermore, it is non-incremental since a large part of the specification may need to be changed when the topology of the network changes or new network elements are introduced. We believe that such solutions are simply inadequate for tomorrow's large-scale, heterogeneous, and highly-dynamic IP networking environments.

Our Proposed Approach. Rather than relying on human operators to "hand-code" the core of the Event-Correlation engine, we propose the use of data-mining techniques to help automate the task of inferring and incrementally maintaining the causal model of network alarm signals (Figure 5). For the inference task (typically performed off-line), our data-mining tool can exploit the database of alarm signals collected and stored over the lifespan of the network along with important "domain-specific knowledge" (e.g., network topology and routing-protocol information) to automatically construct the correct causal model of fault-propagation rules. For the maintenance task (typically performed on-line), our data-mining tool can again exploit such "domain-specific knowledge" along with information on network updates (e.g., topology changes or new additions to the network) and the incoming stream of network alarm signals to automatically effect the appropriate updates to the fault-propagation model.

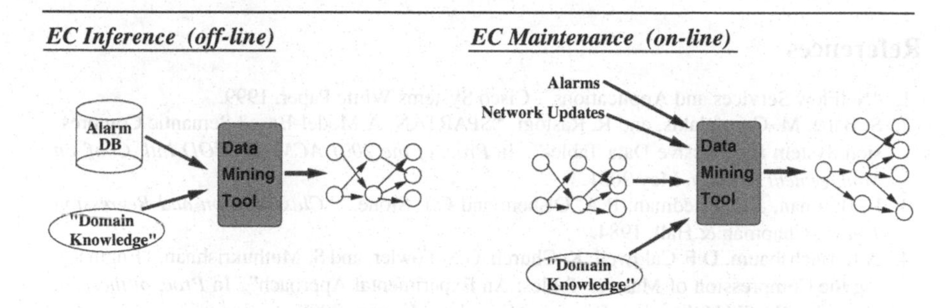

Fig. 5. Exploiting Data Mining for Automated EC Inference and Maintenance.

We should note here that, even though the problem of inferring causal Bayesian models from data has been studied for some time in the data-mining and machine-learning communities [12], the automatic extraction of event-correlation models for communication networks presents a host of new challenges due to several unique characteristics of the problem domain. First, the issue of how to effectively incorporate and exploit important "domain-specific knowledge" (like the network topology or routing-protocol information) in the model-learning algorithm is certainly very challenging and non-trivial. Second, it is important to incorporate the *temporal aspects* of network alarm signals in the data-mining process; for example, alarms that occur within a small time window are more likely to be correlated than alarms separated by larger amounts of time. Finally, the learning algorithm needs to be *robust* to lost or spurious alarm signals, both of which are common phenomena in modern communication networks.

For the Root-Cause Analyzer, data-mining techniques can again be exploited; for example, our tools can use failure data collected from the field to automatically learn failure "signatures" and map them to an associated root cause. Once again, it is crucial to effectively incorporate important "domain-specific knowledge" (like the network topology) in the data-mining process.

4 Conclusions

Operational data collected from modern communication networks is massive and hides "knowledge" that is invaluable to several key network-management tasks. In this short abstract, we have provided an overview of some of our recent and ongoing work in the context of the $\mathcal{NEMESIS}$ project at Bell Labs that aims to develop novel data warehousing and mining technology for the effective storage, exploration, and analysis of massive network-management data sets. We believe that, in years to come, network management will provide an important application domain for innovative, challenging and, at the same time, practically-relevant research in data mining and warehousing.

Acknowledgements: Many thanks to Shivnath Babu (our coauthor in [2]) for his valuable contributions on the $\mathcal{SPARTAN}$ semantic-compression engine.

References

1. "NetFlow Services and Applications". Cisco Systems White Paper, 1999.
2. S. Babu, M. Garofalakis, and R. Rastogi. "SPARTAN: A Model-Based Semantic Compression System for Massive Data Tables". In *Proc. of the 2001 ACM SIGMOD Intl. Conf. on Management of Data*, May 2001.
3. L. Breiman, J.H. Friedman, R.A. Olshen, and C.J. Stone. *"Classification and Regression Trees"*. Chapman & Hall, 1984.
4. A.L. Buchsbaum, D.F. Caldwell, K. Church, G.S. Fowler, and S. Muthukrishnan. "Engineering the Compression of Massive Tables: An Experimental Approach". In *Proc. of the 11th Annual ACM-SIAM Symp. on Discrete Algorithms*, January 2000.
5. K. Chakrabarti, M. Garofalakis, R. Rastogi, and K. Shim. "Approximate Query Processing Using Wavelets". In *Proc. of the 26th Intl. Conf. on Very Large Data Bases*, September 2000.
6. A. Deshpande, M. Garofalakis, and R. Rastogi. "Independence is Good: Dependency-Based Histogram Synopses for High-Dimensional Data". In *Proc. of the 2001 ACM SIGMOD Intl. Conf. on Management of Data*, May 2001.
7. C. Fraleigh, S. Moon, C. Diot, B. Lyles, and F. Tobagi. "Architecture of a Passive Monitoring System for Backbone IP Networks". Technical Report TR00-ATL-101-801, Sprint Advanced Technology Laboratories, October 2000.
8. M.R. Garey and D.S. Johnson. *"Computers and Intractability: A Guide to the Theory of NP-Completeness"*. W.H. Freeman, 1979.
9. M.M. Halldórsson. "Approximations of Weighted Independent Set and Hereditary Subset Problems". *Journal of Graph Algorithms and Applications*, 4(1), 2000.
10. H.V. Jagadish, J. Madar, and R. Ng. "Semantic Compression and Pattern Extraction with Fascicles". In *Proc. of the 25th Intl. Conf. on Very Large Data Bases*, September 1999.
11. G. Jakobson and M.D. Weissman. "Alarm Correlation". *IEEE Network*, November 1993.
12. Judea Pearl. *"Probabilistic Reasoning in Intelligent Systems: Networks of Plausible Inference"*. Morgan Kaufmann Publishers, Inc., 1988.
13. William Stallings. *"SNMP, SNMPv2, SNMPv3, and RMON 1 and 2"*. Addison-Wesley Longman, Inc., 1999. (Third Edition).
14. S. Yemini, S. Kliger, E. Mozes, Y. Yemini, and D. Ohsie. "High Speed & Robust Event Correlation". *IEEE Communications Magazine*, May 1996.

Privacy Preserving Data Mining: Challenges and Opportunities

Ramakrishnan Srikant

IBM Almaden Research Center
650 Harry Road, San Jose, CA 95120
srikant@us.ibm.com

Abstract. The goal of privacy preserving data mining is to develop accurate models without access to precise information in individual data records, thus finessing the conflict between privacy and data mining. In this talk, I will give an introduction to the techniques underlying privacy preserving data mining, and then discuss several application domains. In particular, recent events have led to an increased interest in applying data mining toward security related problems, leading to interesting technical challenges at the intersection of privacy, security and data mining.

M.-S. Chen, P.S. Yu, and B. Liu (Eds.): PAKDD 2002, LNAI 2336, pp. 13–13, 2002.

A Case for Analytical Customer Relationship Management

Jaideep Srivastava[1], Jau-Hwang Wang[2], Ee-Peng Lim[3], and San-Yih Hwang[4]

[1]Computer Science & Engineering
University of Minnesota, Minneapolis, MN 55455, USA
srivasta@cs.umn.edu
[2]Information Management
Central Police University, Taoyuan, ROC
jwang@sun4.cpu.edu.tw
[3]Chinese University of Hong Kong
Hong Kong, PRC
aseplim@ntu.edu.sg
[4]National Sun-Yat Sen University
Kaoshiung, ROC
syhwang@misserv.mis.nsysu.edu.tw

Abstract. The Internet has emerged as a low cost, low latency and high bandwidth customer communication channel. Its interactive nature provides an organization the ability to enter into a close, personalized dialog with individual customers. The simultaneous maturation of data management technologies like data warehousing, and data mining, have created the ideal environment for making customer relationship management (CRM) a much more systematic effort than it has been in the past. In this paper we described how data analytics can be used to make various CRM functions like customer segmentation, communication targeting, retention, and loyalty much more effective. We briefly describe the key technologies needed to implement analytical CRM, and the organizational issues that must be carefully handled to make CRM a reality. Our goal is to illustrate problems that exist with current CRM efforts, and how using data analytics techniques can address them. Our hope is to get the data mining community interested in this important application domain.

1 Introduction

As bandwidth continues to grow, and newer information appliances become available, marketing departments everywhere see this as an opportunity to get in closer touch with potential customers. In addition, with organizations constantly developing more cost-effective means of customer contact, the amount of customer solicitation has been on a steady rise. Today, with Internet as the ultimate low latency, high bandwidth, customer contact channel with practically zero cost, customer solicitation has reached unprecedented levels.

Armed with such tools, every organization has ramped up its marketing effort, and we are witnessing a barrage of solicitations targeted at the ever-shrinking attention span of the same set of customers. Once we consider the fact that potentially

M.-S. Chen, P.S. Yu, and B. Liu (Eds.): PAKDD 2002, LNAI 2336, pp. 14-27, 2002.

good customers, i.e. 'those likely to buy a product', are much more likely to get a solicitation than those who are not so good, the situation for the good customers is even more dire. This is really testing the patience of many customers, and thus we have witnessed a spate of customers signing up to be on 'no solicitation' lists, to avoid being bombarded with unwanted solicitations..

From the viewpoint of the organizations, the situation is no better. Even though the cost of unit customer communication has dropped dramatically; the impact of unit communication has dropped even faster. For example, after a lot of initial enthusiasm, it is now widely accepted that the impact of web page banner advertisements in affecting customer opinion is practically negligible. On the other hand, the impact of targeted e-mails, especially with financial offers, is quite high. In essence, each organization is spinning its wheels in tying to target the same set of good customers, while paying insufficient attention to understanding the needs of the 'not so good customers' of today, and converting them into good customers of tomorrow. A clear example of this mutual cannibalism of customers is the cellular phone industry, where each service provider is constantly trying to outdo the others. "Customer churn" is a well-accepted problem in this industry.

A well-accepted wisdom in the industry is that it costs five to seven times as much to acquire a new customer than to retain an existing one. The reason is that the organization already has the loyalty of existing customers, and all that is required for retention is to meet the customer's expectations. For customer acquisition however, the customer must be weaned away from another organization, which is a much harder task. Given this, it is crucial that the selection of customers to target is done with care, and the right message be sent to each one. Given these needs, it becomes important for an organization to understand its customers well. Thus, one can consider customer relationship management to consist of two parts as follows:

CRM = customer understanding + relationship management

This equation is not new, since in the classical 'neighborhood store' model of doing business, the store had a highly localized audience, and the store owner knew practically everyone in the neighborhood – making it easy for him to meet the needs of his customers. It is the big corporations, serving a mass customer base, that have difficulty in understanding the needs of individual customers. The realization of this gap of knowledge has been one of the driving factors for the rapid adoption of CRM software by many corporations. However, the initial deployment of CRM software has been for the second part of the CRM equation, namely 'relationship management'. As described above, relationship management efforts without an understanding of the customer can be marginally effective at best, and sometimes even counter productive.

The approach that resolves this dilemma is the use of data analytics in CRM, with the goal of obtaining a better understanding of the needs of individual customers. Improved customer understanding drives better customer relationship efforts, which leads to better and more frequent customer response; which in turn leads to more data collection about the customer – from which a more refined customer understanding can be gained. This positive feedback cycle – or 'virtuous loop' as it is often called – is shown in Figure 1.

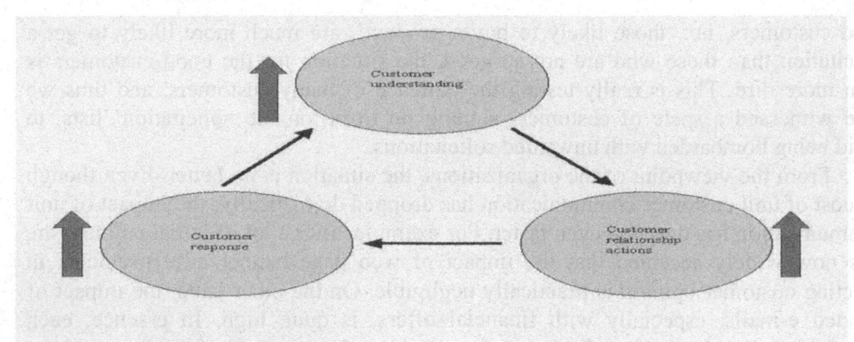

Figure 1. 'Virtuous circle' of CRM.

While this picture is very desirable, unfortunately there are a number of technical and organizational challenges that must be overcome to achieve it. First, much of customer data is collected for operational purposes, and is not organized for ease of analysis. With the advance of data analysis techniques, it is becoming feasible to exploit this data for business management, such as to find existing trends and discover new opportunities. Second, it is critical that this knowledge cover all channels and customer touch points - so that the information base is complete, and delivers a holistic and integrated view of each customer. This includes customer transactions, interactions, customer denials, service history, characteristics and profiles, interactive survey data, click-stream/browsing behavior, references, demographics, psychographics, and all available and useful data surrounding that customer. This may also include data from outside the business as well, for example from third party data providers such as Experian or Axciom. Third, organizational thinking must be changed from the current focus on products to include both customers and products, as illustrated in Figure 2. Successful adoption of CRM requires a change in focus by marketing from "who I can sell this products to?" to "what does this customer need?" It transforms marketing from "tactical considerations, i.e. "how do I get this campaign out of the door" to strategic focus, i.e. "what campaigns will maximize customer value?"

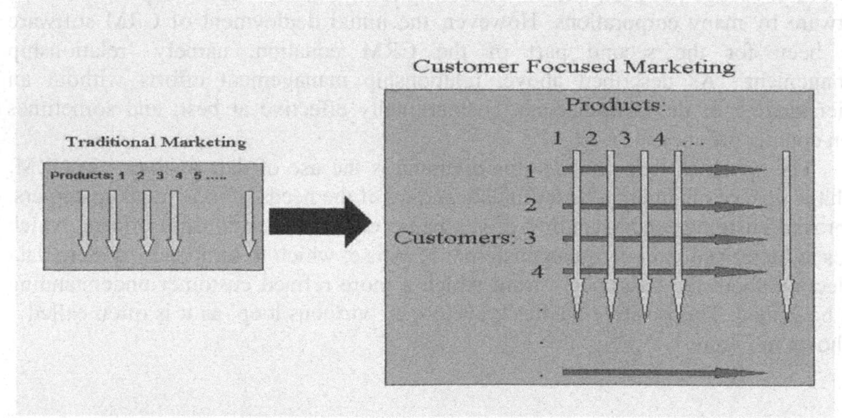

Figure 2. Change of focus from product only to customer+product.

The goal of this paper is to introduce the data mining community to the data analytics opportunities that exist in customer relationship management, especially in the area of customer understanding. As the data collected about customers is becoming more complete, the time is ripe for the application of sophisticated data mining techniques towards better customer understanding. The rest of this paper is organized as follows: in Section 2 we introduce the concept of analytical customer relationship management. Section 3 briefly describes the underlying technologies and tools that are needed, namely data warehousing and data mining. Section 4 describes a number of organizational issues that are critical to successful deployment of CRM in an organization, and Section 5 concludes the paper.

2 Analytical Customer Relationship Management

Significant resources have been spent on CRM, leading to the success of CRM software vendors such as Seibel, Oracle, and Epiphany. However, in the initial stages sufficient attention was not paid to analyzing customer data to target the CRM efforts. Simple heuristics and 'gut-feel' approaches led to profitable customers being bombarded with offers (often turning them off), while there being little attempt to develop today's the 'less valuable' customers into tomorrow's valuable ones. This lack of attention to customer needs is the cause of decreasing customer satisfaction across a wide variety of industries, as illustrated in Figure 3 [Heyg2001].[1]

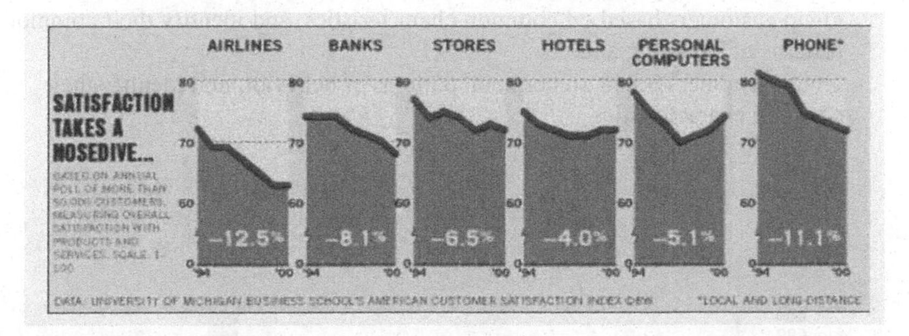

Figure 3. Declining trend in customer satisfaction index.

Fortunately, however, the tremendous advancement in data management and analysis technologies is providing the opportunity to develop fine-grained customer understanding on a mass scale, and use it to better manage the relationship with each customer. It is this approach to developing customer understanding through data analysis, for the purpose of more effective relationship management, that we call *"analytical customer relationship management(ACRM)"*. ACRM can make the customer interaction functions of a company much more effective than they are presently.

[1] Of course, customer expectation keeps rising over time, and the source of dissatisfaction today is very different that that of a few years ago. However, this is a battle that all organizations must constantly fight.

2.1 Customer Segmentation

Customer segmentation is the division of the entire customer population into smaller groups, called *customer segments*. The key idea is that each segment is fairly homogeneous from a certain perspective – though not necessarily from other perspectives. Thus, the customer base is first segmented by the value they represent to an organization, and then by the needs they may have for specified products and services.

The purpose of segmentation is to identify groups of customers with similar needs and behavior patterns, so that they can be offered more tightly focused products, services, and communications. Segments should be identifiable, quantifiable, addressable, and of sufficient size to be worth addressing. For example, a vision products company may segment the customer population into those whose eyesight is perfect and those whose eyesight is not perfect. As far as the company is concerned, everyone whose eyesight is not perfect falls in the same segment, i.e. of potential customers, and hence they are all the same. This segment is certainly not homogeneous from the perspective of a clothing manufacturer, who will perhaps segment on attributes like gender and age.

A company's customer data is organized into *customer profiles*. A customer's profile consists of three categories of data, namely (i) identity, (ii) characteristics, and (iii) behavior. These categories correspond to the questions *Who the person is?, What attributes do they have?*, and *How do they behave?* Two types of segmentation can be performed based on the profile, namely

- group customers based on common characteristics, and identify their common patterns of behavior, and
- group customers based on common patterns of behavior, and identify their common characteristics.

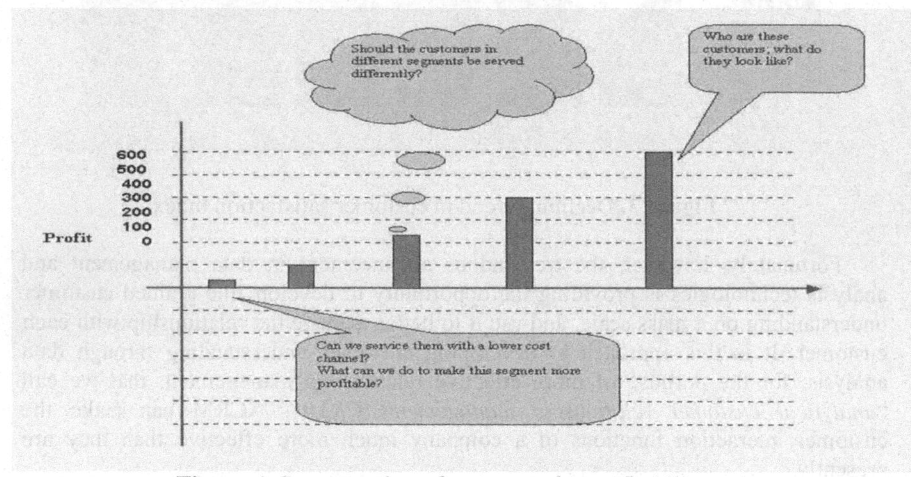

Figure 4. Segmentation of customers by profitability.

As shown in Figure 4, each customer segment represents a different amount of profit per customer; the treatment of each segment can be different. The figure shows examples of the type of questions the company can ask about segments. Also included are some overall strategic questions about which segments to focus on, and how much.

2.2 Customer Communication

A key element of customer relationship management is communicating with the customer. This consists of two components, namely (i) deciding what message to send to each customer segment, and (ii) selecting the channel through which the message must be sent. Message selection for each customer segment depends on the strategy being followed for that segment, as shown in Figure 4. The selection of the communication channel depends on a number of characteristics of each channel, including cost, focus, attention, impact, etc.

Typical communication channels include television, radio, print media, direct mail, and e-mail. Television is a broadcast channel, which is very good at sending a common message to a very large population. While it is very effective in building brand recognition, it is difficult to target a specific segment, as well as to measure response at the individual customer level. Radio, like television is a broadcast medium, and hence difficult to use for targeted communication to individual customers. Some television and radio stations, e.g. public radio and public television, develop a fairly accurate sample of their listener/viewer base through periodic fundraisers. Print media like newspapers and magazines can be used for much more focused communication, since the subscriber's profile is known. However, the readership of print media is usually much larger than the subscription base – a ratio of 1:3 in the US – and hence for a large part of the readership base, no profile is available. Direct mail is a communication channel that enables communicating with individual customers through personalized messages. In addition, it provides the ability of measuring response rates of customers at the individual level, since it enables the contacted customer to immediately respond to the message – if so desired. Finally, given its negligible cost, e-mail is becoming the medium of choice for customer contact for many organizations.

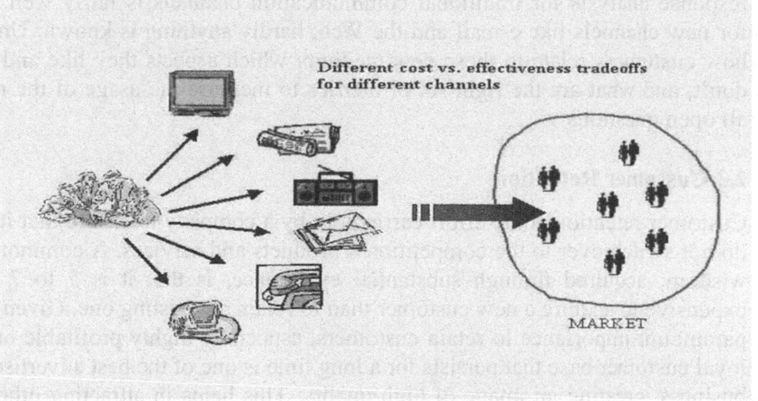

Figure 5. Formulating the optimal customer communication strategy.

Figure 5, courtesy of [Stev1998], illustrates the problem of formulating the customer communication strategy. Each communication channel has its own characteristics in terms of cost, response rate, attention, etc. The goal of communication strategy optimization is to determine the (set of) communication channel(s) for each customer that minimizes cost or maximizes sale, profit, etc. While communication channel optimization has been a well-studied problem in the quantitative marketing literature, characteristics of new channels such as e-mail and the Web are not well understood. Thus, there is a need to revisit these problems.

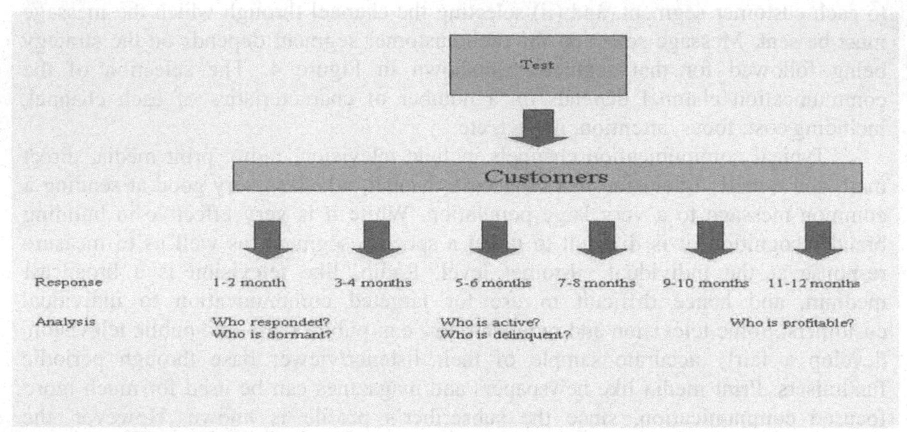

Figure 6. Analyzing the response to customer communications.

Sending the message to each customer through the chosen communication channel is not enough. It is crucial to measure the impact of the communication. This is done by using an approach called *response analysis*. As shown in Figure 6, response analysis metrics, e.g. number of respondents, acquired customers; number of active customers, number of profitable customers, etc. can be calculated. These are analyzed to (i) determine how effective the overall customer communication campaign has been, (ii) validate the goodness of customer segmentation, and (iii) calibrate and refine the models of the various communication channels used. While response analysis for traditional communication channels is fairly well understood, for new channels like e-mail and the Web, hardly anything is known. Understanding how customers relate to these new medium, which aspects they like and which they don't, and what are the right set of metrics to measure the usage of the medium, are all open questions.

2.3 Customer Retention

Customer retention is the effort carried out by a company to ensure that its customers do not switch over to the competition's products and services. A commonly accepted wisdom, acquired through substantial experience, is that it is 5 to 7 times more expensive to acquire a new customer than to retain an existing one. Given this, it is of paramount importance to retain customers, especially highly profitable ones. A good loyal customer base that persists for a long time is one of the best advertisements for a business, creating an image of high quality. This helps in attracting other customers who value long term relationships and high quality products and services.

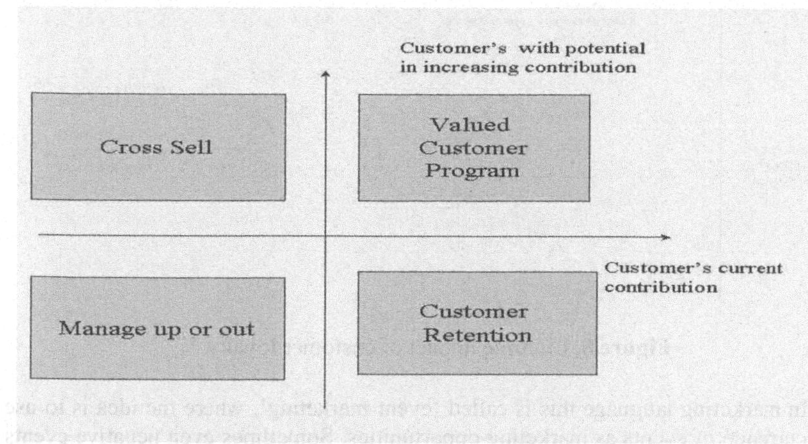

Figure 7. Treatment of various customer segments.

Figure 7 shows how a company thinks of its various customer segments, from a current and future profitability perspective. Clearly, the quadrants on the right bottom and the right top should be targeted for retention. In addition, the right top customer quadrant must be targeted for strengthening the relationship, as there is significant unrealized potential.

A successful customer retention strategy for a company is to identify opportunities to meet the needs of the customer in a timely manner. A specific example is of a bank that used the event "ATM request for cash" is rejected due to lack of funds" to offer unsecured personal loans to credit-worthy customers the next day. This offer was found to have a very high success rate, with the additional advantage of building customer loyalty. Classically, this analysis has been done at an aggregate level, namely for customer segments. Given present day analytic tools, it should be possible to do it at the level of individual customers.

2.4 Customer Loyalty

From a company's perspective, a loyal customer is one who prefers the company's products and services to those of its competition. Loyalty can range from having a mild preference all the way to being a strong advocate for the company. It is well accepted in consumer marketing that an average customer who feels closer to a company (high loyalty) is significantly more profitable than one who feels less close (low loyalty). Thus, ideally a company would like all its customers to become loyal, and then to quickly advance up the loyalty chain.

Figure 8, courtesy of [Heyg2001], illustrates the concept of tracking a customer to identify events in his/her life. Many of these events offer opportunities for strengthening the relationship the company has with this customer. For example, sending a greeting card on a customer's birthday is a valuable relationship building action – with low cost and high effectiveness.

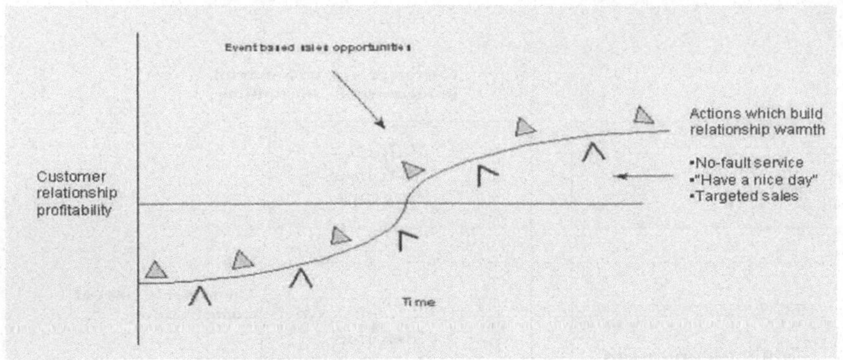

Figure 8. Lifetime impact of customer loyalty.

In marketing language this is called 'event marketing', where the idea is to use the occurrence of events as marketing opportunities. Sometimes even negative events can be used to drive sales. For example, a bank adopted the policy of offering a personal loan to every customer whose check bounced or there were insufficient funds for ATM withdrawal. This program was very successful, and also enhanced the reputation of the bank as being really caring about its customers.
The data mining community has developed many techniques for event and episode identification from sequential data. There is a great opportunity for applying those techniques here, since recognizing a potential marketing event is the biggest problem here.

3 Data Analytics Support for Analytical CRM

In this section we describe the backend support needed for analytical CRM. Specifically, we first outline a generic architecture, and then focus on the two key components, namely data warehousing and data mining.

3.1 Data Analytics Architecture

Figure 9 shows an example architecture needed to support the data analytics needs of analytical CRM. The key components are the data warehouse and the data analysis tools and processes.

3.2 Data Warehouse

Building a data warehouse is a key stepping stone in getting started with analytical CRM. Data sources for the warehouse are often the operational systems, providing the lowest level of data. Data sources are designed for operational use, not for decision support, and the data reflect this fact. Multiple data sources are often from different systems, running on a wide range of hardware, and much of this software is built in-house or highly customized. This causes data from multiple sources to be mismatched. It is important to clean warehouse data since critical CRM decisions will be based on it. The three classes of data extraction tools commonly used are - data migration which allows simple data transformation, data scrubbing which uses domain-specific knowledge to scrub data, and data auditing which discovers rules and relationships by scanning data and detects outliers.

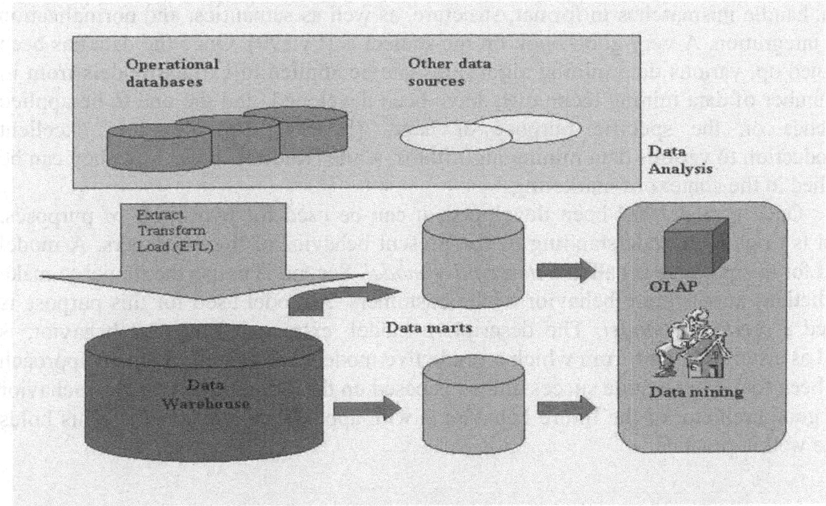

Figure 9. Data analytics architecture.

Loading the warehouse includes some other processing tasks, such as checking integrity constraints, sorting, summarizing, and build indexes, etc. Refreshing a warehouse requires propagating updates on source data to the data stored in the warehouse. The time and frequency to refresh a warehouse is determined by usage, types of data source, etc. The ways to refresh the warehouse includes data shipping, which uses triggers to update snapshot log table and propagate the updated data to the warehouse, and transaction shipping, which ships the updates in the transaction log.

The key entities required for CRM include *Customer, Product, Channel*, etc. Usually information about each of these is scattered across multiple operational databases. In the warehouse these are consolidated into complete entities. For example, the Customer entity in the warehouse provides a full picture of who a customer is from the entire organization's perspective, including all possible interactions, as well as their histories. For smaller organizations the analysis may be done directly on the warehouse, while for larger organizations separate data marts may be created for various CRM functions like customer segmentation, customer communication, customer retention, etc.

3.3 Data Mining

The next generation of analytic CRM requires companies to span the analytical spectrum and focus more effort on looking forward. The 'what has happened' world of report writers and the 'why has it happened' OLAP worlds are not sufficient. Time-to-market pressures, combined with data explosion, are forcing many organizations to struggle to stay competitive in the 'less time, more data' scenario. Coupled with the need to be more proactive, organizations are focusing their analytical efforts to determine what will happen, what they can do to make it happen, and ultimately to automate the entire process. Data mining is now viewed today as an analytical necessity. The primary focus of data mining is to discover knowledge, previously unknown, predict future events and automate the analysis of very large data sets.

The data mining process consist of a number of steps. First the data collected must be processed to make it mine-able. This requires a number of steps to clean the data, handle mismatches in format, structure, as well as semantics, and normalization and integration. A very good book on the subject is [Pyle99]. Once the data has been cleaned up, various data mining algorithms can be applied to extract models from it. A number of data mining techniques have been developed, and the one to be applied depends on the specific purpose at hand. [HMS00] provides and excellent introduction to various data mining algorithms, while [Rud00] shows how they can be applied in the context of marketing.

Once a model has been developed, it can be used for two kinds of purposes. First is to gain an understanding of the present behavior of the customers. A model used for this purpose is called a *descriptive model.* Second is to use the model to make predictions about future behavior of the customers. A model used for this purpose is called a *predictive model.* The descriptive model, extracted from past behavior, is used as a starting point from which a predictive model can be built. Such an approach has been found to be quite successful, as is based on the assumption that past behavior is a good predictor of the future behavior – with appropriate adjustments. This holds quite well in practice.

4 Organizational Issues in Analytical CRM Adoption

While the promise of analytical CRM, both for cost reduction and revenue increase, is significant, this cannot be achieved unless there is successful adoption of it within an organization. In this section we describe some of the key organizational issues in CRM adoption.

4.1 Customer First' Orientation

Companies that offer a number of products and services have traditionally organized their customer facing teams, e.g. sales, marketing, customer service, etc. along product lines, called "Lines of Business (LOB)". The goal of any such product marketing team is to build the next product in this line; the goal of the sales team is to identify the customers who would be likely to buy this product, etc. This product line focus causes customer needs to be treated as secondary.

The customer focusing teams of an organization must be re-oriented to make them focus on customers in addition to product lines. These teams can be organized around well-defined customer segments, e.g. infants, children, teenagers, young professionals, etc., and each given the charter of mapping our product design, marketing, sales, and service strategies that are geared to satisfying the needs of their customer segment. As part of this, some of the activities might be targeted to individual customers.

4.2 Attention to Data Aspects of Analytical CRM

The most sophisticated analytical tool can be rendered ineffective if the appropriate data is not available. To truly excel at CRM, an organization needs detailed information about the needs, values, and wants of its customers. Leading organizations gather data from many customer touch points and external sources, and bring it together in a common, centralized repository; in a form that is available and ready to be analyzed when needed. This helps ensure that the business has a consistent and accurate picture of every customer; and can align its resources according to the highest priorities. Given this observation, it is critical that sufficient attention be paid to the data aspects of the CRM project, in addition to the software.

4.3 Organizational 'Buy In'

While data mining and data warehousing are very powerful technologies, with a proven track record, there are also enough examples of failures when technology is deployed without sufficient organizational 'buy in'. As described earlier, the parts of the organization that will benefit the most from analytical CRM are the business units, i.e. marketing, sales, etc., and not the IT department. Thus, it is crucial to have 'buy in' from the business units to ensure that the results will be used appropriately. A number of steps must be taken to ensure this happens.

First, there needs to be a cross-functional team involved in implementing a CRM project in the organization. While the technical members on the team play an important role, an executive on the business side, who should also be the project owner and sponsor, should head the team. Second, processes need to be adopted, with an appropriate set of measurable metrics, to ensure that all steps for project success are being taken. Finally, incentives for performing well on the project should be included as part of the reward structure to ensure motivation.

4.4 Incremental Introduction of CRM

Introducing CRM into an organization must be managed carefully. Given its high initial cost, and significant change on the organization's processes, it is quite possible that insufficient care in its introduction leads to high expense, seemingly small early benefits, which can lead to low morale and excessive finger-pointing.

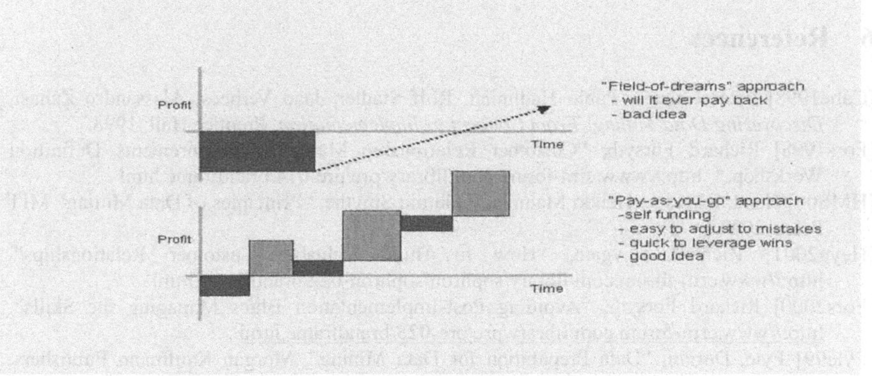

Figure 10. Incremental approach to CRM adoption

As shown in Figure 10, courtesy of [Fors2000], it is better to have an incremental 'pay-as-you-go' approach rather than a 'field-of-dreams' approach. The benefits accrued from the first stage become evident, and act as a catalyst for accelerating the subsequent stages. This makes the choice of the first project and its team very critical. Ideally, the project must be in a potentially high-impact area, where the current process is very ineffective. The ideal (cross-functional) team should have enthusiastic members, who are committed, and are also seen as leaders in their respective parts of the organization. This will make the dissemination of the successes much easier.

5 Conclusion

The Internet has emerged as a low cost, low latency and high bandwidth customer communication channel. In addition, its interactive nature provides an organization the ability to enter into a close, personalized dialog with its individual customers. The simultaneous maturation of data management technologies like data warehousing, and analysis technologies like data mining, have created the ideal environment for making customer relationship management a much more systematic effort than it has been in the past. While there has been a significant growth of software vendors providing CRM software, and of using them, the focus so far has largely been on the 'relationship management' part of CRM rather than on the 'customer understanding' part. Thus, CRM functions such as e-mail based campaigns management; on-line ads, etc. are being adopted quickly. However, ensuring that the right message is being delivered to the right person, that multiple messages being delivered at different times and through different channels are consistent, is still in a nascent stage. This is often leading to a situation where the best customers are being over communicated to, while insufficient attention is being paid to develop new ones into the best customers of the future.

In this paper we have described how Analytical CRM can fill the gap. Specifically, we described how data analytics can be used to make various CRM functions like customer segmentation, communication targeting, retention, and loyalty much more effective. Our hope is that the data mining community will address the analytics problems in this important and interesting application domain.

6 References

[Cabe1998] Peter Cabena, Pablo Hadjinian, Rolf Stadler, Jaap Verhees, Alessandro Zanasi, *Discovering Data Mining: From Concept to Implementation*, Prentice Hall, 1998.

[Fors1998] Richard Forsyth, "Customer Relationship Marketing Requirements Definition Workshop," http://www.crm-forum.com/library/pre/pre-014/brandframe.html

[HMS00] Hand, David J., Heikki Mannila, Padhraic Smythe, "Principles of Data Mining" MIT Press, 2000.

[Heyg2001] Richard Heygate, "How to Build Valuable Customer Relationships", http://www.crm-forum.com/library/sophron/sophron-022/brandframe.html

[Fors2000] Richard Forsyth, "Avoiding Post-Implementation Blues Managing the Skills", http://www.crm-forum.com/library/pre/pre-025/brandframe.html

[Pyle99] Pyle, Dorian, "Data Preparation for Data Mining", Morgan Kaufmann Publishers, 1999, ISBN No. 1558605290

[Rud00] Rud, Olivia C., "Data Mining, Cookbook: Modeling Data for Marketing, Risk and Customer Relationship Management", John Wiley and Sons, 2000.

[Silvon] Silvon Software, "The Bottom Line of CRM: Know your Customer", http://www.crm2day.com/library/wp/wp0032.shtml

[Stev1998] Peter Stevens, "Analysis and Communication - Two Sides of the Same Coin", http://www.crm-forum.com/library/sophron/sophron-003/brandframe.html

[Stev1999] Peter Stevens, and John Hegarty, "CRM and Brand Management - do they fit together?," http://www.crm-forum.com/library/sophron/sophron-002/brandframe.html

[Swif] Ronald S. Swift, "Analytical CRM Powers Profitable Relationships – Creating Success by Letting Customers Guide You", http://www.crm-forum.com/library/ncr/ncr-073/ncr-073.html.

[Think] thinkAnalytics, "The Hidden World of Data Mining", http://www.crm-forum.com/library/ven/ven-051/ven-051.html

[Michigan] University of Michigan Business School Study, "American Customer Satisfaction Index", 2000.

On Data Clustering Analysis: Scalability, Constraints, and Validation

Osmar R. Zaïane, Andrew Foss, Chi-Hoon Lee, and Weinan Wang

University of Alberta, Edmonton, Alberta, Canada

Abstract. Clustering is the problem of grouping data based on similarity. While this problem has attracted the attention of many researchers for many years, we are witnessing a resurgence of interest in new clustering techniques. In this paper we discuss some very recent clustering approaches and recount our experience with some of these algorithms. We also present the problem of clustering in the presence of constraints and discuss the issue of clustering validation.

1 Introduction

Cluster analysis is the automatic identification of groups of similar objects. This analysis is achieved by maximizing inter-group similarity and minimizing intra-group similarity. Clustering is an unsupervised classification process that is fundamental to data mining. Many data mining queries are concerned either with how the data objects are grouped or which objects could be considered remote from natural groupings. There have been many works on cluster analysis, but we are now witnessing a significant resurgence of interest in new clustering techniques. Scalability and high dimensionality are not the only focus of the recent research in clustering analysis. Indeed, it is getting difficult to keep track of all the new clustering strategies, their advantages and shortcomings. The following are the typical requirements for a good clustering technique in data mining [10]:

- **Scalability:** The cluster method should be applicable to huge databases and performance should decrease linearly with data size increase.
- **Versatility:** Clustering objects could be of different types - numerical data, boolean data or categorical data. Ideally a clustering method should be suitable for all different types of data objects.
- **Ability to discover clusters with different shapes:** This is an important requirement for spatial data clustering. Many clustering algorithms can only discover clusters with spherical shapes.
- **Minimal input parameter:** The method should require a minimum amount of domain knowledge for correct clustering. However, most current clustering algorithms have several key parameters and they are thus not practical for use in real world applications.
- **Robust with regard to noise:** This is important because noise exists everywhere in practical problems. A good clustering algorithm should be able to perform successfully even in the presence of a great deal of noise.

M.-S. Chen, P.S. Yu, and B. Liu (Eds.): PAKDD 2002, LNAI 2336, pp. 28–39, 2002.

- **Insensitive to the data input order:** The clustering method should give consistent results irrespective of the order the data is presented.
- **Scaleable to high dimensionality:** The ability to handle high dimensionality is very challenging but real data sets are often multidimensional.

Historically, there is no single algorithm that can fully satisfy all the above requirements. It is important to understand the characteristics of each algorithm so the proper algorithm can be selected for the clustering problem at hand. Recently, there are several new clustering techniques offering useful advances, possibly even complete solutions.

In the next section, we attempt to put various approaches to clustering in perspective and group them by their fundamental approach. We present their basic concepts and principles. In Section 3, we discuss clustering techniques for spatial data in the presence of physical constraints. Finally, in Section 4, we conclude and discuss methods for validating cluster quality.

2 Taxonomy on Clustering Techniques

There exist a large number of clustering algorithms. Generally speaking, these clustering algorithms can be clustered into four groups: partitioning methods, hierarchical methods, density-based methods and grid-based methods. This section gives a taxonomy analysis and an experimental study of representative methods in each group. In order to examine the clustering ability of clustering algorithms, we performed experimental evaluation upon k-means [12], CURE [21], ROCK [8], DBSCAN [2], CHAMELEON [14], WaveCluster [24] and CLIQUE [1]. The DBSCAN code came from its authors while CURE and ROCK codes were kindly supplied by the Department of Computer Science and Engineering, University of Minnesota. k-means, CHAMELEON, WaveCluster, and CLIQUE programs were locally implemented. We evaluate these algorithms by using two dimensional spatial data sets referenced and used in the CHAMELEON paper [14] and data sets referenced and used in the WaveCluster paper [24]. The reason for using two dimensional spatial data is because we can visually evaluate the quality of the clustering result. Often people can intuitively identify clusters on two dimensional spatial data, while this is usually very difficult for high dimensional data sets. We show the experimental results of each algorithm on the t7 data set from the CHAMELEON paper as shown in Figure 1. This data set is a good test because it has various cluster shapes including clusters within clusters and a great deal of noise.

2.1 Partitioning Methods

Suppose there are n objects in the original data set, partitioning methods break the original data set into k partitions. The basic idea of partitioning is very intuitive, and the process of partitioning is typically to achieve certain optimal criterion iteratively. The most classical and popular partitioning methods are k-means [12] and k-medoid [16], where each cluster is represented by the gravity centre of the cluster in k-means method or by one of the "central" objects of the cluster in k-medoid method. Once cluster representatives are selected, data points are assigned to these representatives, and iteratively, new better representatives are selected and points reassigned until no change is made. CLARANS [20] is an improved k-medoid algorithm. Another extension of k-means is the k-modes method [11], which is specially designed for clustering categorical data. Instead of a "mean" in k-means, k-modes defined a "mode" for each cluster. A new development on the standard k-means algorithm is bisecting k-means [26]. Starting with all data points in one cluster, the algorithm proceeds by selecting the largest cluster and splitting it into two using basic k-means. This iterates until the desired number of clusters k is reached. By the nature of the algorithm, bisecting k-means tends to produce clusters of similar sizes unlike k-means, which tends to result in lower entropy as large clusters will often have higher entropy.

All the partitioning methods have a similar clustering quality and the major dificulties with these methods include: (1) The number k of clusters to be found needs to be known prior to clustering requiring at least some domain knowledge which is often not available; (2) it is dificult to identify clusters with large variations in sizes (large genuine clusters tend to be split); (3) the method is only suitable for concave spherical clusters.

Because partitioning algorithms have similar clustering results, we only implemented k-means. K-mean's result on t7 is shown in Figure 1 (A). From here we see k-means tends indeed to find spherical clusters, and is unable to find arbitrary shaped clusters. This is actually a general problem for all the partition methods because they use only one gravity centre to represent a cluster, and clustering of all the other points are decided by their relative closeness to the gravity centres of clusters.

2.2 Hierarchical Methods

A hierarchical clustering algorithm produces a dendogram representing the nested grouping relationship among objects. If the clustering hierarchy is formed from bottom up, at the start each data object is a cluster by itself, then small clusters are merged into bigger clusters at each level of the hierarchy until at the top of the hierarchy all the data objects are in one cluster. This kind of hierarchical method is called agglomerative. The inverse process is called divisive hierarchical clustering. There are many new hierarchical algorithms that have appeared in the past few years. The major difference between all these hierarchical algorithms is how to measure the similarity between each pair of clusters.

BIRCH [33] introduced the concept of clustering features and the CF-tree. It first partitions objects hierarchically using the CF-tree structure. This CF-tree is used as a summarized hierarchical data structure which compresses data while trying to preserve the inherent clustering structure. After the building of the CF-tree, any clustering algorithm can be applied to the leaf nodes. BIRCH is particularly suitable for large data sets, however, it does not perform well if

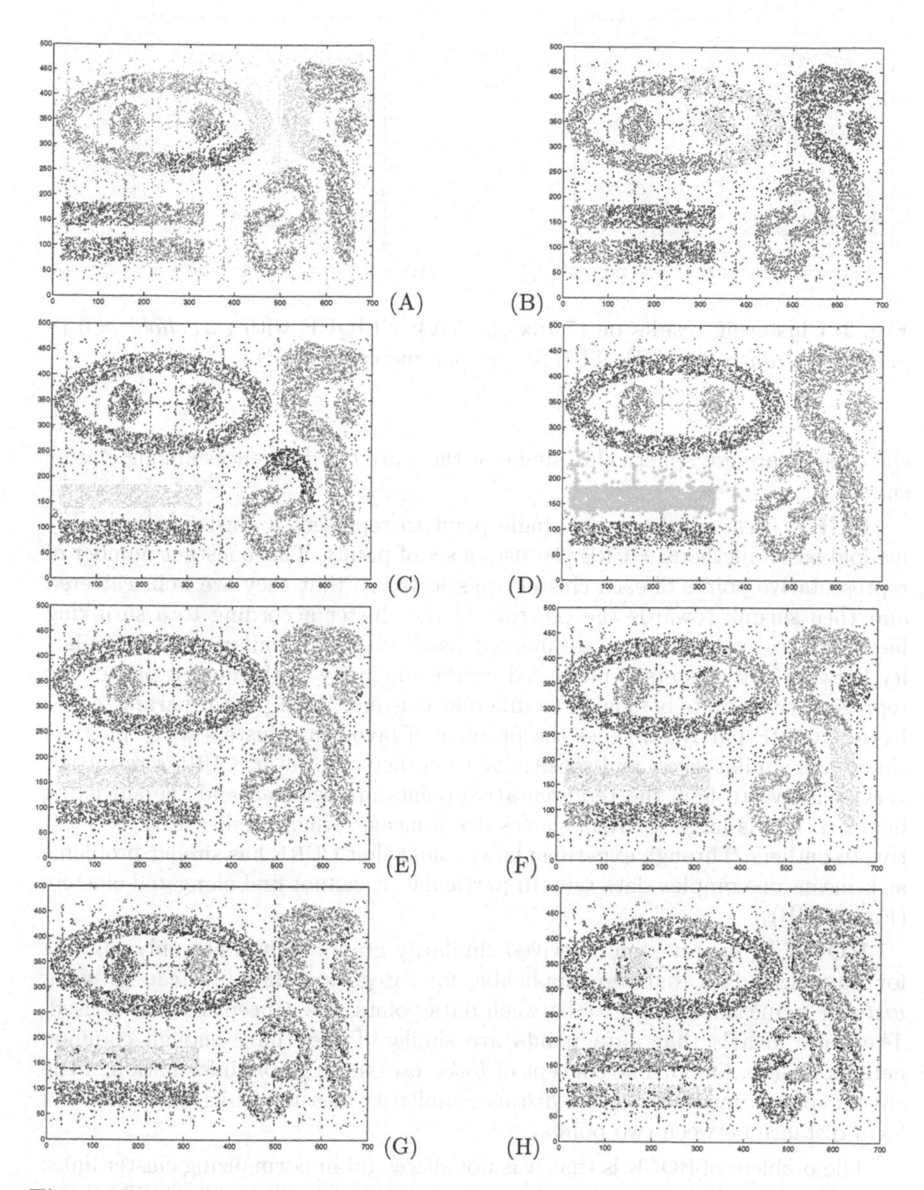

Fig. 1. Clustering results on t7.10k.dat. **(A)**: k-means with k=9; **(B)**: CURE with k=9, α=0.3, and 10 representative points per cluster; **(C)**: ROCK with θ=0.975 and k=1000; **(D)**: CHAMELEON with k-NN=10, MinSize=2.5%, and k=9; **(E)**: DBSCAN with ϵ=5.9 and MinPts=4; **(F)**: DBSCAN with ϵ=5.5 and MinPts=4; **(G)**: WaveCluster with *resolution*=5 and τ=1.5; **(H)**: WaveCluster with *resolution*=5 and τ=1.999397.

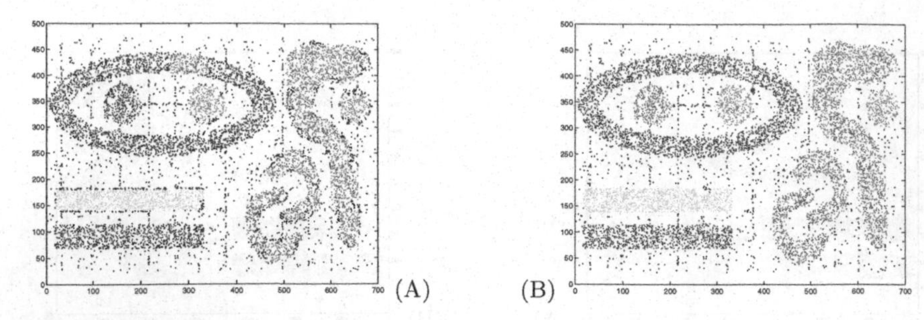

Fig. 2. Clustering results on t7.10k.dat. **(A)**: CLIQUE with *threshold* = 0.18 and *resolution* = 20 ; **(B)**: TURN* (no parameters needed).

the clusters are not spherical in shape or there are big differences among cluster sizes.

CURE: Instead of using a single point to represent a cluster in centroid/ medoid based methods, CURE [21] uses a set of points. This constant number of representative points of each cluster are selected so that they are well scattered and then shrunk towards the centroid of the cluster according to a shrinking factor. Interatively, clusters are merged based on their similarity. The similarity between two clusters is measured by the similarity of the closest pair of the representative points belonging to different clusters. With proper parameter selection, CURE partly remedies the problem of favouring clusters with spherical shape and similar sizes, and sensitivity to outliers. However, CURE's result are very sensitive to how the representative points are selected and the shrinking factor α. If α is large, CURE behaves like k-means, when small, CURE is sensitive to outliers. Through experiments we found that CURE has similar problems as k-means on complex data sets. In particular, it cannot find elongated clusters (Figure 1(B)).

ROCK [8] operates on a derived similarity graph, so it is not only suitable for numerical data, but also applicable for categorical data. Instead of using *distance* to measure similarity between data points, the concept of *links* is used. The basic idea is that data points are similar if they have enough common neighbours (i.e. links). The concept of *links* uses more global information of the cluster space compared with the distance similarity measure which only considers local distance between two points.

The problem of ROCK is that it is not successful in normalizing cluster links: it uses a fixed global parameter to normalize the total number of links. This fixed parameter actually reflects a fixed modeling of clusters, and it is not suitable for clusters with various densities. ROCK's clustering result is not good for complex clusters with various data densities. Also, it is very sensitive to the selection of parameters and sensitive to noise. After adjusting parameters for a long time, the best clustering result on t7 we could find is illustrated in Figure 1(C). Notice that we set the number of clusters to be 1000, then among the resulting 1000 clusters, we got 5 big clusters, all the other 995 are just noise. This is because ROCK does not collect noise in its clustering process. For this data set, if we set cluster number to be 9, then most of the points, 9985 points, are in one cluster, and the other 15 points exists in the 8 noise clusters.

CHAMELEON [14] performs clustering through dynamic modeling: two clusters are merged only if the inter-connectivity and closeness between two clusters are comparable to the internal inter-connectivity and closeness within the clusters. CHAMELEON also operates on a derived similarity graph, so that this algorithm can be applied to both numerical data and categorical data. It operates on a sparse graph in which nodes represent data items, and weighted edges represent similarities among the data items. The sparse graph is formed by keeping *k-nearest neighbour* of each node. A graph partitioning method it used to pre-cluster objects in the sparse graph into a set of small clusters. These small clusters are then merged based on their relative interconnectivity and relative closeness. CHAMELEON has been found to be very effective in clustering, but has significant shortcomings: it cannot handle outliers and has too many parameters such as the number of nearest neighbours in the sparse graph, MINSIZE the stopping condition for the graph partitionning and α for adjusting relative closeness. CHAMELEON's result on the data set t7 is shown in Figure 1(D). Note that all the noise points are included inside neighbouring clusters.

The common disadvantage of hierarchical clustering algorithms is setting a termination condition which requires some domain knowledge for parameter setting. The problem of parameter setting makes them less useful for real world applications. Also typically, hierarchical clustering algorithms have high computational complexity.

2.3 Density-Based Methods

The advantages of density-based methods are that they can discover clusters with arbitrary shapes and they do not need to preset the number of clusters.

DBSCAN [2] connects regions with sufficient high density into clusters. Each cluster is a maximum set of density-connected points. Points not connected are considered outliers. For each object of a cluster, the neighbourhood of a given radius (ϵ) has to contain at least a minimum number of points ($MinPts$). Methaphorically, it is like moving a flashlight of radius ϵ across the data set and connecting data points as long as $MinPts$ points are seen. This is the notion of density-reachability, and it is repeated until all points are labelled. In practice it works very well in spatial clustering. DBSCAN is very sensitive to the selection of ϵ and $MinPts$ and cannot identify clusters efficiently when cluster density varies considerably. When we apply DBSCAN to the data set t7, it gives very good results as illustrated in Figure 1(E). However, if we slightly change ϵ from 5.9 to 5.5, it gives bad results (see Figure 1(F)). If we increase ϵ, the noise creates bridges that cause genuine clusters to merge.

By the same authors, OPTICS [18] is an extension to DBSCAN. Instead of producing one set of clustering results with one pre-setting radius (ϵ), OPTICS produces an augmented ordering of the database representing its density-based clustering structure. This cluster-ordering actually contains the information about every clustering level of the data set. Restrictions of OPTICS are that it is still more suitable for numerical data, and also the user still needs to set one parameter, *MinPts*.

TURN* [6] consists of an overall algorithm and two component algorithms, one, an efficient resolution dependent clustering algorithm TURN-RES which returns both a clustering result and certain global statistics (cluster features) from that result and two, TurnCut, an automatic method for finding the important or "optimum" resolutions from a set of resolution results from TURN-RES. TurnCut uses the core of the TURN algorithm [5] to detect a change in the third differential of a series to identify important areas in a cluster feature across resolution series built by repeated calls to TURN-RES by the *TURN** algorithm. This change is the "*turning*" point in the overall trend of the curve - acceleration or reversal of the rate of change of the clustering feature studied. This is very similar to the concept of finding the "knee" in the cluster feature graph [17] used for cluster validation. Unlike other approaches such as grid-based clustering, a resolution is simply a scale by which all data point values are multiplied, thus all data points are used in the process. At a given resolution, TURN-RES computes how tightly packed the points are around each point and marks points with a high value as "internal". At the same time, those neighbours of each point that are "close" are also marked. These definitions are resolution dependent. Clustering involves combining all close neighbours to an internal point into its cluster, and iterating for all of those points that are internal. Applied to the data set t7, TURN* gives very good results without the need of inputting parameters. Results are illustrated in Figure 2(B).

2.4 Grid-Based Methods

Grid-based methods first quantize the clustering space into a finite number of cells, and then perform clustering on the gridded cells. The main advantage of grid-based methods is that their speed only depends on the resolution of griding, but not on the size of the data set. Grid-based methods are more suitable for high density data sets with a huge number of data objects in limited space.

WaveCluster [24] is a novel clustering approach based on wavelet transforms. WaveCluster first summarizes the data by applying a multi-resolution grid structure on the data space, then the original two-dimensional data space is considered as two-dimensional signals and signal processing techniques, wavelet transforms, are applied to convert the spatial data into the frequency domain. After wavelet transform, the natural clusters in the data become distinguishable in a transformed frequency space. Dense regions, i.e. clusters, are easily captured in the frequency domain. The process of clustering is reduced to finding connected strong signal components in the low pass digital filter along the horizontal and vertical dimensions. Because Wavelet transforms can filter out noise points, clustering in this fequency domain is usually much simpler than clustering in the original 2-dimensional space. In the process of WaveCluster, there are two main parameters to be selected: one is the grid resolution; the other one is the signal threshold τ for deciding whether a cell is a significant cell in the low pass digital filter of both dimensions. WaveCluster's clustering result on t7 is shown in Figure 1(G). Notice that WaveCluster can not separate the two clusters connected by a "bridge". This is because in the convolved and down-sampled image with the low pass filter, the bridge connecting the two clusters is still very a strong signal. To separate the two clusters, other genuine clusters have to be separated as well. Figure 1(H) shows another cluster result of WaveCluster by adjusting signal threshold τ. Now it separates the bridge-connected clusters but breaks other genuine clusters.

CLIQUE [1] is specifically designed for finding subspace clusters in sparce high dimensional data. CLIQUE's clustering process starts with the lower dimensional space. When clustering for the k-dimensional space, CLIQUE makes use of information of the (k-1)-dimension which is already available. For instance, potentially dence cells of the grid in the 2-dimensional space are identified by dense regions in the 1-dimentional space. All other cells are simply disregarded. This is similar to the *a-priori* principle used in mining frequent itemsets for association rules. CLIQUE is, however, unsuitable for very noisy or dense spaces. CLIQUE's clustering result on t7 is shown in Figure 2(A). CLIQUE is not ideal for clustering 2-dimensional data. Its sensitivity to noise makes it merge genuine clusters at high resolution and disregard edge points when the resolution is not high enough.

3 Clustering Spatial Data in Presence of Constraints

So far we have seen algorithms that focus on the efficiency and effectiveness of clustering data. However, none of the algorithms consider possible constraints to the clustering. Examples of this are: (1) Constraints on individual objects; (2) obstacle objects as constraints; (3) clustering parameters as "constraints"; and (4) contraints imposed on each individual cluster [29]. In spatial data in particular, where data clustering has many applications in geograhic information systems, there are physical constaints such as obstacles (rivers, highways, mountain ranges, etc.) and crossings (bridges, pedways, etc.) that can hinder or signicantly alter the clustering process.

3.1 Constraint-Based Clustering

In this section we present three algorithms recently devised that deal with clustering spatial data in the presence of physical constraints: COD-CLARANS [28], AUTOCLUST+ [3] and DBCluC [32].

COD-CLARANS [28] has been derived from CLARANS [20], a variant of the k-medoids approach. COD-CLARANS takes into account the obstacles by integrating an optimization scheme into the distance-error function in the course of the CLARANS algorithm. Obstacles are modeled with *a visibility graph* $VG = (V, E)$ such that each vertex of the obstacle has a corresponding node in V, and two nodes v_1 and v_2 in V are connected by an edge in E if and only if the corresponding vertices they represent are visible to each other. The visibility graph is pre-procossesed in order to compute the obstructed distance between two data objects in a given planar space. The obstructed distance is a detoured distance between two data objects with consideration of the visibility graph. Once the visibility graph is constructed, *Micro-clustering* is applied as pre-processing to group some data points from presumably a same cluster in order to minimize the number of data points to consider during COD-CLARANS clustering and fit the data set into main memory. Along with using the CLARANS clustering algorithm, COD-CLARANS uses a pruning function to reduce the search space by minimizing distance errors when selecting cluster representatives. Unfortunately, COD-CLARANS inherits the problems from CLARANS, a partitioning clustering method. It is sensitive to noise and assumes significant pre-processing of the data space to deal with the obstacles.

AUTOCLUST+ [3] is based on an extension to the clustering algorithm AUTOCLUST [4], a graph partitioning algorithm. The algorithm uses a Delaunay Diagram where all data points are represented and linked by edges based on mean and standard deviations of distances between points. Points linked by short edges, indicating closness, are clustered together. Other edges represent relations between clusters, and between clusters and noise. AUTOCLUST proceeds by eliminating edges from the graph to isolate data points that form clusters. In AUTOCLUST+, an obstacle is modeled by a set of line segments obstructing the edges from the Delaunay Diagram. AUTOCLUST+ then removes the edges from the Delaunay Diagram if they are impeded by a line segment from an obstacle. A removed edge is replaced by a detour path defined as the shortest path between the two data objects.

However, reconstructing the diagram after the removal of edges eventually degrades the performance of the algorithm since the algorithm needs to find a detour path for every intersected line segment with the Delaunay diagram.

DBCluC [32] is a density-based algorithm derived from DBSCAN. It discovers clusters of arbitrary shapes and isolates noise while considering not only disconnectivity dictated by obstacles but also connectivity forced by bridges. Obstacles and bridges are modeled by polygons where bridge-polygons have special edges considered as entry points. The DBCluC algorithm extends the density reachability notion in DBSCAN to take into account visibility spaces created by the polygon edges. To enhance the clustering performance, polygons are reduced to a minimal set of obstruction lines that preserve the integrity of the visibility spaces. These lines stop the propagation of point neighbourhood in DBSCAN when they are from obstruction-polygons, or extend the neighbourhood when they are bridge-polygons.

4 Conclusion: Clustering Validation

One of the main difficulties with clustering algorithms is that, after clustering, how can one assess the quality of the clusters returned? Many of the popular clustering algorithms are known to perform poorly on many types of data sets. In addition, virtually all current clustering algorithms require their parameters to be tweaked for the best results, but this is impossible if one cannot assess the quality of the output. While 2D spatial data allows for assessment by visual inspection, the result is dependent on the resolution presented to the inspector and most clustering tasks are not 2D or even spatial. One solution [22,13,15] is to reduce any output to a 2D spatial presentation. Other solutions are based on 1) external, 2) internal and 3) relative criteria [17]. The External and Internal Criteria approaches use Monte Carlo methods [27] to evaluate whether the clustering is significantly different from chance.

In the External approach, the clustering result C can be compared to an independent partition of the data P built according to our intuition of the structure of the data set or the proximity matrix P is compared to P. The Internal Criteria approach uses some quantities or features inherent in the data set to evaluate the result. If the clustering is hierarchical, a matrix P_c, representing the proximity level at which two vectors are found in the same cluster for the first time, can be compared to P. This is repeated for many synthetic data sets to determine significance. For non-hierarchical methods a cluster membership matrix is compared to P using the same Monte Carlo method to determine significance.

These methods are clearly very expensive in processing time and only tell us that the clustering result is not pure chance. The Relative Criteria does not involve statistical tests but attempts to evaluate among several results arising from different parameter settings. The challenge is to characterize the clustering result in a way that tells us the quality of the clustering. Naturally, there is a grey line between measures used by clustering algorithms to determine where to join or split clusters and indices, e.g. [30,23,19,9], proposed to determine if that was good. Like many clustering algorithms, these indices suffer from problems especially inability to handle non-spherical clusters. Another approach computes several indices such as the Root-Mean-Square Standard Deviation, a measure of homogeneity, and plots them against k [23]. Whatever the index, having created a graph, this is inspected visually for either a minima/maxima or a "knee", being the greatest jump of the index with a change in k. There is, however, no rigorous way of ensuring that this "knee" identifies the correct k. There are different indices defined for evaluating fuzzy clustering, e.g. [7,30]. An evaluation [17] of a number of indices on data that contained only concave but not always circular clusters, found different indices were better on different data sets showing their shape-dependence. In a few cases, Clustering Validation approaches have been integrated into clustering algorithms giving a relatively automatic clustering process. Smyth presented MCCV [25], the Monte Carlo Cross-Validation algorithm though this is intended for data sets where a likelihood function such as Gaussian mixture models can be defined. We have developed TURN [5] and TURN* [6] which handle arbitrary shapes, noise, and very large data sets in a fast and efficient way. TURN is intended for categorical data while TURN* is density based for spatial data.

An extended version of this paper is available in [31].

References

1. Agrawal R., Gehrke J., Gunopulos D. and Raghavan P. (1998) Automatic subspace clustering of high dimensional data for data mining applications. In *Proc. ACM-SIGMOD Int. Conf. Management of Data*, pp 94–105.
2. Ester M., Kriegel H.-P., Sander J. and Xu X. (1996) A density-based algorithm for discovering clusters in large spatial databases with noise. In *Proc. ACM-SIGKDD Int. Conf. Knowledge Discovery and Data Mining*, pp 226–231.
3. Estivill-Castro V. and Lee I. (2000) Autoclust+: Automatic clustering of point-data sets in the presence of obstacles. In *Int. Workshop on Temporal and Spatio-Temporal Data Mining*, pp 133–146.
4. Estivill-Castro V. and Lee I. (2000) Autoclust: Automatic clustering via boundary extraction for mining massive point-data sets. In *Proc. 5th International Conference on Geocomputation.*
5. Foss A., Wang W. and Zaïane O. R. (2001) A non-parametric approach to web log analysis. In *Proc. of Workshop on Web Mining in First International SIAM Conference on Data Mining*, pp 41–50.
6. Foss A. and Zaïane O. R. (2002) TURN* unsupervised clustering of spatial data. submitted to *ACM-SIKDD Intl. Conf. on Knowledge Discovery and Data Mining*, July 2002.
7. Gath I. and Geva A. (1989) Unsupervised optimal fuzzy clustering. *IEEE Transactions on Pattern Analysis and Machine Intelligence*, 11(7).
8. Guha S., Rastogi R. and Shim K. (1999) ROCK: a robust clustering algorithm for categorical attributes. In *15th ICDE Int'l Conf. on Data Engineering.*
9. Halkidi M., Vazirgiannis M. and Batistakis I. (2000) Quality scheme assessment in the clustering process. In *Proc. of PKDD, Lyon, France.*
10. Han J. and Kamber M. (2000) *Data Mining: Concepts and Techniques.* Morgan Kaufmann Publishers.
11. Huang Z. (1998) Extensions to the k-means algorithm for clustering large data sets with categorical values. *Data Mining and Knowledge Discovery*, v2 pp283–304.
12. MacQueen J. (1967) Some methods for classification and analysis of multivariate observations. In *Proc. 5th Berkeley Symp. Math. Statist. Prob.*.
13. Sammon J. Jr. (1969) A non-linear mapping for data structure analysis. *IEEE Trans. Computers*, v18 pp401–409.
14. Karypis G., Han E.-H. and Kumar V. (1999) Chameleon: A hierarchical clustering algorithm using dynamic modeling. *IEEE Computer*, 32(8) pp68–75.
15. Kohonen T. (1995) *Self-Organizing Maps*, Springer-Verlag.
16. Kaufman L. and Rousseeuw P. J. (1990) *Finding Groups in Data: an Introduction to Cluster Analysis.* John Wiley & Sons.
17. Halkidi M., Batistakis Y. and Vazirgiannis M. (2001) On clustering validation techniques. *Journal of Intelligent Information Systems*, 17(2-3) pp 107–145.
18. Ankerst M., Breunig M, Kriegel H.-P. and Sander J. (1999) Optics: Ordering points to identify the clustering structure. In *Proc. ACM-SIGMOD Conf. on Management of Data*, pp 49–60.
19. Pal N. R. and Biswas J. (1997) Cluster validation using graph theoretic concepts. *Pattern Recognition*, 30(6).
20. Ng R. and Han J. (1994) Efficient and effective clustering method for spatial data mining. In *Proc. Conf. on Very Large Data Bases* , pp 144–155.
21. Guha S., Rastogi R. and Shim K. (1998) CURE: An efficient clustering algorithm for large databases. In *Proc. ACM-SIGMOD Conf. on Management of Data.*

22. Schwenker F., Kestler H. and Palm G. (2000) An algorithm for adaptive clustering and visualisation of highdimensional data sets. In H.-J. L. G. della Riccia, R. Kruse, editor, *Computational Intelligence in Data Mining*, pp 127–140. Springer, Wien, New York.
23. Sharma S. (1996) *Applied Multivariate Techniques.* John Willey & Sons.
24. Sheikholeslami G., Chatterjee S. and Zhang A. (1998) Wavecluster: a multi-resolution clustering approach for very large spatial databases. In *Proc. 24th Conf. on Very Large Data Bases.*
25. Smyth P. (1996) Clustering using monte carlo cross-validation. *Proc. ACM-SIGKDD Int. Conf. Knowledge Discovery and Data Mining.*
26. Steinbach M., Karypis G. and Kumar V. (2000) A comparison of document clustering techniques. In *SIGKDD Workshop on Text Mining.*
27. Theodoridis S. and Koutroubas K. (1999) *Pattern recognition*, Academic Press.
28. Tung A. K. H., Hou J. and Han J. (2001) Spatial clustering in the presence of obstacles. In *Proc. ICDE Int. Conf. On Data Engineering.*
29. Tung A. K. H., Ng R., Lakshmanan L. V. S. and Han J. (2001) Constraint-based clustering in large databases. In *Proc. ICDT*, pp 405–419.
30. Xie X. and Beni G. (1991) A validity measure for fuzzy clustering. *IEEE Transactions on Pattern Analysis and Machine Intelligence*, 13(4).
31. Zaïane O. R., Foss A., Lee C.-H. and Wang W. (2002) Data clustering analysis - from simple groupings to scalable clustering with constraints. Technical Report, TR02-03, Department of Computing Science, University of Alberta.
32. Zaïane O. R. and Lee C.-H. (2002) Clustering spatial data in the presence of obstacles and crossings: a density-based approach. submitted to *IDEAS Intl. Database Engineering and Applications Symposium.*
33. Zhang T., Ramakrishnan R. and Livny M. (1996) BIRCH: an efficient data clustering method for very large databases. In *Proc. ACM-SIGKDD Int. Conf. Managament of Data*, pp 103–114.

Discovering Numeric Association Rules via Evolutionary Algorithm

Jacinto Mata[1], José-Luis Alvarez[1], and José-Cristobal Riquelme[2]

[1] Dpto. Ingeniería Electrónica, Sistemas Informáticos y Automática
Universidad de Huelva, Spain
{mata,alvarez}@uhu.es
[2] Dpto. Lenguajes y Sistemas Informáticos
Universidad de Sevilla, Spain
riquelme@lsi.us.es

Abstract. Association rules are one of the most used tools to discover relationships among attributes in a database. Nowadays, there are many efficient techniques to obtain these rules, although most of them require that the values of the attributes be discrete. To solve this problem, these techniques discretize the numeric attributes, but this implies a loss of information. In a general way, these techniques work in two phases: in the first one they try to find the sets of attributes that are, with a determined frequency, within the database (*frequent itemsets*), and in the second one, they extract the association rules departing from these sets. In this paper we present a technique to find the *frequent itemsets* in numeric databases without needing to discretize the attributes. We use an evolutionary algorithm to find the intervals of each attribute that conforms a *frequent itemset*. The evaluation function itself will be the one that decide the amplitude of these intervals. Finally, we evaluate the tool with synthetic and real databases to check the efficiency of our algorithm.

1 Introduction

Association rules were introduced in [1] as a method to find relationships among the attributes of a database. By means of these techniques a very interesting qualitative information with which we can take later decisions can be obtained. In general terms, an association rule is a relationship between attributes in the way $C_1 \Rightarrow C_2$, where C_1 and C_2 are pair conjunctions (attribute-value) in the way $A = v$ if it is a discrete attribute or $A \,\epsilon\, [v_1, v_2]$ if the attribute is continuous or numeric. Generally, the antecedent is formed by a conjunction of pairs, while the consequent usually is a unique attribute-value pair.

In most of databases can appear a rather high number of rules of this kind, so it is essential to define some measures that allow us to filter only the most significant ones. The most used measures to define the *interest* of the rules were described in [1]:

M.-S. Chen, P.S. Yu, and B. Liu (Eds.): PAKDD 2002, LNAI 2336, pp. 40–51, 2002.

- **support.** It is a statistical measure that indicates the ratio of the population that satisfies both the antecedent and the consequent of the rule. A rule $R : C_1 \Rightarrow C_2$ has a support s, if a $s\%$ of the records of the database contain C_1 and C_2.
- **confidence.** This measure indicates the relative frequency of the rule, that is, the frequency with which the consequent is fulfilled when it is also fulfilled the antecedent. A rule $R : C_1 \Rightarrow C_2$ has a confidence c, if the $c\%$ of the records of the database that contain C_1 also contain C_2.

The goal of the techniques that search for association rules is to extract only those that exceed some minimum values of *support* and *confidence* that are defined by the user. The greater part of the algorithms that extract association rules work in two phases: in the first one they try to find the sets of attributes that exceed the minimum value of support and, in the second phase, departing from the sets discovered formerly, they extract the association rules that exceed the minimum value of confidence. Some of these algorithms can be seen on [2, 9, 13, 14, 7, 8, 12].

The first works on association rules were focused on marketing. In them the databases are transactions that represent the purchases made by the customers. Hence, each transaction is formed by a set of elements of variable size. These kind of rules use to be called *classic association rules.* and the nomenclature proposed for them is still being used for the different variants of association rules. The databases with which we will work, unlike these, will be relational tables, that is, will consist of a set of records or tuples formed by a fixed number of continuous attributes, as can be seen in figure 1.

assists	height	time	age	points
0.0888	201	36.02	28	0.5885
0.1399	198	39.32	30	0.8291
0.0747	198	38.8	26	0.4974
...	...	...	...	...
0.1276	196	38.4	28	0.5703

Fig. 1. Basketball database

In this paper we will use the definitions proposed in [1], adapting them to the databases with which we will work.

Definition 1. ***Itemset.*** *It is a set of attributes belonging to the database. Each itemset is formed by a variable number of attributes. An itemset formed by k attributes will be called k-itemset. In our case, an itemset is formed by pair (attribute-range of values)*

Definition 2. ***Frequent itemset.*** *It is that itemset that exceed the minimum value of support.*

Therefore, the problem of mining association rules consists, basically, in finding all the frequent itemsets and obtaining the rules departing from these sets. All the studies and researches are focused on the first phase, which is the most expensive, since the second one can be considered a simple and direct process. Most of the tools cited before work starting with the frequent itemsets of size 1 and joining them to conform frequent itemsets of a greater size in each step.

But in the real world there are numerous databases where the stored information is numeric. In these databases, attributes have thousand of possibilities of taking one value, by this reason the process described above is unthinkable from a computational point of view. Association rules obtained on numeric databases will be called *quantitative association rules*. The problem of mining quantitative association rules was first introduced in [15]. These rules are a variant of classic association rules where the value that the attribute takes in the rule is an interval instead of a discrete value. An example of this kind of rules is: *if height* ϵ $[196, 201]$ *and time* ϵ $[35.3, 37.8]$ *then assist* ϵ $[0.025, 0.076]$.

The basic idea of the algorithm presented in their work consists in dividing the range of each numeric attribute into intervals, treating them, from that moment onwards, as discrete attributes. That strategy is the same that have been followed by the diverse authors that have worked with numeric databases. Each of them uses different methods: clustering techniques, partition of the domain into intervals of the same size, techniques to merge adjacent intervals until reaching a maximum support, discretization by means of fuzzy sets, etc., but all of them have in common the fact that they need information a priori from the user. Some of these techniques can be consulted in [11, 16, 3].

The main problem of all of them lies in the fact that the data must be prepared before applying the tool. This preparation, either by means of the user or by means of an automatic process, conveys a loss of information because the rules will be only generated departing from the partitions previously created.

Our goal is to find association rules in numeric databases without the necessity of preparing previously the data. In order to get this objective we present a tool based in an evolutionary algorithm [4] that discovers the frequent itemsets in numeric databases. We have designed the evolutionary algorithm to find the intervals in each of the attributes that conforms a frequent itemset, in such a way that the fitness function itself is the one that decides the amplitude of the intervals.

2 A Motivation Example

In figure 2 we can see the result obtained by our algorithm for the basketball database. We have only represented two of the frequent itemsets found. The most important of our results with regard to formerly tools is the possibility of obtaining ranges with overlapping in different itemsets. For example, in the first itemset, the best interval for *height* attribute is [179,198], while in the second one, the best interval for this attribute is [175,196]. In the previously referenced techniques, the attributes are discretized before searching the itemsets. So, if the discretization process finds the interval [179,198] for *height* attribute, the interval [175,196] can not appear in any itemset. This fact generates a loss of information. For example, if the minimum support is 30% and the discretization process has created the interval [179,198] for the *height* attribute, the second itemset would never be discovered because, probably, it would not exceed the minimum support or it would be smaller than 36.31%. Nevertheless, if their limits are slightly dynamically modified (we make it by means of mutations), the second itemset can also be discovered.

assist	[0.0721,0.2529]
height	[179,198]
age	[22,32]
Support	=39.45%

assist	[0.0978,0.277]
height	[175,196]
time	[19.02,38.97]
age	[22,31]
points	[0.3071,0.59]
Support	=36.31%

Fig. 2. 2 itemsets discovered in basketball database

3 Preliminaries

The tool presented in this paper is based on the evolutionary algorithm theory (EA). In order to find the optimal itemsets, that is, those with the best support without being their intervals excessively wide, we depart from a population where the individuals are potential itemsets. These individuals will be evolving by means of crossover and mutations, so that, at the end of the process, the individual with the best fitness will correspond with the "best" frequent itemset.

One of the problems we find when we work with EA theory is the convergence of all the individuals towards the same solution. In our case, this means that all the individuals evolve towards the same frequent itemset, that is, the individuals that conform the last generation provide, in practice, the same information. There are many techniques to solve this problem. Among them evolutionary algorithm with niches and iterative rule learning [5], which is the one used in our tool.

In this paper we develop only the first phase of a process of mining association rules, that is, the one that undertakes to find the frequent itemsets, because we use for the second phase some of the algorithm presented in the studies cited before.

4 Practical Implementation

As it was above, the core of this tool is an EA where the individuals are the possible itemsets we want to discover. In the following sections we will see the general structure of the algorithm, the same that the fitness function, representation of the individuals and the meaning of the genetic operators.

4.1 GAR Algorithm

The GAR (Genetic Association Rules) algorithm is based in the theory of evolutionary algorithms and it is an extension of the GENAR algorithm presented in [10], that search directly for the association rules, so it is necessary to prepare the data to indicate to the tool which attributes form part of the antecedent and which one is the consequent. Nevertheless, this process is not necessary in GAR, because the algorithm finds the frequent itemsets and the rules are built departing from them.

```
algorithm GAR
1.  nItemset = 0
2.  while (nItemset < N) do
3.    nGen = 0
4.    generate first population P(nGen)
5.    while (nGen < NGENERATIONS) do
6.      process P(nGen)
7.      P(nGen+1) = select individuals of P(nGen)
8.      complete P(nGen+1) by crossover
9.      make mutations in P(nGen+1)
10.     nGen++
11.   end_while
12.   I[nItemset] = choose the best of P(nGen)
13.   penalize records covered by I[nItemset]
14.   nItemset++
15. end_while
end
```

Fig. 3. GAR algorithm

In figure 3 the structure of the algorithm is shown. The process is repeated until we obtain the desired number of frequent itemsets N. The first step consists in generating the initial population. The evolutionary algorithm takes charge of calculating the fitness of each individual and carries out the processes of selection, crossover and mutation to complete the following generation. At the end of the process, in step 12, the individual with the best fitness is chosen and it will correspond with one of the frequent itemsets that the algorithm returns. The operation made in step 13 is very important. In it, records covered by the obtained itemset in the previous step are penalized. Since this factor affects negatively to the fitness function we achieve that in the following evolutionary process the search space tends to not be repeated.

4.2 Structure of Individuals

Due to the nature itself of the problem to solve, that is, the fact that the value of the attributes are taken from continuous domain, we use real codification to represent the individuals. An individual in GAR is a k-itemset where each gene represents the maximum and minimum values of the intervals of each attribute that belongs to such k-itemset.

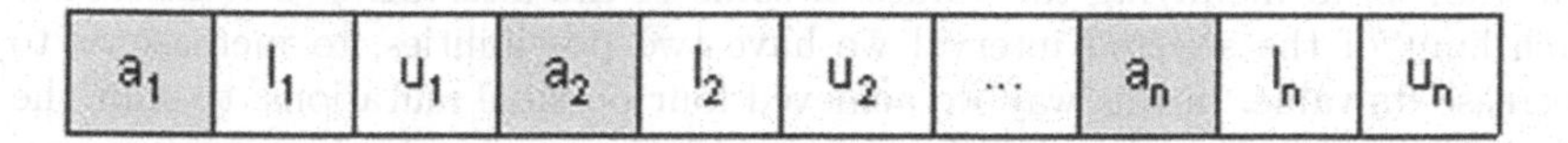

Fig. 4. Representation of an individual (n-itemset)

In general, the frequent itemsets are formed by a variable number of attributes, that is, for a database with n attributes there can be frequent itemsets from size 2 to size n, as can be seen in figure 4, where l_i and u_i are the limits of the intervals corresponding to the attribute a_i.

4.3 Initial Population

The generation of the initial population consists in the random creation of the intervals of each attribute that conforms the itemset. The number of attributes of each itemset is also chosen in a random way between 2 and the maximum number of attributes of the database. We condition the itemesets to cover at least a record of the database and that their intervals have a reduced size.

4.4 Genetic Operators

The genetic operators used in GAR are the usual ones, that is, selection, crossover and mutation. For the selection, we use an elitist strategy to replicate the individual with the best fitness. By means of the crossover operator we complete the rest of the population, choosing randomly, the individuals that will be combined to form new ones. From each crossover between two individuals two new ones are generated, and the best adapted will pass to the next generation. Given two individuals of the population $I = ([l_1, u_1], [l_3, u_3])$ and $I' = ([l'_1, u'_1], [l'_2, u'_2], [l'_3, u'_3])$, that are going to be crossed, the crossover operator generates the following two offspring:

$$O_1 = ([[l_1, u_1] \vee [l'_1, u'_1]], [[l_3, u_3] \vee [l'_3, u'_3]])$$
$$O_2 = ([[l'_1, u'_1] \vee [l_1, u_1]], [l'_2, u'_2], [[l'_3, u'_3] \vee [l_3, u_3]])$$

In figure 5 a possible result of the crossover operator for two itemsets of different size can be seen.

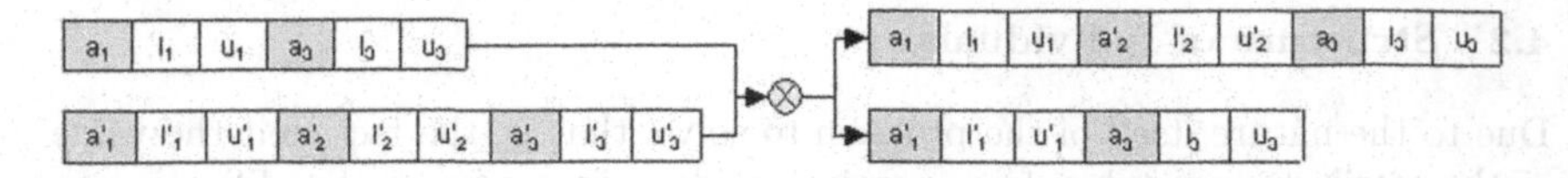

Fig. 5. Example of a crossover operation

The mutation operator consists in altering one or more genes of the individual, that is, in modifying the values of some of the intervals of a itemset. For each limit of the selected interval we have two possibilities, to increase or to decrease its value. In this way we achieved four possible mutations: to shift the whole interval to the left or to the right and to increase or to decrease its size.

Finally, a process of adjusting the chosen individual is carried out. This consists in decreasing the size of its intervals until the number of covered records be smaller than the records covered by the original itemset. Again, the goal of this post processing is to obtain more quality rules.

4.5 Fitness Function

As any evolutionary algorithm, GAR has a function implemented in order to evaluate the fitness of the individuals and to decide which are the best candidates in the following generations.

In our scenery, we look for the frequent itemsets with a larger support, that is, those that cover more records in the database. But, if we use this criterion as the only one to decide the limits of the intervals the algorithm will try to span the complete domain of each attribute. For this reason, it is necessary to include in the fitness function some measure to limit the size of the intervals.

The fitness function f for each individual is:

$$f(i) = covered - (marked * \omega) - (amplitude * \psi) + (nAtr * \mu) \quad (1)$$

The meaning of the parameters of the fitness function is the following:

- **covered.** It indicates the number of records that belong to the itemset that represent to the individual. It is a measure similar to support.
- **marked.** It indicates that a record has been covered previously by a itemset. We achieve with this that the algorithm tend to discover different itemsets in later searches. To penalize the records, we use a value that we call *penalization factor* (ω) to give more or least weight to the marked record, that is, we will permit more or least overlapping between the itemsets found depending on this value. This factor will be defined by the user.
- **amplitude.** This parameter is very important in the fitness function. Its mission is to penalize the amplitude of the intervals that conform the itemset. In this way, between two individuals (itemsets) that cover the same number of records and have the same number of attributes, the best information is given by the one whose intervals are smaller, as we can see in figure 6. By means of the factor ψ it is achieved that the algorithm be more or least

permissive with regard to the growth of the intervals. Within this concept, we penalize both the mean and the maximum amplitude of the intervals.

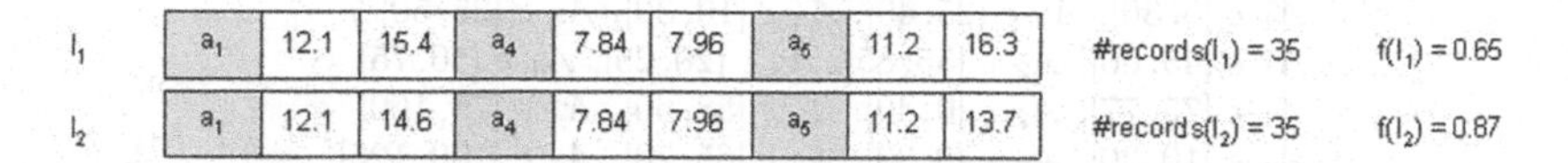

Fig. 6. Amplitude effect

- **number of attributes ($nAtr$).** This parameter rewards the frequent itemsets with a larger number of attributes. We will be able of increasing or decreasing its effect by means of the factor μ.

All the parameters of the fitness function are normalized into the unit interval. In this way all of them have the same weight when obtaining the fitness of each individual.

5 Experimental Results

To test if the developed algorithm finds in a correct way the frequent itemsets, we have generated several synthetic databases. We have used different functions to distribute the values in the records of the database, in such a way that they group on predetermined sets. The goal will be to find, in an accurate way, the intervals of each one of the sets artificially created. Besides, we have tested our tool with numeric databases from the Bilkent University Function Approximation Repository [6].

To carry out the tests, the algorithm was executed with a population of 100 individuals and 200 generations. We have chosen the following parameters in the GAR algorithm: 15% of selected individuals for the selection operator, 50% of crossover probability and 80% of mutation probability.

5.1 Synthetic Databases

A first database formed by four numeric attributes and 1000 records was generated. The values were distributed, by means of a uniform distribution, into 5 sets formed by predetermined intervals. Besides, 500 new records were added with the idea of introducing noise in the data, distributing their values, by means of a uniform distribution, between the minimum and maximum values of the domain of the intervals. In table 1 the 5 sets synthetically created are shown and in table 2 we show the frequent itemsets found by GAR.

The exact support for each of the synthetically defined sets is 13.34%, since each of them cover 200 records. As can be seen in table 2, the support of each of the sets found is quite close to such value, with a suitable size for each interval. The results show that the algorithm behaves in a correct way when the database

Table 1. Sets synthetically created by means of an uniform distribution

sets
$A_1 \in [1,15]$, $A_2 \in [7,35]$, $A_3 \in [60,75]$, $A_4 \in [0,25]$
$A_1 \in [5,30]$, $A_2 \in [25,40]$, $A_3 \in [10,30]$, $A_4 \in [25,50]$
$A_1 \in [45,60]$, $A_2 \in [55,85]$, $A_3 \in [20,25]$, $A_4 \in [50,75]$
$A_1 \in [75,77]$, $A_2 \in [0,40]$, $A_3 \in [58,60]$, $A_4 \in [75,100]$
$A_1 \in [10,30]$, $A_2 \in [0,30]$, $A_3 \in [65,70]$, $A_4 \in [100,125]$

Table 2. Frequent itemsets found by GAR

frequent itemsets	sup(%)	#records
[1, 15], [6, 35], [60, 76], [0, 26]	13.40	201
[5, 30], [24, 40], [10, 30], [26, 51]	13.07	196
[44, 61], [55, 84], [20, 35], [50, 75]	13.34	200
[74, 77], [0, 40], [58, 60], [75, 101]	13.34	200
[9, 29], [0, 30], [62, 71], [102, 125]	12.80	192

contains a set of records that can not be grouped in any frequent itemsets. The values used in the fitness function were: ω=0.7, ψ=0.6 and μ=0.7.

The first experiment was carried out creating sets independent among them, that is, without overlapping. In order to test if the tool works properly when the sets have records in common, a second database was created in the same way that the first one but with overlapping among the sets. In this case 600 records with the values distributed into 3 sets were generated and other 200 records were added to generate noise. In table 3 the three sets synthetically created are shown and in table 4 we show the frequent itemsets found by GAR.

Table 3. Sets synthetically created with overlapping

sets
$A_1 \in [18,33]$, $A_2 \in [40,57]$, $A_3 \in [35,47]$
$A_1 \in [1,15]$, $A_2 \in [7,30]$, $A_3 \in [0,20]$
$A_1 \in [10,25]$, $A_2 \in [20,40]$, $A_3 \in [15,35]$

The penalization factor was decreased to carry out this test in order to permit overlapping among the itemsets. The values used in the fitness function were: $\omega = 0.4$, $\psi = 0.6$ and $\mu = 0.7$. In both examples we can see that the sizes of the intervals have been reduced to discover the smallest intervals that cover the larger number of records.

The next test was carried out to test the behaviour of the tool when the itemsets are of a variable size. For this test we used the first database but distributing the values only among some of the attributes. In table 5 the five sets synthetically created are shown and in table 6 we show the frequent itemsets found by GAR.

Table 4. Frequent itemsets found by GAR

frequent itemsets	sup(%)	#records
[16, 32], [41, 57], [35, 46]	22.12	177
[1, 16], [7, 30], [1, 22]	27.38	219
[11, 25], [19, 41], [13, 35]	23.88	191
[1, 24], [7, 37], [0, 34]	49.50	396

Table 5. Sets variable size

sets
$A_1 \in [1, 15]$, $A_2 \in [7, 35]$, $A_4 \in [0, 25]$
$A_2 \in [25, 40]$, $A_3 \in [10, 30]$, $A_4 \in [25, 50]$
$A_2 \in [55, 85]$, $A_4 \in [50, 75]$
$A_1 \in [75, 77]$, $A_2 \in [0, 40]$, $A_3 \in [58, 60]$, $A_4 \in [75, 100]$
$A_1 \in [10, 30]$, $A_3 \in [65, 70]$

The result of the test shows how the tool found the predefined frequent itemsets. Besides, two new sets appeared as a consequence of the random distribution of the rest of the values. In this test the penalization factor and the number of attributes were loosen to find itemsets of variable size. The values used in the fitness function were: $\omega = 0.5$, $\psi = 0.6$ and $\mu = 0.45$.

5.2 Real-Life Databases

With the idea of evaluating our tool with real databases, we carried out some experiments using the Bilkent University Function Approximation Repository.

Due to the fact that the performance of the tool is based in a EA, we have carried out five times the proofs in the examples and the results fit in with the average values of such proofs. In 7 the results obtained are shown. The first and second column indicate the number of records and the number of numeric attributes of each database respectively. The third column (*#itemsets*) indicates the mean number of frequent itemsets found. The value of the column *support* indicates the mean of support of the found itemsets, while *size* shows the mean number of attributes of the itemsets. The column *%amplitude* indicates the mean size of the intervals that conform the set. This measure is significant to test that the intervals of the sets are not too many ample. The last column (*%records*) shows the percentage of records covered by the found itemsets on the total records.

Due to the fact of not knowing a priori the distribution of the values of the records, we use a minimum support of 20% and thresholds of $\omega = 0.4$, $\psi = 0.7$ and $\mu = 0.5$ to carry out this tests. The tool found frequent itemsets with high values of support but without expanding the intervals in excess (amplitude percentage below 30%).

Table 6. Frequent itemsets found by GAR

frequent itemsets	sup(%)	#records
[1, 15], [8, 34], [0, 24]	10.94	164
[25, 38], [12, 30], [24, 46]	10.20	153
[55, 77], [50, 73]	11.60	174
[75, 78], [1, 37], [58, 61], [75, 100]	12.40	186
[10, 30], [64, 70]	14.07	211
$A_2 \in [0, 40], A_3 \in [13, 70]$	42.74	641
$A_1 \in [0, 31], A_3 \in [9, 73]$	33.47	502

Table 7. Results for real-life databases

Database	records	#att	#itemsets	support	size	%ampl	%records
baskball (BK)	96	5	5.6	36.69	3.38	25	100
bodyfat (FA)	252	18	4.2	65.26	7.45	29	86
bolts (BL)	40	8	5.6	25.97	5.29	34	77.5
pollution (PO)	60	16	4.8	46.55	7.32	15	95
quake (QU)	2178	4	6.9	38.65	2.33	25	87.5
sleep (SL)	62	8	5.2	35.91	4.21	5	79.03
stock price (SP)	950	10	6.8	45.25	5.8	26	99.26
vineyard (VY)	52	4	6.6	36.08	3	17	100

6 Conclusions

We have presented in this paper a tool to discover association rules in numeric databases without the necessity of discretizing a priori, the domain of the attributes. In this way the problem of finding rules only with the intervals created before starting the process is avoided. We have used an evolutionary algorithm to find the most suitable amplitude of the intervals that conform a k-itemset, so that they have a high support value without being the intervals too wide. We have carried out several test to check the tools behaviour in different data distributions, obtaining satisfactory results if the frequent itemsets have no overlapping, if they have overlapping and if they are of a variable size. Nowadays, we are studying new measures to include in the fitness function and to find, with more accuracy, the size of the intervals in a k-itemset.

7 Acknowledgments

This work has been supported by Spanish Research Agency CICYT under grant TIC2001-1143-C03-02

References

[1] Agrawal, R., Imielinski. T., Swami, A.: Mining association rules between sets of items in large databases. Proc. ACM SIGMOD. (1993) 207–216, Washington, D.C.

[2] Agrawal, R., Srikant, R: Fast Algorithms for Mining Association Rules. Proc. of the VLDB Conference (1994) 487–489, Santiago (Chile)
[3] Aumann, Y., Lindell, Y.: A Statistical Theory for Quantitative Association Rules. Proceedings KDD99 (1999) 261–270, San Diego, CA
[4] Goldberg, D.E: Genetic algorithms in search, optimization and machine learning. Addison-Wesley. (1989)
[5] González, A., Herrera, F.: Multi-stage Genetic Fuzzy System Based on the Iterative Rule Learning Approach. Mathware & Soft Computing, 4 (1997)
[6] Guvenir, H. A., Uysal, I.: Bilkent University Function Approximation Repository, http://funapp.cs.bilkent.edu.tr (2000)
[7] Han, J., Pei, J., Yin, Y.: Mining Frequent Patterns without Candidate Generation. Proc. of the ACM SIGMOD Int'l Conf. on Management of Data (2000)
[8] Lin, D-I., Kedem, Z.M.: Pincer Search: A New Algorithm for Discovering the Maximum Frequent Set. In Proc. of the 6th Int'l Conference on Extending Database Technology (EDBT) (1998) 105–119 Valencia
[9] Manila, H., Toivonen, H., Verkamo, A.I.: Efficient algorithms for discovering association rules. KDD-94: AAAI Workshop on Knowledge Discovery in Databases (1994) 181–192 Seatle, Washington
[10] Mata, J., Alvarez, J.L., Riquelme, J.C.: Mining Numeric Association Rules with Genetic Algorithms. 5th Internacional Conference on Artificial Neural Networks and Genetic Algorithms, ICANNGA (2001) 264–267 Praga
[11] Miller, R. J., Yang, Y.: Association Rules over Interval Data. Proceedings of the International ACM SIGMOD Conference (1997) Tucson, Arizona
[12] Pasquier, N., Bastide, Y., Taouil, R., Lakhal, L.: Discovering Frequent Closed Itemsets for Association Rules
[13] Park, J. S., Chen, M. S., Yu. P.S.: An Effective Hash Based Algorithm for Mining Association Rules. Proc. of the ACM SIGMOD Int'l Conf. on Management of Data (1995) San José, CA
[14] Savarese, A., Omiecinski, E., Navathe, S.: An efficient algorithm for mining association rules in large databases. Proc. of the VLDB Conference, Zurich, Switzerland (1995)
[15] Srikant, R, Agrawal, R.: Mining Quantitative Association Rules in Large Relational Tables. Proc. of the ACM SIGMOD (1996) 1–12
[16] Wang, K., Tay. S.H., Liu, B.: Interestingness-Based Interval Merger for Numeric Association Rules. Proc. 4th Int. Conf. KDD (1998) 121–128

Efficient Rule Retrieval and Postponed Restrict Operations for Association Rule Mining

Jochen Hipp[1,3], Christoph Mangold[2,3], Ulrich Güntzer[3], and Gholamreza Nakhaeizadeh[1]

[1] DaimlerChrysler AG, Research & Technology, Ulm, Germany
{jochen.hipp,rheza.nakhaeizadeh}@daimlerchrysler.com

[2] IPVR, University of Stuttgart, Germany
mangold@informatik.uni-stuttgart.de

[3] Wilhelm Schickard-Institute, University of Tübingen, Germany
guentzer@informatik.uni-tuebingen.de

Abstract. Knowledge discovery in databases is a complex, iterative, and highly interactive process. When mining for association rules, typically interactivity is largely smothered by the execution times of the rule generation algorithms. Our approach is to accept a single, possibly expensive run, but all subsequent mining queries are supposed to be answered interactively by accessing a sophisticated rule cache. However there are two critical aspects. First, access to the cache must be efficient and comfortable. Therefore we enrich the basic association mining framework by descriptions of items through application dependent attributes. Furthermore we extend current mining query languages to deal with these attributes through $\exists$ and $\forall$ quantifiers. Second, the cache must be prepared to answer a broad variety of queries without rerunning the mining algorithm. A main contribution of this paper is that we show how to postpone restrict operations on the transactions from rule generation to rule retrieval from the cache. That is, without actually rerunning the algorithm, we efficiently construct those rules from the cache that would have been generated if the mining algorithm were run on only a subset of the transactions. In addition we describe how we implemented our ideas on a conventional relational database system. We evaluate our prototype concerning response times in a pilot application at DaimlerChrysler. It turns out to satisfy easily the demands of interactive data mining.

1 Introduction

1.1 Mining for Association Rules

Association rule mining [1] is one of the fundamental methods for knowledge discovery in databases (KDD). Let the database $\mathcal{D}$ be a multiset of transactions where each transaction $T \in \mathcal{D}$ is a set of items. An association rule $A \to B$ expresses that whenever we find a transaction which contains all items $a \in A$, then this transaction is likely to also contain all items $b \in B$. We call A the body and B the head of the rule. The strength and reliability of such rules are expressed by

M.-S. Chen, P.S. Yu, and B. Liu (Eds.): PAKDD 2002, LNAI 2336, pp. 52–65, 2002.

various rule quality measures [1,3]. The fraction of transactions $T \in \mathcal{D}$ containing an itemset A is called the support of A, $\mathsf{supp}_{\mathcal{D}}(A) = |\{T \in \mathcal{D} \mid A \subseteq T\}|/|\mathcal{D}|$. The rule quality measure support is then defined as $\mathsf{supp}_{\mathcal{D}}(A \to B) = \mathsf{supp}_{\mathcal{D}}(A \cup B)$. In addition the rule confidence is defined as the fraction of transactions containing A that also contain B: $\mathsf{conf}_{\mathcal{D}}(A \to B) = \mathsf{supp}_{\mathcal{D}}(A \cup B)/\mathsf{supp}_{\mathcal{D}}(A)$. These measures are typically supplemented by further rule quality measures. One of these measures that we found very helpful is called lift or interest [3]: $\mathsf{lift}_{\mathcal{D}}(A \to B) = \mathsf{conf}_{\mathcal{D}}(A \to B)/\mathsf{supp}_{\mathcal{D}}(B)$. It expresses in how far the confidence is higher or lower respectively than the a priori probability of the rule head.

For example, DaimlerChrysler might consider vehicles as transactions and the attributes of these vehicles as items. Then one might get rules like

$$\mathsf{Mercedes\ A\text{-}Class, AirCondition} \to \mathsf{BatteryTypeC}.$$

The support of this rule might be about 10% and the confidence about 90%. A lift of about 4 might indicate that the body of the rule actually raises the likelihood of BatteryTypeC clearly over its a priori probability.

1.2 Motivation

Obviously the idea behind association rules is easy to grasp. Even non-experts in the field of data analysis directly understand and can employ such rules for decision support. Unfortunately in practice the generation of valuable rules turns out to imply much more than simply applying a sophisticated mining algorithm to a dataset. In brief, KDD is by no means a push button technology but has to be seen a process that covers several tasks around the actual mining. Although there are different process descriptions, e.g. [2,4,20], KDD is always understood as complex, incremental, and highly iterative. The analyst never walks strictly through the pre-processing tasks, mines the data, and then analyzes and deploys the results. Rather, the whole process has a cyclic character: we often need to iterate and to repeat tasks in order to improve the overall result. It is the human in the loop who, on the basis of current results, decides when to return to previous tasks or proceed with the next steps. In the end, it is the analyst's creativity and experience which determine the success of a KDD project.

When mining for association rules on large datasets, the response times of the algorithms easily range from minutes to hours, even with the fastest hardware and highly optimized algorithms available today [7]. This is problematic because investigating even speculative ideas often requires a rerun of the mining algorithm and possibly of data pre-processing tasks. Yet if every simple and speculative idea implies to be idle for a few minutes, then analysts will – at least in the long run – brake themselves in advance instead of trying out diligently whatever pops into their minds. So, creativity and inspiration are smothered by the annoying inefficiencies of the underlying technology.

1.3 Contribution and Outline of This Paper

Imielinski et al. describe the idea of working on associations stored in a previous algorithm run [10,11]. In Section 2 we put this approach further by explicitly introducing a sophisticated rule cache. Other related work covers the idea of 'rule browsing', e.g. [12,13]. Our idea is to accept a single and possibly expensive algorithm run to fill the cache. After this initial run all mining queries arising during following iterations through the phases of the KDD process are satisfied directly from the cache without touching the original data. The result is interactivity due to short response times that are actually independent from the size of the underlying dataset. However there are two problematic aspects.

First, the analyst must be supported adequately when accessing the rule cache. For this purpose we suggest to employ mining query languages as described in [5,11,14]. We enrich these languages by fundamental extensions. We add language support for dealing with attributes that describe single items and aggregate functions on rules. Furthermore we add explicit $\exists$ and $\forall$ quantifiers on itemsets. Our extensions perfectly supplement today's approaches.

Second, a KDD process starts typically with the general case and comes down to the specifics in later phases. The rule cache can deal easily with restrictions concerning the items. Focusing the analysis on e.g. dependencies between special equipments of vehicles is achieved simply by filtering the cached rules. Unfortunately things get more difficult when restricting the underlying transactions. For example the analyst might decide to focus on a special vehicle model, e.g. asks which rules would have been generated when only E-Class vehicles were taken into account. Today such a restriction is part of the preprocessing and therefore requires a complete regeneration of the associations in the cache. As one of the main contributions of our paper in Section 3 we show how to even answer such queries from the cache without rerunning the rule generation algorithm.

In Section 4 we introduce the SM*ART*SKIP system that efficiently implements the described ideas. The cache structure must be able to store a huge number of rules, occasionally upto several hundreds of thousands. In addition it must be prepared to answer a broad variety of mining queries efficiently. In fact, we show how to realize such a rule cache based on top of a conventional relational database system. The database system stores the rules together with additional information in several relational tables. Moreover we implement an interpreter that translates our enhanced mining language to standard SQL that is directly executed on the database engine. We proof the efficiency of the resulting system by presenting experiences from a pilot application at DaimlerChrysler. Finally, we conclude with a short summary of our results in Section 5.

2 Interactivity Through Caching and Efficient Retrieval

Current approaches focus mainly on speeding up the algorithms. Although there have been significant advances in this area, the response times achieved do not allow true interactivity [7]. One approach is to introduce constraints on the items and to exploit this restriction of the rule set during rule generation, e.g. [15,19]. Yet the resulting response times still are far from the needs of the analyst in an interactive environment. In this section we tackle this problem by rule caching and sophisticated rule retrieval.

2.1 Basic Idea

Instead of speedup through restriction, our own approach does exactly the opposite: we accept that running the mining algorithm implies an interruption of the analyst's work. But if there must be a break, then the result of this interruption should at least be as beneficial as possible. In other words, if running the algorithm is inevitable, then the result should answer as many questions as possible. Hence, the goal of our approach is to broaden the result set by adding any item that might make sense and by lowering the thresholds on the quality measures. Typically response times suffer; but, as this is expected, this should not be a severe problem. In extreme cases, running the mining task overnight is a straightforward and acceptable solution. In general, the number of rules generated is overwhelming and, of course, the result set is full of noise, trivial rules, or otherwise uninteresting associations. Simply to present all rules would hardly make sense, because this would surely overtax the analyst.

Our approach is to store all the generated rules in an appropriate cache and to give the analyst highly sophisticated access to this cache. The goal is to satisfy as many of the mining queries as possible directly from the cache, so that a further mining pass is only needed as an exception. Once the cache is filled, answering mining queries means retrieving the appropriate rules from the cache instead of mining rules from the data. Now interactivity is no longer a problem, because mining queries can be answered quickly without notable response times.

We want to point out that we distinguish strictly between running mining algorithms and retrieving rules from the cache. In other words, we want analysts to be aware of what exactly they are doing: they should either start a mining run explicitly or query the cache. Otherwise, accidently causing a rerun of the rule generation by submitting a query that cannot be satisfied from the cache could interrupt the analysis for hours. Explicitly starting a rerun makes analysts think twice about their query and, moreover, gives them control.

2.2 Enhanced Rule Retrieval

The access to the rule cache must be as flexible as possible in order to be useful for a wide range of mining scenarios. Strict separation between rule mining and querying the cache allows us to also separate the retrieval language from the means to specify the rule generation task. For the latter we want to refer the reader to [5,6,14] where accessing and collecting the underlying transactions is treated exhaustively. The mining language we need is focused on rule retrieval from the cache and is never concerned directly with the mining data itself.

We found languages that cover this aspect [5,11,14]. However for the purpose of demonstrating our new ideas we decided to restrict ourselves to a simple 'core' language. We forego a formal language definition but sketch the concept behind as far as necessary for understanding our ideas. Of course our enhancements are supposed to be integrated into a universal environment, e.g. [5,11,14]. We point out that we do not compete with these systems but see our ideas as supplements arising from experiences during practical mining projects, e.g. [8,9].

A query in our simplified retrieval language always consists of the keyword SelectRulesFrom followed by the name of a rule cache and a sophisticated where-clause that filters the retrieved rules. The basic query restricts the rules by thresholds on the quality measures. For example, we may want to retrieve all rules from cache rulecache that have confidence above 75% and lift of at least 10:

```
SelectRulesFrom rulecache                          (Query 1)
Where conf > 0.75 and lift >= 10;
```

Often we want to restrict rules based on the items they contain respectively not contain. For example we might be interested in rules that 'explain' the existence of a driver airbag, that is rules containing item Airbag in the head:

```
SelectRulesFrom rulecache                          (Query 2)
Where 'Airbag' in head and conf > 0.75 and lift >= 10;
```

At the same time we know that a co driver airbag CoAirbag always implies a driver airbag. So by adding "not 'CoAirbag' in body" to the where-clause we might exclude all trivial rules containing this item in the body.

We think our 'core' language is rather intuitive to use but actually upto this point its capabilities do not go beyond today's mining languages. The first feature we missed are aggregate functions on rule quality measures. In fact, it is often appropriate not to specify thresholds for the quality measures as absolute values but relative to the values reached by the generated rules. For example the following query retrieves rules having relatively low support – less than 1% higher than the lowest support in rulecache – and at the same time having a relatively high confidence – more than 99% of the highest confidence found in the cache. The minimum support respectively the maximum confidence value of all rules in rulecache are denoted by min(supp) and max(conf) (and vice versa). Other aggregate functions like average are also useful extensions:

```
SelectRulesFrom rulecache                          (Query 3)
Where supp < 1.01*min(supp) and conf > 0.99*max(conf);
```

Association mining algorithms treat items as literals and finally map them to integers. Whereas this restriction makes sense during rule generation we found it is not satisfying when retrieving rules from a cache. In brief, although items are literals from the rule generation point of view, in practical applications items normally have structure and we came to the conclusion that rule retrieval can greatly benefit from exploiting this structure. For example the items in a supermarket all have prices and costs associated with them. Similarly production dates, costs, manufacturers etc. are assigned to parts of vehicles. Such attributes can be considered through discretization and taxonomies. But actually the resulting quantitative [18] or generalized [17] rules are typically not what we want.

Formally we extend the basic framework of association mining as follows: let $\mathcal{I} \subseteq \mathrm{N} \times \mathrm{A}_1 \times \cdots \times \mathrm{A}_m$ be a set of items. Each item is uniquely identified by an ID $id \in \mathrm{N}$ and described by attributes $a_1, \ldots a_m \in \mathrm{A}_1 \times \ldots \times \mathrm{A}_m$. For example, one attribute may be the name of the item, another the price of the item, costs associated with it, or other application dependent information. A set of rules is then defined as $\mathcal{R} \subseteq \mathcal{P}(\mathcal{I}) \times \mathcal{P}(\mathcal{I}) \times \mathrm{R} \times \cdots \times \mathrm{R}$. As usually, in addition to the body and the head (subsets of the power set of $\mathcal{I}$), each rule is rated by a fixed number of real valued quality measures. Adding structure to the items in such a way does not affect the mining procedure but nevertheless introduces a new means to formulate practically important mining queries, as we will see.

Let x.attname denote the value of the attribute attname of item x. For example we may want to select all rules satisfying certain thresholds for support and confidence and 'explaining' an item of type special equipment SpEquip which incurs costs above 1000. Such queries now can be expressed through $\exists$ and $\forall$ quantifiers on itemsets:

```
SelectRulesFrom rulecache                                (Query 4)
Where supp > 0.25 and conf > 0.975
and exists x in head (x.type = 'SpEquip' and x.costs > 1000);
```

A more complex and also very useful query is to find all rules with at least one special equipment in the head that originates from a manufacturer who also manufactures at least one special equipment from the body:

```
SelectRulesFrom rulecache                                (Query 5)
Where exists x in head (x.type = 'SpEquip' and
exists y in body (y.type = 'SpEquip' and x.manu = y.manu))
and supp > 0.25 and conf > 0.975;
```

The quantifiers and attributes on the items and the aggregate functions on the rules are both intuitive to use and flexible. In the hands of the analyst a mining language enhanced to express queries considering the structure of the items is a powerful means to efficiently break down the result space. The examples above give a first impression of the potentials.

3 Postponing Restrict Operations on the Mining Data

The basic problem when caching association rules is the validity of the cache when the underlying data changes. In this section, we explain how to circumvent expensive regeneration in the practically very important case of data restrictions.

3.1 Restricting the Mining Data

Restriction in the sense of relational algebra means selecting a subset of the transactions for rule generation. This pre-processing task is quite common and often essential: for example, an analyst might restrict the transactions to a special vehicle model, because he is only interested in dependencies in this subset of the data. Or he might want to analyze each production year separately in order to compare the dependencies over the years. An example from the retail domain are separate mining runs for each day of the week. For example, after generating the rules for all baskets, the analyst might decide to have a closer look at rules that occur if only baskets from Saturdays are taken into account.

The problem with the restrict operation is that the quality measures of the rules change as soon as transactions are removed from the data. For example a rule may hold with low confidence on the set of all vehicles. When restricting the mining data to a subset, e.g. to the vehicles of the A-Class model, we might find the same rule but with much higher confidence. For instance, a stronger battery type in an A-Class vehicle might imply air conditioning with high confidence. In contrast, for the more luxurious Mercedes E-Class, there are many more reasons to implement a stronger battery type. Accordingly the stronger battery type need not imply necessarily air conditioning with high confidence. When postponing restrict operations from pre-processing to post-processing, the adapted values for the quality measures must be derived from the rules in the cache. In the following we show how to do this for fundamental quality measures.

3.2 Postponing Restrict Operations to Retrieval

We presume that attributes being employed for restriction of the mining data are contained as items in the transactions already during the initial rule generation. Such items describe the transactions, e.g. production date or vehicle model. They can be seen as pseudo items and must be distinguished from attributes that are attached to items, e.g. costs or manufacturer of a special equipment.

Let $\mathcal{D}'$ be a subset of $\mathcal{D}$ that is restricted to transactions containing a certain itemset R. The support of the itemset A in $\mathcal{D}'$ can be derived from the support values in $\mathcal{D}$ as follows:

$$\mathsf{supp}_{\mathcal{D}'}(A) = \mathsf{supp}_{\mathcal{D}}(A \cup R) \cdot \frac{|\mathcal{D}|}{|\mathcal{D}'|}$$

Therefore rule quality measures with respect to $\mathcal{D}'$ can be derived from the rule cache generated with respect to $\mathcal{D}$ by the following equations:

$$\mathsf{supp}_{\mathcal{D}'}(A \to B) = \mathsf{supp}_{\mathcal{D}'}(A \cup B) = \mathsf{supp}_{\mathcal{D}}(A \cup B \cup R) \cdot \frac{|\mathcal{D}|}{|\mathcal{D}'|}$$

$$\mathsf{conf}_{\mathcal{D}'}(A \to B) = \frac{\mathsf{supp}_{\mathcal{D}'}(A \cup B)}{\mathsf{supp}_{\mathcal{D}'}(A)} = \frac{\mathsf{supp}_{\mathcal{D}}(A \cup B \cup R)}{\mathsf{supp}_{\mathcal{D}}(A \cup R)}$$

$$\mathsf{lift}_{\mathcal{D}'}(A \to B) = \frac{\mathsf{supp}_{\mathcal{D}'}(A \cup B)}{\mathsf{supp}_{\mathcal{D}'}(A) \cdot \mathsf{supp}_{\mathcal{D}'}(B)} = \frac{\mathsf{supp}_{\mathcal{D}}(A \cup B \cup R)}{\mathsf{supp}_{\mathcal{D}}(A \cup R) \cdot \mathsf{supp}_{\mathcal{D}}(B \cup R)} \cdot \frac{|\mathcal{D}'|}{|\mathcal{D}|}$$

To give an illustrative example: let us restrict the mining data to the vehicles of the model E-Class by setting $R = \{\text{E-Class}\}$. Then the support of the rule

$\mathsf{AirCond} \rightarrow \mathsf{BatteryTypeC}$ that would have been generated when mining only on the E-Class-vehicles can be determined from the rule cache through:

$$\mathsf{supp}_{\mathcal{D}'}(\mathsf{AirCond} \rightarrow \mathsf{BatteryTypeC}) = \mathsf{supp}_{\mathcal{D}}(\{\mathsf{AirCond}, \mathsf{BatteryTypeC}, \mathsf{E\text{-}Class}\}) \cdot \frac{|\mathcal{D}|}{|\mathcal{D}'|}$$

Further quality measures can be derived similarly. But of course the cache contains only information to derive rules r with $\mathsf{supp}_{\mathcal{D}'}(r) \geq \mathsf{minsupp}_{\mathcal{D}'}$ with

$$\mathsf{minsupp}_{\mathcal{D}'} = \frac{|\mathcal{D}|}{|\mathcal{D}'|} \cdot \mathsf{minsupp}_{\mathcal{D}}.$$

This means, to derive rules at a reasonable threshold $\mathsf{minsupp}_{\mathcal{D}'}$, $\mathsf{minsupp}_{\mathcal{D}}$ must be quite low or R must occur comparably often in the data. Low $\mathsf{minsupp}_{\mathcal{D}}$ can easily be achieved in our scenario although drastically lowering this threshold may result in tremendous cache sizes, a potential performance problem concerning rule retrieval, c.f. Section 4.2. Fortunately the other condition turns out to be a minor problem in practical applications. It simply implies that the subset $\mathcal{D}'$ to which we restrict $\mathcal{D}$ must be a reasonable 'portion' of $\mathcal{D}$. For typical subsets we encountered, e.g. restrictions to a special model, production year, or a day of the week, this was always the case. For instance presuming approximately the same sales everyday, lowering $\mathsf{minsupp}_{\mathcal{D}}$ by factor 1/7 is a practical guess for restrictions on 'day of the week'.

4 The SM*ART*SKIP System

SM*ART*SKIP is a prototypical implementation of the ideas described in this paper. As a starting point, we took a collection of mining algorithms called *ART* – Association Rule Toolbox –, which proved their efficiency in several prior research projects, e.g. [7,8]. SM*ART*SKIP implements a comfortable platform for interactive association rule mining. Queries can be submitted on a command shell or through a web browser based GUI. In addition to the features described in Section 2, queries can also contain statements to start mining runs explicitly or to restrict the mining data as suggested in Section 3.

4.1 Implementation

Instead of implementing specialized data structures to hold the cache, we employ a relational database system for this purpose. We did that for two reasons. First, the mining results are typically generated at high cost. Storing them together with the data in the database is natural. Second, the implementation of the mining system benefits directly from the database system. Actually, we translate queries from our mining language to standard SQL and execute these SQL queries directly on the query engine of the database.

Although the generation of association rules from frequent itemsets (sets of items that occur at least with frequency $\mathsf{minsupp}$ in the data) is straightforward [1], we store rules instead of frequent itemsets for a good reason: Typically in our scenario there is no longer a significant difference between the number of frequent itemsets to be cached or the corresponding rules. We experienced that minsupp is rarely employed in the sense of a quality measure but as the only efficient means to reduce the response times of the mining algorithms. The cache releases us from tight runtime restrictions, so we expect the analyst to set minsupp to relatively low values and filter the numerous resulting rules by strict thresholds on the rest of the measures. We learned that as a consequence the number of frequent itemsets and the number of rules get more and more similar. In addition, for such low support thresholds, generating rules from the frequent itemsets can be quite costly. The reason is simply the great number of frequent itemsets that imply an even greater number of rules that need to be checked for satisfying the thresholds on the quality measures during rule generation.

Each rule to be stored consists of body, head and a vector of quality measures. An itemset may occur in several rules as body or head. In order to avoid redundancy, we keep the itemsets in a table separate from the rules and store each itemset only once. This saves a considerable amount of memory, e.g. in practical applications we experienced that memory usage decreased between 50% and upto more than 90%. But of course for each new itemset such a storage approach implies checking whether this itemset already exists or not before adding it to the database. This turned out to be rather inefficient when implemented on the database system. The reason is that itemsets are of arbitrary size and therefore are stored as (ID, Item-ID)-pairs. Fortunately our caches are always filled in a single pass with the complete result of one mining algorithm run. That means duplicates can be eliminated efficiently already outside the database.

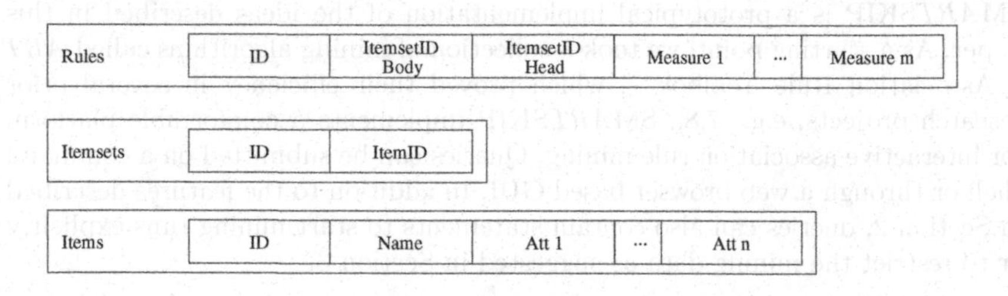

Fig. 1. Database tables that implement the rule cache.

The tables implementing the cache are given in Figure 1. For each rule in the rules table, a rule-ID and two itemset-IDs are stored. The latter identify body and head of the rule in the itemsets table. The values for the rule quality measures are added as floats. The pairs stored in the itemsets table consist of an ID to identify the itemset and an item-ID refering to the IDs in the items table. The latter table is application dependent and stores the attribute values further describing the items. (We regard the name of an item as an attribute.)

Now that the rules are stored in the database system we need to realize the access to the cache. For that purpose we translate the queries from our mining language into SQL queries. As mentioned before, at this point the usage of the database system pays off: implementing a translation unit is rather straight forward and executing the resulting SQL queries is handled entirely by the database query engine. For the realization we employ C++ and a parser generator. As an example the translation of Query 4 from Section 2 is given in Figure 2. The

```
CREATE VIEW extended_itemsets (itemsetid, itemid, type, costs)
AS (    SELECT Itemsets.id, itemid, type, costs
        FROM Itemsets INNER JOIN Items
        ON Itemsets.itemid = Items.id )

SELECT * FROM rules
WHERE   supp > 0.25 AND conf > 0.975
AND     itemsetidhead in (
        SELECT itemsetid FROM extended_itemsets
        WHERE type = 'spec equip' and costs > 1000 )
```

Fig. 2. Translation of Query 4 from Section 2 to SQL.

translated queries may look complicated to human but nevertheless are processed efficiently by the database system.

The number and the types of the attributes linked to the items and stored in the separate items table are of course application dependent. Therefore they are not hard wired into our software. In fact, during the compilation of the translation unit and during the translation of mining queries the types need not to be known. Types are not required until executing the translated queries on the database engine. At this point of course a table describing the attributes with their types must exist.

4.2 Evaluation

For our evaluation, we used the QUIS database – Quality Information System – at DaimlerChrysler. This database is a huge source of interesting mining scenarios, e.g. [8,9]. We consider mining dependencies between the special equipments installed in cars, together with additional attributes like model, production date, etc. We selected a set of 100 million database rows. The generation of association rules took upto more than four hours on a SUN ULTRASPARC-2 clocked at 400 Mhz. As in [6,16], we experienced that the time for rule generation was dominated by the overhead from database access. Obviously a rerun of the algorithm implies a considerable interruption of the analysis. So without rule caching, interactive mining of the data is impossible. We filled four separate caches at different levels of minsupp, containing about 10,000, 100,000, 200,000, and 300,000 rules. We found that rules with too many items hardly make sense in our domain and therefore restricted the maximal number of items per rule to four and to a single item in the head. In addition before submitting the queries from Section 2 we modified the thresholds for the quality measures to restrict the returned rules to less than a thousand rules for each of the caches. We think the retrieval of larger rule sets does not make sense. We employed an item with support of about 33% in the where-clause of Query 2. In Query 5 there are five different manufacturers uniformly distributed over the special equipments. In Figure 3, the response times for the Queries 1-5 from Section 2 on the different rule caches

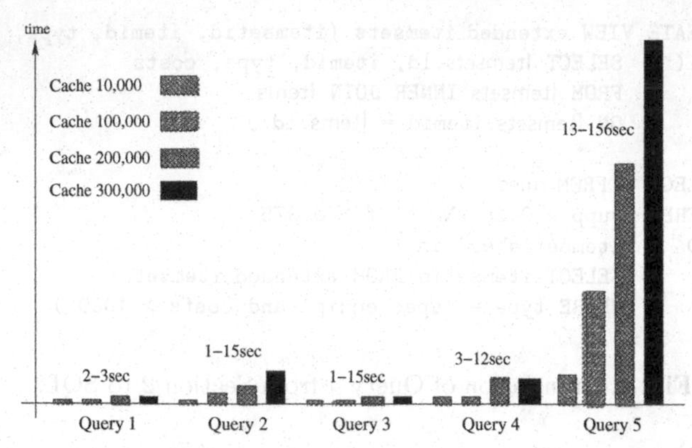

Fig. 3. Response times for Queries 1-5 from Section 2.1 in seconds.

are shown. We employed IBM's DB2 V7.1 running under Linux on a 500Mhz Pentium III for our experiments. The response times clearly satisfy the demands of interactive knowledge discovery.

It is important to note that, when the cache is filled, response times no longer depend on the size of the underlying data. Whereas algorithm runs scale at least linearly with the number of transactions, retrieval from the cache depends only on the number of cached rules. Our experiences show that typically a growing number of transaction does not imply a growing number of rules (constant thresholds presumed). Obviously, in the data we analyzed, frequent patterns are more or less uniformly distributed over all transactions. As a consequence retrieving rules from a cache generated from ten million transactions is not much worse than retrieving rules from a cache generated from one million transactions. The reason is simply that the sizes of the caches do not vary significantly.

For our experiments with postponed restricts we chose three different attributes for restriction that are contained as pseudo items in approximately 25%, 30% and 60% of the vehicles. Actually the achieved response times were nearly the same for all three attributes. Not surprisingly, execution times grow linearly with the number of cached rules. Restricting the four caches took about 20sec, 163sec, 319sec and 497sec respectively.

Although still much faster than a rerun of the algorithm, for very large caches the response times obviously suffer. We therefore explicitly treat the restrict operation separately from the query language. The idea is to always transform a complete rule set and to store the result of this operation also in the database. Then this result is not lost for further mining iterations but accessible through the query language like any other cached rule set.

5 Conclusion

In this paper we set out how to support user interactivity in association rule mining. Our basic idea is exactly the opposite of the common approach taken today: instead of improving response times by restrictions on the result sets, we accept one broad and possibly expensive algorithm run. This initial run fills a sophisticated rule cache. Answering refined search queries by retrieving rules from the cache takes only seconds typically whereas rerunning an algorithm implies several minutes upto hours of idle time for the analyst. In addition response times become independent from the number of underlying transactions.

However there are two critical aspects of rule caching. For both problems we presented a promising approach in this paper. First, the analyst needs a powerful means to be able to navigate in the rather large rule cache. For this purpose, we enhanced the concept of association mining. In brief, we introduced attributes to describe items, quantifiers on itemsets, and aggregate functions on rule sets. The queries becoming possible are powerful and practically relevant.

Second, without rerunning the mining algorithm, the cache must satisfy a broad variety of queries. A KDD process starts typically with the general case and comes down to the specifics in later phases. A common and often employed task is to restrict the mining data to subsets for further investigation, e.g. focus on a special vehicle model or on a special day of the week. Normally this implies a regeneration of all rules in the cache. We solved this problem and showed how to answer even those queries from the cache that specify rules that would have been generated if the mining data were restricted to a subset. For this purpose we do not need to rerun the algorithm or touch the mining data at all.

Finally, we presented the SM*ART*SKIP system, which implements the ideas introduced in this paper. An important aspect of SM*ART*SKIP is that it greatly benefits from its implementation based on a conventional relational database system. We evaluated it on a real database deployed at DaimlerChrysler. We found that our system is scalable and supports interactive data mining efficiently, even on very large databases.

References

1. R. Agrawal, T. Imielinski, and A. Swami. Mining association rules between sets of items in large databases. In *Proc. of the ACM SIGMOD Int'l Conf. on Management of Data (ACM SIGMOD '93)*, pages 207–216, Washington, USA, May 1993.
2. R. J. Brachman and T. Anand. The process of knowledge discovery in databases: A human centered approach. In U. M. Fayyad, G. Piatetsky-Shapiro, P. Smyth, and R. Uthurusamy, editors, *Advances in Knowledge Discovery and Data Mining*, chapter 2, pages 37–57. AAAI/MIT Press, 1996.

3. S. Brin, R. Motwani, J. D. Ullman, and S. Tsur. Dynamic itemset counting and implication rules for market basket data. In *Proc. of the ACM SIGMOD Int'l Conf. on Management of Data (ACM SIGMOD '97)*, pages 265–276, 1997.
4. U. Fayyad, G. Piatetsky-Shapiro, and P. Smyth. The KDD process for extracting useful knowledge from volumes of data. *Communications of the ACM*, 39(11):27–34, November 1996.
5. J. Han, Y. Fu, W. Wang, K. Koperski, and O. Zaiane. DMQL: A data mining query language for relational databases. In *Proc. of the 1996 SIGMOD Workshop on Research Issues on Data Mining and Knowledge Discovery (DMKD '96)*, Montreal, Canada, June 1996.
6. J. Hipp, U. Güntzer, and U. Grimmer. Integrating association rule mining algorithms with relational database systems. In *Proc. of the 3rd Int'l Conf. on Enterprise Information Systems (ICEIS 2001)*, pages 130–137 , Portugal, July 2001.
7. J. Hipp, U. Güntzer, and G. Nakhaeizadeh. Algorithms for association rule mining – a general survey and comparison. *SIGKDD Explorations*, 2(1):58–64, July 2000.
8. J. Hipp and G. Lindner. Analysing warranty claims of automobiles. an application description following the CRISP-DM data mining process. In *Proc. of 5th Int'l Computer Science Conf. (ICSC '99)*, pages 31–40, Hong Kong, China, December 13-15 1999.
9. E. Hotz, G. Nakhaeizadeh, B. Petzsche, and H. Spiegelberger. Waps, a data mining support environment for the planning of warranty and goodwill costs in the automobile industry. In *Proc. of the 5th Int'l Conf. on Knowledge Discovery and Data Mining (KDD '99)*, pages 417–419, San Diego, California, USA, August 1999.
10. T. Imielinski, A. Virmani, and A. Abdulghani. Data mining: Application programming interface and query language for database mining. In *Proc. of the 2nd Int'l Conf. on Knowledge Discovery in Databases and Data Mining (KDD '96)*, pages 256–262, Portland, Oregon, USA, August 1996.
11. T. Imielinski, A. Virmani, and A. Abdulghani. DMajor - application programming interface for database mining. *Data Mining and Knowledge Discovery*, 3(4):347–372, December 1999.
12. M. Klemettinen, H. Mannila, and H. Toivonen. Interactive exploration of discovered knowledge: A methodology for interaction, and usability studies. Technical Report C-1996-3, University Of Helsinki, Department of Computer Science, P.O. 26, 1996
13. B. Liu, M. Hu, and W. Hsu. Multi-level organisation and summarization of the discovered rules. In *Proc. of the 6th ACM SIGKDD Int'l Conf. on Knowledge Discovery and Data Mining (KDD '00)*, pages 208–217, Boston, MA USA, August 20-23 2000.
14. R. Meo, G. Psaila, and S. Ceri. A new sql-like operator for mining association rules. In *Proc. of the 22nd Int'l Conf. on Very Large Databases (VLDB '96)*, Mumbai (Bombay), India, September 1996.
15. R. Ng, L. S. Lakshmanan, J. Han, and T. Mah. Exploratory mining via constrained frequent set queries. In *Proc. of the 1999 ACM-SIGMOD Int'l Conf. on Management of Data (SIGMOD'99)*, pages 556–558, Philadelphia, PA, USA, June 1999.
16. S. Sarawagi, S. Thomas, and R. Agrawal. Integrating association rule mining with relational database systems: Alternatives and implications. *SIGMOD Record (ACM Special Interest Group on Management of Data)*, 27(2):343–355, 1998.
17. R. Srikant and R. Agrawal. Mining generalized association rules. In *Proc. of the 21st Conf. on Very Large Databases (VLDB '95)*, Zürich, Switzerland, Sept. 1995.

18. R. Srikant and R. Agrawal. Mining quantitative association rules in large relational tables. In *Proc. of the 1996 ACM SIGMOD Int'l Conf. on Management of Data (SIGMOD '96)*, Montreal, Canada, June 1996.
19. R. Srikant, Q. Vu, and R. Agrawal. Mining association rules with item constraints. In *Proc. of the 3rd Int'l Conf. on KDD and Data Mining (KDD '97)*, Newport Beach, California, August 1997.
20. R. Wirth and J. Hipp. CRISP-DM: Towards a standard process modell for data mining. In *Proc. of the 4th Int'l Conf. on the Practical Applications of Knowledge Discovery and Data Mining*, pages 29–39, Manchester, UK, April 2000.

Association Rule Mining on Remotely Sensed Images Using P-trees*

Qin Ding, Qiang Ding, and William Perrizo

Department of Computer Science, North Dakota State University,
Fargo, ND 58105-5164, USA
{qin.ding, qiang.ding, william.perrizo}@ndsu.nodak.edu

Abstract. Association Rule Mining, originally proposed for market basket data, has potential applications in many areas. Remote Sensed Imagery (RSI) data is one of the promising application areas. Extracting interesting patterns and rules from datasets composed of images and associated ground data, can be of importance in precision agriculture, community planning, resource discovery and other areas. However, in most cases the image data sizes are too large to be mined in a reasonable amount of time using existing algorithms. In this paper, we propose an approach to derive association rules on RSI data using Peano Count Tree (P-tree) structure. P-tree structure, proposed in our previous work, provides a lossless and compressed representation of image data. Based on P-trees, an efficient association rule mining algorithm P-ARM with fast support calculation and significant pruning techniques are introduced to improve the efficiency of the rule mining process. P-ARM algorithm is implemented and compared with FP-growth and Apriori algorithms. Experimental results showed that our algorithm is superior for association rule mining on RSI spatial data.

1 Introduction

Association rule mining [1,2,3,4,11,12,19] is one of the important advances in the area of data mining. The initial application of association rule mining was on market basket data. Recently study on association rule mining has been extended to more areas, such as multimedia data [8]. An association rule is a relationship of the form X=>Y, where X and Y are sets of items. X is called the antecedent and Y the consequence. An example of the rule can be, "customers who purchase an item X are very likely to purchase another item Y at the same time". There are two primary quality measures for each rule, support and confidence. The rule X=>Y has support s% in the transaction set D if s% of transactions in D contain $X \cup Y$. The rule has confidence c% if c% of transactions in D that contain X also contain Y. The goal of association rule mining is to find all the rules with support and confidence exceeding user specified thresholds.

Remotely Sensed Imagery (RSI) data is one of the promising data areas for application of association rule mining techniques. The quantities of RSI data being collected every day from satellites, aerial sensors, telescopes and other sensor

* Patents are pending on the P-tree technology. This work is partially supported by GSA Grant ACT#: K96130308, NSF Grant OSR-9553368 and DARPA Grant DAAH04-96-1-0329.

M.-S. Chen, P.S. Yu, and B. Liu (Eds.): PAKDD 2002, LNAI 2336, pp. 66-79, 2002.

platforms are so huge that much of this data is archived before its value can be determined. Extracting the interesting rules from these datasets, in combination with other data such as ground and weather data, can provide tremendous benefits. Application areas include precision agriculture; community planning; resource discovery and management; and natural disaster prediction, detection and mitigation to mention just a few. For example, in precision agriculture, association rules can be mined from RSI data to identify crop yield potential, insect and weed infestations, nutrient requirements, and flooding damage and other phenomena. In this paper, we use as an example, the derivation of association rules from RSI data to identify high and low agricultural crop yield potential. In what is called precision agriculture, RSI data is used in mid growing season to determine where additional inputs (fertilizers, herbicides, etc.) can be effectively applied to raise the likelihood of high yield. This application serves as a good example of situations in which RSI data can be mined to alter future outcomes in a timely fashion.

An RSI image can be viewed as a 2-dimensional array of pixels. Associated with each pixel are various descriptive attributes, called "bands" in remote sensing literature [18]. For example, visible reflectance bands (Blue, Green and Red), infrared reflectance bands (e.g., NIR, MIR1, MIR2 and TIR) and possibly some bands of data gathered from ground sensors (e.g., yield quantity, yield quality, and soil attributes such as moisture and nitrate levels, etc.). All the values have been scaled to values between 0 and 255 for simplicity. The pixel coordinates in raster order constitute the key attribute. One can view such data as a relational table where each pixel is a tuple and each band is an attribute. In this paper we focus on the task of deriving association rules in which yield is specified as the rule consequent. The rule, NIR[192,255] ^ Red[0,63] => Yield[128, 255], which is read "Near Infrared reflectance at least 192 and Red reflectance at most 63 implies Yield will be at least 128 (e.g., bushel/acre or some normalized yield measurement)" is the type of rule expected. Such rules are useful to both producers and agribusiness communities. If low yield is predicted early in the growing year, the producer can apply additional inputs (e.g., water and nitrogen) to alter the Red and NIR and produce higher yield potential. For the agribusiness community, wide area yield estimation can improve future price prediction.

Existing algorithms do not scale well to this kind of task due to the amount of data in these images. Therefore, we propose an efficient model to perform association rule mining on RSI data. We use a bit Sequential (bSQ) format [15] to organize images and the Peano Count Tree (P-tree) structure [15] to represent bSQ files in a spatial-data-mining-ready way. P-trees are lossless representation of the image data and its histograms are in a recursive quadrant-by-quadrant arrangement. By using P-trees, association rule mining algorithm with fast support calculation and significant pruning techniques are possible.

The paper is organized as follows. Section 2 summarizes the bSQ data format and the P-tree structure. Section 3 details how to derive association rules using P-trees and related pruning techniques. Experiment results and performance analysis are given in Section 4. Section 5 gives the related work and discussion. The conclusions and future work are given in Section 6.

2 Peano Count Tree (P-tree)

2.1 RSI Data Formats

The concept of remotely sensed imagery covers a broad range of methods to include satellites, aerial photography, and ground sensors. A remotely sensed image typically contains several bands or columns of reflectance intensities. For example, Land satellite Thematic Mapper (TM) scenes contain at least seven bands (Blue, Green, Red, NIR, MIR, TIR and MIR2) while a TIFF image contains three bands (Red, Green and Blue). Each band contains a relative reflectance intensity value in the range 0-to-255 for each pixel. Ground data are collected at the surface of the earth and can be organized into images. For example, yield data can be organized into a yield map. Fig. 1 gives a TIFF image and related yield map.

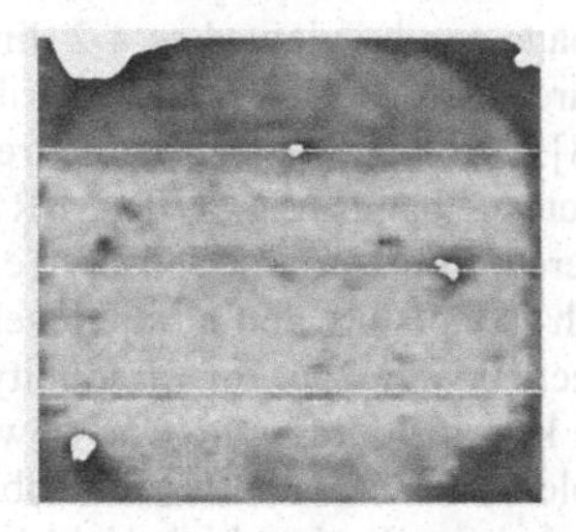

Fig. 1. TIFF image (29NW082598.tiff) and related Yield Map

RSI data are collected in different ways and are organized in different formats. In our previous work [15], we proposed a format, called bit Sequential (bSQ), to organize spatial data. A reflectance value in a band is a number in the range 0-255 and is represented as a byte. We split each band into eight separate files, one for each bit position. There are several advantages of using bSQ format. The bSQ format facilitates the representation of a precision hierarchy (from one bit precision up to eight bit precision). It also facilitates better compression. In image data, close pixels may have similar properties. By using bSQ format, close pixels may share the same bit values in high order bits. This facilitates high compression for high order bit files and brings us the idea of creating P-trees.

2.2 Basic P-trees

In this section, we summarize the P-tree structure and its operations. A P-tree is a quadrant-wise, Peano-order-run-length-compressed, representation of each bSQ file [15]. The idea is to recursively divide the entire image into quadrants and record the count of 1-bits for each quadrant, thus forming a quadrant count tree. For example, given an 8×8 bSQ file, its P-tree is as shown in Fig. 2.

In this example, 39 is the number of 1's in the entire image, called root count. The root level is labeled level 0. The numbers 16, 8, 15, and 0 at the next level (level 1) are the 1-bit counts for the four major quadrants in raster order. Since the first and last level-1 quadrants are composed entirely of 1-bits (called pure-1 quadrant) and 0-

bits (call pure-0 quadrant) respectively, sub-trees are not needed and these branches terminate. This pattern is continued recursively using the Peano (Z-ordering) of the four sub-quadrants at each new level. Eventually, every branch terminates. If we were to expand all sub-trees, including those for pure quadrants, the leaf sequence would be the Peano-ordering of the bSQ file.

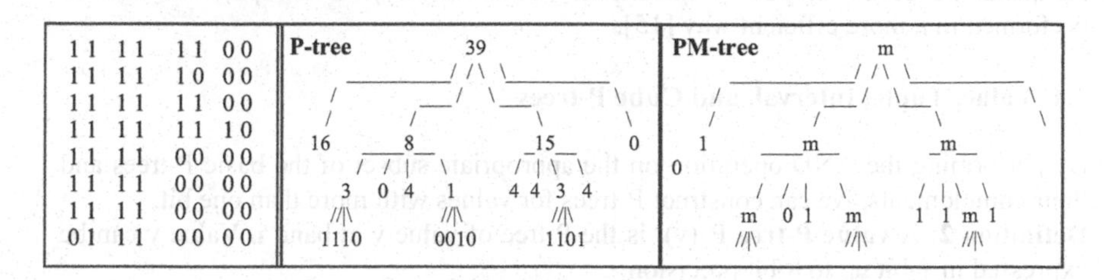

Fig. 2. P-tree and PM-tree for an 8×8 bSQ file

For each band (assuming 8-bit data values, though the model applies to data of any number bits), there are eight P-trees, one for each bit position.

Definition 1: A **basic P-tree** $P_{i,j}$ is a P-tree for the jth bit of the ith band.

A variation of the P-tree data structure, the Peano Mask Tree (PM-tree), is a similar structure in which masks rather than counts are used. In a PM-tree, we use a 3-value logic to represent pure-1, pure-0 and mixed (or called non-pure) quadrants (1 denotes pure-1, 0 denotes pure-0 and m denotes mixed).

2.3 P-tree Operations

There are three basic P-tree operations: complement, AND and OR. The complement of a basic P-tree can be constructed directly from the P-tree by simply complementing the counts at each level, as shown in the example below (Fig. 3). The complement of basic P-tree $P_{i,j}$ is denoted as $P_{i,j}$. Note that the complement of a P-tree provides the 0-bit counts for each quadrant.

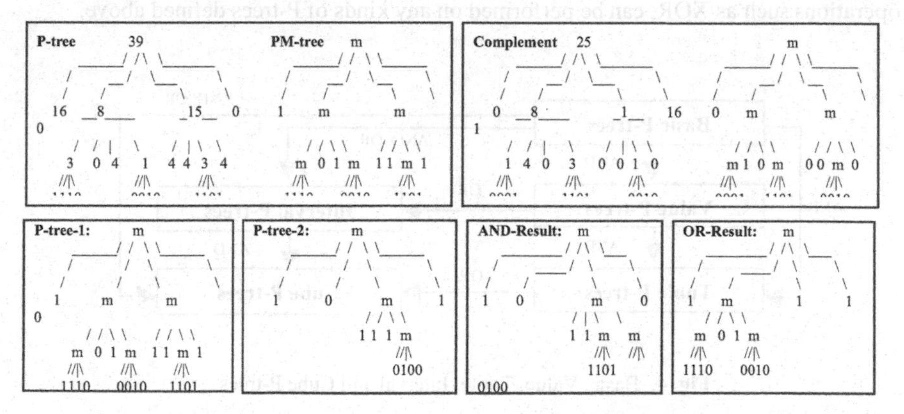

Fig. 3. P-tree Operations (Complement, AND and OR)

P-tree AND/OR operations are also illustrated in Fig. 3. AND is a very important and frequently used operation for P-trees. It can be performed in a very straightforward way. For example, to AND a pure-1 P-tree with any P-tree X will result in X; to AND a pure-0 P-tree with any P-tree will result in a pure-0 P-tree; to AND two non-pure P-trees will result in a non-pure P-tree unless all of the four subquadrants result in pure-0 quadrants. Alternatively, AND operation can be performed in a more efficient way [15].

2.4 Value, Tuple, Interval, and Cube P-trees

By performing the AND operation on the appropriate subset of the basic P-trees and their complements, we can construct P-trees for values with more than one bit.

Definition 2: A **value P-tree** $P_i(v)$, is the P-tree of value v at band i. Value v can be expressed in 1-bit up to 8-bit precision.

Value P-trees can be constructed by ANDing basic P-trees or their complements. For example, value P-tree $P_i(110)$ gives the count of pixels with band-i bit 1 equal to 1, bit 2 equal to 1 and bit 3 equal to 0, i.e., with band-i value in the range of [192, 224). It can be constructed from the basic P-trees as:

$$P_i(110) = P_{i,1} \textbf{ AND } P_{i,2} \textbf{ AND } P_{i,3}'$$

Definition 3: A **tuple P-tree** $P(v_1, v_2, \ldots, v_n)$, is the P-tree of value v_i at band i, for all i from 1 to n. So, we have,

$$P(v_1, v_2, \ldots, v_n) = P_1(v_1) \textbf{ AND } P_2(v_2) \textbf{ AND } \ldots \textbf{ AND } P_n(v_n)$$

If value v_j is not given, it means it could be any value in Band j.

Definition 4: An **interval P-tree** $P_i(v_1, v_2)$, is the P-tree for value in the interval of $[v_1, v_2]$ of band i. Thus, we have,

$$P_i(v_1, v_2) = \textbf{OR } P_i(v), \text{ for all v in } [v_1, v_2].$$

Definition 5: A **cube P-tree** $P([v_{11}, v_{12}], [v_{21}, v_{22}], \ldots, [v_{N1}, v_{N2}])$, is the P-tree for value in the interval of $[v_{i1}, v_{i2}]$ of band i, for all i from 1 to N.

Any value P-tree and tuple P-tree can be constructed by performing ANDing on basic P-trees and their complements. Interval and cube P-trees can be constructed by combining AND and OR operations of basic P-trees (Fig. 4). All the P-tree operations, including basic operations AND, OR, COMPLEMENT and other operations such as XOR, can be performed on any kinds of P-trees defined above.

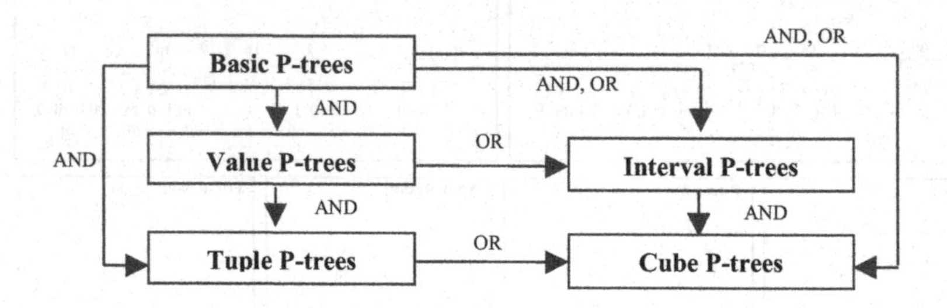

Fig. 4. Basic, Value, Tuple, Interval and Cube P-trees

3 Association Rule Mining on Spatial Data Using P-trees

3.1 Data Partitioning for Numeric Data

To perform association rule mining on image data using the terminology of association rule mining on market basket data ([1,2,3]), each pixel is a transaction while each band is an attribute. Every value from 0 to 255 in each band can be considered as an item (e.g., for a four-band image, the total number of items will be 1024). In this case, the mining process is expensive because of the number of items, itemsets and the database size. In addition, considering every value in each band as an item, we will get rules like "NIR=210 ^ Red=55 => Yield=188" which is more specific rather than general.

To solve this problem, we can partition the numeric data into intervals to reduce the number of items. Partitioning will also produce more interesting and general rules by using intervals instead of single values in the rules. There are several ways to partition the data, such as equi-length, equi-depth, and user-defined partitioning. Equi-length partitioning is a simple but useful method. By truncating some of the right-most bits of the values (low order or least significant bits), we can reduce the size of the itemsets dramatically without losing much information since the low order bits show only subtle differences. For example, we can truncate the right-most six bits, resulting in the set of values {00, 01, 10, 11}. Each of these values represents an equi-length partitioning of the original 8-bit value space (i.e., 00 represents the values in [0,64), 01 represents the values in [64,128), etc.).

In some cases, it would be better to allow users to partition the value space into uneven partitions. Domain knowledge can be applied in partitioning. For example, band B_i can be partitioned into {[0,32), [32,64) [64,96), [96,256)}, if it is known that there will be only a few values between 96 and 255. Applying the user's domain knowledge increases accuracy and data mining efficiency. This type of partitioning will be referred to as user-defined partitioning. Equi-depth partitioning is another type of partitioning in which each partition has approximately the same number of pixels.

Whether partitioning is equi-length, equi-depth or user-defined, appropriate P-trees can be generated as follows. For each band, choose n partition-points, forming n+1 intervals. One way we can do is to replace every value in each band with its interval number in the range [0, n]. Then P-trees are built on these intervalized data. For example, after partitioning data in one band into eight intervals, only three basic P-trees are needed for this band. In the case of equi-length partition, these three basic P-trees are exactly the same as the original basic P-trees for the first three bits.

3.2 Deriving Association Rules Using P-trees

For RSI data, we can formulate the model of association rule mining as follows. Let I be the set of all items and T be the set of all transactions. I = {(b,v) | b=band, v=value (1-bit, 2-bit,..., or 8-bit)}, T = {pixels}.

Definition 6: Admissible Itemsets (Asets) are itemsets of the form, $Int_1 \times Int_2 \times ... \times Int_n = \Pi_{i=1..n} Int_i$, where Int_i is an interval of values in $Band_i$ (some of which may be unrestricted, i.e., [0,255]). A k-band Aset (k-Aset) is an Aset in which k of the Int_i intervals are restricted (i.e., in k of the bands the intervals are not all of [0,255]).

We use the notation $[a, b]_i$ for the interval [a, b] in band i. For example, $[00, 01]_2$ indicates the interval [00, 01] (which is [0, 127] in decimal) in band 2.

The user may be interested in some specific kinds of rules. For an agricultural producer using precision techniques, there is little interest in rules of the type, Red>48 => Green<134. A physicist might be interested in such color relationships (both antecedent and consequent from color bands), but a producer is interested in rules with color antecedents and, e.g., yield consequents (i.e., observed color combinations that predict high yield or foretell low yield). Therefore, for precision agriculture applications, it makes sense to restrict our search to those rules that have a consequent in the yield band. We will refer to rule restrictions in this type as restriction to rules ***of interest***, as distinct from interesting rules. Of-interest rules can be interesting or not interesting, depending on such measures as support and confidence.

Based on the Apriori algorithm [2], we first find all itemsets that are frequent and of-interest (e.g., if B_1 = Yield). The user may wish to restrict attention to those Asets for which Int_1 is not all of B_1, either high-yield or low-yield. For 1-bit data values, this means either yield < 128 or Yield ≥ 128. Other threshold values can be selected using the user-defined partitioning described above.

For a frequent Aset, $B = \Pi_{i=1..n} Int_i$, rules are created by partitioning {1..n} into two disjoint sets, $\hat{A}=\{i_1..i_m\}$ and $\hat{C}=\{j_1..j_q\}$, q+m=n, and then forming the rule, A=>C where $A = \Pi_{k\in \hat{A}} Int_k$ and $C = \Pi_{k\in \hat{C}} Int_k$. As noted above, users may be interested only in rules where q=1 and therefore the consequents come from a specified band (e.g., B_1=Yield). Then there is just one rule of interest for each frequent set found and it needs only be checked for confidence.

We start by finding all frequent 1-Asets. The candidate 2-Asets are those made up of frequent 1-Asets. The candidate k-Asets are those whose (k-1)-Aset subsets are frequent. Various pruning techniques can be applied during this process based on the precision hierarchy and P-trees. Details are given later in this section. Fig. 5 gives the P-ARM algorithm for mining association rules on RSI data using P-trees.

```
Procedure P-ARM
{
    Data Discretization;
    F_1 = {frequent 1-Asets};
    For (k=2; F_k-1 ≠∅) do begin
        C_k = p-gen(F_k-1);
        Forall candidate Asets c ∈ C_k do
            c.count = rootcount(c);
        F_k = {c∈C_k | c.count >= minsup}
    end
    Answer = ∪k F_k
}
```

Fig. 5. P-ARM algorithm

```
insert into C_k
    select p.item_1, p.item_2, …, p.item_k-1,
        q.item_k-1
    from F_k-1 p, F_k-1 q
    where p.item_1 = q.item_1, …, p.item_k-2 =
        q.item_k-2, p.item_k-1 < q.item_k-1,
        p.item_k-1.group <> q.item_k-1.group
```

Fig. 6. Join step in *p-gen* function

The P-ARM algorithm assumes a fixed precision, for example, 3-bit precision in all bands. In the Apriori algorithm, there is a function called "*apriori-gen*" [2] to generate candidate k-itemsets from frequent (k-1)-itemsets. The ***p-gen*** function in the P-ARM algorithm differs from the *apriori-gen* function in the way pruning is done. We use band-based pruning in the *p-gen* function. Since any itemsets consisting of two or more intervals from the same band will have zero support (no value can be in both intervals simultaneously), for example, support ($[00,00]_1 \times [11,11]_1$) = 0, the kind of joining done in [2] is not necessary. We put different items in different groups based on bands. Each item has an associated group ID. *P-gen* only joins items from different groups. The join step in the *p-gen* function is given in Fig. 6.

The ***rootcount*** function is used to calculate Aset counts directly by ANDing the appropriate basic P-trees instead of scanning the transaction databases. For example, in the Asets, {B1[0,64), B2[64,127)}, denoted as $[00,00]_1 \times [01,01]_2$, the count is the root count of $P_1(00)$ AND $P_2(01)$. This provides fast support calculation and is particularly useful for large data sets. It eventually improves the entire mining performance significantly.

3.3 Pruning Techniques

In the above algorithm, several pruning techniques have been applied or provided as options.

- **Band-based pruning**

We already mentioned that band-based pruning is used in *p-gen* function. This avoids unnecessary join operations among intervals from the same band.

- **Consequence constrain pruning**

We are only interested in of-interest rules with specified consequence, such as yield. Therefore, we only consider frequent itemsets with item in yield. This saves time by not considering the itemsets without yield even if the itemsets are frequent.

- **Bit-based pruning for multi-level rules**

There may be interest in multi-level rules [11], which means the different itemsets in the rule can have different precision in our case. A bit-based pruning technique can be applied in this case. The basic idea of bit-based pruning is that, if Aset $[1,1]_2$ (the interval [1,1] in band 2) is not frequent, then the Asets $[10,10]_2$ and $[11,11]_2$ which are covered by $[1,1]_2$ cannot possibly be frequent.

Based on this we start from 1-bit precision, and try to find all the frequent itemsets. Once we find all 1-bit frequent k-Asets, we can use the fact that a 2-bit k-Aset cannot be frequent if its enclosing 1-bit k-Aset is infrequent. A 1-bit Aset encloses a 2-bit Aset if when the endpoints of the 2-bit Aset are shifted right 1-bit position, it is a subset of the 1-bit Aset. This can help to prune out the infrequent items early.

- **Root count boundary pruning**

To determine if a candidate Aset is frequent or not, we need to AND appropriate P-trees to get the root count. In fact, without performing AND operations, we can tell the boundaries for the root count by looking at the root counts of two P-trees. Suppose we have two P-trees for 8×8 bit files, with the first P-tree having root count 28 and the level-1 count 16, 12, 0 and 1, and the second P-tree having root count 36 and the level-1 count 1, 5, 16 and 14. By looking at the root level, we know the root count of ANDing result will be at most 28. If we go one more level, we can say the root count will be at most 7, calculated by min(16,1)+min(12,5)+min(0,16)+min(1,14), where min(x,y) gives the minimum of x and y. If the support threshold is 30%, the corresponding A-set will not be frequent since $7/64 < 0.3$. We provide options to specify the number of levels, from 0 to 3, to check for the boundaries before performing the ANDing.

- **Adjacent interval combination using interval and cube P-trees**

P-ARM algorithm also provides an option to combine the adjacent intervals depending on user's requirements [3]. We can use interval P-trees and cube P-trees to calculate the support after combining the adjacent intervals. Fast algorithms can be used to quickly calculate the root count for interval and cube P-trees from basic P-trees. The basic idea is the root count of the AND of two P-trees with non-overlapping interval values will be the sum of root counts of these two P-trees. This avoids a lot of OR operations.

Max-support measurement proposed in [3] is used here to stop the further combination of adjacent intervals to avoid non-interesting and redundant rules.

3.4 An Example

The following simple example (Table 1) is used to illustrate the method. The data contains four bands with 4-bit precision in the data values, in which B1 is Yield. We are interested to find multi-level rules so bit-based pruning technique will be applied in this example.

Field Coords (X, Y)	Class Label	Remotely Sensed Reflectance Bands		
	Yield	Blue	Green	Red
0,0	0011	0111	1000	1011
0,1	0011	0011	1000	1111
0,2	0111	0011	0100	1011
0,3	0111	0010	0101	1011
1,0	0011	0111	0100	1011
1,1	0011	0011	0100	1011
1,2	0111	0011	1100	1011
1,3	0111	0010	1101	1011
2,0	0010	1011	1000	1111
2,1	0010	1011	1000	1111
2,2	1010	1010	0100	1011
2,3	1111	1010	0100	1011
3,0	0010	1011	1000	1111
3,1	1010	1011	1000	1111
3,2	1111	1010	0100	1011
3,3	1111	1010	0100	1011

Table 1. Sample Data

The data are first converted to bSQ format. The Band 1 bit-bands as well as the basic P-trees are given in Fig. 7.

```
B11    B12    B13    B14       P1,1      P1,2      P1,3   P1,4
0000   0011   1111   1111      5         7         16     11
0000   0011   1111   1111      0 0 1 4   0 4 0 3          4 4 0 3
0011   0001   1111   0001      0001      0111             0111
0111   0011   1111   0011
```

Fig. 7. Basic P-trees for sample data Band 1

Assume the minimum support is 60% (requiring a count of 10) and the minimum confidence is 60%. First, we find all 1-Asets for 1-bit values from B_1. There are two possibilities for Int_1, $[1,1]_1$ and $[0,0]_1$, with support($[1,1]_1$) = 5 (infrequent) while support($[0,0]_1$) = 11 (frequent). Similarly, there are two possibilities for Int_2 with support($[1,1]_2$) = 8 (infrequent) and support($[0,0]_2$) = 8 (infrequent), two possibilities for Int_3 with support($[1,1]_3$) = 8 (infrequent) and support($[0,0]_3$) = 8 (infrequent), and two possibilities for Int_4 with support($[1,1]_4$) = 16 (frequent) and support($[0,0]_4$) = 0 (infrequent).

The set of 1-bit frequent 1-Asets, is $\{[0,0]_1, [1,1]_4\}$. So the set of 1-bit candidate 2-Asets, is $\{[0,0]_1 \times [1,1]_4\}$ with support 11 (root-count of $P_{1,0}$ & $P_{4,1}$) and therefore, the set of 1-bit frequent 2-Asets is $\{[0,0]_1 \times [1,1]_4\}$. The set of 1-bit candidate 3-Asets is empty.

We use Yield (B_1) as the consequent. The rule that can be formed with B_1 as the consequent is $[1,1]_4 \Rightarrow [0,0]_1$ (rule support = 11). The confidence of this rule is 11/16 (68%). Thus, this is a strong rule.

Now consider the 2-bit case. The frequent 1-bit 1-Asets were $[0,0]_1$ and $[1,1]_4$ and the other 1-bit 1-Asets are infrequent. This means all their enclosed 2-bit subintervals are infrequent. The interval $[00,01]_1$ is identical to $[0,0]_1$ in terms of the full 8-bit values that are included, and $[00,10]_1$ is a superset of $[0,0]_1$, so both are frequent. Others in band-1 to consider are: $[00,00]_1$, $[01,01]_1$, $[01,10]_1$ and $[01,11]_1$. $[00,00]_1$ is infrequent (using $P_1(00)$, count=7). $[01,01]_1$ is infrequent (using $P_1(01)$, count=4). For $[01,10]_1$ we use $P_1(01)$ OR $P_1(10)$. The root count of $P_1(01)$ OR $P_1(10)$ is 6 and therefore $[01,10]_1$ is infrequent. The root count of $P_1(01)$ OR $P_1(10)$ OR $P_1(11)$ is 9 and therefore $[01,11]_1$ is infrequent. The only new frequent 2-bit $band_1$ 1-Aset is $[00,10]_1$, which does not form the support set of a rule.

Similarly, we can continue the process to the 3-bit case depending on the user's requirement.

4 Experiment Results and Performance Analysis

In this section, we compare our work with the Apriori algorithm [2], and a recently proposed efficient algorithm, FP-growth [12], which does not have the candidate generation step. The experiments are performed on a 900-MHz PC with 256 megabytes main memory, running Windows 2000. We set our algorithm to find all the frequent itemsets, not just those of interest (e.g., containing Yield) for fairness. We got identical rules by running Apriori, FP-growth and P-ARM algorithms. The images used were actual aerial TIFF images with a synchronized yield band. The data are available on [9]. In our performance study, each dataset has four bands {Blue, Green, Red, Yield}. We used different image sizes up to 1320×1320 pixels.

We only store the basic P-trees for each dataset. All other P-trees (value P-trees and tuple P-trees) are created in real time as needed. This results in a considerable saving of space.

4.1 Comparison of the P-ARM with Apriori

We implemented the Apriori algorithm [2] for the TIFF-Yield datasets using equi-length partitioning. For example, using 3-bit precision, there will be 8 items (intervals) in each band for a total of 32 items. Each pixel is a transaction. For an image size 1320×1320, the total number of transactions will be 1.7M. Experiments have been done on three datasets [9], having the very similar results with regard to performance.

The P-ARM algorithm is more scalable than Apriori in two ways. First, P-ARM is more scalable for lower support thresholds. The reason is that, for low support thresholds, the number of candidate itemsets will be extremely large. Thus, candidate frequent itemset generation performance degrades markedly. Fig. 8 compares the results of the P-ARM algorithm and Apriori for different support thresholds.

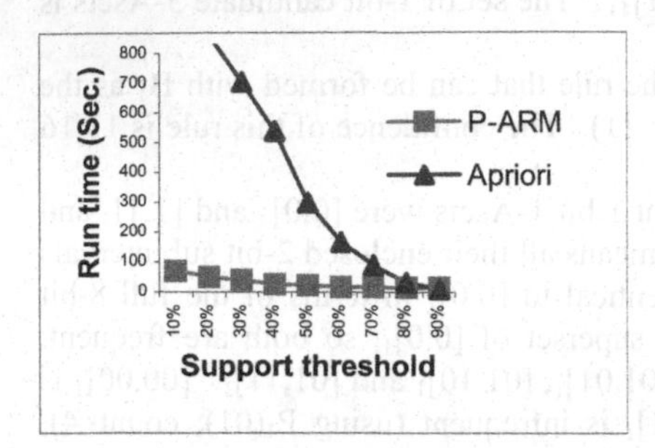

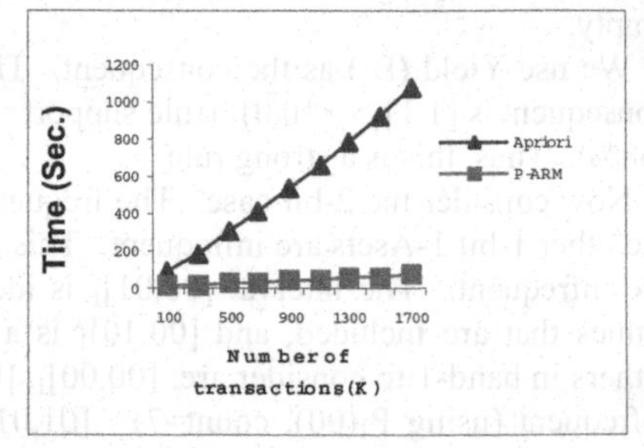

Fig. 8. Scalability with support threshold **Fig. 9.** Scalability with number of transactions

The second conclusion is that the P-ARM algorithm is more scalable for large spatial datasets. In the Apriori algorithm we need to scan the entire database each time a support is to be calculated. This has a high cost for large databases. However, in P-ARM, we calculate the count directly from the root count of a basic P-tree and the AND program. When dataset size is doubled, only one more level is added to each basic P-tree. The additional cost is relatively small compared to the Apriori algorithm as shown in Fig. 9.

4.2 Comparison of the P-ARM Algorithm and the FP-growth Algorithm

FP-growth is an efficient algorithm for association rule mining. It uses a data structure called frequent pattern tree (FP-tree) to store compressed information about frequent patterns. We use the FP-growth object code and convert the image to the required file format in which each item is identified by an item ID. For a dataset of 100K bytes, FP-growth is fast. However, when we run the FP-growth algorithm on the TIFF image of large size, the performance falls off markedly. For large sized datasets and low support thresholds, it takes longer for FP-growth to run than P-ARM.

Fig. 10 shows the experimental result of running the P-ARM and FP-growth algorithms on a 1320×1320 pixel TIFF-Yield dataset (in which the total number of transactions is about 1,700,000). In these experiments, we have used 2-bits precision and equi-length partitioning.

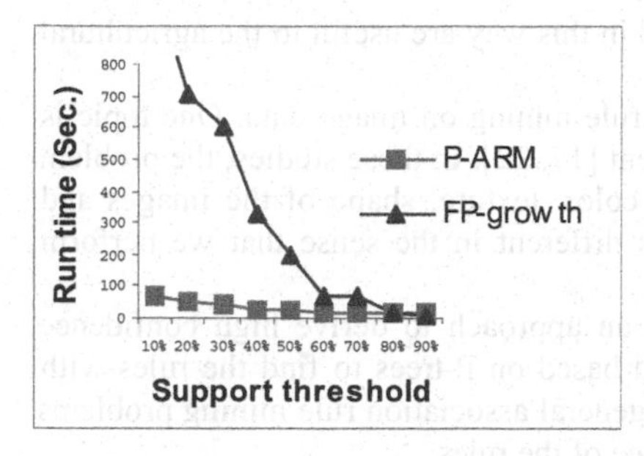

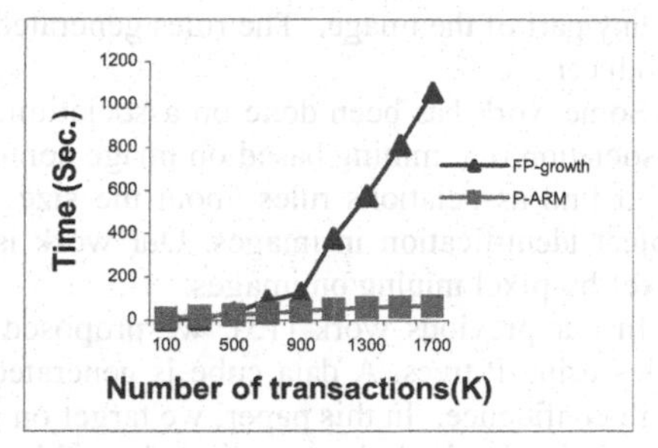

Fig. 10. Scalability with support threshold **Fig. 11.** Scalability with number of transactions

From the above figures we can see that both the P-ARM and the FP-growth algorithms run faster than the Apriori algorithm for most cases. For large image datasets and low support threshold, the P-ARM algorithm runs faster than the FP-tree algorithm.

Our test suggests that the FP-growth algorithm runs quite fast for datasets with transactions numbering less than 500,000 ($|D| < 500K$). For the larger datasets, the P-ARM algorithm gives a better performance. This result is presented in Fig. 11.

Fig. 12 shows how the number of precision bits used affects the performance of the P-ARM algorithm. The more precision bits are used, the greater the number of items, the longer the mining time. By adjusting the number of precision bits, we can balance the trade-off between the number of rules discovered and the quality of the rules.

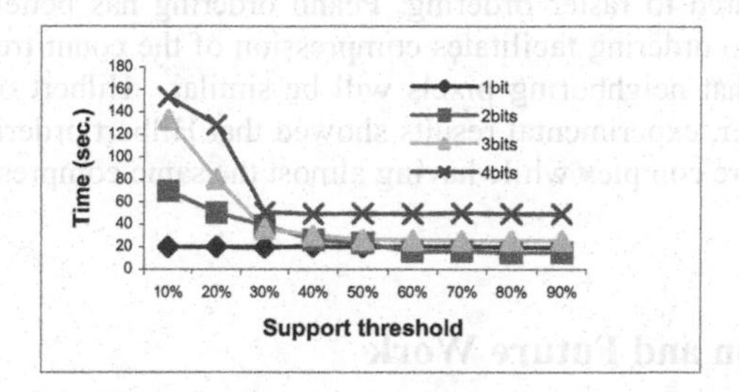

Fig. 12. Performance of the P-ARM algorithm with respect to the number of precision bits

5 Related Work and Discussions

Remotely Sensed Imagery data belongs to the category of spatial data. Some works on spatial data mining are [13,14], including Association Rule Mining on Spatial data [16]. A spatial association rule is a rule indicating that a certain association relationship exists among a set of spatial and possibly some nonspatial predicates [16]. An example rule might be "most big cities in Canada are close to the Canada-U.S. border". In these works, spatial data mining is performed with the perspective of spatial locality. Our work is different from this point of view. The patterns among spectral bands and yield do not necessarily exist on close pixels only. They can exist in any part of the image. The rules generated in this way are useful to the agricultural producer.

Some work has been done on association rule mining on image data. One topic is association rule mining based on image content [17, 22]. In these studies, the problem is to find associations rules about the size, color, texture, shape of the images and object identification in images. Our work is different in the sense that we perform pixel-by-pixel mining on images.

In our previous work [15], we proposed an approach to derive high confidence rules using P-trees. A data cube is generated based on P-trees to find the rules with high confidence. In this paper, we target on general association rule mining problems by considering both the support and confidence of the rules.

The P-tree structure is related to some other structures including quadtrees [6,7] (and its variants point quadtrees [7] and region quadtrees [6]), and HHcode [10]. The similarities between P-trees, quadtrees and HH-Codes are that they are quadrant based. The difference is that P-trees include occurrence counts. P-trees are not indexes, but are representations of the dataset itself. P-trees incorporate inherent data compression. Moreover, P-trees contain useful information to facilitate efficient association rule mining.

The work in [20] presents the ideas of using AD-trees (All-Dimension trees) for machine learning with large datasets. AD-trees are a sparse data structure for representing the cached counting statistics for categorical datasets. The AD-tree records information for all dimensions. P-trees record count information for separate bits and bands. This information is small, simple and complete.

The P-trees are based on Peano ordering. Peano ordering was selected for several reasons. Compared to raster ordering, Peano ordering has better spatial clustering properties. Peano ordering facilitates compression of the count tree structures due to the probability that neighboring pixels will be similar. Hilbert ordering shares this property, however, experimental results showed that Hilbert ordering made the count tree structure more complex while having almost the same compression ratio as Peano ordering.

6 Conclusion and Future Work

In this paper, we propose a new model to derive association rules from Remotely Sensed Imagery. In our model, bSQ format and Peano Count Tree (P-tree) structure are used to organize and represent RSI data. The P-tree structure is a space efficient, lossless, data mining ready structure for spatial datasets. P-trees facilitate advantages, such as fast support calculation and new pruning techniques with early algorithm exit, for association rule mining. Similarly, P-trees can also facilitate the fast calculation of other measurements, such as "interest" and "conviction" defined in [19].

P-trees have potential applications in the areas other than precision agriculture. These areas include flood prediction and monitoring, community and regional planning, virtual archeology, mineral exploration, Bioinformatics, VLSI design and environmental analysis and control. Our future work includes extending P-trees to these and other application areas. Another interesting direction is the application of P-trees on sequential pattern mining.

Acknowledgement

We would like to express our thanks to Dr. Jiawei Han of Simon Fraser University for providing us the FP-growth object code.

References

1. R. Agrawal, T. Imielinski, and A. Swami, "Mining Association Rules Between Sets of Items in Large Database", SIGMOD 93.
2. R. Agrawal and R. Srikant, "Fast Algorithms for Mining Association Rules," VLDB 94.
3. R. Srikant and R. Agrawal, "Mining Quantitative Association Rules in Large Relational Tables", SIGMOD 96.
4. Jong Soo Park, Ming-Syan Chen and Philip S. Yu, "An effective Hash-Based Algorithm for Mining Association Rules," SIGMOD 95.
5. V. Gaede and O. Gunther, "Multidimensional Access Methods", Computing Surveys, 30(2), 1998.
6. H. Samet, "The quadtree and related hierarchical data structure". ACM Computing Survey, 16, 2, 1984.
7. R. A. Finkel and J. L. Bentley, "Quad trees: A data structure for retrieval of composite keys", Acta Informatica, 4, 1, 1974.
8. O. R. Zaiane, J. Han and H. Zhu, "Mining Recurrent Items in Multimedia with Progressive Resolution Refinement", ICDE'2000.
9. TIFF image data sets. Available at http://midas-10.cs.ndsu.nodak.edu/data/images/
10. HH-codes. Available at http://www.statkart.no/nlhdb/iveher/hhtext.htm
11. J. Han and Y. Fu, "Discovery of Multiple-level Association Rules from Large Databases", VLDB 95.
12. J. Han, J. Pei and Y. Yin, "Mining Frequent Patterns without Candidate Generation", SIGMOD 2000.
13. M. Ester, H. P. Kriegel, J. Sander, "Spatial Data Mining: A Database Approach", SSD 1997.
14. K. Koperski, J. Adhikary, J. Han, "Spatial Data Mining: Progress and Challenges", DMKD 1996.
15. William Perrizo, Qin Ding, Qiang Ding and Amalendu Roy, "Deriving High Confidence Rules from Spatial Data using Peano Count Trees", Springer-Verlag, LNCS 2118, July 2001.
16. K. Koperski, J. Han, "Discovery of Spatial Association Rules in Geographic Information Databases", SSD 1995.
17. C. Ordonez and E. Omiecinski, "Discovering Association Rules based on Image Content", Proceedings of the IEEE Advances in Digital Libraries Conference 99.
18. Remote Sensing Tutorial. Available at http://rst.gsfc.nasa.gov/Front/tofc.html
19. S. Brin et al, "Dynamic Itemset Counting and Implication Rules for Market Basket Data", SIGMOD 97.
20. A. Moore, M. Soon Lee, "Cached Sufficient Statistics for Efficient Machine Learning with Large Datasets", Journal of Artificial Intelligence Research, 8 (1998).

On the Efficiency of Association-Rule Mining Algorithms

Vikram Pudi[1] and Jayant R. Haritsa[1]

Database Systems Lab, SERC
Indian Institute of Science
Bangalore 560012, India
{vikram, haritsa}@dsl.serc.iisc.ernet.in

Abstract. In this paper, we first focus our attention on the question of how much space remains for performance improvement over current association rule mining algorithms. Our strategy is to compare their performance against an "Oracle algorithm" that knows in advance the identities of all frequent itemsets in the database and only needs to gather their actual supports to complete the mining process. Our experimental results show that current mining algorithms do not perform uniformly well with respect to the Oracle for all database characteristics and support thresholds. In many cases there is a substantial gap between the Oracle's performance and that of the current mining algorithms. Second, we present a new mining algorithm, called ARMOR, that is constructed by making minimal changes to the Oracle algorithm. ARMOR consistently performs within a factor of two of the Oracle on both real and synthetic datasets over practical ranges of support specifications.

1 Introduction

We focus our attention on the question of how much space remains for performance improvement over current association rule mining algorithms. Our approach is to compare their performance against an **"Oracle algorithm"** that knows *in advance* the identities of all frequent itemsets in the database and only needs to gather the actual supports of these itemsets to complete the mining process. Clearly, *any* practical algorithm will have to do at least this much work in order to generate mining rules. This "Oracle approach" permits us to clearly demarcate the maximal space available for performance improvement over the currently available algorithms. Further, it enables us to construct new mining algorithms from a completely different perspective, namely, as *minimally-altered derivatives* of the Oracle.

First, we show that while the notion of the Oracle is conceptually simple, its *construction* is not equally straightforward. In particular, it is critically dependent on the choice of data structures used during the counting process. We present a carefully engineered implementation of Oracle that makes the best choices for these design parameters at each stage of the counting process. Our experimental results show that there is a considerable gap in the performance between the Oracle and existing mining algorithms.

M.-S. Chen, P.S. Yu, and B. Liu (Eds.): PAKDD 2002, LNAI 2336, pp. 80–91, 2002.

Second, we present a new mining algorithm, called **ARMOR** (Association Rule Mining based on ORacle), whose structure is derived by making minimal changes to the Oracle, and is guaranteed to complete in two passes over the database. Although ARMOR is derived from the Oracle, it may be seen to share the positive features of a variety of previous algorithms such as PARTITION [6], CARMA [2], AS-CPA [3], VIPER [7] and DELTA [4]. Our empirical study shows that ARMOR consistently performs within a factor of two of the Oracle, over both real (BMS-WebView-1 [11] from Blue Martini Software) and synthetic databases (from the IBM Almaden generator [1]) over practical ranges of support specifications.

Problem Scope The environment we consider, similar to the majority of the prior art in the field, is one where the data mining system has a single processor and the pattern lengths in the database are small relative to the number of items in the database. We focus on algorithms that generate *boolean* association rules where the only relevant information in each database transaction is the presence or absence of an item. That is, we restrict our attention to the class of *sequential bottom-up* mining algorithms for generating boolean association rules.

2 The Oracle Algorithm

In this section we present the Oracle algorithm which, as mentioned in the Introduction, "magically" knows in advance the identities of all frequent itemsets in the database and only needs to gather the actual supports of these itemsets. Clearly, *any* practical algorithm will have to do at least this much work in order to generate mining rules. Oracle takes as input the database, $\mathcal{D}$ in item-list format (which is organized as a set of rows with each row storing an ordered list of item-identifiers (IID), representing the items purchased in the transaction), the set of frequent itemsets, F, and its corresponding negative border, N, and outputs the supports of these itemsets by making *one scan* over the database. We first describe the mechanics of the Oracle algorithm below and then move on to discuss the rationale behind its design choices in Section 2.2.

2.1 The Mechanics of Oracle

For ease of exposition, we first present the manner in which Oracle computes the supports of 1-itemsets and 2-itemsets and then move on to longer itemsets. Note, however, that the algorithm actually performs all these computations *concurrently* in one scan over the database.

Counting Singletons and Pairs

Data-Structure Description The counters of singletons (1-itemsets) are maintained in a 1-dimensional lookup array, $\mathcal{A}_1$, and that of pairs (2-itemsets), in a lower triangular 2-dimensional lookup array, $\mathcal{A}_2$. (Similar arrays are also used in Apriori [1, 8] for its first two passes.) The kth entry in the array $\mathcal{A}_1$ contains two fields: (1) *count*, the counter for the itemset X corresponding to the kth item, and (2) *index*, the number of frequent itemsets prior to X in $\mathcal{A}_1$, if X is frequent; **null**, otherwise.

```
ArrayCount (T, A1, A2)
Input: Transaction T, Array for 1-itemsets A1, Array for 2-itemsets A2
Output: Arrays A1 and A2 with their counts updated over T
1.   Itemset T^f = null; // to store frequent items from T in Item-List format
2.   for each item i in transaction T
3.       A1[i.id].count ++;
4.       if A1[i.id].index ≠ null
5.           append i to T^f
6.   for j = 1 to |T^f| // enumerate 2-itemsets
7.       for k = j + 1 to |T^f|
8.           index1 = A1[T^f[j].id].index // row index
9.           index2 = A1[T^f[k].id].index // column index
10.          A2[index1, index2] ++;
```

Fig. 1. Counting Singletons and Pairs in Oracle

Algorithm Description The ArrayCount function shown in Figure 1 takes as inputs, a transaction T along with $\mathcal{A}_1$ and $\mathcal{A}_2$, and updates the counters of these arrays over T. In the ArrayCount function, the individual items in the transaction T are enumerated (lines 2–5) and for each item, its corresponding count in $\mathcal{A}_1$ is incremented (line 3). During this process, the frequent items in T are stored in a separate itemset T^f (line 5). We then enumerate all pairs of items contained in T^f (lines 6–10) and increment the counters of the corresponding 2-itemsets in $\mathcal{A}_2$ (lines 8–10).

Counting k-itemsets, $k > 2$

Data-Structure Description Itemsets in $F \cup N$ of length greater than 2 and their related information (counters, etc.) are stored in a DAG structure $\mathcal{G}$, which is pictorially shown in Figure 2 for a database with items {A, B, C, D}. Although singletons and pairs are stored in lookup arrays, as mentioned before, for expository ease, we assume that they too are stored in $\mathcal{G}$ in the remainder of this discussion.

Each itemset is stored in a separate node of $\mathcal{G}$ and is linked to the first two (in a lexicographic ordering) of its subsets. We use the terms "mother" and "father" of an itemset to refer to the (lexicographically) first and second subsets, respectively. E.g., {A, B} and {A, C} are the mother and father respectively of {A, B, C}. For each itemset X in $\mathcal{G}$, we also store with it links to those supersets of X for which X is a mother. We call this list of links as *childset*. E.g., {BC, BD} is the childset of B.

Since each itemset is stored in a separate node in the DAG, we use the terms "itemset" and "node" interchangeably in the remainder of this discussion. Also, we use $\mathcal{G}$ to denote the set of itemsets that are stored in the DAG structure $\mathcal{G}$.

Algorithm Description We use a *partitioning scheme* [6] wherein the database is logically divided into n disjoint horizontal partitions $P_1, P_2, ..., P_n$. In this scheme, itemsets being counted are enumerated only at the *end of each partition* and not after every tuple. Each partition is as large as can fit in available main memory. For ease of exposition, we

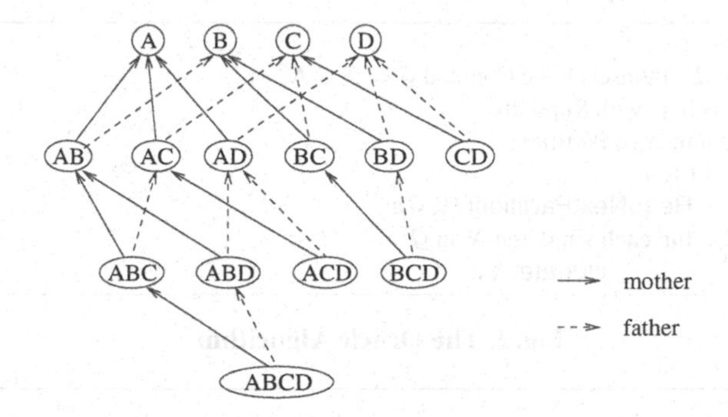

Fig. 2. DAG Structure Containing Power Set of {A,B,C,D}

assume that the partitions are equi-sized. However, the technique is easily extendible to arbitrary partition sizes.

The pseudo-code of Oracle is shown in Figure 3 and operates as follows: The ReadNextPartition function (line 3) reads tuples from the next partition and simultaneously creates tid-lists[1] (within that partition) of singleton itemsets in $\mathcal{G}$. The Update function (line 5) is then applied on each singleton in $\mathcal{G}$. This function takes a node M in $\mathcal{G}$ as input and updates the counts of all descendants of M to reflect their counts over the current partition. The count of any itemset within a partition is equal to the length of its corresponding tidlist (within that partition). The tidlist of an itemset can be obtained as the intersection of the tidlists of its mother and father and this process is started off using the tidlists of frequent 1-itemsets. The exact details of tidlist computation are discussed later.

We now describe the manner in which the itemsets in $\mathcal{G}$ are enumerated after reading in a new partition. The set of links, $\bigcup_{M \in \mathcal{G}} M.childset$, induce a spanning tree of $\mathcal{G}$ (e.g. consider only the solid edges in Figure 2). We perform a *depth first search* on this spanning tree to enumerate all its itemsets. When a node in the tree is visited, we compute the tidlists of all its children. This ensures that when an itemset is visited, the tidlists of its mother and father have already been computed.

The above processing is captured in the function Update whose pseudo-code is shown in Figure 4. Here, the tidlist of a given node M is first converted to the tid-vector format[2] (line 1). Then, tidlists of all children of M are computed (lines 2–4) after which the same children are visited in a depth first search (lines 5–6).

The mechanics of tidlist computation, as promised earlier, are given in Figure 5. The Intersect function shown here takes as input a tid-vector B and a tid-list T. Each tid in T is added to the result if $B[offset]$ is 1 (lines 2–5) where *offset* is defined in line 3 and represents the position of the transaction T relative to the current partition.

[1] A tid-list of an itemset X is an ordered list of TIDs of transactions that contain X.

[2] A tid-vector of an itemset X is a bit-vector of 1's and 0's to represent the presence or absence respectively, of X in the set of customer transactions.

```
Oracle (D, G)
Input: Database D, Itemsets to be Counted G = F ∪ N
Output: Itemsets in G with Supports
1.      n = Number of Partitions
2.      for i = 1 to n
3.          ReadNextPartition(P_i, G);
4.          for each singleton X in G
5.              Update(X);
```

Fig. 3. The Oracle Algorithm

```
Update (M)
Input: DAG Node M
Output: M and its Descendents with Counts Updated
1.      B = convert M.tidlist to Tid-vector format // B is statically allocated
2.      for each node X in M.childset
3.          X.tidlist = Intersect(B, X.father.tidlist);
4.          X.count += |X.tidlist|
5.      for each node X in M.childset
6.      Update(X);
```

Fig. 4. Updating Itemset Counts

```
Intersect (B, T)
Input: Tid-vector B, Tid-list T
Output: B ∩ T
1.      Tid-list result = φ
2.      for each tid in T
3.          offset = tid + 1− (tid of first transaction in current partition)
4.          if B[offset] = 1 then
5.              result = result ∪ tid
6.      return result
```

Fig. 5. Tid-vector and Tid-list Intersection

2.2 Optimality of Oracle

We show that Oracle is optimal in two respects: (1) It enumerates only those itemsets in $\mathcal{G}$ that need to be enumerated, and (2) The enumeration is performed in the most efficient way possible. These results are based on the following two theorems. Due to lack of space we have deferred the proofs of theorems to [5].

Theorem 1. *If the size of each partition is large enough that every itemset in $F \cup N$ of length greater than 2 is present at least once in it, then the only itemsets being enumerated in the Oracle algorithm are those whose counts need to be incremented in that partition.*

Theorem 2. *The cost of enumerating each itemset in Oracle is $\Theta(1)$ with a tight constant factor.*

While Oracle is optimal in these respects, we note that there may remain some scope for improvement in the details of *tidlist computation*. That is, the Intersect function (Figure 5) which computes the intersection of a tid-vector B and a tid-list T requires $\Theta(|T|)$ operations. B itself was originally constructed from a tid-list, although this cost is amortized over many calls to the Intersect function. We plan to investigate in our future work whether the intersection of two sets can, in general, be computed more efficiently – for example, using **diffsets**, a novel and interesting approach suggested in [10]. The diffset of an itemset X is the set-difference of the tid-list of X from that of its mother. Diffsets can be easily incorporated in Oracle – only the Update function in Figure 4 of Section 2 is to be changed to compute diffsets instead of tidlists by following the techniques suggested in [10].

Advantages of Partitioning Schemes Oracle, as discussed above, uses a partitioning scheme. An alternative commonly used in current association rule mining algorithms, especially in hashtree [1] based schemes, is to use a tuple-by-tuple approach. A problem with the tuple-by-tuple approach, however, is that there is considerable wasted enumeration of itemsets. The core operation in these algorithms is to determine all candidates that are subsets of the current transaction. Given that a frequent itemset X is present in the current transaction, we need to determine all candidates that are immediate supersets of X and are also present in the current transaction. In order to achieve this, it is often necessary to enumerate and check for the presence of many more candidates than those that are actually present in the current transaction.

3 The ARMOR Algorithm

As will be shown in our experiments (Section 4), there is a considerable gap in the performance between the Oracle and existing mining algorithms. We now move on to describe our new mining algorithm, ARMOR (Association Rule Mining based on ORacle). In this section, we overview the main features and the flow of execution of ARMOR – the details of candidate generation are deferred to [5] due to lack of space.

The guiding principle in our design of the ARMOR algorithm is that we consciously make an attempt to determine the *minimal amount of change* to Oracle required to result in an online algorithm. This is in marked contrast to the earlier approaches which designed new algorithms by trying to address the limitations of *previous* online algorithms. That is, we approach the association rule mining problem from a completely different perspective.

In ARMOR, as in Oracle, the database is conceptually partitioned into n disjoint blocks $P_1, P_2, ..., P_n$. At most *two* passes are made over the database. In the first pass we form a set of candidate itemsets, $\mathcal{G}$, that is guaranteed to be a superset of the set of frequent itemsets. During the first pass, the counts of candidates in $\mathcal{G}$ are determined over each partition in exactly the same way as in Oracle by maintaining the candidates in a DAG structure. The 1-itemsets and 2-itemsets are stored in lookup arrays as in

```
ARMOR (D, I, minsup)
Input: Database D, Set of Items I, Minimum Support minsup
Output: F ∪ N with Supports
1.      n = Number of Partitions
        //—First Pass —
2.      G = I     // candidate set (in a DAG)
3.      for i = 1 to n
4.          ReadNextPartition(P_i, G);
5.          for each singleton X in G
6.              X.count += |X.tidlist|
7.              Update1(X, minsup);
        //—Second Pass —
8.      RemoveSmall(G, minsup);
9.      OutputFinished(G, minsup);
10.     for i = 1 to n
11.         if (all candidates in G have been output)
12.             exit
13.         ReadNextPartition(P_i, G);
14.         for each singleton X in G
15.             Update2(X, minsup);
```

Fig. 6. The ARMOR Algorithm

Oracle. But unlike in Oracle, candidates are inserted and removed from $\mathcal{G}$ at the end of each partition. Generation and removal of candidates is done *simultaneously* while computing counts. The details of candidate generation and removal during the first pass are described in [5] due to lack of space. For ease of exposition we assume in the remainder of this section that all candidates (including 1-itemsets and 2-itemsets) are stored in the DAG.

Along with each candidate X, we also store the following three integers as in the CARMA algorithm [2]: (1) $X.count$: the number of occurrences of X *since* X was last inserted in $\mathcal{G}$. (2) $X.firstPartition$: the index of the partition *at* which X was inserted in $\mathcal{G}$. (3) $X.maxMissed$: upper bound on the number of occurrences of X *before* X was inserted in $\mathcal{G}$.

While the CARMA algorithm works on a tuple-by-tuple basis, we have adapted the semantics of these fields to suit the partitioning approach. If the database scanned so far is *d*, then the support of any candidate X in $\mathcal{G}$ will lie in the range $[X.count/|d|, (X.maxMissed + X.count)/|d|]$ [2]. These bounds are denoted by minSupport(X) and maxSupport(X), respectively. We define an itemset X to be d-frequent if minSupport(X) $\geq minsup$. Unlike in the CARMA algorithm where only d-frequent itemsets are stored at any stage, the DAG structure in ARMOR contains other candidates, including the *negative border* of the d-frequent itemsets, to ensure efficient candidate generation. The details are given in [5].

At the end of the first pass, the candidate set $\mathcal{G}$ is pruned to include only d-frequent itemsets and their negative border. The counts of itemsets in $\mathcal{G}$ over the entire database are determined during the second pass. The counting process is again identical to that of Oracle. No new candidates are generated during the second pass. However, candidates may be removed. The details of candidate removal in the second pass is deferred to [5].

The pseudo-code of ARMOR is shown in Figure 6 and is explained below.

First Pass At the beginning of the first pass, the set of candidate itemsets $\mathcal{G}$ is initialized to the set of singleton itemsets (line 2). The ReadNextPartition function (line 4) reads tuples from the next partition and simultaneously creates tid-lists of singleton itemsets in $\mathcal{G}$.

After reading in the entire partition, the Update1 function (details in [5]) is applied on each singleton in $\mathcal{G}$ (lines 5–7). It increments the counts of existing candidates by their corresponding counts in the current partition. It is also responsible for generation and removal of candidates.

At the end of the first pass, $\mathcal{G}$ contains a superset of the set of frequent itemsets. For a candidate in $\mathcal{G}$ that has been inserted at partition P_j, its count over the partitions $P_j, ..., P_n$ will be available.

Second Pass At the beginning of the second pass, candidates in $\mathcal{G}$ that are neither d-frequent nor part of the current negative border are removed from $\mathcal{G}$ (line 8). For candidates that have been inserted in $\mathcal{G}$ at the first partition, their counts over the entire database will be available. These itemsets with their counts are output (line 9). The OutputFinished function also performs the following task: If it outputs an itemset X and X has no supersets left in $\mathcal{G}$, X is removed from $\mathcal{G}$.

During the second pass, the ReadNextPartition function (line 13) reads tuples from the next partition and creates tid-lists of singleton itemsets in $\mathcal{G}$. After reading in the entire partition, the Update2 function (details in [5]) is applied on each singleton in $\mathcal{G}$ (lines 14–15). Finally, before reading in the next partition we check to see if there are any more candidates. If not, the mining process terminates.

3.1 Memory Utilization in ARMOR

In the design and implementation of ARMOR, we have opted for speed in most decisions that involve a space-speed tradeoff. Therefore, the main memory utilization in ARMOR is certainly more as compared to algorithms such as Apriori. However, in the following discussion, we show that the memory usage of ARMOR is well within the reaches of current machine configurations. This is also experimentally confirmed in the next section.

The main memory consumption of ARMOR comes from the following sources: (1) The 1-d and 2-d arrays for storing counters of singletons and pairs, respectively; (2) The DAG structure for storing counters of longer itemsets, including tidlists of those itemsets, and (3) The current partition.

The total number of entries in the 1-d and 2-d arrays and in the DAG structure corresponds to the number of candidates in ARMOR, which as we have discussed in [5], is only marginally more than $|F \cup N|$. For the moment, if we disregard the space occupied by tidlists of itemsets, then the amortized amount of space taken by each candidate is a small constant (about 10 integers for the dag and 1 integer for the arrays). E.g., if there are 1 million candidates in the dag and 10 million in the array, the space required is about 80MB. Since the environment we consider is one where the pattern lengths are small, the number of candidates will typically be comparable to or well within the available main memory. [9] discusses alternative approaches when this assumption does not hold.

Regarding the space occupied by tidlists of itemsets, note that ARMOR only needs to store tidlists of d-frequent itemsets. The number of d-frequent itemsets is of the same order as the number of frequent itemsets, $|F|$. The total space occupied by tidlists while processing partition P_i is then bounded by $|F| \times |P_i|$ integers. E.g., if $|F| = 5K$ and $|P_i| = 20K$, then the space occupied by tidlists is bounded by about 400MB. We assume $|F|$ to be in the range of a few thousands at most because otherwise the total number of rules generated would be enormous and the purpose of mining would not be served. Note that the above bound is very pessimistic. Typically, the lengths of tidlists are much smaller than the partition size, especially as the itemset length increases.

Main memory consumed by the current partition is small compared to the above two factors. E.g., If each transaction occupies 1KB, a partition of size 20K would require only 20MB of memory. Even in these extreme examples, the total memory consumption of ARMOR is 500MB, which is acceptable on current machines.

Therefore, *in general we do not expect memory to be an issue* for mining market-basket databases using ARMOR. Further, even if it does happen to be an issue, it is easy to modify ARMOR to free space allocated to tidlists at the expense of time: $M.tidlist$ can be freed after line 3 in the Update function shown in Figure 4.

A final observation to be made from the above discussion is that the main memory consumption of ARMOR is proportional to the size of the *output* and does not "explode" as the input problem size increases.

4 Performance Study

In the previous section, we have described the Oracle and ARMOR algorithms. We have conducted a detailed study to assess the performance of ARMOR with respect to the Oracle algorithm. For completeness and as a reference point, we have also included the classical Apriori in our evaluation suite. The performance of other algorithms including VIPER and FP-growth are not presented here due to lack of space, but are available in [5].

Our experiments cover a range of database and mining workloads, and include the typical and extreme cases considered in previous studies. The performance metric in all the experiments is the *total execution time* taken by the mining operation. Due to space limitations, we show only a few representative experiments here – the others are available in [5].

Our experiments were conducted on a 700-MHz Pentium III workstation running Red Hat Linux 6.2, configured with a 1 GB main memory and three local 18 GB SCSI 10000 rpm disks. All the algorithms in our evaluation suite are written in C++. Finally, the partition size in ARMOR and Oracle was fixed to be 20K tuples.

The real dataset used in our experiments was BMS-WebView-1 [11] from Blue Martini Software, while the synthetic databases were generated using the IBM Almaden

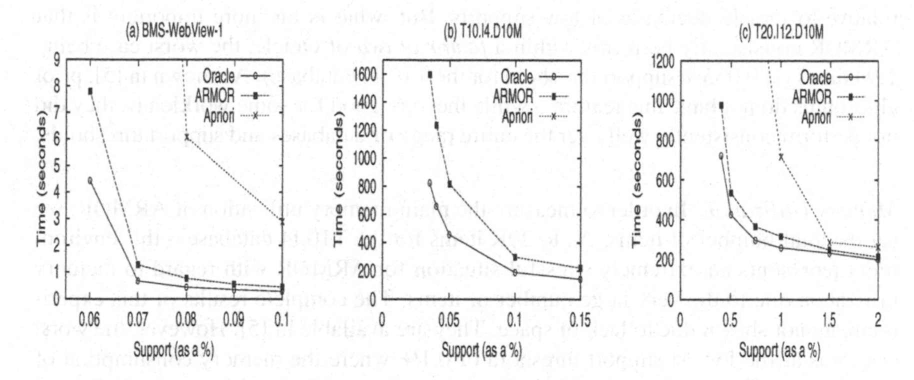

Fig. 7. Performance of ARMOR

generator [1]. Our synthetic databases were generated with parameters T10.I4 (following the standard naming convention introduced in [1]) and T20.I12 with 10M tuples in each of them. The number of items in the supermarket and the number of potentially frequent itemsets was set to 1K and 2K, respectively.

We set the rule support threshold values to as low as was feasible with the available main memory. At these low support values the number of frequent itemsets exceeded sixty thousand! Beyond this, we felt that the number of rules generated would be enormous and the purpose of mining – to find interesting patterns – would not be served. In particular, we set the rule support threshold values for the BMS-WebView-1, T10.I4 and T20.I12 databases to the ranges (0.06%–0.1%), (0.035%–0.1%), and (0.4%–2%), respectively. The results of these experiments are shown in Figures 7a–c. The x-axis in these graphs represent the support threshold values while the y-axis represents the response times of the algorithms being evaluated.

In these graphs, we first see that the response times of all algorithms increase exponentially as the support threshold is reduced. This is only to be expected since the number of itemsets in the output, $F \cup N$, increases exponentially with decrease in the support threshold. We also see that Apriori is uncompetitive with ARMOR and Oracle. In Figure 7b, Apriori did not feature at all since its response time was out of the range of the graph even for the highest support threshold.

Next, we see that ARMOR's performance is close to that of Oracle for high supports. This is because of the following reasons: The density of the frequent itemset distribution is sparse at high supports resulting in only a few frequent itemsets with supports "close" to $minsup$. Hence, frequent itemsets are likely to be locally frequent within most partitions. Even if they are not locally frequent in a few partitions, it is very likely that they are still d-frequent over these partitions. Hence, their counters are updated even over these partitions. Therefore, the complete counts of most candidates would be available at the end of the first pass resulting in a "light and short" second pass. Hence, it is expected that ARMOR's performance will be close to that of Oracle.

Since the frequent itemset distribution becomes dense at low supports, the above argument does not hold in this support region. Hence we see that ARMOR's performance relative to Oracle decreases at low supports. But, what is far more important is that ARMOR consistently performs within a *factor of two* of Oracle, the worst case being 1.94 times (at 0.035% support threshold for the T10.I4 database). As shown in [5], prior algorithms do not have this feature – while they are good for some workloads, they did not perform consistently well over the entire range of databases and support thresholds.

Memory Utilization In order to measure the main memory utilization of ARMOR, we set the total number of items, N, to 20K items for the T10.I4 database – this environment represents an extremely stressful situation for ARMOR with regard to memory utilization due to the very large number of items. The complete results of this experiment are not shown due to lack of space. They are available in [5]. However, the worst case was at the lowest support threshold of 0.1% where the memory consumption of ARMOR for N = 1K items was 104MB while for N = 20K items, it was 143MB – an increase in less than 38% for a 20 times increase in the number of items! The reason for this is that the main memory utilization of ARMOR does not depend directly on the number of items, but only on the size of the output, $F \cup N$, as discussed in Section 3.1.

4.1 Discussion of Experimental Results

We now explain why ARMOR should typically perform within a factor of two of Oracle. First, we notice that the only difference between the single pass of Oracle and the first pass of ARMOR is that ARMOR continuously generates and removes candidates. Since the generation and removal of candidates in ARMOR is dynamic and efficient, this does not result in a significant additional cost for ARMOR.

Since candidates in ARMOR that are neither d-frequent nor part of the current negative border are continuously removed, any itemset that is locally frequent within a partition, but not globally frequent in the entire database is likely to be removed from G during the course of the first pass (unless it belongs to the current negative border). Hence the resulting candidate set in ARMOR is a good approximation of the required mining output. In fact, in our experiments, we found that in the worst case, the number of candidates counted in ARMOR was only about *ten percent* more than the required mining output. The above two reasons indicate that the cost of the first pass of ARMOR is only slightly more than that of (the single pass in) Oracle.

Next, we notice that the only difference between the second pass of ARMOR and (the single pass in) Oracle is that in ARMOR, candidates are continuously removed. Hence the number of itemsets being counted in ARMOR during the second pass quickly reduces to much less than that of Oracle. Moreover, ARMOR does not necessarily perform a complete scan over the database during the second pass since this pass ends when there are no more candidates. Due to these reasons, we would expect that the cost of the second pass of ARMOR is usually less than that of (the single pass in) Oracle.

Since the cost of the first pass of ARMOR is usually only slightly more than that of (the single pass in) Oracle and that of the second pass is usually less than that of (the single pass in) Oracle, it follows that ARMOR will typically perform within a factor of two of Oracle.

5 Conclusions

In this paper, our approach was to quantify the algorithmic performance of association rule mining algorithms with regard to an idealized, but practically infeasible, "Oracle". The Oracle algorithm utilizes a partitioning strategy to determine the supports of itemsets in the required output. It uses direct lookup arrays for counting singletons and pairs and a DAG data-structure for counting longer itemsets. We have shown that these choices are optimal in that only required itemsets are enumerated and that the cost of enumerating each itemset is $\Theta(1)$. Our experimental results showed that there was a substantial gap between the performance of current mining algorithms and that of the Oracle.

We also presented a new online mining algorithm called ARMOR, that was constructed with minimal changes to Oracle to result in an online algorithm. ARMOR utilizes a new method of candidate generation that is dynamic and incremental and is guaranteed to complete in two passes over the database. Our experimental results demonstrate that ARMOR performs within a *factor of two* of Oracle for both real and synthetic databases with acceptable main memory utilization.

Acknowledgments We thank Roberto Bayardo, Mohammed J. Zaki and Shiby Thomas for reading previous drafts of this paper and providing insightful comments and suggestions.

References

[1] R. Agrawal and R. Srikant. Fast algorithms for mining association rules. In *Proc. of Intl. Conf. on Very Large Databases (VLDB)*, September 1994.

[2] C. Hidber. Online association rule mining. In *Proc. of ACM SIGMOD Intl. Conf. on Management of Data*, June 1999.

[3] J. Lin and M. H. Dunham. Mining association rules: Anti-skew algorithms. In *Intl. Conf. on Data Engineering (ICDE)*, 1998.

[4] V. Pudi and J. Haritsa. Quantifying the utility of the past in mining large databases. *Information Systems*, July 2000.

[5] V. Pudi and J. Haritsa. On the optimality of association-rule mining algorithms. Technical Report TR-2001-01, DSL, Indian Institute of Science, 2001.

[6] A. Savasere, E. Omiecinski, and S. Navathe. An efficient algorithm for mining association rules in large databases. In *Proc. of Intl. Conf. on Very Large Databases (VLDB)*, 1995.

[7] P. Shenoy, J. Haritsa, S. Sudarshan, G. Bhalotia, M. Bawa, and D. Shah. Turbo-charging vertical mining of large databases. In *Proc. of ACM SIGMOD Intl. Conf. on Management of Data*, May 2000.

[8] R. Srikant and R. Agrawal. Mining generalized association rules. In *Proc. of Intl. Conf. on Very Large Databases (VLDB)*, September 1995.

[9] Y. Xiao and M. H. Dunham. Considering main memory in mining association rules. In *Intl. Conf. on Data Warehousing and Knowledge Discovery (DAWAK)*, 1999.

[10] M. J. Zaki and K. Gouda. Fast vertical mining using diffsets. Technical Report 01-1, Rensselaer Polytechnic Institute, 2001.

[11] Z. Zheng, R. Kohavi, and L. Mason. Real world performance of association rule algorithms. In *Intl. Conf. on Knowledge Discovery and Data Mining (SIGKDD)*, August 2001.

A Function-Based Classifier Learning Scheme Using Genetic Programming

Jung-Yi Lin, Been-Chian Chien and Tzung-Pei Hong

Institute of Information Engineering
I-Shou University
1, Section 1, Hsueh-Cheng Rd, Ta-Hsu Hsiang,
Kaohsiung County, Taiwan 84008, R.O.C
TEL: +886-7-6577711 ext 6517
FAX: +886-7-6578944
{m893310m, cbc, tphong}@isu.edu.tw

Abstract. Classification is an important research topic in knowledge discovery and data mining. Many different classifiers have been motivated and developed of late years. In this paper, we propose an effective scheme for learning multi-category classifiers based on genetic programming. For a *k*-class classification problem, a training strategy called adaptive incremental learning strategy and a new fitness function are used to generate *k* discriminant functions. We urge the discriminant functions to map the domains of training data into a specified interval, and thus data will be assigned into one of the classes by the values of functions. Furthermore, a *Z*-value measure is developed for resolving the conflicts. The experimental results show that the proposed GP-based classification learning approach is effective and performs a high accuracy of classification.

(Keywords: Knowledge discovery, Data mining, Genetic programming, Classification)

1. Introduction

Machine learning and data mining are important research topics these years. The purpose of machine learning is to try to find human knowledge from a given data set, and the goal of data mining is to turn confused data into valuable information. Since the versatility of human activities and the unpredictability of data, many research issues have been initiated by such challenge. Classification is one of the major issues. The task of classification is to assign an object into a predefined class based on the attributes of the object. Many applications such as pattern recognition, disease diagnosis, and business decision-making can be treated as extensions of classification problem.

In the past decades, many different methods and ideas were proposed for solving classification problem through mathematical models or theories. The well-known classifiers include nearest neighbor classifiers [3][6], Bayesian network [7], neural network [29], and evolutionary approaches (including genetic algorithm [25][26][27] and genetic programming [5][25][26][27][9] [22]). We will review some of the classifiers in Section 2.

M.-S. Chen, P.S. Yu, and B. Liu (Eds.): PAKDD 2002, LNAI 2336, pp. 92-103, 2002.

The technique of genetic programming (GP) was proposed by Koza [11][12]. Genetic programming has been applied to a wide range of areas, such as symbolic regression, the robot controller and classification, etc. Genetic programming can discover underlying data relationships and present these relationships by expressions. The expression may be composed of terminals and functions that include arithmetic operations, trigonometric functions, and conditional operators, etc.

Genetic programming begins with a population. Population is a set of individuals that are randomly created symbolic expressions. Each individual is represented by a binary tree composed of functions and terminals. The fitness value of each individual is evaluated by a predefined function called fitness function. If the fitness value is not satisfied, the genetic operators like reproduction, crossover and mutation are applied to the individuals and a new population will be generated by modifying the individuals with better fitness in next generation. After a number of generations passed, the individual with best fitness will be selected from the population to be the final solution.

In this paper, we propose a new scheme for multi-category classification based on genetic programming. The proposed method will generate the corresponding discriminant functions for each class. We first apply the adaptive incremental learning strategy to handle effectiveness as well as efficiency in learning stage of genetic programming. Next, in order to obtain the corresponding discriminant functions, we design a new fitness function for dealing with the training samples. For solving the problem of conflict, a method called *Z*-value measure is presented. The experiments select two well-known data sets to show the performance of the proposed scheme: the Wisconsin Breast Cancer Diagnostic database (WBCD) [17] and the Fisher's Iris data sets [4]. We also give a comparison of our method with the previous research results. The empirical results show that the discriminant functions obtained by the proposed scheme are efficient for pattern classification.

The remainder of this paper is organized as follows. Section 2 reviews the previous classifiers. In Section 3, we propose a GP-based approach and a resolution method for conflict to accomplish the task of effective classification. Section 4 describes the experimental results and gives a comparison with previous work. Finally, conclusions are made in Section 5.

2. Related Works

In this section, we review the previous proposed classifiers [13] include nearest neighbor classifier, Bayesian network classifier, neural network classifier, histogram classifier, and evolutionary approaches.

The nearest neighbor classifier is based on distances between the input feature vectors and the sample vectors [3][6]. For a given feature vector, the classifier assigns the vector to the class indicated by the nearest vector in the reference sample. In the *k*-nearest neighbor classifier, it maps feature vector to the pattern class that appears the most frequently among the *k*-nearest neighbors. The nearest neighbor classifier is simple and accurate, but there are several disadvantages. For example, the nearest neighbor classifiers can only work well under the data set with normal distribution and they are usually time-consuming.

The Bayesian classifier is based on Bayesian theorem[7]. The Bayesian theorem provides a probability model for constructing a decision network. Each node of the Bayesian network outputs a probability that determines what the next node is. After that, the data is assigned one of the classes with maximum probability. Although the statistical models are soundness, the main limitation of this approach is that users must have a good knowledge about data properties for performing an effective classification.

In neural network method [28][29], a multi-layered network with m inputs and n outputs is trained with a given set of training data. After that, we give an input vector to the neural network and obtain an n-dimensional output vector from the outputs of the network. Then the given vector is assigned the class with the maximum output. The drawbacks of neural network method are that the knowledge representation in the neural network is unclear and the training process is inefficient.

The histogram classifier partitions the feature space into several cells with size h, and the class of each data is assigned according to a majority vote within the cell in which the data falls. There are two requirements in histogram classifiers. First, the size h should be quite small that leads to smaller cells. The small cells are higher specificity may obtain optimal decision. Second, the larger h leads to contain sufficiently many data to average out statistical variation and higher generality. However, in the case of large dimensions, to determine h is difficult, a small h stands for that feature space will be partitioned into too many cells, and then leads to time-consuming. Contrarily, a large h partitions the feature space into few cells may lead to poor accuracy.

The evolutionary approaches include genetic algorithm and genetic programming. For a set of classification rules, genetic algorithm [25][26][27] encodes each rule to be a bit strings called gene. The evolution operations such as reproduction, crossover and mutation then are used to generate the next generation of genes with better fitness. After the specified number of generations is computed or the conditions of fitness functions are satisfied, a set of genes that are classification rules can be obtained usually.

Genetic programming has been applied to classification problem successfully [5][9][22]. Genetic programming can be used to generate classification rules or discriminant functions. Like genetic algorithm, genetic programming uses the evolution operations include reproduction, crossover and mutation to generate better rules or functions. The advantages of classified by functions are more concise, efficient, and accurate than classification rules. However, the main drawback is the situations of conflict. A conflict is that an object is assigned to more than two different classes by distinct discriminant functions at the same time. It needs a good determinant method to solve such problem.

3. The Proposed Learning Scheme and Classifiers

In this section, we present the method to discover discriminant functions for classification using genetic programming. At first, we give a formal description for the classification problem. Then, we describe a training strategy called adaptive incremental learning strategy for training the given samples. We also provide a new fitness function to evaluate the fitness values of individuals. Finally, a conflict resolution, Z-value measure, is contributed to classify the conflicting objects.

3.1 Notations

The notations of used symbols and a formal description for classification problem are described in the following. Consider a given data set S, for each data $x_j \in S$ that having n attributes:

$$x_j=(v_{j1}, v_{j2}, \ldots, v_{jt}, \ldots, v_{jn}),\ 1 \le t \le n,$$

where $v_{jt} \in \mathbf{R}$ stands for the t-th attribute of x_j. Assume that there exists a set C and

$$C = \{C_1, C_2, ..., C_K\}$$

are the set of K predefined classes, we say that $<x_j, c_j>$ is a sample if the data x_j has been assigned to a specified class c_j, $c_j \in C$. We define a training set (TS) to be a set of collections of samples, as follows:

$$TS = \{<x_j, c_j> \mid x_j = (v_{j1}, v_{j2}, ..., v_{jn}),\ c_j \in C,\ 1 \le j \le m\},$$

where m is the number of samples in TS. Let

$$m = \sum_{i=1}^{K} m_i\ ,$$

where m_i is the number of samples belonging to the class C_i, $1 \le i \le K$.

A discriminant function f_i is a function mapping $\mathbf{R}^n$ to $\mathbf{R}$. A classifier is constructed by a set of discriminant functions F such that

$$F = \{f_i \mid f_i : \mathbf{R}^n \to \mathbf{R}, 1 \le i \le K\}.$$

3.2 The Adaptive Incremental Learning Strategy

The first step of the learning procedure is to prepare the training set TS. The samples in TS usually include positive instances and negative instances. Consider a specified class C_i, a sample $<x_j, c_j> \in TS$, $1 \le j \le m$, we say that $<x_j, c_j>$ is a positive instance if $c_j = C_i$; otherwise, $<x_j, c_j>$ is a negative instance.

After TS is prepared, we start the learning procedure using genetic programming. Conventionally, all of the samples in TS are fed to the learning procedure at a time. However, while the size of TS is getting larger, the training steps will relatively spend more time upon learning from samples. Large TS in the genetic programming will causes the number of evolving generations increases rapidly for finding an effective solution. Thus, for obtaining effective solutions efficiently in genetic programming, we develop an adaptive incremental strategy to proceed to learn the training set.

The adaptive incremental learning strategy accomplishes the learning procedure by separating it into several stages. At first, we give an incremental rate, an adaptive factor and a condition of fitness values satisfying effectiveness. In each stage, we evolve the population for a specified number of generations and parts of the training samples. The selected part of training samples in each stage contains the same proportion of positive instances and negative instances from TS. While the current stage is completed, a number of training samples will be added to the selected set by the incremental rate and the adaptive factor. Therefore, the number of training samples in the selected set will increase stage by stage until all of the training samples are selected. Furthermore, if the fitness value of the best individual satisfies the given condition, the adaptive factor will be doubled and the double number of training samples will be added to the selected training set in next stage. Hence, the number of stages is adapted with the obtained best fitness value, and then the training time can be reduced while the given training samples are easy to discriminate. The adaptive incremental learning strategy not only uses incremental concept to generate better discriminant function[9] but also uses an adaptive factor to reduce the training time. The main steps are described in the following.

Assume that g is the specified generations of evolution in each stage of the learning procedure, and m' be the number of training samples in each stage. We further define three parameters ρ, α and ω. The ρ is the basic incremental rate of training samples in each stage. The α is the adaptive factor used to adapt the incremental rate of samples. The ω is the specified criterion used to determine the goodness of fitness values for the obtained discriminant functions, $0 \leq \rho \leq 1$, $0 \leq m' \leq m$, $\alpha \geq 1$.

Step 1: Initially, let α=1 and m'=0. We give the condition ω and specify g and ρ.
Step 2: $m' = m \times \alpha \times \rho + m'$. If $m' \geq m$, then $m' = m$.
Step 3: The population of genetic programming is evolved with the m' training samples for g generations. A function will be obtained after the g generations.
Step 4: If $m'=m$, then stop, else go to next step.
Step 5: If the fitness value satisfies the condition ω, then α=2, else α=1. Go to Step 2.

3.3 The Fitness Function

The evaluation of the fitness in an individual is dependent on a fitness function. The obtained fitness value determines the individual should be selected or not. The proposed fitness function is based on error measure. We specified an interval to determine a data belonging to the class or not. For the positive instances, we urge that the outputs of the individuals should be fallen in the specified interval. On the contrary, the output of the individuals should be out of the interval for negative instances. For this purpose, the proposed fitness function is defined in the following.

Since the individuals are functions, we consider an individual f_i of a class C_i and two specified numbers μ and σ, $\sigma \geq 0$. For the positive instance, we urge that $[f_i(x_j)-\mu]^2 \leq \sigma$; on the contrary, $[f_i(x_j)-\mu]^2 > \sigma$ for negative instance. That is, the fitness values of positive instances should be fallen in the specified interval $[\mu-\sigma, \mu+\sigma]$. We defined a positive constant p, for a given sample $\langle x_j, c_j\rangle \in TS$, if it is a positive instance, we defined a value that denoted by D_p by (1):

$$D_p = \begin{cases} 0 \ \ if \ c_j = C_i \ \ and \ [f_i(x_j) - \mu]^2 \leq \sigma \\ p \ \ if \ c_j = C_i \ \ and \ [f_i(x_j) - \mu]^2 > \sigma \end{cases}, \tag{1}$$

for a negative instance, defined a value that denoted by D_n by (2):

$$D_n = \begin{cases} 0 \ \ if \ c_j \neq C_i \ \ and \ [f_i(x_j) - \mu]^2 > \sigma \\ p \ \ if \ c_j \neq C_i \ \ and \ [f_i(x_j) - \mu]^2 \leq \sigma \end{cases} \tag{2}$$

The fitness value of an individual is defined by

$$-\sum_{j=1}^{m'} (D_p + D_n), \tag{3}$$

where m' is the number of training samples in the current learning stage, $\langle x_j, c_j\rangle \in TS$, $1 \leq j \leq m'$. By the fitness function, the best fitness of an individual is zero. While the fitness value of a function f_i approximates or equals to zero, it means that the individual f_i can discriminate most of the samples of class C_i in *TS*. Hence, the individual f_i can be used to be the discriminant function of class C_i. A data x_j thus can be assigned to a class C_i if it satisfies $[f_i(x_j)-\mu]^2 \le \sigma$.

3.4 *Z*-value Measure

In the real classification applications, classifiers usually cannot correctly recognize all objects completely. Except the case of misclassification, two conflicting situations may occur when discriminant functions are used:

1. An object is recognized by two or more discriminant functions at the same time.
2. An object is rejected by all discriminant functions.

The two above situations are called conflicts. A complete classification scheme should include a resolution for the conflicting objects. In the proposed scheme, we provide the *Z*-value measure to solve the problem.

For a discriminant function f_i and the specified numbers μ and σ, consider a data $x_j \in S$, where S is the given data set. Assume that the discriminant functions are generated by a representative training set of samples. The samples should be viewed as a normal distribution, the confidence interval of $f_i(x_j)$ is denoted as (4):

$$[-Z_{[1-\alpha/2]} \le \frac{f_i(x_j)-\mu}{\sigma/\sqrt{m_i}} \le Z_{[1-\alpha/2]}], \tag{4}$$

where $1 \le j \le |S|$, $1 \le i \le K$ and α is the confidence coefficient, m_i is the number of data. In this paper, since the size of confidence interval is not significant, we don't have to care how α is. Then, we defined the *Z*-value of the data x_j with the discriminant function f_i by (5):

$$Z_i(x_j) = \frac{f_i(x_j)-\mu}{\sigma/\sqrt{m_i}}, \quad 1 \le j \le |S|, 1 \le i \le K. \tag{5}$$

While the $f_i(x_j)$ approximates to μ, the *Z*-value of x_j, $Z_i(x_j)$, is reduced and the confidence interval can be smaller. Furthermore, since a smaller confidence interval indicates that we have more reliance in $f_i(x_j)$ as an estimation of μ, it means that the possibility of x_j belonging to the class C_i is higher if the value of $Z_i(x_j)$ is smaller. Thus, we can assign the conflicting data into a class with less *Z*-value. The procedures of *Z*-value measure is shown as follows:

Step 1: Initially, i=1 and there exists a set Z such that $Z=\Phi$.
Step 2: If $[f_i(x_j)-\mu]^2 \le \sigma$, that is, data x_j is recognized by f_i, then $Z=\{f_i\} \cup Z$, else go to Step 3.
Step 3: If $i<K$, then $i=i+1$, go to Step 2. Else go to Step 4.
Step 4: If $|Z|=1$, then output C_i since $Z=\{f_i\}$ and stop, else go to Step 5.
Step 5: If $|Z|=0$, then $Z=F$.
Step 6: Compute $Z_i(x_j)$ where $f_i \in Z$.
Step 7: Let $Z_i = \arg\min_i_{x_i \in Z}\{Z_i(x_j)\}$, then the data x_j will be assigned to the class C_i.

4. Experimental Results

The experiments are done by using a PC with 866MHz CPU and 128MB RAM. We modified the GP Quick 2.1[24] to fit the requirements of the proposed scheme and performed the experiments by two data sets: the Wisconsin Breast Cancer Diagnostic Dataset [17] and Iris dataset [4]. The two data sets are both well-known benchmark for evaluating the performance of classifiers.

In the proposed scheme, the results of classification are accomplished by different discriminant functions, except the traditional measures: average accuracy and overall accuracy [28], we evaluate the accuracy of each discriminant function by two additional criterions: precision and recall. Let N_i be the number of objects belonging to class C_i, N_{fi} be the number of objects recognized by function f_i and N^i_{fi} be the number of objects belonging to class C_i and recognized by function f_i. For a specific class C_i, the precision and recall for its corresponding discriminant function f_i are defined as equations (6) and (7), respectively.

$$Precision = \frac{N^i_{f_i}}{N_{f_i}}, \tag{6}$$

$$Recall = \frac{N^i_{f_i}}{N_i}. \tag{7}$$

Assume that $|T|$ is the number of the test data. K is the number of classes. The definitions of average accuracy and overall accuracy [28] are given in equations (8) and (9).

$$Average\ accuracy = \frac{1}{K}\sum_{i=1}^{K}\frac{N^i f_i}{N_i}, \tag{8}$$

$$Overall\ accuracy = \frac{\sum_{i=1}^{K} N^i_{f_i}}{|T|}. \tag{9}$$

A discriminant function f_i with higher precision has lower misclassification rate for class C_i. A discriminant function f_i with higher recall means that the f_i behaves higher recognition rate for class C_i. The average accuracy stands for the average recognition rate of discriminant functions. The overall accuracy presents the performance of recognition for a classifier.

I. Wisconsin Breast Cancer Diagnostic Data Set

The first experiment uses Wisconsin Breast Cancer Database (WBCD) data set [17]. The data set contains 699 data that separated into two classes called "malignant" and "benign". There are 241 data in the "malignant" and 458 data in the "benign". Each data in WBCD data set contains nine numerical attributes. However, there are 16 data of WBCD have missing values, we use the remaining 683 data to be the final test date set after deleting the data with missing values. The final "malignant" remains 239 data and the final "benign" remains 444 data.

The training set *TS* is constructed by 119 data from "malignant" and 222 data from "benign". Therefore, the total number of training set *TS* is 341 samples. The other 342 data are used to be the test set. The parameters in adaptive incremental training strategy are g=1000, ρ=0.1, ω=0 and α=2. The parameters in fitness function are set to be p=100, μ=0 and σ=100. The 9 attributes of a sample are denoted as $F1$, $F2$, ..., $F9$, which are the terminals in GP. The related parameters used in GP program are shown in Table 1.

The experimental results for WBCD data are shown in Table 2 and Table 3. The Table 2 is the results without Z-value measure and the Table 3 is the results with Z-value measure.

Actually, a two-class classification problem also can be finished by a single discriminant function only. While a data object was not recognized by the discriminant function, it would be assigned to the other one class. The advantage of using single discriminant function is that it is never have the problem of conflicts. The accuracies of using single discriminant function are listed in Table 4.

We found that either the average accuracy or the overall accuracy of the results without Z-measure can be improved by Z-value measure. Furthermore, after resolving the conflicts by the Z-value measure, we found that neither the average accuracy nor the overall accuracy of the results with single discriminant function is better than the results with Z-measure.

Table 5 is the comparison of other methods and the proposed method. The comparison shows that the proposed method performs better accuracy than other methods perform.

Table 1: Parameters used in the experiment of WBCD.

Parameter	Value	Parameter	Value
Node mutate weight	43.5%	Mutation weight annealing	40%
Mutate constant weight	43.5%	Population size	1000
Mutate shrink weight	13%	Generations per stage g	1000
Selection method	Tournament	Incremental rate ρ	0.1
Tournament size	7	Adaptive factor α	2
Crossover weight	28%	Criterion number ω	0
Mutation weight	8%	Terminal set	$F1, F2 \ldots F9$
Specified constant μ	0	Function set	$+, -, \times, \div$
		Specified constant σ	100

Table 2: The results without Z-value measure

	$f_{malignant}$	f_{benign}
Malignant	118	3
Benign	6	204
Precision	95.16%	98.55%
Recall	98.33%	91.89%

Table 3: The results with Z-value measure

	$f_{malignant}$	f_{benign}
Malignant	120	0
Benign	6	216
Precision	95.24%	100%
Recall	100%	97.30%
Average	98.65%	
Overall	98.25%	

Table 4: The results with single discriminant function in test set

	Single $f_{malignant}$		Single f_{benign}	
Malignant	118	2	117	3
Benign	6	216	18	204
Precision	95.16%	99.08%	86.67%	98.55%
Recall	98.33%	97.30%	97.50%	91.89%
Average	97.82%		94.70%	
Overall	97.66%		93.86%	

Table 5: A comparison of classification results on WBCD

Models or methods	Testing recognition rate
MSC [16]	94.9%
C4.5*	93.1%
NEFCLASS [18]	92.7%
NNFS with 9 features [21]	93.94%
NNFS with avg. 2.73 features [21]	94.15%
FEBFC with 9 features [15]	94.67%
FEBFC with selected 6 features [15]	95.14%
Our method	98.65%

* The result is available in [18]

II. Fisher's Iris

The second experiment uses Fisher's Iris data set [4]. There are 150 data separated into three classes: "setosa", "versicolor" and "virginica". Each class contains 50 data. There are four numerical attributes namely sepal length, sepal width, petal length and petal width, respectively. We denote the four attributes as *SL*, *SW*, *PL* and *PW*, respectively.

We randomly select 25 data from each class to construct the training set. The parameters in the adaptive incremental learning strategy are g=1000, ρ=0.2, ω=0 and α=2. The parameters in fitness function are p=100, μ=0 and σ=100. The parameters used in GP program are shown in Table 6.

After the learning procedure, three discriminant functions are obtained:

f_{setosa}: $SW\times PL^2-SW^2+SW\times PL$,

$f_{versicolor}$: $((-110/PL)-SW+36)$,

$f_{virginica}$: $((-23+SL)/PW)$.

Furthermore, we can translate the three functions into the following rules.

IF ($SW \times PL^2 - SW^2 + SW \times PL \leq 10$) **THEN** "setosa"

IF ($((-110/PL) - SW + 36) \leq 10$) **THEN** "versicolor"

IF ($((-23 + SL)/PW) \leq 10$) **THEN** "virginica"

The performance of f_{setosa}, $f_{versicolor}$ and $f_{virginica}$ are shown in Table 7 and Table 8. The Table 7 is the result without *Z*-value measure. The Table 8 is the results with *Z*-value measure. We found that either the three discriminant functions or the three rules are concise, but still accurate and effective. We also found that the method of *Z*-value measure really improves the accuracy either in precision or in recall. Thus, the average accuracy and the overall accuracy are also improved. Finally, we compare the proposed method with the results of other classifiers in Table 9. The comparison shows that the proposed method behaves better accuracy than other methods behave.

Table 6: Parameters used in the experiment of Iris

Parameter	Value	Parameter	Value
Node mutate weight	43.5%	Mutation weight annealing	40%
Mutate constant weight	43.5%	Population size	1000
Mutate shrink weight	13%	Generations per stage g	1000
Selection method	Tournament	Incremental rate ρ	0.2
Tournament size	7	Adaptive factor α	2
Crossover weight	28%	Criterion number ω	0
Mutation weight	8%	Terminal set	*SL, SW, PL, PW*
Specified constant μ	0	Function set	$+, -, \times, \div$
		Specified constant σ	100

Table 7: The results of test set without *Z*-value measure

	f_{setosa}	$f_{versicolor}$	$f_{virginica}$
Setosa	25	0	0
Versicolor	0	22	1
Virginica	0	0	23
Precision	100%	100%	95.83%
Recall	100%	88.0%	92.0%

Table 8: The results of test set with *Z*-value measure

	f_{setosa}	$f_{versicolor}$	$f_{virginica}$
Setosa	25	0	0
Versicolor	0	23	2
Virginica	0	0	25
Precision	100%	100.0%	92.6%
Recall	100%	92.0%	100.0%
Average		97.3%	
Overall		97.3%	

Table 9: The comparison on classification results on Fisher's Iris data set

Models or Methods	Testing recognition rate
FMMC [23]	97.3%
FUNLVQ-FENCE [2]	95.7%
FUNLVQ+GFENCE [14]	96.3%
FEBFC with 4 features [15]	96.7%
FEBFC with 2 selected features [15]	97.1%
Our method	97.3%

5. Conclusions

This paper presents an efficient scheme to learn discriminant functions for classification based on genetic programming. The proposed approaches include the adaptive incremental learning strategy, a new fitness function that constrains the outputs of the discriminant functions, and the *Z*-value measure that is a conflict resolution method. The experimental results of the two test datasets, WBCD and Iris, show that the proposed scheme obtains high accuracy on both. However, the attributes in the two datasets are all numerical. The categorical attribute seems to have no simple way to be applied for classification by discriminant functions. The categorical classification based on genetic programming is worth to be future investigated.

References

1. Bojarczuk, C. C., Lopes, H. S., Freitas, A. A.: Discovering comprehensible classification rules using genetic programming: a case study in a medical domain. Proc. Genetic and Evolutionary Computation Conf. (GECCO-99). Orlando, FL, USA (1999) 953-958
2. Chen, K. H. et al.: A multiclass neural network classifier with fuzzy teaching inputs. Fuzzy Sets System, Vol. 91, no. 1 (1997) 15-35
3. Duda, R. O., Hart, P. E.: Pattern classification and scene analysis, John Wiley & Sons (1973)
4. Fisher, R. A.: The use of multiple measurements in taxonomic problems. Ann. Eugenics, pt. II, Vol. 7 (1936) 179-188
5. Freitas, A. A.: A genetic programming framework for two data mining tasks: classification and generalized rule induction. Proc. 2nd Annual Conf. Morgan Kaufmann (1997) 96-101
6. Han, E. H., Karypis, G., Kumar, V.: Text categorization using weight adjusted k-nearest neighbor classification. PhD thesis, University of Minnesota (1999)
7. Heckerman, D., Wellman, M. P.: Bayesian networks. Communications of the ACM, Vol. 38, No. 3 (1995)
8. Hunt, E. B., Marin, J., Stone, P. J.: Experiments in induction. Academic Press (1966)
9. Kishore, J. K., Patnaik, L. M., Mani, V., Agrawal, V. K.: Application of genetic programming for multicategory pattern classification. IEEE Trans. on Evolutionary Computation, Vol. 4, No. 3 (2000) 242-258
10. Kotani, M., Ozawa, S., Nakai, M., Akazawa, K.: Emergence of feature extraction function using genetic programming. Proc. Third Int. Conf. on Knowledge-Based Intelligent Information Engineering System (1999) 149-152

11. Koza, J. R.: Genetic Programming: On the programming of computers by means of Natural Selection. MIT Press (1992)
12. Koza, J. R.: Introductory genetic programming tutorial. Genetic Programming 1996 Conf. Stanford University (1996)
13. Kulkarni, S. R., Lugosi, G., Venkatesh, S. S.: Learning pattern classification – a survey. IEEE Trans. on Information Theory, Vol. 44, No. 6 (1998) 2178-2206
14. Lee, H. M.: A neural network classifier with disjunctive fuzzy information. Neural Networks, Vol. 11, No. 6 (1998) 1113-1125
15. Lee, H. M., Chen, C. M., Chen, J. M., Jou, Y. L.: An efficient fuzzy classifier with feature selection based on fuzzy entropy. IEEE Trans. on Systems, Man, and Cybernetics-part b: Cybernetics, Vol. 31, No. 3 (2001) 426-432
16. Lovel, B. C., Bradley, A. P.: The multiscale classifier. IEEE Trans. on Pattern Analysis Machine Intelligence, Vol. 18 (1996) 124-137
17. Mangasarian, O. L., Wolberg, W. H.: Cancer diagnosis via linear programming. SIAM News, Vol. 23, No. 5 (1990) 1-18
18. Nauck, D., Kruse, R.: A neuro-fuzzy method to learn fuzzy classification rules from data. Fuzzy Sets System, Vol. 89, No. 3 (1997) 277-288
19. Quinlan, J. R.: Induction of decision trees. Machine Learning, 1: (1986) 81-106
20. Ross, K. A., Wright, C. R. B.: Discrete Mathematics. Prentice-Hall, Inc. (1992)
21. Setiono, R. et al., "Neural-network feature selector", IEEE Trans. on Neural Networks, Vol. 8 (1997) 654-662
22. Sherrah, J., Bogner, R. E., Bouzerdoum, A.: Automatic selection of features for classification using genetic programming. in Proc. Australian New Zealand Conf. On Intelligent Information Systems (1996) 284-287
23. Simpson, P. K.: Fuzzy Min–Max neural networks—Part 1: Classification. IEEE Trans. on Neural Networks, Vol. 3 (1992) 776-786
24. Singleton, A.: Genetic programming with C++. Byte, Feb (1994) 171-176
25. Wang, C. H., Liu, J. F., Hong, T. P., Tseng, S. S.: A fuzzy inductive learning strategy for modular rules. Fuzzy Set and Systems (1999) 91-105
26. Wang, C. H., Hong, T. P., Tseng, S. S.: Integrating fuzzy knowledge by genetic algorithms. IEEE Trans. on Evolutionary Computation, Vol. 2, No. 4 (1998) 138-149
27. Wang, C. H., Hong, T. P., Tseng, S. S., Liao, C. M.: Automatically integrating multiple rule sets in a distributed-knowledge environment. IEEE Trans. On Systems, Man, And Cybernetics Part C: Applications and Reviews, Vol. 28, No. 3 (1998) 471-476
28. Yoshida, T., Omatu, S.: Neural network approach to land cover mapping. IEEE Trans. on Geosciences Remote Sensing, Vol. 32, No. 5 (1994) 1103-1108
29. Zhang, G. P.: Neural networks for classification: a survey. IEEE Trans. on Systems, Man, and Cybernetics-Part C: Applications and Reviews, Vol. 30, No. 4 (2000) 451-462

SNNB: A Selective Neighborhood Based Naïve Bayes for Lazy Learning

Zhipeng Xie[1] Wynne Hsu[1] Zongtian Liu[2] Mong Li Lee[1]

[1]School of Computing
National University of Singapore
Lower Kent Ridge Road, Singapore, 119260
{xiezp, whsu, leeml}@comp.nus.edu.sg
[2]School of Computing
Shanghai University of China
Shanghai, P.R.China, 200072
ztliu@mail.shu.edu.cn

Abstract. Naïve Bayes is a probability-based classification method which is based on the assumption that attributes are conditionally mutually independent given the class label. Much research has been focused on improving the accuracy of Naïve Bayes via eager learning. In this paper, we propose a novel lazy learning algorithm, Selective Neighbourhood based Naïve Bayes (SNNB). SNNB computes different distance neighborhoods of the input new object, lazily learns multiple Naïve Bayes classifiers, and uses the classifier with the highest estimated accuracy to make decision. The results of our experiments on 26 datasets show that our proposed SNNB algorithm is efficient and it outperforms Naïve Bayes, and state-of-the-art classification methods NBTree, CBA, and C4.5 in terms of accuracy.

Key words: Naïve Bayes, Classification, Lazy Learning

1 Introduction

Naive Bayes [5] is a probability-based classification method which is based on the assumption that attributes are conditionally mutually independent given the class label. Although simple, Naïve Bayes has surprisingly good performance in a wide variety of domains, including many domains where there are clear dependencies between the attributes. Naïve Bayes is also robust to noise and irrelevant attributes and the learnt theories are easy for domain experts to understand. As a result, Naïve Bayes has attracted much attention from researchers. Research work to extend the Naïve Bayes can be broadly divided into three main categories.

The first category aims to improve Naïve Bayes by transforming the feature space such as feature subset selection and constructive feature. Kononenko's semi-naïve Bayesian classifier [10] performs exhaustive search by iteratively joining pairs of attribute values to generate constructive features based on statistical tests for independence. The constructive Bayesian classifier [14] employs a wrapper model to find the best Cartesian product attributes from existing nominal attributes, and possible deletion of existing attributes. Langley and Sage [11] use the Forward Sequential Selection (FSS) method to select a subset of the available attributes, with

M.-S. Chen, P.S. Yu, and B. Liu (Eds.): PAKDD 2002, LNAI 2336, pp. 104-114, 2002.

which to build a Naïve Bayes classifier. It is shown that such attribute selection can improve upon the performance of the Naïve Bayes classifier when attributes are interdependent, especially when some attributes are redundant.

The second category of research extends Naïve Bayes by relaxing the attribute independence assumption. This covers many classification methods based on Bayesian network [2]. Friedman and Goldszmidt [8] explore the Tree Augmented Naïve Bayes (TAN) model for classifier learning, which belongs to a restricted subclass of Bayesian network by inducing a tree-structure network structure.

The third category employs the principle of local learning to extend Naïve Bayes. It is well-established that large, complex databases are not always amenable to a unique global approach to generalization. This is because there may exist different models specific to a data point. A typical example in this category is the Naïve Bayes tree, NBTree [9], which uses decision tree techniques to partition the whole instance space (root node) into several subspaces (leaf nodes), and then trains a Naïve Bayes classifier for each leaf node. [17] presents the Lazy Bayesian Rule (LBR) classification method to solve the small disjunct problem of NBTree.

In addition, Zheng Zijian [16] presents a method to generate Naïve Bayes Classifier Committees by building individual naïve Bayes classifiers using different attribute subsets in sequential trials. Majority vote of the committees was applied in the classification stage. It has been shown that this method is able to improve the accuracy of Naïve Bayes by a wide margin.

On the other hand, we can divide classification methods into two types: eager learning and lazy learning, depending on when the major computation occurs [1]. Lazy learning is distinguished by spending little or no effort during training and delaying computation until classification time. On the other hand, eager learning replaces the training inputs with an abstraction expression such as rule set, decision tree, concept lattice or neural network) and use it to process queries. The majority of the methods to extend Naïve Bayes are eager, except for LBR. We observe that most existing techniques for improving the performance of the Naïve Bayesian classifier require complex induction processes.

In this paper, we propose a novel Naïve Bayes classifier, the Selective Neighborhood Naïve Bayes (SNNB), for lazy classification. SNNB constructs multiple Naïve Bayes classifiers on multiple neighborhoods by using different radius values for an input new object. It then selects the most accurate one to classifier the new object. Experimental results shows that SNNB outperforms not only the Naïve Bayes and NBTree, but also several other state-of-art classification methods.

The rest of the paper is organized as follows. Section 2 briefly reviews the Naïve Bayes method. Section 3 gives an example to motivate the work of SNNB. A detailed description and analysis of SNNB is given in Section 4. Section 5 gives the results from the performance study of SNNB. Section 6 discuss some related work and we conclude in Section 7 by highlighting our contributions.

2 Naïve Bayes and Accuracy Estimation

For simplicity, we shall assume that the dataset is a relational table with only nominal attributes and consists of the descriptions of n objects in the form of tuples. These n

objects have been classified into q known classes, C_1, C_2, ..., C_q. Each object in the database is described by m distinct attributes, $Attr_1$, ..., $Attr_i$, ..., $Attr_m$. In an instantiation of object description, an attribute $Attr_i$ takes on the value $v_{ij} \in domain(Attr_i)$. Let $T=\{x_1, x_2, ..., x_n\}$ denote the set of objects and $A=\{Attr_1, ..., Attr_m\}$ denote the set of attributes. Various kinds of classification method have been developed to induce classifiers on a dataset, and the classifier can be thought as a function assigning a class label to an unclassified object.

Naïve Bayes is a probability-based classification method, built on the assumption that all the attributes are mutually independent within each class. Given an unlabelled instance $x=<v_1, ..., v_m>$ consisting of m attribute values, the classification technique of Naïve Bayes classifier will assign the object x to the class C_i such that the value of $P(C_i|x)$ is maximal. $P(C_i|x)$ can be calculated via Bayesian Theorem and the independence assumption as follows:

$$P(C_i|x)=\frac{P(x \mid C_i)\times P(C_i)}{P(x)} \propto P(x|C_i)\times P(C_i)=\prod_{j=1}^{m} P(A_j = v_j \mid C_i)\times P(C_i). \quad (1)$$

Note that $P(x)$ is fixed for a given x. To estimate $p(C_i)$, the simplest probability estimate, or occurrence frequency, is used. That is,

$$p(C_i) = N(C_i)/N$$

where N is the number of the training examples, and $N(C_i)$ is the number of the training examples with class C_i. To estimate the conditional probability $p(Attr_k=v_k|C_i)$, we adopt the Laplace-corrected estimate, which leads to

$$p(Attr_k=v_k|C_i)=(N(Attr_k=v_k, C_i)+f)/(N(C_i)+fn_j)$$

where n_j is the number of values of the k-th attribute, and f is a multipicative factor (default value as $1/N$) [4]

For Naïve Bayes classifier, Leave-One-Out is used to get the accuracy on training set. This can be implemented efficiently, and is linear to the number of objects, number of attributes, and number of label values [9]. For Naive Bayes classifier cls_{NB}, $acc(cls_{NB})$ is used to denote the accuracy computed through Leave-One-Out method.

3 Motivating Example

Before presenting the details of our SNNB algorithm, let us look at a simple example, for which Naïve Bayes fails while one possible solution succeeds.

Example: Suppose we are given a small dataset that comprises of 300 objects which are described by two conditional attributes, A and B. These objects are divided into two classes, d=0 and d=1. Table 1 shows that among the 50 objects with description (A=0, B=0), 45 are classified as (d=0), while 5 are classified as (d=1). Obviously, an ideal classifier should be able to classify an object (A=0, B=0) to class (d=0), object (A=0, B=1) to class (d=1), object (A=1, B=0) to class (d=1), and object (A=1, B=1) to class (d=0). Note that for simplicity, all the probabilities are estimated with occurrence frequencies.

A	B	d=0	d=1
0	0	45	5
0	1	5	95
1	0	5	45
1	1	95	5

Table 1. Example of a dataset

Let us first consider the classifier generated by Naïve Bayes. We have the following probabilities:

p(A=0|d=0)=1/3, p(A=1|d=0)=2/3, p(B=0|d=0)=1/3, p(B=1|d=0)=2/3
p(A=0|d=1)=2/3, p(A=1|d=1)=1/3, p(B=0|d=1)=1/3, p(B=1|d=1)=2/3
p(d=0)=p(d=1)=1/2

According to Equation (1) (in Section 2), the following results will be obtained:

(A=0, B=0) will be classified as d=1,
(A=0, B=1) will be classified as d=1,
(A=1, B=0) will be classified as d=0, and
(A=1, B=1) will be classified as d=0,

Clearly, Naïve Bayes can not produce the ideal results.

Now, let us consider the situation where we construct a naïve bayes trained for an input test example on its neighborhood including all the objects with distance no larger than 1. Thus, for an test object with description (A=0, B=0), only (A=0, B=0), (A=0, B=1), (A=1, B=0) in Table 1 will be considered as being in the neighborhood. The following probabilities can be computed on these training objects:

p(A=0|d=0)=50/55, p(A=1|d=0)=5/55
p(B=0|d=0)=50/55, p(B=1|d=0)=5/55
p(A=0|d=1)=100/145,p(A=1|d=1)=45/145
p(B=0|d=1)=50/145, (B=1|d=1)=95/145
p(d=0)=55/200, p(d=1)=145/200

Hence,

p(A=0|d=0)*p(B=0|d=0)*p(d=0) = (50/55)*(50/55)*(55/200) = 0.227
p(A=0|d=1)*p(B=0|d=1)*p(d=1) = (100/145)*(50/145)*(145/200) = 0.172

Therefore, (A=0, B=0) will be classified as d=0.

Similarly, for each other input object, through constructing a naïve bayes classifier on its 1-neighborhood, and applying this classifier to make decision for the input object, an ideal classification result can be achieved.

4 Selective Neighborhood Based Naïve Bayes

We will now present the details of SNNB, a Selective Neighborhood based Naïve Bayesian classifier. For a given new object, the basic idea is to construct multiple classifiers on its multiple neighborhoods with different radius, then select out the classifier with the highest estimated accuracy to classify it.

For any two objects x and y, the distance between them normally can be defined as the number of the attributes on which x and y take on the different values, that is,

$$distance(x, y)=|\{Attr_i \in A | Attr_i(x) \neq Attr_i(y)\}|.$$

For an input new object x, its k-Neighborhood consists of all the objects in T with the distance to x not larger than k, denoted as

$$NH_k(x)=\{x_i \in T \mid distance(x_i, x) \leq k\}$$

We also call the Naïve Bayes classifier, $k\text{-}NB_x$, trained on $NH_k(x)$ as the k-th local Naïve classifier. Clearly, for any input object x, $m\text{-}NB_x$ is trained on the whole object set T, so it is also called the global Naïve Bayes classifier. The pseudo-code for SNNB is given below, where OWD is an array of sets of objects, and OWD[*dist*], $0 \leq dist \leq m$, stores all the objects in T that have distance *dist* to the input object x.

```
Algorithm SNNB
input: Training set T,
       the trained global NB classifier CLS_global,
       unknown new object x;
output: the predicted class of x;

begin
1  Add t into OWD[distance(t,x)] for each t∈T;
2  k=|A|; j=0;
3  total=|T|;
4  k-NB=CLS_global; NH_k=T;
5  Candidates={k-NB};
6  while (true)
7    count=0;
8    while(count<(1-θ)×total)//θ is set as 0.5 defaultly
9      count+=|OWD[k]|;
10     k--;
11   endwhile;
12   if (count<ϕ×|T|) break; endif;
13   j++;
14   NH_k=OWD[0]∪OWD[1]∪...∪OWD[k];
15   total=|NH_k|;
16   Train a Naïve Bayes classifier k-NB on NH_k;
17   Add k-NB into Candidates;
18 endwhile;
19 Select out the classifier q-NB with the maximal
   value of acc(q-NB) from Candidates;
20 return q-NB(x);
end;
```

The algorithm SNNB consists of three main steps. The first step calculates the distance between the input new object x and each training object t in training set T, and stores all the training objects according to their distances (line 1). The second step constructs a series of NB classifiers on different subsets of training objects (line 2-line 18). The last step classifies the new object with the most accurate NB classifier (line 19-line 20). There are two parameters in the algorithm: one is the support difference threshold, θ, with 0.5 as the default value; the other is the support threshold ϕ, with 0.03 as the default value. The support threshold is to ensure the generalizing ability of learnt model, while the support difference threshold is mainly for controlling the speed of the algorithm.

Complexity Analysis

We will now give an analysis of the complexity of the algorithm. Let m be the number of attributes and n be the number of objects. It is obvious that the complexities of the first step and the third step are O(m×n) and O(m) respectively. We will now examine the complexity of the second step.

Fact: Given a set TS of training objects, the complexity of inducing a NB classifier CLS and estimating its accuracy with Leave-1-out is O(|TS|×m).

Suppose the generated series of NB classifiers is CLS_1, CLS_2, …, CLS_p, which are trained on TS_1, TS_2, …, TS_p respectively. From lines 8-10, we know that

(1) $|TS_1| \leq \theta \times n$, and

(2) $|TS_{i+1}| \leq \theta \times |TS_i|$ for i = 1, 2, …, p−1

That is, $|TS_i| \leq \theta^i \times n$ for i=1, 2, …, p−1. According to line 12, we also have $|TS_p| = \theta^p \times n \geq \phi \times n$, that is $p \leq \log_\theta \phi$. Given the values of θ and ϕ, $\log_\theta \phi$ is a constant value. Hence, the complexity of the second step is also O(m×n).

5 Experimental Results

Dataset	No. Attrs	No. Classes	Size	Dataset	No. Attrs	No. Classes	Size
anneal	38	6	798	australian	14	2	690
auto	25	7	205	breast-w	10	2	699
cleve	13	2	303	crx	15	2	690
diabetes	8	2	768	german	20	2	1000
glass	9	7	214	heart	13	2	270
hepatitis	19	2	155	horse	22	2	368
hypo	25	2	3163	ionosphere	34	2	351
iris	4	3	150	labor	16	2	57
led7	7	10	3200	lymph	18	4	148
pima	8	2	768	sick	29	2	2800
sonar	60	2	229	tic-tac-toe	9	2	958
vehicle	18	4	846	waveform	21	3	5000
wine	13	3	178	zoo	16	7	101

Table 2. Datasets used in the experiments

We carried out an empirical comparison of the algorithm SNNB by using the 26 datasets from UCI Machine Learning Repository [13]. The characteristics of these datasets are listed in Table 2. Since the current version of SNNB can only deal with nominal attributes, the entropy-based discretization algorithm [6] is used for preprocessing.

5.1 Error-Rate Comparison

Dataset	NBTree	CBA	C4.5Rules	NB	SNNB
anneal	**1.0 (1)**	2.1 (4)	5.2 (5)	1.6 (3)	1.4 (2)
australian	14.5 (2)	14.6(3)	15.3 (5)	**14.1 (1)**	14.8 (4)
auto	22.8 (4)	**19.9 (1)**	**19.9 (1)**	27.7 (5)	22.5 (3)
breast-w	2.6 (2)	3.7 (4)	5.0 (5)	**2.4 (1)**	3.0 (3)
cleve	19.1 (4)	**17.1 (1)**	21.8 (5)	18.1 (2)	18.5 (3)
crx	14.2 (2)	14.6 (4)	15.1 (5)	14.5 (3)	**13.9 (1)**
diabetes	**24.1 (1)**	25.5 (4)	25.8 (5)	**24.1 (1)**	**24.1(1)**
german	**24.5 (1)**	26.5 (4)	27.7 (5)	**24.5 (1)**	26.2 (3)
glass	28.0 (2)	**26.1 (1)**	31.3 (5)	28.5 (4)	28.0 (2)
heart	**17.4 (1)**	18.1 (2)	19.2 (5)	18.1 (2)	18.9 (4)
hepatitis	**11.7 (1)**	18.9 (4)	19.4 (5)	15.6 (3)	14.3 (2)
horse	18.7 (4)	17.6 (3)	**17.4 (1)**	21.7 (5)	**17.4 (1)**
hypo	1.0 (2)	1.0 (2)	**0.8 (1)**	1.8 (4)	1.8 (4)
ionosphere	12.0 (5)	**7.7 (1)**	10.0 (2)	10.5 (3)	10.5 (3)
iris	7.3 (5)	5.3 (2)	**4.7 (1)**	5.3 (2)	5.3 (2)
labor	12.3 (3)	13.7 (4)	20.7 (5)	5.0 (2)	**3.3 (1)**
led7	26.7 (3)	28.1 (5)	**26.5 (1)**	26.7 (3)	**26.5 (1)**
lymph	17.6 (2)	22.1 (4)	26.5 (5)	19.0 (3)	**17.0 (1)**
pima	24.9 (3)	27.1 (5)	**24.5 (1)**	**24.5 (1)**	25.1 (4)
sick	22.1 (5)	2.8 (2)	**1.5 (1)**	4.2 (4)	3.8 (3)
sonar	22.6 (4)	22.5 (3)	29.8 (5)	21.6 (2)	**16.8 (1)**
tic-tac-toe	17.0 (4)	**0.4 (1)**	0.6 (2)	30.1 (5)	15.4 (3)
vehicle	29.5 (3)	31 (4)	**27.4 (1)**	40.0 (5)	28.4 (2)
waveform	**16.1 (1)**	20.3 (4)	21.9 (5)	19.3 (3)	17.4 (2)
wine	2.8 (3)	5.0 (4)	7.3 (5)	**1.7 (1)**	**1.7 (1)**
zoo	5.9 (4)	3.2 (2)	7.8 (5)	3.9 (3)	**2.9 (1)**
Average	16.02	15.19	16.67	16.33	**14.57**

Table 3. Experiment results on the error rates of classifiers

We first compare the accuracy results of SNNB with Naïve Bayes, and three other state-of-art classification methods[1]:

- NBTree in [9] (a state-of-art hybrid classification method to improves the accuracy of Naïve Bayes),
- CBA in [12] (a classification method based on association rules), and
- C 4.5 Rules (Release 8) [15].

The error rates of the different algorithms on the experimental domains are listed in Table 3. All the error rates are obtained through 10-fold cross validation. We use the same train/test set split for different classification methods in the experiments. Throughout the experiment, the parameters of SNNB are set as default values without adjusting.

---

[1] These classification systems are all available from Web, where NBTree is implemented in the MLC Utilities available from http://www.sgi.com/tech/mlc/, CBA is downloadable from http://www.comp.nus.edu.sg/~dm2/, and C4.5 from http://www.cse.unsw.edu.au/~quinlan/.

From table 3, we have observed the following facts:

(1) SNNB obtains lower error rates than Naïve Bayes in 15 out of the 26 domains, and higher error rates in 6 domains. It also obtains lower error rates than CBA in 16 domain and higher error rates in 9 domains. When compared with NBTree, SNNB gets the similar sore, winning in 15 domains and losing in 9 domains, while compared with C4.5Rules, SNNB wins in 16 domains and loses in 8 domains.

(2) To give further insight into the experimental results, for each dataset, all the classification methods are sorted from the lowest error rate to the highest error rate. The ranking of each method is recorded in the parentheses, where number i ($1 \le i \le 5$) means that the method gets the i-th lowest error rate among the five methods. Such information has been summarized into the following table. SNNB gets the lowest error rates on 9 domains, the sencond lowest error rates on 6 domains, the third lowest error rates on 7 domains, the fourth lowest error rates on 4 domain, and doesn't get the worst error rate on any domain. We can also conclude that SNNB is better than NB, NBTee, C4.5Rules and CBA. A interesting fact is that, C4.5Rules gets the lowest error rates on 8 domains, which is better than NB, NBTree and CBA, but it also gets the worst error rates on 16 domains.

	1st	2nd	3rd	4th	5th
NBTree	6	6	5	6	3
CBA	5	5	3	11	2
C4.5Rules	8	2	0	0	16
NB	6	5	8	3	4
SNNB	9	6	7	4	0

Table 4. Summary of ranking information

(3) With the comparison of the average accuracy, SNNB also produces more accurate classifiers than any other method.

5.2 Computational Requirements

At the end of section 4, we have already shown the time complexity of running SNNB to classify a new object is linear with the number of training objects and the number of attributes. To get an intuitionistic idea of the computational requirements of SNNB, the table 5 records the average time of SNNB in CPU seconds on the personal computer (Pentium III 900Mhz with 128M memory) for classifying each object. We have not recorded the training time used to induce a global Naive Bayes classifier, because this process is always very fast.

In table 5, when θ set as 0.5, the three worst cases are 0.1477 seconds, 0.1471 seconds, and 0.1402 seconds in the hypo domain, sick domain and waveform domain. These three domains have 3163, 2800, and 5000 objects respectively, and they also have 25, 29, and 21 attributes respectively.

Dataset	Support-Difference Threshold θ			
	θ=0.3	θ=0.5	θ=0.7	θ=0.9
anneal	0.0496	0.0607	0.0868	0.1475
australian	0.0092	0.0110	0.0148	0.0223
auto	0.0074	0.0101	0.0156	0.0324
breast-w	0.0063	0.0073	0.0102	0.0157
cleve	0.0051	0.0058	0.0070	0.0098
crx	0.0098	0.0120	0.0159	0.0249
diabetes	0.0075	0.0087	0.0105	0.0138
german	0.0203	0.0240	0.0310	0.0429
glass	0.0033	0.0042	0.0048	0.0092
heart	0.0046	0.0053	0.0070	0.0091
hepatitis	0.0038	0.0046	0.0058	0.0099
horse	0.0070	0.0088	0.0124	0.0206
hypo	0.1315	0.1477	0.1798	0.2734
ionosphere	0.0104	0.0131	0.0184	0.0347
iris	0.0007	0.0007	0.0012	0.0015
labor	0.0011	0.0019	0.0021	0.0021
led7	0.0371	0.0520	0.0812	0.1274
lymph	0.0028	0.0047	0.0081	0.0104
pima	0.0068	0.0081	0.0094	0.0123
sick	0.1315	0.1471	0.1735	0.2396
sonar	0.0164	0.0190	0.0235	0.0357
tic-tac-toe	0.0066	0.0089	0.0125	0.0176
vehicle	0.0141	0.0199	0.0312	0.0646
waveform	0.1103	0.1402	0.2057	0.3432
wine	0.0025	0.0037	0.0052	0.0067
zoo	0.0018	0.0028	0.0060	0.0113
Average	0.0234	0.0282	0.0377	0.0592

Table 5. Average time to classify a new object (in seconds)

6 Related Work

Although the underlying mechanism in SNNB is lazy, SNNB is closely related to the Naïve Bayes Tree method. Let us consider two kinds of neighborhood,

(1) Feature-based: For any object x and a feature subset F of its description, the F-neighborhood of x consists of all the objects containing the feature subset F.

(2) Distance-based: For any object x and a distance d, the d-Neighborhood of x consists of all the objects whose distances are less than d.

Now, it is clear that SNNB uses the Naïve Bayes classifier trained on a selective distance-based neighborhood of the input test object to make classification. On the other hand, NBTree uses the Naïve Bayes classifier trained on a selective feature-based neighborhood of the input test object to make classification.

Furthermore, our method is also related to the k-nearest neighbor (k-NN) algorithm [3], which is one of the most venerable algorithms in machine learning. Both SNNB and k-NN are based on distance (or similarity). The k-NN algorithm can be decomposed into two phases: Phase 1 determines a neighborhood which is formed by the nearest k neighbors, while Phase 2 applies a simple classifier (Majority Voting) to classify the new object. If we replace the simple classifier (Majority voting in original k-NN algorithm) with Naïve Bayes, we can find that SNNB is also similar with the k-NN algorithm, and SNNB can be viewed as a hybrid method of Naïve Bayes and Nearest Neighborhood.

7 Conclusion

In this paper, we have developed a selective neighborhood-based Naive Bayes algorithm, SNNB. The contribution of SNNB is that it lazily constructs a set of Naïve Bayes classifiers, and chooses one of them to make decisions. Experimental results demonstrate that our proposed algorithm is not only able to improve the accuracy of Naïve Bayes, but it is also able to outperform existing state-of-art classification methods such as NBTree, CBA and C4.5. We have also shown that SNNB is computationally efficient.

Acknowledgements
We thank the anonymous reviewers for their suggestions. This work is supported by the NSTB-NUS research project R-252-000-102-112 and R-252-000-102-303. This work is also supported by National Natural Science Fund of China (No. 69985004).

References

[1] Aha, D.W. Lazy Learning. Dordrecht: Kluwer Academic, 1997
[2] Cheng, J., & Greiner, R. Comparing Bayesian network classifiers. in Proceedings of the fifteenth conference on uncertainty in artificial intelligence, 1999
[3] Cover, T.M., & Hart, P.E. Nearest neighbor pattern classification. IEEE Transactions on Information Theory, 1996, vol. 13, pp. 21-27
[4] Domingos, P., & Pazzani, M. On the optimality of the simple bayesian classifier under zero-one loss. Machine Learning, 1997, Vol. 29, pp. 103-130
[5] Duda, R.O., & Hart, P.E. Pattern Classification and Scene Analysis. New Yaork: John Wiley, 1973
[6] Fayyad, U.M, & Irani, K.B. Multi-interval discretization of continuous-valued attributes for classification learning. IJCAI-93, pp. 1022-1027
[7] Friedman, J.H., Kohavi, R., and Yun, Y. Lazy decision trees. In Proceedings of the Thirteenth National Conference on Artificial Intelligence, 1996, pp. 717—724
[8] Friedman, N., & Goldszmidt, M. Building classifiers using Bayesian networks. Proceedings of the Thirteenth National Conference on Artificial Intelligence, 1996, pp. 1277-1284
[9] Kohavi R. Scaling Up the Accuracy of Naïve-Bayes Classifiers: a Decision-Tree Hybrid. in Simoudis E. & Han J. (eds.), KDD-96: Proceedings Second International Conference on Knowledge Discovery & Data Mining, AAAI Press/MIT press, Cambridge/Menlo Park, pp. 202-207, 1996.
[10] Kononenko, I. Semi-naïve Bayesian classifier. in Proceedings of European Conference on Artificial Intelligence, 1991, pp. 206-219

[11] Langley, P., & Sage, S. Induction of selective Bayesian classifiers. in Proceedings of the Tenth Conference on Uncertainty in Artificial Intelligence, 1994, pp. 339-406
[12] Liu, B., Hsu, W., and Ma, Y. Integrating Classification and Association Rule Mining. Proceedings of the Fourth International Conference on Knowledge Discovery and Data Mining. New York, USA, 1998.
[13] Merz, C.J., and Murphy, P. UCI repository of machine learning database [http://www.cs.uci.edu/~mlearn/MLRepository.html], 1996
[14] Pazzani, M. Constructive induction of Cartesian product attributes. in Proceedings of the Conference ISIS96: Information, Statistics and Induction in Science, 1996, pp. 66-77
[15] Quinlan, J.R. C4.5: Programs for Machine Learning. Morgan Kaufmann, 1993
[16] Zheng, Z. Naïve Bayesian classifier committees. in Proceedings of European Conference on Machine Learning, 1998, pp. 196-207
[17] Zheng, Z. and Webb, G.I. Lazy Learning of Bayesian Rules. Machine Learning, 2000, Vol. 41(1), Kluwer Academic Publishers, pp. 53-84

A Method to Boost Naïve Bayesian Classifiers

Lili Diao, Keyun Hu, Yuchang Lu, and Chunyi Shi

The State Key Laboratory of Intelligent Technology and System, Dept. of Computer Science and Technology, Tsinghua University, Beijing 100084, China
diaolili@mails.tsinghua.edu.cn
http://www.cs.tsinghua.edu.cn

Abstract. In this paper, we introduce a new method to improve the performance of combining boosting and naïve Bayesian. Instead of combining boosting and Naïve Bayesian learning directly, which was proved to be unsatisfactory to improve performance, we select the training samples dynamically by bootstrap method for the construction of naïve Bayesian classifiers, and hence generate very different or unstable base classifiers for boosting. Besides, we devise a modification for the weight adjusting of boosting algorithm in order to achieve this goal: minimizing the overlapping errors of its constituent classifiers. We conducted series of experiments, which show that the new method not only has performance much better than naïve Bayesian classifiers or directly boosted naïve Bayesian ones, but also much quicker to obtain optimal performance than boosting stumps and boosting decision trees incorporated with naïve Bayesian learning.

1 Introduction

Boosting is an iterative machine learning procedure that successively classifies a weighted version of the sample, and then re-weights the sample dependent on how successful the classification was. Its purpose is to find a highly accurate classification rule by combining many weak or base hypotheses (classifiers), many of which may be only moderately accurate. As boosting progresses, training examples and their corresponding labels that are easy to classify get lower weights. The intended effect is to force the base classifier to concentrate on "difficult to identify" samples with high weights, which will be beneficial to the overall goal of finding a highly accurate classification rule [1]. Naïve Bayesian classifier is an effective classification technique, though its performance is always influenced by its attribute independency assumption and the incompleteness of training examples and hence usually not optimal. Boosting decision trees has been proved to be very successful for many machine-learning problems, but some experiments show that, Naïve Bayesian classifiers cannot be effectively improved by boosting [2]. The most possible reason is that Naïve Bayesian classifier is quite stable with respect to small disturbances of training data, since these

---

Supported by the National Grand Fundamental Research 973 Program of China under Grant No.G1998030414 and the National Natural Science Foundation of China under Grant No.79990580.

M.-S. Chen, P.S. Yu, and B. Liu (Eds.): PAKDD 2002, LNAI 2336, pp. 115-122, 2002.

disturbances will not result in important changes of the estimated probabilities. Therefore, boosting Naïve Bayesian classifiers may not be able to generate multiple models with sufficient diversity. These models cannot effectively correct each other's errors during classification by majority voting or weighted combination. In such a situation, boosting cannot greatly reduce the error of naïve Bayesian classification. In the sense, the stability of naïve Bayesian classifiers becomes a real problem when employing naïve Bayesian classifiers as the base classifiers of boosting. Kai Ming Ting and Zijian Zheng proposed to introduce tree structures into naïve Bayesian classification to increase its instability, and expect that this can improve the success of boosting for naïve Bayesian classification [2]. This method provides a classifier with lower error rates than the naïve Bayesian one. However, to achieve optimal results, those naïve Bayesian trees need to be large, say, many of them need to have depth over 5. Since the leaves of those trees are all naïve Bayesian classifiers using many training samples, its converging speed becomes very slow, and its computational cost is also expensive, especially for the task of text categorization that always have huge number of training documents with very high dimensions. In this paper we propose a method quite different with that of naïve Bayesian trees to obtain "instable" naïve Bayesian classifiers for boosting: using bootstrap methods to establish different training sample sets, based on which we can obtain different naïve Bayesian Classifiers. We modify the weight adjusting criteria of boosting to place extra emphasis on the sample documents that would be helpful in minimizing the overlapping errors among the constituent classifiers. Besides, we employ un-weighted majority vote for combining base naïve Bayesian classifiers because we expect the classification results would not be influenced by the order of different training sample sets.

Section 2 introduce the basic ideas of Boosting, naive Bayesian learning respectively. Section 3 and 4 illustrates our new algorithm in detail. Section 5 presents the corresponding experimental results of compared performance with regarding to this new algorithm and other related methods. Section 6 concludes our work.

2 Boosting and Naïve Bayesian Classifier

The Bayesian approach to classification estimates the (posterior) probability that an instance V belongs to a class, given the observed attribute values for the instance. Suppose there are L possible class labels $\{C_j\}$, $j = 1,\ldots,L$. When making a categorical rather than probabilistic classification, the class with the highest estimated posterior probability is selected. The posterior probability, $P(C_j \mid V)$, of an instance being class C_j, given the observed attribute values $V = \langle v_1, v_2, \ldots, v_n \rangle$, can be computed using the apriori probability of an instance V being class C_j, $P(C_j)$; the probability of an instance of class C_j having the observed attribute values, $P(V \mid C_j)$; and the prior probability of observing the attribute values, $P(V)$. Here n is the number of possible attributes. In the naïve Bayesian approach, the likelihood of the observed attribute values are assumed to be mutually independent for each class. With this attribute independence assumption, the probability $P(C_j \mid V)$ can be re-expressed as:

$$P(C_j \mid V) \propto P(C_j)P(V \mid C_j) \propto P(C_j)\prod_{i=1}^{n} P(v_i \mid C_j).$$

The description of boosting follows Freund and Schapire's AdaBoost.M1 [1]. We assume a set of training cases V_i, $i = 1,2,\ldots,N$. At each repetition or trial $s = 1,2,\ldots,T$, case V_i has weight $w_s[i]$, where $w_1[i] = 1/N$ for all i. The process at trial s can be summarized as follows:

- The base learning system constructs classifier H_s from the training cases using the weights w_s.
- The error rate ε_s of this classifier on the training data is determined as the sum of the weights $w_s[i]$ for each misclassified case V_i.
- Update the weight of each training case $w_s[i]$ for all i:

$$w_{s+1}[i] = w_s[i] \div \begin{cases} 2\varepsilon_s & \textit{if} \quad H_s \quad \textit{misclassifies} \quad \textit{case} \quad V_i \\ 2(1-\varepsilon_s) & \textit{otherwise} \end{cases}$$

The composite classifier H_B is obtained by voting each of the component classifiers $\{H_s\}$. If H_s classifies some case V as belonging to class C_j, the total vote for C_j is incremented by $\log((1-\varepsilon_s)/\varepsilon_s)$. H_B then assigns V to the class with the greatest total vote. Provided the learning system can reliably generate from weighted cases a hypothesis that has less than 50% error on those cases, a sequence of "weak" classifiers $\{H_s\}$ can be boosted to a "strong" classifier H_B that is at least as, and usually more accurate than, the best weak classifier.

In quite a number of experiments, boosting cannot increase, or even can decrease, the accuracy of naïve Bayesian classifier learning, which is in sharp contrast to the success of boosting decision trees. One key difference between naïve Bayesian classifiers and decision trees, with respect to boosting, is the stability of the classifiers – naïve Bayesian classifier learning is relatively stable with respect to small changes to training data, but decision tree learning is not. How to make the naïve Bayesian classifiers unstable (say, quite different with each other) becomes very urgent a problem to solve for incorporating Bayesian ideas into boosting method. Fortunately, we observed that, when the training cases change, the naïve Bayesian classifier constructed from them would be quite "instable", hence solve the problem. An idea to realize this should be selecting training sample set for constructing each base naïve Bayesian classifier dynamically and randomly.

3 Improved Boosting Naïve Bayesian Learning

Here we introduce our newly devised method for boosting naïve Bayesian classifiers.

Before generate each naïve Bayesian classifier as one of the base learners of boosting, we need to establish different training sample sets, one for each naïve Bayesian classifier. Naturally "Bootstrap" is a good solution, which establishes the training set that consists of the training samples drawn randomly with replacement from the original training set of the same size. Such a training set is called a *bootstrap replicate* of the original training set [3]. Each bootstrap replicate contains, on the average, 63.2% of the original training set, with several training samples appearing multiple times. For each round of boosting iteration, this method provide a distinct training set for constructing a distinct base classifier, say, the naïve Bayesian learner. Thus generated sample sets are only used for constructing base classifiers. For adjusting of boosting weights, the original training set should be used.

Other than employing the original form, the naïve Bayesian classifier needs to be updated for meeting the requirements of Boosting as its base classifier.

Suppose there is a training set D that consists of samples $\{V_i\}$, $i=1,...,N$. As introduced in section 2, the naïve Bayesian classification rule is: finding the maximum of $P(C_j \mid V) \propto P(C_j)\prod_{k=1}^{n} P(v_k \mid C_j)$, and choosing the corresponding C_j as its prediction. C_j represents the *j*-th class label in the possible label space, and L is the volume of this space.

$P(C_j)$ can be estimated by $P(C_j) = \frac{1+\sum_{i=1}^{N} P(C_j \mid V_i)}{L+N}$. $P(v_k \mid C_j)$ can be estimated by $P(v \mid C_j) = \frac{1+\sum_{i=1}^{N} F(v,V_i)P(C_j \mid V_i)}{n+\sum_{k=1}^{n}\sum_{i=1}^{N} F(v_k,V_i)P(C_j \mid V_i)}$. Here $F(v,V)$ is a function to caculate the frequency of attribute v appears in a certain sample V. $P(V \mid C_j)$ is defined as 1 when $V \in C_j$, otherwise, 0. Because boosting requires its base classifiers to work with a weight distribution over training documents, we use $P(V \mid C_j) \cdot w[V]$ to replace $P(V \mid C_j)$. Here $w[V]$ is the weight (maintained by boosting) of training sample V.

Because we expect the classification results would not be influenced by the order of training sets selected for the base classifiers, the uniform weights suggested by Breiman will be used instead of the weights defined for AdaBoost that depend on the values of ε_s [4][5].

Suppose we settle the boosting algorithm to T rounds. For 2 class problems, T should be odd number to avoid ties. If there are more than two classes, ties are possible if some classes have the same number of votes. In this situation, the class predicted by the first classifier with the maximum votes will be preferred. Let H_k be the classifier generated in the s-th boosting round, $s=1,2,...,T$, and H_B the composite classifier. Let $P(e_k)$ and $P(e_B)$ denote their respective probabilities of error under the original (uniform) weight distribution for the training cases. The same notation will be extended to more complex events; for instance, $P(e_1e_2)$ will denote the probability of simultaneous errors by H_1 and H_2. It is clear that boosted classifier H_B will misclassify a document only when at least $\lceil T/2 \rceil$ constituent classifiers are in error, so,

$$\begin{aligned}
&P(e_B) \le \\
&P(e_1 e_2 \ldots e_{\lceil T/2 \rceil} \bar{e}_{\lceil T/2 \rceil+1} \ldots \bar{e}_T) + P(e_1 \bar{e}_2 e_3 \ldots e_{\lceil T/2 \rceil+1} \bar{e}_{\lceil T/2 \rceil+2} \ldots \bar{e}_T) + \cdots + P(\bar{e}_1 \bar{e}_2 \ldots \bar{e}_{T-\lfloor T/2 \rfloor-1} e_{T-\lfloor T/2 \rfloor} \ldots e_T) \\
&\{\lceil T/2 \rceil \ \textit{classifiers misclassify}\} \\
&+ P(e_1 e_2 \ldots e_{\lceil T/2 \rceil+1} \bar{e}_{\lceil T/2 \rceil+2} \ldots \bar{e}_T) + P(e_1 \bar{e}_2 e_3 \ldots e_{\lceil T/2 \rceil+2} \bar{e}_{\lceil T/2 \rceil+3} \ldots \bar{e}_T) + \cdots + P(\bar{e}_1 \bar{e}_2 \ldots \bar{e}_{T-\lfloor T/2 \rfloor} e_{T-\lfloor T/2 \rfloor+1} \ldots e_T) \\
&\{\lceil T/2 \rceil + 1 \ \textit{classifiers misclassify}\} \\
&+ \ldots\ldots \\
&+ P(e_1 e_2 \ldots e_T) \qquad (\textit{all classifiers misclassify})
\end{aligned}$$

Discussions of ensemble classifiers often focus on independence of the constituents [1][6], then,

$$\begin{aligned}
&P(e_1 \ldots e_s \bar{e}_{s+1} \ldots \bar{e}_{s+\lceil T/2 \rceil} e_{s+\lceil T/2 \rceil+1} \ldots e_T) = \\
&P(e_1) \times \cdots \times P(e_s) \times (1 - P(e_{s+1})) \times \cdots \times (1 - P(e_{s+\lceil T/2 \rceil})) \times P(e_{s+\lceil T/2 \rceil+1}) \times \cdots \times P(e_T)
\end{aligned}$$

Unless $P(e_1), P(e_2), \cdots, P(e_T)$ are all zero, $P(e_B)$ will remain positive. However, consider the situation in which at least $\lceil T/2 \rceil$ classifiers have non-overlapping errors. Then $P(e_B)$ becomes error-free. The goal of boosting should thus be generating constituent classifiers with minimal overlapping errors, rather than independent classifiers. This idea came from the experiments of Quinlan [5]. To achieve this goal, we modify the weight adjusting method for the next boosting round. In the problem of multi-class learning, a sample still may not be misclassified even if it has already been misclassified by more than $\lceil T/2 \rceil$ base classifiers. Only the sample with the largest number of votes for any class which this sample does not belong to will be misclassified. In another word, at any boosting round s, if a sample has already been misclassified by $\lceil T/2 \rceil$ classifiers, it still have the chance to be correctly classified if, in the remained rounds, it can receive enough correct votes to defeat the combined votes of any incorrect class. Therefore, in the remained rounds, whether we can give enough correct votes for these samples will be very crucial for us to minimize the overlapping errors of the base classifiers. In the sense, we increase the weight of such sample, no matter in this round it is classified correctly or incorrectly. There is an adaptive coefficient to determine the size of incensement: the closer the sample is to being misclassified, the larger the coefficient is (say, the more important the sample is). As for the samples already having no chance to receive correct classification even if all the remained $T-s$ base classifiers correctly classify it, their weight can be set to 0 from round $s+1$, since they will have no help for minimizing the overlapping errors. If in round s, a sample has already been misclassified by less than $\lceil T/2 \rceil$, we employ the weight adjusting criteria of boosting: if it is misclassified at this round, then its weight will increase; otherwise, decrease. To facilitate the convergence of boosting, we also introduce a coefficient to enlarge the increment of the weight of the document misclassified by H_s. This coefficient is also adaptive: the closer the number of previous classifiers misclassifying this sample is to $\lceil T/2 \rceil$, the larger the coefficient is. If in round s a sample is correctly classified, the reduction of weight is not changed.

4 Algorithm Description

Given training set:

$D=\{(V_1,Y(V_1)),(V_2,Y(V_2)),\cdots,(V_N,Y(V_N))\}$, where V_i is a training sample, $Y(V_i)\in y=\{C_1,C_2,\cdots,C_L\}$ denotes the classes V_i belongs to, and L is the number of different classes. N is the number of training samples. $i=1,2,\cdots,N$.

Given a settled number of T, the iteration times of boosting.

This algorithm maintains a set of weights as a distribution W over samples and labels (classes), i.e., for each $V_i\in D$, and at each boosting round s, there is an associated real value $w_s[i]$.

Boosting maintains an error counter for each training sample. For example, the error counter of V_i is denoted by $ec[i]$. Boosting also maintains an array cc for each training sample to record the votes of all possible classes. Each class has an element of cc. $cc[i,j]$ denotes how many base classifiers predict that V_i belongs to C_j.

Step #1:

- Initialize $w_1[i]=1/N$ for all $i=1,2,\cdots,N$; set $ec[i]=0$ for all $i=1,2,\cdots,N$;
- Set $cc[i,j]=0$ for all $i=1,2,\cdots,N$ and for all $j=1,2,\cdots,L$;

Step#2: For $s=1,2,\cdots,T$:

- Using Bootstrap to establish a new training set D_s for constructing base classifiers. Its size is still N.
- Now we can start to construct weighted naïve Bayesian classifier H_s according to D_s. Its classification rule is:

$$H_s(V)=C_j=\arg\max_j P(C_j\mid V)=\arg\max_j P(C_j)\prod_{k=1}^{n}P(v_k\mid C_j)$$

$$=\arg\max_j \frac{1+\sum_{i=1}^{|D_s|}P(C_j\mid V_i)}{L+|D|}\prod_{m=1}^{n}\frac{1+\sum_{i=1}^{|D_s|}F(v_m\in V,V_i)P(C_j\mid V_i)w_i[i,j]}{L+\sum_{k=1}^{n}\sum_{i=1}^{|D_s|}F(v_k,V_i)P(C_j\mid V_i)w_i[i,j]}$$

for any sample V.

- Using H_s to classify all the training samples in the original training sample set D. For all $i=1,2,\cdots,N$ and for all $j=1,2,\cdots,L$, if any sample V_i is predicted by H_s as class C_j, then $cc[i,j]=cc[i,j]+1$; if this prediction is incorrect, then $ec[i]=ec[i]+1$.
- Compute the weighted error $\varepsilon_s=\sum_{i=1}^{N}w[i]\cdot[\![V_i \textit{ is misclassifed by } H_s]\!]$, considering all samples in D. $[\![\bullet]\!]$ is a function that maps its content to 1 if it is true, otherwise, maps to 0.

- If $\varepsilon_s > 1/2$ break;
- For all $i = 1,2,\cdots,N$, do:

If $ec[i] < \lceil T/2 \rceil$, then:

$$w_{s+1}[i] = \exp\left(\frac{1}{\lceil T/2 \rceil - ec[i]}\right) \times w_s[i] \div \begin{cases} 2\varepsilon_s & H_s \;\; misclassify \;\; V_i \\ 2(1-\varepsilon_s) & otherwise \end{cases};$$

If $ec[i] \geq \lceil T/2 \rceil$ then:

Compute:

a) $CMEV = \max(cc[i,j] : C_j \neq Y(V_i))$; $CCV = cc[i,j] : C_j = Y(V_i)$;

b) If $CMEV - CCV < T - s$ then:

$$w_{s+1}[i] = w_s[i] \times \exp\left(\frac{1}{(t-s)-(CMEV-CCV)}\right) \times \frac{1}{\varepsilon_s};$$

Else: $w_{s+1}[i] = 0$;

Step #3: $H_B(V) = C_j =$ The C_j with the maximum un-weighted votes from $H_s(V)$ for all $s = 1,2,\cdots,T$.

5 Experimental Results

We choose other 4 algorithms to execute the same task: text categorization, and conducted a number of experiments to compare their performance with that of our newly devised algorithm, which is denoted as BNBX. They are: 1) Naïve Bayesian classifier [7] (for simplicity, we denote it as NB), 2) Boosted naïve Bayesian classifier (for simplicity, we denote it as BNB), 3) Boosted Leveled naïve Bayesian Trees (for simplicity, we denote it as BLNBT), a method of boosting decision trees incorporated with naïve Bayesian learning [2], and 4) ADABOOST.MH [8] (for simplicity, we denote it as ABM), a boosting-based text categorization method whose performance are better than many traditional text categorization algorithms. For these experiments we have used the Yahoo! Chinese News (news.cn.yahoo.com), which consists of a set of 1041 news stories. The documents have an average length of 577 Chinese characters. We choose "Precision" as the main measure for assessing and comparing the performances of different text categorization methods. Table 1 presents the precision for 5 selected topics.

From table 1 we can find that BNBX has the best precision in 3 out of 5 topics. Directly boosted naïve Bayesian classifiers (BNB) cannot really improve the performance of naïve Bayesian learning, but BLNBT and our BNBX can do that. Generally, the classification performance of BNBX with $T = 20$ is much better than all the other boosting algorithms compared here with the same T, and is nearly equal to the performance of other boosting algorithms with large T. Because other boosting algorithms require many rounds to achieve optimal performance, our method BNBX provides a cheaper way to obtain the same goal. BNBX also obviously enhances the learning ability of naïve Bayesian learning, which is reflected by the gap of precision between BNBX and NB.

Table 1. The Precision Of 5 Text Categorization Methods. Here T is the iteration times of boosting algorithms. Naive Bayesian classifiers (NBs) do not need this parameter

Algorithm	T	Topic				
		Econ.	Tech.	Liv.	Sprt.	Cult.
BNBX	20	0.831	0.868	0.798	0.882	0.894
NB	-	0.701	0.765	0.724	0.711	0.793
BNB	100	0.686	0.693	0.729	0.730	0.753
BLNBT	20	0.622	0.704	0.747	0.736	0.760
BLNBT	100	0.852	0.842	0.839	0.838	0.877
ABM	100	0.700	0.756	0.703	0.723	0.714
ABM	200	0.821	0.855	0.772	0.865	0.889

6 Conclusion

In this paper, we introduce a new method to improve the performance of combining boosting and naïve Bayesian learning, which not only has performance much better than naïve Bayesian classifiers or directly boosted naïve Bayesian ones, but also much quicker to obtain optimal performance than boosting stumps and boosting decision trees incorporated with naïve Bayesian learning. How to improve its generalization error bounds should be further explored in our following work.

References

1. Freund, Y., Schapire, R.: A Decision-theoretic Generalization of On-line Learning and an Application to Boosting. Journal of Computer and System Sciences (1997), 55(1), 119-139
2. Ting, K., Zheng, Z.: Improving the Performance of Boosting for Naïve Bayesian Classification. School of Computing and Mathematics, Deakin University (2000)
3. Breiman, L.: Bagging Predictors. Machine Learning (1996), 24(2), 123-140
4. Bremain, L.: Bias, Variance, and Arcing Classifiers. Machine Learning (2000)
5. Quinlan, J.R.: Mini-Boosting Decision Trees. AI Access Foundation and Morgan Kaufmann Publishers (1998)
6. Dietterich, T.G.: Machine Learning Research. AI Magazine (1997), 18(4), 97-136
7. McCallum, A., Nigam, K.: A Comparison of Event Models for Naive Bayesian Text Classification. Just Research. 4616 Henry Street Pittsburgh, PA 15213 (1999)
8. Schapire, R., Singer, Y.: BoosTexter: A Boosting-based System for Text Categorization. Machine Learning (2000), 39(2/3), 135-168

Toward Bayesian Classifiers with Accurate Probabilities

Charles X. Ling[1] and Huajie Zhang[1]

Department of Computer Science,
The University of Western Ontario
London, Ontario, Canada N6A 5B7
{ling, hzhang}@csd.uwo.ca

Abstract. In most data mining applications, accurate ranking and probability estimation are essential. However, many traditional classifiers aim at a high classification accuracy (or low error rate) only, even though they also produce probability estimates. Does high predictive accuracy imply a better ranking and probability estimation? Is there any better evaluation method for those classifiers than the classification accuracy, for the purpose of data mining applications? The answer is the area under the ROC (Receiver Operating Characteristics) curve, or simply AUC. We show that AUC provides a more discriminating evaluation for the ranking and probability estimation than the accuracy does. Further, we show that classifiers constructed to maximise the AUC score produce not only higher AUC values, but also higher classification accuracies. Our results are based on experimental comparison between error-based and AUC-based learning algorithms for TAN (Tree-Augmented Naive Bayes).

1 Introduction

Classification is the most important task in machine learning. In classification, a classifier is built from a set of training examples with class labels. A key performance measure of a classifier is its predictive accuracy (or error rate, which is one (1) minus the accuracy) on the training and testing examples. Predictive error rate is simply the percentage of the number of incorrectly classified examples versus the total number of examples. Many classifiers can also produce probability estimation or confidence of the prediction. However, the error rate does not consider how "far-off" (be it 0.45 or 0.01) the prediction of each example is from its target, but only the class with the largest probability estimation. Such error-based classifiers are optimal only under the following two assumptions: First, we care nothing more about the classification results; and second, different types of errors, such as false positive and false negative, are treated as equivalent.

In data mining applications, however, neither of those two assumptions is often true. For example, in direct marketing, we often need to promote the top X% of customers during gradual roll-out, or we often deploy different promotion strategies to customers with different likelihoods of buying some products. To

M.-S. Chen, P.S. Yu, and B. Liu (Eds.): PAKDD 2002, LNAI 2336, pp. 123–134, 2002.

accomplish these tasks, we need more than a classification of buyers and non-buyers. We need at least a ranking of customers in terms of their likelihoods of buying. Thus, a ranking is much more desirable than just a classification. Is it true, however, that a classifier with a smaller error rate always has a better ranking?

In addition, the costs of errors, such as false positive and false negative, can be quite different. In direct marketing, for example, the cost of missing a valuable buyer is often much higher than the cost of promoting a non-buyer. In this case, classifiers that produce small error rates do not optimise the business goal: which customers shall be promoted for the maximum profit?

Last, the probabilities on the prediction are often crucial for optimal decision making. For example, if for an example e, the cost of predicting class i when the true class of e is j is $C(i, j, e)$, the Bayes optimal prediction for e is the class i that minimises the expected loss:

$$R(i|e) = \sum_j P(j|e)C(i, j, e). \tag{1}$$

$R(i|e)$ is the expected cost of predicting e to be class i. Clearly, Bayes optimal prediction requires accurate probability estimation, more than ranking and classification alone.

If we are aiming at accurate ranking or probability estimation from a classifier, one might naturally think that we must need true ranking or true probabilities in the training examples. In most scenarios, however, that is not possible. Most likely, what we are given is a dataset of examples with class labels. Thus, given only classification labels in training and testing sets, are there better methods than the error rate to evaluate classifiers that also produce probability estimates or ranking, for the purpose of data mining applications?

The answer is the ROC curve! ROC (Receiver Operating Characteristics) curve [11,10] compares the classifiers' performance cross the entire range of class distributions and error costs. Details of ROC curve and related calculations are in the Appendix, and we only give an intuitive explanation here. Figure 1 shows a plot of four ROC curves, each representing one of the four classifiers, A through D. A ROC curve X is said to *dominate* another ROC curve Y if X is always above and to the left of Y. This means that the classifier of X always has a lower expected cost than that of Y, over all possible error costs and class distributions. In this example, A and B dominate D.

However, often there is no clear dominating relation between two ROC curves. For example, curves A and B are not dominating each other in the whole range. In those situations, the area under the ROC curve, or simply AUC, is a good "summary" for comparing the two ROC curves. Clearly, if one ROC curve dominates the other, its AUC must be larger. Intuitively, if the AUC of one ROC curve is larger than the AUC of the other ROC curve, its probability estimation would likely to be better too. Bradley [1] shows that AUC is a proper metric for the quality of classifiers averaged across all possible probability thresholds.

We believe that AUC should always be used as a more discriminating evaluation method than the error rate for classifiers that produce probabilities. For

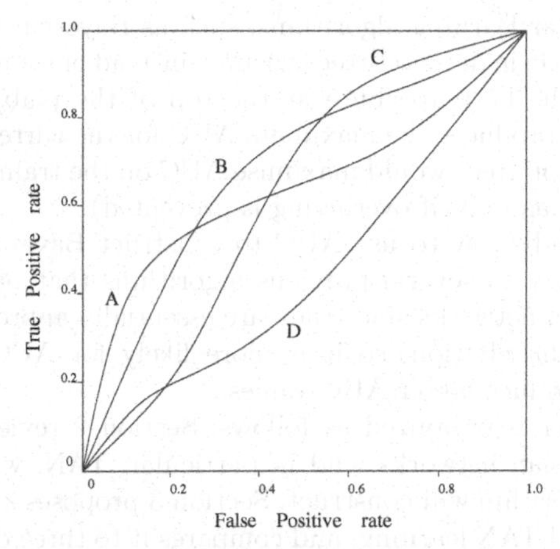

Fig. 1. An example of four ROC curves.

example, if we have two classifiers, 1 and 2, both producing probabilities for a set of 10 testing examples. Assume that both classifiers classify 5 of the 10 examples as positive, and the other 5 negative. If we rank the testing examples according to increasing probabilities, we get the two rank lists as in Table 1.

Table 1. An example in which two classifiers have the same classification accuracy, but different AUC values.

Classifier 1	$- - - - +\|- + + + +$
Classifier 2	$+ - - - -\|+ + + + -$

Clearly, both classifiers produce an error rate of 20% (one false positive and one false negative), thus the two classifiers are equivalent in terms of the error rate. However, intuition tells us that Classifier 1 is better than Classifier 2, since overall positive examples are ranked higher in Classifier 1 than 2. If we calculate the AUCs, we obtain that the AUC of Classifier 1 is $\frac{24}{25}$, and the AUC of Classifier 2 is $\frac{16}{25}$. Clearly, AUC tells us that Classifier 1 is indeed better than 2.

For a perfect classifier, which ranks all positive examples with higher probability than any negative examples, the AUC would be 1. In the worst case (the reverse of above), the AUC is equal to 0. If the ranking is random, then the AUC is 0.5. Note, however, that AUC is a global measure of a classifier, and is not always better than error rate. Counter examples can be found in Section 3.1.

Since AUC is a better evaluation method than the error rate for data mining algorithms that also produce probabilities, the next natural question is: Can we construct classifiers using AUC directly, rather than using error based matrix (such as entropy, error rate in cross-validation)? Clearly, we can use AUC in constructing most popular learning algorithms, such as Bayesian networks and decision trees. For example, in decision-tree learning, instead of using information gain (ratio) to choose the best attribute at the top of the (sub)tree, we can choose an attribute that produces the maximum AUC for the current tree under construction. Such decision trees would maximise AUC on the training data, and hopefully, on the testing as well (if overfitting is prevented).

In this paper, we study how to use AUC to construct Bayesian classifiers, and compare this strategy to several previous algorithms that based on error rate. We choose Bayesian networks since they are essentially approximating the underlying probability distribution, so it is more likely for AUC-constructed Bayesian networks to produce larger AUC values.

The rest of the paper is organized as follows: Section 2 reviews necessary background about Bayesian networks, and in particular, TAN, which our new AUC-based learning algorithm will construct. Section 3 proposes a new learning algorithm for AUC-based TAN learning, and compares it to three other learning algorithms by empirical experiments.

2 Review of Learning Simple Bayesian Networks

Bayesian networks (BNs) are probabilistic models that combine the probability theory and the graph theory. They represent causal and probabilistic relations (by arcs and conditional probability tables) among random variables (nodes) that are governed by the probability theory. Probabilistic inference can be made directly from Bayesian networks. Bayesian networks have been widely used in many applications, because they provide intuitive and causal representations of our real-world applications, and they are supported by a rigorous theoretical foundation.

Bayesian networks have often been used in classification. Assume A_1, A_2,$\cdots$, A_n are n attributes. An example e is represented by a vector $(a_1, a_2, , \cdots, a_n)$, where a_i is the value of A_i. Let C represent the classification variable that corresponds to the class, and c represent the value that C takes.

From the Bayes Rule, the probability of an example $e = (a_1, a_2, \cdots, a_n)$ being class c is

$$p(c|e) = \frac{p(a_1, a_2, \cdots, a_n|c)p(c)}{p(a_1, a_2, \cdots, a_n)}.$$

e is classified into the most probable class c; i.e.,

$$g(e) = \max_c p(c|a_1, a_2, \cdots, a_n), \tag{2}$$

where $g(e)$ is called a Bayesian classifier.

The term $p(c|a_1, a_2, \cdots, a_n)$ is difficult to estimate. Assume that all attributes are independent given the value of the class variable; that is,

$$p(a_1, a_2, \cdots, a_n|c) = \prod_{i=1}^{n} p(a_i|c),$$

the resulting $g(e)$ is then:

$$g(e) = \max_{c} p(c) \prod_{i=1}^{n} p(a_i|c). \tag{3}$$

$g(e)$ is called a Naive Bayesian classifier, or simply Naive Bayes (NB). Figure 2 (a) shows an example of Naive Bayes.

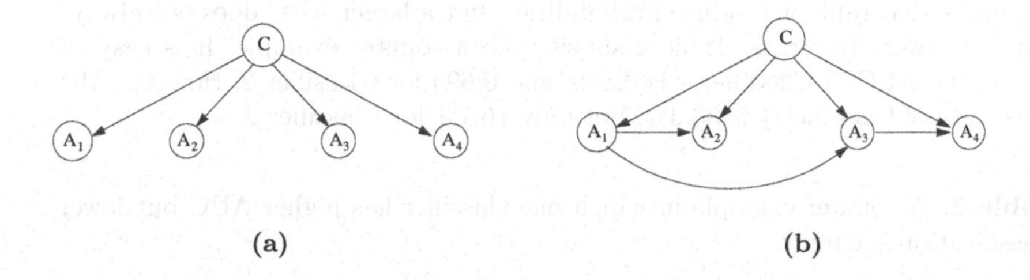

Fig. 2. (a) an example of Naive Bayes (b) an example of TAN

Because values of $p(a_i|c)$ can be estimated easily from the training examples, Naive Bayes is easy to construct. It is also, however, surprisingly effective in classification. Unfortunately, the independence assumption is rarely true in real-world applications. Indeed, Naive Bayes is found to work poorly for regression problems [3], and produces poor probability estimates [8]. Therefore, researchers have extended the structure of Naive Bayes to represent dependencies among attributes[1]. Tree Augmented Naive Bayes (TAN) is such an extension, in which the classification node points directly to all attributes (as in Naive Bayes), but an attribute can have one parent from another attribute. Figure 2 (b) shows an example of TAN. TAN is a specific case of general Augmented Naive Bayesian networks (or simply ANB), where the classification node also points directly to all attributes, but where there is no limitation on the arcs among attributes (except that they do not form any directed cycle). The general ANB is as powerful as general Bayesian networks.

TAN is a nice trade-off between complexity of learning and representational power. In the past, a number of learning algorithms have been published on learning TAN. Some algorithms, such as [2,4], are based on conditional independencies among attributes (CI-based). Others, such as [6], are based on minimising the error rate (error-based). There is no previous work that learns TAN based on AUC. The questions we will answer in this paper are: Do error-based learning algorithms also produce accurate probability estimates? What about learning

[1] Most of the extensions, however, aim at improving the predictive accuracy, not at better probability estimations.

TAN by maximising directly AUC? How do we compare these three learning algorithms (CI-based, error-based, and AUC-based) in terms of the error rate and the AUC?

3 Learning TAN with Accurate Probability Estimates

3.1 Error Rate Vs AUC

As we have seen, AUC can be viewed as a better measure than the error rate for classifiers that rank or produce probabilities, but a larger AUC does not always imply a lower error rate. Table 2 shows such a counter example. It is easy to obtain the AUC of Classifier 1 is 0.889, and 0.694 for Classifier 2. However, the error rate of Classifier 1 is 33.3%, but only 16.7% for Classifier 2.

Table 2. A counter example in which one classifier has higher AUC but lower classification accuracy.

Classifier 1	− − − − + + | − − + + + +
Classifier 2	+ − − − − − | + + + + + −

We will compare three different TAN learning algorithms in terms of the error rate and AUC values: one is CI-based, the second one is error-based, and the third is AUC-based. We will discuss each of these algorithms below.

3.2 CI-based TAN Learning Algorithms

CI (conditional independence) based learning algorithms have been studied by many researchers [4]. The basic idea is to detect the conditional dependence between two attributes by performing conditional independence tests based on the data, and then search for a network using the detected dependencies. Essentially, CI-based learning algorithms attempt to directly approximate the underlying probability distribution. Intuitively, they may tend to produce more accurate probability estimates.

Chow and Liu's TAN learning algorithm [2] is a typical CI-based algorithm. The idea here is to estimate dependencies between each pair of attributes, and find a maximum spanning tree based on the estimation in building the TAN. Friedman et al. [4] extend Chow and Liu's algorithm [2] by using conditional mutual information between two attributes given the class variable. This function is defined as

$$I_P(X;Y|Z) = \sum_{x,y,z} P(x,y,z) log \frac{P(x,y|z)}{P(x|z)P(y|z)}.$$

Friedman et al.'s algorithm [4] is described below, and is used in our comparison. We refer to this algorithm as CI-TAN in our paper.

1. Compute $I_P(A_i, A_j|C)$ between each pair of attributes, $i \neq j$.
2. Build a complete undirected graph in which the nodes are the attributes A_1, $\cdots$, A_n. Annotate the weight of an edge connecting A_i to A_j by $I_P(A_i; A_j|C)$.
3. Build a maximum weighted spanning tree.
4. Transform the resulting undirected tree to a directed one by choosing a root variable and setting the direction of all edges to be outward from it.
5. Construct a TAN model by adding a node labelled by C and adding an arc from C to each A_i.

3.3 Error-Based TAN Learning Algorithms

Error-based algorithms of learning Bayesian networks search for the network that minimises the classification error (or maximises the accuracy). They belong to scored-based algorithms since the accuracy is the score to maximise. In learning TAN, the SuperParent algorithm is a recent algorithm that searches for arcs by maximising the accuracy via cross validation [6]. A node is called a SuperParent if we extend arcs from it to every orphan (nodes without parent). The algorithm SuperParent is depicted below.

1. Initialise network to Naive Bayes.
2. Evaluate the current classifier by its classification accuracy.
3. Consider making each node a SuperParent. Let A_{sp} be the SuperParent which increases accuracy the most.
4. Consider an arc from A_{sp} to each orphan. If the best such arc improves accuracy, then keep it and go to 2; else return the current classifier.

Since error-based algorithms aim directly at higher classification accuracy, intuitively, it may tend to produce a model with higher classification accuracy than CI-based algorithms. This has been verified by results in [6], as well as in our experiment (see Section 3.5). However, the most intriguing question is: Would an AUC-based SuperParent algorithm produce better probability estimation measured by AUC, compared to error-based SuperParent algorithm?

3.4 AUC-based SuperParent Algorithm

AUC-based TAN learning algorithms simply use AUC as the score to maximise via cross-validation during the search for arcs in learning TAN. Intuitively, since we search the best TAN structure based on higher AUC values, the resulting TAN would tend to produce higher AUC, or more accurate probability estimates.

We extend SuperParent algorithm in which AUC is used for evaluating current network instead of classification accuracy. This algorithm is called AUC-SuperParent in our paper. The algorithm is described below.

1. Initialise network to Naive Bayes.
2. Evaluate the current classifier in terms of its AUC.
3. Consider making each node a SuperParent. Let A_{sp} be the SuperParent which increases AUC the most.
4. Consider an arc from A_{sp} to each orphan. If the best such arc improves AUC, then keep it and go to 2; else return the current classifier.

3.5 Empirical Comparisons

The questions we are interested in are: Do AUC-based algorithms really result in classifiers with higher AUC? If AUC is a better and more discriminating measure than the error rate, would AUC-based learning algorithms also produce a lower error rate?

We will answer those questions by empirical experiments. We use twelve datasets from the UCI repository [7] to conduct our experiments. Table 3 lists the properties of the datasets we use in our experiments. Most of these datasets were also used in comparing SuperParent and TAN published by Keogh and Pazzani [6].

Table 3. Descriptions of the datasets used in our experiments.

Dataset	Attributes	Class	Instances
Australia	14	2	690
breast	10	10	683
cars	7	2	700
dermatology	34	6	366
ecoli	7	8	336
hepatitis	4	2	320
import	24	2	204
iris	5	3	150
pima	8	2	392
segment	19	7	2310
vehicle	18	4	846
vote	16	2	232

We have also included Naive Bayes in our experiments, since other algorithms are derived from it. Our experiments follow the procedure below:

1. The continuous attributes in the dataset are discretized.
2. For each dataset, run Naive Bayes, CI-based TAN [4], SuperParent [6], and our AUC-SuperParent with the 5-fold cross-validation, and obtain the AUC and classification accuracy on the testing set unused in the training.
3. Repeat 2 above 20 times and obtain the average AUC and classification accuracy on the testing data.

Table 4 shows the experimental results of AUC values of four learning algorithms: Naive Bayes, CI-based TAN, SuperParent, and AUC-SuperParent. They are represented as NB, CI-TAN, SP, and AUC-SP respectively in the table.

We use a simple threshold, 1%, on average AUC values to judge if an algorithm outperforms another. The comparison of the four algorithms on these datasets is presented in Table 5. In the table, i-j-k means that the algorithm at the corresponding row wins in i datasets, loses in j datasets, and ties in k datasets, compared to the algorithm at the corresponding column.

Table 4. Experimental results of the AUCs for Naive Bayes, CI-TAN, SuperParent and AUC-SuperParent.

Dataset	**NB**	**CI-TAN**	**SP**	**AUC-SP**
Australia	75.2±0.30	75.2±0.48	74.7±0.44	76.2±0.48
breast	47.6±1.19	43.3±1.11	42.9±1.10	42.2±0.92
cars	93.7±0.17	93.4±0.24	96.1±0.16	97.1±0.12
dermatology	99.9±0.01	99.2±0.47	99.3±0.46	99.6±0.33
ecoli	99.3±0.06	98.9±0.08	99.2±0.07	99.2±0.08
hepatitis	62.5±0.70	62.4±0.60	61.3±0.73	62.5±0.70
import	99.2±0.10	97.8±0.19	99.0±0.14	99.2±0.11
iris	96.5±0.67	97.1±0.18	97.3±0.19	97.9±0.18
pima	77.4±0.44	75.8±0.51	77.2±0.46	76.7±0.50
segment	95.3±0.05	97.0±0.05	97.2±0.06	97.0±0.05
vehicle	91.1±0.49	92.4±0.71	93.3±0.39	93.2±0.42
vote	86.2±0.54	86.0±0.56	86.4±0.55	85.4±0.55

We can see that the average AUC of AUC-SuperParent is slightly better (3 wins, 1 loss, and 8 ties) than error-based SuperParent and CI-TAN, and significantly better than Naive Bayes. There are five datasets in which AUC-SuperParent is higher than that of Naive Bayes for more than one percent, but only one dataset in which the reverse happens. Overall, AUC-SuperParent is best regarding the AUC score. That confirms that using AUC directly to build Bayesian networks will result in a network with more accurate ranking or probability estimation. The tables also show that CI-based TAN does not perform

Table 5. Comparison of the algorithms in terms of AUC.

Algorithms	**NB**	**CI-TAN**	**SP**	**AUC-SP**
NB				
CI-TAN	2-3-7			
SP	3-2-7	3-1-8		
AUC-SP	5-1-6	3-1-8	3-1-8	

well; it is even slightly worse than Naive Bayes. CI-based TAN is constructed with the goal of a good fit for conditional independencies of all attributes with the data, not necessarily a good fit of the classification accuracy or its probability estimation, which is what we actually care to measure. This shows indirectly that if we want to learn a Bayesian network for a certain goal, the best bet is to search the network that maximises a score of that goal. Ranking and probability estimation are important for data mining, thus, data mining models should be constructed to maximise the AUC value, not the predictive accuracy.

Since AUC is a more discriminating evaluation method compared to the accuracy, one may expect that the AUC-SuperParent would also have a higher predictive accuracy compared to error-based SuperParent. Table 6 shows the experimental results of the four learning algorithms on the classification accuracy on the testing set.

Table 6. Experimental results of the accuracies of Naive Bayes, CI-TAN, SuperParent and AUC-SuperParent.

Dataset	**NB**	**CI-TAN**	**SP**	**AUC-SP**
Australia	76.1±0.39	76.7±0.32	76.0±0.30	76.6±0.39
breast	68.3±0.36	73.3±0.37	74.8±0.34	74.6±0.33
cars	86.1±0.29	85.4±0.37	90.0±0.27	90.8±0.25
dermatology	98.3±0.14	97.7±0.17	98.5±0.13	98.2±0.13
ecoli	96.9±0.20	96.1±0.23	96.8±0.21	96.7±0.21
hepatitis	71.0±0.48	70.5±0.42	70.3±0.48	71.0±0.48
import	97.0±0.24	93.6±0.37	96.7±0.28	96.7±0.23
iris	91.4±0.45	91.2±0.48	91.6±0.47	91.3±0.49
pima	71.9±0.40	70.5±0.46	71.5±0.39	71.0±0.44
segment	73.1±0.21	82.3±0.17	82.6±0.18	81.9±0.19
vehicle	82.0±0.26	89.3±0.23	89.4±0.22	90.0±0.23
vote	76.0±0.55	78.6±0.61	77.0±0.60	79.9±0.52

When we set the threshold to be 0.5%, the comparison of the four algorithms in terms of their classification accuracies is shown in Table 7. It indicates that AUC-SuperParent is better, in terms of predictive accuracy, than Naive Bayes (6 wins, 1 loss, and 5 ties), CI-TAN (9 wins, 0 loss, and 3 ties), and SuperParent (5 wins, 2 losses, and 5 ties).

Table 7. Comparison of the algorithms in terms of classification accuracy.

Algorithms	**NB**	**CI-TAN**	**SP**	**AUC-SP**
NB				
CI-TAN	5-5-2			
SP	5-1-6	6-2-4		
AUC-SP	6-1-5	9-0-3	5-2-5	

As we discussed earlier, AUC is a more discriminating evaluation criterion than the accuracy, thus, to build a TAN for the purpose of high classification accuracy, we should probably still maximise AUC during the search of network structures.

4 Conclusion

In this paper, we investigate TAN learning algorithms for the purpose of accurate probability estimation, often required in many applications of data mining. We show that AUC is a more discriminating measure of the quality of ranking or probability estimation. We also propose a new algorithm, AUC-SuperParent, for learning TAN by directly using AUC as the search criterion. By empirical experiments, we have obtained the following interesting results:

- AUC-based Bayesian network learning algorithms tend to produce more accurate ranking and probability estimation, than the error-based algorithms.
- AUC-based Bayesian network learning algorithms tend to produce higher classification accuracy than the error-based algorithms.

We can thus conclude that AUC should be used both as an evaluation criterion, and as a scoring function, for data mining algorithms.

In our future research, we will study other methods for improving probability estimation such as smoothing, binning, and bagging. Another direction we are working on is to study other learning algorithms using AUC, such as AUC-based decision-tree learning algorithms.

References

1. Bradley, A. P.: The Use of the Area under the ROC Curve in the Evaluation of Machine Learning Algorithms. Patten Recognition, Vol. 30 (1997), 1145–1159.
2. Chow, C. K., Liu, C. N.: Approximating Discrete Probability Distributions with Dependence Trees. IEEE Trans. on Information Theory, Vol. 14 (1968), 462–467.
3. Frank, E., Trigg, L., Holmes, G., Witten, I. H.: Naive Bayes for Regression. Machine Learning, Vol. 41 (2000), 5–15.
4. Friedman, N., Geiger, D., Goldszmidt, M.: Bayesian Network Classifiers. Machine Learning, Vol: 29 (1997), 131–163.
5. Hand, D. J., Till, R. J.: A Simple Generalisation of the Area under the ROC Curve for Multiple Class Classification Problems. Machine Learning, Vol. 45 (2001), 171–186.
6. Keogh, E. J., Pazzani, M. J.: Learning Augmented Naive Bayes Classifiers. In: Proceedings of the Seventh International Workshop on AI and Statistics, Ft. Lauderdale (1999).
7. Merz, C., Murphy, P., Aha, D.: UCI Repository of Machine Learning Databases. In: Dept of ICS, University of California, Irvine (1997). http://www.www.ics.uci.edu/mlearn/MLRepository.html.
8. Monti, S., Cooper, G. F.: A Bayesian Network Classifier That Combines a Finite Mixture Model and a Naive Bayes Model. In: Proceedings of the 15th Conference on Uncertainty in Artificial Intelligence. Morgan Kaufmann (1999) 447–456.
9. Provost, F., Fawcett, T.: Analysis and Visualization of Classifier Performance: Comparison under Imprecise Class and Cost Distribution. In: Proceedings of the Third International Conference on Knowledge Discovery and Data Mining. AAAI Press (1997) 43–48.

10. Provost, F., Fawcett, T., Kohavi, R.: The Case Against Accuracy Estimation for Comparing Induction Algorithms. In: Proceedings of the Fifteenth International Conference on Machine Learning. Morgan Kaufmann (1998) 445–453.
11. Swets, J.: Measuring the Accuracy of Diagnostic Systems. Science, Vol. 240 (1988), 1285–1293.

Appendix

We review the basics of ROC and AUC here. See [9,10] for more details.

Let $\{P, N\}$ be the positive and negative instance classes, and let $\{\tilde{P}, \tilde{N}\}$ be the classifications produced by a classifier. Let $P(P|I)$ be the posterior probability that an instance I is positive. The true positive rate, TP, of a classifier is:

$$TP = P(\tilde{P}|P) \approx \frac{\text{positives correctly classified}}{\text{total positives}}.$$

The false positive rate, FP, of a classifier is:

$$FP = P(\tilde{P}|N) \approx \frac{\text{negatives incorrectly classified}}{\text{total negatives}}.$$

On a ROC graph, TP is plotted on the Y axis and FP is plotted on the X axis. In the ROC space, each classifier with a given class distribution and cost matrix is represented by a point (FP, TP). For a model that produces a continuous output, TP and FP can vary as the threshold on the output varies between its extremes (0 and 1). The resulting curve is called the ROC curve. A ROC curve illustrates the tradeoff available with a given model, in which a point is recorded for a different cost and class distribution. When a ROC dominates another, then its classifier always has a lower expected cost than the other over all possible error costs and class distributions. However, if two ROCs do not dominate each other, then AUC can be used as a rough measure for the expected cost. Hand and Till [5] showed that AUC is equivalent to the probability that a randomly chosen negative example will have a smaller estimated probability of belonging to the positive class than a randomly chosen positive example. Therefore, the larger the AUC, the more likely that a negative example will not be misclassified. They give a simple formula for calculating AUC [5].

Pruning Redundant Association Rules Using Maximum Entropy Principle

Szymon Jaroszewicz and Dan A. Simovici

University of Massachusetts at Boston,
Department of Mathematics and Computer Science,
Boston, Massachusetts 02125, USA
{sj,dsim}@cs.umb.edu

Abstract. Data mining algorithms produce huge sets of rules, practically impossible to analyze manually. It is thus important to develop methods for removing redundant rules from those sets. We present a solution to the problem using the Maximum Entropy approach. The problem of efficiency of Maximum Entropy computations is addressed by using closed form solutions for the most frequent cases. Analytical and experimental evaluation of the proposed technique indicates that it efficiently produces small sets of interesting association rules.

Keywords: association rule, rule interestingness, rule pruning, maximum entropy

1 Introduction

Many data mining algorithms produce huge sets of rules, practically impossible to analyze manually. Typically, those sets are highly redundant and, so, it is important to develop methods for removing redundant rules and for helping the user select from thousands of discovered rules those which are the most interesting from his point of view.

Our goal is to identify a reasonably small, nonredundant set of interesting association rules describing well (and as completely as possible) the relationships within the data. The paper presents a solution to this problem using the maximum entropy approach. A subrule of an association rule $I \to J$ is a rule $K \to J$, such that $K \subset I$ (see [AIS93] or further sections for a detailed discussion of association rules). In [LHM99, AL99] a rule is considered not interesting if its confidence is close to that of one of its subrules. A similar approach (although in a slightly generalized setting) is used in [PT00] to prune the discovered rules. Also, in [PT00] a rule is considered interesting with respect to some set of beliefs if it contradicts at least one of the rules in the beliefs under the so called monotonicity assumption. A detailed statistical analysis of interestingness of a rule with respect to a single subrule, and algorithms for finding rules interesting in this setting can be found in [Suz97, SK98].

M.-S. Chen, P.S. Yu, and B. Liu (Eds.): PAKDD 2002, LNAI 2336, pp. 135–147, 2002.

The current work on evaluation of interestingness considers the influence of each subrule separately, while in our approach we take into account the combined influence of all the subrules of a rule. Examples illustrating the advantages (in our opinion) of our approach are given in Section 3.

In [LHM99], apart from pruning, the authors also find so called *direction setting* rules which summarize the dataset. This procedure takes into account many subrules of a rule and is thus similar to our approach. However, our approach has the advantage of giving a more precise, probabilistic quantification of the influence of subrules on the interestingness of a rule.

Another approach to pruning discovered rules is based on selecting a minimal set of rules covering the dataset [TKR+95, BVW00]. Again, those methods do not take into consideration probabilistic interactions between rules in the cover. Also, they may prune many interesting rules if they cover instances already covered by other rules.

A general study of measures of rule interestingness can be found in [BA99, JS01, HH99].

An overview of the interestingness of a rule with respect to a set of constraints can be found in [GHK94]. In [GHK94] the authors propose the method of *random worlds* and prove that in many important cases it is equivalent to the principle of maximum entropy.

Maximum entropy principle and other probability models have been also used in datamining in query selectivity estimation [PMS01]. There has also been work in applying MaxENT in speech processing [Rat96]

Let us now introduce notation used throughout the paper. If A is an attribute of a table we denote its domain by $\mathrm{Dom}(A)$. When $\mathrm{Dom}(A) = \{0, 1\}$ we say that A is a binary attribute. In this note we use tables whose headings have the form $H = \{A_1, A_2, \ldots, A_m\}$ and consist of binary attributes. The heading H will be written, as usual as $A_1 \cdots A_m$.

Subsets of H, referred to as *itemsets*, will be denoted using uppercase Roman letters $I, J, K, L, \ldots$. Single attributes will be denoted by uppercase letters $A, B, C, \ldots$

The domain of a set of attributes $I \subseteq H$, where $I = A_{i_1} A_{i_2} \ldots A_{i_r}$ is defined as

$$\mathrm{Dom}(I) = \mathrm{Dom}(A_{i_1}) \times \mathrm{Dom}(A_{i_2}) \times \cdots \times \mathrm{Dom}(A_{i_r}) = \{0, 1\}^r.$$

Values from domains of attributes will be denoted by corresponding boldface lowercase letters, e.g. $\mathbf{i} \in \mathrm{Dom}(I)$.

For $\mathbf{h} \in \mathrm{Dom}(H)$ and $I \subseteq H$, we denote the projection of $\mathbf{h}$ on I by $\mathbf{h}_I$.

For a probability distribution P on $\mathrm{Dom}(H)$ let P_I be the marginal probability distribution on $\mathrm{Dom}(I)$, where $I \subseteq H$, obtained by marginalizing the distribution P. In other words, we have

$$P_I(\mathbf{i}) = \sum \{P(\mathbf{h}) : \mathbf{h}_I = \mathbf{i}\}$$

for $\mathbf{i} \in \mathrm{Dom}(I)$. The joint distribution of H estimated from the data will be denoted by $\hat{P}$.

Let P_I and P'_I be two probability distributions over an itemset I. The *Kullback-Leibler divergence* and the *chi-squared divergence* [KK92] between P_I and P'_I are defined respectively as

$$D_{KL}(P_I : P'_I) = \sum_{\mathbf{i} \in \mathrm{Dom}(I)} P_I(\mathbf{i}) \log \frac{P_I(\mathbf{i})}{P'_I(\mathbf{i})},$$

$$D_{\chi^2}(P_I : P'_I) = \sum_{\mathbf{i} \in \mathrm{Dom}(I)} \frac{(P_I(\mathbf{i}) - P'_I(\mathbf{i}))^2}{P'_I(\mathbf{i})}.$$

Intuitively, the divergence represents how much distribution P_I differs from P'_I. Since the choice of divergence is immaterial for the rest of the paper we will simply denote the divergence by D meaning that either Kullback-Leibler or chi-squared divergence can be used.

A *constraint* C on the set of attributes H is a pair $\mathsf{C} = (I, p)$ where $I \subseteq H$, $p \in [0,1]$. A probability distribution P *satisfies* a constraint $\mathsf{C} = (I, p)$ if $P_I(\mathbf{1}_I) = p$, where $\mathbf{1}_I = (1, 1, \ldots, 1) \in \mathrm{Dom}(I)$. Usually the attribute set will be clear from context, so we will just write $\mathbf{1}$.

To remove redundancies in the rule set we need to define how interesting a rule is with respect to a set of constraints introduced by other rules.

Definition 1. *A set of constraints $\mathcal{C}$ is* consistent *if there exists a joint probability distribution over H which satisfies all the constraints in $\mathcal{C}$. Otherwise, $\mathcal{C}$ is* inconsistent.

In this paper we will only be concerned with consistent sets of constraints. Dealing with inconsistent sets of constraints is an interesting topic of future research.

While determining interestingness of rules with respect to a consistent set of constraints $\mathcal{C}$ we will associate with $\mathcal{C}$ some joint probability distribution $P^{\mathcal{C}}$ over H.

Note that a set of constraints does not have to determine the joint probability distribution uniquely, and we have to choose one of the conforming distributions. The three main approaches to this problem are the maximum entropy principle (MaxENT), the minimum interdependence principle, and the maximum likelihood (see [KK92, Adw97]). We use MaxENT, but it can be shown [KK92, Adw97], that in most cases all three approaches are equivalent. Philosophical justifications of the principles can be found in [KK92, GHK94].

Definition 2. *Let $\mathcal{C}$ be a consistent set of constraints. A* probability distribution $P^{\mathcal{C}}$ over H is induced by $\mathcal{C}$ *if it satisfies the following conditions:*

1. *$P^{\mathcal{C}}$ satisfies all the constraints in $\mathcal{C}$.*
2. *Of all probability distributions over H satisfying $\mathcal{C}$, $P^{\mathcal{C}}$ has the largest entropy.*

It can be shown [Adw97] that $P^{\mathcal{C}}$ is unique.

2 Interestingness of a Rule with Respect to a Set of Constraints

We are now ready to define the interestingness of an association rule with respect to some set of constraints $\mathcal{C}$. For the definition of association rules see [AIS93].

The support of an itemset I is $\text{supp}(I) = \hat{P}_I(\mathbf{1})$. Rules with empty antecedents are allowed and the support and confidence of such rules are defined to be equal to the support of their consequents.

The set of constraints generated by an association rule $I \to J$ is defined as

$$\mathsf{C}(I \to J) = \{(I, \text{supp}(I)), (I \cup J, \text{supp}(I \cup J))\}.$$

We introduce two interestingness measures for association rules: the active and passive interestingness. The active interestingness reflects the impact of adding to the current set of constraints the set of constraints generated by the rule itself. The passive interestingness is determined by the difference between the confidence estimated from the data and the confidence estimated starting from the probability distribution induced by the constraints.

Definition 3. *Let $\mathcal{C}$ be a consistent set of constraints, $I \to J$ be a rule and D some measure of distribution divergence. Denote by $Q^{\mathcal{K}}$ the probability distribution over $I \cup J$ induced by the set of constraints $\mathcal{K}$.*

The active interestingness *of $I \to J$ with respect to $\mathcal{C}$ is defined as:*

$$\mathsf{I}^{\text{act}}(\mathcal{C}, I \to J) = D(Q^{\mathcal{C} \cup \mathsf{C}(I \to J)}, Q^{\mathcal{C}}).$$

The passive interestingness *of $I \to J$ with respect to $\mathcal{C}$ is defined as:*

$$\mathsf{I}^{\text{pass}}(\mathcal{C}, I \to J) = \left| \text{conf}(I \to J) - \frac{Q^{\mathcal{C}}(\mathbf{1})}{Q_I^{\mathcal{C}}(\mathbf{1})} \right|,$$

where $\text{conf}(I \to J)$ *denotes the confidence of rule* $I \to J$.

Whenever we state facts that hold for either of these measures we simply talk about rule interestingness I.

3 Pruning Redundant Association Rules

Definition 4. *Let $\mathcal{R}$ be a set of association rules. Consider an association rule $I \to J$, where $I, J \subseteq H$. The rule $I \to J$ is* I-nonredundant with respect to $\mathcal{R}$, *if $I = \emptyset$ or $\mathsf{I}(\mathcal{C}^{I,J}(\mathcal{R}), I \to J)$ is significantly greater than* 0, *where $\mathcal{C}^{I,J}(\mathcal{R}) = \{\mathsf{C}(K \to J) : K \to J \in \mathcal{R}, K \subset I\}$.*

Note that we do not specify precisely what 'significantly greater' means. This may mean 'greater than some threshold' or 'statistically significant at some confidence level' or some combination of both.

A feature of our definition of redundancy is that it is not influenced by rules involving attributes not in $I \cup J$. For example, suppose that the joint distribution of attributes ABC is fully explained by rules $A \to B$ and $B \to C$. The rule $A \to C$ may still be considered I-nonredundant, even though it does not introduce any new information.

We believe this is the correct behavior. In general, if we have a long chain of rules $A \to B \to C \to \ldots \to Y \to Z$, the rule $A \to Z$ might not be easy to see and thus be interesting. Furthermore, the discovered rules do not necessarily correspond to true causality relations, and it might be better, at least until the user develops a better understanding of the dataset, to present him/her also rules indirectly implied by some other rules.

Another important advantage of our method is that single rules usually involve very few attributes, and thus local interestingness can be efficiently determined, even by direct application of the Generalized Iterative Scaling algorithm, see later in this section.

An algorithm for producing a set of I-nonredundant rules with a single attribute in the consequent is given below:

Input: A set $\mathcal{S}$ of association rules.
Output: Set $\mathcal{R}$ of I-nonredundant association rules of $\mathcal{S}$.

1. For each $A_i \in H$
2. $\quad \mathcal{R}_i = \{\emptyset \to A_i\}$
3. $\quad k = 1$
4. $\quad$ For each rule $I \to A_i \in \mathcal{S}$, $|I| = k$ do
5. $\quad\quad$ If $I \to A_i$ is I-nonredundant with respect to $\mathcal{R}_i$ then
6. $\quad\quad\quad$ Let $\mathcal{R}_i = \mathcal{R}_i \cup \{I \to A_i\}$
7. $\quad k = k + 1$
8. $\quad$ Goto 4
9. $\mathcal{R} = \bigcup_{A_i \in H} \mathcal{R}_i$

Examples below show how our method compares with other work in certain situations. Passive interestingness measure I^{pass} is used, but it is easy to see that the statements remain valid also for the active interestingness measure I^{act}. See discussion later in this section for details on how the maximum entropy distributions can be computed.

Example 1. Let A, B, C be binary attributes, $P_A(1) = P_B(1) = 0.5$. The attribute C depends on A, B according to the following association rules:

assoc. rule	confidence
$\emptyset \to C$	0.5
$A \to C$	0.3
$B \to C$	0.7
$AB \to C$	0.3

Using the approach from [PT00, LHM99, AL99, SLRS99] rules $\emptyset \to C$, $A \to C$ and $B \to C$ are interesting but $AB \to C$ is not, since it is explained by the rule $A \to C$. We claim however that the rule $AB \to C$ is interesting, since it tells us that when both A and B are 'present' it is A that influences C stronger.

Consider rules $\emptyset \to C$, $A \to C$, and $B \to C$. The set of constraints corresponding to them is $\mathcal{C} = \{(A, 0.5), (B, 0.5), (C, 0.5), (AC, 0.15), (BC, 0.35)\}$. The MaxENT distribution in this case is

$$P^{\mathcal{C}} = \begin{pmatrix} 000 & 001 & 010 & 011 & 100 & 101 & 110 & 111 \\ 0.105 & 0.105 & 0.045 & 0.245 & 0.245 & 0.045 & 0.105 & 0.105 \end{pmatrix},$$

and $P^{\mathcal{C}}_{ABC}(\mathbf{1})/P^{\mathcal{C}}_{AB}(\mathbf{1}) = 0.5$, different from $\mathrm{conf}(AB \to C) = 0.3$, making the rule $AB \to C$ interesting.

Example 2. Assume now that the confidences of the rules in the example above are

assoc. rule	confidence
$\emptyset \to C$	0.5
$A \to C$	0.3
$B \to C$	0.7
$AB \to C$	0.5

Using methods given in [LHM99, AL99] the rule $AB \to C$ is interesting, since its confidence differs from $\mathrm{conf}(A \to C)$ and $\mathrm{conf}(B \to C)$.

However, as seen above the maximum entropy distribution induced by rules $\emptyset \to C$, $A \to C$ and $B \to C$ gives $P^{\mathcal{C}}_{ABC}(\mathbf{1})/P^{\mathcal{C}}_{AB}(\mathbf{1}) = 0.5$, and the rule $AB \to C$ is considered uninteresting. In other words, knowing the joint influence of AB on C does not give us any more information over what we have already know from other rules, since A and B are conditionally independent given C. The above result is intuitive since when both A and B influence C we would expect their joint influence to be an 'average' between the influences of A and B alone.

Example 3. Suppose that attribute A is independent of B, C, and jointly of BC. Then, $P^{\mathcal{C}}_{ABC}(\mathbf{1})/P^{\mathcal{C}}_{AB}(\mathbf{1}) = P^{\mathcal{C}}_{BC}(\mathbf{1})/P^{\mathcal{C}}_{B}(\mathbf{1}) = \mathrm{conf}(B \to C)$, and the rule $AB \to C$ is considered not interesting using our approach, but also using methods from [PT00, LHM99, SLRS99, AL99] which explains their good behavior in practice. However as the examples above show, those methods can filter out certain interesting rules, and include some uninteresting ones.

To compute the maximum entropy distribution we can use the Generalized Iterative Scaling (GIS) Algorithm [Adw97, Bad95, DR72, Csi89].

Let $\mathcal{C} = \{\mathsf{C}_1, \mathsf{C}_2, \ldots, \mathsf{C}_n\}$ be a set of constraints, where $\mathsf{C}_k = (I_k, p_k)$. GIS proceeds by assigning some initial values to each probability in $P^{\mathcal{C}}$, and iteratively updating them until all the constraints are satisfied. Let $P^{\mathcal{C}(i)}$ denote the distribution after i iterations. Updating in each iteration is performed according to the formula

$$P^{\mathcal{C}(i+1)}(\mathbf{h}) = P^{\mathcal{C}(i)}(\mathbf{h}) \prod_{\mathbf{h}_{I_k}=\mathbf{1}} \left[\frac{p_k}{P_{I_k}^{\mathcal{C}(i)}(\mathbf{h}_{I_k})} \right]^{\frac{1}{c}},$$

for every $\mathbf{h} \in \mathrm{Dom}(H)$, assuming that $\frac{0}{0} = 0$. The algorithm is guaranteed to converge if $\sum_{k=1}^{n} f_k(\mathbf{h}) = c$ is a constant independent of $\mathbf{h}$. In practice, this condition can always be satisfied by adding an additional constraint. See [Adw97, DR72, Csi89] for details and proof of convergence. The version of the algorithm presented in [Csi89] has the advantage of being able to cope with distributions with zero probabilities, and this is the one we use in our implementation.

The disadvantage of the GIS algorithm is its high computational cost caused by the necessity of computing the marginal probabilities, and in some cases by the large number of iterations required.

One of the main techniques for speeding up MaxENT computations is *decomposition* [Bad95, And74, DLS80]. However, in our case, we will only use maximum entropy distributions in few variables, and our experiments showed that decomposition does not give real improvement in efficiency. We noticed however that the number of rules considered interesting is small and thus constraints are usually simple. Closed form solutions are used for a few common cases; in every other situation we use the GIS algorithm.

Below we describe closed form solutions used in this paper. For attribute set I denote $N_I = |\{\mathbf{x} \in \mathrm{Dom}(H) : \mathbf{x}_I = \mathbf{1}\}|$.

Theorem 1. *Let* $\mathcal{C} = \{(J, \hat{P}_J(\mathbf{1})), (K, \hat{P}_K(\mathbf{1})), (K \cup J, \hat{P}_{K\cup J}(\mathbf{1}))\}$, $J, K \subset H$, $K \cap J = \emptyset$ *be a set of constraints. The MaxENT distribution induced by* $\mathcal{C}$ *is*

$$P^{\mathcal{C}}(\mathbf{x}) = \begin{cases} \frac{\hat{P}_{K\cup J}(\mathbf{1})}{N_{K\cup J}}, & \text{if } \mathbf{x}_J = \mathbf{1} \wedge \mathbf{x}_K = \mathbf{1} \\ \frac{\hat{P}_J(\mathbf{1}) - \hat{P}_{K\cup J}(\mathbf{1})}{N_J - N_{K\cup J}}, & \text{if } \mathbf{x}_J = \mathbf{1} \wedge \mathbf{x}_K \neq \mathbf{1} \\ \frac{\hat{P}_K(\mathbf{1}) - \hat{P}_{K\cup J}(\mathbf{1})}{N_K - N_{K\cup J}}, & \text{if } \mathbf{x}_J \neq \mathbf{1} \wedge \mathbf{x}_K = \mathbf{1} \\ \frac{1 - \hat{P}_K(\mathbf{1}) - \hat{P}_J(\mathbf{1}) + \hat{P}_{K\cup J}(\mathbf{1})}{|\mathrm{Dom}(H)| - N_K - N_J + N_{K\cup J}}, & \text{if } \mathbf{x}_J \neq \mathbf{1} \wedge \mathbf{x}_K \neq \mathbf{1}, \end{cases}$$

for $\mathbf{x} \in \mathrm{Dom}(H)$.

Proof. For every $R \subseteq \{K, J\}$ denote X_R the set of all $\mathbf{x} \in \mathrm{Dom}(H)$ such that $\mathbf{x}_I = \mathbf{1}$ if $I \in R$ and $\mathbf{x}_I \neq \mathbf{1}$ otherwise, for all $I \in \{J, K\}$. Note that $|X_{\{K,J\}}| = N_{K\cup J}$, $|X_{\{J\}}| = N_J - N_{K\cup J}$, $|X_{\{K\}}| = N_K - N_{K\cup J}$, and $|X_\emptyset| = |\mathrm{Dom}(H)| - N_K - N_J + N_{K\cup J}$. Also denote $P_R^* = \sum_{\mathbf{x}\in X_R} \hat{P}(\mathbf{x})$ for all $R \subseteq \{K, J\}$. Note that $P_{\{K,J\}}^* = \hat{P}_{K\cup J}(\mathbf{1})$, $P_{\{J\}}^* = \hat{P}_J(\mathbf{1}) - \hat{P}_{K\cup J}(\mathbf{1})$, $P_{\{K\}}^* = \hat{P}_K(\mathbf{1}) - \hat{P}_{K\cup J}(\mathbf{1})$, and $P_\emptyset^* = 1 - \hat{P}_K(\mathbf{1}) - \hat{P}_J(\mathbf{1}) + \hat{P}_{K\cup J}(\mathbf{1})$.

For a probability distribution P on H that satisfies the the set of constraints $\mathcal{C}$ we have:

$$\begin{aligned} H(P) &= -\sum_{R\subseteq\{K,J\}}\sum_{\mathbf{x}\in X_R} P(\mathbf{x})\log P(\mathbf{x}) \\ &= -\sum_{R\subseteq\{K,J\}} P_R^*\sum_{\mathbf{x}\in X_R}\frac{P(\mathbf{x})}{P_R^*}\log\frac{P(\mathbf{x})}{P_R^*} - \sum_{R\subseteq\{K,J\}} P_R^*\log P_R^*. \end{aligned}$$

It suffices to maximize the first term. Notice that for every $R \subseteq \{K, J\}$, $\sum_{\mathbf{x}\in X_R}\frac{P(\mathbf{x})}{P_R^*} = 1$ and thus P/P_R^* is a probability distribution over X_R, and its entropy $-\sum_{\mathbf{x}\in X_R}\frac{P(\mathbf{x})}{P_R^*}\log\frac{P(\mathbf{x})}{P_R^*}$ is maximized when $\frac{P(\mathbf{x})}{P_R^*} = 1/|X_R|$ for every $\mathbf{x} \in X_R$. This gives $P(\mathbf{x}) = P_R^*/|X_R|$ for every $\mathbf{x} \in X_R$ and completes the proof since every $\mathbf{x} \in \mathrm{Dom}(H)$ belongs to exactly one of the X_R's. □

Notice that when $J = K$, the above result reduces to

$$P^{\mathcal{C}}(\mathbf{x}) = \begin{cases} \frac{\hat{P}_J(\mathbf{1})}{N_J}, & \text{if } \mathbf{x}_J = \mathbf{1} \\ \frac{1-\hat{P}_J(\mathbf{1})}{|\mathrm{Dom}(H)|-N_J}, & \text{if } \mathbf{x}_J \neq \mathbf{1}, \end{cases} \qquad (1)$$

for $\mathbf{x} \in \mathrm{Dom}(H)$.

Frequently the only subrules of a rule $I \to J$ are $\emptyset \to J$, and $K \to J$, where $K \subset I$. In this case the MaxENT distribution induced by the subrules can be found by application of Theorem 1. If the only subrule is $\emptyset \to J$, then we can use Equality (1). Our experiments revealed that using the above theorem reduces pruning time up to a factor of 10. See [Bad95, PMS01] for a more detailed discussion of methods of speeding up MaxENT computations.

4 Experimental Evaluation of the Pruning Algorithm

In this section we present an experimental evaluation of our pruning algorithm. We used passive interestingness I^{pass}, and considered a rule I^{pass}-nonredundant if its passive interestingness was greater than some threshold. Our experiments have shown that the passive measure of interestingness performed better than the active one I^{act}, which often pruned interesting rules with small support. The reason for that is that rules with small support usually have many attributes in the antecedent, and thus adding them as constraints affects only very few values in the joint probability distribution, while active interestingness depends on the whole distribution. Also, we did not use any minimum confidence threshold, because pruning provided a sufficient reduction in the number of rules, and setting a minimum confidence threshold occasionally pruned some of the interesting rules.

We first present the result of running the algorithm on the `lenses` database from the UCI machine learning archive [BM98]. The database has the advantage of being very small thus allowing manual selection of rules. Table 1 shows the rules having the `lenses` attribute as consequent, selected manually by the

antecedent→lenses	conf. [%]	supp. [%]
∅→soft	20.8	20.8
∅→hard	16.6	16.6
∅→none	62.5	62.5
tears=reduced→none	100	50
astigmatism=no,tears=normal→soft	83.3	20.8
astigmatism=yes,tears=normal→hard	66.6	16.6
age=pre-presbyopic,prescription=hypermetrope,astigmatism=yes→none	100	8.3
age=presbyopic,prescription=myope,astigmatism=no→none	100	8.3
age=presbyopic,prescription=hypermetrope,astigmatism=yes→none	100	8.3

Table 1. Rules manually selected from the `lenses` database

antecedent→lenses	I^P [%]	conf. [%]	supp. [%]
∅→soft	0	20.8	20.8
∅→hard	0	16.6	16.6
∅→none	0	62.5	62.5
tears=reduced→none	37.5	100	50
astigmatism=no,tears=normal→soft	62.5	83.3	20.8
astigmatism=yes,tears=normal→hard	50	66.6	16.6
tears=normal→none	-37.5	25	12.5
prescription=myope,astigmatism=yes→hard	33.3	50	12.5
prescription=myope,tears=normal→hard	33.3	50	12.5
prescription=hypermetrope,astigmatism=yes,tears=normal→none	41.4	66.6	8.3
age=pre-presbyopic,prescription=hypermetrope,astigmatism=yes→none	37.5	100	8.3
age=presbyopic,prescription=myope,astigmatism=no→none	37.5	100	8.3
age=presbyopic,prescription=hypermetrope,astigmatism=yes→none	37.5	100	8.3
age=young,astigmatism=yes→hard	33.3	50	8.3
age=young,tears=normal→hard	33.3	50	8.3
age=presbyopic,astigmatism=no,tears=normal→soft	-32.9	50	4.1
age=presbyopic,prescription=hypermetrope,astigmatism=no,tears=normal→soft	49.3	100	4.1
prescription=hypermetrope,astigmatism=yes,tears=normal→hard	-32.9	33.3	4.1
age=young,prescription=hypermetrope,astigmatism=yes,tears=normal→hard	39.3	100	4.1

Table 2. Rules selected from the `lenses` database

authors, providing a complete description of the dataset. Table 2 shows rules involving `lenses` attribute as consequent generated by the Apriori algorithm with minimum support 1 (1 record), no minimum confidence, post-processed with our pruning algorithm using passive interestingness with interestingness threshold 0.3. Negative values of interestingness mean that the presence of the antecedent decreases the probability of presence of the consequent.

Rules have been sorted based on the product of support and interestingness, with an extra condition, that a rule cannot be printed until all its subrules have been printed. Also, note that the `lenses` dataset contains multivalued attributes. Since our method only handles boolean attributes we encode each original attribute with a number of boolean attributes, one for each possible value of the original attribute.

The Apriori algorithm produced 113 rules having `lenses` attribute as the consequent. After pruning, 16 nonredundant rules were left with a nonempty antecedent. This is a significant reduction.

When rules with all possible consequents are considered, our method outputs 40 rules out of 890 produced by Apriori. Also, note that all rules selected manually are also considered interesting by our pruning algorithm, and the top three rules are indeed identical in both cases, which suggests that really interesting rules are indeed retained by our algorithm.

We also applied our method to a dataset of census data of elderly people obtained from The University of Massachusetts at Boston Gerontology Cen-

antecedent→lenses	I^P [%]	conf. [%]	supp. [%]
∅→urban=no	0	22.4	22.4
∅→urban=yes	0	77.5	77.5
immigr=no,region=south→urban=yes	-11.8	65.7	26.2
race=white→urban=yes	-10.6	66.8	22.5
region=west→urban=yes	12.8	90.3	16.9
race=hisp→urban=yes	12.4	89.9	15.4
region=south,race=black→urban=yes	-10.6	66.8	17.8
immigr=no,region=south→urban=no	11.8	34.2	13.7
alone=yes,region=south→urban=yes	-10.5	66.9	15
immigr=before75→urban=yes	15.9	93.4	9.7
region=neast,race=black→urban=yes	19.7	97.2	6.7
region=midw,race=black→urban=yes	18.9	96.5	6.9
age=below75,region=neast→urban=yes	10.5	88	11.3
race=white→urban=no	10.6	33.1	11.1

Table 3. Top 12 rules involving `urban` attribute generated from the elderly people census data

ter. The dataset contains about 330 thousand records, 11 attributes with up to five values, and is available at `http://www.cs.umb.edu/ ~sj /datasets /census.arff.gz`. We used 1% minimum support and no minimum confidence. The Apriori algorithm produced 247476 rules practically impossible to analyze by hand. After pruning with 10% interestingness threshold only 2056 were considered nonredundant, and after further restricting this set to rules with a given consequent attribute we were able to obtain easily manageable sets of interesting association rules. Some of them, concerning the `urban` (whether a person lives in a city or not) attribute are given in Table 3. Although the pruning time was quite long (over 4 hours on a 100MHz Pentium machine), it was still much easier to use our method than to handle hundreds of thousands of rules manually. See Table 4 for further details.

Table 4 shows the number of rules generated by Apriori compared with the number of rules considered interesting by our algorithm, as well as pruning time, for various datasets from the UCI Machine Learning Archive [BM98]. All datasets have been mined with 0 minimum confidence. The interestingness thresholds and minimum supports have been chosen manually by trial and error such that the unpruned rules provide a lot of interesting information while keeping their number reasonably small. For some datasets values for a few different thresholds are given for comparison. All experiments have been performed on a 100MHz Pentium machine with 64MB of memory.

5 Conclusions and Further Research

A method for pruning redundant association rules using the Maximum Entropy approach has been presented along with methods of speeding up MaxENT computations by using closed form formulas. The method has been experimentally shown to produce relatively small sets of interesting rules in a reasonable amount of time. A detailed analytical comparison of our method with other approaches has also been presented.

We plan to concentrate our further efforts on including background knowledge in the mining/pruning process. In most real applications the user already

dataset	min. support	interestingness threshold	number of rules		pruning time [s]
			Apriori	after pruning	
lenses	1(4%)	0.3	890	40	1.3
mushroom*	500(16%)	0.2	164125	5141	418
breast-cancer	30(10%)	0.15	2128	74	2.8
primary-tumor*	30(9%)	0.3	43561	67	21.8
primary-tumor*	30(9%)	0.2	43561	432	24
car	10(0.5%)	0.3	20669	293	11.1
car	10(0.5%)	0.15	20669	580	30.2
splice*	300(9%)	0.5	4847	24	3.0
splice*	300(9%)	0.3	4847	95	5.6
splice*	300(9%)	0.15	4847	290	7.2
splice*	200(6%)	0.3	35705	463	33.8
census(elderly people)	3000(1%)	0.3	247476	194	4801
census(elderly people)	3000(1%)	0.2	247476	621	5683
census(elderly people)	3000(1%)	0.1	247476	2056	15480

* itemsets of up to 4 attributes

Table 4. Numbers of rules and computation times for various datasets

has a lot of domain knowledge about the dataset and is only interested in rules which are not implied by what is already known. We believe that this is the approach one should use to achieve further reductions in the number of rules and make them more applicable for the user.

Currently, the selection of interestingness threshold is made by trial and error. A more formal procedure, possibly based on statistical tests and confidence levels would be very useful.

Even though our method was fast enough to apply it to real datasets, we plan to further improve its performance by analyzing which configurations of constraints occur most frequently and provide closed form solutions for them.

6 Acknowledgments

The authors would like to thank Prof. Jeffrey Burr from the University of Massachusetts at Boston Gerontology Center for providing us with the elderly people census data.

References

[Adw97] Ratnaparkhi Adwait. A simple introduction to maximum entropy models for natural language processing. IRCS Report 97–08, University of Pennsylvania, 3401 Walnut Street, Suite 400A, Philadelphia, PA, May 1997. ftp://www.cis.upenn.edu/pub/ircs/tr/97-08.ps.Z.

[AIS93] R. Agrawal, T. Imielinski, and A. Swami. Mining association rules between sets of items in large databases. In *Proc. ACM SIGMOD Conference on Management of Data*, pages 207–216, Washington, D.C., 1993.

[AL99] Y. Aumann and Y. Lindell. A statistical theory for quantitative association rules. In *Knowledge Discovery and Data Mining*, pages 261–270, 1999.

[And74] A. H. Andersen. Multidimensional contigency tables. *Scand. J. Statist*, 1:115–127, 1974.

[BA99] R. J. Bayardo and R. Agrawal. Mining the most interesting rules. In *Proc. of the 5th ACM SIGKDD Int'l Conf. on Knowledge Discovery and Data Mining*, pages 145–154, August 1999.

[Bad95] J. Badsberg. *An Environment for Graphical Models.* PhD thesis, Aalborg University, 1995.

[BM98] C.L. Blake and C.J. Merz. UCI repository of machine learning databases. University of California, Irvine, Dept. of Information and Computer Sciences, 1998. http://www.ics.uci.edu/~mlearn/MLRepository.html.

[BVW00] T. Brijs, K. Vanhoof, and G. Wets. Reducing redundancy in characteristic rule discovery by using integer programming techniques. *Intelligent Data Analysis Journal*, 4(3), 2000.

[Csi89] I. Csiszar. A geometric interpretation of Darroch and Ratcliff's generalized iterative scaling. *The Annals of Statistics*, 17(3):1409–1413, 1989.

[DLS80] J. N. Darroch, S. L. Lauritzen, and T. P. Speed. Markov fields and log-linear interaction models for contingency tables. *Annals of Statistics*, 8:522–539, 1980.

[DR72] J. N. Darroch and D. Ratcliff. Generalized iterative scaling for log-linear models. *Annals of Mathematical Statistics*, 43:1470–1480, 1972.

[GHK94] A. J. Grove, J. Y. Halpern, and D. Koller. Random worlds and maximum entropy. *Journal of Artificial Intelligence Research*, 2:33–88, 1994.

[HH99] R. Hilderman and H. Hamilton. Knowledge discovery and interestingness measures: A survey. Technical Report CS 99-04, Department of Computer Science, University of Regina, 1999.

[JS01] S. Jaroszewicz and D. A. Simovici. A general measure of rule interestingness. In *Proc of PKDD 2001, Freiburg, Germany*, volume 2168 of *Lecture Notes in Computer Science*, pages 253–265. Springer, September 2001.

[KK92] J. N. Kapur and H. K. Kesavan. *Entropy Optimization Principles with Applications.* Academic Press, San Diego, 1992.

[LHM99] Bing Liu, Wynne Hsu, and Yiming Ma. Pruning and summarizing the discovered associations. In Surajit Chaudhuri and David Madigan, editors, *Proceedings of the Fifth ACM SIGKDD International Conference on Knowledge Discovery and Data Mining*, pages 125–134, N.Y., August 15–18 1999. ACM Press.

[PMS01] D. Pavlov, H. Mannila, and P. Smyth. Beyond independence: Probabilistic models for query approximation on binary transaction data. Technical Report ICS TR-01-09, Information and Computer Science Department, UC Irvine, 2001.

[PT00] B. Padmanabhan and A. Tuzhilin. Small is beautiful: discovering the minimal set of unexpected patterns. In Raghu Ramakrishnan, Sal Stolfo, Roberto Bayardo, and Ismail Parsa, editors, *Proceedinmgs of the 6th ACM SIGKDD International Conference on Knowledge Discovery and Data Mining (KDD-00)*, pages 54–63, N. Y., August 2000. ACM Press.

[Rat96] Adwait Ratnaparkhi. A maximum entropy model for part-of-speech tagging. In Eric Brill and Kenneth Church, editors, *Proceedings of the Conference on Empirical Methods in Natural Language Processing*, pages 133–142. Association for Computational Linguistics, Somerset, New Jersey, 1996.

[SK98] E. Suzuki and Y. Kodratoff. Discovery of surprising exception rules based on intensity of implication. In *Proc of PKDD-98, Nantes, France*, pages 10–18, 1998.

[SLRS99] D. Shah, L. V. S. Lakshmanan, K. Ramamritham, and S. Sudarshan. Interestingness and pruning of mined patterns. In *1999 ACM SIGMOD Workshop on Research Issues in Data Mining and Knowledge Discovery*, 1999.

[Suz97] E. Suzuki. Autonomous discovery of reliable exception rules. In David Heckerman, Heikki Mannila, Daryl Pregibon, and Ramasamy Uthurusamy, editors, *Proceedings of the Third International Conference on Knowledge Discovery and Data Mining (KDD-97)*, page 259. AAAI Press, 1997.

[TKR+95] H. Toivonen, M. Klemettinen, P. Ronkainen, K. Hätönen, and H. Mannila. Pruning and grouping discovered association rules. In *MLnet Workshop on Statistics, Machine Learning, and Discovery in Databases*, pages 47–52, Heraklion, Crete, Greece, April 1995.

A Confidence-Lift Support Specification for Interesting Associations Mining

Wen-Yang Lin[1], Ming-Cheng Tseng[2], and Ja-Hwung Su[3]

[1] Department of Information Management, I-Shou University,
Kaohsiung 84008, Taiwan
wylin@isu.edu.tw

[2] Institute of Information Engineering, I-Shou University,
Kaohsiung 84008, Taiwan
tmc001@ksts.seed.net.tw

[3] Institute of Information Engineering, I-Shou University,
Kaohsiung 84008, Taiwan
m893324m@isu.edu.tw

Abstract. Recently, the weakness of the canonical support-confidence framework for associations mining has been widely studied in the literature. One of the difficulties in applying association rules mining to real world applications is the setting of support constraint. A high support constraint avoids the combinatorial explosion in discovering frequent itemsets, but at the expense of missing interesting patterns of low support. Instead of seeking the way for setting the appropriate support constraint, all current approaches leave the users in charge of the support setting, which, however, puts the users in a dilemma. This paper is an effort to answer this long-standing open question. Based on the notion of confidence and lift measures, we propose an automatic support specification for mining high confidence and positive lift associations without consulting the users. Experimental results show that this specification is good at discovering the low support, but high confidence and positive lift associations, and is effective in reducing the spurious frequent itemsets.

1 Introduction

Mining association rules from a large database of business data, such as transaction records, has been a hot topic within the area of data mining. This problem is motivated by applications known as market basket analysis to find relationships between items purchased by customers, that is, what kinds of products tend to be purchased together [1]. An association rule is an expression of the form $A \Rightarrow B$, where A and B are sets of items. Such a rule reveals that transactions in the database containing items in A tend to contain items in B, and the conditional probability, measured as the fraction of transactions containing A also containing B, i.e., $P(B|A) = P(A \cup B)/P(A)$, is called the *confidence* of the rule. The *support* of the rule is the fraction of the transactions that contain all items both in A and B, i.e., $sup(A \Rightarrow B) = P(A \cup B)$. For an association rule to hold, the

M.-S. Chen, P.S. Yu, and B. Liu (Eds.): PAKDD 2002, LNAI 2336, pp. 148–158, 2002.

support and the confidence of the rule should satisfy a user-specified minimum support called *minsup* and minimum confidence called *minconf*, respectively.

One of the difficulties in applying association rules mining to real world applications is the setting of support constraint. A high support constraint avoids the combinatorial explosion in discovering frequent itemsets, but at the expense of missing interesting patterns of low support. However, most rules with high support are obvious and well-known, and it is the rules with low-support that provide interesting new insight, such as deviations or exceptions.

Instead of seeking the way for setting the appropriate support constraint, all current approaches leave the users in charge of the support setting, which, however, puts the users in a dilemma: how to specify the most appropriate support constraint, either uniform or non-uniform, to discover interesting patterns without suffering from combinatorial explosion and missing some less-support but perceptive rules. The best one can do is either setting the support at the lowest value ever specified or performing a consecutive sequence of mining processes with various constraints to extract the right patterns.

Our intent is to seek all rules of high confidence without the need for user specified support constraint. To this end, we proposed an automatic support specification without consulting the users. The idea is based on the notion of *lift* measure [3] (also called *interest* [5]) and confidence measure. Experimental results show that this specification is good at discovering the low support, but high confidence and positive lift associations, and is effective in reducing the spurious frequent itemsets.

The remaining of the paper is organized as follows. The problem of support-confidence framework for associations mining and the related work are presented in Section 2. In Section 3, we provide a modified association framework and explain the support specification and mining process for this model. An evaluation of the proposed specification on IBM synthetic data is described in Section 4. Finally, our conclusions are stated in Section 5.

2 Problem of Support-Confidence Framework and Related Work

In the past few years, there has been work on challenging the canonical support-confidence framework for associations mining. These efforts can be categorized into two paradigms: extending the constant support constraint and/or seeking substitutes for confidence measure.

The uniform support constraint was first argued by [7] while generalizing the association model into mutiple-level associations on account of the item hierarchy. In [7], Han and Fu extended the uniform support constraint to a form of level-by-level, decreasing assignment. That is, items at the same level receive the same minimum support and higher level items have larger support constraint. This level-wise support specification accounts for their progressive mining approach: A apriori-like algorithm is performed progressively from the top level to the bottom, and stop at the very level when no frequent itemset is generated.

Another form of association rules mining with non-uniform minimum supports was proposed by Liu et al [9]. Their method allows the users to specify different minimum supports to different items and the support constraint of an itemset is defined as the lowest minimum item support among the items in the itemset. The motivation behind this approach is that in natural the supports of items are non-uniform and high profit items (e.g., TV) usually occurs less frequently than low value items (e.g., toothpaste). The multi-supported model was then extended to generalized associations with taxonomy information in [11]. The problem thus remains tangling.

Wang et al. [12] proposed a bin-oriented, non-uniform support constraint: Items are grouped into disjoint sets, called *bins*, and items within the same bin are regarded as non-distinguished with respect to the specification of minimum support. In particular, each support constraint specifies a set of bins $B_1, B_2, \ldots, B_s$ of the form $SC_i(B_1, B_2, \ldots, B_s) \geq \theta_i$, where $s \geq 0$ and θ_i is a minimum support. The problem of their approach is that the specification is complex and hard to follow. Furthermore, the end-user needs to determine the appropriate bins and θ_i prior to the mining.

In general, the support of an itemset is decreasing as its length increases. The uniform support constraint thus may hinder the discovery of long itemsets. In some applications, however, it may be interesting to discover associations between long itemsets. To solve this problem, Seno and Karypis [10] used a support constraint that decreases with the length of the itemset, which helps to find long itemsets without generating a large number of shorter itemsets.

There is also work on mining high confidence associations without support constraints [8]. The proposed method, however, is restricted to discover all top rules ($conf = 100\%$) with the consequent being given.

The primary criticism on confidence-based association is the poor predictive ability, which means the measure is unable to capture the real implication [2][4][5]. As an illustration, consider the example transaction database in Table 1. For a minimum support of 30% and minimum confidence of 60%, the following association rule is discoveried:

$$\text{Scanner} \Rightarrow \text{Printer}\ (sup = 44.4\%, conf = 66.7\%),$$

One may conclude that this rule is strong and interesting because of its high confidence. However, note that the support of Printer is 77.8%, which means a priori probability that customers purchase Printer is 77.8% . The rule is misleading since it does not increase the sales of Printer through promotion on Scanner. To remedy this deficiency, two additional measures have been proposed. They are *lift* [3] (also known as *interest* [5]) and *conviction* [5].

For an association rule $A \Rightarrow B$, the lift is defined as

$$lift(A \Rightarrow B) = \frac{P(A \cup B)}{P(A)P(B)} = \frac{sup(A \cup B)}{sup(A)sup(B)} = \frac{conf(A \Rightarrow B)}{sup(B)}$$

Table 1. A transaction database($\mathcal{D}$)

tid	Items Purchased
11	PC, Printer, PDA
12	Printer, Notebook
13	Printer, Scanner
14	PC, Printer, Notebook
15	PC, Scaner
16	Printer, Scanner
17	PC, Scanner
18	PC, Printer, Scanner, PDA
19	PC, Printer, Scanner

and the conviction is

$$conv(A \Rightarrow B) = \frac{P(A)P(\neg B)}{P(A \cup \neg B)} = \frac{1 - sup(B)}{1 - conf(A \Rightarrow B)}.$$

There is also work on investigating alternatives to the association model for attribute set mining [2][4][6]. The models proposed in [2] and [4] aim at mining strong correlated attribute sets. What differentiates them is the way for measuring correlation. In [4], the correlation is measured using the well-known *chi-squared test* from classical statistics, while in [2], a new criterion called *collective strength* is proposed.

In [6], instead of searching for high confidence associations, they focus on identifying similar itemsets (column pairs) without any support threshold. The confidence measure is replaced by a similarity measure.

3 Interesting Association Rules and Support Specification

3.1 Interesting Association Rules

As concluded from the previous discussions, the primary problems of support-confidence framework are poor predictive ability and uniform support constraint. Though previous work has provided solutions to these problems, such as adding the lift or conviction measure and using non-uniform support constraint, there remains no guideline in the support specification. Without such instruction, the users may set an inappropriate support constraint and suffer from the combinatorial explosion or missing new insight patterns. Our view for solving this problem is to discharge the end-users from specifying the support constraint. That is, we would like to seek all interesting association rules, without the need of user specified support constraint. To this end, we first refine the canonical association rule model.

Definition 1. *Let $\mathcal{I}$ be a set of items and $ms(a)$ denote the minimum support of an item a, $a \in \mathcal{I}$. An itemset $A = \{a_1, a_2, ..., a_k\}$, where $a_i \in \mathcal{I}$, is frequent if* *the support of A is larger than the lowest value of minimum support of items in A, i.e.,* $sup(A) \geq \min_{a_i \in A} ms(a_i)$.

Definition 2. *An association rule* $A \Rightarrow B$ *is strong if* $sup(A \Rightarrow B) \geq \min_{a_i \in A \cup B} ms(a_i)$ *and* $conf(A \Rightarrow B) \geq minconf$.

Definition 3. *An association rule* $A \Rightarrow B$ *is interesting if it is strong and* $lift(A \Rightarrow B) > 1$. [1]

3.2 The Confidence-Lift Based Support Specification

The basic idea of our approach is to "push" the confidence and lift measure into the support constraint to prune the spurious frequent itemsets that fail in generating interesting associations as early as possible. First let us show how to specify the constraint to reduce the frequent itemsets that fail in generating strong associations.

Lemma 1. *Let* $A \cup B$ *be a frequent itemset,* $A \cap B = \emptyset$, $a = \min_{a_i \in A \cup B} sup(a_i)$ *and* $a \in A$. *If the minimum support of a is set as* $ms(a) = sup(a) \cdot minconf$, *then the association rule* $A \Rightarrow B$ *is strong, i.e.,* $sup(A \cup B)/sup(A) \geq minconf$.

Proof. Since $A \cup B$ is frequent, $sup(A \cup B) \geq ms(a) = sup(a) \cdot minconf$. Furthermore, $sup(A) \leq sup(a)$ since $a \in A$. It follows that $sup(A \cup B)/sup(A) \geq minconf$.

According to Lemma 1, the minimum support of an item a, for $a \in \mathcal{I}$, is specified as follows:

$$ms(a) = sup(a) \cdot minconf. \tag{1}$$

Note that Lemma 1 does not imply that the rule $B \Rightarrow A$ is strong. This is because the confidence measure is not symmetric over the antecedence and consequence. Therefore, Eq. 1 does not guarantee all rules generated from the frequent itemsets are strong.

Example 1. Consider Table 1. Let $minconf = 50\%$ and the minimum supports of items be set according to Eq. 1. Then we have $ms(\text{PC}) = 6/9 \times 1/2 = 33\%$, $ms(\text{Printer}) = 7/9 \times 1/2 = 39\%$, and $ms(\text{Notebook}) = 2/9 \times 1/2 = 11\%$. It can be verified that {PC, Printer, Notebook} is a frequent itemset. Consider the following two rules generated from this itemset

$$r1 : \text{Printer} \Rightarrow \text{PC, Notebook}\ (sup = 11\%, conf = 14\%),$$
$$r2 : \text{PC, Notebook} \Rightarrow \text{Printer}\ (sup = 11\%, conf = 50\%).$$

Clearly, rule $r1$ is not strong while $r2$ is.

---

[1] The lift measure can be replaced by conviction; indeed, $lift(A \Rightarrow B) > 1$ if and only if $conv(A \Rightarrow B) > 1$.

Now that we have known how to specify the support constraint to obtain strong associations, our next step is to consider how to specify the constraint to generate interesting associations.

Consider any association rule derived from a frequent 2-itemset, say $r : a \Rightarrow b$. Assume that r is strong. Rule r is interesting if

$$\frac{sup(a,b)}{sup(a) \cdot sup(b)} > 1.$$

Since a, b is frequent, it is true that

$$sup(a,b) \geq \min\{ms(a), ms(b)\}.$$

Without loss of generality, let $\min\{ms(a), ms(b)\} = ms(a)$. Hence, $sup(a,b) \geq ms(a)$. To make r be an interesting association rule, the minimum support of a should be set at least of the value $sup(a)sup(b)$, i.e., $ms(a) > sup(a)sup(b)$. Since our intent is to make all generated rules interesting to the users, the minimum support for any item $a_i \in \mathcal{I}$ can be set as

$$ms(a_i) = sup(a_i) \cdot \max_{a_j \in \mathcal{I} - \{a_i\}} sup(a_j). \qquad (2)$$

Though Eq. 2 is derived from rules consisting of only two items, the following lemma shows that this setting suffices for all strong association rules.

Lemma 2. *Let $\mathcal{I}$ be a set of items and the minimum support of each item is specified according to Eq. 2. Then any association rule $A \Rightarrow B$, for $A, B \subset \mathcal{I}$ and $A \cap B = \emptyset$, is interesting, i.e.,*

$$\frac{sup(A \cup B)}{sup(A) \cdot sup(B)} > 1.$$

Proof. Since $A \Rightarrow B$ is strong, $sup(A \cup B) \geq \min_{a \in A \cup B} ms(a)$. Specifically, let $a = a_i$. It follows that

$$sup(A \cup B) \geq ms(a_i) = sup(a_i) \cdot \max_{a_j \in \mathcal{I} - a_i} sup(a_j).$$

Since $A \cap B = \emptyset$, a_i belongs to either A or B. Without loss of generality, let $a_i \in A$. It is easy to show that

$$sup(A) \leq sup(a_i) \text{ and } sup(B) \leq \max_{a_j \in \mathcal{I} - a_i} sup(a_j).$$

The lemma then follows.

Note that the support constraint specified in Eq. 2 only provides a sufficient condition for obtaining interesting association rules from frequent itemsets. That is, it is an upper bound in guaranteeing that all frequent itemsets will not generate associations having negative lift ($lift < 1$). There may exist some itemsets that are infrequent with respect to this constraint but can generate positive lift associations.

Example 2. Let us consider Table 1 again. The minimum supports of items PC, Printer, and Notebook, according to Eq. 2, will be set as $ms(\mathsf{PC}) = 6/9 \times 7/9 = 52\%$, $ms(\mathsf{Printer}) = 7/9 \times 6/9 = 52\%$, and $ms(\mathsf{Notebook}) = 2/9 \times 7/9 = 17\%$, respectively. Clearly, the itemset $\{\mathsf{PC}, \mathsf{Printer}, \mathsf{Notebook}\}$ is not frequent for $sup(\{\mathsf{PC}, \mathsf{Printer}, \mathsf{Notebook}\}) = 11\% < ms(\mathsf{Notebook})$ and so rule $r1$ will not be generated. But it can be verified that $lift(r1) > 1$.

Now, if our purpose is to construct all associations without missing any positive lift rule, we should seek other form of support specification. The intuition is to set the minimum support of an item a to accommodate all frequent itemsets consisting of a as the smallest supported item capable of generating at least one nonnegative lift association rule.

Consider an association rule $A \Rightarrow B$. Without loss of generality, we assume that $a \in A$, $a = \min_{a_i \in A \cup B} sup(a_i)$ and $b = \min_{a_i \in B} sup(a_i)$. The following conditions hold

$$sup(A \cup B) \geq ms(a), sup(A) \leq sup(a) \text{ and } sup(B) \leq sup(b),$$

Thus, to make $lift(A \Rightarrow B) \geq 1$, we should specify $ms(a) \geq sup(a) \cdot sup(b)$. Note that b can be any item in the item set $\mathcal{I}$ except a, and $sup(b) \geq sup(a)$. What we need is the smallest qualified item, i.e., $b = \min\{a_i | a_i \in \mathcal{I} - \{a\} \text{ and } sup(a_i) \geq sup(a)\}$. Let $\mathcal{I} = \{a_1, a_2, \ldots, a_n\}$ and $sup(a_i) \geq sup(a_{i+1})$, $1 \leq i \leq n-1$. The minimum item support with respect to nonnegative lift (LS) can be specified as follows:

$$ms(a_i) = \begin{cases} sup(a_i) \cdot sup(a_{i+1}), & \text{if } 1 \leq i \leq n-1 \\ sup(a_i), & \text{if } i = n \end{cases} \tag{3}$$

Now we have two separate support settings: One is based on the confidence measure and the other is on lift. To prune the spurious frequent itemsets to make most of the generated rules be interesting, we combine these two specifications as shown below, which we call the confidence-lift support constraint (CLS).

$$ms(a_i) = \begin{cases} sup(a_i) \cdot \max\{minconf, sup(a_{i+1})\}, & \text{if } 1 \leq i \leq n-1 \\ sup(a_i), & \text{if } i = n \end{cases} \tag{4}$$

Example 3. Let $minconf = 50\%$ and the sorted set of items in Table 1 be $\{\mathsf{Notebook}, \mathsf{PDA}, \mathsf{PC}, \mathsf{Scanner}, \mathsf{Printer}\}$. The minimum item supports will be $ms(\mathsf{Notebook}) = 2/9 \times 1/2 = 11\%$, $ms(\mathsf{PDA}) = 2/9 \times 6/9 = 15\%$, $ms(\mathsf{PC}) = 6/9 \times 6/9 = 44\%$, $ms(\mathsf{Scanner}) = 6/9 \times 7/9 = 52\%$, and $ms(\mathsf{Printer}) = 7/9 = 78\%$.

3.3 Methods for Mining Interesting Associations Using CLS

Using Eq. 4, the traditional associations mining process can be refined like this: The user specify the minimum confidence ($minconf$) and wait for the results; users no longer have to specify the minimum support. The only focus is on how strong the rules they expect to see and whether these rules are interesting.

The task of associations mining remains the same. First, the set of frequent itemsets are generated and then from which the interesting association rules are constructed. Since the support constraint we adopt in this paper follows the specification in [9], one might expect to use the MSapriori algorithm [9] to accomplish the first subtask. However, MSapriori is not the only candidate. Indeed, it is not suitable under the support constraint as specified in Eq. 4. The reason is that the MSapriori algorithm was developed to resolve the sorted closure problem [9][11], which however does not exist when $L_1 = \mathcal{I}$, i.e., all items are frequent 1- itemset. As a consequence, any effective Apriori-like algorithm can be adapted to meet this purpose, with the following modification: Remove the step for generating L_1 and modify the initial steps as setting $L_1 = \mathcal{I}$, scanning the database $\mathcal{D}$ to obtain the support of each item, and setting the minimum item supports according to Eq. 4. Finally, for the second subtask, all proposed methods for constructing associations from the frequent itemsets can be used, with an additional pruning of the nonpositive lift rules.

4 Experiments

In this section, we evaluate the proposed confidence-lift support specification as opposed to the standard uniform specification (US) and the varied item support specification (VIS) [9]. The test set is generated using IBM synthetic data generator [1]. Parameter settings are shown in Table 2.

Table 2. Parameter settings for synthetic data generation

Parameter		Default value
$\lvert\mathcal{D}\rvert$	Number of transactions	100,000
$\lvert t\rvert$	Average size of transactions	5
$\lvert I\rvert$	Average size of maximal potentially frequent itemsets	2
N	Number of items	200

The evaluation is examined from two aspects: The number of frequent itemsets and the ratio of generated rules that are interesting in terms of high confidence and positive lift (conviction). Both aspects are inspected under different support specifications, including US (with support = 1%), VIS (randomly set between 1% and 10%), and our CLS. The result is shown in Table 3, where $|\mathcal{F}|$, $|\mathcal{F}_e|$, $|\mathcal{R}|$, $|\mathcal{R}_c|$, $|\mathcal{R}_l|$, and $|\mathcal{R}_i|$ denote the number of frequent itemsets, number of effective frequent itemsets (itemsets that can generate at least one interesting rules), number of rules, number of rules with $conf \geq minconf$, number of rules with positive lift, and the number of interesting rules, respectively. It can be observed that CLS yields less number of frequent itemsets but more number of interesting rules than US and VIS. This is because all frequent itemsets generated by CLS are effective while the other two generates far more spurious frequent itemsets. Furthermore, note that for US and VIS, the number of frequent itemsets are the same under different confidence thresholds. This means that as the confidence becomes higher, the ratio of spurious frequent itemsets increases as well and more generated rules are uninteresting. This phenomenon is depicted in Figures 1 and 2.

Table 3. Statistics of frequent itemsets and association rules.

	minconf = 50%				*minconf* = 80%		
	US	VIS	CLS		US	VIS	CLS
$\|\mathcal{F}\|$	395	310	271	$\|\mathcal{F}\|$	395	310	214
$\|\mathcal{F}_e\|$	232	216	271	$\|\mathcal{F}_e\|$	190	177	214
$\|\mathcal{R}\|$	1130	896	834	$\|\mathcal{R}\|$	1130	896	482
$\|\mathcal{R}_c\|$	499	465	634	$\|\mathcal{R}_c\|$	303	319	338
$\|\mathcal{R}_l\|$	862	762	834	$\|\mathcal{R}_l\|$	862	762	482
$\|\mathcal{R}_i\|$	499	465	634	$\|\mathcal{R}_i\|$	303	319	338

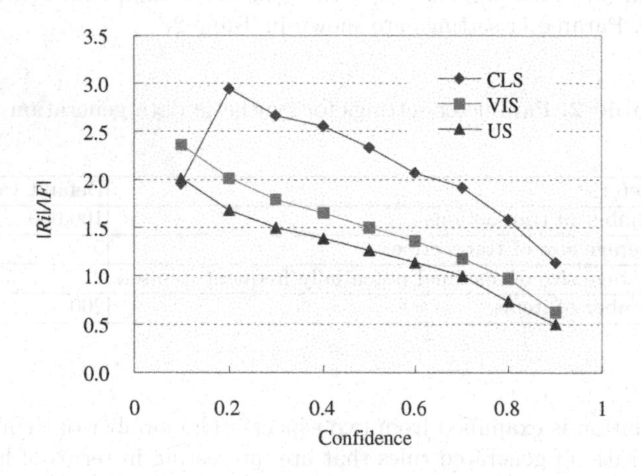

Fig. 1. Ratios of $|\mathcal{R}_i|$ to $|\mathcal{F}|$ under various confidences ranging from 0.1 to 0.9

5 Conclusions

We have investigated in this paper the problem of mining interesting association rules without the user-specified support constraint. We proposed a confidence-

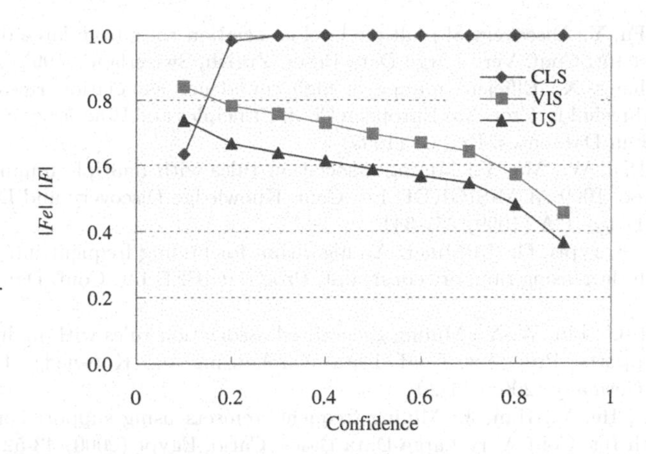

Fig. 2. Ratios of $|\mathcal{F}_e|$ to $|\mathcal{F}|$ under various confidences ranging from 0.1 to 0.9

lift based support constraint which can be automatically derived from the item support. Empirical evaluation has showed that the proposed support specification is good at discovering the low support, high confidence and positive lift associations, and is effective in reducing the spurious frequent itemsets.

Finally, we should point out that the proposed support specification still can not find all interesting association rules. In fact, it may miss some interesting rules composed of more than two items because the specification is derived from frequent 2-itemsets. Our future work is to investigate methods for remedying this deficiency.

References

1. Agrawal, R., Imielinski T., Swami, A.: Mining association rules between sets of items in large databases. Proc. 1993 ACM-SIGMOD Int. Conf. Management of Data. Washington, D.C. (1993) 207-216
2. Aggarwal, C. C., Yu, P. S.: A new framework for itemset generation. Proc. 17th ACM Symp. Principles of Database Systems. Seattle, WA (1998) 18-24
3. Berry, J. A., Linoff, G. S.: Data Mining Techniques for Marketing, Sales and Customer Support. John Wiley & Sons, Inc. (1997)
4. Brin, S., Motwani, R., Silverstein, C.: Beyond market baskets: generalizing association rules to correlations. Proc. 1997 ACM-SIGMOD Int. Conf. Management of Data. (1997) 265-276
5. Brin, S., Motwani, R., Ullman, J.D., Tsur, S.: Dynamic itemset counting and implication rules for market-basket data. Proc. 1997 ACM-SIGMOD Int. Conf. Management of Data. (1997) 207-216
6. Cohen, E., Datar, M., Fujiwara, S., Gionis, A., Indyk, P., Motwani, R., Ullman, J. D., Yang, C.: Finding Interesting associations without support pruning. Proc. IEEE Int. Conf. Data Engineering. (2000) 489-499

7. Han, J., Fu, Y.: Discovery of multiple-level association rules from large databases. Proc. 21st Int. Conf. Very Large Data Bases. Zurich, Swizerland (1995) 420-431
8. Li, J., Zhang, X.: Efficient mining of high confidence association rules without support thresholds. Proc. 3rd European Conf. Principles and Practice of Knowledge Discovery in Databases. Prague (1999)
9. Liu, B., Hsu, W., Ma, Y.: Mining association rules with multiple minimum supports. Proc. 1999 ACM-SIGKDD Int. Conf. Knowledge Discovery and Data Mining. San Deigo, CA (1999) 337-341
10. Seno, M., Karypis, G.: LPMiner: An algorithm for finding frequent intemsets using length-decreasing Support constraint. Proc. 1st IEEE Int. Conf. Data Mining. (2001)
11. Tseng, M.-C., Lin, W.-Y.: Mining generalized association rules with multiple minimum supports. Proc. Int. Conf. Data Warehousing and Knowledge Discovery. Munich, Germany (2001) 11-20
12. Wang, K., He, Y., Han, J.: Mining frequent itemsets using support constraints. Proc. 26th Int. Conf. Very Large Data Bases. Cario, Egypt (2000) 43-52

Concise Representation of Frequent Patterns Based on Generalized Disjunction-Free Generators

Marzena Kryszkiewicz and Marcin Gajek

Institute of Computer Science, Warsaw University of Technology
Nowowiejska 15/19, 00-665 Warsaw, Poland
mkr, gajek@ii.pw.edu.pl

Abstract. Frequent patterns are often used for solving data mining problems. They are applied e.g. in discovery of association rules, episode rules, sequential patterns and clusters. Nevertheless, the number of frequent itemsets is usually huge. In the paper, we overview briefly four lossless representations of frequent itemsets proposed recently and offer a new lossless one that is based on generalized disjunction-free generators. We prove on theoretical basis that the new representation is more concise than three of four preceding representations. In practice it is much more concise than the fourth representation too. An algorithm determining the new representation is proposed.

1 Introduction

Frequent patterns are often used for solving data mining problems. They are useful in the discovery of association rules, episode rules, sequential patterns and clusters etc. (see [4] for overview). Nevertheless, the number of frequent itemsets is usually huge. Therefore, it is important to work out concise, preferably lossless, representations of frequent itemsets. By lossless we mean representations that allow derivation and support determination of all frequent itemsets without accessing the database. Four such lossless representations have been investigated recently. They are based on the following families of itemsets: closed itemsets (see e.g. [2,8-9]), generators (see e.g. [2,6,8]), disjunction-free sets [3] and disjunction-free generators [6]. Unlike the closed itemsets representation, the others utilize a concept of a negative border [7]. Applications of closed itemsets and generators have been demonstrated in the case of the discovery of association rules and their essential subsets (see e.g. [5,8-9]). In particular, in the case of representative rules [5] as well as generic basis and informative basis [8], rule's antecedent is a generator, while its consequent is a closed itemset diminished by the items present in the antecedent. As proved on theoretical basis, the disjunction-free generators representation is more concise than the other representations except for closed itemsets [6], in which case it may be more or less concise depending on data. In the case of large highly correlated data sets, it is much more concise than the closed itemsets representation too.

In this paper, we introduce yet another lossless representation of frequent itemsets, which is a generalization of the disjunction-free generators representation (Section 4). We prove that the new representation constitutes a subset of the disjunction-free generators representation, and by this also a subset of the generators and the disjunction-free sets representations. An algorithm for extracting the new

M.-S. Chen, P.S. Yu, and B. Liu (Eds.): PAKDD 2002, LNAI 2336, pp. 159-171, 2002.

representation is offered (Section 5). There are provided selected experimental results related to the extraction time and cardinality of all five representations (Section 6).

2 Basic Notions and Properties

2.1. Itemsets, Frequent Itemsets

Let $I = \{i_1, i_2, ..., i_m\}$, $I \neq \emptyset$, be a set of distinct literals, called *items*. In the case of a transactional database, a notion of an item corresponds to a sold product, while in the case of a relational database an item will be an (*attribute,value*) pair. Any non-empty set of items is called an *itemset*. An itemset consisting of k items will be called *k-itemset*. Let D be a set of transactions (or tuples, respectively), where each transaction (tuple) T is a subset of I. (Without any loss of generality, we will restrict further considerations to transactional databases.) *Support* of an itemset X is denoted by $sup(X)$ and defined as the number of transactions in D that contain X. The itemset X is called *frequent* if its support is greater than some user-defined threshold *minSup*. The set of all frequent itemsets will be denoted by F:

$$F = \{X \subseteq I \mid sup(X) > minSup\}.$$

Property 2.1.1 [1]. If $X \in F$, then $\forall Y \subset X, Y \in F$.

2.2 Closures, Closed Itemsets, and Generators

Closure of an itemset X is denoted by $\gamma(X)$ and is defined as the greatest (w.r.t. set inclusion) itemset that occurs in all transactions in D in which X occurs. Clearly, $sup(X) = sup(\gamma(X))$. The itemset X is defined *closed* iff $\gamma(X) = X$. The set of all closed itemsets will be denoted by C, i.e.

$$C = \{X \subseteq I \mid \gamma(X) = X\}.$$

Let X be a closed itemset. A minimal itemset Y satisfying $\gamma(Y) = X$ is called a *generator* of X. By $G(X)$ we will denote the set of *all generators of X*. The *union of generators of all closed itemsets* will be denoted by G, i.e.

$$G = \bigcup\{G(X) \mid X \in C\}.$$

Example 2.2.1. Let D be the database from Table 1. To make the notation brief, we will write itemsets without brackets and commas (e.g. ABC instead of $\{A,B,C\}$).

Table 1. Example database D

Id	*Transaction*
T_1	$\{A,B,C,D,E,G\}$
T_2	$\{A,B,C,D,E,F\}$
T_3	$\{A,B,C,D,E,H,I\}$,
T_4	$\{A,B,D,E\}$
T_5	$\{A,C,D,E,H,I\}$
T_6	$\{B,C,E\}$

The itemset $ABCDE$ is closed since $\gamma(ABCDE) = ABCDE$. The itemset $ABCD$ is not closed as $\gamma(ABCD) = ABCDE \neq ABCD$. Clearly, $sup(ABCD) = sup(ABCDE) = 3$. The itemset ABC is a minimal subset the closure of which equals to $ABCDE$. Hence, $ABC \in G(ABCDE)$.

Property below states that the closure of an itemset X can be computed: 1) as the intersection of the transactions in D that are supersets of X, or 2) as the intersection of the closed itemsets that are supersets of X.

Property 2.2.1 [8]. Let $X \subseteq I$. $\gamma(X) = \bigcap\{T \in D \mid T \supseteq X\} = \bigcap\{Y \subseteq I \mid Y \in C \wedge Y \supseteq X\}$.

Property 2.2.2 [2]. Let $X \subseteq I$. $X \in G$ iff $sup(X) \neq \min\{sup(X \setminus \{A\}) | A \in X\}$.

The next property states that all subsets of a generator are generators.

Property 2.2.3 [2,6]. Let $X \in G$. Then, $\forall Y \subset X$, $Y \in G$.

The property beneath states that for any itemset its support can be computed if we know supports of either all closed itemsets or all generators.

Property 2.2.4 [6]. Let $X \subseteq I$.

a) $sup(X) = \max\{sup(Y) | Y \in C \wedge Y \supseteq X\}$.
b) $sup(X) = \min\{sup(Y) | Y \in G \wedge Y \subseteq X\}$.

2.3 Disjunctive Rules and Disjunction-Free Sets

The notion of *disjunction-free sets* was introduced in [3]. Let us present this concept by means of an auxiliary notion called a *2-disjunctive rule*.

Let $Z \subseteq I$. $X \Rightarrow A_1 \vee A_2$ is defined a *2-disjunctive rule based on Z* (and Z is *the base of* $X \Rightarrow A_1 \vee A_2$) if $X \subset Z$ and $A_1, A_2 \in Z \setminus X$. Observe, that a 2-disjunctive rule $X \Rightarrow A_1 \vee A_2$ can have an empty antecedent ($X = \emptyset$) and its consequents can be equal ($A_1 = A_2$). *Support of* $X \Rightarrow A_1 \vee A_2$, denoted by $sup(X \Rightarrow A_1 \vee A_2)$, is defined as the number of transactions in D in which X occurs together with A_1 or A_2, that is:

$$sup(X \Rightarrow A_1 \vee A_2) = sup(X \cup \{A_1\}) + sup(X \cup \{A_2\}) - sup(X \cup \{A_1, A_2\}).$$

Confidence of $X \Rightarrow A_1 \vee A_2$, denoted by $conf(X \Rightarrow A_1 \vee A_2)$, is defined as follows:

$$conf(X \Rightarrow A_1 \vee A_2) = sup(X \Rightarrow A_1 \vee A_2) / sup(X).$$

$X \Rightarrow A_1 \vee A_2$ is defined a *certain rule* if $conf(X \Rightarrow A_1 \vee Y_2) = 1$. Thus, $X \Rightarrow A_1 \vee A_2$ is certain if each transaction containing X contains also A_1 or A_2.

Example 2.3.1. Let us consider the database D from Table 1. The rule $\emptyset \Rightarrow A \vee A$ is not certain in D, since there is a transaction (T_6) that contains $\emptyset$, and does not contain A. On the other hand, $\emptyset \Rightarrow A \vee C$ is certain because each transaction in D contains A or C. Similarly, $C \Rightarrow D \vee E$ is certain as each transaction containing C contains also D or E.

Property 2.3.1 [3].

$X \Rightarrow A_1 \vee A_2$ is certain iff $sup(X) = sup(X \cup \{A_1\}) + sup(X \cup \{A_2\}) - sup(X \cup \{A_1, A_2\})$.

Property 2.3.2 [3]. If $X \Rightarrow A_1 \vee A_2$ is certain, then $\forall Z \supset X$, $Z \Rightarrow A_1 \vee A_2$ is certain.

Example 2.3.2. Let us consider the database D from Table 1. The rule $C \Rightarrow D \vee E$ is certain, so $AC \Rightarrow D \vee E$ and $ABC \Rightarrow D \vee E$ (and so forth) are also certain rules.

An itemset X is defined *disjunctive* if there are $A, B \in X$ such that $X \setminus \{A, B\} \Rightarrow A \vee B$ is a certain rule. Otherwise, the itemset is called *disjunction-free* (see [3] for the original definition of a disjunction-free set). The set of all disjunction-free sets will be denoted by *DFree*, i.e.:

$$DFree = \{X \in I | \neg \exists A_1, A_2 \in X, conf(X \setminus \{A_1, A_2\} \Rightarrow A_1 \vee A_2) = 1\}.$$

Example 2.3.3. Let us consider the database D from Table 1 and the itemset DE. The only 2-disjunctive rules based on DE are: $\emptyset \Rightarrow D \vee E$, $D \Rightarrow E \vee E$, $E \Rightarrow D \vee D$. The rule $E \Rightarrow D \vee D$ is not certain, however $\emptyset \Rightarrow D \vee E$ and $D \Rightarrow E \vee E$ are certain. Hence, DE is a disjunctive set (i.e. $DE \notin DFree$). Now, since $\emptyset \Rightarrow D \vee E$ is certain in D, then by Property 2.3.2, $A \Rightarrow D \vee E$ is also certain. Therefore, $ADE \notin DFree$.

Property 2.3.3 [3].

a) If $X \notin DFree$, then $\forall Y \supset X$, $Y \notin DFree$.

b) If $X \in DFree$, then $\forall Y \subset X$, $Y \in DFree$.

Property 2.3.4. [6].

a) Let $X \subseteq I$. If $X \in DFree$, then $X \in G$.

b) $X \in G$ iff $\neg \exists A \in X$ such that $X \backslash \{A\} \Rightarrow A$ is a certain rule.

3 Overview of Concise Lossless Representations

3.1 Closed Itemsets Representation

The majority of research on concise representations of frequent itemsets was devoted to closed itemsets. Here we will present this representation.

An itemset X is defined to be *frequent closed* if X is closed and frequent. In the sequel, the set of *all frequent closed itemsets* will be denoted by FC, i.e.

$$FC = F \cap C.$$

Closed itemsets representation (CR) is defined as the set FC enriched by the information on support for each $X \in FC$.

CR is sufficient to determine all frequent itemsets and their supports (see e.g. [6]).

3.2 Generators Representation

Generators are commonly used as an intermediate step for the discovery of closed itemsets. However, the generators themselves can constitute a concise lossless representation of frequent itemsets [6]:

Frequent generators, denoted by FG, are defined as:

$$FG = F \cap G.$$

Negative generators border, denoted by GBd^-, is defined as follows:

$$GBd^- = \{X \in G \mid X \notin F \wedge (\forall Y \subset X, Y \in FG)\}.$$

GBd^- consists of all minimal (w.r.t. set inclusion) infrequent generators.

Generators representation (GR) is defined as:

- the set FG enriched by the information on support for each $X \in FG$,
- the border set GBd^-.

GR is sufficient to determine all frequent itemsets and their supports [6].

3.3 Disjunction-Free Sets Representation

The *disjunction-free sets representation* was introduced in [3].

Frequent disjunction-free itemsets, denoted by $FDFree$, are defined as:

$$FDFree = DFree \cap F.$$

Negative border of FDFree is denoted by $DFreeBd^-$ and defined as follows:

$$DFreeBd^- = \{X \subseteq I \mid X \notin FDFree \wedge (\forall Y \subset X, Y \in FDFree)\}.$$

Disjunction-free sets representation (DFSR) is defined as below:

- $FDFree$ enriched by the information on support for each $X \in FDFree$,
- $DFreeBd^-$ enriched by the information on support for each $X \in DFreeBd^-$.

DFSR is sufficient to determine all frequent itemsets and their supports [3].

3.4 Disjunction-Free Generators Representation

The *disjunction-free generators representation* was introduced in [6] as the combination of the generators representation and disjunction-free sets representation:

Disjunction-free generators, denoted by *DFreeG*, are defined as follows:

$$DFreeG = DFree \cap G.$$

Frequent disjunction-free generators, denoted by *FDFreeG*, are defined as:

$$FDFreeG = F \cap DFreeG.$$

Property 3.4.1 [6]. If $X \in FDFreeG$, then $\forall Y \subset X$, $Y \in FDFreeG$.

Negative infrequent generators border, denoted by $IDFreeGBd^-$, is defined as:

$$IDFreeGBd^- = \{X \in G \mid X \notin F \land (\forall Y \subset X,\ Y \in FDFreeG)\}.$$

$IDFreeGBd^-$ consists of all minimal (w.r.t. set inclusion) infrequent generators the proper subsets of which are disjunction-free generators.

Negative frequent generators border, denoted by $FDFreeGBd^-$, is defined beneath:

$$FDFreeGBd^- = \{X \in G \mid X \in F \land X \notin DFreeG \land (\forall Y \subset X,\ Y \in FDFreeG)\}.$$

$DFreeGBd^-$ consists of all minimal frequent disjunctive generators.

Disjunction-free generators representation (DFGR) is defined as follows:

- *FDFreeG* enriched by the information on support for each $X \in FDFreeG$,
- $FDFreeGBd^-$ enriched by the information on support for each $X \in FDFreeGBd^-$
- $IDFreeGBd^-$.

DFGR constitutes a subset of both GR and DFSR, but still it is sufficient to determine all frequent itemsets and their supports [6].

4 New Representation of Frequent Itemsets Based on Generalized Disjunction-Free Generators

In this section we will introduce a new representation of frequent itemsets that will be based on an extended concept of a disjunction-free generator. We will prove that the new representation is sufficient to derive all frequent itemsets and constitutes a subset of the disjunction-free generators representation.

4.1 Generalized Disjunction-Free Sets

Let $Z \subseteq I$. $X \Rightarrow A_1 \lor ... \lor A_n$ is defined as a *generalized disjunctive rule based on Z* (and Z is *the base of* $X \Rightarrow A_1 \lor ... \lor A_n$) if $X \subset Z$ and $A_i \in Z \backslash X$ for $i=1..n$.

Support of $X \Rightarrow A_1 \lor ... \lor A_n$, denoted by $sup(X \Rightarrow A_1 \lor ... \lor A_n)$, is defined as the number of transactions in D in which X occurs together with A_1 or A_2, or...or A_n.

Confidence of $X \Rightarrow A_1 \lor ... \lor A_n$, denoted by *conf*, is defined in usual way:

$$conf(X \Rightarrow A_1 \lor ... \lor A_n) = sup(X \Rightarrow A_1 \lor ... \lor A_n) / sup(X).$$

$X \Rightarrow A_1 \lor ... \lor A_n$ is defined a *certain rule* if $conf(X \Rightarrow A_1 \lor ... \lor A_n) = 1$.

Thus $X \Rightarrow A_1 \lor ... \lor A_n$ is certain if each transaction containing X contains also A_1 or A_2, or ... or A_n.

An itemset X is defined *generalized disjunctive* if there are $A_1,\dots,A_n \in X$ such that $X\setminus\{A_1,\dots,A_n\} \Rightarrow A_1\vee\dots\vee A_n$ is a certain rule. Otherwise, the itemset is called *generalized disjunction-free*. The set of all generalized disjunction-free sets will be denoted by *GDFree*, i.e.

$$GDFree = \{X\in I \mid \neg\exists A_1,\dots,A_n \in X,\ conf(X\setminus\{A_1,\dots,A_n\} \Rightarrow A_1\vee\dots\vee A_n) = 1,\ n \geq 1\}.$$

Example 4.1.1. Let us consider the database D from Table 2. Below we list all itemsets followed by subscript informing on their support: $\emptyset_7$, A_4, B_4, C_4, AB_2, AC_2, BC_2, ABC_1. Let us consider the itemset *ABC*. Table 3 presents all generalized disjunctive rules based on *ABC*. The set *ABC* is not disjunction-free since no rule with at most two disjuncts is certain. However, it is a generalized disjunction-free set since there is a generalized disjunctive rule based on *ABC* (namely, $\emptyset \Rightarrow A\vee B\vee C$), which is certain.

Table 2. Example database D.

Id	*Transaction*
T_1	$\{A,B,C\}$
T_2	$\{A,B\}$
T_3	$\{A,C\}$
T_4	$\{A\}$
T_5	$\{B,C\}$
T_6	$\{B\}$
T_7	$\{C\}$

Table 3. Generalized disjunctive rules based on *ABC*.

$r: X\Rightarrow A_1\vee\dots\vee A_n$	$sup(X)$	$sup(r)$	certain?
$AB \Rightarrow C$	2	1	no
$AC \Rightarrow B$	2	1	no
$BC \Rightarrow A$	2	1	no
$A \Rightarrow B\vee C$	4	3	no
$B \Rightarrow A\vee C$	4	3	no
$C \Rightarrow A\vee B$	4	3	no
$\emptyset \Rightarrow A\vee B\vee C$	7	7	yes

Property 4.1.1. If $X \Rightarrow A_1\vee\dots\vee A_n$ is certain, then $\forall Z\supset X$, $Z \Rightarrow A_1\vee\dots\vee A_n$ is certain.

The next property states that supersets of a generalized disjunctive set are generalized disjunctive and subsets of a generalized disjunction-free set are generalized disjunction-free.

Property 4.1.2. Let $X\subseteq I$.

a) If $X\notin GDFree$, then $\forall Y\supset X$, $Y\notin GDFree$.

b) If $X\in GDFree$, then $\forall Y\subset X$, $Y\in GDFree$.

Proof: Ad. a) Follows immediately from Property 4.1.1.

Ad. b) Follows from Property 4.1.2a.

Property 4.1.3. $GDFree \subseteq DFree$.

Proof: $GDFree = \{X\in I \mid \neg\exists A_1,\dots,A_n \in X,\ conf(X\setminus\{A_1,\dots,A_n\} \Rightarrow A_1\vee\dots\vee A_n) = 1,\ n \geq 1\} \subseteq \{X\in I \mid \neg\exists A_1,A_2 \in X,\ conf(X\setminus\{A_1,A_2\} \Rightarrow A_1\vee A_2) = 1\} = DFree$.

Property 4.1.4. Let $X\subseteq I$.

a) If $X\in GDFree$, then $X\in G$.

b) $\emptyset\in GDFree$.

c) Let $A\in I$. $\{A\}\in GDFree$ iff $\{A\}\in G$.

Proof: Ad. a) Since $GDFree \subseteq DFree$ (by Property 4.1.3) and $DFree=DFree\cap G$ (by Property 2.3.4a), then $GDFree \subseteq DFree\cap G$, thus $GDFree \subseteq G$.

Ad. b) Immediate by definition of a generalized disjunction-free set.

Ad.c) By definition of a generalized disjunction-free set and Property 2.3.4b.

Let Y be an itemset such that $Y = \{A_1,\dots,A_n\}$. In the sequel, the disjunction $A_1\vee\dots\vee A_n$ will be denoted by $\vee Y$. Now, we will investigate how to compute support of a generalized disjunctive rule, say $X\Rightarrow\vee Y$, based on the information on supports of rules with smaller number of disjuncts or alternatively on supports of subsets of $X\cup Y$.

Property 4.1.5. Let $Y \subset I$ and $A \in I \setminus Y$.

$$sup(X \Rightarrow \vee Y \vee A) = sup(X \Rightarrow \vee Y) + sup(X \Rightarrow A) - sup(X \cup \{A\} \Rightarrow \vee Y).$$

Property 4.1.6. Let $X, Y \subset I$ and $X \Rightarrow \vee Y$ be a generalized disjunctive rule. Then:

$$sup(X \Rightarrow \vee Y) = \{\Sigma_{i=1..|Y|} (-1)^{i-1} \times [\Sigma_{i\text{-itemsets } Z \subseteq Y} sup(X \cup Z)]\}.$$

Proof (by induction w.r.t. $|Y|$): Let $A \in I \setminus Y$ and $Y' = Y \cup \{A\}$.

Induction hypothesis: $sup(X \Rightarrow \vee Y) = \{\Sigma_{i=1..|Y|} (-1)^{i-1} \times [\Sigma_{i\text{-itemsets } Z \subseteq Y} sup(X \cup Z)]\}$.

We are to prove that $sup(X \Rightarrow \vee Y') = \{\Sigma_{i=1..|Y'|} (-1)^{i-1} \times [\Sigma_{i\text{-itemsets } Z \subseteq Y'} sup(X \cup Z)]\}$.

$sup(X \Rightarrow \vee Y') = sup(X \Rightarrow \vee Y \vee A) =$ /* by Property 4.1.5 */ $= sup(X \Rightarrow \vee Y) + sup(X \Rightarrow A) - sup(X \cup \{A\} \Rightarrow \vee Y) =$ /* by induction hypothesis */ $= \{\Sigma_{i=1..|Y|} (-1)^{i-1} \times [\Sigma_{i\text{-itemsets } Z \subseteq Y} sup(X \cup Z)]\} + sup(X \cup \{A\}) - \{\Sigma_{i=1..|Y|} (-1)^{i-1} \times [\Sigma_{i\text{-itemsets } Z \subseteq Y} sup(X \cup \{A\} \cup Z)]\} = \{\Sigma_{i=1..|Y|} (-1)^{i-1} \times [\Sigma_{i\text{-itemsets } Z \subseteq Y} sup(X \cup Z)]\} + sup(X \cup \{A\}) + \{\Sigma_{i=2..|Y'|} (-1)^{i-1} \times [\Sigma_{i\text{-itemsets } Z' \subseteq Y'\text{, such that } A \in Z'} sup(X \cup Z')]\} = \{\Sigma_{i=1..|Y|} (-1)^{i-1} \times [\Sigma_{i\text{-itemsets } Z \subseteq Y} sup(X \cup Z)]\} + \{\Sigma_{i=1..|Y'|} (-1)^{i-1} \times [\Sigma_{i\text{-itemsets } Z' \subseteq Y'\text{, such that } A \in Z'} sup(X \cup Z')]\} = \{\Sigma_{i=1..|Y|} (-1)^{i-1} \times [\Sigma_{i\text{-itemsets } Z \subseteq Y'} sup(X \cup Z)\} + \{(-1)^{|Y'|-1} \times sup(X \cup Y')\} = \{\Sigma_{i=1..|Y'|} (-1)^{i-1} \times [\Sigma_{i\text{-itemsets } Z \subseteq Y'} sup(X \cup Z)]\}$.

It follows from the presented property that the support of $X \Rightarrow \vee Y$ depends on supports of all itemsets Z such that $X \subset Z \subseteq X \cup Y$. Alternatively, one can calculate $sup(X \cup Y)$ based on $sup(X \Rightarrow \vee Y)$ and supports of itemsets Z such that $X \subset Z \subset X \cup Y$:

Corollary 4.1.1. Let $X, Y \subset I$ and $X \Rightarrow \vee Y$ be a generalized rule. Then:

$$sup(X \cup Y) = (-1)^{|Y|} \times \{-sup(X \Rightarrow \vee Y) + \Sigma_{i=1..|Y|-1} (-1)^{i-1} \times [\Sigma_{i\text{-itemsets } Z \subset Y} sup(X \cup Z)]\}.$$

Proof: By Property 4.1.6, $sup(X \Rightarrow \vee Y) = \{\Sigma_{i=1..|Y|} (-1)^{i-1} \times [\Sigma_{i\text{-itemsets } Z \subseteq Y} sup(X \cup Z)]\} = \{\Sigma_{i=1..|Y|-1} (-1)^{i-1} \times [\Sigma_{i\text{-itemsets } Z \subset Y} sup(X \cup Z)]\} + (-1)^{|Y|-1} \times sup(X \cup Y)$. Hence: $sup(X \cup Y) = (-1)^{|Y|} \times \{-sup(X \Rightarrow \vee Y) + \Sigma_{i=1..|Y|-1} (-1)^{i-1} \times [\Sigma_{i\text{-itemsets } Z \subseteq Y} sup(X \cup Z)]\}$.

Clearly, if the rule $X \Rightarrow \vee Y$ is certain, then $sup(X \Rightarrow \vee Y) = sup(X)$ and the equation from Corollary 4.1.1, can be rewritten as follows:

$$sup(X \cup Y) = (-1)^{|Y|} \times \{-sup(X) + \Sigma_{i=1..|Y|-1} (-1)^{i-1} \times [\Sigma_{i\text{-itemsets } Z \subset Y} sup(X \cup Z)]\}.$$

In the sequel, the formula on the right-hand side of the equation above will be treated as a *hypothetical support* of $X \cup Y$ driven by the rule $X \Rightarrow \vee Y$:

Hypothetical support of base $X \cup Y$ with respect to a generalized rule $X \Rightarrow \vee Y$ (also called shortly: *hypothetical base support* w.r.t. $X \Rightarrow \vee Y$) is denoted by $HBSup(X \Rightarrow \vee Y)$ and defined as follows:

$$HBSup(X \Rightarrow \vee Y) = (-1)^{|Y|} \times \{-sup(X) + \Sigma_{i=1..|Y|-1} (-1)^{i-1} \times [\Sigma_{i\text{-itemsets } Z \subset Y} sup(X \cup Z)]\}.$$

Let us note that $HBSup(X \Rightarrow \vee Y)$ is computable from the supports of the itemsets Z such that $X \subseteq Z \subset X \cup Y$.

Property 4.1.7. Let $X, Y \subset I$.

$X \Rightarrow \vee Y$ is a certain generalized rule iff $sup(X \cup Y) = HBSup(X \Rightarrow \vee Y)$.

Proof: Comparing the formula for calculating $sup(X \cup Y)$ in Corollary 4.1.1 with the definition of $HBSup(X \Rightarrow \vee Y)$ we observe: $sup(X \cup Y) = HBSup(X \Rightarrow \vee Y)$ iff $sup(X \Rightarrow \vee Y) = sup(X)$ iff $X \Rightarrow \vee Y$ is a certain generalized rule.

Thus, the rule $X \Rightarrow \vee Y$ is certain if and only if the support of its base $X \cup Y$ equals to the hypothetical support of $X \cup Y$ w.r.t. $X \Rightarrow \vee Y$.

4.2 Generalized Disjunction-Free Generators Representation

Generalized disjunction-free generators are a key concept for a new lossless representation, we will introduce in this section.

Generalized disjunction-free generators, denoted by *GDFreeG*, are defined as:

$$GDFreeG = GDFree \cap G.$$

Frequent generalized disjunction-free generators, denoted by *FGDFreeG*, are defined as:

$$FGDFreeG = F \cap GDFreeG.$$

Property 4.1.4a implies: $GDFreeG = GDFree$ and $FGDFreeG = F \cap GDFree$. As in the case of *F*, *GDFree* and *G*, subsets of a frequent generalized disjunction-free generator are frequent generalized disjunction-free generators:

Property 4.2.1. Let $X \in FGDFreeG$. Then, $\forall Y \subset X$, $Y \in FGDFreeG$.

Proof: Follows immediately from Properties 2.1.1, 2.2.3, and 4.1.2b.

Generalized infrequent generators border, denoted by $IGDFreeGBd^-$, is defined as:

$$IGDFreeGBd^- = \{X \in G \mid X \notin F \wedge (\forall Y \subset X, Y \in FGDFreeG)\}.$$

$IGDFreeGBd^-$ consists of all minimal (w.r.t. subset inclusion) infrequent generators the proper subsets of which are generalized disjunction-free generators.

Generalized frequent generators border, denoted by $FGDFreeGBd^-$, is defined as:

$$FGDFreeGBd^- = \{X \in G \mid X \in F \wedge X \notin GDFreeG \wedge (\forall Y \subset X, Y \in FGDFreeG)\}.$$

$FGDFreeGBd^-$ consists of all minimal (w.r.t. subset inclusion) frequent generalized disjunctive generators. Let us note that $IGDFreeGBd^- \cap FGDFreeGBd^- = \emptyset$.

Generalized disjunction-free generators representation (GDFGR) is defined as:

- *FGDFreeG* enriched by the information on support for each $X \in FGDFreeG$,
- the border set $FGDFreeGBd^-$ enriched by the information on support for each $X \in FGDFreeGBd^-$,
- the border set $IGDFreeGBd^-$.

Theorem 4.2.1. GDFGR is capable of determining for any itemset if it is frequent and if so, its support is computable from GDFGR.

Proof (constructive): The proof will be made by induction on $|X|$.

Induction hypothesis. For every itemset $V \subset X$, we can determine if it is frequent or not, and if *V* is frequent then we can determine its support.

One can distinguish the following cases:

- If $X \in FGDFreeG$ or $FGDFreeGBd^-$, then *X* is frequent and its support is known.
- If $\exists Y \in IGDFreeGBd^-$, $Y \subseteq X$, then *X* is not frequent.
- If $\neg\exists Z \in IGDFreeGBd^-$, $Z \subseteq X$, and $\exists Y \in FGDFreeGBd^-$, $Y \subset X$, then *X* is a generalized disjunctive set as a superset of some generalized disjunctive itemset in $FGDFreeGBd^-$ (by Property 4.1.2a). Let $Y \in FGDFreeGBd^-$, $Y \subset X$. Hence, there are some items $A_1,\ldots,A_n \in Y$ such that the rule $Y \setminus \{A_1,\ldots,A_n\} \Rightarrow A_1 \vee \ldots \vee A_n$ is certain. Let $A_1,\ldots,A_n$ be such items. Then, by Property 4.1.1, $X \setminus \{A_1,\ldots,A_n\} \Rightarrow A_1 \vee \ldots \vee A_n$ is also certain and by Property 4.1.7 *sup*(*X*) can be computed based on the information on supports of proper subsets of *X*. By induction hypothesis, we can determine if all subsets of *X* are frequent, and if so we can determine their supports. Let us assume that some of these itemsets is not

frequent, then X is not frequent either. Otherwise, X is frequent only if $sup(X) > minSup$.

- Let $X \notin FGDFreeG$ and $\neg\exists Z \in FGDFreeGBd^- \cup IGDFreeGBd^-$, $Z \subseteq X$. Then no generator being a subset of X is a superset of any $Z \in FGDFreeGBd^- \cup IGDFreeGBd^-$. Hence, all generators being subsets of X are contained in $FGDFreeG$. By Property 2.2.4b, $sup(X) = \min(\{sup(Y) | Y \in G \wedge Y \subseteq X\})$. In our case, this equation is equivalent to: $sup(X) = \min(\{sup(Y) | Y \in FGDFreeG \wedge Y \subseteq X\})$. Clearly, X is frequent as $sup(X)$ is equal to the support of some frequent generalized disjunction-free generator.

The proof of Theorem 4.2.1 can be treated as a naive algorithm for determining frequent itemsets and their supports. GDFGR can be further reduced by discarding all infrequent singleton itemsets from the negative border $IGDFreeGBd^-$ and instead storing the itemset $I_F = \{A \in I | \{A\} \in F\}$, whose elements are all frequent items. When applying such a reduced representation, the usual procedure of frequent itemsets derivation should be preceded with checking whether an itemset is contained in I_F.

Similarly, one could reduce the infrequent parts of the borders of the previously described GR, DFSR, and GDFGR representations and store I_F instead.

Proposition 4.2.1.

a) $FGDFreeG \subseteq FDFreeG$,

b) $IGDFreeGBd^- \subseteq IDFreeGBd^-$,

c) $FGDFreeGBd^- \cup FGDFreeG \subseteq FDFreeGBd^- \cup FDFreeG$,

d) GDFGR $\subseteq$ DFGR.

Proof: Ad. a) $FGDFreeG = GDFree \cap FG \subseteq$ /*by Prop. 4.1.3*/ $DFree \cap FG = FDFreeG$.

Ad. b) $IGDFreeGBd^- = \{X \in G \setminus F | \forall Y \subset X, Y \in FGDFreeG\} \subseteq$ /* by Proposition 4.2.1a */ $\{X \in G \setminus F | \forall Y \subset X, Y \in FDFreeG\} = IDFreeGBd^-$.

Ad. c) Applying Property 4.2.1, we can write $FGDFreeG = \{X \in FG | X \in GDFreeG \wedge (\forall Y \subset X, Y \in FGDFreeG)\}$ (*). Similarly, by Property 3.4.1, $FDFreeG = \{X \in FG | X \in DFreeG \wedge (\forall Y \subset X, Y \in FDFreeG)\}$ (**). Now, $FGDFreeGBd^- \cup FGDFreeG =$ /* by definition of $FGDFreeGBd^-$ & (*) */ $= \{X \in FG | \forall Y \subset X, Y \in FGDFreeG\} \subseteq$ /* by Proposition 4.2.1a */ $\{X \in FG | \forall Y \subset X, Y \in FDFreeG\}$ = /* by definition of $FDFreeGBd^-$ & (**) */ $= FDFreeGBd^- \cup FDFreeG$.

Ad. d) Immediate conclusion from Proposition 4.2.1b-c.

5 Algorithm for Computing New Representation

The *GDFGR-Apriori* algorithm, we propose, determines the newly introduced generalized disjunction-free representation. *GDFGR-Apriori* is analogous to *DFreeGenApriori* (also called briefly: *DFGR-Apriori*) we offered for computing DFGR in [6]. The main difference consists in computing support of (generalized) disjunctive rules and testing if itemsets are (generalized) disjunctive. In the algorithm we apply the following notation:

- $FGDFreeG_k$, $FGDFreeGBd^-_k$, $IGDFreeGBd^-_k$ – k-itemsets in the respective components of the generalized disjunction-free generators representation;
- C_k – candidate frequent generalized disjunction-free k-generators.

The itemsets are assumed kept in an ascending order. With each itemset c there are associated the following fields:

- *sup* – support of c;
- *minSubSup* – minimum of the supports of the proper subsets of c.

GDFGR-Apriori starts with checking if the number of transactions in D is greater than *minSup*. If so, then $\emptyset$ is frequent. By Property 4.1.4a-b, $\emptyset$ is a generalized disjunction-free generator. Hence, $\emptyset$ is included in $FGDFreeG_0$ provided $\emptyset \in F$. Next, all items in I are stored as 1-candidates in C_1. By Property 2.2.2, each itemset in C_1 is a generator if its support differs from $sup(\emptyset)$. In addition, Property 4.1.4c guarantees that each generator in C_1 is a generalized disjunction-free set. Hence, each generator in C_1 is added to the set of frequent generalized disjunction-free generators $FGDFreeG_1$, if its support is sufficiently high. Otherwise, it is included in the negative infrequent generators border $IGDFreeGBd^-_1$. Then, 2-candidates C_2 are created from $FGDFreeG_1$ by the *AprioriGGen* function (described later).

```
Algorithm GDFGR-Apriori;
FGDFreeG={}; FGDFreeGBd⁻={}; IGDFreeGBd⁻
={∅};
if |D| > minSup then begin
  ∅.sup = |D|;
  move ∅ from IGDFreeGBd⁻_0 to FGDFreeG_0;
  C_1 = {1-itemsets in D with minSubSup
    initialized to ∅.sup};
  forall candidates c∈ C_1 do begin
    SupportCount(C_1);
    if c.sup ≠ ∅.sup then                  // c is a generator
      if c.sup ≤ minSup then
        add c to IGDFreeGBd⁻_1              // c is infrequent
    else add c to FGDFreeG_1;
  endfor;
  C_2 = AprioriGGen(FGDFreeG_1);
  for (k = 2; C_k ≠ ∅; k++) do begin
    SupportCount(C_k);
    forall candidates c∈ C_k do
      if c.sup ≠ c.minSubSup then           // c is a generator
        if c.sup ≤ minSup then
          add c to IGDFreeGBd⁻_k            // c is infrequent
        elseif IsGDis(c, ∪_{l<k} FGDFreeG_l) then
          add c to FGDFreeGBd⁻_k
        else add c to FGDFreeG_k;
    endfor;
    C_{k+1} = AprioriGGen(FGDFreeG_k);
  endfor;
endif;
return <∪_k FGDFreeG_k, ∪_k FGDFreeGBd⁻_k,
  ∪_k IGDFreeGBd⁻_k>;
```

```
function AprioriGGen
  (FGDFreeG_k)
forall f, h ∈ FGDFreeG_k do
  if f[1]=h[1] ∧ ... ∧ f[k-1]=h[k-1]
    ∧ f[k]<h[k] then begin
      c = f[1]•f[2]•...•f[k]•h[k];
      add c to C_{k+1} endif;
endfor;
/* Pruning */
forall c∈ C_{k+1} do
  forall k-itemsets s ⊂ c do
    if s ∉ FGDFreeG_k then
      delete c from C_{k+1}
    else c.minSubSup =
          min(c.minSubSup, s.sup);
  endfor;
endfor;
return C_{k+1};
```

```
procedure SupportCount(var C_k)
forall transactions t∈ D do
  forall candidates c∈ C_k do
    if c ⊆ t then c.count++;
  endfor;
endfor;
endproc;
```

The following steps are performed level-wise for all k-candidates, $k \geq 2$:

1. Supports of the candidate k-itemsets C_k are determined by a pass over the database (see the *SupportCount* procedure).
2. The k-candidates C_k the supports of which differ from the supports of their proper (k-1)-subsets (*c.sup* $\neq$ *c.minSubSup*) are found generators (by Property 2.2.2).
3. Infrequent k-generators in C_k are added to the negative infrequent generators border $IGDFreeGBd^-_k$. The *IsGDis* function (described later) determines for each frequent k-generator if it is generalized disjunctive. Frequent generalized disjunctive k-generators are added to the negative frequent generators border $FGDFreeGBd^-_k$. The remaining frequent k-generators are generalized disjunction-free and hence, they are added to $GDFreeG_k$.
4. The *AprioriGGen* function is called to generate the candidate (k+1)-itemsets C_{k+1} from the frequent generalized disjunction-free k-generators $FGDFreeG_k$ and to initialize the *minSubSup* field for each new candidate. *AprioriGGen* follows Property 4.2.1 to guarantee that the (k+1)-candidates include only itemsets having all their subsets in $FGDFreeG_k$.

The algorithm ends when there are no more candidates. Let us note that an algorithm for computing the reduced disjunction-free generators representation would differ only slightly from the presented algorithm: it would return $<\cup_k FGDFreeG_k$, $\cup_k FGDFreeGBd^-_k$, $\cup_{k\neq 1} IGDFreeGBd^-_k$, $I_F>$, where $IGDFreeGBd^-$ would not contain infrequent 1-itemsets and I_F (set of frequent items) could be calculated as $I \setminus \cup IGDFreeGBd^-_1$.

Generating Candidates. Candidates are generated by the *AprioriGGen* function, which is similar to *AprioriGen* (see [1] for details). The difference consists in additional computing the value of ***minSubSup*** field. For each new candidate c, *minSubSup* is assigned the minimum from the supports of the proper subsets of c.

Checking whether Generator is Generalized Disjunctive. The purpose of the *IsGDis* function is to check whether a generator X of size 2 or greater is generalized disjunctive or not. In general, in order to state if X is generalized disjunctive, one should evaluate (possibly all) rules of the form $X\setminus Y \Rightarrow \vee Y$, where $Y \sqsubseteq X$. If there is such certain rule, then X is generalized disjunctive. However, when X is a generator, no rule with one disjunct is certain (Property 2.3.4b). Thus, we should restrict examining rules to those that have two or more disjuncts (at most $|c|$) in the consequent.

The *IsGDis* function takes two arguments: a generator X of size 2 or greater, the support of which is known, and all frequent disjunction-free generators of all sizes up to k-1 ($FGDFreeG' = \cup_{l<k} FGDFreeG_l$), the supports of which are also known. (Hence, supports of all proper subsets of X are known too.) The function creates and evaluates rules based on X. Each rule $X\setminus Y \Rightarrow \vee Y$ is evaluated as follows:

- The *HBSupCount* function (described later) calculates the hypothetical base support $HBSup(X\setminus Y \Rightarrow \vee Y)$ of X from supports of X's subsets in $FGDFreeG'$.
- The hypothetical support $HBSup(X\setminus Y \Rightarrow \vee Y)$ of X is compared with $sup(X)$. If these values are equal, then $X\setminus Y \Rightarrow \vee Y$ is found certain (by Property 4.1.7), and thus X is found generalized disjunctive (and the function returns the value **true**).

If none of the generalized disjunctive rules based on X is certain, then X is found generalized disjunction-free (and the function returns the value **false**).

```
function IsGDis(k-itemset c, FGDFreeG');
/* Assert: c is a frequent generator */
/* Assert: k ≥ 2                      */
forall i=2 to k do begin
  forall i-itemsets s ⊆ c do
    if c.sup =
      HBSupCount(c\s⇒∨s, FGDFreeG')
    then  // c is generalized disjunctive
      return true;
  endfor;
endfor;
return false;
```

```
function HBSupCount(X⇒∨Y, FGDFreeG');
sign = -1;
find X in FGDFreeG'; sup = X.sup;
forall i=1 to |Y|-1 do begin
  sign = sign × (-1); sup_i = 0;
  forall i-itemsets Z ⊂ Y do begin
    find X∪Z in FGDFreeG';
    sup_i = sup_i + (X∪Z).sup;
  endfor;
  sup = sup + (sign × sup_i)
endfor;
sup = sign × sup;
return sup;
```

Calculating Hypothetical Base Support. The *HBSupCount* function takes two arguments: the rule $X \Rightarrow \vee Y$ and the family *FGDFreeG'* including (at least) all proper subsets of $X \cup Y$ that contain X. The supports of all itemsets in *FGDFreeG'* are assumed known. *HBSupCount* calculates the hypothetical support of $X \cup Y$ w.r.t. $X \Rightarrow \vee Y$ according to the definition of $HBSup(X \Rightarrow \vee Y)$.

6 Experimental Results

Because of lack of space, we briefly report the performance of the new algorithm and the algorithms extracting the other concise representations. All the algorithms are modifications of *Apriori*, except for *CR* (*CR* is found as the closures of *GR* [8]). The presented results were obtained for the *Connect-4* data set (highly correlated data from UCI Machine Learning DB Repository). The experiments confirm that GDFGR outperforms the other four representations w.r.t. extraction time and cardinality (see Fig. 1-2, Table 4).

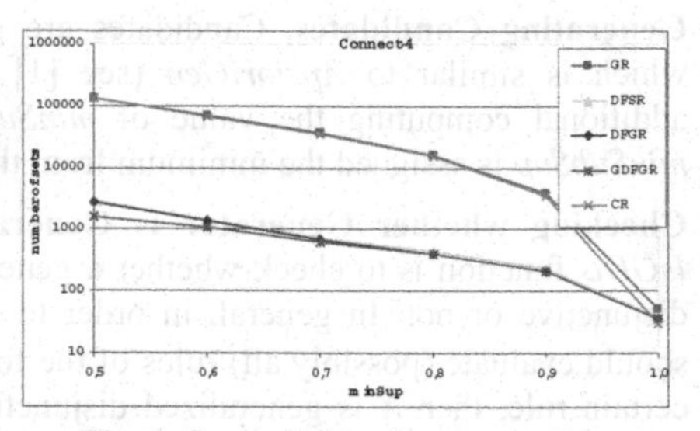

Fig. 1. Cardinalities of representations

Table 4. GDFGR compared with other representations XR (time(GDFGR) / time(XR) and |GDFGR| / |XR|), where XR∈ {DFGR, DFSR, GR, CR} for *minSup* = 50%

GDFGR / XR	DFGR	DFSR	GR	CR
time	56,12%	56,22%	0,74%	0,30%
cardinality	58,15%	56,51%	1,18%	1,23%

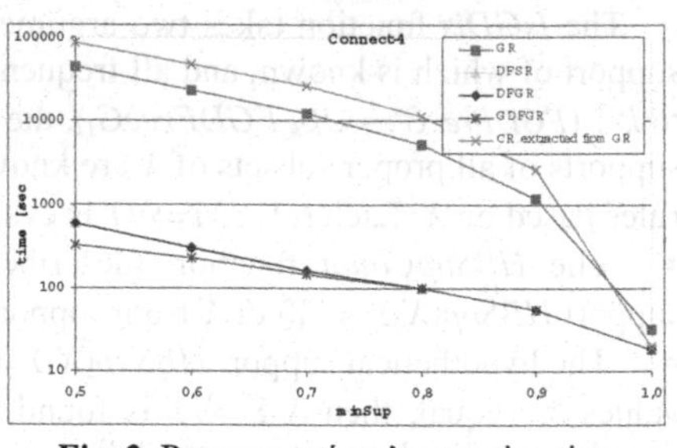

Fig. 2. Representations' extraction time

7 Conclusions

The new representation GDFGR of frequent patterns was offered as well as the algorithm for its extraction. It was proved theoretically that the new representation is more concise than DFGR, which in turn is more concise than DFSR and GR. In the case of *Connect-4* data, the new representation is up to 2 orders of magnitude more concise than CR and GR, and up to 50% smaller than DFGR and DFSR

References

[1] Agrawal, R., Mannila, H., Srikant, R., Toivonen, H., Verkamo, A.I.: Fast Discovery of Association Rules. In: Advances in KDD. AAAI, Menlo Park, California (1996) 307-328

[2] Bastide, Y., Taouil, R., Pasquier, N., Stumme, G., Lakhal, L.: Mining Frequent Patterns with Counting Inference. ACM SIGKDD Explorations, Vol. 2(2). (2000) 66-75

[3] Bykowski, A., Rigotti, C.: A Condensed Representation to Find Frequent Patterns. In: Proc. of the 12th ACM SIGACT-SIGMOD-SIGART PODS' 01 (2001)

[4] Han, J., Kamber, M.: Data Mining: Concepts and Techniques. Morgan Kaufmann (2000)

[5] Kryszkiewicz, M.: Closed Set based Discovery of Representative Association Rules. In: Proc. of IDA '01. Springer (2001) 350-359

[6] Kryszkiewicz, M.: Concise Representation of Frequent Patterns based on Disjunction-Free Generators. In: Proc. of ICDM '01. IEEE (2001) 305-312

[7] Mannila, H., Toivonen., H.: Levelwise Search and Borders of Theories in Knowledge Discovery. In: Data Mining and Knowledge Discovery 1(3). (1997) 241-258

[8] Pasquier, N.: DM: Algorithmes d'extraction et de réduction des règles d'association dans les bases de données. Ph.D. Thesis. Univ. Pascal-Clermont-Ferrand II (2000)

[1] Zaki, M.J.: Generating Non-redundant Association Rules. In: 6th ACM SIGKDD (2000)

Mining Interesting Association Rules: A Data Mining Language

Show-Jane Yen* and Yue-Shi Lee**

*Dept. of Computer Science and Info. Engineering, Fu Jen Catholic University, Taiwan, R.O.C.
sjyen@csie.fju.edu.tw

**Dept. of Information Management, Ming Chuan University, Taiwan, R.O.C.
leeys@mcu.edu.tw

Abstract. *Mining association rules* is to discover customer purchasing behaviors from a transaction database, such that the quality of business decision can be improved. However, the size of the transaction database can be very large. It is very time consuming to find all the association rules from a large database, and users may be only interested in some information. Hence, a data mining language needs to be provided such that users can query only interesting knowledge to them from a large database of customer transactions. In this paper, a data mining language is presented. From the data mining language, users can specify the interested items and the criteria of the association rules to be discovered. Also, the efficient data mining techniques are proposed to extract the association rules according to the user requirements.

1 Introduction

An association rule [1, 2, 3, 4] describes the association among items in which when some items are purchased in a transaction, others are purchased too. It can be represented as X ⇒ Y, in which X is an antecedent and Y is a consequent of this rule, and X and Y are two *itemsets*.

A transaction *t* supports an itemset *i* if *i* is contained in *t*. The *support* for an itemset *i* is defined as the ratio of the number of transactions that supports the itemset *i* to the total number of transactions. If the support for an itemset *i* satisfies the user-specified *minimum support* threshold, then *i* is called *frequent itemset.* The confidence of a rule X ⇒ Y is defined as the ratio of the support for the itemsets X∪Y to the support for the itemset X. If itemset Z=X∪Y is a frequent itemset and the confidence of X ⇒ Y is no less than the user-specified *minimum confidence*, then the rule X ⇒ Y is an association rule.

We present a data mining language, from which users only need to specify the criteria and the items for discovering the association rules.

2 Data Mining Language and Database Transformation

In this section, we propose a data mining query language and transform the original transaction data into another type to improve the efficiency of query processing.

M.-S. Chen, P.S. Yu, and B. Liu (Eds.): PAKDD 2002, LNAI 2336, pp. 172-176, 2002.

Our proposed data mining language is defined as follows:

Mining <Association Rules>
From <CSD>
With <{D_1},{D_2}>
  Support <s%>
  Confidence <c%>

1. In the **From** clause, <CSD> is used to specify the database name to which users query the association rules.
2. In the **With** clause, D_1 and D_2 are the itemsets in the antecedent and consequent, respectively, of the discovered rules. Besides, {D_i} and the items in D_i can be the notation "*" which represents any items.
3. **Support** clause is followed by the user-specified minimum support *s%.*
4. **Confidence** clause is followed by the user-specified minimum confidence c%.

Table 1. Customer Sequence Database (CSD)

CID	Customer Sequence
1	{C}{AC}{ACE}
2	{AE}{A}{ACE}{CE}
3	{C}{E}{E}{CE}
4	{BD}{AE}{BC}{AE}{ABE}{F}
5	{D}{DEF}{CEF}{AD}{BD}{DF}

In order to find the interesting association rules efficiently, we need to transform the original transaction data into another type. A *customer sequence* is the list of all the transactions of a customer, which is ordered by increasing transaction-time. Each item in each customer sequence is transformed into a bit string. The length of a bit string is the number of the transactions in the customer sequence. If the *i*th transaction of the customer sequence contains an item, then the *i*th bit in the bit string for this item is set to 1. Otherwise, the *i*th bit is set to 0. Hence, we can transform the customer sequence database (Table 1) into the *bit-string database* (Table 2). From the bit-string database, we can easily compute the number of the transactions in a customer sequence, which contain an itemset.

Table 2. Bit-String Database

CID	Transaction Items	Bit String for Each Item
1	A, C, E	011,111,001
2	A, C, E	1110,0011,1011
3	C, E	1001,0111
4	A, B, C, D, E, F	010110,101010,001000,100000,010110,000001
5	A, B, C, D, E, F	000100,000010,001000,110111,011000,011001

3 Mining Interesting Association Rules

For a user's query, if there is no notation "*" specified in the **With** clause, that is, the items in the antecedent and the consequent of the rule are fully specified, then this query is to check if the rule is an association rule. We call this type of users' queries the Type I query. If the user would like to extract the association rules whose antecedent or consequent contains other items except the items specified in the **With** clause, then the notation "*"s have to be specified in the **With** clause. We call this type of users' queries the Type II query.

3.1 Query Processing for Type I Query

Suppose that the two itemsets specified in the antecedent and the consequent in the **With** clause are X and Y, respectively. The method to check if this rule is an association rule is described as follows:

Step 1: Scan each record in the bit-string database. For the record in CID i, if all items in X are contained in this record, then perform logical AND operations on the bit strings for all the items in X. The resultant bit string is the bit string for itemset X in CID i. The number m_i of 1's in the resultant bit string is the number of the transactions which contain the itemset X for the record in CID i. If the itemset X is not contained in CID i, then m_i is equal to 0. If m_i is not zero and both itemsets X and Y are contained in CID i, then perform logical AND operations on the bit strings for itemset X and itemset Y. The number n_i of 1's in the resultant bit string is the number of the transactions which contain both itemsets X and Y for the record in CID i.

Step 2: Suppose there are p customers and q transactions in the bit-string database. The numbers of the transactions which contain the itemset X and itemset X∪Y are $m = \sum_{i=1}^{p} m_i$ and $n = \sum_{i=1}^{p} n_i$, respectively. The support for the itemset X∪Y can be obtained by n/q, and the confidence for the rule X⇒Y is n/m. If the support and the confidence both are no less than the user-specified minimum support and minimum confidence, respectively, then this rule is an association rule.

For example, consider Query 1, the user would like to check if {A}⇒{C, E} satisfies the user-specified minimum support 5%, and minimum confidence 70% in the customer sequence database in Table 1.

Query 1:

Mining <Association Rules>
From <CSD>
With <{A},{C,E}>
Support <5%>
Confidence <70%>

3.2 Query Processing for Type II Query

For Type II query, we divide it into three cases. Case 1: there are items in the antecedent of the discovered rules specified. Case 2: there are items in the consequent of the discovered rules specified, but the items in the antecedent are not specified, which can be any items. Case 3: there is no any specified items in the **With** clause, that is, all rules which satisfy the user-specified minimum support and minimum confidence will be discovered. Suppose that the itemset specified in the antecedent of the discovered rules in Case 1 is X, and the itemset specified in the consequent of the discovered rules in Case 2 is Y. The algorithm to find all the association rules, which satisfy the user requirements, is described as follows:

Step 1: Scan the bit-string database to compute the support for the specified itemset, and then find all the frequent 1-itemsets. If the specified itemset is not a frequent itemset, then terminate this algorithm.

Step 2: Generate candidate (k+1)-itemsets (k is the length of itemset X for Case 1, and k=1 for Cases 2 and 3), scan the bit-string database to find the frequent (k+1)-itemsets, which contain the itemset X for Case 1, and generate the (k+1)-itemset database.

For Case 1, the method to generate (k+1)-itemsets is as follows: For each frequent 1-itemset f, the candidate (k+1)-itemset $X \cup f$ can be generated. For the record in CID *i* in the bit-string database, if the record contains the itemset X and a frequent 1-itemset g, then generate the itemset $X \cup g$, and perform logical AND operation on the bit strings for itemset X and the item g. For Cases 2 and 3, the method to generate candidate 2-itemsets is as follows: For every two frequent 1-itemsets u and v ($u \neq v$), the candidate 2-itemset $u \cup v$ can be generated. For each record in CID *i* in the bit-string database, if the record contains two frequent 1-itemsets x and y, then generate the itemset $x \cup y$, and perform logical AND operation on the bit strings for items x and y. If the resultant bit string is not zero, then output the itemset $X \cup g$ (or $x \cup y$) and the resultant bit string to the (k+1)-itemset database, and accumulate the number of the transactions which contain the candidate itemset $X \cup g$ (or $x \cup y$) by counting the number of 1's in the resultant bit string. After scanning each record in the bit-string database, the (k+1)-itemset database can be generated and the support for each candidate (k+1)-itemset can be obtained.

Step 3: The frequent itemsets are generated for each iteration. In the (h-1)th iteration ($h \geq 2$) (the (h-k)th iteration ($h \geq k+1$) for Case 1), generate candidate (h+1)-itemsets, scan the h-itemset database to generate (h+1)-itemset database and find all the frequent (h+1)-itemsets).

In this step, we first generate candidate (h+1)-itemsets. For every two frequent h-itemsets $\{a_1, a_2, \ldots, a_{h-1}, b\}$ and $\{a_1, a_2, \ldots, a_{h-1}, c\}$ (b>c), the candidate (h+1)-itemset $\{a_1, a_2, \ldots, a_{h-1}, b, c\}$ can be generated.

Step 4: Generate all the association rules, which satisfy the user requirements

For Case 1, if there is a specified itemset *Y* in the consequent in the **With** clause, then find the frequent itemset *Z* which contains itemset *Y* from all the frequent itemsets. For every two sub-itemsets X_1 and Y_1 of the frequent itemset *Z*, where $X \subseteq X_1$, $Y \subseteq Y_1$, $X_1 \cap Y_1 = \phi$ and $X_1 \cup Y_1 = Z$, if the confidence of the rule $X_1 => Y_1$ is no less than the user-specified minimum confidence, then the rule $X_1 => Y_1$ is an association rule for user requirements. If there is no items specified in the consequent in the **With** clause, then for every two sub-itemsets X_2 and Y_2 of each frequent itemset V, where $X \subseteq X_2$, $X_2 \cap Y_2 = \phi$ and $X_2 \cup Y_2 = V$, if the confidence of the rule $X_2 => Y_2$ is no less than the user-specified minimum confidence, then the rule $X_2 => Y_2$ is an association rule for user requirements.

In Case 2, for every two sub-itemsets X_3 and Y_3 of each frequent itemset Z, where $Y \subseteq Y_3$, $X_3 \cap Y_3 = \phi$ and $X_3 \cup Y_3 = Z$, if the confidence of the rule $X_3 => Y_3$ is no less than the user-specified minimum confidence, then the rule $X_3 => Y_3$ is an association rule for user requirements.

In Case 3, for every two sub-itemsets X_4 and Y_4 of each frequent itemset Z, where $X_4 \cap Y_4 = \phi$ and $X_4 \cup Y_4 = Z$, if the confidence of the rule $X_3 => Y_3$ is no less than the user-specified minimum confidence, then the rule $X_4 => Y_4$ is an association rule for user requirements.

Query 2 is an example for Case 1 query. This query means that the user would like to find all the association rules whose antecedent and consequent contain items A and C, respectively, from the customer sequence database CSD (Table 1). The minimum support and the minimum confidence are set to 5% and 20%, respectively. We can find three association rules: {A}⇒{C, E}, {A, E}⇒{C} and {A}⇒{C}.

Query 2:

Mining <Association Rules>
From <CSD>
With <{A,*},{C,*}>
Support <5%>
Confidence <20%>

References

1. Rakesh Agrawal and et al. Fast Algorithm for Mining Association Rules. In *Proceedings of International Conference on Very Large Data Bases*, pages 487-499, 1994.
2. J.S. Park, M.S. Chen, and P.S. Yu. An Effective Hash-Based Algorithm for Mining Association Rules. In *Proceedings of ACM SIGMOD*, 24(2):175-186, 1995.
3. Show-Jane Yen and Arbee L.P. Chen. An Efficient Approach to Discovering Knowledge from Large Databases. In *Priceedings of the International Conference on Parallel and Distributed Information Systems*, pages 8-18,1996.
4. Show-Jane Yen and A.L.P. Chen. "A Graph-Based Approach for Discovering Various Types of Association Rules," *IEEE Transactions on Knowledge and Data Engineering*, Vol.13, No.5, pp. 839-845, 2001.

The Lorenz Dominance Order as a Measure of Interestingness in KDD

Robert J. Hilderman

Department of Computer Science
University of Regina
Regina, Saskatchewan, Canada S4S 0A2
robert.hilderman@uregina.ca

Abstract. Ranking summaries generated from databases is useful within the context of descriptive data mining tasks where a single data set can be generalized in many different ways and to many levels of granularity. Our approach to generating summaries is based upon a data structure, associated with an attribute, called a *domain generalization graph* (DGG). A DGG for an attribute is a directed graph where each node represents a domain of values created by partitioning the original domain for the attribute, and each edge represents a generalization relation between these domains. Given a set of DGGs associated with a set of attributes, a *generalization space* can be defined as all possible combinations of domains, where one domain is selected from each DGG for each combination. This generalization space describes, then, all possible summaries consistent with the DGGs that can be generated from the selected attributes. When thc number of attributes to be generalized is large or the DGGs associated with the attributes are complex, the generalization space can be very large, resulting in the generation of many summaries. The number of summaries can easily exceed the capabilities of a domain expert to identify interesting results. In this paper, we show that the Lorenz dominance order can be used to rank the summaries prior to presentation to the domain expert. The *Lorenz dominance order* defines a partial order on the summaries, in most cases, and in some cases, defines a total order. The rank order of the summaries represents an objective evaluation of their relative interestingness and provides the domain expert with a starting point for further subjective evaluation of the summaries.

1 Introduction

We describe a data mining problem where the task is description by summarization, the representation language is generalized relations, and the method for searching is the Multi-Attribute Generalization algorithm [5, 6]. Let a *summary* S be a relation defined on the columns $\{(A_1, D_1), (A_2, D_2), \ldots, (A_n, D_n)\}$, where each (A_i, D_i) is an attribute-domain pair. Also, let $\{(A_1, v_{i1}), (A_2, v_{i2}), \ldots, (A_n, v_{in})\}$, $i = 1, 2, \ldots, m$, be a set of m unique tuples, where each (A_j, v_{ij}) is an attribute-value pair and each v_{ij} is a value from the domain D_j associated with

M.-S. Chen, P.S. Yu, and B. Liu (Eds.): PAKDD 2002, LNAI 2336, pp. 177–185, 2002.

attribute A_j. One attribute A_k is a derived attribute, called *Count*, whose domain D_k is the set of positive integers, and whose value v_{ik} for each attribute-value pair (A_k, v_{ik}) is equal to the number of tuples which have been aggregated from the base relation (i.e., the unconditioned data present in the original database). A sample summary is shown in Table 1.

Table 1. A sample summary

Office	*Quantity*	*Amount*	*Count*
West	8	$200.00	4
East	11	$275.00	3

A summary, such as the one shown in Table 1, can be generated using *attribute-oriented generalization* (AOG) [3] and *domain generalization graphs* (DGGs) [5, 6]. For example, a sample DGG for the *Office* attribute is shown in Figure 1. In Figure 1, each node in the DGG is a partition of the domain values that can be used to describe the attribute, and the arcs connecting each pair of adjacent nodes defines a generalization relation based upon AOG.

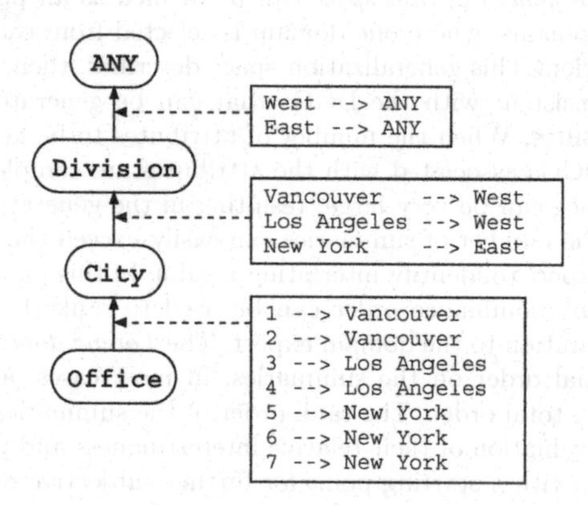

Fig. 1. A DGG for the *Office* attribute

AOG summarizes the information in a database by replacing specific attribute values with more general concepts according to user-defined taxonomies. For example, the domain for the *Office* attribute shown in the sales transaction database of Table 2 is represented by the *Office* node in Figure 1. Increasingly general descriptions of the domain values are represented by the *City*, *Division*, and *ANY* nodes. In Figure 1, the generalization relations consists of table lookups (other generalization relations besides table lookups are possible, but we restrict our discussion for the sake of simplicity and clarity). The table associated with the arc between the *Office* and *City* nodes defines the mapping of the domain values of the *Office* node to the domain values of the *City* node. The table associated with the arc between the *City* and *Division* nodes can be described similarly. The table associated with the arc between the *Division* and *ANY* nodes maps all values in the *Division* domain to the special value *ANY*. The summary shown in Table 1 corresponds to the *Division* node of the DGG in Figure 1, and was generated from Table 2. That is, the tuples aggregated in the *West* tuple of Table 1 include the first, third, fifth, and seventh tuples from Table 2, for a *Count* attribute value of four. Similarly, the *East* tuple includes the second, fourth, and sixth tuples from Table 2, for a *Count* attribute value of three. The corresponding values in the *Quantity* and *Amount* attributes from Table 2 are also aggregated in Table 1.

Table 2. A sales transaction database

Office	*Quantity*	*Amount*
2	2	$50.00
5	3	$75.00
3	1	$25.00
7	4	$100.00
1	3	$75.00
6	4	$100.00
4	2	$50.00

When there are multiple DGGs associated with an attribute, meaning knowledge about the attribute can be expressed in different ways, a multi-path DGG can be constructed from the single-path DGGs. Similarly, when there are DGGs associated with multiple attributes, more complex summaries can be generated using a process known as *multi-attribute generalization*. Briefly, in multi-attribute generalization, a set of individual attributes can be considered to be a single attribute (called a *compound attribute*) whose domain is the cross-product of the individual attribute domains. A summary generated from the cross-product domain for the compound attribute corresponds to a unique combination of nodes from the DGGs associated with the individual attributes, where one node is selected from the DGG associated with each attribute.

When the number of attributes to be generalized is large or the DGGs associated with the attributes are complex, the generalization space can be very large, resulting in the generation of many summaries. The number of summaries can easily exceed the capabilities of a domain expert to identify interesting results. What is needed is an effective measure of interestingness to assist in the preliminary interpretation and evaluation of the discovered knowledge. The values in the *Count* attribute of the summaries under consideration (which we refer to as *count vectors*, or simply *vectors*) can provide a basis for this rank ordering. We have previously described heuristics, based upon measures from information theory, statistics, ecology, and economics, that consider these distributions to assign a single real-valued index that represents the interestingness of a summary relative to the other summaries [5, 6]. And in [4], we developed the foundation for an intuitive understanding of the term "interestingness" when used within the context of ranking summaries generated from databases.

In this paper, we describe a technique for ranking summaries based upon the Lorenz dominance order [1, 7]. The *Lorenz dominance order* is used to determine the relationship between every possible pair of count vectors (i.e., summaries). The relationship that we seek to discover is whether the vectors are comparable. If vector X is more diverse than Y according to the Lorenz dominance order (the criteria to be discussed in the next section), then the vectors are comparable and we can say that X *majorizes* Y. For our purposes, we consider majorization and interestingness to be equivalent, so we say that X is more interesting than Y, and the Lorenz dominance order provides the basis for an objective measure of interestingness. If X and Y are not comparable according to the Lorenz dominance order, then the relationship between the two vectors remains undefined.

2 Theoretical Results

An important property of the Lorenz dominance order is that it defines a partial order on the set of all possible vectors, a property useful and important for ranking summaries. We formalize this property as it relates to ranking summaries in the series of principles and theorems that follow.

Transfer Principle (P1). Given a vector $(n_1, \ldots, n_m)$, $n_i \geq n_j$, $i < j$, and $0 < c \leq n_j$, $f(n_1, \ldots, n_i + c, \ldots, n_j - c, \ldots, n_m) > f(n_1, \ldots, n_i, \ldots, n_j, \ldots, n_m)$.

P1, adopted from [2], specifies that when a strictly positive transfer is made from the count of one tuple to another tuple whose count is greater, then interestingness increases. For example, given the vectors $X = (10, 7, 5, 4)$ and $Y = (10, 9, 5, 2)$, where Y is derived from X via a positive transfer of 2 units from the fourth tuple of X to the second tuple of Y, then we require that $f(10, 9, 5, 2) > f(10, 7, 5, 4)$.

Majorization Principle (P2). Given vectors $(n_1, \ldots, n_m)$ and $(n_1', \ldots, n_m')$, whenever $f(n_1', \ldots, n_m') > f(n_1, \ldots, n_m)$, then $(n_1', \ldots, n_m') \succ (n_1, \ldots, n_m)$, which we read as $(n_1', \ldots, n_m')$ *majorizes* $(n_1, \ldots, n_m)$.

The majorization operator, $\succ$, is based upon the Lorenz dominance order [1, 7]. The *Lorenze dominance order* compares vectors with different distributions and says for any two vectors $(n_1, \ldots, n_m)$ and $(n_1', \ldots, n_m')$, that $(n_1', \ldots, n_m') \succ (n_1, \ldots, n_m)$ if the following fours conditions hold:

1. $n_1 \geq \ldots \geq n_m$.
2. $n_1' \geq \ldots \geq n_m'$.
3. $\sum_{i=1}^{j} n_i' \geq \sum_{i=1}^{j} n_i$, for every $j = 1, \ldots, m$.
4. $\sum_{i=1}^{m} n_i' = \sum_{i=1}^{m} n_i$.

In other words, $(n_1',\ldots,n_m') \succ (n_1,\ldots,n_m)$ if and only if the Lorenz curve of $(n_1,\ldots,n_m)$ is completely nested within that of $(n_1',\ldots,n_m')$.

Theorem 1. P1 and P2 define a partial order on ranked summaries.

Proof. We need to show, in general, that for any two vectors $(n_1,\ldots,n_i+c,\ldots,n_j-c,\ldots,n_m)$ and $(n_1,\ldots,n_i,\ldots,n_j,\ldots,n_m)$, where $f(n_1,\ldots,n_i+c,\ldots,n_j-c,\ldots,n_m) > f(n_1,\ldots,n_i,\ldots,n_j,\ldots,n_m)$ according to P1, that $(n_1,\ldots,n_i+c,\ldots,n_j-c,\ldots,n_m) \succ (n_1,\ldots,n_i,\ldots,n_j,\ldots,n_m)$.

Since we assume the vectors are arranged in descending order, then conditions 1 and 2 of the Lorenz dominance order are obviously true. Also, since N is fixed, condition 4 is obviously true. It remains to be shown that condition 3 is true.

Suppose that n_i+c and n_j-c occupy positions $i-r$, $r \geq 0$, and $j+s$, $s \geq 0$, respectively, after reordering (if necessary) the vector $(n_1,\ldots,n_i+c,\ldots,n_j-c,\ldots,n_m)$. This implies that the elements in positions 1 to $i-r-1$ and $j+s+1$ to m did not require reordering, so we have

$$\sum_{k=1}^{i-r-1} n_k' = \sum_{k=1}^{i-r-1} n_k$$

and

$$\sum_{k=j+s+1}^{m} n_k' = \sum_{k=j+s+1}^{m} n_k.$$

We need to show that for the remaining elements

$$\sum_{k=i-r}^{j+s} n_k' \geq \sum_{k=i-r}^{j+s} n_k,$$

for every $k = i-r,\ldots,j+s$.

First, we consider the situation where n_{i-r} and n_{j+s} are adjacent. That is, $i = i-r$ and $j = j+s$, meaning $r = s = 0$, so the vectors that we compare are $(n_{i-r}', n_{j+s}') = (n_i+c, n_j-c)$ and $(n_{i-r}, n_{j+s}) = (n_i, n_j)$. For $k = i$, we have

$$n_i + c > n_i,$$

which is obviously true, and for $k = j$, we have

$$(n_i + c) + (n_j - c) = n_i + n_j,$$

so condition 3 is shown to be true.

We now consider the situation where n_{i-r} and n_{j+s} are not adjacent. The vectors that we compare are $(n_{i-r}', n_{i-r+1}', \ldots, n_{i-1}', n_i', n_{i+1}', \ldots, n_{j-1}', n_j', n_{j+1}', \ldots, n_{j+s}')$ and $(n_{i-r}, n_{i-r+1}, \ldots, n_{i-1}, n_i, n_{i+1}, \ldots, n_{j-1}, n_j, n_{j+1}, \ldots, n_{j+s})$, where the first vector contains the reordered elements of the second vector, as follows:

1. $n_{i-r}' = n_i + c$. That is, $n_i + c$ is now the greatest element in positions $i-r$ to $j+s$.

2. $n'_{i-r+1} = n_{i-r}$, $n'_{i-r+2} = n_{i-r+1}$, ..., $n'_{i-1} = n_{i-2}$, $n'_i = n_{i-1}$. That is, all the elements in positions $i - r$ to $i - 1$ in the second vector are shifted one position to the right into positions $i - r + 1$ to i in the first vector.
3. $n'_k = n_k$, for $k = i+1, i+2, \ldots, j-1$. That is, all elements in positions $i+1$ to $j - 1$ of the second vector are identical to those in the same positions of the first vector and are not affected by the reordering.
4. $n'_j = n_{j+1}$, $n'_{j+1} = n_{j+2}$, ..., $n'_{j+s-2} = n_{j+s-1}$, $n'_{j+s-1} = n_{j+s}$. That is, all elements in positions $j + 1$ to $j + s$ in the second vector are shifted one position to the left into positions j to $j + s - 1$ in the first vector.
5. $n'_{j+s} = n_j - c$. That is, $n_j - c$ is now the least element in positions $i - r$ to $j + s$.

Therefore, the first vector can be written as $(n_i + c, n_{i-r}, n_{i-r+1}, \ldots, n_{i-1}, n_{i+1}, \ldots, n_{j-1}, n_{j+1}, \ldots, n_j - c)$. So, we have

$$\sum_{k=i-r}^{i} n'_k > \sum_{k=i-r}^{i} n_k,$$

for every $k = i - r, \ldots, i$ because $n'_k > n_k$, for all $k = i - r, \ldots, i$. Similarly, we have

$$\sum_{k=i+1}^{j-1} n'_k = \sum_{k=i+1}^{j-1} n_k,$$

for every $k = i + 1, \ldots, j - 1$ because $n'_k = n_k$, for all $k = i + 1, \ldots, j - 1$. Now

$$\sum_{k=i-r}^{j-1} n'_k - \sum_{k=i-r}^{j-1} n_k =$$

$$(n_i + c + n_{i-r} + n_{i-r+1} + \ldots + n_{i-1} + n_{i+1}, \ldots, n_{j-2} + n_{j-1}) -$$
$$(n_{i-r} + n_{i-r+1} + \ldots + n_{i-1} + n_i + n_{i+1}, \ldots, n_{j-2} + n_{j-1}),$$

which after canceling terms yields

$$\sum_{k=i-r}^{j-1} n'_k - \sum_{k=i-r}^{j-1} n_k = c.$$

Similarly,

$$\sum_{k=j}^{j+s} n_k - \sum_{k=j}^{j+s} n'_k = (n_j + n_{j+1} + \ldots + n_{j+s-1} + n_{j+s}) - (n_{j+1} + nj+2 + \ldots + n_{j+s} + n_j - c),$$

which after canceling terms yields

$$\sum_{k=j}^{j+s} n_k - \sum_{k=j}^{j+s} n'_k = c.$$

So, for every $k = j, \ldots, j+s$, we have

$$\sum_{k=j}^{j+s} n_k - \sum_{k=j}^{j+s} n_k^{'} \leq c.$$

Since

$$\sum_{k=i-r}^{j-1} n_k^{'} - \sum_{k=i-r}^{j-1} n_k = c,$$

then for every $k = j, \ldots, j+s$, we have

$$\sum_{k=i-r}^{j-1} n_k^{'} - \sum_{k=i-r}^{j-1} n_k - \left(\sum_{k=j}^{j+s} n_k - \sum_{k=j}^{j+s} n_k^{'} \right) \geq 0.$$

The above inequality can be written as

$$\sum_{k=i-r}^{j-1} n_k^{'} + \sum_{k=j}^{j+s} n_k^{'} \geq \sum_{k=i-r}^{j-1} n_k + \sum_{k=j}^{j+s} n_k,$$

so condition 3 is shown to be true. Since $f(n_1, \ldots, n_i + c, \ldots, n_j - c, \ldots, n_m) > f(n_1, \ldots, n_i, \ldots, n_j, \ldots, n_m)$ satisfies P1 is given, and $(n_1, \ldots, n_i + c, \ldots, n_j - c, \ldots, n_m) \succ (n_1, \ldots, n_i, \ldots, n_j, \ldots, n_m)$, then P2 is also satisfied, and it is proved. □

Definition. Let $(n_1^{'}, \ldots, n_m^{'})$ be a vector derived from $(n_1, \ldots, n_m)$ according to P1. That is, for some $n_i \geq n_j$, $i < j$, $0 < c \leq n_j$, we have $(n_1^{'}, \ldots, n_m^{'}) = (n_1, \ldots, n_i + c, \ldots, n_j - c, \ldots, n_m)$. The transfer from n_j to n_i is called *one elementary transfer*.

Theorem 2. Whenever a vector $(n_1^{'}, \ldots, n_m^{'})$ can be derived from a vector $(n_1, \ldots, n_m)$ via a finite series of elementary transfers, then $(n_1^{'}, \ldots, n_m^{'}) \succ (n_1, \ldots, n_m)$.

Sketch of Proof. We need to show that after every elementary transfer, $(n_1^{'}, \ldots, n_m^{'}) \succ (n_1, \ldots, n_m)$. That is, $(n_1, \ldots, n_i{+}c, \ldots, n_j{-}c, \ldots, n_m) \succ (n_1, \ldots, n_i, \ldots, n_j, \ldots, n_m)$. For one elementary transfer, the proof is similar to that for Theorem 1. For multiple elementary transfers, the proof relies on the transitive property of vectors in the Lorenz dominance order, where each succeeding vector dominates the one that preceded it such that $(n_1^a, \ldots, n_m^a) \succ (n_1^b, \ldots, n_m^b) \succ \cdots \succ (n_1^x, \ldots, n_m^x)$. □

Theorem 3. For a summary whose distribution of tuples corresponds to the vector $(n_1, \ldots, n_m)$, if a more general summary resides along the same path in a DGG whose distribution of tuples corresponds to the vector $(n_1^{'}, \ldots, n_m^{'})$, then $(n_1^{'}, \ldots, n_m^{'})$ can be derived from $(n_1, \ldots, n_m)$ via a finite series of elementary transfers.

Proof. Given a summary whose distribution of tuples corresponds to the vector $(n_1, \ldots, n_m)$, any more general summary along the same path whose distribution of tuples corresponds to the vector $(n_1', \ldots, n_m')$ is derived by aggregating elements. Say the r elements in positions $i, \ldots, j$ (not necessarily contiguous) are to be aggregated into position k. That is, $n_k = n_i + \ldots + n_j$. Then each element in positions $i, \ldots, j$ can be transferred via an elementary transfer. The number of elementary transfers required is equal to r, one for each element $n_i, \ldots, n_j$, and it is proved. □

Theorem 4. P1 and P2 define a total order for the summaries along a single path in a DGG.

Proof. According to Theorem 2, whenever a vector $(n_1', \ldots, n_m')$ can be derived from a vector $(n_1, \ldots, n_m)$ via a finite series of elementary transfers, then $(n_1', \ldots, n_m') \succ (n_1, \ldots, n_m)$. Also, according to Theorem 3, for any vector $(n_1, \ldots, n_m)$, any vector $(n_1', \ldots, n_m')$ along the same path that is more general can be derived from $(n_1, \ldots, n_m)$ via a finite series of elementary transfers. Now each summary along a single path is derived via a finite series of elementary transfers, so due to the transitive property of vectors in the Lorenz dominance order $(n_1^a, \ldots, n_m^a) \succ (n_1^b, \ldots, n_m^b) \succ \ldots \succ (n_1^x, \ldots, n_m^x)$. Also, each summary is unique, so the relation defines a total order on the summaries. Since each pair of vectors $(n_1', \ldots, n_m')$ and $(n_1, \ldots, n_m)$ satisfies P1, then P2 is also satisfied, and it is proved. □

3 Conclusion

The Lorenz dominance order compares vectors with different distributions. Here we have presented principles and theorems describing the Lorenz dominance order as a measure of interestingness for ranking summaries generated from databases. Future research will focus on evaluating other diversity measures to determine their suitability for ranking the interestingness of summaries generated from databases.

References

[1] B.C. Arnold. *Majorization and the Lorenz Order: A Brief Introduction.* Springer-Verlag, 1987.

[2] H. Dalton. The measurement of the inequality of incomes. *Economic Journal*, 30:348–361, 1920.

[3] J. Han, Y. Cai, and N. Cercone. Data-driven discovery of quantitative rules in relational databases. *IEEE Transactions on Knowledge and Data Engineering*, 5(1):29–40, February 1993.

[4] R.J. Hilderman and H.J. Hamilton. Principles for mining summaries using objective measures of interestingness. In *Proceedings of the 12th IEEE International Conference on Tools with Artificial Intelligence (ICTAI'00)*, pages 72–81, Vancouver, BC, November 2000.

[5] R.J. Hilderman and H.J. Hamilton. *Knowledge Discovery and Measures of Interest.* Kluwer Academic Publishers, 2002.
[6] R.J. Hilderman, Liangchun Li, and H.J. Hamilton. Visualizing data mining results with domain generalization graphs. In U. Fayyad, G.G. Grinstein, and A. Wierse, editors, *Information Visualization in Data Mining and Knowledge Discovery*, pages 251–270. Morgan Kaufmann Publishers, 2002.
[7] A.W. Marshall and I. Olkin. *Inequalities: Theory of Majorization and its Applications.* Academic Press, 1979.

Efficient Algorithms for Incremental Update of Frequent Sequences*

Minghua Zhang, Ben Kao, David Cheung, and Chi-Lap Yip

Department of Computer Science and Information Systems
The University of Hong Kong
{mhzhang, kao, dcheung, clyip}@csis.hku.hk

Abstract. Most of the works proposed so far on mining frequent sequences assume that the underlying database is static. However, in real life, the database is modified from time to time. This paper studies the problem of incremental update of frequent sequences when the database changes. We propose two efficient *incremental* algorithms `GSP+` and `MFS+`. Throught experimetns, we compare the performance of `GSP+` and `MFS+` with `GSP` and `MFS` — two efficient algorithms for mining frequent sequences. We show that `GSP+` and `MFS+` effectively reduce the CPU costs of their counterparts with only a small or even negative additional expense on I/O cost.
keywords: data mining, sequence, incremental update

1 Introduction

One of the many data mining problems is mining frequent sequences from transactional databases. The goal is to discover frequent sequences of events. The problem was first introduced by Agrawal and Srikant [1]. In their model, a database is a collection of transactions. Each transaction is a set of items (or an itemset) and is associated with a customer ID and a time ID. If one groups the transactions by their customer IDs, and then sorts the transactions of each group by their time IDs in increasing value, the database is transformed into a number of customer sequences. Each customer sequence shows the order of transactions a customer has conducted. Roughly speaking, the problem of mining frequent sequences is to discover "subsequences" (of itemsets) that occur frequently enough among all the customer sequences.

A few efficient algorithms for mining frequent sequences have been proposed, notably, `AprioriAll` [1], `GSP` [13], `SPADE` [15], `MFS` [16] and `PrefixSpan` [10]. The above studies assume the database is static, and even a small change in the database will require the algorithms to run again to get the updated frequent sequences. In practice, the content of a database changes continuously, and data mining has to be performed repeatedly. If each time we have to rerun the mining algorithms from scratch, it will be very inefficient.

* This research is supported by Hong Kong Research Grants Council grant HKU 7035/99E

M.-S. Chen, P.S. Yu, and B. Liu (Eds.): PAKDD 2002, LNAI 2336, pp. 186–197, 2002.

In this paper we study the problem of incremental maintenance of frequent sequences. We assume that a mining exercise has been performed on an *old* database to obtain a set of frequent sequences. The old database is then updated by inserting new sequences and/or by deleting some old sequences. Our objective is to discover the set of frequent sequences in the *new* database efficiently, taking advantage of information that was obtained in a previous mining exercise. We propose two new algorithms GSP+ and MFS+ to solve the problem. The two algorithms are modified versions of GSP [13] and MFS [16]. The modification is made based on the following observations:

- If one knows about the support of a *frequent* sequence in the old database, then the sequence's support w.r.t. the new database can be deduced by scanning the inserted customer sequences and the deleted customer sequences. The portion (typically the majority) of the database that has not been changed needs not be processed.
- Given that a sequence is infrequent w.r.t. the old database, the sequence cannot become frequent unless its support in the inserted customer sequences is *large enough* and that its support in the deleted customer sequences is *small enough*. This observation allows us to determine whether a candidate sequence should be considered by "looking" at the *small* portion of the database that has been changed.
- If the old database and the new database share a non-trivial set of common sequences, then the set of frequent sequences found in the old database gives a good indication of which sequences in the new database are *likely* to be frequent. As we will see later, this allows us to generate long candidate sequences fairly early on in the mining algorithm. The effect is that fewer passes over the data are required compared with the mine-from-scratch approach. We can thus effectively reduce the I/O cost.

The rest of this paper is organized as follows. Section 2 gives a formal definition of the maintenance problem. In Section 3, we review some related works, in particular, GSP and MFS. The two new algorithms GSP+ and MFS+ are presented in Section 4. Experiment results comparing the performance of the algorithms are shown in Section 5. Finally, we conclude the paper in Section 6.

2 Model

In this section, we give a formal statement of the maintenance problem.

Let $I = \{i_1, i_2, \ldots, i_m\}$ be a set of literals called items. An itemset X is a set of items (hence, $X \subseteq I$). A sequence $s = \langle t_1, t_2, \ldots, t_n \rangle$ is an ordered set of transactions, where each transaction t_i $(i = 1, 2, \ldots, n)$ is an itemset.

The length of a sequence s is defined as the number of items contained in s. (If an item occurs several times in different itemsets of a sequence, the item is counted for each occurrence.) We use $|s|$ to represent the length of s.

Given two sequences $s_1 = \langle a_1, a_2, \ldots, a_n \rangle$ and $s_2 = \langle b_1, b_2, \ldots, b_l \rangle$, we say s_1 contains s_2 (or equivalently s_2 is a subsequence of s_1) if there exist integers $j_1, j_2, \ldots, j_l$, such that $1 \leq j_1 < j_2 < \ldots < j_l \leq n$ and $b_1 \subseteq a_{j_1}, b_2 \subseteq a_{j_2}, \ldots, b_l \subseteq a_{j_l}$. We represent this relationship by $s_2 \sqsubseteq s_1$.

In a sequence set V, a sequence $s \in V$ is *maximal* if s is not a subsequence of any other sequence in V.

Given a sequence set V and a sequence s, if there exists a sequence $s' \in V$ such that $s \sqsubseteq s'$, we write $s \vdash V$. Given a database D of sequences, the *support count* of a sequence s, denoted by δ_D^s, is defined as the number of sequences in D that contain s. The *fraction* of sequences in D that contain s is called the *support* of s. If we use the symbol $|D|$ to denote the number of sequences in D (or the size of D), we have: support of $s = \delta_D^s / |D|$.

If the support of s is no less than a user specified support threshold ρ_s, s is a frequent sequence. The problem of mining frequent sequences is to find all *maximal* frequent sequences in a database D. We use symbol L_i to denote the set of all length-i frequent sequences, and L to denote the set of all frequent sequences.

Given a database D, we assume that a previous mining exercise has been executed to obtain the supports of the frequent sequences. The database D is then updated by deleting a set of sequences Δ^- followed by inserting a set of sequences Δ^+. Let us denote the updated database D'. Note that $D' = (D - \Delta^-) \cup \Delta^+$. We denote the set of unchanged sequences by $D^- = D - \Delta^- = D' - \Delta^+$. Since the relative order of the sequences within a database does not affect the mining results, we may assume (without loss of generality) that all of the deleted sequences are located at the beginning of the database and all of the new sequences are appended at the end, as illustrated in Figure 1.

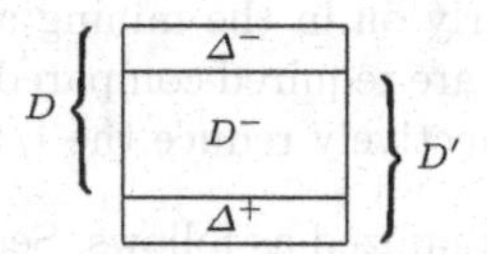

Fig. 1. Definitions of D, D', Δ^-, D^- and Δ^+

Our objective is to find all maximal frequent sequences in the database D' given Δ^-, D^-, Δ^+, and the result of mining D.

3 Related Works

Agrawal and Srikant [1] first studied the problem of mining frequent sequences, and proposed 3 algorithms. Later, the same authors proposed a more efficient algorithm GSP [13]. Similar to the structure of the Apriori algorithm [11] for mining association rules, GSP starts by finding all frequent 1-sequences from the database. A set of candidate 2-sequences are then generated. The support counts of the candidate sequences are then counted by scanning the database once. Those frequent 2-sequences are then used to generate candidate 3-sequences, and so on. In general, GSP uses a function GGen to generate candidate $(k+1)$-sequences given the set of all frequent k-sequences. The algorithm terminates when no more frequent sequences are discovered during a database scan. Figure 2 shows the GSP algorithm (In the algorithm, C_i means the set of candidate sequences of length-i). Details of the candidate generation function GGen is omitted due to space limitation. Readers are referred to [13] for more information. We note that if the database is huge and if it contains very long frequent sequences, the I/O cost of GSP is high.

```
1  Algorithm GSP(D, ρ_s, I)
2    C_1 := {⟨{i}⟩ | i ∈ I}
3    Scan D to compute δ^s_D for every sequence s in C_1
4    L_1 := {s | s ∈ C_1, δ^s_D ≥ ρ_s × |D|}
5    i := 1
6    while (L_i ≠ ∅)
7      C_{i+1} := GGen(L_i)
8      Scan D to compute δ^s_D for every sequence s in C_{i+1}
9      L_{i+1} := {s | s ∈ C_{i+1}, δ^s_D ≥ ρ_s × |D|}
10     i := i + 1
11   Return L_1 ∪ L_2 ∪ ... ∪ L_{i−1}
```

Fig. 2. Algorithm GSP

An interesting I/O-efficient algorithm, SPADE, was proposed by Zaki [15]. The idea is to first transform a sequence database from a "horizontal" representation to a "vertical" representation. SPADE works on the transformed database, and requires three database scans to find frequent sequences. While SPADE is an efficient algorithm, it requires the availability of the "vertical" database. Such a transformation requires either a high I/O cost or a lot of memory.

Based on SPADE, ISM [9] was put forward to address the maintenance problem of frequent sequences. By making use of a *sequence lattice*, which contains information about the old database, ISM can determine frequent sequences in the new database efficiently. Our algorithms differ from ISM in the following aspects. First, we consider both sequence insertions and deletions, while ISM handles insertion only. Second, similar to SPADE, ISM works on the "vertical" representation of the database. Our algorithms do not require the transformation.

PrefixSpan [10] is a newly devised efficient algorithm for mining frequent sequences. It needs to generate a number of intermediate databases during the process. If main memory is large enough, PrefixSpan is very efficient; otherwise it will require a high cost.

Another I/O-efficient algorithm MFS was proposed in [16]. MFS achieves I/O efficiency by making use of a candidate generation function MGen to generate sequences of various lengths given a set of frequent sequences of various lengths. This allows long sequences to be generated early, reducing the number of iterations and hence the I/O cost. For MFS to be effective, it requires an initial estimate (S_{est}) of the set of frequent sequences be available. S_{est} could be obtained by mining a small sample of the database. For the maintenance problem, MFS can use the frequent sequences in the old database as S_{est} directly. Hence, MFS is potentially efficient for the maintenance problem. Figure 3 shows the MFS algorithm. Details of the candidate generation function MGen is omitted due to space limitation. Readers are referred to [16] for more information.

1 Algorithm MFS(D, ρ_s, I, S_{est})
2 $MFSS := \emptyset$
3 $CandidateSet := \{\langle\{i\}\rangle | i \in I\} \cup \{s | s \vdash S_{est}, |s| > 1\}$
4 Scan D to get δ_D^s for every sequence s in $CandidateSet$
5 $NewFrequentSequences := \{s | s \in CandidateSet, \delta_D^s \geq \rho_s \times |D|\}$
6 $AlreadyCounted := \{s | s \vdash S_{est}, |s| > 1\}$
7 $Iteration := 2$
8 while ($NewFrequentSequences \neq \emptyset$)
9 $//Max(S)$ returns the set of all maximal sequences is S
10 $MFSS := Max(MFSS \cup NewFrequentSequences)$
11 $CandidateSet :=$ MGen($MFSS$, $Iteration$, $AlreadyCounted$)
12 Scan D to get δ_D^s for every sequence s in $CandidateSet$
13 $NewFrequentSequences := \{s | s \in CandidateSet, \delta_D^s \geq \rho_s \times |D|\}$
14 $Iteration := Iteration + 1$
15 Return $MFSS$

Fig. 3. Algorithm MFS

Finally, a number of studies have been done on the problem of maintaining discovered association rules including [2, 3, 4, 5, 7, 8, 12, 14].

4 Algorithms

In this section we describe our incremental update algorithms GSP+ and MFS+. The idea is that, given a sequence s, we use the support count of s in D (if available) to deduce whether s could have enough support in D'. In the deduction process, the portion of the database that has been changed, namely, Δ^- and Δ^+, might have to be scanned. If we deduce that s cannot be frequent in D', s's support in D^- (the portion of the database that has not been changed) is not counted. If D^- is large comparing with Δ^- and Δ^+, the pruning technique saves much CPU cost.

Before we present the algorithms, let us consider a few mathematical equations that allow us to perform the pruning deductions.

First of all, since $D' = D - \Delta^- \cup \Delta^+ = D^- \cup \Delta^+$, we have, $\forall s$,

$$\delta_D^s = \delta_{D^-}^s + \delta_{\Delta^-}^s, \quad (1)$$

$$\delta_{D'}^s = \delta_{D^-}^s + \delta_{\Delta^+}^s, \quad (2)$$

$$\delta_{D'}^s = \delta_D^s + \delta_{\Delta^+}^s - \delta_{\Delta^-}^s. \quad (3)$$

Let us define $b_X^s = \min_{s'} \delta_X^{s'}$ for any sequence s and database X, where $(s' \sqsubseteq s) \wedge (|s'| = |s| - 1)$. That is to say, if s is a k-sequence, b_X^s is the smallest support count of the $(k-1)$-subsequences of s in the database X. Since the support count of a sequence s must not be larger than the support count of any subsequence of s, b_X^s is an *upper bound* of δ_X^s.

The reason for considering b_X^s is to allow us to estimate δ_X^s without counting it. As we will see later, under both GSP+ and MFS+, a candidate sequence s is considered (and may have its support counted) only if all of s's subsequences are frequent. Since frequent sequences would already have their supports counted (in order to conclude that they are frequent), we would have the necessary information to deduce b^s_X when we consider s.

To illustrate how the bound is used in the deduction, let us consider the following simple Lemma:

Lemma 1 *If a sequence s is frequent in D', then $\delta^s_D + b^s_{\Delta^+} \geq \delta^s_D + b^s_{\Delta^+} - \delta^s_{\Delta^-} \geq |D'| \times \rho_s$.*

Proof: If s is frequent in D', we have,

$$\begin{aligned} |D'| \times \rho_s &\leq \delta^s_{D'} && \text{(by definition)} \\ &= \delta^s_D + \delta^s_{\Delta^+} - \delta^s_{\Delta^-} && \text{(by Equation 3)} \\ &\leq \delta^s_D + b^s_{\Delta^+} - \delta^s_{\Delta^-} \\ &\leq \delta^s_D + b^s_{\Delta^+}. \end{aligned}$$

Given a sequence s, if s is frequent in D, we know δ^s_D. If $b^s_{\Delta^+}$ is available, we can compute $\delta^s_D + b^s_{\Delta^+}$ and conclude that s is infrequent in D' if the quantity is less than $|D'| \times \rho_s$. Otherwise, we scan Δ^- to find $\delta^s_{\Delta^-}$. We conclude that s is infrequent in D' if $\delta^s_D + b^s_{\Delta^+} - \delta^s_{\Delta^-}$ is less than the required support count ($|D'| \times \rho_s$). Note that in the above cases, the deduction is made without processing D^- or Δ^+.

If a sequence s is not frequent in D, δ^s_D is unavailable. The pruning tricks derived from Lemma 1 are thus not applicable. However, being infrequent in D means that the support of s (in D) is *small*. The following Lemma allows us to prune those sequences.

Lemma 2 *If a sequence s is frequent in D' but not in D, then $b^s_{\Delta^+} \geq b^s_{\Delta^+} - \delta^s_{\Delta^-} \geq \delta^s_{\Delta^+} - \delta^s_{\Delta^-} > (|\Delta^+| - |\Delta^-|) \times \rho_s$.*

Proof: If s is frequent in D' but not in D, we have, by definition:

$$\begin{aligned} \delta^s_{D'} &\geq |D'| \times \rho_s = (|D^-| + |\Delta^+|) \times \rho_s, \\ \delta^s_D &< |D| \times \rho_s = (|D^-| + |\Delta^-|) \times \rho_s. \end{aligned}$$

Hence,

$$\begin{aligned} \delta^s_{D'} - \delta^s_D &> (|D^-| + |\Delta^+|) \times \rho_s - (|D^-| + |\Delta^-|) \times \rho_s \\ \delta^s_{\Delta^+} - \delta^s_{\Delta^-} &> (|\Delta^+| - |\Delta^-|) \times \rho_s \quad \text{(by Equation 3)}. \end{aligned}$$

Also,

$$b^s_{\Delta^+} \geq b^s_{\Delta^+} - \delta^s_{\Delta^-} \geq \delta^s_{\Delta^+} - \delta^s_{\Delta^-}.$$

Lemma 2 thus follows.

Given a candidate sequence s that is not frequent in the old database D, we first compare $b^s_{\Delta^+}$ against $(|\Delta^+| - |\Delta^-|) \times \rho_s$. If $b^s_{\Delta^+}$ is not large enough, s cannot be frequent in D', and hence s can be pruned. Otherwise, we scan Δ^- to find $b^s_{\Delta^+} - \delta^s_{\Delta^-}$ and see if s can be pruned. If not, we scan Δ^+ and consider $\delta^s_{\Delta^+} - \delta^s_{\Delta^-}$.

Similar to Lemma 1, Lemma 2 allows us to prune some candidate sequences without completely counting their supports in the new database.

4.1 GSP+

Based on the Lemmas, we modify GSP to incorporate the pruning techniques mentioned. The new algorithm, GSP+, is shown in Figure 4.

1 Algorithm GSP+(Δ^-, D^-, Δ^+, ρ_s, I)
2 $C_1 := \{\langle\{i\}\rangle | i \in I\}$
3 Scan Δ^-, D^-, Δ^+ to get $\delta^s_{\Delta^-}$, $\delta^s_{D^-}$, $\delta^s_{\Delta^+}$ for each $s \in C_1$
4 $L_1 := \{s | s \in C_1, \delta^s_{D^-} + \delta^s_{\Delta^+} \geq |D'| \times \rho_s\}$
5 $i := 2$
6 $C_i :=$ GGen(L_{i-1})
7 while ($C_i \neq \emptyset$)
8 calculate $b^s_{\Delta^+}$ for each $s \in C_i$
9 $candp = \{s | s \in C_i, s\ is\ frequent\ in\ D\}$, $candq = C_i - candp$
10 $\forall s \in candp$, if $\delta^s_D + b^s_{\Delta^+} < |D'| \times \rho_s$, delete s from $candp$
11 $\forall s \in candq$, if $b^s_{\Delta^+} \leq (|\Delta^+| - |\Delta^-|) \times \rho_s$, delete s from $candq$
12 scan Δ^- to count $\delta^s_{\Delta^-}$ for each s in $candp$ and $candq$
13 $\forall s \in candp$, if $\delta^s_D + b^s_{\Delta^+} - \delta^s_{\Delta^-} < |D'| \times \rho_s$, delete s from $candp$
14 $\forall s \in candq$, if $b^s_{\Delta^+} - \delta^s_{\Delta^-} \leq (|\Delta^+| - |\Delta^-|) \times \rho_s$, delete s from $candq$
15 scan Δ^+ to count $\delta^s_{\Delta^+}$ for each s in $candp$ and $candq$
16 $\forall s \in candq$, if $\delta^s_{\Delta^+} - \delta^s_{\Delta^-} \leq (|\Delta^+| - |\Delta^-|) \times \rho_s$, delete s from $candq$
17 scan D^- to count $\delta^s_{D^-}$ for each s in $candq$
18 $L_i = \emptyset$
19 $\forall s \in candp$, if $\delta^s_D + \delta^s_{\Delta^+} - \delta^s_{\Delta^-} \geq |D'| \times \rho_s$, insert s into L_i
20 $\forall s \in candq$, if $\delta^s_{D^-} + \delta^s_{\Delta^+} \geq |D'| \times \rho_s$, insert s into L_i
21 if ($|L_i| > i$), $C_{i+1} :=$ GGen(L_i)
22 $i := i + 1$
23 Return $L_1 \cup L_2 \cup \ldots \cup L_{i-1}$

Fig. 4. Algorithm GSP+

GSP+ shares the same structure with GSP. GSP+ is an iterative algorithm. During each iteration, GSP+ also uses GGen to generate C_i (set of candidate sequences of length-i) based on L_{i-1}. Before the database is scanned to count the support of the candidate sequences, the pruning tests derived from Lemmas 1 and 2 are applied. Depending on the test results, the datasets Δ^- and/or Δ^+ may have to be processed to count the support of a candidate sequence. GSP+ carefully controls when such countings are necessary. If all pruning tests fail on a candidate sequence s, GSP+ checks whether s is frequent in D. If so, δ^s_D is available. Hence, $\delta^s_{D'}$ can be computed by $\delta^s_D + \delta^s_{\Delta^+} - \delta^s_{\Delta^-}$. Finally, if s is not frequent in D, the unchanged part of the database, D^-, is scanned to find out the actual support of s. Since D^- is typically much larger than Δ^+ and Δ^-, saving is achieved by avoiding processing D^- for certain candidate sequences. As we will see later in Section 5, the pruning tests can prune up to 60% of the candidate sequences in our experiment setting. The tests are thus quite effective.

4.2 MFS+

The pruning tests can also be applied to MFS. We call the resulting algorithm MFS+ (see Figure 5). The interesting thing about MFS+ is that it uses the set of frequent sequences (L_{old}) of the old database D as an initial estimate of the set of frequent sequences of the new database D'. These sequences together with all possible 1-sequences are put into a candidate set *CandidateSet*. It then scans Δ^-, D^-, and Δ^+ to obtain $\delta^s_{\Delta^-}$, $\delta^s_{D^-}$, and $\delta^s_{\Delta^+}$ for each sequence s in *CandidateSet*. From these counts, we can deduce which sequences in *CandidateSet* are frequent in D'. The *maximals* of such frequent sequences are put into the set *MFSS*. We can consider the set *MFSS* as the set of maximal frequent sequences that `MFS+` knows given the information that `MFS+` has obtained so far. `MFS+` then executes a loop, trying to refine *MFSS*. During each iteration, `MFS+` uses `MGen` to generate a set of candidate sequences *CandidateSet* from *MFSS*. `MFS+` then deduces which candidate sequences must not be frequent by applying the pruning tests. In the process, the datasets Δ^- and Δ^+ may have to be scanned. For those sequences that are not pruned by the tests, D^- is scanned to obtain their exact support counts. Since sequences that are originally frequent in the old database D have already had their supports (w.r.t D') counted in the initial part of `MFS+`, all candidate sequences considered by `MFS+` in the loop section are infrequent in D. Hence, only those pruning tests resulting from Lemma 2 are used. `MFS+` terminates when no refinement is made to *MFSS* during an iteration.

1 Algorithm `MFS+`(Δ^-, D^-, Δ^+, ρ_s, I, L_{old})
2   $MFSS := \emptyset$
3   $CandidateSet := \{\langle\{i\}\rangle | i \in I\} \cup \{s | s \vdash L_{old}, |s| > 1\}$
4   Scan Δ^-, D^-, Δ^+ to get $\delta^s_{\Delta^-}$, $\delta^s_{D^-}$, $\delta^s_{\Delta^+}$ for each $s \in CandidateSet$
5   $NewFrequentSequences := \{s | s \in CandidateSet, \delta^s_{D^-} + \delta^s_{\Delta^+} \geq |D'| \times \rho_s\}$
6   $AlreadyCounted := \{s | s \vdash L_{old}, |s| > 1\}$
7   $Iteration := 2$
8   while ($NewFrequentSequences \neq \emptyset$)
9   //$Max(S)$ returns the set of all maximal sequences is S
10     $MFSS := Max(MFSS \cup NewFrequentSequences)$
11     $CandidateSet :=$ `MGen`($MFSS$, $Iteration$, $AlreadyCounted$)
12   calculate $b^s_{\Delta^+}$ for each $s \in CandidateSet$
13     $\forall s \in CandidateSet$, if $b^s_{\Delta^+} \leq (|\Delta^+| - |\Delta^-|) \times \rho_s$, delete s from *CandidateSet*
14   scan Δ^- to count $\delta^s_{\Delta^-}$ for each $s \in CandidateSet$
15     $\forall s \in CandidateSet$,
      if $b^s_{\Delta^+} - \delta^s_{\Delta^-} \leq (|\Delta^+| - |\Delta^-|) \times \rho_s$, delete s from *CandidateSet*
16   scan Δ^+ to count $\delta^s_{\Delta^+}$ for each $s \in CandidateSet$
17     $\forall s \in CandidateSet$,
      if $\delta^s_{\Delta^+} - \delta^s_{\Delta^-} \leq (|\Delta^+| - |\Delta^-|) \times \rho_s$, delete s from *CandidateSet*
18   scan D^- to count $\delta^s_{D^-}$ for each $s \in CandidateSet$
19     $NewFrequentSequences := \{s | s \in CandidateSet, \delta^s_{D^-} + \delta^s_{\Delta^+} \geq |D'| \times \rho_s\}$
20     $Iteration := Iteration + 1$
21   Return *MFSS*

Fig. 5. Algorithm `MFS+`

5 Results

We performed a number of experiments comparing the performance of `GSP+` and `MFS+` with `GSP` and `MFS`. In this section we present some representative results from our experiments. The test databases are synthetic data generated by a generator provided in the IBM Quest data mining project. Readers are referred to [6] for the details of the data generator. The experiments were performed on a 700MHz PIII Xeon machine with 4GB main memory running Solaris 8.

5.1 Performance

In our first experiment, we used a database D of 1,500,000 sequences. We first executed a sequence mining program on D to obtain all frequent sequences and their support counts. Then, 10% (150,000) of the sequences in D are deleted. These sequences form the dataset Δ^-. Another 10% (150,000) sequences, which form the set Δ^+, are added into D to form the updated database D'. After that, four algorithms GSP, MFS, GSP+, and MFS+ were executed to mine D'. We did the experiment using different values of support thresholds ($0.35\% \leq \rho_s \leq 0.65\%$).

We compared the CPU costs and I/O costs of the four algorithms. I/O cost is measured in terms of database scans normalized by the size of D'. For example, if GSP scans D' 8 times, then the I/O cost of GSP is 8. For an incremental algorithm, if it reads Δ^- n_1 times, D^- n_2 times, and Δ^+ n_3 times, then the I/O cost is $(n_1|\Delta^-| + n_2|D^-| + n_3|\Delta^+|)/|D'|$. We notice that while the I/O costs of GSP and MFS are integral numbers (because D' is the only dataset they read), those of GSP+ and MFS+ could be fractional. Figure 6 shows the experiment results.

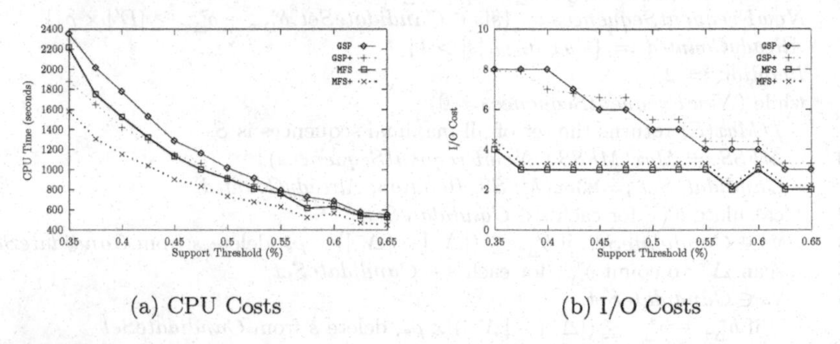

(a) CPU Costs (b) I/O Costs

Fig. 6. Comparison of four algorithms under different support thresholds

Figure 6(a) shows that as ρ_s increases, the CPU costs of all four algorithms decrease. This is because a larger ρ_s means fewer frequent sequences, and thus fewer candidate sequences whose supports need to be counted.

Among the four algorithms, GSP has the highest CPU cost. Using the set of frequent sequences in D as an initial estimate allows MFS to disover the long frequent sequences early. As we will see shortly, this results in fewer iterations for MFS. During each iteration, each sequence s in the database is matched against a set of candidate sequences to see which candidates are contained in s, and to increase their support counts. Since MFS performs fewer database scans compared with GSP, the database sequences are matched against candidate sequences fewer times. The overall effect is a lower CPU cost for MFS compared with GSP.

From the figure, we also see that GSP+ and MFS+ need less CPU time than their counterparts — GSP and MFS. The saving is obtained by the pruning effect of the incremental algorithms. Some candidate sequences are pruned by processing only Δ^- and/or Δ^+; the set D^- is avoided. In our experiment setting, the size of D^- is 9 times that of Δ^+ and Δ^-. Avoid counting the supports of the pruned candidate sequences in D^- results in a significant CPU cost reduction.

Table 1 shows the effectiveness of the pruning tests of GSP+. The total number of candidate sequences processed by pruning tests is shown in the second row, and the number of them requiring the scanning of D^- is shown in the third row. We can see that GSP+ needs to process D^- to obtain the support counts of only about 40% of all candidate sequences. This accounts for the saving in CPU cost achieved by GSP+ over GSP. Similarly, MFS+ outperforms MFS, mainly due to candidate pruning.

ρ_s	0.35%	0.4%	0.45%	0.5%	0.55%	0.6%	0.65%
Total # of candidates	34,065	18,356	10,024	5,812	3,365	2,053	1,160
Those requiring $\delta^s_{D^-}$	13,042	7,161	3,966	2,313	1,353	867	509
Percentage (row3/row2)	38%	39%	40%	40%	40%	42%	44%

Table 1. Effectiveness of prunning tests

Finally, when ρ_s becomes large, the CPU times of the algorithms are roughly the same. This is because a large support threshold implies short and few frequent sequences. In such a case, GSP and MFS take similar number of iterations, and hence the CPU saving of MFS over GSP is diminished. Moreover, there are much fewer candidate sequences when ρ_s is large, and there are much fewer opportunities for the incremental algorithms to prune candidate sequences.

Figure 6(b) shows the I/O costs of the four algorithms. We see that as ρ_s increases, the I/O costs of the algorithms decrease. This is due to the fact that a larger ρ_s leads to shorter frequent sequences. Hence, the number of iterations (and database scans) needed is small.

Comparing GSP and MFS, we see that MFS is a very I/O-efficient algorithm. Because MFS uses the frequent sequences in D as an estimate of those in D', which gives MFS a head start and allows MFS to discover all maximal frequent sequences in much fewer database scans.

The incremental algorithms generally require a slightly higher I/O cost than their non-incremental counterparts. The reason is that the incremental algorithms scan and process Δ^- to make pruning deductions, which is not needed by GSP and MFS. However, in some cases, the pruning tests remove all candidate sequences during an iteration of the algorithm. In such cases, incremental algorithms save some database passes. For example, when $\rho_s < 0.45\%$, GSP+ has a smaller I/O cost than GSP (see Figure 6(b)).

From Figure 6 we can conclude that GSP has a high CPU cost and a high I/O cost. GSP+ reduces the CPU requirement but does not help in terms of I/O. MFS is very I/O-efficient and it also performs better than GSP in terms of CPU cost. MFS+ is the overall winner. It requires the least amount of CPU time and its I/O cost is comparable to that of MFS.

5.2 Varying $|\Delta^+|$ and $|\Delta^-|$

As we have discussed, GSP+ and MFS+ achieve efficiency by processing Δ^+ and Δ^- to prune candidate sequences. The performance of GSP+ and MFS+ is thus dependent on the sizes of Δ^+ and Δ^-. we ran an experiment to study how $|\Delta^+|$ and $|\Delta^-|$ affect the performance of GSP+ and MFS+. In the experiment, we set $|D| = |D'| = 1,500,000$ sequences, $\rho_s = 0.5\%$, and varied $|\Delta^+|$ and $|\Delta^-|$ from 15,000 to 600,000 (1% – 40% of $|D|$). Figure 7 shows the experiment results.

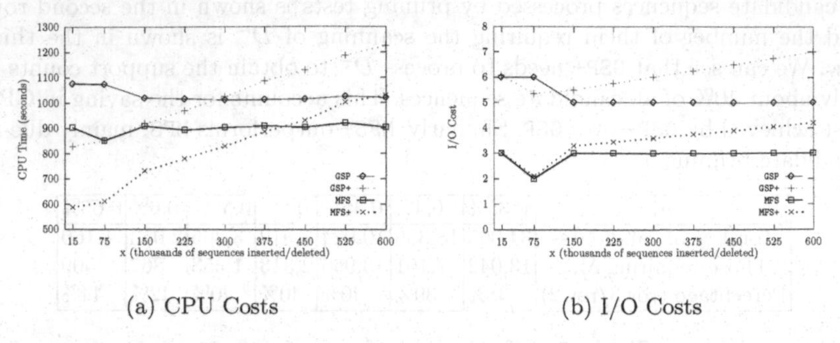

(a) CPU Costs (b) I/O Costs

Fig. 7. Comparison of four algorithms under different insertion and deletion sizes

Figure 7(a) shows that the CPU costs of GSP and MFS stay relatively steady. This is expected since $|D'|$ does not change. The CPU costs of GSP+ and MFS+, on the other hand, increase linearly with the size of Δ^+ and Δ^-. This increase is due to a longer processing time taken by GSP+ and MFS+ to deal with Δ^+ and Δ^-. From the figure, we see that MFS+ outperforms others even when the database is changed substantially. In particular, if $|\Delta^+|$ and $|\Delta^-|$ are less than 375,000 sequences (or 25% of $|D|$), MFS+ is the most CPU-efficient algorithm in our experiment. For cases in which only a small fraction of the database is changed, the incremental algorithms can achieve significant performance gains.

As we have discussed previously, GSP+ and MFS+ usually have slightly higher I/O costs than their non-incremental counterparts, since they have to read Δ^-. Therefore the gap between GSP and GSP+ (and also that between MFS and MFS+) widens as $|\Delta^-|$ increases (see Figure 7(b)).

6 Conclusions

This paper studies the maintenance problem of frequent sequences. We proposed two incremental algorithms, GSP+ and MFS+. We analyzed their performance in terms of CPU efficiency and I/O efficiency, and compared them with GSP and MFS. Our experiment results show that the incremental algorithms outperform their non-incremental counterparts in CPU cost at the expense of a small penalty in I/O cost. The performance gain is obtained by pruning candidate sequences to avoid processing the bulk of the database. The gain is thus most prominent when the database contains long and numerous frequent sequences, since in such cases there are many candidate sequences to consider and the impact of the pruning technique is most significant.

References

[1] Rakesh Agrawal and Ramakrishnan Srikant. Mining sequential patterns. In *Proc. of the 11th Int'l Conference on Data Engineering*, Taipei, Taiwan, March 1995.

[2] Necip Fazil Ayan, Abdullah Uz Tansel, and Erol Arkun. An efficient algorithm to update large itemsets with early pruning. In *Proc. 5th ACM SIGKDD International Conference on Knowledge Discovery and Data Mining*, San Diego, CA USA, August 1999.

[3] D. W. Cheung, J. Han, V. Ng, and C. Y. Wong. Maintenance of discovered association rules in large databases: An incremental updating techniques. In *Proc. 12th IEEE International Conference on Data Engineering (ICDE)*, New Orleans, Louisiana, U.S.A, March 1996.

[4] D. W. Cheung, S. D. Lee, and B. Kao. A general incremental technique for maintaining discovered association rules. In *Proc. International Conference On Database Systems For Advanced Applications (DASFAA)*, Melbourne, Australia, April 1997.

[5] D. W. Cheung, V. Ng, and B. W. Tam. Maintenance of discovered knowledge: A case in multi-level association rules. In *Proc. Second International Conference on Knowledge Discovery and Data Mining (KDD)*, Portland, Oregon, August 1996.

[6] http://www.almaden.ibm.com/cs/quest/.

[7] S. D. Lee and D. W. Cheung. Maintenance of discovered association rules: When to update? In *Proc. 1997 ACM-SIGMOD Workshop on Data Mining and Knowledge Discovery (DMKD)*, Tucson, Arizona, May 1997.

[8] E. Omiecinski and A. Savasere. Efficient mining of association rules in large dynamic databases. In *Proc. BNCOD'98*, 1998.

[9] S. Parthasarathy, M. J. Zaki, M. Ogihara, and S. Dwarkadas. Incremental and interactive sequence mining. In *Proceedings of the 1999 ACM 8th International Conference on Information and Knowledge Management (CIKM'99)*, Kansas City, MO USA, November 1999.

[10] Jian Pei, Jiawei Han, Behzad Mortazavi-Asl, Helen Pinto, Qiming Chen, Umeshwar Dayal, and Mei-Chun Hsu. Prefixspan: Mining sequential patterns by prefix-projected growth. In *Proc. 17th IEEE International Conference on Data Engineering (ICDE)*, Heidelberg, Germany, April 2001.

[11] T. Imielinski R. Agrawal and A. Swami. Mining association rules between sets of items in large databases. In *Proc. ACM SIGMOD International Conference on Management of Data*, page 207, Washington, D.C., May 1993.

[12] N. L. Sarda and N. V. Srinivas. An adaptive algorithm for incremental mining of association rules. In *Proc. DEXA Workshop'98*, 1998.

[13] Ramakrishnan Srikant and Rakesh Agrawal. Mining sequential patterns: Generalizations and performance improvements. In *Proc. of the 5th Conference on Extending Database Technology (EDBT)*, Avignion, France, March 1996.

[14] S. Thomas, S. Bodagala, K. Alsabti, and S. Ranka. An efficient algorithm for the incremental updation of association rules in large databases. In *Proc. KDD'97*, 1997.

[15] Mohammed J. Zaki. Efficient enumeration of frequent sequences. In *Proceedings of the 1998 ACM 7th International Conference on Information and Knowledge Management(CIKM'98)*, Washington, United States, November 1998.

[16] Minghua Zhang, Ben Kao, C.L. Yip, and David Cheung. A GSP-based efficient algorithm for mining frequent sequences. In *Proc. of IC-AI'2001*, Las Vegas, Nevada, USA, June 2001.

DELISP: Efficient Discovery of Generalized Sequential Patterns by Delimited Pattern-Growth Technology

Ming-Yen Lin, Suh-Yin Lee*, and Sheng-Shun Wang

Department of Computer Science and Information Engineering
National Chiao Tung University, Taiwan 30050, R.O.C.
{mylin, sylee, sswang}@csie.nctu.edu.tw

Abstract. An active research in data mining is the discovery of sequential patterns, which finds all frequent sub-sequences in a sequence database. Most of the studies specify no time constraints such as maximum/minimum gaps between adjacent elements of a pattern in the mining so that the resultant patterns may be uninteresting. In addition, a data sequence containing a pattern is rigidly defined as only when each element of the pattern is contained in a distinct element of the sequence. This limitation might lose useful patterns for some applications because sometimes items of an element might be spread across adjoining elements within a specified time period or time window. Therefore, we propose a pattern-growth approach for mining the generalized sequential patterns. Our approach features in reducing the size of sub-databases by bounded and windowed projection techniques. Bounded projections keep only time-gap valid sub-sequences and windowed projections save non-redundant sub-sequences satisfying the sliding time window constraint. Furthermore, the delimited growth technique directly generates constraint-satisfactory patterns and speeds up the growing process. The empirical evaluations show that the proposed approach has good linear scalability and outperforms the well-known *GSP* algorithm in the discovery of generalized sequential patterns.

1 Introduction

Sequential pattern mining is one of the active researches in data mining [1, 2, 3, 7, 8, 12]. The focus has been on improving the efficiency of the mining process owing to the huge number of potential patterns [1, 4, 14]. Typically, a user specifies a minimum support threshold to locate all the frequent sub-sequences, called *sequential patterns*, from a set of data sequences. Nevertheless, the time constraints between elements of a sequential pattern are not specified so that some uninteresting patterns may appear. For example, without specifying the maximum time gap, one may find a pattern <(*b*, *d*, *e*)(*a*, *f*)>, which means an item-set having *a* and *f* will occur after the occurrence of an item-set having *b*, *d*, and *e*. However, the pattern could be insignificant if the time interval between the two item-sets is too long such as over months. Therefore, time constraints including *maximum gap* and *minimum gap* should be incorporated in the mining to reinforce the accuracy and significance of mining results.

Moreover, common definition of an element of a sequential pattern is too rigid for some applications. Essentially, a data sequence is defined to support a pattern if each

M.-S. Chen, P.S. Yu, and B. Liu (Eds.): PAKDD 2002, LNAI 2336, pp. 198-209, 2002.

element of the pattern is contained in an individual element of the sequence, in the occurrence order of elements. However, a user may not care whether the items in an element (of the pattern) are from a single element or adjoining elements of a data sequence if the adjoining elements appear within a specified time interval. The specified interval is named *sliding (time) window* [11]. For instance, given a sliding window of *7*, a data sequence $<_{t1}(d, e)_{t2}(b)_{t3}(a, f)>$ can support the pattern $<(b, d, e)(a, f)>$ if the difference between time *t1* and time *t2* is no greater than *7*. Relaxing the definition of an element with sliding window will broaden the applications of sequential patterns.

Many algorithms were proposed for sequence mining without time constraints [1, 3, 6, 7, 12]. The Apriori-like algorithms [1, 5, 10] iteratively generate candidate patterns and determine true ones in multiple database scans, employing the Apriori heuristic (*all sub-patterns of a frequent pattern must be frequent*) in candidate pruning. The *SPADE* algorithm [14] quickly searches in a lattice constructed from the database of vertical-layout for the sequential patterns. The *FreeSpan* [4] projects sequences into sub-databases to confine the search and growth of sub-sequences. The *PrefixSpan* [9], representing a new methodology for the discovery of sequential patterns, explores prefix-projection and efficiently mines the complete set of sequential patterns. Nevertheless, these algorithms discover only patterns without the time constraints.

The *GSP* (**G**eneralized **S**equential **P**attern) algorithm, also an Apriori-like algorithm, was introduced to solve the problem of generalized sequential patterns [11]. In every database scan, each data sequence is transformed into time-lists of items for fast finding of certain element with a time tag. Since the start-time and end-time of a transaction (may comprise several elements) must be considered, *GSP* defined 'contiguous sub-sequence' for candidate generation, and moved between 'forward phase' and 'backward phase' for checking whether a data sequence contains a certain candidate [11].

The *cSPADE* [13] algorithm extended the *SPADE* [14] algorithm and modified the 'idlists' in *SPADE* to incorporate *min_gap*, *max_gap*, *time window*, and some other constraints. However, the time window refers to the window of occurrence of the whole sequence, rather than the sliding time window described here.

In this paper, we extend the pattern-growth technique [4] and propose a new algorithm, the *DELISP* (**Deli**mited **S**equential **P**attern) mining algorithm, for the discovery of sequential patterns generalized with constraints of maximum/minimum gap and sliding window. The *DELISP* grows sequential patterns in each sub-database generated by projecting each unfinished sub-sequence according to certain frequent prefixed elements. In *DELISP*, the *bounded projection* technique eliminates invalid sub-sequence projections caused by unqualified maximum/minimum gaps. The *windowed projection* technique reduces redundant projections for adjacent elements satisfying the sliding window constraint. The *delimited growth* technique grows only patterns satisfying constraints. The conducted experiments show that the *DELISP* outperforms the *GSP* algorithm. The scale-up experiments also confirm that the *DELISP* has good linear scalability with the number of data sequences.

The remaining paper is organized as follows. We formulate the problem and show an example in Section 2. Section 3 presents the *DELISP* algorithm. The experimental evaluation is described in Section 4. We discuss the performance factors of *DELISP* in Section 5. Section 6 concludes our study.

2 Problem Statement

Let $\Psi = \{\alpha_1, \alpha_2, \ldots, \alpha_z\}$ be a set of literals, called *items*. An *itemset* $I = (\beta_1, \beta_2, \ldots, \beta_m)$ is a set of m items, such that $I \subseteq \Psi$. A *sequence* x, denoted by $<a_1a_2\ldots a_n>$, is an ordered list of n *elements* where each *element* a_j is an itemset. Without loss of generality, we assume the items in an element are in lexicographic order. A sequence $\omega = <b_1b_2\ldots b_w>$ is a *subsequence* of another sequence $\varpi = <a_1a_2\ldots a_n>$ if there exist $1 \leq i_1 < i_2 < \ldots < i_w \leq n$ such that $b_1 \subseteq a_{i_1}$, $b_2 \subseteq a_{i_2}$, …, and $b_w \subseteq a_{i_w}$.

A user specifies four parameters, *min_gap* (minimum gap), *max_gap* (maximum gap), *t_win* (sliding time window), and *minsup* (minimum support), for mining generalized sequential patterns in the database *DB* of data sequences. A *data sequence* is represented by $sid/<_{t_1}a_1\ {}_{t_2}a_2\ldots {}_{t_n}a_n>$, where element a_j occurred at time t_j, $t_1 < t_2 <\ldots< t_n$, and *sid* is a unique identifier. A data sequence $ds=sid/<_{t_1}a_1\ {}_{t_2}a_2\ldots {}_{t_n}a_n>$ *contains* a sequence $x = <b_1b_2\ldots b_w>$ if there exist integers $1 \leq l_1 \leq u_1 < l_2 \leq u_2 < \ldots < l_w \leq u_w \leq n$ such that the four conditions hold: (1) $b_i \subseteq a_{l_i} \cup \ldots \cup a_{u_i}$, $1 \leq i \leq w$ (2) $t_{u_i} - t_{l_i} \leq t\_win$, $1 \leq i \leq w$ (3) $t_{u_i} - t_{l_{i-1}} \leq max\_gap$, $2 \leq i \leq w$ (4) $t_{l_i} - t_{u_{i-1}} > min\_gap$, $2 \leq i \leq w$. Assume that t_j, *min_gap*, *max_gap*, and *t_win* are all positive integers and *min_gap* < *max_gap*.

The *minsup* is the user specified minimum support threshold. The *support* of a sequence x, denoted by $x.sup$, is the number of data sequences containing x divided by the total number of data sequences in *DB*. A sequence x is a *frequent sequence* if $x.sup \geq minsup$. The sequence x is also called a **(generalized)** ***sequential pattern***. Given the *minsup*, the three constraints *min_gap*, *max_gap*, *t_win*, and the database *DB*, the problem of generalized sequential pattern mining is to discover the set of all (generalized) sequential patterns. Note that the mining of sequential patterns without generalized constrains is a special case with *min_gap*=0, *max_gap*=∞, and *t_win*=0 here.

Example 1: Generalized sequential patterns in an example database

An example sequence database having 5 sequences is listed in the first column in Table 1. Take the first sequence for instance. It has two itemsets, one having a single item *c* occurring at time *1* and the other having items *b* and *f* occurring at time *35*. The sequential patterns with and without time constraints in the example database are illustrated as follows.

(i) Given the *minsup* = 40%. The <(a)>, occurring at *C3*, *C4*, and *C5*, is a sequential pattern. The <(a)(a)> (contained in *C3* and *C4*) and <(a)(d)> (contained in *C4* and *C5*) are both sequential patterns. The set of all sequential patterns are listed in the second column in Table 1.

(ii) Given the *minsup* = 40%, *min_gap* = 2, *max_gap* = 30, *t_win* = 2. The *C1* does not contain <(c)(b)> or <(c)(f)> now since the time gap (35-1) > *max_gap* (30). The *C2* does not contain <(b)(d)> any more since the time gap is not bigger than *min_gap*. However, <(a,d)> becomes a (generalized) sequential pattern because, with *t_win*=2, *C4* may contain <(a,d)> now. Likewise, *C5* may contain <(b,f)> due to *t_win*=2 so that <(b,f)> becomes a new pattern. The set of all generalized sequential patterns are listed in the third column in Table 1. We also underline patterns that turn out to be invalid and italicize the new sequential patterns, due to the specified constraints.

Table 1. Example sequence database *DB* and the sequential patterns

Sequence	Sequential patterns	Generalized sequential patterns
$C1/<_1(c)_{35}(b,f)>$ $C2/<_2(b)_4(d)>$ $C3/<_1(a,d)_5(c)_6(c)_8(b)_{35}(a,f)>$ $C4/<_2(a)_4(d)_{30}(f)_{33}(a)_{61}(f)>$ $C5/<_1(a,b,e)_4(e)_7(f)_8(d)_9(b)>$	***(minsup=40%)*** <(a)>, <(a)(a)>, <(a)(b)>, <(a)(d)>, <(a)(f)>, <(b)>, <(b)(d)>, <(b)(f)>, <(c)>, <(c)(b)>, <(c)(f)>, <(d)>, <(d)(a)>, <(d)(b)>, <(d)(f)>, <(f)>	***(minsup=40%, t_win=2, min_gap=2, max_gap=30)*** <(a)>,<(a)(b)>, <(*a,d*)>, <(a)(f)>, <(b)>, <(*b,d*)>, <(*b,f*)>, <(b)(f)>, <(c)>, <(d)>, <(f)>

3 DELISP: Delimited Sequential Pattern Mining

In this section, we propose a new pattern-growth method for mining generalized sequential patterns, called *DELISP*. Analogous to the pattern-growth approaches [4, 9], the main idea is finding the frequent items, and then 'growing' potential patterns in the sub-databases constructed by projecting sub-sequences corresponding to the frequent items. However, *DELISP* projects fewer but complete combinations by windowed and bounded projections, and grows potential patterns effectively by delimited growth. In Section 3.1, we demonstrate the method by mining the example database in Table 1. Section 3.2 describes the proposed algorithm.

3.1 Mining Generalized Sequential Patterns by *DELISP*: An Example

Definition 1 (Prefix) Given a sequence $\varpi = <a_1a_2...a_n>$, a sequence $\omega = <a_1'a_2'...a_m'>$ $(m \le n)$ is called a *prefix* of ϖ if and only if (1) $a_i' = a_i$, $1 \le i \le m-1$; (2) $a_m' \subseteq a_m$; and (3) for each item $\beta_j \in (a_m - a_m')$, $\beta_j > \beta_j'$, for all $\beta_j' \in a_m'$.
Definition 2 (Stem, type-1 growth, and type-2 growth) Given a sequential pattern ρ and a frequent item x in the database *DB*, x is called the *stem* of the sequential pattern ρ' if ρ' can be formed by (1) appending (x) as a new element to ρ or (2) extending the last element of ρ with x. The formation of ρ' is a *type-1 growth* if it is formed by appending (x), and a *type-2 growth* if it is formed by extending with x.
Example 2: Given *minsup*=40%, *min_gap*=2, *max_gap*=30, *t_win*=2, and the *DB* in Table 1. *DELISP* mines the patterns by the following steps.
Step 1. Find frequent items. By scanning *DB* once, we have <(a)>:3, <(b)>:4, <(c)>:2, <(d)>:4, <(e)>:1, and <(f)>:4. The notation *sequence_x:count* means that the support count of the *sequence_x* is *count*. Non-frequent item '*e*' is omitted from mining afterward.
Step 2. Project corresponding sub-sequences to sub-databases. Considering the generalized sequential patterns having $\rho = <(x)>$ as the prefix, each can be found in the sub-database, named ρ*-DB*, generated by projecting all data sequences having item x in *DB*. Assume that ${}_{t_\rho}a_\rho$ is the first element containing the stem of ρ in a data sequence *ds*. We omit non-frequent items and project each element ${}_{t_j}a_j$ of *ds* into ρ*-DB* sequentially, according to the following rules.

(R0) $t_j < t_\rho - t\_win$: the a_j is dropped.

(R1) (a) $t_\rho - t\_win \le t_j \le t_\rho$: the items in a_j ordered after ρ are projected. (b) $t_\rho < t_j \le t_\rho + t\_win$: project a_j.

(R2) $t_\rho + t\_win < t_j$: let the first element in this range be ${}_{t_k}a_k$. The a_j having $t_j > t_\rho + min\_gap$ is projected if $t_k \le t_\rho + max\_gap$.

Note that for each new data sequence, we mark t_ρ as the start-time (abbreviated as *st*) and the end-time (abbreviated as *et*) for use later. We tabulate the sub-databases <(a)>-*DB*, <(b)>-*DB*, <(c)>-*DB*, <(d)>-*DB*, and <(f)>-*DB* in part 1 of Table 2.

Table 2. The projected sub-sequences in the *ρ-DB* sub-databases

ρ-DB	**Projected sub-sequences**
part 1: sub-databases of *DB*	
<(a)>-*DB*	$1{:}1/C3/<_{1}(d)_{5}(c)_{6}(c)_{8}(b)_{35}(a,f)>$; $2{:}2/C4/<_{4}(d)_{30}(f)_{33}(a)_{61}(f)>$; $1{:}1/C5/<_{1}(b)_{7}(f)_{8}(d)_{9}(b)>$
<(b)>-*DB*	$35{:}35/C1/<_{35}(f)>$; $2{:}2/C2/<_{4}(d)>$; $8{:}8/C3/<_{6}(c)_{35}(a,f)>$; $1{:}1/C5/<_{7}(f)_{8}(d)_{9}(b)>$
<(c)>-*DB*	$5{:}5/C3/<_{6}(c)_{8}(b)_{35}(a,f)>$
<(d)>-*DB*	$1{:}1/C3/<_{5}(c)_{6}(c)_{8}(b)_{35}(a,f)>$; $4{:}4/C4/<_{30}(f)_{33}(a)_{61}(f)>$; $8{:}8/C5/<_{7}(f)>$
<(f)>-*DB*	$30{:}30/C4/<_{33}(a)_{61}(f)>$
part 2: sub-databases of <(a)>-*DB*	
<(a)(b)>-*DB*	$8{:}8/C3/<_{35}(f)>$
<(a,d)>-*DB*	$1{:}1/C3/<_{8}(b)_{35}(f)>$; $2{:}4/C4/<_{30}(f)_{61}(f)>$
<(a)(f)>-*DB*	None

Note: the notation '*st*:*et*' prior to a data sequence denotes the *start-time* and the *end-time* (lists) of the data sequence with respect to ρ projection.

Take <(a)>-*DB* for instance. For $C3/<_{1}(\mathbf{a},d)_{5}(c)_{6}(c)_{8}(b)_{35}(a,f)>$, we mark *st*:*et* as 1:1, and project the remaining item 'd' as the first element (rule/R1). The time of the next element is 5, hence we project all the remaining elements since the *min_gap*/*max_gap* constraints are satisfied (rule/R2). For $C4/<_{2}(\mathbf{a})_{4}(d)_{30}(f)_{33}(a)_{61}(f)>$, the *st*:*et* is marked as 2:2, ${}_{4}(d)$ is projected (rule/R1), and then all the remaining elements after time 2+min(2,2) are projected (rule/R2). Similar to the projection of *C3*, we have $1{:}1/C5/<_{1}(b)_{7}(f)_{8}(d)_{9}(b)>$ in <(a)>-*DB* from the projection of $C5/<_{1}(\mathbf{a},b)_{4}(e)_{7}(f)_{8}(d)_{9}(b)>$.

Note that while projecting *C3* to <(b)>-*DB*, *st*:*et*=8:8 and all the elements before time (8-2) are dropped (rule/R0). The projection completes after including ${}_{6}(c)$ by rule/R1 and ${}_{35}(a,f)$ by rule/R2. The rule/R1 also includes ${}_{7}(f)$ in the projection of *C5* in <(d)>-*DB* (*st*:*et*=8:8). However, the ${}_{9}(b)$ of *C5* is not projected into <(d)>-*DB* owing to b < d. In other words, the case that *C5* contains (b,d) with sliding-window is considered in <(b)>-*DB* instead of <(d)>-*DB*. Similarly, the ${}_{9}(b)$ of *C5* in *DB* need not be kept in <(f)>-*DB* (*st*:*et*=7:7) for b < f. Moreover, the ${}_{6}(c)$ in <(c)>-*DB* for the projection of *C3* (*st*:*et*=5:5) and the ${}_{8}(d)$ in <(f)>-*DB* (*st*:*et*=7:7) for the projection of *C5* are dropped due to the unqualified *min_gap* constraint (rule/R2).

Step 3. Mine each sub-database for the subsets of generalized sequential patterns. In each sub-database, we grow the patterns in each sequence according to the time constraints, and determine which pattern is a valid sequential pattern. Assume that we are growing patterns from a sequence with *st*:*et* mark. The type-1 growth appends a new element with the item within the *(et+min_gap, st+max_gap]* range, and the type-2 puts the item within the *[et-t_win, st+t_win]* range in the same element having the ρ prefix. We mine <(a)>*-DB* as follows.

We grow the type-1 patterns <(a)(c)> and <(a)(b)> from the sub-sequence of *C3* using items from (1+2, 1+30]. In addition to the <(a,f)> at time 35, we also have pattern <(a,d)> from [1-2, 1+2]. The sub-sequence of *C4* generates <(a)(f)> from range (2+2, 2+30], and <(a,d)> from [2-2,2+2]. For the case of *C5*, we have <(a)(f)>, <(a)(d)>, and <(a)(b)> from the range (1+2, 1+30] and <(a,b)> from [1-2,1+2]. Consequently, the generalized sequential patterns are <(a)(b)>:2, <(a,d)>:2, and <(a)(f)>:2 by mining <(a)>*-DB*.

Step 4. Find all patterns by applying step 2 and step 3 on the sub-databases recursively. Next, we can apply the rules in step 2 to project the sub-sequences in <(a)>*-DB* further into sub-databases <(a)(b)>*-DB*, <(a,d)>*-DB*, and <(a)(f)>*-DB*. In step 2, the occurrence-time of the first element containing the stem determines how elements are projected. Assume that we are projecting a sequence with *st*:*et* mark. The occurrence-time of the searching element must be after *et* for sub-databases of type-1 growth (e.g. <(a)(b)>*-DB*), and within *[et-t_win, st+t_win]* for the other (e.g. <(a,d)>*-DB*). With ${}_{t_\rho}a_\rho$ being the first element found, we mark t_ρ:t_ρ as the *st*:*et* of the new sequence in type-1 growth. However, we use t_ρ to update either *st* or *et* of the new sequence of the other type. The projected sub-databases of <(a)>*-DB* are shown in part 2 of Table 2. We proceed to the mining of these sub-databases.

First on <(a)(b)>*-DB*, we mine the generalized sequential patterns having prefix (a)(b) by applying step 3. Since <(a)(b)(f)> has insufficient support, the mining stops. We then recursively apply the above steps on <(a,d)>*-DB* and on <(a)(f)>*-DB*. Since no patterns are found so that the mining for patterns having prefix 'a' in the 'a' prefixed sub-databases completely stops. We then recursively apply the steps on <(b)>*-DB* for 'b' prefixed patterns, on <(c)>*-DB* for 'c' prefixed patterns, ..., and on <(f)>*-DB* for 'f' prefixed patterns.

By collecting the patterns found in the above process, *DELISP* efficiently discovers all the sequential patterns satisfying the generalized constraints.

3.2 The *DELISP* Algorithm

Intuitively, one may modify the *PrefixSpan* Algorithm [4] to handle the *min_gap* and *max_gap* constraints. The modification requires recording the time-difference for each growing pattern, and invalidating patterns with insufficient/excess time gaps. However, the extra effort for verifying patterns will slow the whole process. In addition, modifying *PrefixSpan* algorithm to incorporate the sliding window constraint is a demanding task since *t_win* allows combining bi-directional elements. Growing patterns in either way would destroy the correctness of the pattern-growth framework.

Analogous to *PrefixSpan* algorithm, *DELISP* decomposes the mining problem by recursively growing patterns, one item longer than the current patterns, in the projected sub-databases. However, the potential items used to grow are subjected to *min_gap* and *max_gap* constraints, called *de-limited growth*. Therefore, we apply type-1 growth with item within range [et+*min_gap*, st+*max_gap*], and apply type-2

Algorithm *DELISP*

Input: *DB* = a sequence database; *minsup* = minimum support; *min_gap* = minimum time gap; *max_gap* = maximum time gap; *t_win* = sliding time window.

Output: the set of all generalized sequential patterns.

1. Scan *DB* once, find the set of all frequent items.
2. For each frequent item *x*,
   (i) form a sequential pattern $\rho = <(x)>$ and output ρ.
   (ii) call *ProjectDB*(ρ, $<>$-*DB*) to construct ρ-projected database ρ-*DB*.
   (iii) call *Mine*(ρ-*DB*).

Subroutine *ProjectDB*(ρ', ρ-*DB*)

Parameters: ρ' = a sequential pattern; ρ-*DB* = the sequence database *DB* if $\rho = <>$, or the ρ-projected database.

Output: the ρ'-projected database ρ'-*DB*.

1. If $\rho = <>$ then execute (1a), otherwise execute (1b).
   (1a) For each sequence $ds = sid/<_{t_1}a_1\ _{t_2}a_2\ \ldots\ _{t_n}a_n>$ in *DB*, find the first element $_{t_\rho}a_\rho$ which contains *x*, where *x* is the stem of ρ'. Mark the *st*:*et* of a new sequence *ds'* as t_ρ:t_ρ.
   (1b) For each sequence $ds = st{:}et/sid/<_{t_1}a_1\ _{t_2}a_2\ \ldots\ _{t_n}a_n>$ in ρ-*DB*, find the first element $_{t_\rho}a_\rho$ which contains *x*, where *x* is the stem of ρ' and
   (i) $et+min\_gap < t_\rho \le st+max\_gap$ if ρ' is a type-1 growth pattern. Mark the *st*:*et* of a new sequence *ds'* as t_ρ:t_ρ.
   (ii) $et-t\_win \le t_\rho \le st+t\_win$ if ρ' is a type-2 growth pattern. Mark the *start-time*:*end-time* of a new sequence *ds'* as min(*st*, t_ρ):max(*et*, t_ρ).
2. **(Windowed-projection)** For each element a_j such that $t_\rho - t\_win \le t_j \le t_\rho$, insert $_{t_j}a_j'$ to *ds'*, where a_j' is the set of all frequent items α_i in a_j satisfying $\alpha_i > x$. For each a_j such that $t_\rho < t_j \le t_\rho + t\_win$, insert $_{t_j}a_j$ to *ds'*.
3. **(Bounded-projection)** Find the first a_k such that $t_k > t_\rho + \max(t\_win, min\_gap)$. Insert $_{t_k}a_k$, $_{t_{k+1}}a_{k+1}$, …, and $_{t_n}a_n$ to *ds'* if $t_k \le t_\rho + max\_gap$.
4. Insert *ds'* to ρ'-*DB* if *ds'* is not empty.

Subroutine *Mine*(ρ-*DB*)

Parameter: ρ-*DB* = the ρ-projected database.

1. **(Delimited-growth)** For each data sequence $ds = st{:}et/sid/<_{t_1}a_1\ _{t_2}a_2\ \ldots\ _{t_n}a_n>$ in ρ-*DB*,
   (i) for each item *x* in a_i such that $et+min\_gap < t_i \le st+max\_gap$, increase the support count of potential type-1 growth pattern ρ' having the stem *x*.
   (ii) for each item *x* in a_i such that $et-t\_win \le t_i \le st+t\_win$, increase the support count of potential type-2 growth pattern ρ' having the stem *x*.
2. Validate the supports of the potential patterns in 1. Find the set of frequent items *x* which forms a type-1 or a type-2 growth pattern.
3. For each frequent item *x*,
   (i) form a sequential pattern ρ' and output ρ'.
   (ii) call *ProjectDB*(ρ', ρ-*DB*) to construct ρ'-projected database ρ'-*DB*.
   (iii) call *Mine*(ρ'-*DB*).

Fig. 1. Algorithm *DELISP*

growth with item from range [et-*t_win*, st+*t_win*], where st and et denotes the time span for the current pattern. On projecting sub-databases, we avoid the bi-directional growth by imposing the item-order to type-2 growth. We always add a new item lexicographically ordered after all the items existing in the type-2 growth pattern. Considering an example element (b, d, e) formed by combining t_1(d, e) and t_2(b). When the time t_2 is earlier than time t_1, (b, d, e) will be discovered in the projected b-*DB* if $(t_1-t_2) \leq t\_win$. In case $t_1 < t_2$, we keep (d, e) in b-*DB* while projecting by allowing elements having time tag later than (t_2-t_win) be projected. We refer such projection to as *windowed-projection*. Fig. 1 presents the proposed *DELISP* algorithm.

4 Experimental Results

We conducted several experiments to access the performance of the *DELISP* algorithm. The experiments used an 866 MHz Pentium-III PC with 1024MB memory running the Windows NT. Like most studies on sequential pattern mining [1, 4, 11, 14], we generated the synthetic datasets for these experiments using the procedure described in [11]. Due to space limit, we only report the results on dataset *C10*–T*2.5*–S*4*–I*2.5* having 100,000 sequences and *N*=1000, N_S=5000, N_I=25000. We refer readers to [11] for the details of the parameters.

We compare the execution times of *GSP* with *DELISP* in mining generalized sequential patterns with various values of *minsup*, *min_gap*, *max_gap*, and *t_win*. In these experiments, *DELISP* is about 3 times faster than *GSP*. The result of varying *min_gap* is shown in Fig. 2 and Fig. 3 shows the result of varying *max_gap*. When we increase the *min_gap* or decrease the *max_gap* value, the number of generalized sequential patterns will decrease since both changes restrict some data sequences to contain certain patterns. Fig. 4 displays the effect on performance when constraint *t_win* is increased. More patterns are discovered, and more execution time is required, with the increased *t_win*. The results of varying *minsup* (2%, 1.5%, 1%, 0.75%, 0.5%) are consistent. Finally, Fig. 5 shows that the execution time of *DELISP* scales up linearly with the number of customer sequences.

5 Discussion

In this section, we summarize the factors contributing to the efficiency of the proposed *DELISP* algorithms, by comparison with the well-known *GSP* algorithm.

- **No candidate generation.** *DELISP* generates no candidates and saves the time not only in candidate generation but also in candidate testing. Moreover, the huge space required for candidate hash-tree is eliminated thoroughly. When we decreased the amount of installed memory (e.g. from 1024M down to 256M), the candidate *2*-sequences no longer fit into the memory at the same time. *GSP* needs to partition the candidates and scan the database more than once for support counting. *DELISP* grows patterns directly and frees from such limitation.

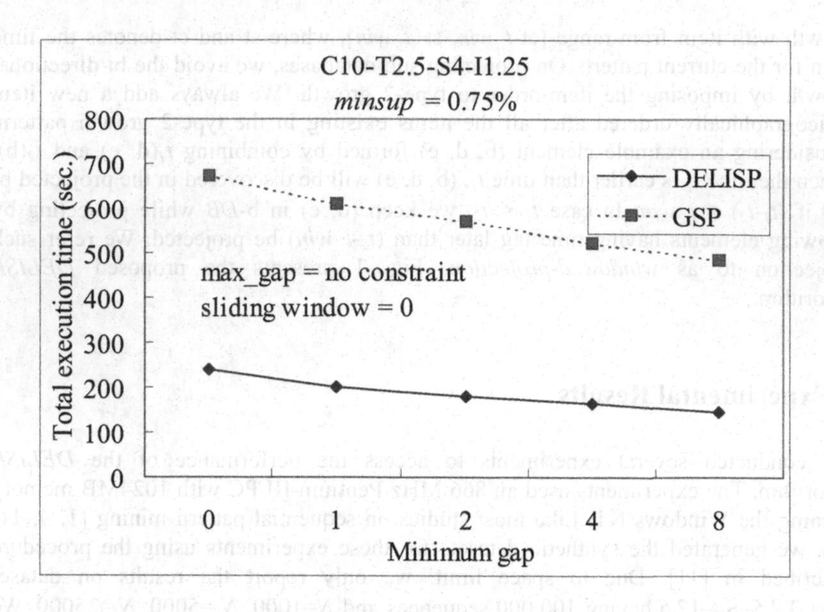

Fig.2. Effect of the *min_gap* constraint

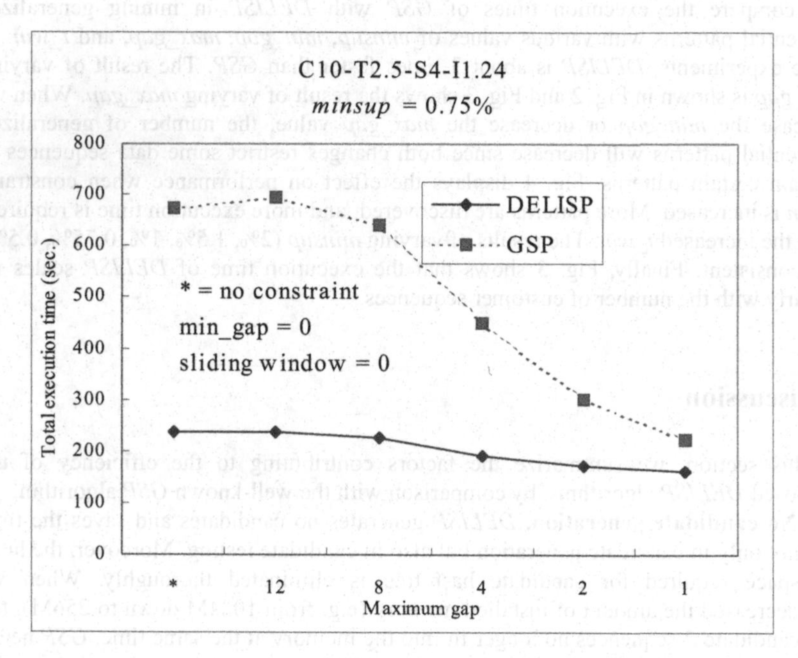

Fig. 3. Effect of the *max_gap* constraint

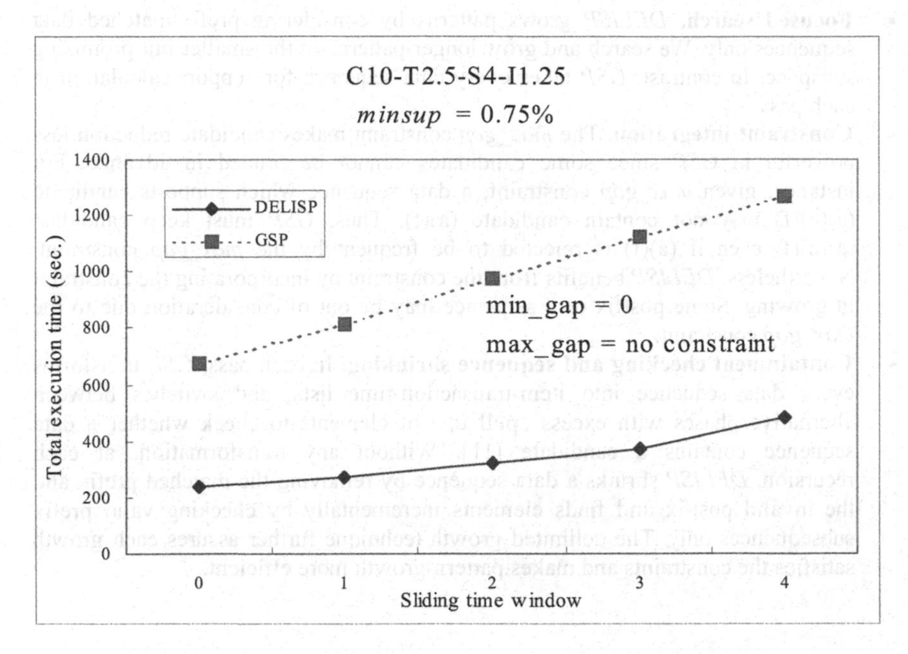

Fig. 4. Effect of the *t_win* constraint

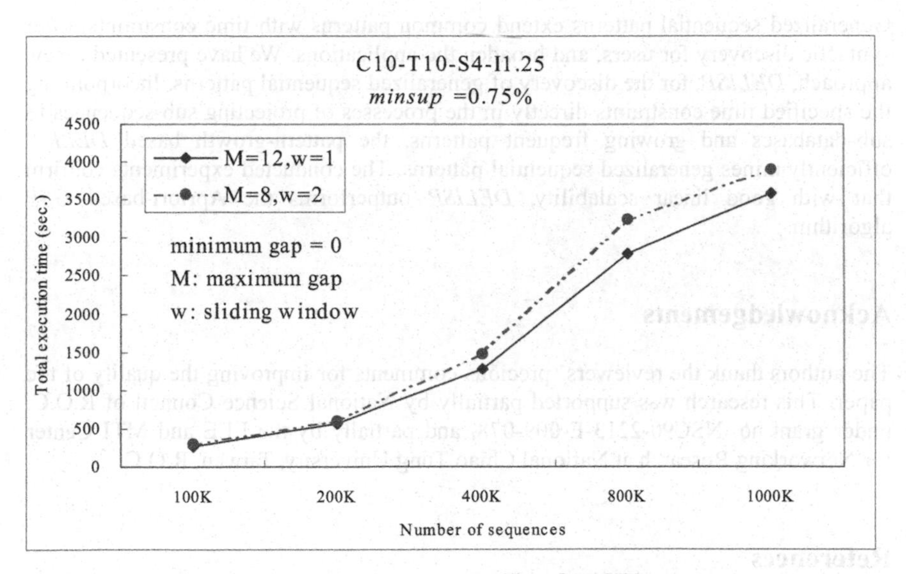

Fig. 5. Linear scalability of *DELISP*

- **Focused search.** *DELISP* grows patterns by considering prefix-matched data sequences only. We search and grow longer patterns in the smaller but promising subspace. In contrast, *GSP* takes every data sequence for support calculation in each pass.
- **Constraint integration.** The *max_gap* constraint makes candidate reduction less powerful in *GSP* since some candidates cannot be pruned in advance. For instance, given *max_gap* constraint, a data sequence which supports candidate (a)(e)(f) may not contain candidate (a)(f). Thus, *GSP* must keep candidate (a)(e)(f) even if (a)(f) is rejected to be frequent by the *max_gap* constraint. Nevertheless, *DELISP* benefits from the constraint by incorporating the constraint in growing. Some postfix of a sequence may be out of consideration due to the *max_gap* constraint.
- **Containment checking and sequence shrinking.** In each pass, *GSP* transforms every data sequence into item-transaction-time lists, and switches between alternative phases with excess „pull up“ of elements to check whether a data sequence contains a candidate [11]. Without any transformation, at each recursion, *DELISP* shrinks a data sequence by removing the matched prefix and the invalid postfix and finds elements incrementally by checking valid prefix subsequences only. The delimited growth technique further assures each growth satisfies the constraints and makes pattern-growth more efficient.

6 Conclusion

Generalized sequential patterns extend common patterns with time constraints, offer syntactic discovery for users, and broaden the applications. We have presented a new approach, *DELISP*, for the discovery of generalized sequential patterns. Incorporating the specified time constraints directly in the processes of projecting sub-sequences to sub-databases and growing frequent patterns, the pattern-growth based *DELISP* efficiently mines generalized sequential patterns. The conducted experiments confirm that with good linear scalability, *DELISP* outperforms the Apriori-based *GSP* algorithm.

Acknowledgements

The authors thank the reviewers' precious comments for improving the quality of the paper. This research was supported partially by National Science Council of R.O.C. under grant no. NSC90-2213-E-009-078, and partially by the LEE and MTI Center for Networking Research at National Chiao Tung University, Taiwan, R.O.C.

References

[1] R. Agrawal and R. Srikant, „Mining Sequential Patterns,“ *Proceedings of the 11th International Conference on Data Engineering*, Taipei, Taiwan, pp. 3-14, 1995.

[2] C. Bettini, X. S. Wang, and S. Jajodia, „Mining Temporal Relationships with Multiple Granularities in Time Sequences," *Data Engineering Bulletin*, Vol. 21, pp. 32-38, 1998.
[3] M. N. Garofalakis, R. Rastogi, and K. Shim, „SPIRIT: Sequential Pattern Mining with Regular Expression Constraints," *Proceedings of the 25th International Conference on Very Large Data Bases*, Edinburgh, Scotland, pp. 223-234, 1999.
[4] J. Han, J. Pei, B. Mortazavi-Asl, Q. Chen, U. Dayal and M.-C. Hsu, „FreeSpan: Frequent pattern-projected sequential pattern mining," *Proceedings of the 6th ACM SIGKDD international conference on Knowledge discovery and data mining*, pp. 355-359, 2000.
[5] M. Y. Lin and S. Y. Lee, „Incremental Update on Sequential Patterns in Large Databases," *Proceedings of 10th IEEE International Conference on Tools with Artificial Intelligence*, Taipei, Taiwan, pp. 24-31, 1998.
[6] H. Mannila, H. Toivonen and A. I. Verkamo, „Discovery of Frequent Episodes in Event Sequences," *Data Mining and Knowledge Discovery*, Vol. 1, Issue 3, pp. 259-289, 1997.
[7] F. Masseglia, F. Cathala, and P. Poncelet, „The PSP Approach for Mining Sequential Patterns," *Proceedings of 1998 2nd European Symposium on Principles of Data Mining and Knowledge Discovery*, Vol. 1510, Nantes, France, pp. 176-184, Sep. 1998.
[8] T. Oates, M. D. Schmill, D. Jensen, and P. R. Cohen, „A Family of Algorithms for Finding Temporal Structure in Data," *Proceedings of the 6th International Workshop on AI and Statistics*, Fort Lauderdale, Florida, pp. 371-378, 1997.
[9] J. Pei, J. Han, H. Pinto, Q. Chen, U. Dayal and M.-C. Hsu, „PrefixSpan: Mining sequential patterns efficiently by prefix-projected pattern growth," *Proceedings of 2001 International Conference on Data Engineering*, pp. 215-224, 2001.
[10] T. Shintani and M. Kitsuregawa, „Mining algorithms for sequential patterns in parallel: Hash based approach," *Proceedings of the Second Pacific–Asia Conference on Knowledge Discovery and Data mining*, pp. 283-294, 1998.
[11] R. Srikant and R. Agrawal, „Mining Sequential Patterns: Generalizations and Performance Improvements," *Proceedings* of the *5th International Conference on Extending Database Technology*, Avignon, France, pp. 3-17, 1996. (An extended version is the IBM Research Report RJ 9994)
[12] K. Wang, „Discovering patterns from large and dynamic sequential data," *Journal of Intelligent Information Systems*, Vol. 9, No. 1, pp. 33-56, 1997.
[13] M. J. Zaki, „Sequence Mining in Categorical Domains: Incorporating Constraints," *Proceedings of the 9th International Conference on Information and Knowledge Management*, Washington D.C., pp. 422-429, 2000.
[14] M. J. Zaki, „SPADE: An Efficient Algorithm for Mining Frequent Sequences," *Machine Learning Journal*, Vol. 42, No. 1/2, pp. 31-60, 2001.

Self-Similarity for Data Mining and Predictive Modeling A Case Study for Network Data

Jafar Adibi[1], Wei-Min Shen[1], Eaman Noorbakhsh[2]

[1]Information Sciences Institute, Computer Science Department,
University of Southern California
4676 Admiralty Way, Marina del Ray, CA 90292
{adibi,shen}@isi.edu

[2] Electrical Engineering Department, University of Southern California
University Campus Park, Los Angeles, CA 90089
noorbakh@usc.edu

Abstract. Recently there are a handful study and research on observing self-similarity and fractals in natural structures and scientific database such as traffic data from networks. However, there are few works on employing such information for predictive modeling, data mining and knowledge discovery. In this paper we study, analyze our experiments and observation of self-similar structure embedded in Network data for prediction through Self Similar Layered Hidden Markov Model (SSLHMM). SSLHMM is a novel alternative of Hidden Markov Models (HMM) which proven to be useful in a variety of real world applications. SSLHMM leverage HMM power and extend such capability to self-similar structures and exploit this property to reduce the complexity of predictive modeling process. We show that SSLHMM approach can captures self-similar information and provides more accurate and interpretable model comparing to conventional techniques.

1 Introduction

In recent years there have been interest in research and development for traffic modeling and forecasting the Network data. Recent measurements of local-area and wide-area traffic [16, 22, 23] have shown that network traffic exhibits variability at a wide range of time scales. For instance, analysis of traffic data from networks and services such as ISDN traffic, Ethernet LAN's, Common Channel Signaling Network (CCNS) and Variable Bit Rate (VBR) video have all convincingly demonstrated the presence of features such as self-similarity, long range dependence, slowly decaying variances, heavy-tailed distributions and fractal dimensions [23]. Through past decade, different models suggested for Network behavior analysis. Although several works attempt to analyze, model and predict Network data such as Markov Model, but there are only a few remarkable works, which have shown successful, result and the problem of modeling Network data still is an open issue [5, 9, 15, 19] . Recently there have been several attempts to apply data mining techniques through fractal dimensions and self-similarity. Among those, using fractal dimension, using fractal dimension for dimension reduction [21], learning association rules[3] and application in spatial joint [10] are considerable.

M.-S. Chen, P.S. Yu, and B. Liu (Eds.): PAKDD 2002, LNAI 2336, pp. 210-217, 2002.

Self Similar Layered Hidden Markov Model (SSLHMM) has been introduced by Adibi et al in [1] with application in Network data. In this paper we only study and analyze our experiments and observations of such data and the relation among SLLHMM structure, fractal dimension and self-similarity. We would like to show if we can use the fractal dimension and self-similarity of a given data for a better estimation of SSLHMM structure.

2 Related Work and Background

The convectional methods for analyzing the network data was Poisson arrivals, in which the number of arrivals in the time interval T follows the exponential distribution with parameter λT. This model works well in case of traditional telephone network, but not for the internet traffic data. The failure of the Poisson model is explained in [19] by Paxson and Floyd. The second method which used for traffic modeling was Autoregressive type traffic models. These models define the next variates in the sequence as an explicit function of previous variates from the same time series within a time window stretching from the present into the past. These models are Auto Regressive (AR), Moving Average model (MA), ARMA model and also ARIMA model. This method worked well in the early years of developing Internet, because the change in the amount of traffic was not abrupt in those days. But recent studies have indicated that this method also fails for highly volatile traffic[4].

Recent measurements of local-area and wide-area traffic have shown that network traffic exhibits variability at a wide range of time scales. What is striking is the ubiquitousness of the phenomenon which has been observed in diverse networking contexts, from Ethernet to ATM, compressed VBR video, and HTTP-based WWW traffic [5, 9, 16, 22-24]. A number of performance studies have shown that self-similarity can have a detrimental on network performance leading to increased queuing delay and packet loss rate which implied that they also exhibited long range dependency (LRD). Recent research suggests that not only packet traffic, but also the TCP session arrival pattern is self-similar. This means that if a switched virtual circuit network is to be substituted for a connectionless TCP/IP, the network will have to cope with periodical overloads of control units of its switches. Since then, this feature has been discovered in many other traffic traces, such as Transmission Control Protocol (TCP), Motion Pictures Experts Group (MPEG) video, World Wide Web, and Signaling System traffic[6, 9, 15, 22, 24]. The importance of this discovery becomes apparent when it is observed that Poisson; ARMA and Markov processes are unable to exhibit LRD. In fact they are short-range dependent (SRD) processes. The major flaw with the traditional traffic models is that they do not model the burstiness of the Internet traffic correctly. The burstiness exists in every time-scale while with traditional models it disappears in the long time intervals.

The mathematical study of self-similar shapes and their relationship to natural shapes was first presented by Benoit Mandelbrot. Self-similar stochastic processes were introduced by Kolmogorov in a theoretical context and brought to the attention of probabilists and statisticians by Mandelbrot and his co-workers and have been used in hydrology, geophysics, biophysics, and biology and communication systems [17]. Among different alternative to test the self-similarity of a sequence we used variance-

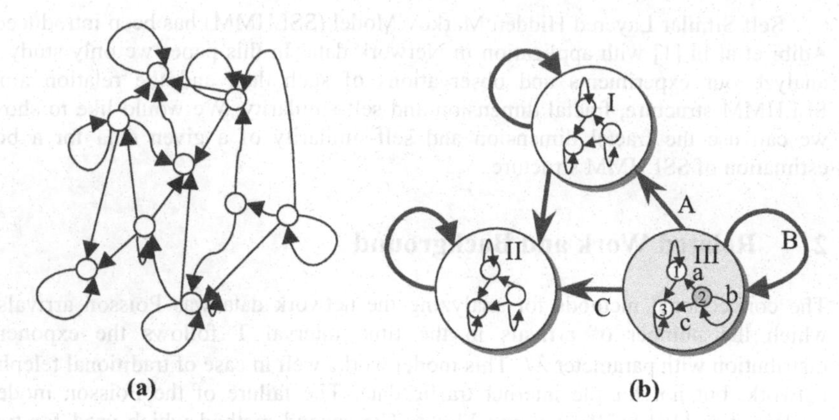

Fig. 1. **(a)** A normal Hidden Markov Model with 9 states, **(b)** Self-Similar Layered Hidden Markov Model with 9 states and 3 phases. As it shows each *phase* contains similar structure

time plot[16]. In this method the estimated variance of $X^{(m)}$ is plotted against m on log-log scale. A straight line with $(-\beta)$ slope indicates self-similarity. The discovery of self-similarity in computer networks led researcher to use of fractional Brownian motion (fBm) - a more general class of self-similar processes of which Brownian motion is a special case - and fractional ARIMA process for modeling the self-similarity in the network data traffic. In general the ARIMA trace is often obtained by generating a fractional Gaussian noise trace with suitable H and filtering this noise with the ARMA coefficients. Although the above-mentioned models have been employed by several groups but the capability of these models for predicting self-similar processes is not clear and prediction of Network data seems inevitable [15]. For example Bates and McLaughlin [5] showed some evidence, both qualitatively and quantitatively, to suggest that Ethernet data does not conform to popular self-similar models. The evidence suggests that Ethernet is more impulsive than the Gaussian case, which these models assume. There are also some other methods for self-similar data modeling which are less popular than the above models [16][21]. These models are M/G/∞ process, aggregated AR process; heavy tailed on-off process, stable self-similar process, fractal shot noise and renewal process and stochastic difference equations.

3 Self-Similar Layered Hidden Markov Model (SSLHMM)

Hidden Markov Models (HMM) have proven to be useful in a variety of real world applications where considerations for uncertainty are crucial. Such an advantage can be more leveraged if HMM can be scaled up to deal with complex problems. However, despite the broad range of application areas shown for HMM, they do have limitations and do not easily handle problems with certain characteristics. For example Markov Model has not reported as a successful alternative for analyze the Network behavior. To extend HMM we only focus on complexity of HMM for a certain category of problems with the following characteristics: 1) The uncertainty and complexity embedded in these application makes it difficult and impractical to construct the model in one step. 2) Systems are self-similar, contain self-similar structures and have been generated through recurrent processes. In the modeling of complex processes, when the number of states goes high, the maximization process gets more difficult. A solution provided in other literature is to use of a Layered HMM instead [2, 12]. Layered HMM has the capability to model more than one process. Hence, it provides an easier platform for modeling complex processes. Layered HMM is a combination of two or more HMM processes in a hierarchy. Fig. 1(b) shows a Layered HMM with 9 states and 3 super-states, or macro-states, which we refer to them as p*hases*. As we can see, each phase is a collection of states bounded to each other. The real model transition happens among the states. However, there is another transition process in upper layer among phases. The comprehensive transition model is a function of transition among states and transition among phases.

SSLHMM is a special form of Layered HMM in which there are some constraints on state layer transition, phase layer transition, and observation distribution. There are also some extensions to HMM such as [12], [11] which through limited space we only refer to them for further study. The advantage of SSLHMM is that like any other self-similar model it is possible to learn the whole model having any part of the model. Although there are a couple of assumptions to hold such properties but fortunately for a large group of systems in nature self-similarity is one of their characteristics

4 Experimental Result

We have applied SSLHMM approach to a real Network database collected during 18 weeks, from October 16 1994 to February 12 1995,on Cabletron corporate network. There are 16849 entries, representing measurements roughly every 10 minutes for 18 weeks. This network has a router with 16 ports connected to 16 links. The packet traffic of each port is investigated independently. There are 16 ports router that connect to 16 links, which in turn connect to 16 Ethernet subnets. Note that the traffic has to flow through the router ports in order to reach the 16 subnets. There are three independent variables:

- *Bandwidth*: the percentage of bandwidth utilization of a port during a 10 minute period.
- *Packet Rate:* the rate at which packets are moving through a port per minute.
- *Collision Rate:* the number of collided packets during a 10 minute period

Fig. 2 shows the variance-time plot for the data. Here we measured the Hurst parameter by fitting the least square line through the resulting points in the plane, ignoring the small values of m. Values of the estimate β of the asymptotic slope between –1 and 0 suggest self similarity is given by $H = 1 - \beta / 2$. Fig.2 shows the data is indeed self-similar.

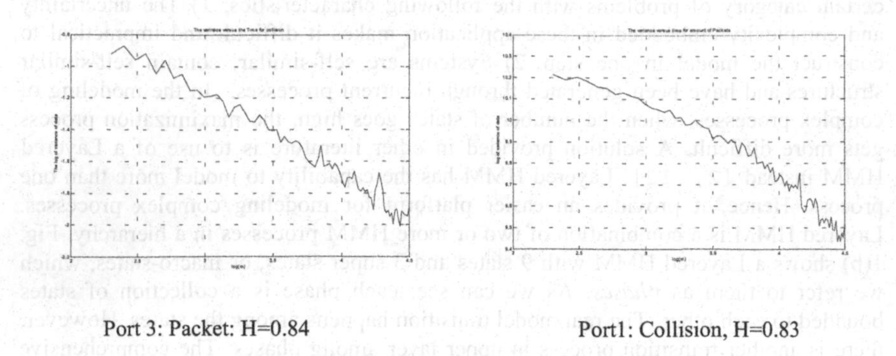

Fig. 2: variance-time plot for tow given ports (No. 1 and 3) illustrate slowly decaying variance

Table 1 shows the statistical information of *self-similarity* and *fractal dimension* for all 16 ports. We use of well-known methods in [13] to compute fractal dimension and similar technique in [16] for computing the self-similarity.

	Self-Similarity				Fractal Dimension			
	Min	Max	Mean	STDV	Min	Max	Mean	STDV
Bandwidth	0.66	0.83	0.78	.0755	1.63	1.73	1.56	.0603
Packet	0.66	0.85	0.77	.0528	1.68	1.80	1.50	.0435
Collision	0.62	0.83	0.70	.0640	1.78	1.92	1.65	.0630

Table 1 : Statistical Analysis of fractal dimension and self similarity for all 16 ports

There are some amazing observations, which we found out through our experiments. The average fractal dimension for all three variables: Bandwidth, Packet and Collision is from 1.5 to 1.65. This may indicates that all three variables might have similar characteristics. We generated a synthetic data similar to Network data. User had the capability to define the number of sequence in experimental pool, length of each sequence, number of layers, number of states in each phase, number of phases and observation set for discrete environment or a range for continuous observation. We observed that the fractal dimension for the synthetic data for a SSLHMM with N=4 and n = 4 (number of states in each phase) is close to 1.5-1.6 in different experiments. This may indicates that this model, N and n are the good estimates for a SSLHMM or at least a good initial point to start. The number of states in HMM is hidden and has been a recurrent issue in past decade. The brute forth way to find the best number of states may not be suitable in some cases in which a model desired to be obtained in real time. In addition, a better starting point in a blind search decrease the search time in general.

To compare SSLHMM with HMM and conventional techniques, we employed a SSLHMM model in which user has the capability to define the number of *states* and number of *phases*. In this paper we only report the comparison of *Baum-Welch forward algorithm* for HMM with n_{HMM} states and a 2-layer *strong* SSLHMM with *N* phases and *n* states. In addition we assume a one-to-one relationship among states and phases in two layers. In addition we ran a synthetic data generator and we found a SSLHMM with 4 phases and 4 states in each phase generate a sequence with fractal dimension about 0.6. We did use such number as a estimation for number of phases and states. The main purpose of this experiment is built on the following chain of principles:

- Assume there is a sequence of observation $O = \{o_1, o_2, \cdots, o_T\}$ as Network data.
- We would like to construct a model λ for such data.
- We would like to illustrate that for $O = \{o_1, o_2, \cdots, o_T\}$, $P(O \mid SSLHMM)$ is higher than $P(O \mid HMM)$, and other techniques (the probability of the observation given each model).

We applied the HMM and SSLHMM to a given port of database with the purpose of modeling the Network data. We did test our technique through cross validation and in each round we trained the data with a random half of the data and test over the rest. We repeat the procedure for Bandwidth , Packet Rate and Collision Rate. Table 2. illustrates the comparison of HMM and SSLHMM for Bandwidth, Packet Rate and Collision Rate on 3 randomly selected ports. Respectively, we ran HMM with prior number of states equal to 2, 3, 4, 9 and 16, and SSLHMM with number of phases equal to 2, 3 and 4 (shown as 2-s, 3-s and 4-s in the Table 2). As it shows in Table 2 the SSLHMM model with *N=4* outperforms other competitors in all series of experiments. *-log(likelihood)* increases by increasing the number of states more than 16 as it over fits the data. The best SSLHMM performance beats the best HMM by 23%, 41% and 38% for Collision Rate, Bandwidth and Packets Rate respectively. For ARMA, AR and RW techniques we used the error as the measure for likelihood with the assumption of a normal distribution for the trained data. Hence, the likelihood became the probability of error.

Technique	Bandwidth	Packet	Collision
HMM- 2	382.05	403.32	284.71
HMM- 3	206.86	228.68	183.74
HMM- 4	213.05	220.17	176.86
HMM- 9	169.23	176.86	176.61
HMM- 16	151.72	170.12	163.86
SSLHMM 2-S	355.77	381.93	294.08
SSLHMM - 3-S	97.49	166.78	130.12
SSLHMM - 4-S	105.19	102.90	100.20
AR	227.74	300.50	267.35
ARMA	230.74	297.78	239.25
Random Walk	232.99	285.51	197.92

Table 2: Negative log likelihood of different techniques for Network data

Our experiments show SSLHMM approach behave properly and does not perform worse than HMM even when the data is not self similar or when we do not have enough information. However due to space limitation we do not report the result in this paper. The SSLHMM provides a more satisfactory model of the network data from three point of views. First, the time complexity is such that it is possible to consider model with a large number of states in a hierarchy. Second, these larger number of states do not require excessively large numbers of parameters relative to the number of states. Learning a certain part of the whole structure is enough to extend to the rest of the structure. Finally SSLHMM resulted in significantly better predictors; the test set likelihood for the best SSLHMM was about 100 percent better than the best ARIMA, ARMA and Random Walk.

While the SSLHMM is clearly better predictor than HMM, it is easily interpretable than an HMM as well. The notion of phase may be considered as a collection of locally connected sets, groups, levels, categories, objects, states or behaviors and it comes with the idea of granularity, organization and hierarchy. As it mentioned before in Network application domain a phase could define as "congestion" or "stable". This characteristics is the main advantage of SSLHMM over other approaches such as FHMM [12]. SSLHMM is designed toward better interpretation as one the main goal of data mining approaches in general.

5 Acknowledgement

This work was partially supported by the National Science Foundation Grant: 9529615.

6 References

1. Adibi, J., Shen, W-M. *Self Similar Layered Hidden Markov Model*. in *5th European Conference on Principles and Practice of Knowledge Discovery in Databases (PKDD'01)*. (2001). Freiburg, Germany.
2. Adibi, J., Shen, W-M. *General structure of mining through layered phases*. in *International Conference on Data Engineering*. (2002). San Jose, California, USA: IEEE.
3. Barbara, D. *Chaotic Mining: Knowledge discovery using the fractal dimension*. in *ACM SIGMOD Workshop on Research Issues in Data Mining and Knowledge Discovery (DMKD)*. (1999). Philadelphia, USA,.
4. Basu, S., Mukherjee, A. and Klivansky, S. *Time series models for Internet Traffic*. in *IEEE Infocom Conference*. (1996). San Francisco.
5. Bates, S., and McLaughlin, S. *Testing the gaussian assumption for Self-Similar Teletraffic models*. in *IEEE Signal Processing*. (1997).
6. Beran, J., Sherman, R., Taqqu M. S., Willinger, W.,, *Long-range dependence in Variable-Bit-Rate Video Traffic*. IEEE Transactions on Communications, (1995). **43**.
7. Chang, J.W., Glass, J, *Segmentation and modeling in segment-based recognition*. (1997).
8. Cohen, J., *Segmentation speech using dynamic programming*. ASA, (1981). **69**(5): p. 1430-1438.
9. Erramilli, A., Pruthi, P. and Willinger, W. *Self-similarity in high-speed network traffic measurements: Fact or artifact?* in *12th Nordic Teletraffic Seminar NTS 12 (VTT Symposium 154)*. (1995).

10. Faloutsos, C., Seeger, B., Traina, A. and Traina Jr., C. *Spatial Join selectivity using power law*. in *SIGMOD*. (2000). Dallas, TX.
11. Fine, S., Singer Y, Tishby N, *The hierarchical Hidden Markov Model: Analysis and applications*. Machine Learning, (1998). **32**(1): p. 41-62.
12. Ghahramani, Z., Jordan, M., *Factorial Hidden Markov Models*. Machine Learning, (1997). **2**: p. 1-31.
13. Higuchi, T., *Approach to an irregular time series on the basis of the fractal theory*. Physica D, (1998). **31**.
14. Holmes, W.J., Russell, M. L., *Probabilistic-trajectory segmental HMMs*. Computer Speech and Languages, (1999). **13**: p. 3-37.
15. Jagerman, J.D., Melamed, B., Willinger, W., ed. *Stochastic modeling of traffic processes*. Frontiers in Queuing: Models, Methods and Problems, ed. J. Dshalalow. (1996).
16. Leland, W., Taqqu, M., Willinger, W., Wilson, D. *On the self-similar nature of Ethernet traffic*. in *ACM SIGComm*. (1993). San Francisco, CA.
17. Mandelbrot, B., Van Ness, J. W., *Brownian motion fractional noises and applications*. SIAM review, (1968). **422**(437).
18. Oates, T. *Identifying distinctive subsequences in multivariate time series by clustering*. in *KDD-99*. (1999). San Diego, CA: ACM.
19. Paxon, J.V., and Floyd, S., *Wide area traffic : the failure of Poisson modeling*. IEEE/ACM Transactions on Networking, (1995). **3**: p. 226-244.
20. Rabiner, L.R., *A tutorial on hidden markov models and selected applications in speech recognition*. IEEE, (1989). **7**(2): p. 257-286.
21. Traina, C., Traina, A., Wu, L., and Faloutsos, C. *Fast feature selection using the fractal dimension*. in *XV Brazilian Symposium on Databases (SBBD)*. (2000). Paraiba, Brazil.
22. Willinger, W., Taqqu M. S., Leland, W. and Wilson, D., *Self-similarity in high speed packet traffic: Analysis and modeling of Ethernet traffic measurements*. Statistical science, (1995). **10**(1): p. 67-85.
23. Willinger, W., Taqqu M. S., Erramilli, A., ed. *A bibliographical guide to self-similar trace and performance modeling for modern high-speed networks*. Stochastic Networks: Theory and Applications, ed. F.P. Kelly, Zachary, S. and Ziedins, I. (1996), Clarendon Press, Oxford University Press: Oxford. 339-366.
24. Willinger, W., Taqqu M. S., Sherman, W. and Wilson, D., *Self-similarity through high variability: statistical analysis of Ethernet LAN traffic at the source level*. IEEE/ACM Transactions on Networking, (1997). **5**(1): p. 71-86.

A New Mechanism of Mining Network Behavior

Shun-Chieh Lin, Shian-Shyong Tseng, and Yao-Tsung Lin

Department of Computer and Information Science, National Chiao Tung University, Hsinchu, Taiwan 300, R.O.C.
{jielin, sstseng, gis88801}@cis.nctu.edu.tw

Abstract. In this work, a new mechanism, which consists of Preprocessing Phase, Two-Layer Pattern Discovering Phase (2LPD), and Pattern Explanation Phase, is proposed to discover unknown patterns. Two heuristics are proposed to detect outlier users in 2LPD Phase. Next, we are also concerned about subsequences of user's behaviors in this phase. As the patterns which are previously unknown have been discovered in 2LPD Phase, they will be incrementally feedbacked to knowledge base for further detection. Through this incremental learning mechanism, the known patterns can be increased.

1 Introduction

In recent years, many network-based applications stored and processed confidential and important information. In [6], IDML-based intrusion detection model has been designed. Although it is good to detect known intrusions, but is still weak to detect unknown intrusions. According to previous researches [3][5], the input of the intrusion behaviors mining includes system log, packet information, connection, and network information. The mining techniques include statistic, classification, association rule, and sequential pattern analysis. The output includes notice report, rule, and pattern.

In this paper, we will propose a new mechanism and the corresponding algorithms to detect unknown network behaviors. This mechanism consists of Preprocessing Phase, Two-Layer Pattern Discovering Phase (2LPD), and Pattern Explanation Phase In Preprocessing Phase, the packets with the same identifier during a period of time will be aggregated into a feature vector. In 2LPD Phase, these feature vectors are first grouped into several clusters, and then to mine the patterns of user's behaviors. Finally, these patterns will be determined in Pattern Explanation Phase. Therefore, the known patterns can be increased by this incremental learning mechanism.

2 Our Method

The input of our method is standard packet database sorted by arrival time. It should be noted here that the packet database is stored only information about suspected users, which come from distrust IP address. The concept of our method is shown in Fig. 1. At first, the Preprocessing Phase could select packets from database and

M.-S. Chen, P.S. Yu, and B. Liu (Eds.): PAKDD 2002, LNAI 2336, pp. 218-223, 2002.

aggregated these packets into a feature vector. Furthermore, each user's behavior can be presented as a sequence of feature vectors. In 2LPD Phase, there may be millions of distinct feature vectors, which will be first clustered into several clusters. Two heuristics are proposed to detect outliers in this phase. Accordingly, each user's behavior can be transformed into a sequence of cluster labels. Next, we are also concerned about subsequence of user's behaviors to mine the patterns of user's behaviors in 2LPD Phase. Finally, the pattern discovered in previous phase can be represented as a sequence of property sets, can be determined to be normal or abnormal, and can be feedbacked into knowledge base in Pattern Explanation Phase.

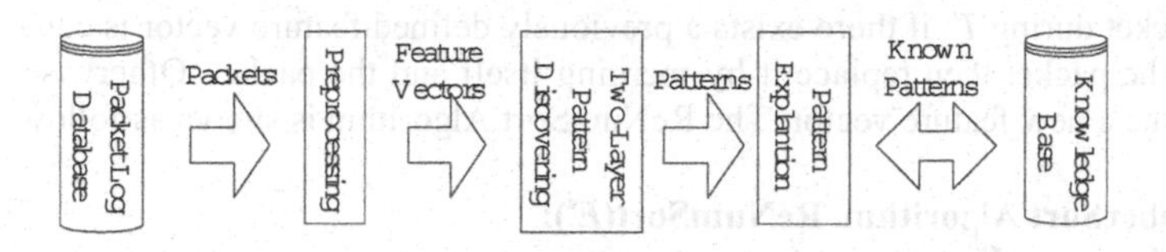

Fig. 1. The Concept Diagram of Our Method

2.1 Preprocessing Phase

Assume there are n senders $u_1, u_2, \ldots, u_n$. Each u_q can be represented by source IP address and let $U = \{u_1, u_2, \ldots, u_n\}$. $T = [t_0, t_0+wc]$ is the time interval concerned to collect packet logs where c is a constant, $t_0 = 0$, $t_i = t_{i-1} + c$, and $T^i = (t_{i-1}, t_i]$, $1 \le i \le w$. $E^i = <e_1^i, e_2^i, \ldots, e_{\alpha_i}^i>$ is a sorted sequence of packet logs in time order during T^i and we assume $|E^i| = \alpha_i \le \alpha$, for each i. '•' is a concatenation operator, i.e., $E^1 \bullet E^2 = <e_1^1, e_2^1, \ldots, e_{\alpha_1}^1, e_1^2, e_2^2, \ldots, e_{\alpha_2}^2>$. $E = E^1 \bullet E^2 \bullet \ldots \bullet E^w$ is the whole packet logs we are concerned in T. Let *e-id* is the event identifier which is defined by the triple fields <*source IP, destination IP, destination port*>, where *source IP* $\in U$. And $ID(e_j^i)$ is an extracting function to extract the *e-id* of e_j^i.

By the experiment of [1], the windows size c is suggested to be 2 seconds. And w $= \lceil T/c \rceil$. If c is very small, information collected seems too little to process in the period. On the contrary, if c is very large, some features may be blurred.

2.1.2 Renumber Sort Algorithm

Since the information of single packet is not sufficient enough to represent the network behavior, several packets with same *e-id* selected from E^i are first aggregated during T^i and then transformed into a feature vector. Let f_j^i = ReNumSort(E^i) is the jth distinct *e-id* during T^i. $S^i = <f_1^i, f_2^i, \ldots, f_{\beta_i}^i>$ is a sequence of feature vectors during T^i, where $\beta_i \le \alpha_i$. $F^i = \{ f_j^i \mid \text{for } 1 \le j \le \beta_i\}$, and $F = \bigcup_{i=1}^{w} F^i$. S_q^i is a subsequence of S^i for $q \in U$. $v_q = <S_q^1, S_q^2, \ldots, S_q^w>$ is a behavior vector of u_q. $V = \{v_q \mid \text{for } q \in U\}$.

Table 1 presents the format of standard packet. The *Time* field indicates the occurred time of packet. There are also other useful packet information will not be discussed in this paper.

Table 1. The Format of Standard Packet

Time	SIP	DIP	DPort	SPort	Protocol	Flag	Length	...

In aggregating the packets into feature vector, we first sort the original packet database by ReNumSort Algorithm to get the distinct *e-id* during T^i, saying f_j^i. For each packet during T^i, if there exists a previously defined feature vector is equal to the *e-id* of the packet then replace it by merging itself and the packet. Otherwise, create and define a new feature vector. The ReNumSort Algorithm is shown as follows.

ReNumberSort Algorithm, ReNumSort(E^i):

Input: E^i

Output: F^i, S^i

Step1. $F^i = \phi$, $S^i = <>$, $DistinctFlag = True$, $\beta_i = 0$.

Step2. For $j = 1$ to α_i,

Step2.1. If $DistinctFlag = True$, β_i ++, $f_{\beta_i}^i = ID(e_j^i)$, Set $DistinctFlag = False$.

Step2.2. For $k = 1$ to β_i, If $ID(e_j^i) \neq f_k^i$, Set $DistinctFlag = True$.

Else Replace f_k^i by merging e_j^i and f_k^i,

Set $DistinctFlag = False$, EXIT.

Step3. For $j = 1$ to β_i, Put f_j^i into F^i, $S^i = S^i \bullet f_j^i$.

Step4. Return F^i, S^i

2.1.3 Preprocessing Phase

As defined above, the feature vector is aggregated from the selected packets with same *e-id* during T^i, so the feature vector is also identified by the *e-id*. The feature vector can be treated as a network event, which represents the user's behavior during T^i. Therefore, the behavior of the sender u_q during T can be represented by a sequence of feature vectors with time order. The format of feature vector shown in Table 2 refers to the experimental data format in [1] and with slight modification.

Table 2. The Format of Feature Vector.

Time	Duration	SIP	DIP	DPort	SPort	Protocol	Flag	Traffic	Packet No.	...

In Table 2, the Time field indicates the starting time of the aggregated feature vector. The *Traffic* field indicates the total packet length of feature vectors during T^i. The *PacketNo.* field indicates the number of packets with feature vector during T^i. Besides, other useful statistical information, which can also be calculated by aggregation algorithm is omitted here.

The preprocessing phase has two major stages: the first stage is to select the packets from packet log database during time window T^i and second stage is to calculate the feature vectors F^i during T^i by aggregating the selected packets. Thus, we can have the sequence of feature vectors S^i and each user's behavior during T^i. Therefore, each user's behaviors during T can be represented as $v_q = <S_q^1, S_q^2, ..., S_q^w>$, for each $q \in U$.

Preprocessing Algorithm, Preprocess(E):
Input: E
Output: F, V

Step1. $F = \phi$, $V = \phi$, $v_q = <>$.
Step2. For $i = 1$ to w,
Select E^i from E, $(F^i, S^i) = \text{ReNumSort}(E^i)$, $F = F \cup F^i$, $v_q = v_q \bullet S_q^i$.
Step3. For $q = 1$ to n, $V = V \cup \{v_q\}$.

2.2 Two-Layer Pattern Discovering Phase (2LPD)

After the preprocessing phase, the original packet logs are already transformed into feature vectors F, and user behaviors have been represented by V. All of them will be treated as input in 2LPD Phase to provide three detection strategies. Without loss of generality, we assume there are at most m clusters. Let $C = \{C_1, C_2, ..., C_m\}$ is a set of clusters where C_i is a subset of F and $1 \le i \le m$. and OC indicates the outlier cluster set. $SEL_q(C_i)$ is a selecting function to select the feature vectors of u_q from C_i. $M(f_j^i) = C_k$ if $f_j^i \in C_k$. $M(S^i) = <M(f_1^i), M(f_2^i), ..., M(f_{\beta_i}^i)>$. $M(S_q^i)$ is a subsequence of $M(S^i)$ for $q \in U$. $M(v_q) = <M(S_q^1), M(S_q^2), ..., M(S_q^w)>$ is a sequence of u_q during T. $M(V) = \{M(v_q) | \text{ for } q \in U\}$.

2.2.1 Behavior Clustering Stage

There are millions of feature vectors with different values. The Behavior Clustering Stage is then proposed to group the similar feature vectors for mining. Since the number of clusters cannot be predicted in advance, a clustering alorithm with the capability of dynamic adjusting the number of clusters is used, e.g., ISODATA [4].

As the intrusions do not frequently happen and are usually performed by few intruders, two heuristics of outlier clusters are proposed as follows:

Heuristic$_1$: A cluster is treated as an outlier cluster if the number of its members is smaller than a threshold θ_1.

Heuristic$_2$: A cluster is treated as an outlier cluster if the ratio of $|SEL_q(C_i)|$ and $|C_i|$ is greater than a threshold θ_2.

Since the system is starting with no priori knowledge about intrusion, thresholds θ_1, θ_2 are set loose; e.g., $\theta_1 = \alpha w/m$, $\theta_2 = 0.5$. θ_1 will gradually decrease and θ_2 will gradually increase according to the patterns discovered in knowledge base.

After the execution of this phase, there may exist some outlier clusters containing information about outlier behaviors or outlier users. All of these discovered outlier clusters can be further analyzed in following phase.

Behavior Clustering Algorithm, Cluster(F, k, θ_1, θ_2):

Input: F, k, θ_1, θ_2
Output: C, OC
Step1. Randomly choose k initial seeds as cluster centers.
Step2. Run ISODATA to generate a number of cluster C= $\{C_1, C_2, \ldots, C_m\}$.
Step3. For $i = 1$ to m,
If $|C_i| \le \theta_1$, put C_i into OC.
If $|SEL_q(C_i)|/|C_i| \ge \theta_2$, put C_i into OC.
Step4. Return (C, OC).

2.2.2 User's Sequence Transforming Stage

As mentioned above, each user's behavior during T can be represented as a sequence of features vectors v_q. Moreover, these feature vectors are grouped into several groups and each feature vector belongs to a unique cluster. Therefore, each user's behaviors can be transformed into a sequence of cluster labels $M(v_q)$.

2.2.3 Sequential Pattern Mining Stage

Since some intrusions can be accomplished by single intruder, the subsequence of each sender should be mined and concerned. On the other hand, some intrusion can be generated by huge amount malicious packet to denial the important services (DoS); the most frequent pattern is also to be considered. Therefore, all the patterns of embedded users' behaviors will be mined in this stage. As each user has a sequence $M(v_q)$, a symbolic sequential mining algorithm, e.g., Agrawal and Strikant's mining algorithm [2] will be used to mine patterns from all users' sequence of behaviors.

Sequential Pattern Mining Algorithm, *SEQUENTIAL($M(V)$, d)*:

Input: $M(V), d$
Output: The subsequences of single user's behaviors and all users' behaviors
Step1. For $q = 1$ to n,
generate $<M(S_q^1), M(S_q^2), \ldots, M(S_q^d)>$, $<M(S_q^2), M(S_q^3), \ldots, M(S_q^{d+1})>$, …, $<M(S_q^{w-d+1}), M(S_q^{w-d+2}), \ldots, M(S_q^w)>$, Run [2] to obtain the subsequence of single user's behavior with $M(v_q)$.
Step2. Run [2] to obtain the subsequences of all users' behaviors with $M(V)$.
Step3. Return the all subsequences of user's behaviors.

d is set to be large if we address long terms sequence of a user's behavior. Otherwise, we chose a small d.

Algorithm of Two-Layer Pattern Discovering: 2LPD($F, M(V), k, \theta_1, \theta_2, d$)

Input: $F, M(V), k, \theta_1, \theta_2, d$
Output: C, OC, and the subsequences of users' behaviors
Step1. $M(V) = \phi$.
Step2. (C, OC) = Cluster $(F, k, \theta_1, \theta_2)$.
Step3. For $q = 1$ to n, Transform v_q into $M(v_q)$, Put $M(v_q)$ into $M(V)$.
Step4. Obtain all subsequences = *SEQUENTIAL*($M(V)$, d).

2.3 Pattern Explanation Phase

The goal of the Pattern Explanation Phase is to explain the meaning of the discovered pattern. Since the heuristic used in behavior cluster is to cluster the similar behaviors, each cluster may have some properties, which can be extracted by analyzing the feature vector space related to each dimension. Using the property of standard derivation evaluation, the most significant attributes of the cluster can be obtained. Therefore, each cluster may be represented as a set of properties and domain expert can explain the meaning of the pattern. These discovered patterns can then be incrementally feedbacked to knowledge base. Hence, with this incremental learning and feedback mechanism the well-known patterns can be increased.

3 Conclusions

In this paper, we proposed a new mechanism to discovery unknown patterns. The corresponding algorithms were also designed. Two heuristics were proposed to detect outlier users to discover intrusion behaviors and the subsequences of single user's behaviors and all users' behaviors were concerned to discover unknown patterns in 2LPD Phase. All discovered patterns can be explained in Pattern Explanation Phase. The results of our method can feedback to knowledge base with known patterns and the numbers of known patterns are gradually increasing. As the discovered patterns can be described in IDML format, they can be further used to detect intrusions online.

Acknowledgement: This work was partially supported by Ministry of Education and National Science Council of the Republic of China under Grand No. 90-E-FA04-1-4, High Confidence Information Systems.

References

1. ACM: KDD cup 1999 data. http://kdd.ics.uci.edu/databases/kddcup99/kddcup99. html (2000)
2. R. Agrawal and R. Srikant: Mining sequential patterns. Proc. of 7th IEEE International Conference on Data Engineering, (1995) 3-14
3. CERT: http://www.cert.org/ (2001)
4. A. K. Jain and R. C. Dubes (eds.): Algorithms for Clustering Data. Prentice-Hill, Englewood Cliffs, N.J. (1988)
5. W. Lee, S. J. Stolfo, and K. W. Mok: A data mining framework for building intrusion detection models. Proc. of 1999 IEEE Symposium on Security and Privacy, (1999)
6. Y. T. Lin, S. S. Tseng, and S. C. Lin: An intrusion detection model based upon intrusion detection markup language (IDML). Journal of Information Science and Engineering Vol. 17, No.6, (2001) 899-919

M-FastMap: A Modified FastMap Algorithm for Visual Cluster Validation in Data Mining*

Michael Ng[1] and Joshua Huang[2]

[1] Department of Mathematics, The University of Hong Kong
Pokfulam Road, Hong Kong
mng@maths.hku.hk

[2] E-Business Technology Institute, The University of Hong Kong
Pokfulam Road, Hong Kong.
jhuang@eti.hku.hk

Abstract. This paper presents M-FastMap, a modified FastMap algorithm for visual cluster validation in data mining. In the visual cluster validation with FastMap, clusters are first generated with a clustering algorithm from a database. Then, the FastMap algorithm is used to project the clusters onto a 2-dimensional (2D) or 3-dimensional (3D) space and the clusters are visualized with different colors and/or symbols on a 2D (or 3D) display. From the display a human can visually examine the separation of clusters. This method follows the principle that *if a cluster is separate from others in the projected 2D (or 3D) space, it is also separate from others in the original high dimensional space* (the opposite is not true). The modified FastMap algorithm improves the quality of visual cluster validation by optimizing the separation of clusters on the 2D or (3D) space in the selection of pivot objects (or projection axis). The comparison study has shown that the modified FastMap algorithm can produce better visualization results than the original FastMap algorithm.

1 Introduction

Cluster validation refers to the procedures that are used to evaluate clusters generated from a data set by a clustering algorithm [16]. Cluster validation is required due to the facts that no clustering algorithm can guarantee the discovery of genuine clusters from real data sets and that different clustering algorithms often impose different cluster structures on a data set even if there is no cluster structure present in it [11] [18].

In statistics, cluster validation is treated as a hypothesis test problem [11] [16] [18]. More precisely, let S be a statistic and H_0 a null hypothesis stating that no cluster structure exists in data set $\mathcal{X}$. Let $\text{Prob}(\mathbf{B}|H_0)$ be the baseline distribution $\mathbf{B}$ under H_0. The event $\mathbf{B}$ could be either $S \geq s_\alpha$ or $S < s_\alpha$, where s_α is a fixed number called a threshold at significance level α and $\text{Prob}(S \geq s_\alpha) = \alpha$. Suppose that s_* is the value of S calculated from a clustering result of data set $\mathcal{X}$.

* supported in part by RGC Grant No. 7132/00P and HKU CRCG Grant Nos 10203501, 10203907 and 10203408.

M.-S. Chen, P.S. Yu, and B. Liu (Eds.): PAKDD 2002, LNAI 2336, pp. 224–236, 2002.

If $s_* \geq s_\alpha$, then we reject hypothesis H_0. Cluster validation methods have been well developed in statistics and other disciplines [5][16][19][21]. Recent surveys are given in [11][18]. Most methods are statistical-based. A few employ graphical displays to visually verify the validity of a clustering [23]. The problem of using these cluster validation methods in data mining is that the computational cost is very high when the data sets are large and complex.

Recent work on clustering in data mining has been focused on the development of fast clustering algorithms to deal with large data sets. Interesting results include CLIQUE [1], CLARANS [22], BIRCH [26], DBSCAN [6] and the k-means extension algorithms [13]. These progresses are extremely important because without fast clustering algorithms one cannot conduct any thorough cluster analysis on large data sets. However, without effective cluster validation tools the problem of cluster analysis on large data sets is only partially solved. Unfortunately, this problem has not been well studied in the data mining community.

In [14], we proposed a visual method for cluster validation in data mining. The visual method uses the FastMap algorithm [8] to project objects and candidate clusters onto a two-dimensional space and allows a human to visually examine the clusters created with a clustering algorithm and determine the genuine clusters found. The visual method is based on the principle that *if a cluster is separate from others in the 2D (or 3D) space, it is also separate from others in the original high dimensional space* (the opposite is not true). We have used this method in a real case study to interactively cluster a mobile service marketing data set and discover a few interesting clusters of customers [14]. In that case study, we used the visual method to solve two common clustering problems, (1) to verify the separations of clusters created by a clustering algorithm and (2) to determine the number of clusters to be produced. Because the Fastmap algorithm is efficient in processing large data sets, the visual method in combination with a fast clustering algorithm provides a complete solution to the clustering problem in data mining.

In [15], we reported our empirical study on comparison of the visual cluster validation method with two statistical indices, C-Index [19] and U-Statistics [11] on a series of constructed artificial data sets with controlled cluster structures and dimensions. The Monte Carlo evaluation has shown that there is a high degree of agreement between the visual method and the statistical methods. These numerical results showed that the visual method can produce validation results equivalent to those of statistical methods. In the comparison study, we have also found that the visualization results of the FastMap projection can be improved by selecting proper pivot objects for projection through the optimization of the separation of discovered cluster centers in the projected space. This observation motivated us to conduct this research.

In this paper, we present a modified FastMap algorithm, called M-FastMap, for data projection in visual cluster validation. M-FastMap improves the quality of visual cluster validation by optimizing the separation of clusters on the 2D (or 3D) space in the selection of pivot objects (or the projection axis). The original

FastMap algorithm [8] has no such objective in selection of the pivot objects. Therefore, it leaves more chances for cluster overlapping on the projected 2D (or 3D) display even if the clusters may be separate in the original space. To make the FastMap algorithm more effective for cluster visualization, our approach is to select the pivot objects that can maximize the total dissimilarity among a small set of objects from each cluster or among the cluster centers in the projected space. This projection will enable the objects in different clusters to be separate on the 2D (or 3D) display as much as possible. Since the number of objects considered in the maximization problem is small, the proposed M-FastMap algorithm does not suffer its efficiency when applied to large data sets. Our numerical experiments have shown that the M-FastMap algorithm can produce better visualization results than the original FastMap algorithm.

This paper is organized as follows: In Section 2, we review the FastMap algorithm [8]. In Section 3, we present our new algorithm M-FastMap. The synthetic data generation and experimental results on comparison of M-FastMap and its original version in visual cluster validation are presented in Section 4. We draw some concluding remarks on this research in Section 5.

2 Visual Cluster Validation with FastMap

Let $\mathcal{O}$ be a set of N objects in an n-dimensional space and d_n a dissimilarity measure between objects in $\mathcal{O}$ in the n-dimensional space. The FastMap projection of the N objects onto an m-dimensional space ($n \gg m$) uses the following formula to calculate the coordinate $x_{m,i}$ of object o_i on the mth axis of the m-dimensional space.

$$x_{m,i} = \frac{d_{n-m+1}(o_a, o_i)^2 + d_{n-m+1}(o_a, o_b)^2 - d_{n-m+1}(o_b, o_i)^2}{2d_{n-m+1}(o_a, o_b)} \tag{1}$$

where objects o_a and o_b are two pivot objects from $\mathcal{O}$ and $d_{n-m+1}(o_i, o_j)^2$ is calculated as $d_{n-m+1}(o_i, o_j)^2 = d_n(o_i, o_j)^2 - d_{m-1}(o_i, o_j)^2$ where $d_{n-m+1}(o_i, o_j)$ is the distance in the $(n - m + 1)$-dimensional space, $d_n(o_i, o_j)$ the distance in the n-dimensional space and $d_{m-1}(o_i, o_j)^2 = \sum_{\ell=1}^{m-1}(x_{\ell,i} - x_{\ell,j})^2$ is the square distance in the $(m - 1)$-dimensional space. Proofs of these formulas are given in [8]. Objects o_a and o_b are referred to as *pivot objects.* The line passing the pivot objects determines a projection axis. Essentially, there are m pairs of pivot objects required. The selection of the pivot objects can be arbitrary provided that the projection lines are not parallel in the original space. Faloutsos and Lin proposed a method to choose the pivot objects effectively and efficiently. Their idea is to find a line which the projections are as far apart from each other as possible. To achieve that, they choose o_a and o_b such that the distance $d_n(o_a, o_b)$ is maximized. However, this would require $O(N^2)$ distance computations. Thus, they considered the linear heuristic algorithm *choose-distant-objects()*:

Algorithm 1 *choose-distant-objects()*:
begin
1. Choose arbitrarily an object, and declare it to be the second pivot object o_b.
2. set o_a (the object that is farthest apart from o_b).
3. set o_b (the object that is farthest apart from o_a).
4. output o_a and o_b.

end

All the steps in the above algorithm are linear on N. The middle two steps can be repeated in a constant number of times (Faloutsos and Lin suggested to use 5 iterations [8]) but still maintain the linearity of the heuristic. To project the objects into an m-dimensional space using **Algorithm 1**, the computational complexity is $O(mN)$. FastMap is suitable for data mining applications.

Faloutsos and Lin [8] has shown that the Fastmap projection has a property to reveal clusters of data existing in the original (sometimes unknown) high dimensional space. This property can be used for data clustering [8] [10] and cluster validation [14]. For clustering, we use Fastmap to project objects into a k-dimensional space and then apply a clustering algorithm to cluster data in the low dimensional space [10]. This method is useful when the dimensions of the original space are unknown. However, the clusters found in the projected space cannot be guaranteed being clusters in the original space.

For cluster validation, we apply a clustering algorithm to cluster data in the original space and use the Fastmap algorithm to project clusters into a 2D (or 3D) space and visualize them. If a cluster is observed to be separate from other objects on the 2D (or 3D) plots, we can claim that it is also separated from other objects in the original space. Let T be the minimal distance between an object o_{in} in the cluster and an object o_{out} outside the cluster. Assume o and o' are identifiable on the 2D (or 3D) plot if $d_2(o, o') \geq T$. The two objects are also identifiable in the original space because $d_n(o, o') \geq d_2(o, o') \geq T$.

Based on the above principle, we can use the Fastmap display to visually validate clusters generated from high dimensional data using a fast clustering algorithm. In data mining, clustering and cluster validation can be conducted interactively. Given a real large data set, we first use a clustering algorithm to partition it into k clusters. Then, we use the Fastmap algorithm to project the clusters into a 2D (or 3D) space and visualize objects of different clusters in different colors and/or symbols. In this way, we can visually identify some clusters which are separate from other objects. These clusters are also separate from other objects in the original high dimensional space. In analysis, we can extract these clusters from the data set and continue the clustering process on the remaining data.

In [14], we described an interactive approach to clustering and cluster validation with Fastmap for data mining. We use a top-down approach to interactively building cluster trees from data. Starting with the whole data set that is considered as a cluster on its own right, we stepwise decompose the data and grow a tree of clusters. In the tree, a node containing children is a composite cluster while all leaves are atomic clusters.

3 M-FastMap Algorithm

The objective of object projection in visual cluster validation is to make clusters as apart as possible in the projected space so genuine clusters can be visually identified. However, the pivot object selection method of **Algorithm 1** tries to make the distances between individual objects and the pivot objects as apart as possible in the projected space. In general, the two objectives are not equivalent.

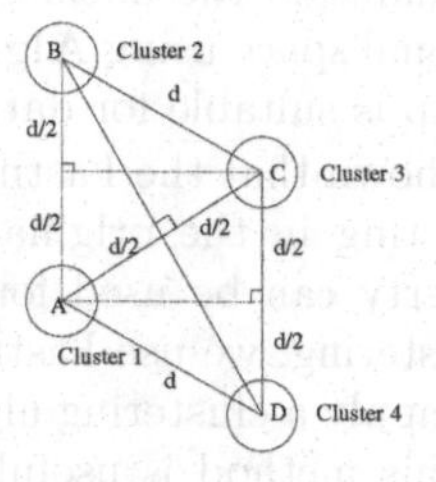

Fig. 1. Proper selection of projection axis for cluster visualization.

For example, Figure 1 contains four clusters in a 2D space, each being bounded by a circle of the equal radius r. The points A, B, C and D are the centers of the clusters. Assume $d >> 2r$. The pivot objects selected from objects in different clusters can be approximated by the cluster centers. To project the objects into a 1D space, **Algorithm 1** will choose points B, D as the pivot objects because $BD = \sqrt{3}d > AB = AC = AD = BC = BD = CD = d$. In this case, the coordinates x_A, x_C of points A, C calculated by the FastMap algorithm on the projection axis passing BD will be the same. That is, $x_A = x_C = \sqrt{3}d/2$. This projection will make Clusters 1 and Cluster 3 overlap in the projected space so the two clusters will not be visually identifiable. This example shows that the FastMap projection based on **Algorithm 1** is not optimal in terms of cluster separation in the projected space. Visualization of clusters in the projected space can be improved if we choose the pivot objects that maximize the sum of the minimum distances among the cluster centers A, B, C and D in the projected space, i.e.,

$$\max_{ij \in \{AB,AC,AD,BC,BD,CD\}} \sum_{X \in \{A,B,C,D\}} \min_{Y \in \{A,B,C,D\} \wedge (Y \neq X)} \{(XY)_{ij}\},$$

where $(XY)_{ij}$ is the distance between points X and Y on the projection axis ij. This approach enables us to find the projection axis such that the clusters in the projected space are separate as much as possible. In this example, if we choose A, B as pivot objects, the minimum projected distances are calculated as follows

$$\min\{(AB)_{AB}, (AC)_{AB}, (AD)_{AB}\} = \min\{d, d/2, d/2\} = d/2;$$
$$\min\{(BA)_{AB}, (BC)_{AB}, (BD)_{AB}\} = \min\{d, d/2, 3d/2\} = d/2;$$
$$\min\{(CA)_{AB}, (CB)_{AB}, (CD)_{AB}\} = \min\{d/2, d/2, d\} = d/2;$$
$$\min\{(DA)_{AB}, (DB)_{AB}, (DC)_{AB}\} = \min\{d/2, 3d/2, d\} = d/2.$$

where $(XY)_{AB}$ is the projected distance of objects X and Y on line AB. The sum of the minimum distances is $2d$. If we choose BD as pivot objects, with the same calculation method, we can find that the sum of the minimum distances is $\sqrt{3}d$. The sum becomes d if we choose AC as pivot objects. Among the three pairs of pivot objects, objects AB are clearly the best selection to make the four clusters well-separate in the projected space. This selection avoids the overlapping of two clusters on the projected space, which occurs when AC or BD are selected. Therefore, we can visually identify all four clusters in the projected space. Other best selections are objects AD, BC or CD in this example.

M-FastMap is a modification of Faloutsos and Lin's FastMap algorithm [8] with a new method for pivot object selection. M-FastMap is sketched in **Algorithm 2**. In order to make clusters separate as much as possible in the projected space, the new method selects the pivot objects o, o' from a set of candidate objects $\mathcal{S}$ that maximize the following cost function:

$$(P) \qquad \max_{o,o' \in \mathcal{S}} \sum_{o_i \in \mathcal{S}} \min_{o_j \in \mathcal{S} \wedge (o_j \neq o_i)} \{|x_{m,i} - x_{m,j}|\}, \tag{2}$$

where $\mathcal{S}$ contains objects randomly selected from each cluster, and $x_{m,\cdot}$ is defined as in (1). If the number of clusters is greater than the number of the projected dimensions, we can simply use all cluster centers for $\mathcal{S}$. This can be explained as follows. Suppose $\mathcal{C}_1$ and $\mathcal{C}_2$ are two clusters in an n-dimensional space and their cluster centers are c_1 and c_2. Let $r_1 = \max_{o \in \mathcal{C}_1} d_n(c_1, o)$ and $r_2 = \max_{o' \in \mathcal{C}_2} d_n(c_2, o')$. We note that if the distance between c_1 and c_2 in the projected space is greater than $r_1 + r_2$, i.e., $d_k(c_1, c_2) > r_1 + r_2$, then $\min_{o \in \mathcal{C}_1, o' \in \mathcal{C}_2} d_k(o, o')$ is greater than zero. This indicates that the two clusters are well separated in the projected space. From this result, we find that if the boundaries of two clusters do not intersect in the original space, the two clusters do not overlap in the projected space. When more clusters are involved, this result still holds if for all clusters $\mathcal{C}_i$ and $\mathcal{C}_j$, we have $d_k(c_i, c_j) > r_i + r_j$ where c_i and c_j are cluster centers, $r_i = \max_{o \in \mathcal{C}_i} d_n(c_i, o)$ and $r_2 = \max_{o' \in \mathcal{C}_j} d_n(c_j, o')$, i.e., the distances among cluster centers in the projected space are greater than their maximum distances among clusters in the original space. This result leads us to take cluster centers as a set of candidates of the pivots objects for the projection. We note that our M-FastMap algorithm is to make cluster centers separate as much as possible. Therefore, the projection determined by solving the optimization problem (P) in (2) will achieve to make the cluster centers separate as much as possible. Because objects in a cluster usually have high density near the cluster center, the projection also makes clusters separate on the 2D (or 3D) display and easily identifiable visually.

Algorithm 2 *M-FastMap*:
begin
1. Randomly choose a set $\mathcal{S}$ of objects from each cluster and $N \gg |S|$.
2. Obtain the dissimilarity matrix D among the chosen objects.
3. For $\ell = 1$ to m
   Solve problem (P) to determine the pivot objects o_a, o_b.
   Use formula (1) to compute the coordinate $x_{k,i}$ of object o_i.
   Update the dissimilarity matrix D.
4. End For

end

Algorithm 2 does not suffer its efficiency if cluster centers are selected as candidate pivot objects because the number of clusters is usually small. Let s be the number of objects in S. Step 2 of **Algorithm 2** requires $O(s^2)$ operations to compute the dissimilarity matrix D from objects in $\mathcal{S}$. The optimization problem (P) in step 3 is solved in $O(s^4)$ operations since there are $s(s-1)/2$ pairs of pivot objects to choose and the evaluation of each objective function value requires $O(s^2)$ operations. The coordinates $x_{k,i}$ for all objects can be obtained in $O(N)$ operations, and dissimilarity matrix D can also be updated in $O(s^2)$ operations. To map the objects into points in m-dimensional space using **Algorithm 2**, the M-FastMap algorithm requires $O(ms^4 + mN)$ operations. Because the number of clusters is usually sufficiently less than the number of objects in the practical application, i.e., $s << N$, the M-FastMap algorithm is still very efficient.

4 Experiments

Data Generation: Two types of synthetic data sets were generated. The data sets of the first type contained well-formed clusters distributed in the specified region in an n-dimensional space. The clusters were generated by a multidimensional normal distribution random number generator.

Table 1 lists the control parameters used to generate these data sets. The first three parameters were randomly generated for each data set. However, we restricted the number of objects in each cluster to the range of $0.8 \times N/K \leq N_k \leq 1.2 \times N/K$ where $N = 100$ for all data sets. Both the number of dimensions n and the number of clusters K were within the range between 3 and 5. We generated 5 different configurations of data sets. For each configuration, 20 data sets were generated. Totally, we generated 100 synthetic data sets containing randomly distributed clusters within a region of unit hyper-boxes. In each data set, we randomly generated K cluster centers. The distances between cluster centers were first randomly set between 1.5 and 3, and then the minimum distances between cluster centers were computed. Based on the minimum distances, we randomly generated the covariance Σ_k for the kth cluster. N_k objects were generated by using the multi-normal distribution generator with the cluster center as the mean and Σ_k as the covariance. Finally, we re-scaled all the data points to [-1,1] in each dimension. The data sets of the second type contained randomly generated objects which were uniformly distributed in the same region of unit hyper-boxes. For each specific n, we generated 100 data sets to calculate the baseline distribution of C-index and U-statistics. Totally, 500 random data sets were generated using a uniform distribution random generator. Each data set contained 100 objects.

4.1 Experimental Results

Performance: The visual cluster validation method was designed to validate partitions generated from a data set by a clustering algorithm and to identify genuine clusters from a partition. In this experiment, we used the k-means algorithm to generate a partition from a synthetic data set because all the data sets were numeric. In dealing with real data sets, we use the k-prototypes algorithm that can process both numeric and categorical data [13].

The purpose of this experiment was to compare M-FastMap (**Algorithm 2**) with FastMap in validation of clusters generated from the synthetic data sets. As we did in our previous study [15], we used the two algorithms to validate both partitions and clusters. To visually validate a partition, we used these two algorithms to project the clusters into 2D spaces and displayed the clusters in different symbols. If we saw well-separated clusters on a display, we considered the partition was validated. If we saw any overlapping between clusters, we considered the partition was not validated by the visual method. In such a case, we used the statistical C-index to evaluate the validity of the partition. Therefore, for each partition, we could get different validation results from the two algorithms and C-index. We were interested in the partitions which were validated by C-index. For these partitions, we investigated whether they could be validated by FastMap or M-FastMap or both or none. In such a way, we could compare the performances of the two algorithms in validating partitions. For example, Figures 2 shows the visual validations of the same partition generated by the k-means algorithm from the same synthetic data set. The C-index value for this partition was -1.4102. In comparison with the baseline distribution, this partition was statistically valid. The validity was confirmed from the display projected by M-FastMap (Figure 2(right)) but could not be confirmed from the display projected by FastMap (Figure 2(left)). This example shows that M-FastMap produced a better validation result than FastMap.

Table 2 shows the summary of validation results of 60 data sets with three different configurations, each with 20 randomly generated data sets. Each data set contained five clusters. The results were created as follows. For each data set, we used the k-means algorithm to generate a partition of five clusters and calculated the C-index value for the partition. The partition was considered valid if the C-index value is less than -1. Then, we used FastMap and M-FastMap to project the same partition into 2D spaces and visualized the clusters in different symbols. Finally, we visually investigated the displays to see whether clusters were overlapped. A partition was not validated if the projection of the partition resulted in overlapping clusters. We can see from Table 2 that M-FastMap produced more validated results than FastMap. In the 3-dimensional configuration, M-FastMap produced validation results equivalent to the C-index. The average increase over FastMap is more than 20%. To further evaluate the separation of objects in the 2D spaces projected by the two algorithms, we calculated the sum of minimum distances among objects in the projected space, i.e., $\sum_{o_i \in \mathcal{S}} \min_{o_j \in \mathcal{S} \wedge (o_j \neq o_i)} \{|x_{m,i} - x_{m,j}|\}$, and then took the sum. The sum of minimum distances among objects in the projected space is used as an objective function to maximize in selection of pivot objects. The larger the sum, the more separate the objects in the projected space and the better the visual results. Table 3 lists the minimum, maximum and average of these distances between objects in the projected spaces. We can see that in average the M-FastMap projection resulted larger total distance values than the FastMap projection. This implies the M-FastMap projection often produces better visualization than the FastMap projection.

Efficiency: Next, we compared the efficiency of FastMap and M-FastMap in projecting large data sets with different numbers of clusters. The data sets consist of $N (= 400, 800, 1600, 3200, 6400, 12800, 25600)$ objects, each being described by 5 attributes. We tested two scalabilities of the FastMap and M-FastMap algorithms using these data sets. The first is the scalability of the two algorithms against the number of objects for different given numbers of clusters ($s = 5, 10, 15, 20$). The second is the scalability of the two algorithms against the number of clusters for different given numbers of objects ($N = 3200, 6400, 12800, 25600$). Figure 3 shows the results of these two types of tests. The times reported in the figures only include the pivot objects selection times because the coordinates calculation times for both algorithms are the same. These results are very encouraging. Figure 3(a) shows clearly that for a fixed number of clusters, the pivot objects selection time of the M-FastMap algorithm is constant as the number of objects increases. This is because the response time for selecting pivot objects is fixed when the number of clusters is fixed. The selection time of the FastMap algorithm increases linearly as the number of objects increases. From Figure 3(b), we can see that when the number of objects is fixed, the selection time of the M-FastMap algorithm increases as the number of clusters increases. In case of the FastMap algorithm, the selection time remains constant. In the complexity analysis of the M-FastMap algorithm in Section 3.2, we know that the M-FastMap algorithm requires $O(s^4)$ operations to solve the optimization problem (P) in (2). When the number of clusters is small, the M-FastMap algorithm is still faster than the FastMap algorithm.

To further check the efficiency of the M-FastMap algorithm over the original FastMap algorithm, in Figure 3(c), we list the largest number $s_{\max}(N)$ of clusters such that the selection time of the M-FastMap algorithm is still faster than that of the FastMap algorithm for different given numbers of objects. For the reference, we also display the curve for $3 \log N$ in the same figure. We see from the figure that the form of the curve $s_{\max}(N)$ is similar to the form of $3 \log N$. This implies that if the number of objects to be clustered is N and the number of clusters considered is $O(\log N)$, then the M-FastMap algorithm will be more efficient than the FastMap algorithm.

Finally, we use a simple example to show how to use the visual cluster validation method to identify genuine clusters interactively in multi-steps clustering. Figure 4(left) shows a projected clustering result generated by the k-means algorithm. This is an invalid partition. However, among the five generated clusters, two clusters in pluses and solid boxes are clearly identifiable. They can also be validated by their U-statistic values (see Table 4). Other clusters in crosses, boxes and stars are not valid visually and statistically because their U-statistic values are quite large compared with the baseline distribution (cf. Table 4). In this step, we cannot identify all genuine clusters but two. In stead of re-clustering the data with other parameters of the clustering algorithm, we remove the two identifiable clusters in pluses and solid boxes from the data and use the k-means algorithm to further cluster the rest of the data set and visually validate the clustering result again. From the new projection in Figure 4(right), we can visually identify all three clusters generated in this step. The C-index value for this partition is -2.5078 (significant for the partition). The U-statistic values also validate the three clusters (see Table 5). This example shows that, by combining the k-means and visual cluster validation methods, genuine clusters can be identified from multiple levels of clustering started from the same data set.

No.	Parameter	Definition
1	μ_k	Mean vector of a cluster
2	Σ_k	Covariance of a cluster
3	N_k	Number of objects in the kth cluster
4	N	Number of objects in a data set
5	n	Number of dimensions
6	K	Number of clusters in a data set

Table 1. Control parameters

Dimensions	Invalid (C-index) Stat. significant		Valid (C-index) FastMap		Valid (C-index) M-FastMap	
	percentage (%)	mean C-index	Validated (%)	Not Validated (%)	Validated (%)	Not Validated (%)
3	10	0.1894	65	25	90	0
4	25	0.0374	50	25	70	5
5	25	0.2548	40	35	60	15

Table 2. Validation results of 60 data sets in three different configurations, each with 20 randomly generated data sets. Each data set contains five clusters.

Dimensions	Total mutual distances FastMap			Total mutual distances M-FastMap		
	minimum	maximum	mean	minimum	maximum	mean
3	521.6	576.9	553.9	619.5	923.2	746.2
4	476.5	553.7	518.2	544.0	726.8	602.3
5	534.3	595.0	569.4	565.4	688.1	628.8

Table 3. Comparison of total mutual distances between FastMap and M-FastMap.

Clusters	M	U	Baseline Distribution minimum	maximum	mean
1 (+)	21	33	173890	252540	211150
2 (×)	9	1823	25273	42379	54265
3 (∗)	37	205283	725700	889710	807870
4 (□)	14	2337	74819	118600	95471
5 (■)	19	9	144430	210920	171740

Clusters	M	U	Baseline Distribution minimum	maximum	mean
1 (+)	19	4	144430	210920	171740
2 (×)	23	0	204300	310920	254300
3 (∗)	18	6	117910	186730	152760

Table 4. U-statistic values (U) for (left) an invalid partition shown as Figure 4(a) and (right) an valid partition shown as Figure 4(b), where M is the number of objects.

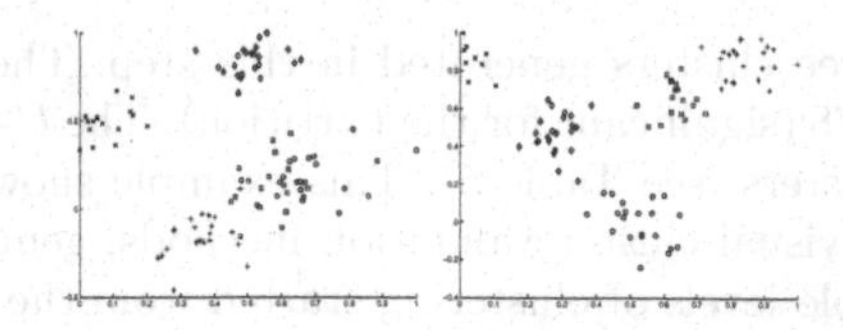

Fig. 2. Invalid partition using FastMap (left) and valid partition using M-FastMap (right).

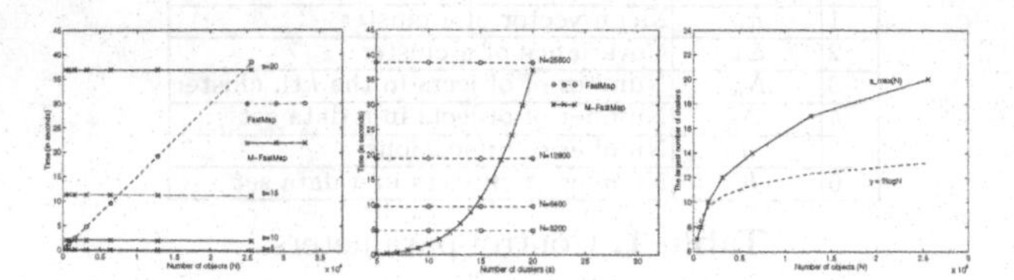

Fig. 3. Pivot objects selection time vs. (left) database size N for the synthetic datasets, without including the coordinates calculation time, (middle) the number of clusters s for the synthetic datasets, without including the coordinates calculation time, and (right) The largest number $s_{\max}$ of clusters vs. number of objects for the synthetic datasets.

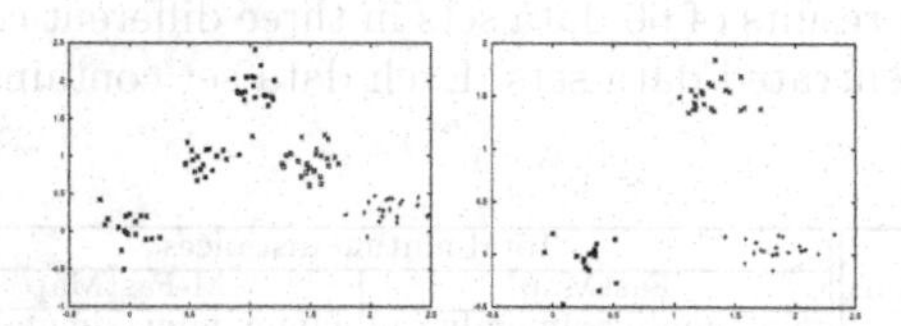

Fig. 4. An invalid partition with two identifiable clusters in pluses and filled boxes (left) and The partition of the data set in Figure 4 after removing two identified genuine clusters and re-clustering (right).

5 Conclusions

We have presented M-FastMap, a modification of the FastMap algorithm for visual cluster validation in data mining. In contrast to other FastMap applications [8][10], the purpose here is to use FastMap to project a clustering result obtained from the original high dimensional space into a 2D (or 3D) space so a human can visually validate the clustering result and identify genuine clusters. As such, the original FastMap algorithm has a drawback in selection of pivot objects for cluster visualization. The M-FastMap algorithm has removed this drawback by using a new pivot object selection method to maximize the separation of clusters in the projected 2D (or 3D) space. Our experimental results have shown that M-FastMap can produce more accurate cluster validation results than FastMap. The average improvement in this study was over 20%. Therefore, using M-FastMap, the user has much more chance to identify genuine clusters through visual cluster validation. On the other hand, unless the number of clusters is large, M-FastMap does not suffer its efficiency when applied to a large data set because the time for selecting pivot objects is independent of the number of objects. In real applications, the number of clusters is usually much less than the number of objects. In our previous papers [14][15], we have shown that visual cluster validation with FastMap is an effective method for cluster validation in data mining because the existing statistical validation methods are unable to handle large data sets. Without cluster validation, the clustering problem in data mining is only partially solved. The M-FastMap algorithm further improves this visual cluster validation method. In combination with a fast clustering algorithm, it offers a flexible way to identify genuine clusters in multiple clustering steps. In this approach, the user is involved in making decisions, which is critical in data mining tasks.

References

1. Agrawal, R., Gehrke, J, Gunopulos, D. and Raghavan, P. (1998) Automatic subspace clustering of high dimensional data for data mining applications. In Proceedings of SIGMOD Conference.
2. Cormack, R. (1971) A review of classification. Journal of Royal Statistical Society, Series A, Vol. 134, pp. 321-367.
3. Cox, T and Cox, M (1994) Multidimensional Scaling. Chapman & Hall.
4. Dubes, R. C. (1987) How many clusters are best? - an experiment. Pattern Recognition, Vol. 20, No. 6, pp.645-663.
5. Dubes, R. and Jain, A. K. (1979) Validity studies in clustering methodologies. Pattern Recognition, Vol. 11, pp. 235-254.
6. Ester, M., Kriegel, H.-P., Sander, J. and Xu, X. (1996) A density-based algorithm for discovering clusters in large spatial databases with noise. In Proceedings of the 2nd International Conference on Knowledge Discovery in Databases and Data Mining, Portland, Oregon, USA.
7. Everitt, B. (1974) Cluster Analysis. Heinemann Educational Books Ltd.
8. Faloutsos, C. and Lin, K., (1995) Fastmap: a fast algorithm for indexing, data-mining and visualization of traditional and multimedia datasets. In Proceedings of ACM-SIGMOD, pp. 163-174.
9. Fukunaga, K. (1990) Introduction to Statistical Pattern Recognition. Academic Press.
10. Ganti, V., Ramakrishnan, R., Gehrke, J, Powell, A. L. and French, J. C. (1999) Clustering large datasets in arbitrary metric spaces. ICDE 1999, pp. 502-511.
11. Gordon, A. D. (1998) Cluster validation, In Data Science, Classification, and Related Methods, ed. C Hayashi, N Ohsumi, K Yajima, Y Tanaka, H-H Bock and Y Baba, Springer, Tokyo, pp 22-39.
12. Gordon, A. D. (1994) Identifying genuine clusters in a classification. Computational Statistics and Data Analysis 18, pp.516-581.

13. Huang, Z. (1998) Extensions to the k-means algorithm for clustering large data sets with categorical values. Data Mining and Knowledge Discovery, Vol. 2, No. 3, pp. 283-304.
14. Huang, Z. and Lin, T. (2000) A visual method of cluster validation with Fastmap. In Proceedings of PAKDD2000, Kyoto, Japan.
15. Huang, Z., Ng, M. K. and Cheung, D. W. (2001) An empirical study on the visual cluster validation method with Fastmap. In Proceedings of DASFAA2001, Hong Kong.
16. Jain, A. K. and Dubes, R. C. (1988) Algorithms for Clustering Data. Prentice Hall.
17. Kruskal, J. B. and Carroll, J. D. (1969) Geometrical models and badness-of-fit functions. in Multivariate Analysis II, ed. P. R. Krishnaiah, Academic Press, pp.639-670.
18. Milligan, G. W. (1996) Clustering validation: results and implications for applied analysis. in Clustering and Classification, ed. P. Arabie, L. J. Hubert and G. De Soete, World Scientific, pp.341-375.
19. Milligan, G. W. (1981) A Monte Carlo study of thirty internal criterion measures for cluster analysis. Psychometrika, Vol. 46, No. 2, pp.187-199.
20. Milligan, G. W. and Cooper, M. C. (1985) An examination of procedures for determining the number of clusters in a data set. Psychometrika, Vol. 50, No. 2, pp.159-179.
21. Milligan, G. W. and Isaac, P. D. (1980) The validation of four ultrametric clustering algorithms. Pattern Recognition, Vol. 12, pp.41-50.
22. Ng, R. and Han, J. (1994) Efficient and effective clustering methods for spatial data mining. In Proceedings of VLDB, 1994.
23. Rousseeuw, P. J. (1987) Silhouettes: a graphical aid to the interpretation and validation of cluster analysis. Journal of Computational and Applied Mathematics, Vol. 20, pp.53-65.
24. Theodoridis, S. and Koutroumbas, K. (1999) Pattern Recognition. Academic Press.
25. Young, F. W. (1987) Multidimensional scaling: history, theory and applications. Lawrence Erlbaum Associates.
26. Zhang, T. and Ramakrishnan, R. (1997) BIRCH: A new data clustering algorithm and its applications. Data Mining and Knowledge Discovery, Vol. 1, No. 2, pp. 141-182.

An Incremental Hierarchical Data Clustering Algorithm Based on Gravity Theory

Chien-Yu Chen, Shien-Ching Hwang, and Yen-Jen Oyang

Department of Computer Science and Information Engineering
National Taiwan University, Taipei, Taiwan
cychen@mars.csie.ntu.edu.tw
schwang@mars.csie.ntu.edu.tw
yjoyang@csie.ntu.edu.tw

Abstract. One of the main challenges in the design of modern clustering algorithms is that, in many applications, new data sets are continuously added into an already huge database. As a result, it is impractical to carry out data clustering from scratch whenever there are new data instances added into the database. One way to tackle this challenge is to incorporate a clustering algorithm that operates incrementally. Another desirable feature of clustering algorithms is that a clustering dendrogram is generated. This feature is crucial for many applications in biological, social, and behavior studies, due to the need to construct taxonomies. This paper presents the GRIN algorithm, an incremental hierarchical clustering algorithm for numerical data sets based on gravity theory in physics. The GRIN algorithm delivers favorite clustering quality and generally features $O(n)$ time complexity. One main factor that makes the GRIN algorithm be able to deliver favorite clustering quality is that the optimal parameters settings in the GRIN algorithm are not sensitive to the distribution of the data set. On the other hand, many modern clustering algorithms suffer unreliable or poor clustering quality when the data set contains highly skewed local distributions so that no optimal values can be found for some global parameters. This paper also reports the experiments conducted to study the characteristics of the GRIN algorithm.

Keyword: data clustering, hierarchical clustering, incremental clustering, gravity theory.

Section 1. Introduction

Data clustering is one of the most traditional and important issues in computer science [3, 8, 9]. In recent years, due to emerging applications such as data mining and document clustering, data clustering has attracted a new round of attention [6, 14]. One of the main challenges in the design of modern clustering algorithms is that, in many applications, new data sets are continuously added into an already huge database. As a result, it is impractical to carry out data clustering from scratch whenever there are new data instances added into the database. One way to tackle this challenge is to incorporate a clustering algorithm that operates incrementally.

The development of incremental clustering algorithms can be traced back to 1980s [4, 5]. In 1989, Fisher proposed CLASSIT [5], which is an alternative version of COBWEB [4] and was designed for handling numerical data sets. However,

M.-S. Chen, P.S. Yu, and B. Liu (Eds.): PAKDD 2002, LNAI 2336, pp. 237-250, 2002.

CLASSIT assumes that the attribute values of the clusters are normally distributed. As a result, its application is limited. In recent years, several incremental clustering algorithms have been proposed, including BIRCH [15], the clustering algorithm proposed by Charikar et al. in 1997 [1], Incremental DBSCAN [2], and the clustering algorithm proposed by Ribert et al. in 1999 [12]. However, the algorithm proposed by Charikar et al. employs a clustering quality measure that may not be appropriate for some real data sets, especially when the data set contains mainly arbitrarily shaped clusters. Incremental DBSCAN lacks a desirable feature for some applications in biological, social, and behavior studies, because it does not output a clustering dendrogram. For such applications, creating a clustering dendrogram is an essential work due to the need to construct taxonomies [9]. The algorithm proposed by Ribert et al. suffers higher time complexity and, therefore, is not suitable for handling large databases. BIRCH is an incremental hierarchical clustering algorithm with $O(n)$ time complexity. However, as will be elaborated in the following, BIRCH may fail to deliver satisfactory clustering quality in some cases.

This paper presents the GRIN algorithm, a novel incremental hierarchical clustering algorithm for numerical data sets based on gravity theory in physics. As the experiments conducted in this study reveal, the GRIN algorithm delivers favorite clustering quality in comparison with the BIRCH algorithm and generally features $O(n)$ time complexity. One main factor that makes the GRIN algorithm able to deliver favorite clustering quality is that the optimal parameters settings in the GRIN algorithm are not sensitive to the distribution of the data set. On the other hand, parameter setting is a main problem for many modern clustering algorithms and could lead to unreliable or poor clustering quality. As Han and Kamber summarized in [6], "Such parameter settings are usually empirically set and difficult to determine, especially for real-world, high-dimensional data sets. Most algorithms are very sensitive to such parameter values: slightly different settings may lead to very different clusterings of the data. Moreover, high-dimensional real data sets often have very skewed distributions such that their intrinsic clustering structure may not be characterized by global density parameters." In the BIRCH algorithm, there is a global parameter that controls the diameter the leaf subclusters. As the experiments conducted in this study reveal, this factor along with dependence on input data ordering cause the BIRCH algorithm unable to deliver satisfactory clustering quality in some cases. Fig. 1 presents an example. In this example, θ denotes the threshold imposed on the diameters of leaf subclusters and it is assumed that distance(instance1, instance2) $< \theta$, distance(instance2, instance3) $< \theta$, and distance(instance1, instance3) $> \theta$. As the example shows, if the data is inputted in the following order $1 \rightarrow 2 \rightarrow 3$ …, then data instance 2 will be clustered with data instance 1 instead of with data instance 3. Such a clustering result is not in conformity with what we consider

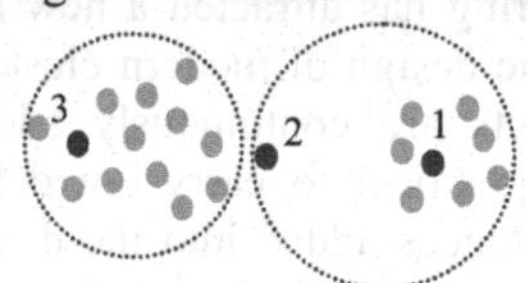

Fig. 1. An case in which BIRCH may fail to deliver satisfactory clustering quality.

natural clusters. It is certain that this problem can be resolved by setting θ to a smaller value. However, since θ is a global parameter automatically set based on some system metrics, it could occur that the setting is not appropriate for some leaf subclusters. Though the optional phase 4 of the BIRCH algorithm can be carried out to remove some of the flaws like the one shown in Fig. 1, this option is not applicable, when BIRCH is invoked to perform incremental clustering.

As mentioned above, the time complexity of the GRIN algorithm is generally $O(n)$. This argument holds, provided that the spatial distribution of the new data instances that are continuously added into the data set is similar to that of the data instances already in the data set. Even if the spatial distributions of the new data set and the existing data set are very different, it does not imply that the linearity of time complexity no longer holds, because the GRIN algorithm keeps flattening and pruning the clustering dendrogram as new data instances continue to arrive. Only when the incoming data instances continue to form brand new clusters far away from existing clusters, will the time complexity of the GRIN algorithm approaches $O(n^2)$.

In the following part of this paper, section 2 discusses how the GRIN algorithm works. Section 3 describes the agglomerative hierarchical clustering algorithm that the GRIN algorithm invokes to construct the clustering dendrogram. Section 4 reports the experiments conducted to study the characteristics of the GRIN algorithm. Finally, concluding remarks are given in section 5.

Section 2. The GRIN Algorithm

This section describes how the GRIN algorithm works. The GRIN algorithm operates in two phases. In both phases, it invokes the gravity-based agglomerative hierarchical clustering algorithm presented in next section to construct clustering dendrograms. One key idea behind the development of the GRIN algorithm is that any arbitrarily shaped cluster can be represented by a set of spherical clusters as exemplified in Fig. 2. Therefore, spherical clusters are the primitive building blocks of the clustering dendrogram derived. Note that, though any two clusters contain disjoint sets of data instances, the spheres defined by their respective centroids and radii may overlap. The centroid of a cluster is defined to be the geometric center of the data instances in the

(a) A cluster of arbitrary shape.

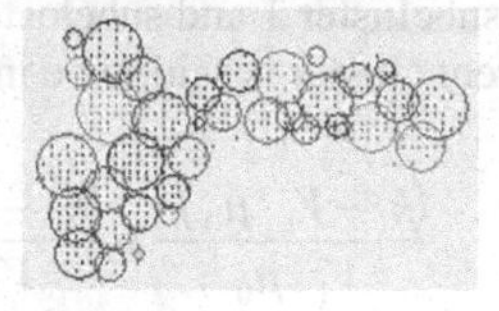

(b) Representation of the arbitrarily shaped cluster by a set of spherical clusters.

Fig. 2. An example demonstrating how an arbitrarily shaped cluster can be represented by a set of spherical clusters.

cluster and the radius is defined to be the maximum distance between the centroid and the data instances.

In the GRIN algorithm, it is assumed that all the incoming data instances are first buffered in an *incoming data pool.* In the first phase of the algorithm, a number of samples are taken from the incoming data pool and the GRACE algorithm, the gravity-based agglomerative hierarchical clustering algorithm described in section 3, is invoked to build a clustering dendrogram for these samples. Actually, how sampling is carried out is not a concern with respect to clustering quality, because, as will be shown in the later part of this paper, the clustering quality of the GRIN algorithm is immune from how incoming data instances are ordered. However, the order of incoming data instances may impact the execution time of the second phase of the GRIN algorithm.

Fig. 3 presents an example that illustrates the operations carried out by the GRIN algorithm. Fig. 3(a) shows the data set and Fig. 3(b) shows 100 random samples taken from the data set. Fig. 3(c) depicts the dendrogram built based on the samples. Each node in the clustering dendrogram corresponds to a cluster of data samples. A cluster is said to be in the spherical shape, if the cluster satisfies either one of the following two conditions:

(1) The cluster contains less than *Min* data instances, where *Min* is a parameter to be set based on the statistical sense discussed in the following.
(2) The cluster contains *Min* or more data instances and passes the statistical test described in the following.

A cluster containing less than *Min* data instances is considered as a spherical cluster by default, because such a cluster does not contain sufficient number samples for any meaningful statistical test to be conducted. Therefore, we just trust GRACE, the gravity- based clustering algorithm, for its capability of identifying spherical clusters of small size. Concerning a cluster containing *Min* or more data instances, the chi-square goodness of fit test [7] is conducted. Fig. 4 presents an example that illustrates the statistical test. The hypothesis of the statistical test is that the data instances of the cluster are uniformly distributed in the sphere defined by the centroid and the radius of the cluster. Accordingly, for the case shown in Fig. 4, the chi-square test of goodness of fit is applied to determine whether the distributions of the data instances in the following 3 subspaces conform with the hypothesis or not:

(1) The subspace enclosed by the sphere of subcluster 1;
(2) The subspace enclosed by the sphere of subcluster 2;
(3) The subspace enclosed by the sphere of the parent cluster but outside subcluster 1 and subcluster 2.

The parent cluster is said to be in the spherical shape, if

$$\frac{(k_1 - V_1 \cdot \mu_0)^2}{V_1 \cdot \mu_0} + \frac{(k_2 - V_2 \cdot \mu_0)^2}{V_2 \cdot \mu_0} + \frac{[(V_0 - V_1 - V_2) \cdot \mu_0]^2}{(V_0 - V_1 - V_2) \cdot \mu_0} \le \chi_\alpha^2 \quad (1)$$

, where $\mu_0 = (k_1 + k_2)/V_0$ and χ_α^2 is a threshold for the chi-square distribution of 2 degrees of freedom. In general, if the parent cluster of concern contains m subclusters, then the chi-square test with m degrees of freedom is conducted. In the statistical test, if one of the child subcluster contains only one single data instance, then the radius of the child cluster is defined to be one half of the distance between

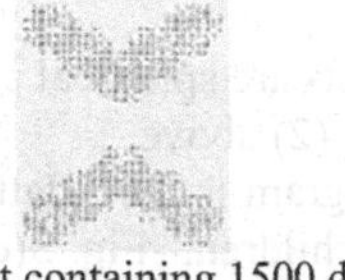

(a) The data set containing 1500 data instances

(b) 100 random samples of the data set in (a).

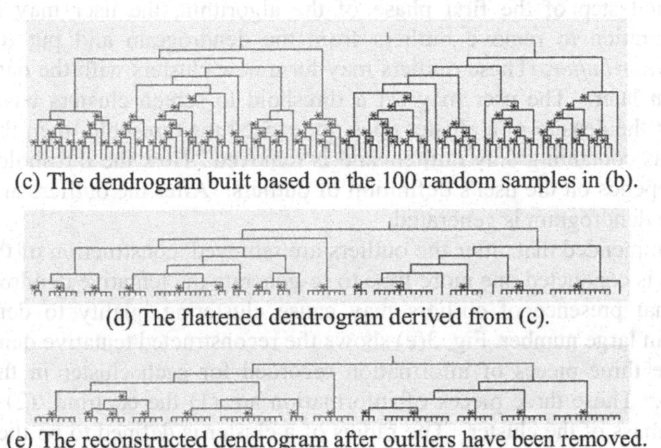

(c) The dendrogram built based on the 100 random samples in (b).

(d) The flattened dendrogram derived from (c).

(e) The reconstructed dendrogram after outliers have been removed.

Fig. 3. An example employed to demonstrate the operations of the GRIN algorithm.

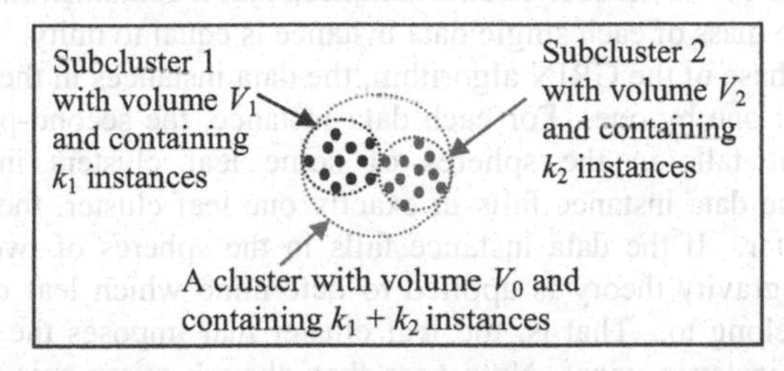

Fig. 4. An example illustrating the statistical test of spherical clusters.

the instance and its nearest neighbor. Note that applying the chi-square goodness of fit in identifying spherical clusters is an empirical mechanism in some sense, as inequality (1) above assumes that the spheres of subclusters 1 and 2 do not overlap and both are completely enclosed by the sphere of the parent cluster.

In Fig. 3(c), the clusters that pass the spherical cluster criterion described above are marked by "*". Note that, the test described here is not applicable to the so-called primitive clusters, i.e. those clusters that contain single data instance. In the following discussion, such clusters are treated as primitive spherical clusters.

The next operation performed in the first phase of the GRIN algorithm is to flatten and prune the bottom levels of the clustering dendrogram in order to derive the so-called *tentative dendrogram*. It is called the tentative dendrogram, because its structure may be modified repetitively in the second phase of the GRIN algorithm. In the flattening process, a spherical cluster in the original dendrogram will become a *leaf cluster* in the tentative dendrogram, if it satisfies all of the following three conditions:

(1) The cluster is of spherical shape;
(2) If the cluster has descendents, all of its descendents are spherical clusters;
(3) Its parent does not satisfy both conditions (1) and (2) above.

The structure under a leaf cluster in the original dendrogram is then flattened so that all the data instances under the leaf cluster become its children. Fig. 3(d) shows the dendrogram derived from flattening the dendrogram depicted in Fig. 3(c).

In the final step of the first phase of the algorithm, the user may conduct an optional operation to remove outliers from the dendrogram and put them into a *tentative outlier buffer*. These outliers may form new clusters with the data instances that come in later. The user may set a threshold to screen clusters based on their densities. If the density of a cluster does not exceed the threshold, then the cluster is considered as containing only outliers and is removed. How the threshold should be set really depends on the user's definition of outliers. After the outliers are removed, the tentative dendrogram is generated.

It is recommended that, after the outliers are removed, construction of the tentative dendrogram is conducted one more time to re-generate the tentative dendrogram. The reason is that presence of outliers may cause clustering quality to deteriorate, if outliers are in large number. Fig. 3(e) shows the reconstructed tentative dendrogram.

There are three pieces of information recorded for each cluster in the tentative dendrogram. These three pieces of information are (1) the centroid, (2) the radius, and (3) the mass of the cluster. The radius of a cluster is defined to be the maximum distance between the centroid and the data instances in this cluster. The mass of a cluster is defined to be the number of data instances that it contains. In other words, it is assumed that the mass of each single data instance is equal to unity.

In the second phase of the GRIN algorithm, the data instances in the incoming data pool are examined one by one. For each data instance, the second-phase algorithm checks whether it falls in the spheres of some leaf clusters in the tentative dendrogram. If the data instance falls in exactly one leaf cluster, then it is inserted into that leaf cluster. If the data instance falls in the spheres of two or more leaf clusters, then the gravity theory is applied to determine which leaf cluster the data instance should belong to. That is, the leaf cluster that imposes the largest gravity force on the data instance wins. Note here that, though every pair of leaf clusters contain disjoint sets of data instances, their spheres defined by their respective centroids and radii may overlap. The third possibility is that the input data instance does not fall in any leaf cluster. In this case, the data instance may be an outlier and a test is conducted to check that. The gravity theory is first applied to determine which leaf cluster imposes the largest gravity force on the data instance. Then, a test is conducted to determine whether this leaf cluster would still satisfy the criterion of being a spherical cluster, if the data instance were added into the cluster. If yes, then the data instance is added into that leaf cluster. If no, the data instance is currently an outlier to the tentative dendrogram and is therefore put into the tentative outlier buffer temporarily. The data instance, however, may form a cluster with other data instances that are already in the tentative outlier buffer or that come in later.

Once the number of data instances in the tentative outlier buffer exceeds a threshold, the gravity-based agglomerative hierarchical clustering algorithm described in section 3 is invoked to construct a new tentative dendrogram. In this reconstruction process, the primitive objects are the leaf clusters in the current tentative dendrogram and the data instances in the tentative outlier buffer. When a new tentative dendrogram has been generated, the same flattening and pruning process invoked in the first phase is conducted and the same criteria is applied to remove the outliers from the new tentative dendrogram. It could occur that the tentative outlier buffer overflows. Should this situation occurs, those data instances that have been in the tentative outlier buffer longest can be treated as outliers and removed from the buffer.

As far as time complexity of the GRIN algorithm is concerned, the time complexity of the first-phase is a constant, as long as the number of samples taken in the first phase is not a function of the size of the data set. It has been shown in [10] that the time complexity of the GRACE algorithm, the hierarchical clustering algorithm invoked by the GRIN algorithm to construct dendrograms, is $O(n^2)$. However, as long as the number of samples is constant, then the time taken by the first phase is a constant.

The analysis of the time complexity of the second phase is a little bit complicated, because it depends on whether the clustering dendrogram will grow indefinitely or not, as new data instances are continuously added into the data set. One important observation regarding the operations of the second phase algorithm is that, if the samples taken in the first phase are good representatives of the entire data set, then most data instances processed in the second phase will fall into one of the leaf clusters in the tentative dendrogram and the structure of the tentative dendrogram will remain mostly unchanged. Even if the spatial distribution of the new data instances that will be continuously added into the data set is very different from that of the data instances that are already in the data set, it does not imply that the dendrogram will continue to grow indefinitely, because the GRIN algorithm keeps flattening and pruning the dendrogram in the second phase. As long as the dendrogram does not grow indefinitely, then the time complexity of the second phase is $O(n)$. The reason is that the time taken to examine each incoming data instance is constant and the time taken to reconstruct the dendrogram in the second phase is bounded, since the size of the tentative outlier buffer is fixed and the number of leaf clusters in the dendrogram is bounded. Only when the incoming data instances keep forming new clusters far away from existing clusters, will the dendrogram continues to grow indefinitely. In this case, the time complexity of the GRIN algorithm is $O(n^2)$.

Section 3. The Gravity-Based Hierarchical Clustering Algorithm

This section discusses the gravity-based hierarchical clustering algorithm that is invoked by the GRIN algorithm for constructing the clustering dendrogram. This algorithm is called the GRACE algorithm, which stands for **GRA**vity-based **C**lustering algorithm for the **E**uclidean space. The GRACE algorithm simulates how a number of water drops move and interact with each other in the cabin of a spacecraft. Due to the gravity force, the water drops in the cabin of a spacecraft will move toward each other. When these water drops move, they will also experience resistance due to the air in the cabin. Whenever two water drops hit, which means that the distance between these two drops is less than the lumped sum of their radii, they merge to form one new and larger water drop. In the simulation model, the merge of water drops corresponds to forming a new, one-level higher cluster that contains two existing clusters. The air resistance is intentionally included in the simulation model in order to guarantee that all these water drops eventually merge into one big drop regardless of how these water drops spread in the space initially. An analysis of why the GRACE algorithm is guaranteed to terminate and its main characteristics can be found in [10, 11].

Fig. 5 shows the pseudo-code of the GRACE algorithm. Basically, the GRACE algorithm iteratively simulates the movement of each node during a time interval and check for possible merge. One key operation in the GRACE algorithm is to compute the velocity of each disjoint node remaining in the system. In the GRACE algorithm, equation (2) below is employed to compute the velocity of a node during one time interval. The derivation of equation (2) involves solving a differential equation under several pragmatical assumptions and is elaborated in [11].

$$v_j = \sqrt{\frac{\left\| \sum_{\text{node } n_i} \vec{F}_{g_i} \right\|}{C_r}} \tag{2}$$

, where $\sum_{\text{node } n_i} \vec{F}_{g_i}$ is the vector sum of the gravity forces that node n_j experiences from all the other disjoint nodes remaining in the physical system at the beginning of the time interval, and C_r is the coefficient of air resistance. According to gravity theory,

$$\left\| \vec{F}_{g_i} \right\| = C_g \cdot \frac{(\text{mass of } n_i)(\text{mass of } n_j)}{\text{distance}^k (n_i, n_j)},$$

where C_g is a coefficient. In the physical world, $k = 2$. However, in the GRACE algorithm, k can be any positive integer number.

Section 4. Experiments

This section reports the experiments conducted to study the following 4 issues concerning the GRIN algorithm.

(1) How the clustering quality of the GRIN algorithm compares with the BIRCH algorithm.
(2) Whether the GRIN algorithm is immune from the order of input data.
(3) Whether the optimal settings of the parameters in the GRIN algorithm are sensitive to the distribution of the input data set.
(4) How the GRIN algorithm performs in terms of execution time in real applications.

Table 1 shows how the parameters in the GRIN and GRACE algorithms are set in the experiments.

```
W : the set containing all disjoint nodes remaining in the system. At the beginning, W contains all
initial nodes.
R : Resolution of time interval. (R = 100)
Repeat
    min_D = MAX;
    max_V = 0;
    For every pair of nodes n_i, n_j ∈ W {
        calculate the distance D_ij between n_i and n_j;
        if (min_D > D_ij) min_D = D_ij;
    };
    For every n_i ∈ W {
        calculate the new velocity V_i of n_i according to equation (2);
        if (max_V < V_i) max_V = V_i;
    };
    time interval T = ( min_D / R ) / max_V;
    For every n_i ∈ W {
        calculate the new position of n_i based on V_i and T;
    }
    For every pair of nodes n_i, n_j ∈ W {
        if (n_i and n_j hit during the time interval T){
            create a new cluster containing the clusters represented by n_i and n_j;
            merge n_i and n_j to form a new node n_h with lumped masses and merged
                momentum;
            delete n_i and n_j from W;
            add n_h to W;
        };
    };
Until (W contains only one node);
```

Fig. 5. The pseudo-code of the GRACE algorithm.

Fig. 6(a) shows a data set used in the experiments. In Fig. 6(a), natural clusters are identified according to human's intuition and are numbered. Fig. 6(b) depicts the clusters identified by the GRIN algorithm and the dendrogram constructed. In this experiment, data is fed to the GRIN algorithm one natural cluster by another natural cluster in the following order $1 \rightarrow 2 \rightarrow 3 \rightarrow \ldots$ For providing better visualization

GRACE	
k : Order of the distance term gravity force formula	10
C_g : Coefficient of gravity force	1
C_r : Coefficient of air resistance	100
M_0 : Initial mass of each node	1
D_0 : Material density of the node	1

GRIN	
sample size	500
size of tentitive outlier buffer	500
Min	3
Significance of the χ^2 test	0.01

Table 1. Parameter settings of the GRIN and GRACE algorithms in the experiments.

quality, only the clusters at the top few levels of the dendrogram are plotted. The result in Fig. 6(b) shows that the GRIN algorithm is able to identify natural clusters flawlessly. Several runs of the GRIN algorithm were executed with the data fed to the algorithm in different orders. In one particular run, the data was fed to the algorithm also one natural cluster by another natural cluster but in the reverse order.

In the remaining runs, data was fed to the algorithm in a random order. The outputs of all these separate runs of the GRIN algorithm are basically identical to what is depicted in Fig. 6(b). The experiment results show that the clustering quality of the GRIN algorithm is immune from the order of input data.

Fig. 6(c) and (d) depict the data clusters identified by the BIRCH algorithm with different hierarchical clustering algorithms incorporated. The BIRCH algorithm is employed for comparison, because it is a well-known incremental hierarchical clustering algorithm that features *O(n)* time complexity. In Fig. 6(c), the portion of the data set in which the BIRCH algorithm with the complete-link algorithm incorporated fails to deliver reasonable clustering quality is enlarged and data instances belonging to different clusters are marked by different symbols. In this case, the BIRCH algorithm mixes the data instances from natural clusters 2, 3, and 4. BIRCH's failure is due to two factors. First, BIRCH uses a global parameter to control the diameter of leaf subclusters. Therefore, as exemplified in Fig. 1, it could occur that no optimal value for this parameter can be found when the local distributions of the data set are highly skewed. Second, the complete-link algorithm itself suffers bias towards spherical clusters [10, 11]. Fig. 6(d) reveals that the clustering quality of BIRCH is improved when the GRACE algorithm is incorporated instead of the complete-link algorithm. The GRACE algorithm contributes to improvement of clustering quality, because the complete-link algorithm suffers bias towards spherical clusters in a much higher degree than the GRACE algorithm [10, 11]. Nevertheless, there are still a few flaws in Fig. 6(d) due to the parameter setting problem with the BIRCH algorithm.

Fig. 7(a) depicts the clusters identified by the GRIN algorithm and the dendrogram constructed for another data set. Fig. 7(b) shows the clusters outputted by the BIRCH algorithm with the GRACE algorithm incorporated. Again, several flaws are observed as marked by the squares.

In the third experiment, a subset of the Sequoia 2000 benchmark [13] is used to test how the GRIN algorithm performs in dealing with a real data set. The subset contains the locations of all the high schools in California. Fig 8(a) plots the 989 location instances in the subset. Fig. 8(b) depicts the outlook of the data set after outliers are removed by GRIN algorithm. In this case, the threshold is set to 1% of the average density of all the leaf clusters. After the outliers are removed, the remaining data set contains 946 location instances. Fig. 8(c) shows the clusters outputted by the GRIN algorithm. In Fig. 8(c), different clusters are plotted using different symbols. We also use the data set shown in Fig. 8(b) to test the how the BIRCH algorithm performs when operating with the complete-link algorithm and the GRACE algorithm, respectively. Due to the limited space, we only show the clustering results inside the rectangle box in Fig. 8(c). As shown in Fig. 8(d) and 8(e), in both cases, the largest natural cluster in the rectangle region, the cluster marked by X in Fig. 8(c) is divided into two parts and the left part is clustered with the location instances further to the left.

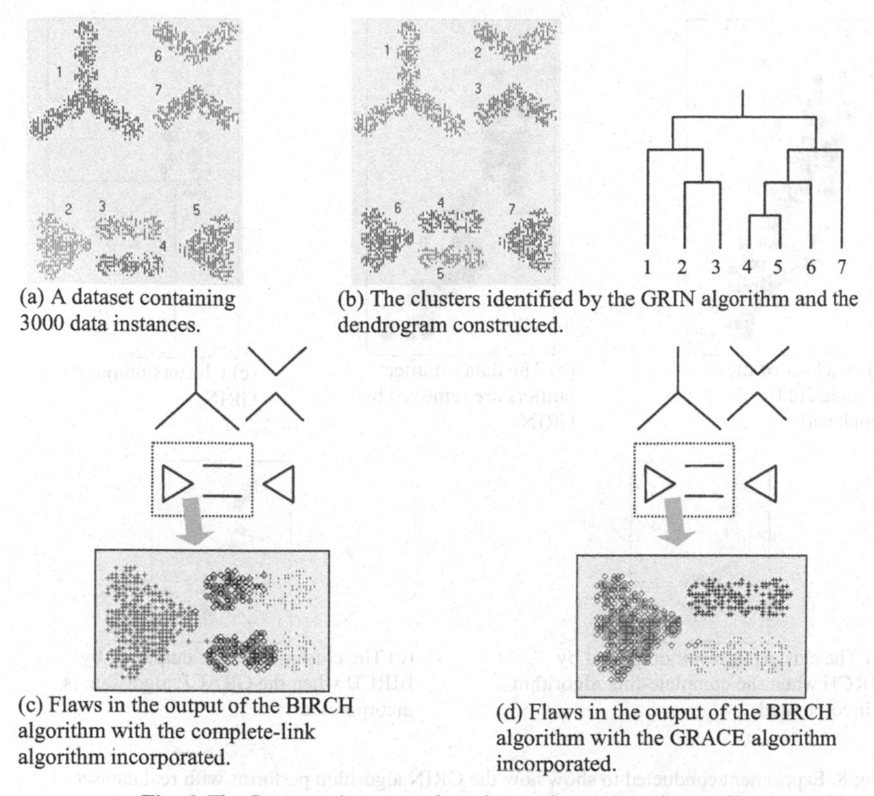

(a) A dataset containing 3000 data instances.

(b) The clusters identified by the GRIN algorithm and the dendrogram constructed.

(c) Flaws in the output of the BIRCH algorithm with the complete-link algorithm incorporated.

(d) Flaws in the output of the BIRCH algorithm with the GRACE algorithm incorporated.

Fig. 6. The first experiment conducted to evaluate clustering quality.

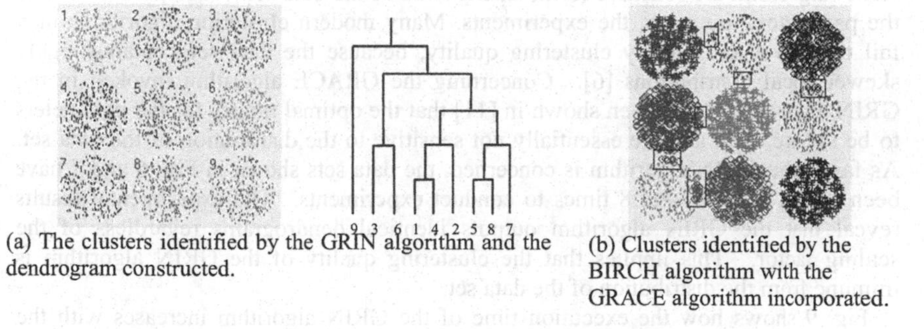

(a) The clusters identified by the GRIN algorithm and the dendrogram constructed.

(b) Clusters identified by the BIRCH algorithm with the GRACE algorithm incorporated.

Fig. 7. The second experiment conducted to evaluate clustering quality.

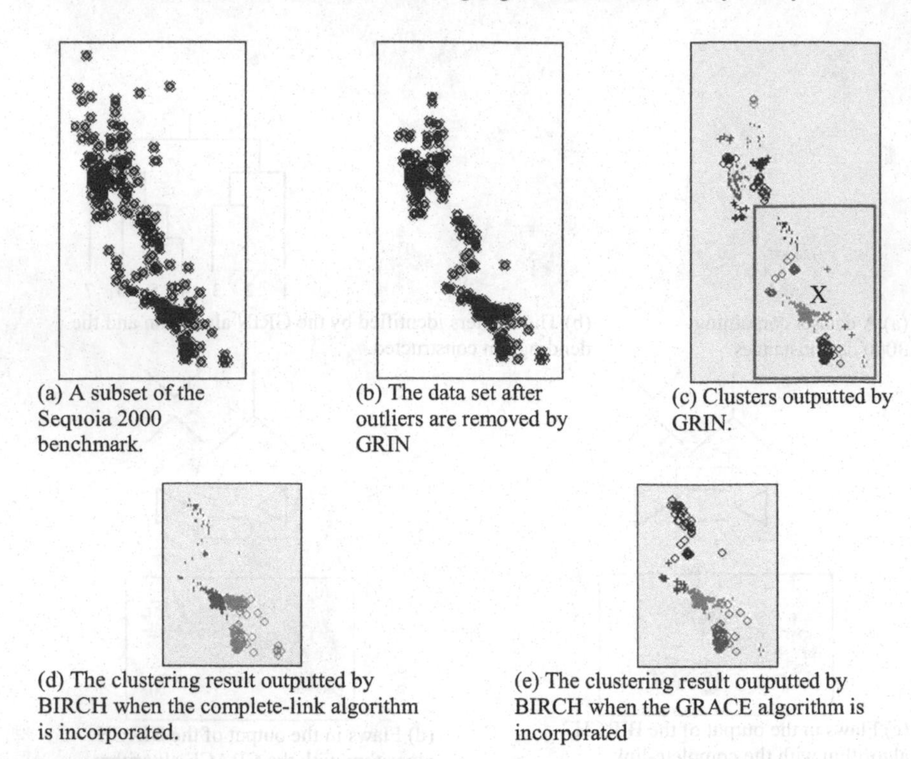

(a) A subset of the Sequoia 2000 benchmark.

(b) The data set after outliers are removed by GRIN

(c) Clusters outputted by GRIN.

(d) The clustering result outputted by BIRCH when the complete-link algorithm is incorporated.

(e) The clustering result outputted by BIRCH when the GRACE algorithm is incorporated

Fig. 8. Experiment conducted to show how the GRIN algorithm performs with real datasets.

Experiments have been conducted to check whether the parameter settings in the GRIN algorithm are sensitive to the distribution of the data set. Table. 1 shows how the parameters are set in the experiments. Many modern clustering algorithms may fail to deliver satisfactory clustering quality, because the data set contains highly skewed local distributions [6]. Concerning the GRACE algorithm invoked in the GRIN algorithm, it has been shown in [11] that the optimal ranges for the parameters to be set are wide and are essentially not sensitive to the distribution of the data set. As far as the GRIN algorithm is concerned, the data sets shown in Fig. 6 and 7 have been scaled up 2, 4, and 8 times to conduct experiments. The experimental results reveal that the GRIN algorithm outputs identical dendrograms regardless of the scaling factor. This implies that the clustering quality of the GRIN algorithm is immune from the distribution of the data set.

Fig. 9 shows how the execution time of the GRIN algorithm increases with the number of data instances in the data set. The experiment was conducted on a machine equipped with a 600-MHz Intel Pentium-III CPU and 328 Mbytes main memory and running Microsoft Window 2000 operating system. The dataset used is the point data

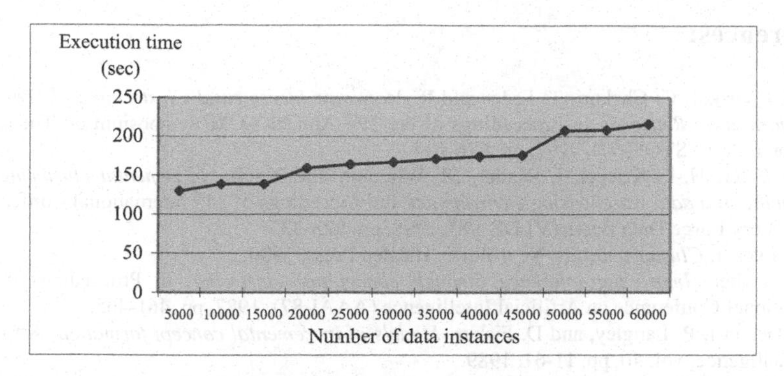

Fig. 9. Experiment conducted to test the performance of the GRIN algorithm.

in Sequoia 2000 Earth benchmark [13], which contains 62556 data instances in total. In this experiment, we ran the GRIN algorithm 3 times with different initial samples randomly selected from the benchmark dataset. The result reveals that the GRIN algorithm generally features $O(n)$ time complexity. One may observe that there are two sections with steeper slopes, one is around 20000 and another one is around 50000. The abrupt rises of execution time are due to the operation to reconstruct the dendrogram in the second phase of the GRIN algorithm.

Section 5. Conclusions

This paper presents the GRIN algorithm, an incremental hierarchical clustering algorithm based on gravity theory in physics. The incremental nature of the GRIN algorithm implies that it is particularly suitable for handling the already huge and still growing databases in modern environments. Its hierarchical nature provides a highly desirable feature for many applications in biological, social, and behavior studies due to the need to construct taxonomies. In addition, the GRIN algorithm delivers favorite clustering quality and generally features $O(n)$ time complexity. The experiments conducted in this study reveal that the clustering quality of the GRIN algorithm is immune from the order of input data and the optimal parameter settings are not sensitive to the distribution of the data set.

There are some issues regarding the GRIN algorithm that deserve further study. One interesting issue is the approach to identify outliers. Due to different natures of applications, some may require more strict criteria for identifying outliers, while the other may require less strict criteria. Therefore, alternative approaches may be employed for achieving different goals. Another interesting issue is the approach to prune the dendrogram. If a more aggressive approach was employed, then the efficiency of the GRIN algorithm would be upgraded, because the dendrogram would be smaller in terms of the number of nodes. However, clustering quality may be traded. Again, different applications may impose different criteria. Therefore, this issue deserves further study.

References:

1 M. Charikar, C. Chekuri, T. Feder and R. *Motwani: Incremental Clustering and Dynamic Information Retrieval.* In Proceedings of the 29th Annual ACM Symposium on Theory of Computing (STOC-97), 1997, pp. 626-634.

2 M. Ester, H.-P. Kriegel, J. Sander, M. Wimmer, and X. Xu. *Incremental clustering for mining in a data warehousing environment.* In Proceedings of 24th International Conference on Very Large Data Bases (VLDB-98), 1998, pp. 323-333.

3. B. Everitt, *Cluster analysis*, New York : Halsted Press, 1980.

4 D. Fisher, *Improving inference through conceptual clustering*, In Proceedings of 6th National Conference on Artificial Intelligence (AAAI-87), 1987, pp. 461-465.

5 J. Gennari, P. Langley, and D. Fisher, *Models of incremental concept formation*, Artificial Intelligence, vol. 40, pp. 11-61, 1989.

6. J. Han, M. Kamber, *Data Mining: Concepts and Techniques*, San Francisco : Morgan Kaufmann Publishers, 2000.

7 R. V. Hogg and E. A. Tanis, *Probability and statistical inference*, New Jersey : Prentice-Hall, 2001.

8. A.K. Jain, R.C. Dubes, *Algorithms for clustering data*, Englewood Cliffs, N.J. : Prentice Hall, 1988.

9. A.K. Jain, M.N. Murty, P.J. Flynn, *Data Clustering: A Review*, ACM Computing Surveys, vol. 31, no. 3, pp. 264-323, 1999.

10 Yen-Jen Oyang, Chien-Yu Chen, and Tsui-Wei Yang, *A Study on the Hierarchical Data Clustering Algorithm Based on Gravity Theory*, In Proceedings of 5th European Conference on Principles and Practice of Knowledge Discovery in Databases (PKDD-01), 2001, pp. 350-361.

11 Yen-Jen Oyang, Chien-Yu Chen, Shien-Ching Hwang, and Cheng-Fang Lin, *Characteristics of a Hierarchical Data Clustering Algorithm Based on Gravity Theory*, Technical Report of NTUCSIE 02-01.
(Available at *http://mars.csie.ntu.edu.tw/~cychen/publications_on_dm.htm*)

12 A. Ribert, A. Ennaji, and Y. Lecourtier, *An incremental Hierarchical Clustering*, In Proceedings of 1999 Vision Interface Conference, 1999, pp. 586-591.

13. M. Stonebraker, J. Frew, K. Gardels and J. Meredith, *The Sequoia 2000 Storage Benchmark,* In Proceedings of 1993 ACM-SIGMOD International Conference on Management of Data (SIGMOD-93), 1993, pp. 2-11.

14 I. H. Witten, *Data mining: practical machine learning tools and techniques with Java implementations*, San Francisco, Califonia : Morgan Kaufmann, 2000.

15. T. Zhang, R. Ramakrishnan, M. Livny, *BIRCH: An Efficient Data Clustering Method for Very Large Databases*, In Proceedings of the 1996 ACM-SIGMOD International Conference on Management of Data (SOGMOD-96), Jun. 1996, pp. 103-114.

Adding Personality to Information Clustering

Ah-Hwee Tan and Hong Pan

Laboratories for Information Technology
21 Heng Mui Keng Terrace, Singapore 119613
{ahhwee,panhong}@lit.org.sg

Abstract. This article presents a new information management method called user-configurable clustering that integrates the flexibility of clustering systems in handling novel data and the ease of use of categorization systems in providing structure. Based on a predictive self-organizing network that performs synchronized clustering of information and preference vectors, a user can influence the clustering of information vectors by encoding his/her preferences as preference vectors. We illustrate a sample session to show how a user may create and personalize an information portfolio according to his/her preferences and how the system discovers novel information groupings while organizing familiar information according to user-defined themes.

1 Introduction

Categorization and clustering have been two fundamentally distinct approaches to information management. Categorization is supervised in nature. It provides good control in the sense that information is organized according to the structure defined by a user. However, due to the predefined structure, categorization is not well suited to handle novel data. In addition, much effort is needed to build a categorization system beforehand. It is necessary to specify classification knowledge in terms of some classification rules/keywords [3] or to construct a categorization system through some supervised learning algorithms [5].

Clustering, on the other hand, is unsupervised in nature. For unsupervised systems such as k-means [2], Scatter/Gather [1], and Self-Organizing Map (SOM) [4], there is no need to train or construct a classifier as information is organized automatically into groups based on their similarities. However, a user has very little control on how the information is grouped together. Although it is possible to fine tune the parameter values of similarity measures to control the degree of coarseness, the clustering structure is affected globally. In addition, the structure uncovered through the clustering process can be quite unpredictable.

In this paper, we present a novel information management method known as user-configurable clustering that integrates the complementary strengths of clustering and categorization. Using user-configurable clustering, information is first organized through automatic clustering to derive the natural information groupings. A user, upon inspecting the information groupings, can then modify the structure according to his/her requirement and preferences through a suite of

M.-S. Chen, P.S. Yu, and B. Liu (Eds.): PAKDD 2002, LNAI 2336, pp. 251–256, 2002.

cluster manipulation functions. It is an interactive process of clustering, personalization, and discovery through which a user turns an automatically generated cluster structure into his/her preferred organization.

The rest of this article is organized as follows. Section 2 presents the user-configurable clustering system architecture. Section 3 summarizes the key clustering and personalization algorithms. Section 4 illustrates how user-configurable clustering can be used for personalized content management through a sample session.

2 System Architecture

A user-configurable information clustering system (Figure 1) comprises an information clustering engine for clustering of information based on similarities, a user interface module for displaying the information groupings and obtaining user preferences, a personalization module for defining, labelling, and modifying cluster structure, and a knowledge base for storing user-defined cluster structures.

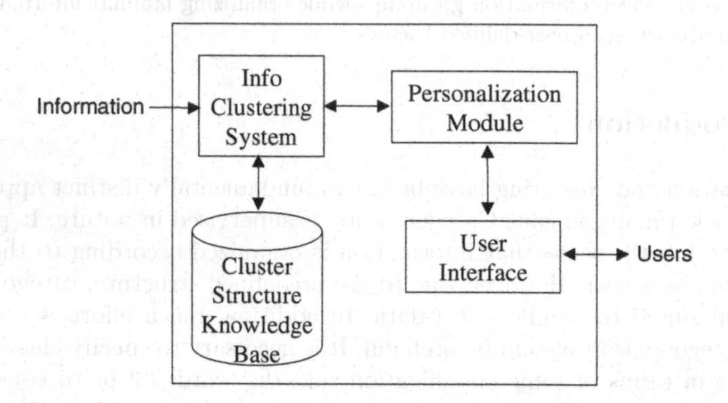

Fig. 1. The personalized information clustering system architecture.

The information clustering engine is based on a class of predictive self-organizing networks known as fuzzy Adaptive Resonance Associative Map (fuzzy ARAM) [6] that learns information groupings or clusters dynamically on-the-fly to encode pairs of information and preference vectors. ARAM is a natural extension of Adaptive Resonance Theory (ART) networks to incorporate supervisory preference signals.

The personalization module works in conjunction with the information clustering engine to incorporate user preferences to modify the automatically generated cluster structure. Through the user interface module and the personalization module, a user is able to perform a wide range of cluster manipulation functions including labelling, adding, deleting, merging, and splitting of clusters through the use of labels or themes. The customized cluster structure, in the form of fuzzy ARAM network, can be stored in the cluster structure knowledge base and retrieved at a later stage for processing new information. Based on the personalized cluster structure, new information can be organized according to the user's preferences captured over the previous sessions.

3 Algorithms

3.1 Clustering

If a predefined cluster structure exists, the information clustering engine loads the network from the cluster structure knowledge base before clustering. Otherwise, a new network is created which contains zero cluster. During clustering, for each unit of information ($\mathcal{I}$), a pair of vectors $(\mathbf{A}, \mathbf{E})$ is presented to the system under a *learning* mode, where $\mathbf{A}$ is the information vector of $\mathcal{I}$ and $\mathbf{E}$ is a null vector such that $E_i = 0$ for $i = 1, \ldots, N$. Clustering is completed when the network is stable, in the sense that for a given set of information, no new cluster is created and changes in template vectors are below a specific threshold.

With a predefined cluster structure, the information clustering engine organizes the information according to the cluster structure. Without predefined network structure, it reduces to a pure clustering system that self-organizes the information based on the similarities among the information vectors only.

3.2 Personalization

The information clustering engine can also operate in an *insertion* mode whereby a pair of information and preference vectors can be inserted directly into a fuzzy ARAM network. Whereas the *learning* mode is used for the discovery of the information clusters and obtaining the cluster assignments of the information vectors, the *insertion* mode enables a user to influence the clusters created through indicating his/her own preferences in the forms of preference vectors. Due to the space constraint, we summarize the main cluster personalization functions below.

Labelling Information Clusters: Associating clusters with labels or themes allows a user to "mark" specific information groupings that are of interest to the user so that the information can be found readily in the future and new information can be organized according to such information groupings.

Adding Information Clusters: A user can define and insert his/her own clusters into an ARAM network so that the information can be organized according to such information groupings. The inserted clusters reflect the user's preferred way of grouping information and are used as the default slots of organizing information.

Merging Information Clusters: Merging of clusters allows a user to combine two or more information groupings generated by clustering into a common theme.

Splitting Information Clusters: Splitting of clusters allows a user to reorganize an information group, that he/she deems containing diverse content, into smaller clusters of specific themes.

4 Experiments

An information portfolio on "text mining" was created by integrating search results of four internet search engines. For illustration purpose, we constrained the size of the portfolio by selecting only top 25 hits from each search engine. After removing duplicated links, there were a total of 71 hits. Figure 2 depicts the clustering results based on a combination of URL and content-based keyword features. There are 17 clusters, each characterized by one to three keywords listed in decreasing order of importance. Three clusters, namely *fortune*, *data*, and *information* [1] are the most prominent ones with 20, 17, and 7 documents respectively.

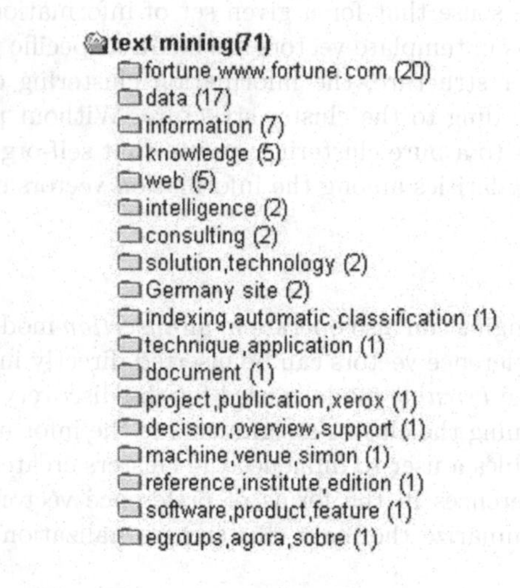

Fig. 2. Clusters created based on the 71 documents collected through four internet search engines.

Based on the raw cluster structure generated, we illustrate how a user may use the various cluster manipulation functions, namely labelling, inserting, merging, and splitting, to personalize his/her portfolios.

Figure 3 shows an exemplary personalized portfolio. The *fortune* cluster containing news articles from the Fortune news site has been labelled under the theme of *Fortune News*. In addition, a number of user-defined cluster have been created under the theme of *Technology*. The documents in these user-defined

[1] For convenience, we refer to a cluster by the first keyword in its keyword list.

clusters are mainly from the original *data*, *information*, and *knowledge* clusters in figure 2. A user-defined cluster with a keyword *IBM* under *Company/Product* manages to pull out a link to a IBM Business Intelligence/Text Mining page which was buried somewhere previously. With these user-defined clusters, new clusters have emerged. The most interesting grouping is the *websom* cluster (under Company/Product) containing three links related to WEBSOM, a well known text mining technique.

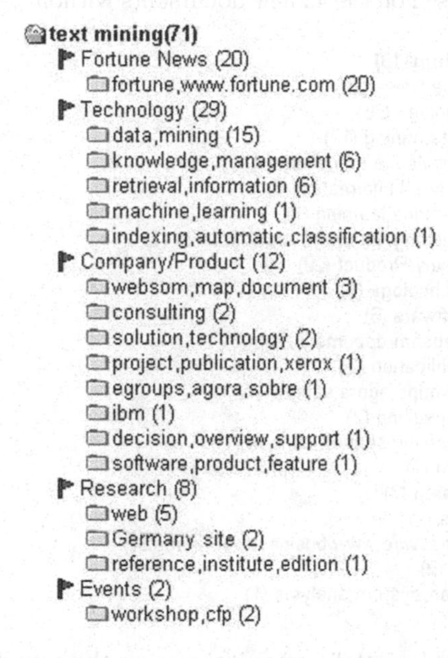

Fig. 3. A personalized portfolio on text mining.

A new set of 42 documents was collected through three additional search engines. Without prior structure, the documents would be organized into the clusters as shown in figure 4. In contrast, figure 5 shows the clustering result when the new documents are organized based on the personalized cluster structure. There are 113 documents in the combined portfolio. A significant portion of the new information, especially those in the *search*, *software*, *information*, and *knowledge* clusters (figure 4), have been organized under the themes of *technology* and *Company/Product*. Some of the other clusters remain, highlighting information that do not fit into the personalized portfolio. The most prominent group is the *businesswire* cluster that contains news articles from the BusinessWire news site. This indicates that the system is able to discover novel information groupings while organizing familiar information into the user's personalized structure.

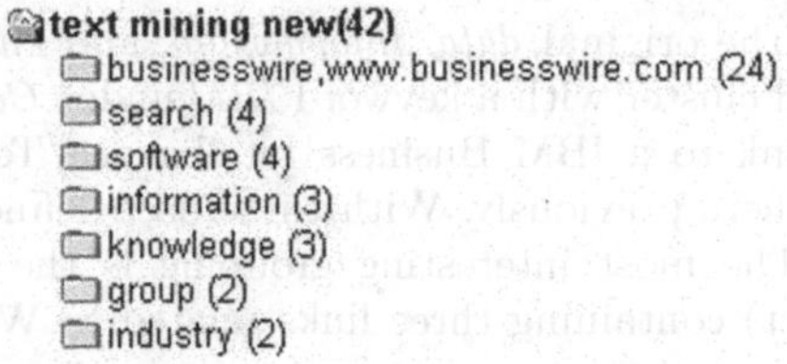

text mining new(42)
businesswire,www.businesswire.com (24)
search (4)
software (4)
information (3)
knowledge (3)
group (2)
industry (2)

Fig. 4. Clusters created based on the 42 new documents without personalization.

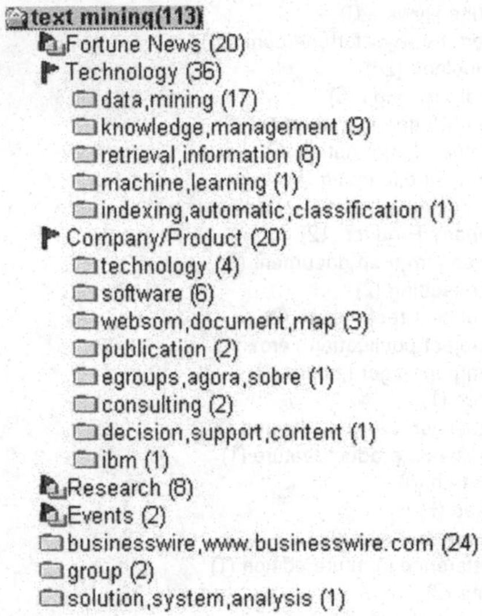

text mining(113)
Fortune News (20)
Technology (36)
data,mining (17)
knowledge,management (9)
retrieval,information (8)
machine,learning (1)
indexing,automatic,classification (1)
Company/Product (20)
technology (4)
software (6)
websom,document,map (3)
publication (2)
egroups,agora,sobre (1)
consulting (2)
decision,support,content (1)
ibm (1)
Research (8)
Events (2)
businesswire,www.businesswire.com (24)
group (2)
solution,system,analysis (1)

Fig. 5. The personalized portfolio after integrating the new documents.

References

1. D. Cutting, D. Karger, J. Pedersen, and J. Tukey. Scatter/gather: a cluster-based approach to browsing large document collections. In *Proceedings, 15th ACM SIGIR*, 1992.
2. V. Faber. *Clustering and the Continuous k-Means Algorithm.* Los Alamos Science, 1994.
3. P. J. Hayes, P. M Andersen, I. B. Nirenburg, and L. M. Schmandt. Tcs: A shell for content-based text categorization. In *Proceedings, Sixth IEEE Conference on Artificial Intelligence Applications*, pages 320–326, 1990.
4. S. Kaski, T. Honkela, K. Lagus, and T. Kohonen. Creating an order in digital libraries with self-organizing maps. In *Proceedings, WCNN'96, San Diego*, 1996.
5. D. D. Lewis, R. E. Schapire, J. P. Callan, and R. Papka. Training algorithms for linear text classifiers. In *Proceedings, SIGIR'96*, pages 298–306, 1996.
6. A.-H. Tan. Adaptive Resonance Associative Map. *Neural Networks*, 8(3):437–446, 1995.

Clustering Large Categorical Data

François-Xavier Jollois and Mohamed Nadif

Université de Metz,
Laboratoire d'Informatique Théorique et Appliquée,
Ile du Saulcy, 57045 Metz Cedex, France,
jollois@lita.univ-metz.fr, nadif@iut.univ-metz.fr

Abstract. Clustering methods often come down to the optimization of a numeric criterion defined from a distance or from a dissimilarity measure. It is possible to show that this problem is often equivalent to the estimation of the parameters of a probabilistic model under the classification likelihood approach. For instance, we know that the inertia criterion optimized under the k-means algorithm corresponds to the hypothesis of a population arising from a Gaussian mixture. In this paper, we propose an adapted mixture model for categorical data. Using the classification likelihood approach, we develop the Classification EM algorithm (CEM) to estimate the parameters of the mixture model. With our probabilistic model, the data are not denatured and the estimated parameters readily indicate the characteristics of the clusters. This probabilistic approach gives an interpretation of the criterion optimized by the k-modes algorithm which is an extension of k-means to categorical attributes and allows us to study the behavior of this algorithm.

1 Introduction

When partitioning the data is the main concern, it is implicitly assumed that each cluster can be approximately viewed as a sample from one of the mixture components. Thus, the clustering problem can be regarded as a problem of estimating of the mixture parameters (see for instance [1] and [9]). Setting this problem under the classification maximum likelihood ([13]), it has been shown that some of most popular heuristic clustering methods are approximate estimation methods for particular probability models. For instance, the inertia criterion optimized by the well known k-means clustering algorithm corresponds to a Gaussian mixture with variances and proportions fixed.

For categorical data, Ralambondrainy ([12]) has used the k-means algorithm on binary data obtained after conversion of multiple category attributes into binary attributes. In contrast, Nadif and Marchetti ([11]) and Huang ([7]) directly worked on the categorical data and proposed the same criterion to be optimized, but carried out the optimization using two different algorithms. Huang ([7]) used the version of k-means proposed by MacQueen ([10]) and adapted it into the k-modes algorithm. Nadif and Marchetti ([11]) used dynamic cluster analysis ([3]). The aim of our contribution is to prove that the criterion optimized by these

M.-S. Chen, P.S. Yu, and B. Liu (Eds.): PAKDD 2002, LNAI 2336, pp. 257–263, 2002.

authors corresponds to a *restricted* probabilistic model. Moreover, we generalize this criterion and we present an algorithm based on the Classification EM algorithm, noted CEM ([1]).

The paper is organized as follows. Section 2 begins with a brief description of the k-modes algorithm. In Section 3, we will review the definition of the mixture model and the classification likelihood approach in general context then we will focus on the case of categorical data. We will propose the probabilistic model and the criterion to be optimized by the Classification EM algorithm. Section 4 will devoted to the comparison between these algorithms on simulated data. Finally, Section 5 summarizes the main points of this work.

2 k-modes Algorithm

Let be a set of n instances $\mathbf{x}_1, \ldots, \mathbf{x}_n$ described by q categorical attributes $\mathbf{x}^1, \ldots, \mathbf{x}^q$. The data matrix is noted $\mathbf{x}$ and defined by $\mathbf{x} = \{(x_i^j); i = 1, \ldots, n; j = 1, \ldots, q\}$. Instance $\mathbf{x}_i$ is represented as $[x_i^1, \ldots, x_i^q]$ and for each attribute $\mathbf{x}^j$, we note c^j the number of categories. For finding the optimal partition $\mathbf{z} = (z_1, \ldots, z_k)$ where k is supposed to be known, Huang ([7]) proposed to minimize the following criterion

$$E = \sum_{\ell=1}^{k} \sum_{i \in z_\ell} d(\mathbf{x}_i, \mathbf{m}_\ell),$$

where $\mathbf{m}_\ell$ is the center of the ℓth cluster z_ℓ and d is a dissimilarity measure. The author proposed to use the measure d defined in the simplest case by $d(\mathbf{x}_i, \mathbf{m}_\ell) = \sum_{j=1}^{q} \delta(x_i^j, m_\ell^j)$ where $\delta(x,y) = 0$ if $x = y$ and $\delta(x,y) = 1$ otherwise. Then, $d(\mathbf{x}_i, \mathbf{x}_h)$ gives the total mismatches of the corresponding attribute categories of the two instances $\mathbf{x}_i$ and $\mathbf{x}_h$. Hence, the criterion is simply the total number of differences between the instances of each cluster z_ℓ and their cluster representation $\mathbf{m}_\ell$.

With the k-modes algorithm which aims to minimize the criterion E, each cluster z_ℓ is characterized by a center $\mathbf{m}_\ell = (m_\ell^1, \ldots, m_\ell^q)$ where m_ℓ^j is the mode of the attribute $\mathbf{x}^j$ in the cluster z_ℓ. For convenience, the vectors $\mathbf{m}_\ell$ are called the modes. Like the k-means algorithm, k-modes also produces local optimal solutions which are affected by the initialization step. Even if the major advantages of this algorithm are simplicity and fast convergence, k-modes presents limitations in some situations that we will study later.

3 Mixture Model and Classification Likelihood Approach

3.1 General Context

In the mixture model, the data matrix $\mathbf{x}$ is assumed to be a sample $\mathbf{x}_1, \ldots, \mathbf{x}_n$ from a probability distribution with the density $f(\mathbf{x}_i; \theta) = \sum_{\ell=1}^{k} p_\ell f_\ell(\mathbf{x}_i; \alpha_\ell)$ where the p_ℓ's are the mixing proportions ($0 < p_\ell < 1$ for all $\ell = 1, \ldots, k$ and $\sum_{\ell=1}^{k} p_\ell = 1$) and the $f_\ell(\mathbf{x}_i; \alpha_l)$ are the density from the same parametric family.

The parameter θ of this model is the vector $(p_1, \ldots, p_k; \alpha_1, \ldots, \alpha_k)$. And the parameter k is assumed to be known.

For convenience, we represent the partition of the sample into k clusters by $\mathbf{z} = (z_1, \ldots, z_k)$. With the classification likelihood approach ([13]), the problem of clustering reduces to maximizing of the complete data log-likelihood also called the classification log-likelihood, which can be written

$$L_c(\mathbf{x}, \mathbf{z}, \theta) = \sum_{\ell=1}^{k} \sum_{\mathbf{x}_i \in z_\ell} \log(f_\ell(\mathbf{x}_i; \alpha_\ell)) + \sum_{\ell=1}^{k} \#z_\ell \log(p_\ell),$$

where $\#z_\ell$ denotes the cardinality of the cluster z_ℓ. The complete data log-likelihood $L_c(\mathbf{x}, \mathbf{z}, \theta)$ is maximized using generally the Classification EM (CEM) algorithm ([1]) which is a version of EM algorithm ([2]). The CEM algorithm proceeds very much like the EM algorithm, with an additional C(Classification)-step between the E(Estimation)-step and the M(Maximization)-step. In the C-step, we assign each $\mathbf{x}_i$ to the cluster which provides the maximum posterior probability

$$t_\ell^{(c)}(\mathbf{x}_i) = \frac{p_\ell f_\ell(\mathbf{x}_i; \alpha_\ell)}{\sum_{\ell'=1}^{k} p_{\ell'} f_{\ell'}(\mathbf{x}_i; \alpha_{\ell'})}.$$

We repeat these three steps until the convergence of the algorithm and we obtain a local maximum of $L_c(\mathbf{x}, \mathbf{z}, \theta)$. Like the k-means, the CEM algorithm depends on initialization step. For example, it suffices to start with random instances at each time. Note that the computational complexity of this algorithm is $O(nkqi)$ where i is the number of iterations, unknown and independent of the number of instances n. Usually, k, q and i are clearly less than n. In particular, for each of our runs of CEM, we note that CEM converges generally in less than 25 iterations. Then, the computational complexity is linear, in $O(n)$. Later in this paper, we present a practical study of the computational complexity of CEM.

3.2 Categorical Data

For the categorical data, we consider a restricted latent class model ([4]) noted $[p_\ell, \varepsilon_\ell^j]$ ([11]) where the density (frequency function) of an instance $\mathbf{x}_i$ can be expressed as

$$f(\mathbf{x}_i; \theta) = \sum_{\ell=1}^{k} p_\ell f_\ell(\mathbf{x}_i; \alpha_\ell), \tag{1}$$

with the ℓth component density given by

$$f_\ell(\mathbf{x}_i; \alpha_\ell) = \prod_{j=1}^{q} (1 - \varepsilon_\ell^j)^{1-\delta(x_i^j, m_\ell^j)} \left(\frac{\varepsilon_\ell^j}{c^j - 1} \right)^{\delta(x_i^j, m_\ell^j)}. \tag{2}$$

Here, the parameter α_ℓ consists of $(\mathbf{m}_\ell, \varepsilon_\ell)$ with $\mathbf{m}_\ell$ is the mode of the ℓth component and ε_ℓ is a q-dimensional vector of probabilities. The density (2) expresses that, for the ℓth component, the attribute $\mathbf{x}^j$ takes category m_ℓ^j with the greatest probability $1 - \varepsilon_\ell^j$ and takes each other category with the same probability $\varepsilon_\ell^j/(c^j - 1)$. Thus, this mixture model is *parsimonious*. Indeed, the number of parameters to be estimated is $k(2q+1)$ instead of $k(\sum_{j=1}^q c^j - q + 1) - 1$ when we consider the latent class model. From (1), the associated classification log-likelihood can be written

$$L_c(\mathbf{x}, \mathbf{z}, \theta) = \sum_{\ell=1}^{k} \sum_{\mathbf{x}_i \in z_\ell} \sum_{j=1}^{q} \log\left(\frac{\varepsilon_\ell^j}{(1-\varepsilon_\ell^j)(c^j-1)}\right) \delta(x_i^j, m_\ell^j)$$
$$+ \sum_{\ell=1}^{k} \sum_{j=1}^{q} \#z_\ell \log(1-\varepsilon_\ell^j) + \sum_{\ell=1}^{k} \#z_\ell \log(p_\ell).$$

Now, we describe the different steps of the CEM algorithm.

- *E-step*: For $\ell = 1, \ldots, k$ and for $i = 1, \ldots, n$, compute the current posterior probabilities: $t_\ell(\mathbf{x}_i) \propto p_\ell f_\ell(\mathbf{x}_i; (\mathbf{m}_\ell, \varepsilon_\ell))$,
- *C-step*: For $\ell = 1, \ldots, k$, the cluster $\mathbf{z}_\ell$ is defined by

$$\mathbf{z}_\ell = \{\mathbf{x}_i / t_\ell(\mathbf{x}_i) = max[t_h(\mathbf{x}_i); h = 1, \ldots, k]\},$$

- *M-step*: For $\ell = 1, \ldots, k$, compute the maximum likelihood (p_ℓ, α_ℓ). For $j = 1, \ldots, q$, it leads to

$$\begin{cases} p_\ell = \frac{\#\mathbf{z}_\ell}{n}, \\ m_\ell^j = \text{ mode of } \{x_i^j, \mathbf{x}_i \in \mathbf{z}_\ell\}, \\ \varepsilon_\ell^j = \frac{e_\ell^j}{n_\ell} \text{ where } e_\ell^j = \sum_{\mathbf{x}_i \in \mathbf{z}_\ell} \delta(x_i^j, m_\ell^j). \end{cases}$$

At the convergence of the CEM algorithm, the obtained parameters of the model readily indicate the characteristics of the clusters: proportions p_ℓ, modes $\mathbf{m}_\ell$ and degree of homogeneity for each class and for each attribute ε_ℓ^j. Next, we will study on the link between $L_c(\mathbf{x}, \mathbf{z}, \theta)$ and k-modes criterion.

3.3 *k*-modes Criterion

It is straightforward to derive different models by imposing constraints on the parameters of the model. For instance, we suppose here that the proportions are equal and the $\varepsilon_\ell^j = \varepsilon$ for $\ell = 1, \ldots, k$ and $j = 1, \ldots, q$. When all the attributes have the same number of categories $\mathbf{c}$, maximizing $L_c(\mathbf{x}, \mathbf{z}, \theta)$ reduces to minimizing of $W = A \sum_{\ell=1}^k \sum_{\mathbf{x}_i \in z_\ell} d(\mathbf{x}_i, \mathbf{m}_\ell)$ where $A = \log\left(\frac{(1-\varepsilon)(\mathbf{c}-1)}{\varepsilon}\right)$. Hence minimizing the criterion W is equivalent to minimizing the k-modes criterion $E = \sum_{\ell=1}^k \sum_{\mathbf{x}_i \in z_\ell} d(\mathbf{x}_i, \mathbf{m}_\ell)$. And, our probabilistic model allows us to explain the criterion E which appears as a too restricted model that we note $[p, \varepsilon]$. To minimize the criterion W, the CEM algorithm reduces to a version called k^*-modes. With this version, the proportions do not affect the *E-step* and *C-step*. It is straightforward that these steps can be reduced to one *Assignment-step* where each instance is allocated to the nearest cluster, according to the dissimilarity d between an instance and the mode of the cluster. Unlike to k-modes, k^*-modes starts from an initial random instances and it is based on the process where the update of the modes are realized after all instances are assigned. This process is introduced with the k-means of Forgy ([5]) and extended to the dynamic cluster analysis ([3]). Through experiments on real data, it appeared preferable to us to initialize the k-modes algorithm with k random instances.

4 Simulated Data Sets

Through simulated data according the model $[p, \varepsilon]$ we have noted that from initial random instances, k-modes and k^*-modes are equivalent. In this section, to illustrate the effects of the parameters of the model we have performed some simulations to compare k^*-modes and CEM. The process of simulation is the following. We generated 30 two-components samples with the same ones $n = 1000$, $q = 10$, $\mathbf{m}_1$, $\mathbf{m}_2$, $p_1 = 1 - p_2$, ε_1 and ε_2. For each generated sample and for each experiment, we ran both algorithms 20 times from random initial position and we selected the solution out of 20 runs which provided the best value of the optimized associated criterion. We evaluated the performances of the algorithms with the proportions of misallocated instances by comparing the partition we obtained with the simulated one. We close this section with a practical study of linearity of the CEM algorithm.

Effect of the Proportions. We generated two-components according the model $[p_\ell, \varepsilon]$ where $\varepsilon = 0.2$. As the degree of separation of the two components depends on $\Delta = d(\mathbf{m}_1, \mathbf{m}_2) = \sum_{j=1}^{q} \delta(m_1^j, m_2^j)$ ([6]), we took $\mathbf{m}_1$=(1,1,1,1,1,1,1,1,1,1) and $\mathbf{m}_2$=(1,1,1,2,2,2,2,2,2,2) and we consider that the components are well separated ($\Delta = 7$) then $\mathbf{m}_2$=(1,1,1,1,1,1,1,2,2,2) and we consider that the components are ill-separated ($\Delta = 3$). Also, we took for all the q-attributes $c^j = 4$. Results are shown in Fig. 1. Note that k^*-modes, contrary with CEM, was unable to provide two clusters when p_1 is close to 0.25, for ill-separated components. For each run, it puts all instances in a single cluster. This is one of the limitations of algorithm k^*-modes. Moreover, CEM is better than k^*-modes, particularly for ill-separated components.

Effect of the Parameters ε_ℓ. To study the effect of the ε_ℓ's, we took $p_1 = p_2 = 0.5$, we fixed $\varepsilon_1^j = 0.1$ for $j = 1, \ldots, q$ and varied ε_2^j (all equal for $j = 1, \ldots, q$). For $j = 1, \ldots, q$, the value of ε_2^j is increased from 0.10 to 0.40 by 0.01. According to the same process of simulation previously described, the results are presented in Fig. 2. For each situation, CEM is clearly better than k^*-modes. When simulated data are close to the model $\lfloor p, \varepsilon \rfloor$ (i.e. when values of ε_2 are close to 0.10), CEM and k^*-modes have the same behavior, and give the same results. But, when the data are far from the model $\lfloor p, \varepsilon \rfloor$, CEM works very much better than k^*-modes.

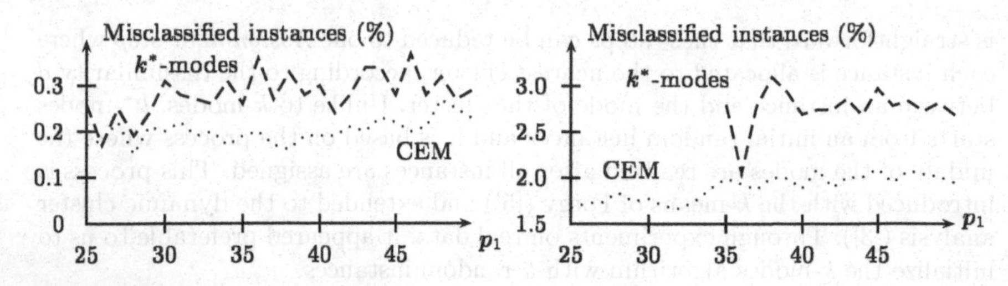

Fig. 1. k^*-modes vs. CEM with simulated categorical data ($\Delta = 7$, then $\Delta = 3$) with varying p_1.

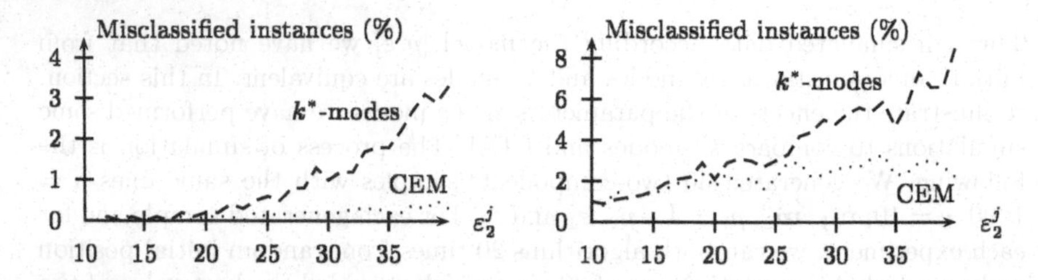

Fig. 2. k^*-modes vs. CEM with simulated categorical data ($\Delta = 7$ and $\Delta = 3$) with varying ε_2^j for $j = 1, \ldots, q$.

Large Database. In order to show the linearity of complexity of CEM, we run it on simulated data sets, where sized varying from $n = 50000$ to $n = 500000$, with $q = 50$, $k = 5$ and well separated clusters. We took the value mean time of 5 runs of CEM for each n. From results shown in Fig. 3, we see that the running time of CEM increases linearly when n increases. Then, this algorithm is particularly adapted to very large database.

5 Summary

We have considered the problem to cluster the categorical data under the classification maximum likelihood approach. In this setting, with a probabilistic model, we have defined a generalization of k-modes criterion and we have presented the CEM algorithm. Also, we have proved that the k-modes algorithm is a particular version of CEM. Taking advantage of this approach, we have studied the behaviors of both algorithms on simulated data. It appears clearly that CEM is better than k-modes and k^*-modes, even on real data (not reported here). Indeed, the criterion used by CEM takes account of the proportions, the number of categories and the degree of homogeneity of the clusters. The CEM algorithm is robust even when proportions are not equal and clusters are ill-separated, unlike to k-modes and k^*-modes. Then, the CEM algorithm has the following qualities: simplicity, fast convergence and processing large data sets.

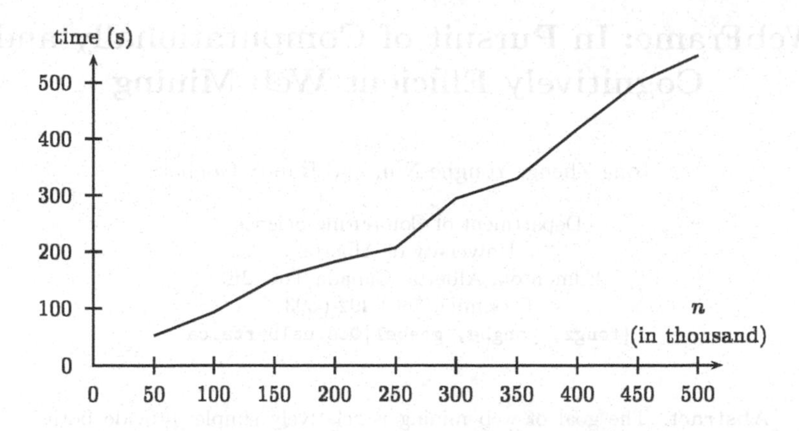

Fig. 3. Scalability of CEM when n increases

References

1. Celeux, G. and Govaert, G. (1992): A Classification EM Algorithm for Clustering and two Stochastic Versions. *Computational Statistics & Data Analysis, 14*, 315–332.
2. Dempster, A., Laird, N. and Rubin, D. (1977): Mixture Densities, Maximum Likelihood from incomplete data via the EM Algorithm. *Journal of the Royal Statistical Society, 39*, 1, 1–38.
3. Diday, E., Bochi, S., Brossier, G. and Celeux, G. (1980): *Optimisation en Classification Automatique*, Le Chesnay, INRIA.
4. Everitt, B. (1984): *An introduction to Latent Variables Models*, Chapman and Hall.
5. Forgy, E. W. (1965): Cluster Analysis of Multivariate Data: Efficiency versus Interpretability of Classification. *Biometrics, 21*, 3, 768.
6. Govaert, G. and Nadif, M. (1996): Comparison of the Mixture and the Classification Maximum Likelihood in Cluster Analysis with binary data. *Comput. Statis. and Data Analysis, 23*, 65–81.
7. Huang, Z. (1997): A Fast Clustering Algorithm to Cluster very large categorical data sets in Data Mining. *SIGMOD Workshop on Research Issues on Data Mining and Knowledge Discovery (SIGMOD-DMKD'97)*.
8. Huang, Z. (1998): Extensions to the k-means Algorithm for Clustering Large Data Sets with Categorical Values. *Data Mining and Knowledge Discovery, 2*, 283–304.
9. Mc Lachlan, G. J. and Basford, K. E. (1989): *Mixture Models, Inference and Applications to Clustering*, Marcel Dekker.
10. Mac Queen, J. B. (1967): Some Methods for Classification and Analysis of multivariate observations. *In Proceedings of the 5th Berkeley Symposium on Mathematical Statistics and Probability*, 281–297.
11. Nadif, M. and Marchetti, F. (1993): Classification de Données Qualitatives et Modèles. *Revue de Statistique Appliquée, XLI*, 1, 55–69.
12. Ralambondrainy H. (1995): A Conceptual Version of the k-means Algorithm, *Pattern Recognition Letters*, **16**, pp. 1147-1157.
13. Symons M. J. (1981): Clustering Criteria and Multivariate Normal Mixture, *Biometrics*, **27**, pp 387-397.

WebFrame: In Pursuit of Computationally and Cognitively Efficient Web Mining

Tong Zheng, Yonghe Niu, and Randy Goebel

Department of Computing Science
University of Alberta
Edmonton, Alberta, Canada T6G 2E8
Facsimile (780) 492-6393
{tongz, yonghe, goebel}@cs.ualberta.ca

Abstract. The goal of web mining is relatively simple: provide both computationally *and* cognitively efficient methods for improving the value of information to users of the WWW. The need for computational efficiency is well-recognized by the data mining community, which sprung from the database community concern for efficient manipulation of large datasets. The motivation for cognitive efficiency is more elusive but at least as important. In as much as cognitive efficiency can be informally construed as ease of understanding, then what is important is any tool or technique that presents cognitively manageable abstractions of large datasets.

We present our initial development of a framework for gathering, analyzing, and redeploying web data. Not dissimilar to conventional data mining, the general idea is that good use of web data first requires the careful selection of data (both usage and content data), the deployment of appropriate learning methods, and the evaluation of the results of applying the results of learning in a web application. Our framework includes tools for building, using, and visualizing web abstractions.

We present an example of the deployment of our framework to navigation improvement. The abstractions we develop are called Navigation Compression Models (NCMs), and we show a method for creating them, using them, and visualizing them to aid in their understanding.

Keywords: data mining, web mining, navigation compression, visualization

1 Introduction

After only half a decade, the world wide web (WWW) has become an information playground where every possible learning technique is of potential value for improving web usage — if only one could match application performance goals with appropriate learning technologies. Given enough resources, one can typically find almost anything on the web. In fact, at the estimated growth rate of about 14 million pages per day[1], it is a practical tautology that we can't find value without creating relevant human-oriented abstractions.

[1] Whatever a *page* is?

M.-S. Chen, P.S. Yu, and B. Liu (Eds.): PAKDD 2002, LNAI 2336, pp. 264–275, 2002.

The process of web mining is to create abstractions, with the overall goal of providing both computationally *and* cognitively efficient methods for improving the value of information for WWW users. The need for computational efficiency is well-recognized by the data mining community, which sprung from the database community concern for efficient manipulation of large datasets.

In as much as cognitive efficiency can be informally construed as ease of understanding, then what is important is any tool or technique that presents cognitively manageable abstractions of large datasets. The visualization of web space is based on exactly this idea: that some abstracted form of a large data set can provide insight into some important attributes of that space (e.g., see [6]).

The idea of *web mining* is to apply the tools and techniques of data mining to world wide web (WWW or web) data to induce "interesting" hypotheses that can be used to improve various web usage applications. So the only realistic research direction is to develop web mining software architectures that explicitly address both aspects: computational efficiency in order to provide access to large volumes of data, and cognitive efficiency in enabling users to guide learning processes to information abstractions of appropriate relevance.

The most common instance of this combination is the application of learning to user generated web usage data, sometimes referred to as *web usage mining* (WUM). Here we use WUM as a specific instance of a web mining task, to illustrate the development of a general framework for web mining.

The biggest challenge is to provide a "mining" software architecture that not only provides a harness for efficient learning methods, but also aids in the incremental user formulation of mining goals. This is important because humans are the ultimate determiner of what "relevance" means.

Everyone has their idea of what an abstraction should be. For example, web search engines are a dynamically created operational abstraction that continually update indices, which are then coupled with user query systems to identify relevant web information. Similarly, meta search engines provide another level of abstraction, working at a granularity above search engines by transforming single user queries into several queries to regular search engines. In the other direction, corporate, intranet, and e-business search engines provide local indexing structure, imposing more rigid abstractions that are targeted to circumscribe corporate policies, workflow constraints, and sales strategies.

This is why the notion of *web data* is as broad as the potential applications of its mining. Most current applications of web mining (e.g., [3, 7, 9, 10, 11, 14]) have concentrated on data that is created by the browsing user, beginning with the usage logs produced by web servers (e.g., [12]). Of course there is a broad spectrum of such user "web data," including ordinary web logs, cookies, page exit surveys, search query collections, and even hand collected user surveys. Even though this spectrum of information can itself be incredibly broad, there is another aspect of web data that is even broader and deeper: the web content itself.

The *Web Mining Framework* proposed here is designed to facilitate all aspects of web mining we can currently envisage, including the use of browsing data, web content, and web meta content. The framework consists of three broad components: 1) data capture tools, 2) learning tools, and 3) evaluation tools.

Our development of this framework is itself an experiment, based on our belief that we need such a framework to assess the various combinations of data, learning, and application evaluation methods. We hope to incrementally improve our framework, and develop answers to questions like "What are the tradeoffs between intrusive data gathering and navigation improvement?" or "Can we measure how the analysis of search queries be focused on specific demographic groups to increase relevance?"

In what follows, we provide a more detailed description of the current status of each of our components, together with examples of our preliminary experiments. In all cases we attempt to be as general as possible in identifying the "inductive opportunities" that arise within web data, and anticipate their ultimate role in improving the value of a range of web activities.

2 System Architecture for Web Mining Framework

Like existing data mining architectural proposals (e.g., [3, 7, 11]), the gross level component architecture will require a module to support data capture, a module to support the specification and deployment of a repertoire of learning methods, and, perhaps less common, an explicit module designed to support evaluation of any combination of the first two.

In our particular instance, we have already extended the simple three component architecture into something more elaborate, as depicted by the diagram of Figure 1. One simple way to understand this instance of our architecture is to consider a high-level description of the process control within it. With respect to the diagram of Figure 1, our current web mining procedure can be described as follows:

1. Determine what data can be used, and obtain the data through *Data Acquisition* modules.
2. Convert the original data into specific formats that can be used by various data mining methods. This work is done through *Data Preparation* modules.
3. Determine what data mining algorithms are appropriate for the task. Then, apply the algorithms to the formatted data to obtain corresponding knowledge. This work is done through *Data Mining* modules.
4. The knowledge can be queried or evaluated through various methods. This work is done through *Knowledge Analysis* modules. The evaluation of the knowledge can then be used as a feedback for steps 1-3.
5. We can make use of the knowledge to achieve various tasks in certain applications, and evaluate the performance improvement. This work is done through *Applications & Evaluation* modules. The evaluation of the performance improvement can be used as a feedback for steps 1-3.
6. Original data, formatted data, and even the knowledge can all be visualized through *Visualization* modules, to provide feedback for steps 1-3.

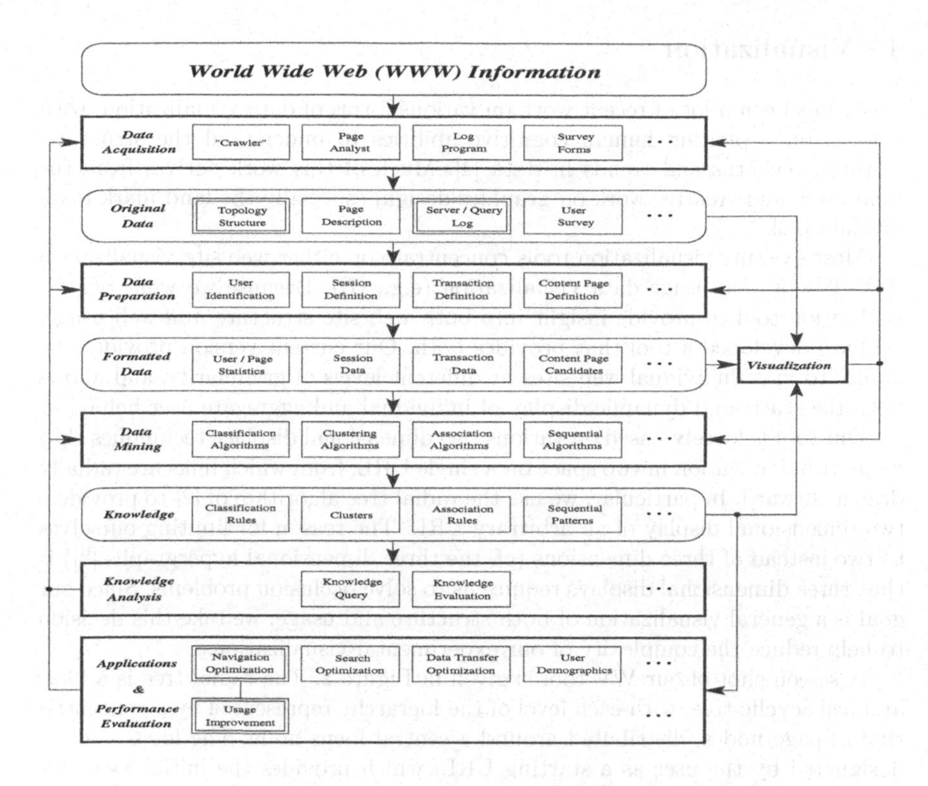

Fig. 1. System Architecture for Web Mining Framework

3 An Architecture Instance: Creating NCMs

Our first experiments with our web mining framework have the goal of improving user navigation. Our input is web usage data, e.g., web logs, and our evaluation methods (described below) help indicate whether we are actually "improving" a user's navigation, e.g., by helping reduce the number of links traversed to find "interesting" pages.

The missing middle component is that object to be created by various learning algorithms, and then tested to see whether the learning algorithm has found something "interesting" which can provide navigation improvements, as measured by the evaluation methods. We call the objects created by the application of learning methods navigation compression models (NCMs). NCMs are simply some representation of an abstract navigation space, determined as a function of a collection of user navigation paths on actual websites.

4 Visualization

There has been a lot of recent work on various forms of data visualization, with a focus on exploiting human cognitive abilities to understand the structure, features, patterns and trends in data [4]. Much of this work derives from the innovative and creative work on graphics design, especially the land mark texts of Tufte [13].

Most existing visualization tools concentrate on either web site visualization (e.g., [8]) or web usage data visualization (e.g., [1]). Because we want our visualization tool to provide insight into both web site structure *and* web usage, we have developed a tool that provides both. Our current version provides the ability to view individual web sites at different levels of granularity, and allows both the static and dynamic display of individual and aggregate user behavior.

Our tool is loosely based on various two dimensional displace techniques that focus visual attention in two space on a single URL, from which links are radially drawn outward. In particular, we use the radial tree algorithm of [2] to provide a two dimensional display of an arbitrary URL. The reason for limiting ourselves to two instead of three dimensions (cf. the three dimensional hypergraphs [8]) is that three dimensional displays require us to solve occlusion problems. Since our goal is a general visualization of both structure and usage, we take this decision to help reduce the complexity of our experimental visualizations.

A screen shot of our WV tool is given in Figure 2. The radial tree is a hierarchical acyclic tree, with each level of the hierarchy represented by a concentric ring of page nodes, distributed around a central focus node. The focus node is designated by the user as a starting URL, which provides the initial focus for any visualization display.

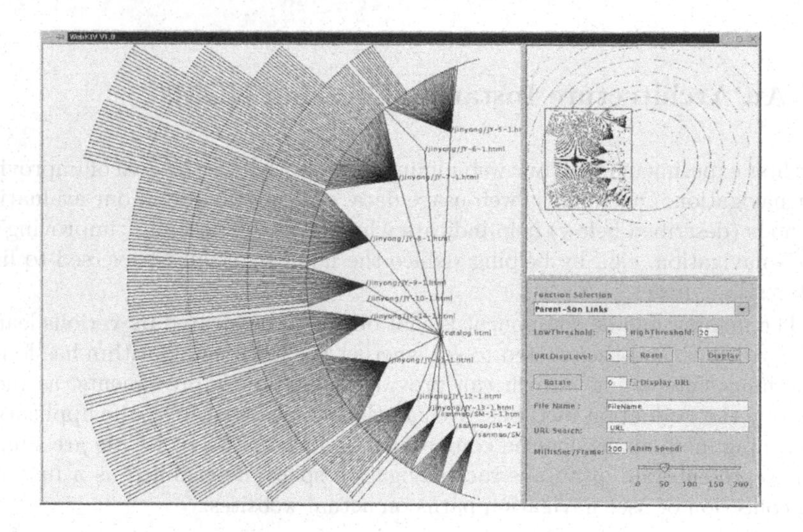

Fig. 2. Web Visualization Tool

Similar to heights on a topographic map, nodes linked from the focus URL are displayed on different concentric circles according to their levels in the hierarchical tree. A breadth-first search that avoids cycles is used to traverse the web site, to build the hierarchical structure. The radial pattern of links from a given node is represented by a radial distribution of nodes on a concentric circle, proportional to the number of nodes. Note, for example the concentric rings of linked nodes in the diagram of Figure 2.

This two dimensional display based on concentric link "isobars" doesn't provide any insight to web structure, without an ability to shift the focus and level of detail. With this in mind, our visualization also provides the user with the ability to drag a rectangle over the visualization, then zoom on that rectangle. As shown in Figure 2, a smaller window (context window) always provides a context for the web site structure, while the other larger window (focus window) displays the detail of choice, which is a subset of the larger context. Whenever there is a need to drill down, the selected sub area is enlarged, and both windows are appropriately modified. If the user "zooms in" to sufficient detail then individual URLs are used to label each node.

If we can visualize *individual* user navigation paths superimposed on top of the web site visualization described above, we can begin to recognize well-traversed paths. These might be links that are popular over some particular time period, or trajectories of heavily visited web pages which help us understand how users arrived at web site "hot spots."

In addition to visualizing the navigation paths of individual users, we can also visualize aggregate paths. For example, when an individual traversal of a hyperlink is indicated by drawing a line from one node to another, the aggregate behavior of two users can simply annotate that link for each traversal. There are an arbitrary number of ways to visualize aggregate link traversal (e.g., line width, color, annotation). But when we can visualize *aggregate* user navigation pathers superimposed on top of the web site visualization, we can begin to recognize web navigation clusters in order to understand aggregate user behavior. For example, this is useful to validate web site design (cf. [5]), by statically viewing the distribution of aggregate navigation paths on a web site. And with appropriate navigation path annotation, we can dynamically observe aggregate behavior, e.g., aggregate navigation path changes over different time periods.

5 Experiments with Navigation Compression Models

Navigation optimization means improving the ease with which users can reach the contents of interest more quickly. The basic idea is that we can discover navigation patterns from previous visitations, and then use these patterns to provide guidance for new users. As explained above, we call our "patterns" navigation compression models or "NCMs."

We can evaluate our NCMs statically by measuring overall navigation improvements. To actually use such information, we can use one of two approaches: one is using synthetic static index pages, each of which contains indices to a set of pages belonging to similar or related topics; another is using dynamic recommendations, which use NCMs to make recommendations to a user based on the pages the user has already visited. In our preliminary experiments, we use our NCM to implement dynamic recommendations.

Note that there is a significant problem with user navigation patterns: a traveled path is not necessarily a desired path. Therefore we propose a recommendation mechanism which ignores those auxiliary pages and makes recommendations only on "relevant" pages. Of course the identification of "relevant" pages is a very difficult problem, but we can make assumptions based on certain heuristics. Though this definition is heuristic, we can still compare NCMs for how well they help users find the relevant pages more quickly.

The inputs to create an NCM include knowledge of page visiting patterns and the path of the user sessions; the outputs are recommendations for useful links. As mentioned above, a NCM can be created from the results of various learning methods, such as association rules, sequential patterns, and clusters. Moreover, a NCM can also be created with some pre-defined navigation templates.

5.1 Data Preparation

Our preliminary experiment uses only the server-produced access logs. Moreover, we experiment on two different original data sets so that we can make sure the result is not exceptional, and some comparisons can be done when necessary. Both data sets are the server access logs of our academic department: one is in the period of September and October, 2001 (*DS1*); another one is in the period of August and September, 2001 (*DS2*). In both data sets, we use the previous one-month logs as training set, and the later one-month logs as test set.

Our data preparation consists of breaking the access log into user sessions. Currently we use two different user identification methods: (a) we exclude all records without both authentication and cookie, and identify users just by their authentication or cookie value; (b) we identify users with all the information we have (including all log information as well as topology structure), and no record is excluded.

To compare these two user identification methods, we generated sessions with each of them on all the three-month logs. The result shows that with method (b), we can identify a lot more users and sessions. But interestingly, after we removed those useless sessions (with $length = 1$), the results from method (a) and (b) became quite similar (with a difference less than 2%). This implies that: (i) most users choose to accept cookies, not rejecting them; (ii) the large amount of "$length = 1$" sessions might come from some robots or web crawling programs.

In what follows, we use only method (a) for user identification, because it is more accurate and runs much faster than method (b). Another important parameter is the session timeout, for which we arbitrarily select 30 minutes.

5.2 Evaluation of Usage Improvement

Exercising our general framework, our experiment took three steps:

1. Convert original log data into sessions (data preparation).
2. Apply learning algorithms to the sessions to obtain NCMs. In our preliminary experiment, we only generate association-rule-based NCMs.
3. Apply the NCMs to the sessions in the test set, and generate a new set of shortened navigation paths.

The specific measure we use for evaluation is called *Usage Improvement (UI)*, which is computed based on the number of hyperlinks traversed:

$$UI = \frac{N_{org} - N_{com}}{N_{org}} \tag{1}$$

Here, N_{org} is the number of requests in the original sessions, and N_{com} is the number of requests in the compressed sessions.

For example, suppose we obtain an association rule $A \longrightarrow D$, where D is a content page. Then an original session $S_{org} = \{A, B, C, D, E\}$ (where B and C are not content pages) can be compressed by removing B and C to obtain $S_{com} = \{A, D, E\}$. The usage improvement for this session would be $UI = \frac{5-3}{5} = 40\%$.

As mentioned above, our NCM mechanism only makes recommendations for relevant pages. Correspondingly, an association rule is applied only when it is used to shorten the path to a relevant content page. We determine relevant content pages using three different approaches: (a) *Maximal Forward Reference* (MFR) [3], which assumes that *maximal forward references are content pages*, and the pages leading up to the maximal forward references are auxiliary pages. Here, a maximal forward reference is defined to be the last page requested (before session timeout) by a user before backtracking; (b) *Reference Length* (RL) [3], which assumes that *a user generally spends more time on content pages than auxiliary pages*, therefore identifies content pages based on a cutoff viewing time. In our experiment, we set the cutoff time to an empirical value — 1 minute; and (c) *Visit Count* (VC), which simply assumes that those pages mostly visited are content pages.

Here we generated three kinds of Usage Improvement *UI*:

- UI — usage improvement on all sessions.
- UI_c — usage improvement on those sessions each of which has at least one content page.
- UI_r — usage improvement on those sessions for each of which at least one rule is applied.

5.3 Experimental Results

We report two experiments. In the first, we tested the usage improvement with varying numbers of association rules and a fixed number of content pages. To simplify the problem, we set the minimal *confidence* of rules to a fixed value — 75%, and only adjusted the *support* of rules. The content pages were selected based on a restriction on their visit counts. So a web page is determined to be content page if: (a) it is classified to be a content page by one of the content page identification methods (MFR-based, RL-based, or VC-based); and (b) it has a visit count no less than a threshold vc. In this experiment, we set vc to $\frac{S}{1000}$, where S is the total amount of sessions. The number of content pages obtained with this approach may not be the same for different content page identification methods. In that case, we simply use the minimum among those numbers, and remove those less-visited pages when necessary.

In the second experiment, we tested the usage improvement with varying numbers of content pages and a fixed number of association rules. The support and confidence were set to 0.75% and 75% respectively. This is because smaller support and confidence values will generate a lot more rules without bringing apparent improvement to the performance.

The experiment results for *DS1* and *DS2* are shown exclusively in Figure 3.

First, we found that the total usage improvement (UI) we can expect is relatively small, though the usage improvement on those sessions where rules took effect (UI_r) is quite large. This means that only a very small part (less than 1% in our experiment) of the sessions were compressed with the association rules we obtained. A possible reason might be that, users' interest in this web site is too diverse, and the content pages we selected (1000 at most) are only a very small part of that. Moreover, among those three content page identification approaches, *reference length* seems to have the potential for the best Usage Improvement (*UI*).

Secondly, with a given set of content pages, the usage improvement is improved when rule number gets larger (i.e., when the *support* gets smaller). However, this also means that users will have more recommendations to choose from in the real world, which can be seen as an extra cost.

Thirdly, we found that more content pages can have two effects: a positive one is that more paths can be compressed (no recommendation will be made if it is not for a content page, therefore no compression); however, there are more pages in the paths we can not skip. Our experiment showed that, with a given set of association rules, the usage improvement is impaired when content page number gets larger.

In those $UI = f(Support)$ figures, we can see that UI can change a lot in some area, and changes very little in some other area. This means that not all rules have the same effect on UI: some rules are very useful, while some other rules can be completely useless. There might be three reasons behind this: (a) an association rule is possibly useful only when some content pages appear at its right side; (b) an association rule may not be useful even it has a content page at right side, because users might have already taken the path it suggests; (c) a lot of rules could relate to very few content pages.

5.4 Visualization of Navigation Compression Modules

Our visualization tool WV provides a way of comparing the navigation behavior of two different navigation strategies, by superimposing two sets of user navigation paths (in this case web logs). This is shown in Figure 4, where part (a) is the visualization of a user web log before applying a navigation learning method,

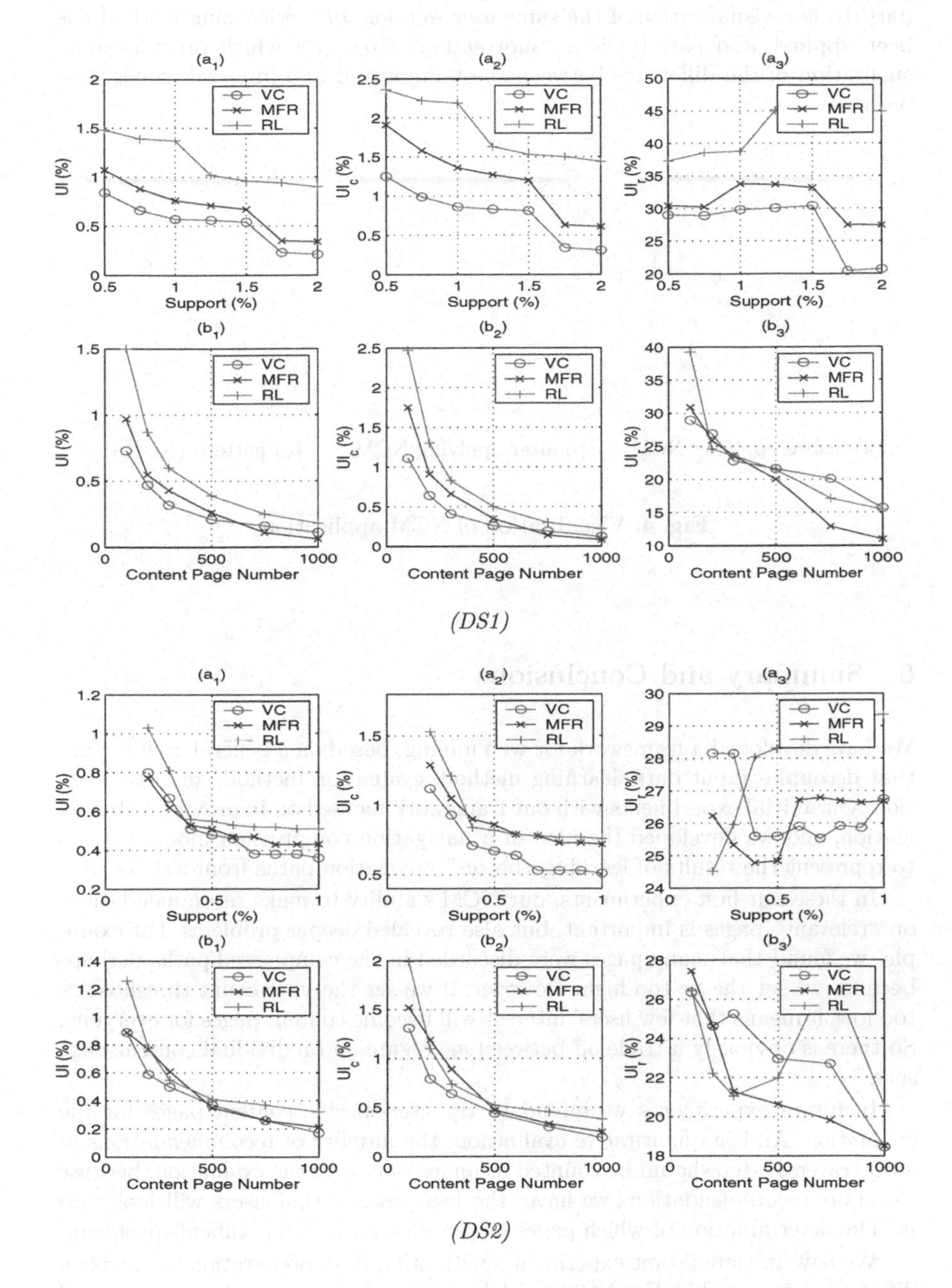

Fig. 3. Experiment Results on *DS1* & *DS2*

part (b) is a visualization of the same user web log *after* a learning method has been applied, and part (c) is a "subtraction" (b) - (a), which provides a visualization of the difference between the unimproved and improved navigation paths.

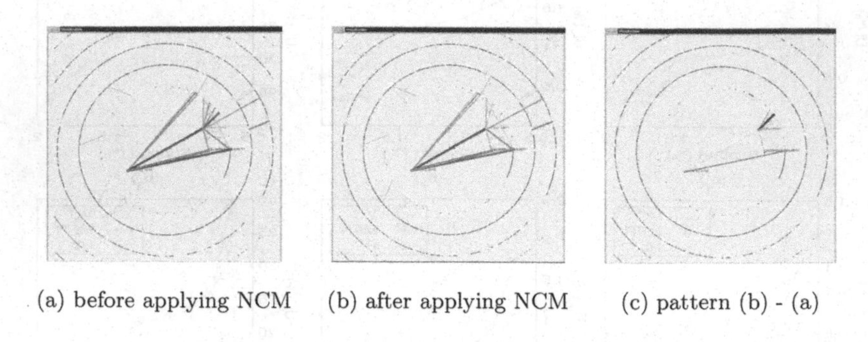

(a) before applying NCM (b) after applying NCM (c) pattern (b) - (a)

Fig. 4. Visualization of NCM application

6 Summary and Conclusions

We have developed a framework for web mining, based on a general architecture that decouples input data, learning method, evaluation method, and visualization. Our initial experiments with our framework focused on improving web navigation, and we developed the idea of a navigation compression model (NCM) to represent the results of learning "better" navigation paths from web logs.

In these our first experiments, our NCM's ability to make recommendations on "relevant" pages is important, but also revealed deeper problems. For example, we found that many pages were discarded in the compressed paths perhaps because we set the vc too high. However, if we set the visit count threshold vc too low, it means that few users' interest will become content pages for everyone. So there is obviously a trade off between aggregate and individual content page sets.

In future experiments we intend to try user-specific content pages for the evaluation. And in quantitative evaluation, the number of recommendations at each traversal step should be counted as an extra cost in the evaluation, because the more recommendations we have, the less possible that users will look into it. The determination of which pages to recommend is still a difficult problem.

We have initiated some experiments with other data preparation methods on different data sets but the NCM models differ only in size and so far have all been created with association rules. Our future experiments will create NCMs from other learning methods, and make comparisons among them.

Acknowledgements

Our work is supported by the Canadian Natural Sciences and Engineering Research Council (NSERC), and by the Institute for Robotics and Intelligent Systems (IRIS) Networks of Centres of Excellence.

References

[1] Igor Cadez, David Heckerman, and Christopher Meek. Visualization of navigation patterns on a web site using model_based clustering. In *Proceedings of KDD 2000*, August 2000.

[2] E.H. Chi, J. Pitkow, J. Mackinlay, P. pirolli, and J. Konstan. Visualizing the evolution of web ecologies. In *proceedings of CHI'98*, 1998.

[3] R. Cooley, B. Mobasher, and J. Srivastava. Data preparation for mining world wide web browsing patterns. *Journal of Knowledge and Information Systems*, 1(1), 1999.

[4] Usama Fayyad, Georges G. Grinstein, and Andreas Wierse. *Information Visualization in Data Mining and Knowledge Discovery.* Morgan Kaufmann, 2001.

[5] Ed Huai hsin Chi and Stuart K. Card. Sensemaking of evolving web sites using visualization spreadsheets. In *proceedings of the Symposium on Information Visualization*, 1999.

[6] Paul Kahn and Krzysztof Lenk. *mapping web sites.* RotoVision SA, 2001.

[7] Bamshad Mobasher, Robert Cooley, and Jaideep Srivastava. Automatic personalization based on web usage mining. Technical Report TR99-010, Department of Computer Science, Depaul University, 1999.

[8] Tamara Munzner and Paul Burchard. Visualizing the structure of the world wide web in 3d hyperbolic space. In *VRML'95*, 1995.

[9] M. Perkowitz and O. Etzioni. Adaptive web sites: Automatically synthesizing web pages. In *Proceedings of the Fifteenth National Conference on Artificial Intelligence (AAAI-98)*, 1998.

[10] S. Schechter, M. Krishnan, and M.D. Smith. Using path profiles to predict http requests. In *Proceedings of the Seventh International World Wide Web Conference*, Brisbane, Australia, April 1998.

[11] J. Srivastava, R. Cooley, M. Deshpande, and P.N. Tan. Web usage mining: Discovery and applications of usage patterns from web data. In *SIGKDD Explorations*, volume 1, Issue 2, January 2000.

[12] R. Stout. *Web Site Stats: Tracking Hits and Analyzing Traffic.* McGraw-Hill, 1997.

[13] Edward R. Tufte. *The Visual Display of Quantitative Information.* Graphics Press, 1982.

[14] I. Zukerman, D.W. Albrecht, and A.E. Nicholson. Predicting users' requests on the www. In *User Modeling: Proceedings of the Seventh International Conference (UM-99)*, pages 275–284, June 1999.

Naviz: Website Navigational Behavior Visualizer

Bowo Prasetyo[1], Iko Pramudiono[1], Katsumi Takahashi[1,2], Masaru Kitsuregawa[1]

[1]Institute of Industrial Science, University of Tokyo
4-6-1 Komaba, Meguro-ku, Tokyo 153-8505, Japan
{praz,iko,katsumi,kitsure}@tkl.iis.u-tokyo.ac.jp
[2]NTT Information Sharing Platform Laboratories Midori-cho 3-9-11,
Musashino-shi, Tokyo 180-8585, Japan

Abstract. Navigational behavior of website visitors can be extracted from web access log files with data mining techniques such as sequential pattern mining. Visualization of the discovered patterns is very helpful to understand how visitors navigate over the various pages on the site. Currently several web log visualization tools have been developed. However those tools are far from satisfactory. They do not provide global view of visitor access as well as individual traversal path effectively. Here we introduce Naviz, a system of interactive web log visualization that is designed to overcome those drawbacks. It combines two-dimensional graph of visitor access traversals that considers appropriate web traversal properties, i.e. hierarchization regarding traversal traffic and grouping of related pages, and facilities for filtering traversal paths by specifying visited pages and path attributes, such as number of hops, support and confidence. The tool also provides support for modern dynamic web pages. We apply the tool to visualize results of data mining study on web log data of Mobile Townpage, a directory service of phone numbers in Japan for i-Mode mobile internet users. The results indicate that our system can easily handle thousands of discovered patterns to discover interesting navigational behavior such as success paths, exit paths and lost paths.

Keywords. navigational behavior visualization, sequential pattern mining, web traversal property, class-instance view, mobile internet user

1. Introduction

Website developers, designers, and maintainers are having difficulties to analyze the efficiency of their website. Two of the major problems with current web log analysis are difficulty to understand *what* visitors are trying to do on a website and *how* they are doing it. This requires analysis of the visitor navigational behavior on the website, which gives webmasters know how to improve structure of the hyperlinks as well as the content of the documents in the website. Some data mining techniques to discover navigational patterns have been proposed. However due to the large number of discovered patterns, effective method to visualize the results is needed. Towards this

M.-S. Chen, P.S. Yu, and B. Liu (Eds.): PAKDD 2002, LNAI 2336, pp. 276-289, 2002.

end, we develop a Java JApplet based navigational behavior visualization tool (running on almost all modern browsers support Java 2 plugin), called Naviz.

Web log visualization firstly should give understanding of global view of visitor access. To achieve this goal, it is important to consider about inherent *web traversal topology*, i.e. WWW visitor access traversals are intrinsically directed cyclic graphs. This can be thought of as a hypermedia-like structure, with nodes represent pages and edges represent traversals between pages. We found that it is very helpful to introduce two *appropriate web traversal properties* below when visualizing web log access:

1. Hierarchical structure regarding traversal traffic, i.e. more traversed edges are placed at the higher level and less traversed edges at the lower level position. This would give an intuitive description of visitors that are traversing over the site, which enter from top of the graph and then exit to bottom.
2. Grouping of related pages, for example, pages that have high degree of transitional probability among them are better to be placed together near to each other. This pages grouping would be useful for webmasters to easily find what pages are related each other.

Another important factor regarding web log visualization is the nature of modern web pages. Today HTML pages are often dynamic pages that automatically created and destroyed whenever needed. A dynamic page can be thought of as a page class, which will create different instances for every access depending on input parameters. Dynamic page brings information for example by retrieving data from the back-end database. Hence, the different access to the same page may bring different information. This kind of page requires different visualization technique comparing to traditional static HTML pages. Class-instance view technique, i.e. a technique to visualize dynamic page either as a node of representative class or as many nodes of its instances, is one of the solutions for this problem. A class of dynamic page can be expanded to show information of some of its instances, and instances can be merged back to a class to simplify the global view.

The other difficulty is how to visualize individual traversal path while keeping the global view of navigational behavior. We use the shortest unit of traversal paths, that is traversal path between two consecutive page classes, as the underlying unit to draw traversal diagram. Such path can be treated as the edge of directed graph.

By interactively visualizing the traversal paths, webmasters should be able to reveal *what* their visitors are trying to do and *how* they are doing it. For those purposes, Naviz uses several strategies to form effective traversal diagram and to visualize traversal paths. These include considerations of web traversal property, class-instance view, traversal path filtering by specifying visited pages and path attributes such as hop number, navigational behavior comparison and interactive environment. Naviz try to take into account the appropriate web traversal properties by hierarchization and pages grouping. For website which contains modern dynamic pages, class-instance view can greatly simplify the graph layout while still provide the information needed. Naviz utilizes particular attribute on web log data to do navigational behavior comparison: i.e day behavior vs. night behavior. Traversal paths can be filtered by specifying visited pages and number of hops (path length). Furthermore Naviz utilizes the interactive environment to provide more capabilities to explore navigational behavior from the data.

We use Naviz to visualize results from data mining study on web log data of Mobile Townpage, i-Mode version of NTT i-Townpage, a directory service of phone numbers in Japan. As far as we know, this is the first experiment to visualize navigational behavior of mobile phone users. Due to the limited display and communication ability of mobile phones, a web site for mobile phone has to be carefully designed so that the visitors can reach their goal with least clicks. Using Naviz we discover some interesting visitor behavior such as success paths, exit paths and lost paths, which are useful for webmasters to get the idea how to improve structure of the hyperlinks as well as the content of the documents in the website.

The rest of the paper is organized as follows. Chapter 2 will overview about related works such as currently available web log analysis visualizers. In chapter 3 we will introduce Naviz, website navigational behavior visualizer, which combines the power of log miner and intuitive-looking of aesthetic graph. Our experiment on NTT i-Townpage served on i-Mode will be discussed in chapter 4, while chapter 5 will give the conclusion and future works regarding Naviz development.

2. Related Works

Pitkow et. al. in 1994 proposed WebViz [4] as a tool for web log analysis and provides graphical view of website's local documents and access patterns. By incorporating the Web-Path paradigm [4], i.e. establishment of relationship between documents in database and web access log, into an interactive tool, webmasters can see the documents in their website as well as the hyperlinks travelled (represented visually as links) by visitors. WebViz also enables webmasters to filter the access log by domain names or DSN numbers, directory names, and start and stop times, and play back the events in the access log. The drawback of WebViz is that it was designed to visualize the statistical property of the data only (i.e. frequency and recency information), it can not handle sequential patterns (i.e. has no related data mining tools). It displayed web log access in two-dimensional directed graph, but without enough considerations about appropriate web traversal properties. Also WebViz did not visualize modern dynamic page effectively.

In 1998, Spiliopoulou et. al. presented Web Utilization Miner [5], WUM as a mining system for the discovery of interesting navigation patterns. One of the most important features of WUM is that using WUM's mining language MINT, human expert can dynamically specify the interestingness criteria for navigation patterns. This includes specification of criteria of statistical, structural and textual nature. To discover the navigation patterns satisfying the expert's criteria, WUM exploits Aggregation Service that extracts information on web access log and retains aggregated statistical information. Although WUM provides an integrated and robust environment to do web log analysis job, but since it focuses mainly on data mining and its mining language, it lacks aesthetic visualization of the data as well as considerations of appropriate web traversal properties and dynamic page.

Cugini et. al. in 1999 proposed VISVIP [6], a tool which allows web site developers and usability engineers to visualize the paths taken through the site. The graphical layout of the web site can be dynamically customized and simplified, and which subjects paths to view can be dynamically selected. VISVIP also provide an animated representation of traversal along the path through the web site. The time spent on each page visit can be represented using the third dimension of the 3D display. VISVIP gives good visualization regarding paths taken through the site, but it has drawbacks such as it needs client customization to instrument a web site so as to record the activity of a subject navigating the site. As almost currently available visualizations, VISVIP also did not consider appropriate web traversal properties as well as dynamic page.

The recent work done by Hong et. al. in 2001, proposed WebQuilt [7] as a web logging and visualization system that helps web design teams run usability tests and analyze the collected data. To overcome many of the problems with server-side and client-side logging, WebQuilt uses a proxy to log the activity. It aggregates logged usage traces and visualize in a zooming interface that shows the web pages viewed. Also it shows the most common paths taken through the website for a given task, as well as the optimal path for that task. WebQuilt as well as almost currently available visualizations, also has drawbacks that it did not consider appropriate web traversal properties and dynamic page.

The most recent work in visual mining of web log data has been done by Zhao et. al. in 2001 [8]. They proposed a technique to track the behavior of discovered rules over the time, and then based on it they try to predict the future behavior of the rules. First, the support or confidence value of rules is plotted against the time; this will show the rule behavior over the time. Using the similarity function based on time behavior of support or confidence value, rules will be clustered to the "similar rule" groups. The trend of "similar rule" group in turn will be used to predict the future behavior of individual rule that belongs to the group. The idea to use support or confidence time behavior for clustering the discovered rules is very interesting.

3. Naviz: Navigational Behavior Visualizer

3.1 Known Problems in Previous Works

Agrawal et. al. [1] in 1995 has given the basic of sequential pattern data mining, which later also widely used to discover traversal paths from web log data. But finding interesting knowledge from the discovered patterns is not an easy task. Since the number of discovered patterns is generally large, and there is no metric that well represents their "usefulness". Thus visualization tool is needed to help human users to interpret them.

On the other hand, construction of an effective visualization is a challenging task. As stated in works of Mukherjea et. al [2]. there are various problems involved: i.e. the navigational views are two or three dimensional projections of generally multidimensional hypermedia networks; even if such a view is developed, the resulting structure would be very complex for any non trivial hypermedia system; as the size of the underlying information space increases, it becomes very difficult to fit the whole information structure on a screen; and yet the user should be able to get an idea of not only the structure but the actual contents of the nodes and links just by looking at the navigational views.

As mentioned in the previous chapter, the drawbacks of current visualization tools includes lack of capability to handle sequential pattern, lack of aesthetic visualization, necessity of client-side customization and the common drawbacks are that most of them do not consider about modern dynamic page and appropriate web traversal properties as described in introduction, i.e. hierarchical structure regarding traversal traffic; and related pages grouping. Naviz was designed to overcome these drawbacks by combining two-dimensional graph of traversal diagram that considers appropriate web traversal properties as well as modern dynamic web pages, and facilities for visualize interesting traversal paths by specifying certain visited pages and path attributes such as number of hops, support and confidence. It combines the power of sequential pattern mining tools and intuitive-looking of graph drawing tools, to create interactive visualization of traversal paths, in an aesthetic layout of graph. Since it uses log miner and graph layout producer separately from visualization part, Naviz may easily adapt to the latest technology of this two methods whenever its newer version is available, while we can continue to explore the future development of its visualization part.

3.2 Traversal Diagram

First we will explain the notation page class and instance we use here. A page class is an abstraction of web pages that offers the same navigational functions in the web traversal topology. Instance is a member of page class that has certain semantics. Instance membership is user defined. Usually an instance is defined by parameters of its page class, however we can ignore some parameters so that an instance may include some different representations of web access logs. For example, logs often include visitor ID as a CGI parameter. But user can define logs with certain CGI parameters as an instance although they have different visitor ID, since this CGI parameter does not affect the semantics of the instance. The page class is a generalization of the web page definition so far since a static web page can be seen as a page class with a single instance.

To represent navigational behavior we use traversal path, a sequence of page classes traversed by visitors. Following association rule, we define some parameters to assess the strength of the traversal path:

1. Support of a traversal path A→B is the percentage of sessions that contain the sequence of A→B against the total number of sessions.
2. Users of a traversal path A→B is the percentage of visitors that traversed between A→B against the total number of visitors.
3. The confidence of a traversal path is the probability of a visitor to visit the last page in the sequence of traversal path. For example the confidence of traversal path A→B→C→D is the probability of a visitor to visit page D after he/she visited pages A→B→C consecutively. It can be obtained by dividing the number of sessions that contain sequence A→B→C→D by the number of sessions that contain page A→B→C. When the sequence only contains two page classes A→B, the confidence can also be interpreted as page transition probability from A to B.

We use GSP algorithm [1] to derive traversal paths from web access logs. Note that Naviz is not limited to this algorithm only. We also add the capability to set constraints such as deriving only the traversal paths of visitor with certain attributes. For example we filter web logs whose accesses between 6:00 until 18:00 to extract traversal paths of daytime visitors.

To acquire a global view of navigational behavior, we need to draw all traversal paths above certain parameter thresholds. As mentioned before, we also have to be able to draw directed cyclic graphs, hierarchical structure and nodes grouping. We use the shortest traversal paths with a pair of page classes as the underlying unit of drawing since they also convey the relationships between two adjacent page classes. We draw traversal diagram as a directed graph with page classes as the nodes and the traversal paths between any pair of page classes as the weighted edges.

Gansner et. al. in their paper [3] in 1993, described a four-pass algorithm for drawing directed graphs. The first pass assigns discrete ranks to nodes. In a top to bottom drawing, ranks determine Y coordinates. Edges that span more than one rank are broken into chains of "virtual" nodes and unit-length edges. The second pass orders nodes within ranks to minimize crossings. The third pass sets X coordinates of nodes to keep edges short. The last pass routes edge splines. Furthermore, as secondary role in the algorithm, Gansner et. al. propose the next aesthetic principles: 1) expose hierarchical structure in the graph, 2) avoid visual anomalies that do not convey information about the underlying graph, 3) keep edges short, and 4) favor symmetry and balance. This algorithm is proven to be able to draw graph layout very well and very fast, while still maintaining the aesthetic principles. Furthermore, this algorithm satisfies the requirements of appropriate website traversal properties that are needed by Naviz to draw traversal diagram.

3.3 System Overview

Naviz was designed to visualize traversal paths discovered from data mining on web log file. Naviz was developed as a Java JApplet program, which run on almost modern browsers with Java 2 plugin installed. It combines a separated log miner that utilize the traversal path mining [1], and a graph layout producer called Graphviz [9], an open source graph drawing software from AT&T that implemented algorithm in [3].

As showed in **Fig. 1** Naviz is a three-tier client-server application, consists of:

1. A java applet as user interface in the client side.
   Naviz applet displays the visualization to the user, interactively responds to user's request and passes it to the server through servlets. Users can set the threshold of support value, confidence degree to control the number of nodes and edges that will be displayed. For traversal path visualization, they can specify number of hops and visited pages to find interesting paths. Furthermore, they can change interactively strategy of hiearchization and grouping to explore the best possible structure of traversal diagram.

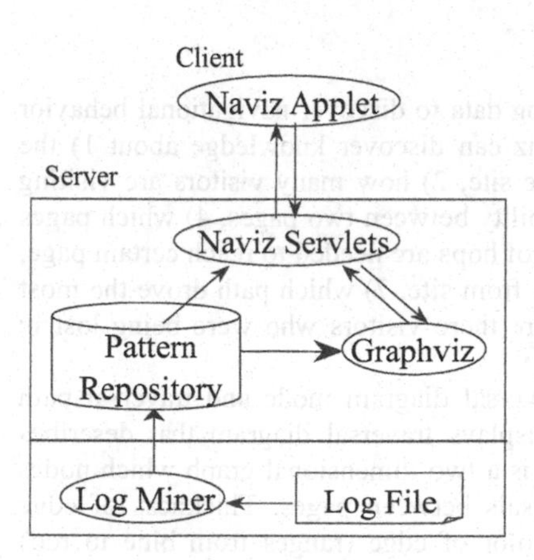

Fig. 1. *Naviz* Architecture

2. Java servlets in the server side, as interface that allows communication between client and server.
   Due to security restrictions that are applied to applet, Naviz applet cannot directly write-access file system in the server, nor running the program. Hence, Naviz servlets acts as interface between applet and server-side programs, by opening HTTP connections between client and server.
3. Pattern repository in the server containing discovered patterns.
   Currently Naviz did not implement interactive communication between Naviz applet and log miner. Instead it uses a pattern repository that manages the miner output files such as traversals paths discovered from web log data. Currently repository contains traversals paths only. As Naviz visualization capability will improve, in the future we will expand repository to handle other pattern also such as clustering results.
4. Graphviz as server-side application that is responsible for drawing the graph layout.
   Graphviz that is responsible for drawing graph layout, allows us to specify detail parameters to draw graph regarding appropriate web traversal properties. Moreover, it is able to give the output as coordinate and size information of the graph in easily readable format for Naviz to use.
5. Log miner to discover traversals/paths from web log data in the server.
   Our mining engine currently implemented association rule mining, sequential pattern mining (traversals paths), and clustering. We have also a group of tools to preprocess web access logs. In the future we will add the capability that allows Naviz to communicate with these tools in order to improve its performance.

Naviz has been programmed to operate on internal representations of graph, and to allow the user to work with the graph. Naviz which run as an applet in the client browser, start up Graphviz on server through the servlet, to compute the graph layout. Whenever a user asks for new layout, Naviz sends graph to Graphviz. Graphviz computes the layout and sends the graph back to Naviz along with coordinate and size information as graph attributes. Naviz then redraws the graph using the new layout.

3.4 Features

Using Naviz webmasters can analyze web log data to discover navigational behavior of their visitors. From our experiment Naviz can discover knowledge about 1) the traversal diagram of visitor traversals on the site, 2) how many visitors are visiting which page, 3) how high is transition probability between two pages, 4) which pages are related each other, 5) how many number of hops are needed to reach certain page, 6) which path drove the most visitors to exit from site, 7) which path drove the most visitors to successfully find their goal, 8) are there visitors who were being lost in website etc.

Naviz has two operation modes, i.e. traversal diagram mode and traversal path mode. In traversal diagram mode Naviz displays traversal diagram that describes global visitor traversals on the entire site. It is a two-dimensional graph which nodes represent pages and edges represent traversals between pages. Thickness of edge represents support value of traversal and color of edge (ranges from blue to red) represents confidence degree (low to high). To form traversal diagram Naviz take into account considerations about appropriate web traversal properties and modern dynamic page, which will be explained below.

In traversal path mode, Naviz displays traversal path on top of traversal diagram, in such a manner that only one path is showed at one time. Which paths are showed can be filtered by specifying the pages that are visited and optionally by the number of hops required to reach those pages. Found paths are then displayed in traversal diagram one after another and can be ordered by support, confidence, or number of hops.

Naviz uses several strategies to form effective view of traversal diagram and traversal paths:

1. Consideration of appropriate web traversal properties.
   - Hierarchization
     Default strategy of node placing in hierarchical position is that more traversed edges are placed in upper level position and less traversed ones in lower level. This would intuitively describe visitor traversals on a website from top to bottom of graph and give better understanding of the traversal diagram. This strategy can be changed interactively, i.e. higher confidence edges in upper level and lower confidence edges in lower level, although it may not suitable to visualize the traversal diagram. Moreover, graph orientation can be changed either vertically or horizontally.
   - Grouping
     In visualizing the traversal diagram, it is often useful to group pages that related each other. This is done by give weight to indicate the importance of edges, such as more important edge will have shorter length. By default Naviz binds weight to confidence degree of edges, such that pages with high degree of transition probability will be grouped together. It can be changed interactively to be support value, so pages with heavy traffic of traversal among them will be grouped together.
2. Class-instance view
   Class-instance view is a technique to visualize a dynamic page either as a node of representative class or as many nodes of its instances. Modern dynamic page is a class that has many instances. Every instance will bring different information on different access. Visualization of this kind of dynamic pages requires different technique from those of traditional static pages. By class-instance view technique, a class of dynamic page can be expanded to show information of some of instances it may have, and they can be merged back to a class to simplify the global view.
3. Navigational behavior comparison
   Web log data contains the information about various attributes of the visitor visit such as time, place etc. Utilizing this information Naviz can display the same traversal diagram of two different attributes in two different windows side by side. This is useful particularly to compare the visitor behavior of the same website regarding the different attribute, i.e. day behavior vs. night behavior, Tokyo people behavior vs. Osaka people behavior, in which Tokyo represents east of Japan while Osaka represents west of Japan.
4. Traversal path filtering
   Traversal paths is visualized over the underlying traversal diagram in such a manner that only one path will be showed in one time. As we slide path number

slide bar, the corresponding path is showed. For traversal path visualizations, Naviz implements two filtering mechanisms i.e. filtering by number of hops (path length) and by visited pages.

- The number of displayed paths will decrease/increase correspondingly as we specify the number of hops.
- We can choose which pages are visited, and find the paths of visitors that traverse over those pages. There is no limit in the number of pages to choose. In general there are four cases to find the path:
  - Find paths that begin exactly at first chosen page, traverse over consecutive chosen pages and finish exactly at last chosen page.
  - Find paths that begin exactly at first chosen page, traverse over consecutive chosen pages and finish at any page.
  - Find paths that begin at any page, traverse over consecutive chosen pages and finish exactly at last chosen page.
  - Find paths that begin at any page, traverse over consecutive chosen pages and finish at any page.

5. Interactive environment.
   Furthermore Naviz utilizes the interactive environment to provide more capabilities to explore navigational behavior from the data. Layout of the traversal diagram, strategy of hierarchization, grouping, and filtering all can be changed interactively to select the best structure representing the web log data. Users use mouse to point to a particular node/edge and Naviz will show detail explanations about the node/edge below the graph such as page's title and url, edge's support value and confidence degree etc. Clicking on the node will either bring the corresponding page in the browser (in traversal diagram mode), or selecting the pages for path searching (in traversal path mode).

4. Visualization of NTT i-Townpage Served on i-Mode

Townpage is the name of ``yellow pages", a directory service of phone numbers in Japan. It consists 11 million listings under 2000 categories. At 1995 it started internet version namely i-Townpage whose URL http://itp.ne.jp/. The visitors of i-Townpage can specify the location and some other search conditions such as business category or any free keywords and get the list of companies or shops that matched, as well as their phone number and address. Visitors can input the location by browsing the address hierarchy or from the nearest station or landmark. Currently i-Townpage records about 40 million page views monthly.

At this moment i-Townpage has four versions:

1. Standard version: for access with ordinary web browser, which is also equipped with some features like online maps.
2. Lite version: simplified version for device with limited display capability such as PDA.
3. Mobile Townpage: a version for i-Mode users. An illustration of its usage is shown in **Fig. 2** i-mode users can directly make a call from the search results.

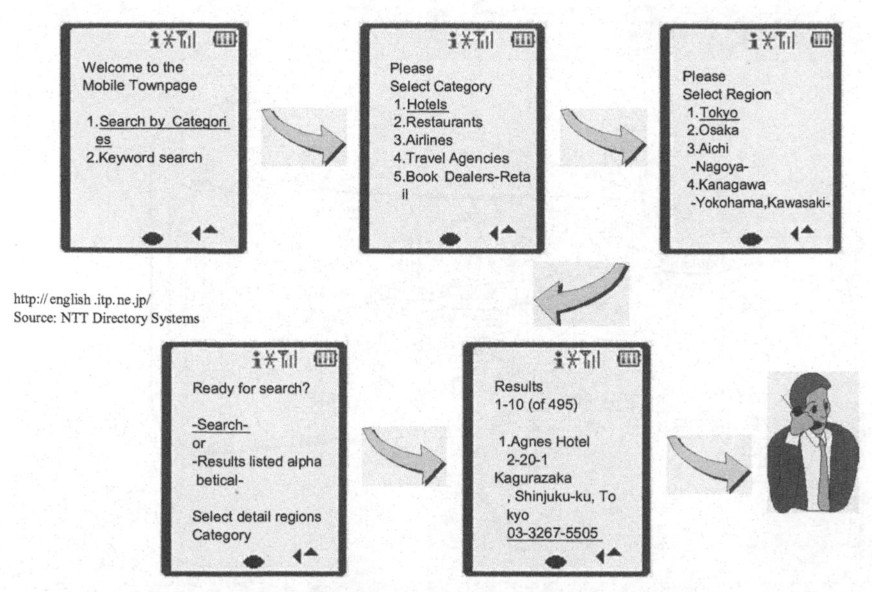

Fig. 2. Mobile Townpage typical usage

4. L-mode version: a version for L-mode access, a new service from NTT that enables internet access from stationary/fixed phone.

Because the limited display and communication ability of mobile phones, a web site for mobile phone has to be carefully designed so that the visitors can reach their goal with least clicks. The Mobile Townpage is not simply a reduced version of standard one but it is completely redesigned to meet the demand of i-Mode users. Nearly 40% visitors of i-Townpage accesses from mobile phones. Figure gives a simple illustration of Mobile Townpage typical usage. A visitor first inputs industry category by choosing from prearranged "Category List" or entering free keywords in "Input Form". Then he/she decides the location. Afterward he/she can begin the search or browse more detailed location, and then get the "Search Result".Data mining performed on the web log data of Mobile Townpage site from 1 to 7 May 2000 with size of 15 GB, using minimum support threshold of 0.1% resulted a set of traversal paths containing 1116 traversal between two pages and 4595 traversal paths among many pages (greater than two). **Fig. 3** is the visualization result of traversal path between two pages by Naviz on traversal diagram mode, it gives the global view of visitor traversal on the entire site of Mobile Townpage. Here, we set minimum-maximum support to 0.7%-100% and minimum-maximum confidence to 0%-100%. The "Start" and "End" are not actual pages belong to the site, they are actually another sites placed somewhere on the internet, and indicate the entry and exit door to and from the site. The important pages include "Top" as the site's top page, "Category List" as visitors choose their search here, "Input Form" where visitors fill in the free keyword, and "Search Result" which indicates that visitors found the answer and thus reached their goal.

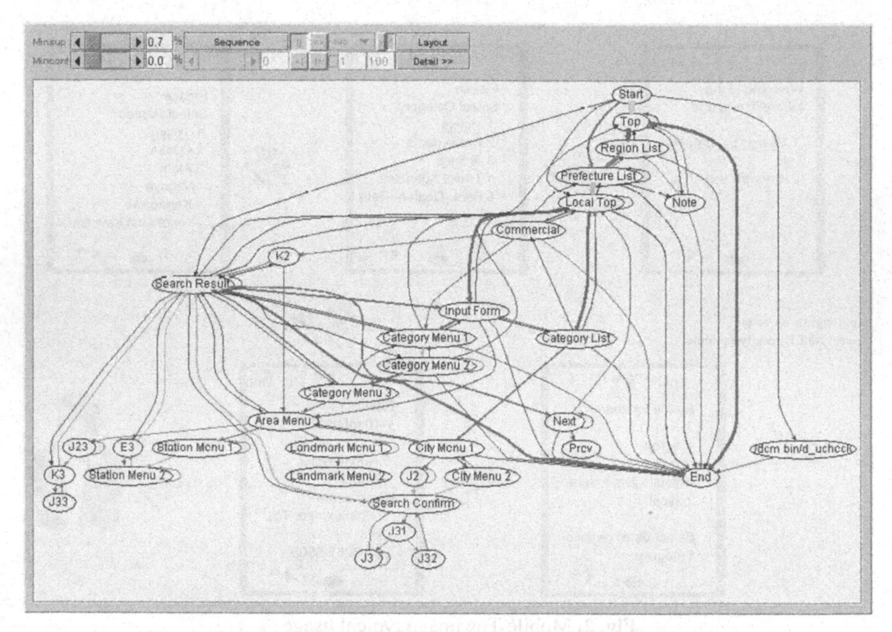

Fig. 3. Traversal diagram of visitor behavior of NTT Mobile Townpage.

In traversal diagram mode, Naviz considers appropriate web traversal properties to form view, so that we can think of visitors came into the site from top of the graph and went out from the bottom. We can see the most traversed edges, the thick ones, that are connecting pages "Top"→"Region List"→"Prefecture List"→"Local Top" etc. are placed in the upper position of the graph. While the less traversed edges that are connecting "City Menu 1"→"City Menu 2", Station Menu 1"→"Station Menu 2" etc. are placed in the lower position of the graph. Naviz also allows us to find related pages easily. As we can see there are several groups of related pages, which indicates that those pages have high degree of transition probability among them: i.e. group ("Top"→"Region List"→"Prefecture List"→"Local Top"→"Commercial"), group ("Input Form"→"Category Menu 1"→"Category Menu 2"→"Category Menu 3"), group ("Area Menu 1"→"Landmark Menu 1"→"Landmark Menu 2"), group ("Station Menu 1"→"Station Menu 2"→"E3") etc.

Switching to traversal path mode, using Naviz we discovered some interesting user behavior on Mobile Townpage site. First, we try to search paths in which visitors are successful finding their goal, i.e. which traverse from "Start" →... →"Search Result"→...→"End". We start by selecting pages, "Start", "Search Result", "End" consecutively, optionally we can specify number of hops too, and then do searching. As the result **Fig. 4** shows one of the success paths: i.e. 1.89 % of visitors successfully reached the "Search Result" page through the path "Top"→"Region List" →"Prefecture List"→"Local Top"→"Input Form"→"Search Result" and then exit from the site. Next is exit path as showed in **Fig. 5**: i.e. 26.33 % of visitors left the site as soon as they came into the "Top" page. Finally **Fig. 6** shows that 0.21 % of visitors were being lost in the website; i.e. once they have tried to choose from "Category

List", but failed and went back to "Local Top", then filled in some keywords in "Input Form", failed again and eventually exit the site without getting the answer. Due to limited space, we cannot include examples of other features such as navigational behavior comparison and class-instance view.

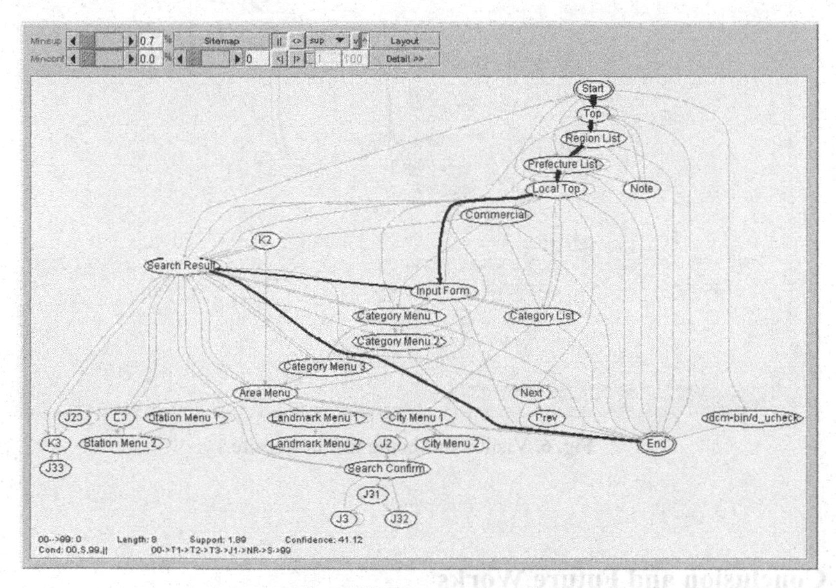

Fig. 4. Visitor success path

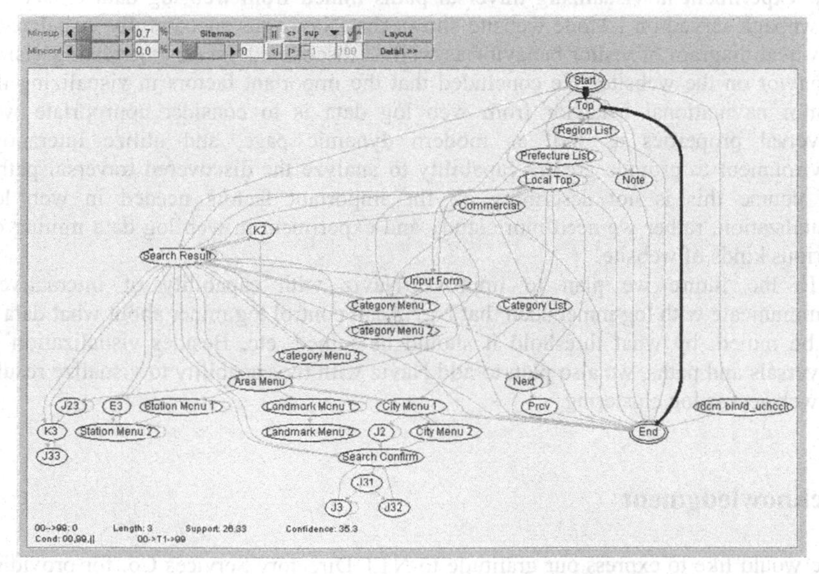

Fig. 5. Visitor exit path

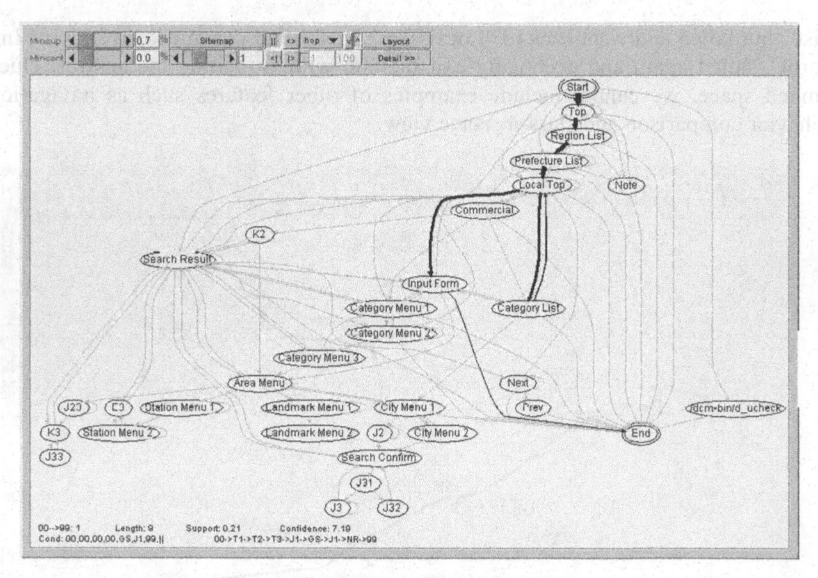

Fig. 6. Visitor was being lost in website.

5. Conclusion and Future Works

Our experiment in visualizing traversal paths mined from web log data of NTT i-Townpage served on i-Mode website shows that Naviz has successfully visualized a traversal diagram of visitor behavior as well as discovered various interesting visitor behavior on the website. We concluded that the important factors in visualizing the visitor navigational behavior from web log data is to consider appropriate web traversal properties as well as modern dynamic page, and utilize interactive environment to provide greater capability to analyze the discovered traversal paths. Of course this is not describing all the important factors needed in web log visualization, rather we need more study and experiment in web log data mining on various kinds of website.

In the future we plan to improve Naviz with capability of interactively communicate with log miner such that Naviz can control log miner about what data is to be mined, by what threshold it should be mined, etc. Besides visualization of traversals and paths, we also plan to add Naviz with the capability to visualize results of web access log clustering.

Acknowledgment

We would like to express our gratitude to NTT Directory Services Co. for providing us with web access log data of NTT i-Townpage served on i-Mode website.

References

[1] R. Srikant, and R. Agrawal. Mining Sequential Patterns: Generalizations and Performance Improvements. In Fifth Int'l Conference on Extending Database Technology (EDBT'96), pages 3-17, Avignon, France, March 1996.
[2] Mukherjea, S & Foley, J., D. (1995). Visualizing the World-Wide Web with the Navigational View Builder. Computer Networks and ISDN Systems, 27 (6), 1075-1087
[3] E. Gansner, E. Koutsofios, S. North, and K. Vo. A technique for drawing directed graphs. Transactions on Software Engineering, 19(3):214–230, March 1993.
[4] Pitkow, J. and K. Bharat. WebViz: A Tool for World-Wide Web Access Log Analysis. In Proceedings of First International Conference on the World-Wide Web 1994.
[5] M. Spiliopoulou and L.C. Faulstich. WUM : A Web Utilization Miner. EDBT Workshop WebDB98, Valencia, Spain, 1998. Springer Verlag.
[6] Cugini, J. and J. Scholtz. VISVIP: 3D Visualization of Paths through Web Sites. In Proceedings of International Workshop on Web-Based Information Visualization (WebVis'99). Florence, Italy. pp. 259-263. IEEE Computer Society, September 1-3 1999.
[7] Jason I. Hong, and James A. Landay, "WebQuilt: A Framework for Capturing and Visualizing the Web Experience." In Proceedings of The Tenth International World Wide Web Conference (WWW10), Hong Kong, May 2001.
[8] Kaidi Zhao, Bing Liu. "Visual Analysis of The Behavior of Discovered Rules." To appear in *Workshop Notes in ACM SIGKDD-2001 Workshop on Visual Data Mining*, San Francisco, CA; Aug 20, 2001
[9] http://www.research.att.com/sw/tools/graphviz/

Optimal Algorithms for Finding User Access Sessions from Very Large Web Logs

Zhixiang Chen[1], Ada Wai-Chee Fu[2], and Frank Chi-Hung Tong[3]

[1] Department of Computer Science, University of Texas-Pan American, Edinburg TX 78539 USA.
chen@cs.panam.edu

[2] Department of Computer Science, Chinese University of Hong Kong, Shatin, N.T., Hong Kong.
adafu@cse.cuhk.edu.hk

[3] Department of Computer Science and Information Systems, The University of Hong Kong, Pokfulam Road, Hong Kong.
ftong@eti.hku.hk

Abstract. Although efficient identification of user access sessions from very large web logs is an unavoidable data preparation task for the success of higher level web log mining, little attention has been paid to algorithmic study of this problem. In this paper we consider two types of user access sessions, *interval sessions* and *gap sessions*. We design two efficient algorithms for finding respectively those two types of sessions with the help of new data structures. We present both theoretical and empirical analysis of the algorithms and prove that both algorithms have optimal time complexity.

1 Introduction

Web log mining has been receiving extensive attention recently (e.g., [1,3,4,5,13]) because of its significant theoretical challenges and great application and commercial potentials. The goal of web log mining is to discover user access behaviors and interests that are buried in vast web logs that are being accumulated every day, every minute and every second. The discovered knowledge of user access behaviors and interests will certainly help the construction and maintenance of real-time intelligent web servers that are able to dynamically tailor their designs to satisfy users' needs [7,10,11]. The discovered knowledge will also help the administrative personnel to predict the trends of the users' needs so that they can adjust their products to attract more users (and customers) now and in the future [3].

The *definition of* **session** *or* **visit** *is the group of activities performed by a user from the moment she enters a server site to the moment she leaves the site* [12]. A user access session determines the set of all the web pages a user accessed from the beginning of the session to the end. This kind of understanding is similar to the *"market basket"* concept in data mining and knowledge discovery [5], the set of all items in the market basket a customer buys at a retail store. As such,

M.-S. Chen, P.S. Yu, and B. Liu (Eds.): PAKDD 2002, LNAI 2336, pp. 290–296, 2002.

the existing data mining techniques, such as the popular association rule mining, may be applied to web log mining [13]. Some researchers [6] have suggested that user access sessions are too coarse grained for mining tasks such as discovery of association rules so that it is necessary to refine single user access sessions into smaller transactions. But such refining process is also based on identification of user access sessions. The difficulty of finding users and user access sessions from web logs have been addressed in [6,4,8,9]. Researchers have proposed cut-off thresholds to approximate the start and the end of a session (e.g., [6]). One may also propose a 30-minutes gap limit between any two pages accessed by the user. If the gap limit is exceeded, then a session boundary should be inserted between the two pages [2,10,11].

There is little algorithmic study of efficient identification of user access sessions in literature. Web logs are typically large, from hundreds of mega bytes to tens of giga bytes of user access records. The logs are usually stored in hard disks. Certainly, after eliminating irrelevant records from the raw web log, sorting algorithms can be applied to group the same user's access records together. After that a one-pass sequential reading process can be used to cut each group of the same user's access records into sessions according to some threshold. This sorting approach may be efficient for smaller web logs, but it is not practically efficient for finding sessions from very large web logs, say, logs with hundreds of mega bytes or tens of giga bytes of records. When real-time applications based on dynamic discovery of the users' hidden access patterns is concerned, faster algorithms that have the ability to find user access sessions in several seconds or several minutes is highly preferred. Without such faster algorithms, any higher level mining tasks would be slowed down.

This paper is organized as follows. In section 2, we give formal definitions of α-interval sessions and β-gap sessions. In section 3, we design data structures and two session finding algorithms. We given theoretical and empirical analysis about the two algorithms in sections 4 and 5, and conclude the paper in section 6.

2 Web Logs and User Access Sessions

The web log for any given web server is a sequential file with one user access record per line. Each user access record consists of the following fields: (1) User's IP address or host name, (2) access time, (3) request method (e.g., "GET", "POST"), (4) URL of the page accessed, (5) Protocol (e.g., HTTP/1.0), (6) return code, and (7) number of bytes transmitted. Other fields may also be included when the server is specially configured. One should note that user access records in a web log are ordered according to record access time with the oldest access record placed on the top and the newest at the bottom.

Given any user access record r, let $t(r)$ be the number of seconds starting at `01/Jan/1900:00:00:00` and ending at the time when r was performed. Let $\mathcal{L}$ be a web log and α and β two positive values. We propose the following two ways to define user access sessions. Recent studies show that the two ways lead to high accuracy in web usage analysis [2].

α-interval sessions: the duration of a session may not exceed a threshold of α. Given any user u, let $r_1, \ldots, r_m$ be the ordered list of u's access records in $\mathcal{L}$. The α-interval sessions for u is a list of subsets $s_1 = \{r_{i_0+1}, \ldots, r_{i_1}\}, s_2 = \{r_{i_1+1}, \ldots, r_{i_2}\}, \ldots, s_k = \{r_{i_{k-1}+1}, \ldots, r_{i_k}\}$ such that for $1 \leq j \leq k$ we have

$$0 \leq t(r_{i_j}) - t(r_{i_{j-1}+1}) \leq \alpha,\ t(r_{i_j+1}) - t(r_{i_{j-1}+1}) > \alpha,$$

where $r_{i_0+1} = r_1$ and $r_{i_k} = r_m$.

β-gap sessions: the time between any two consecutively assessed pages may not exceed a threshold of β. Given any user u, let $r_1, \ldots, r_m$ be the ordered list of u's access records in $\mathcal{L}$. The β-gap sessions for u is a list of subsets $s_1 = \{r_{i_0+1}, \ldots, r_{i_1}\}, s_2 = \{r_{i_1+1}, \ldots, r_{i_2}\}, \ldots, s_k = \{r_{i_{k-1}+1}, \ldots, r_{i_k}\}$ such that for $1 \leq j \leq k$ and for $i_{j_1} + 1 \leq l < i_j$ we have

$$0 \leq t(r_{l+1}) - t(r_l) \leq \beta,\ t(r_{i_j+1}) - t(r_{i_j}) > \beta,$$

where $r_{i_0+1} = r_1$ and $r_{i_k} = r_m$.

3 Data Structures and Algorithms

Our algorithms maintain a data structure as follows. We define a URL node structure to store the URL and the access time of a user access record and a pointer to point to the next URL node. We define a user node to store the following information for a user: the user ID that is usually the hostname or IP address in the access record; the start time that is the time of the first access record entered into the user node; the current time that is the time of last access record entered into the user node; the URL node pointer that points to the linked list of URL nodes; the counter to record the total number of URL nodes added; and two user node pointers to respectively point to the previous user node and the next. Finally, we define a structure with two members, one pointing to the beginning of the two-way linked list of user nodes and the other storing the total number of relevant records that have been processed so far.

The interval session finding algorithm, called ISessionizer, is given in Figure 1. Definitions of the functions used in the algorithm should be inferred from their names. E.g., `purgeNodes(outfile, S, t(r))` searches every user node n in the linked list *S.head*. If it finds that $t(r) - n.startTime > \alpha$, then it outputs the session consisting of the URLs in *n.listPtr* to *outfile*, and then deletes the URL linked list *n.listPtr* and the user node n via memory deallocation.

The gap session finding algorithm, called GSessionizer, is shown in Figure 2. Function *purge2Nodes(outfile, S, t(r))* checks, for every user node in *S.head*, whether the "time gap" between the current record and the last record of the user node is beyond the threshold β or not. If so, it purges the user node in a similar manner as algorithm ISessionizer does.

4 Theoretical Analysis

We present two theorems for algorithm ISessionizer only and the proofs are omitted due to space limit but will be included in the full version of the paper. Similar theorems can be obtained for algorithm GSessionizer. Several notations are introduced: T_1 = the time to read one record from the web log $\mathcal{L}$. T_2 = the time to check whether a record is relevant or not. T_3 = the time to write the URL and the access time of a relevant record from RAM to an output file. and T_4 = the time to search once the user linked list *S.head* and the URL linked list at each user node.

Algorithm ISessionizer (Interval Sessionizer):

```
    input:
        infile: input web log file
        outfile: output user access session file
        α > 0: threshold for defining interval sessions
        γ > 0: threshold for removing old user nodes
    begin
1.      open infile and outfile
2.      createHeadNode(S); S.head = null; S.counter=0
3.      while (infile is not empty)
4.              readRecord(infile, r)
5.              if (isRecordRelevant(r))
6.                      n = findRecord(S, r)
7.                      if (n is null)
8.                              addRecord(S, r)
9.                      else if (t(r) − n.startTime ≤ α)
10.                             n.currentTime = t(r)
11.                             addURL(n.urlListPtr, r)
12.                     else
13.                             writeSessionAndReset(outfile, n, r)
14.             S.counter = S.counter + 1
15.             if (S.counter > γ)
16.                     purgeNodes(outfile, S, t(r)); S.counter=1
17.     cleanList(outfile,S);
18.     close infile and outfile
    end
```

Fig. 1. Algorithm ISessionizer

Algorithm GSessionizer (Gap Sessionizer):

```
    input:
        infile: input web log file
        outfile: output user access session file
        β > 0: threshold for defining gap sessions
        γ > 0: threshold for removing old user nodes
    This algoprithm is the same as algorithm ISessionier except that
    Lines 9 and 16 are replaced respectively by
9.                      else if (t(r) − n.currentTime ≤ β)
16.                     purge2Nodes(outfile, S, t(r)); S.counter=1
```

Fig. 2. Algorithm GSessionizer

Theorem 1. *Given $\alpha > 0$, let $\mathcal{S}$ denote the set of all the α-interval sessions in $\mathcal{L}$. Then, algorithm ISessionizer finds exactly the set $\mathcal{S}$.*

Theorem 2. *Assume that the web log $\mathcal{L}$ have N user access records with M relevant ones. Then, the time complexity of algorithm ISessionizer with a γ-threshold for purging user nodes is at most $(T_1 + T_2)N + T_3M + T_4(1 + \frac{1}{\gamma}))M$.*

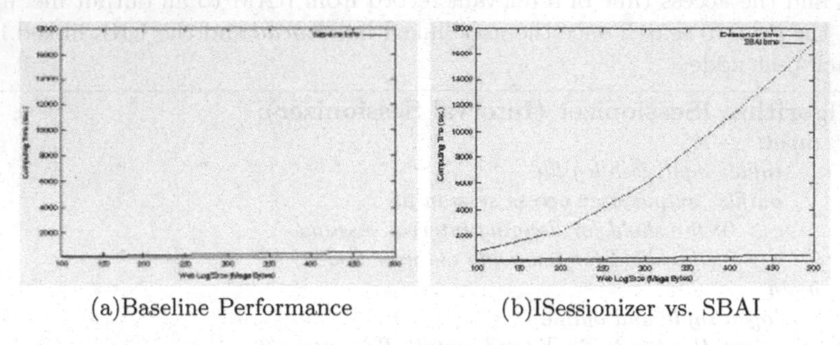

(a)Baseline Performance (b)ISessionizer vs. SBAI

Fig. 3. Performances of Algorithm ISessionizer vs. SBAI vs. Baseline Time

5 Empirical Analysis

We report empirical performance analysis of algorithms ISessionizer and GSessionizer. We use five web logs with respectively 100, 200, 300, 400, and 500 mega bytes of user access records that were collected from the web server of the Department of Computer Science, the University of Texas - Pan American. The computing environment is a Dell OptiPlex GX1 PC with a 300 MHz processor, 512 mega bytes of RAM and 8 Giga bytes of hard disk. The baseline time is the time needed for reading a Web log once sequentially from disk to RAM, testing whether each record is pertinent or not, and writing each pertinent record back to disk. The sorting based algorithm, denoted by *SBAI*, for finding α-interval sessions is the one that sorts the web log to group the same user's access records together and then reads the sorted log sequentially once from disk to RAM to divide the relevant access records in each group into α-interval sessions. The sorting based algorithm, denoted by *SBAG*, for finding β-gap sessions works similarly as algorithm SBAI, but it divides the relevant access records in each group into β-sessions. We choose $\beta = \alpha = 30$ minutes and $\gamma = 500$. The same *isRecordRelevant( )* function was used for the four algorithms and the baseline time testing. Performance comparisons are illustrated in Figures 3 and 4.

6 Conclusions

We have settled the problem of efficient identification of interval sessions and gap sessions from very large Web logs. However, we feel that more robust session definitions may be needed because of the vast diversity of the web users and wide applications of cashing and proxy servers, and that machine learning may help the design of new algorithms.

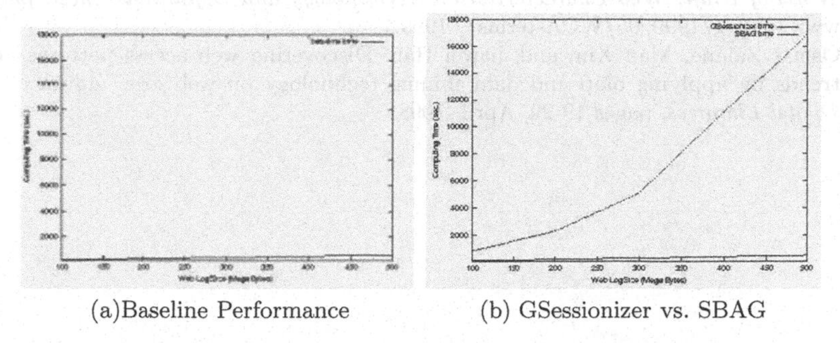

(a)Baseline Performance (b) GSessionizer vs. SBAG

Fig. 4. Performances of Algorithm GSessionizer vs. SBAG vs. Baseline Time

References

1. B. Berendt and M. Spiliopoulou. Analysis of navigation behavior in web sites integrating multiple information systems. *The VLDB Journal*, 9:56-75, 2000.
2. Bettina Berendt, Bamshad Mobasher, Myra Spiliopoulou, and Jim Wiltshire. Measuring the accuracy of sessionizers for web usage analysis. *Proceedings of the Workshop on Web Mining at the First SIAM International Conference on Data Mining*, pages 7-14, April 2001.
3. Alex G. Buchner and Maurice D. Mulvenna. Discovering internet marketing intelligence through online analytical web usage mining. *ACM SIGMOD RECORD*, pages 54-61, Dec. 1998.
4. L. Catledge and J. Pitkow. Characterizing browsing behaviors on the world wide web. *Computer Networks and ISDN Systems*, 27, 1995.
5. M.S. Chen, J.S. Park, and P.S. Yu. Efficient data mining for path traversal patterns. *IEEE Transactions on Knowledge and Data Engineering*, 10:2:209-221, 1998.
6. Robert Cooley, Bamshad Mobasher, and Jaidep Srivastava. Data preparation for mining world wide web browsing patterns. *Journal of Knowledge and Information Systems*, 1:1, 1999.
7. M. Perkowitz and O. Etzioni. Adaptive web pages: Automatically synthesizing web pages. *Proceedings of AAAI/IAAI'98*, pages 727-732, 1998.
8. J. Pitkow. In search of reliable usage data on the WWW. *Proceedings of the Sixth World Wide Web Conference.* pages 451-463, Santa Clara, CA, 1997.
9. P. Pirolli, J. Pitkow, and R. Rao. Silk from sow's ear: Extracting usable structures from the Web. *Proceedings of the 1996 Conference on Human Factors in Computing Systems (CHI'96).* Vancouver, British Columbia, Canada, 1996.
10. Myra Spiliopoulou and Lukas C. Faulstich. Wum: A tool for web utilization analysis. *Proceedings of EDBT Workshop WebDB'98*, LNCS1590, pages 184-203. Springer Verlag, 1999.

11. Myra Spiliopoulou, Carsten Pohle, and Lukas C. Faulstich. Improving the effectiveness of a web site with web usage mining. *KDD'99 Workshop on Web Usage Analysis and User Profiling WEBKDD'99*, Aug, 1999.
12. W3C. World wide web committee web usage characterization activity. *W3C Working Draft: Web Characterization Terminology and Definitions Sheet*, pages www.w3.org/1999/05/WCA-terms/, 1999.
13. Osmar Zaïane, Man Xin, and Jiawei Han. Discovering web access patterns and trends by applying olap and data mining technology on web logs. *Advances in Digital Libraries*, pages 19-29, April, 1998.

Automatic Information Extraction for Multiple Singular Web Pages

Chia-Hui Chang, Shih-Chien Kuo, Kuo-Yu Hwang, Tsung-Hsin Ho, and Chih-Lung Lin

Dept. of Computer Science and Information Engineering
National Central University, Chung-Li 320, Taiwan
chia@csie.ncu.edu.tw, {bruce, want, windson, nono}@db.csie.ncu.edu.tw

Abstract. The World Wide Web is now undeniably the richest and most dense source of information, yet its structure makes it difficult to make use of that information in a systematic way. This paper extends a pattern discovery approach called IEPAD to the rapid generation of information extractors that can extract structured data from semi-structured Web documents. IEPAD is proposed to automate wrapper generation from a multiple-record Web page without user-labeled examples. In this paper, we consider another case when multiple Web pages are available but each input Web page contains only one record (called *singular* Web pages). To solve this case, a hierarchical multiple string alignment is proposed to allow wrapper induction for multiple singular Web pages.

1 Introduction

The problem of information extraction is to transform the contents of input documents into structured data. Unlike information retrieval, which concerns how to identify relevant documents from a collection, information extraction produces structured data ready for post-processing, which is crucial to many applications of text mining. Therefore, information extraction from Web pages is a crucial step enabling content mining and many other intelligent applications of the Web. Examples include meta-search engines [6], which organize search results from multiple search engines for users, and shopping agents [2], which compare prices of products from multiple Web merchants.

Information extraction has been studied for years, but mostly concentrated on free-text documents. In this case, linguistic knowledge such as lexicons and parsers can be useful. However, a huge volume of information on the Web is rendered in a *semi-structured* manner, where linguistic knowledge only provides limited hints. Another difference is that different Web sites have their own unique layout and format. Virtually no general "grammar rule" can describe all possible layouts and formats so that we can not have one extractor for all Web pages. As a result, each Web site may require a specific extractor, which makes it impractical to program extractors by hand.

In recent years, several research efforts have focused on learning the extraction rules to automate wrapper generation. For example, WIEN, Softmealy,

M.-S. Chen, P.S. Yu, and B. Liu (Eds.): PAKDD 2002, LNAI 2336, pp. 297–303, 2002.

STALKER, and IEPAD, etc. [7, 4, 5, 1]. The former three are machine learning based approaches and rely on user-labeled training examples indicating what information to be extracted. The last one, on the other hand, uses a pattern mining approach to discover display template for data records. Therefore, it saves manual effort for labeling examples. Such a property is especially useful when dealing with Web pages containing too many variations such that a lot of user-labeled examples are required.

However, the input to IEPAD is limited to one Web page containing multiple records (called *plural pages*). When a Web page contains only one record (called *singular pages*), IEPAD fails to work since IEPAD relies on the discovery of regular and contiguous patterns to "estimate" the display template of these records. This multiple singular pages input is a contrast to the one plural page input to IEPAD, where each containing several records. In fact, IEPAD's approach is to solve the extraction of the record boundary not the extraction of individual features. In this paper, we consider the case when multiple Web pages are available but each input Web page contains only one record.

To extract meaningful information from multiple pages with similar structure, one thought is to align these multiple pages and sifting representing features for users. However, alignment of long pages is time consuming and error-prone. Therefore, we propose a hierarchical approach to solve the problem layer by layer. Such an approach is similar to the idea of building the parsing trees of these pages and then compare the them level by level. The main problem is how to correctly align multiple pages. If the alignment is done, what information should be extracted? In the next section, we describe a hierarchical multiple string alignment to show how data are divided for alignment. Section 3 describes the criteria for feature selection from the divided blocks and the logic of the extractor. Section 4 shows the preliminary experiment results to demonstrate the effect. Finally, section 5 concludes the paper.

2 Alignment of Multiple Singular Pages

In this section, we describe a hierarchical approach to align the input pages. The key idea is to divide each input page into a sequence of blocks and align these block sequences to understand the page structure for a Web site. The division is based on an abstraction/encoding function of the input pages. In this paper, we use a layered encoding function which utilize the HTML tags to translate each page into tag sequence and divide the page content at the same time. The alignment is based on both the markups in the HTML pages and the structure similarities between two blocks.

2.1 Layered Encoding

Layered encoding is a tag-based encoding scheme but considers only tags at the same layer. The so-called *layer* refers to the depth from the root in the HTML parse tree. For each tag, there is an associated layer. The layered encoding,

Layer_Encoding, transforms an information block to a token sequence with all tags of the smallest layer (or the outermost layer tags) and denotes any other text between two such tags as the <TEXT> token. For example,

Layer_Encoding("SOME TEXT<P><A>LINK</A></P>")
="<TEXT><P><TEXT></P>".

The encoding scheme used here is different from block-level encoding or text-level encoding used for IEPAD. Layered encoding does not need prior knowledge about the characteristics of individual tags, but depends on the layer positions of the tags in the parsing tree. Remember that the encoding scheme is not only used for data abstraction (such that the alignment program can recognize the structures), it is also a segmentation of the input text. For a token string with k tokens, there is a natural section of the page into k pieces. For m multiple singular pages, we will have m token strings for multiple string alignment.

2.2 Multiple String Alignment

Multiple string alignment is a technique used to find a general presentation of the critical common features for multiple strings. It has been successfully applied in IEPAD for constructing patterns which can tolerate missing attributes [1]. In this paper, we use multiple string alignment to find the general expression for the m token sequences.

Multiple string alignment is a generalization of *alignment* for two strings which can be solved in $O(s * t)$ by *dynamic programming* to obtain optimal *edit distance*, where s and t are string lengths. As an example of *two string alignment*, consider the alignment of two strings "*dtbtbt*" and "*dtbt*" shown below:

$$\begin{array}{l} d\,t\,b\,t\,b\,t \\ d\,t\,b\,t\,-\,- \end{array}$$

In this alignment, the first four tokens match their counterparts in the opposite string and the last two tokens are opposite two hyphens. The edit distance between two strings are defined as the summation of the matching score between two aligned tokens. For example, the value for matching two same tokens is zero, while matching a token against a hyphen charges 1; a value of 3 is given for matching two different tokens to avoid such alignment. However, such matching values can not distinguish the above alignment from the following alignment, because both alignments have the same edit distance 2.

$$\begin{array}{l} d\,t\,b\,t\,b\,t \\ d\,t\,-\,-\,b\,t \end{array}$$

In order to find which alignment is correct, the match function needs to consider more than just facile tokens, especially on matching two <TEXT> tokens since <TEXT> tokens actually represent some text contents that have been abstracted. There can be many ways to do the computations. For example, we can represent a text string as a vector $(t_1, t_2, ..., t_n)$ of n tags and compute the cosine of the angle between two text vectors for the distance.

$$dis(\boldsymbol{t}, \boldsymbol{s}) = 1 - \frac{\sum_{i=1}^{n} t_i \cdot s_i}{\sqrt{\sum_{i=1}^{n} s_i}\sqrt{\sum_{i=1}^{n} t_i}}$$

Once the correct alignment for two token strings is given, we can extend the idea to multiple string alignment. The score or goodness of a multiple string alignment is not as easily generalized. A common used scores is the *sum of pairs* score which is the sum of the scores of pairwise edit distances. To avoid $O(n^m)$ complexity by dynamic programming for m strings of length n, an bounded-error approximation algorithm, called *center star*, is used such that the score of the alignment is no greater than twice the score of optimal alignment. The center star algorithm first computes the center string S_c which minimizes $\sum_{S_i \in S} D(S_c, S_i)$ over all strings in S. Then, align each string to the center string to construct multiple alignment [3].

From the alignment of multiple token strings, we can summarize the *signature representation* for them as a regular expression. For example, suppose we have constructed the alignment for three token strings "`dtbtbt`", "`dtbt`" and "`dtbat`" as follows:

$$\begin{array}{l} d\,t\,b - t\;b\;t \\ d\,t\,b - t - - \\ d\,t\,b\;a\;t - - \end{array}$$

The signature representation will be expressed by "`dtb[a|-]t[b|-][t|-]`", which divides each text information into 7 blocks. Among them, some blocks are of particular interests. Notice that token strings are composed of tag tokens and `<TEXT>` token; and only `<TEXT>` and some special tag tokens (which contain hyperlinks such as <A> and <IMG> tags) might contain useful information. Therefore, whenever there is a `<TEXT>` token or special tag tokens in the pattern, the corresponding information should be retained. To summarize, the process for block division and extraction proceeds as follows:

1. First, apply layered encoding for each of the m contents.
2. Second, align m token strings by center star algorithm.
3. Finally, extract those blocks denoted by `<TEXT>` and hyperlink tag tokens.

2.3 The Hierarchical Approach

From the $m \times n_1$ matrix of the multiple string alignment, we are especially interested in columns that are denoted by `<A>` tags and `<IMG>` that contain link information and `TEXT` tags that needs to be further divided. For each of the k `TEXT` columns, we apply the above procedure iteratively and get a result of $L_2 = n_{2,1} + n_{2,2} + \ldots + n_{2,k}$ blocks, where $n_{2,i}$ denotes the number of divided blocks for ith `TEXT` column. Then, for each of the m `TEXT` columns in the L_2 blocks, we can apply the block extraction procedure and get a result of $L_3 = n_{3,1} + n_{3,2} + \ldots + n_{3,m}$ subblocks, where $n_{3,i}$ denotes the number of divided blocks for the ith `TEXT` block. The same process can be applied level by level until no column can be divided anymore.

3 Feature Extraction

Through the hierarchical multiple string alignment approach, the input data can be divided as detailed as any tags can separate. Often this can divide as many blocks from hundreds to thousands. Therefore, the system should also decide what columns to be presented for users' selection. For example, if the texts of a column contain only string "`Author`", such a column might be the name describing the following column. If a good judgement is available, the system can always present users with information that might be useful and greatly reduces the number of blocks to be presented.

To extract useful information, a simple judgement is to examine the varieties among the blocks in a column. If all m blocks contains the same text string, it will be unnecessary to extract sun a column. On the other hand, if large varieties are presented among these blocks, the column might also be useless. After all, whether information is useful or not is quite objective. In this paper, we use edit distance to measure the dis-similarities between two text blocks, where the contents are compared literally byte by byte. A threshold is given for the decision of feature selection as shown in experiments. With the sifting process, the system presents only useful information for users and make this approach practical.

4 Experimental Results

Since most integration systems have to deal with CGI generated pages, we therefore focus ourselves on such data sources. The experiments here contain five commercial Web sites including Barnes&Noble, YahooKimo, NewWorld, E-Family, and DVDmall, each containing 50 pages. The average document size is 6KB~57KB bytes for the data set. Comparing to the average number of 306~1694 tags in a page, it's about 3.7% the page size. The number of columns segmented from a page is about 40~393 for a depth of 20 levels.

4.1 Generalizing over Unseen Pages

Although the process for the pattern discovery is not like typical machine learning process, the goal is the same to generalize the extraction over unseen pages. Therefore, we have designed the experiments in a way similar to that used in machine learning. For each Web site, we randomly select m pages as the training pages and use remaining $N - m$ pages as validation set. For each Web page, the m training Web pages are referenced to get the alignment. To measure the correctness of the data alignment, we compute the ratio of correctly aligned blocks to the number of expected attributes. If any page has accuracy less than 1, such a page is added to the training pages such that $m + 1$ training pages is used as baseline for the extractor. As shown in Figure 1, the accuracy rate is pretty high for the first four data sources. Barnes&Noble has accuracy 85% for two training pages and increase to 90% for 5 training pages. The average accuracy is about 95% with 2 training example, and increased to 97% for two training examples.

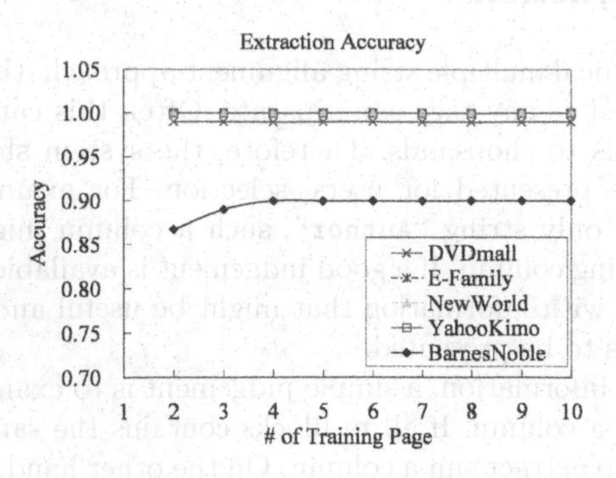

Fig. 1. The accuracy for different training example.

4.2 Feature Selection

As mentioned before, feature selection is designed to do the filtering of unnecessary information. The selection retains those columns with column variety greater than a threshold ϵ. With this mechanism, half the columns are filtered when a value of 0 is used for many Web site. The number of columns retained for Barnes&Noble is 67 as shown in Figure 2. As the threshold increased, more blocks are filtered. Suppose the number of desired attributes are 20 for this Web site, we can choose larger threshold ϵ to filter more columns. However, there is a risk that the desired blocks are filtered if the threshold is set too large.

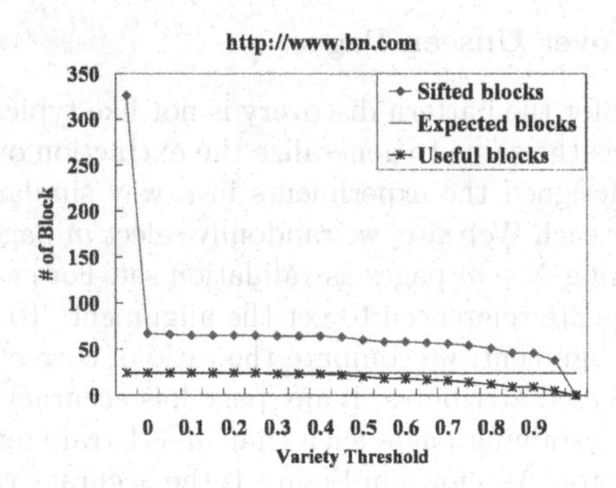

Fig. 2. The number of columns retained versus the threshold.

5 Conclusion and Future Work

In this paper, we presented a novel approach to the extraction of multiple one-record pages. The approach of hierarchical multiple string alignment to automatic feature extraction follows the track of IEPAD to save the efforts of labeling for machine-learning based approach. By taking the advantages of HTML tags, we can divide the Web content as detail as those tags can separate. Meanwhile, the variety measurement of strings in the same column is used to sift interesting features successfully. The result of experiments shows that the alignment can achieve 96% with two training pages. An average of 30 features are selected from 10KB Web pages, where most useful information are preserved.

Acknowledgement

This work is sponsored by National Science Council, Taiwan under grant NSC90-2213-E-008-042.

References

[1] C.-H. Chang and S.-C. Lui. Iepad: Information extraction based on pattern discovery. In *Proceedings of the 10th International Conference on World Wide Web*, pages 223–231, Hong-Kong, 2001.

[2] R.B. Doorenbos, O. Etzioni, and D.S. Weld. A scalable comparison-shopping agent for the world-wide web. In *Proceedings of the 1st International Conference on Autonomous Agents*, pages 39–48, NewYork, USA, 1997.

[3] D. Gusfield. *Algorithms on Strings, Trees, and Sequences*. Cambridge, 1997.

[4] C.-N. Hsu and M.-T. Dung. Generating finite-state transducers for semi-structured data extraction from the web. *Information Systems*, 23(8):521–538, 1998.

[5] I. Muslea, S. Minton, and C. Knoblock. A hierarchical approach to wrapper induction. In *Proceedings of the 3rd International Conference on Autonomous Agents*, pages 190–197, Seattle, WA, 1999.

[6] E. Selberg and O. Etzioni. Multi-engine search and comparison using the metacrawler. In *Proc. of the Fourth Intl. WWW Conference*, Boston, USA, 1995.

[7] N. Kushmerick D. Weld and R. Doorenbos. Wrapper induction for information extraction. In *Proceedings of the 15th International Joint Conference on Artificial Intelligence (IJCAI)*, pages 729–737, Japan, 1997.

An Improved Approach for the Discovery of Causal Models via MML

Honghua Dai and Gang Li

School of Computing and Mathematics, Deakin University
Melbourne Campus, Burwood, Vic 3125, AUSTRALIA
hdai@deakin.edu.au

Abstract. Discovering a precise causal structure accurately reflecting the given data is one of the most essential tasks in the area of data mining and machine learning. One of the successful causal discovery approaches is the information-theoretic approach using the Minimum Message Length Principle[19]. This paper presents an improved and further experimental results of the MML discovery algorithm. We introduced a new encoding scheme for measuring the cost of describing the causal structure. Stiring function is also applied to further simplify the computational complexity and thus works more efficiently. The experimental results of the current version of the discovery system show that: (1) the current version is capable of discovering what discovered by previous system; (2) current system is capable of discovering more complicated causal models with large number of variables; (3) the new version works more efficiently compared with the previous version in terms of time complexity.

Keywords: Minimum message length, MML, causal discovery, causal modeling, inductive inference, machine learning, Bayesian networks.

1 Introduction

With the increasing demanding of using data mining techniques in solving real world application problems, there is a great interest in having more efficient mining algorithms and more accurate discovered rules. To achieve this goal, further refinement and improvement of the existing successfully implemented and tested discovering algorithms becomes critical.

Discovery of Bayesian Network representations for the wide applications of reasoning under uncertainty[10,8] has resulted in the invention of several successful algorithm for the discovery of Bayesian Networks [1,2,5,6,19,3]. In this area, there has been a substantial research program aimed at the automated learning of the linear causal models employed in the social sciences (such as structural equation models or path models; cf. [7]), natural sciences and in medicine, which is that of Clark Glymour and his company at the University of Pittsburgh, underway for the past decade [4,13]. This has lead to a successful delivery of a

M.-S. Chen, P.S. Yu, and B. Liu (Eds.): PAKDD 2002, LNAI 2336, pp. 304–315, 2002.

commercially available program TETRAD II [15], and its successive versions TETRAD III and IV[14].

The methods employed in TETRAD II incorporate a number of principles based upon Judea Pearl's work including what they call Principles I and II [12], otherwise rely upon finding models which minimize residual variance in the data.

In parallel in 1996, Wallace and his fellows, Korb and Dai successfully introduced an information-theoretic approach to the discovery of causal models. The approach uses Wallace's Minimum Message Length Principle (MML) [16,18,17] to search for and evaluate linear causal models given sample data. The experimental results of the comparisons with TETRAD II was reported in [19].

It has been revealed in the article [19] that the MML induction approach is capable of recovering causal models from generated data which are quite accurate reflections of the original models and compare favourably with those of TETRAD II. However, the system's capability in discovering complicated causal models with large number of variables is yet to be improved. This paper will report the improvement we have made to the discovery algorithm and the further experimental results.

In section 2, we will provide a brief introduction of the original algorithm and presents our improved algorithm. Section 3 describes our searching strategy and the implementation. Section 4 reports the experimental results.

2 Encoding Schemes and Model Discovery

The basic idea of the original algorithm published in 1996 is that an encoding scheme based on the minimum message length principle was provided to describe (1) the causal structure, (2) the strength (model parameters) of the causality of the DAG and (3) the data given the model is true. For each model we could find from the model space, we calculate the total message length based on the data provided. The best model we will choose among all the models considered is the one with the shortest total message length. According to information theory, the total message length is given by,

$$L = -log_2 P(M) - log_2 P(D|M) = L(M) + L(D|M) \tag{1}$$

where $L(M) = -log_2 P(M)$ is the cost (in number of bits) of encoding the causal model M, and $L(D|M) = -log_2 P(D|M)$ is the cost of encoding the sample data D given the model M. $L(M)$ is itself composed of two main parts: the cost of encoding the causal structure (the DAG), $L^{(s)}$, and the cost of encoding the model parameters, $L^{(p)}$, i.e.,

$$L(M) = L^{(s)} + L^{(p)} \tag{2}$$

Therefore, the total cost L of encoding a causal model is:

$$L = L(M) + L(D|M) = L^{(s)} + L^{(p)} + L(D|M). \tag{3}$$

2.1 Improved Encoding Schemes

An Improved DAG Encoding Scheme. Initially two ways of encoding the directed acyclic graph were proposed for measuring $L^{(s)}$.

Method 1: Assume that the DAG is encoded by specifying a total ordering (requiring $\log K!$ bits) and then specifying which pairs of nodes are connected; this requires $\frac{K(K-1)}{2}$ bits on the assumption that the probability that a link is present is 1/2. It corresponds to maximal ignorance about the degree of connectedness of the graph (i.e., once again we avoid the use of explicit prior information about the causal models we are looking for). It is enough to specify the presence or absence of arcs, since directionality is implied by the ordering already provided. Since more than one ordering is consistent with the DAG, actually specifying a *particular* ordering is inefficient, so we reduce the message length by the number of bits needed to select among the ϕ total orderings consistent with the DAG (where we count ϕ). Hence,

$$L_1^{(s)} = log\ K! + \frac{K(K-1)}{2} - log\ \phi. \quad (4)$$

Algorithm 21 Counting the total orderings ϕ consistent with the DAG.
Given K variables and n_l links, start from any one possible graph which is composed of the n_l links, the following algorithm is applied for finding out the number of total cyclic graphs composed of the n_l links.

1. *$doms[i] \leftarrow 0$ for $i = 1, 2, \cdots, K$.*
2. *If there is a link directly from i to j, insert 1 to the $i-1$ bit of $doms[j]$, for $i = 1, \cdots, K$.*
3. *If there is a path indirectly link the node i to j, insert 1 to the ith bit of $doms[j]$, for $i = 1, \ldots, K$.*
4. *$nper \leftarrow 0$ and $lev \leftarrow 0$.*
5. *Start permutation at level 1; call subroutine* perms(lev) *recursively until $lev \geq K$.*
6. *Output $M \leftarrow nper$, exit.*

Algorithm 22 Recursive Algorithm. *Subroutine* perms(lev)
This assumes all nodes in perm[1..lev] are in place and entered in Dset. For each node in perm[lev+1...K] it tries to see if the node can be placed in perm[lev+1] and entered in Dset. This can be done only if Dset contains all nodes dominating the node to be placed.

1. *$np \leftarrow 0; perm[i] \leftarrow i$.*
2. *$lev \leftarrow lev + 1$.*
3. *$k \leftarrow perm[lev]$;*
4. *If ($lev < K$)*
   $i \leftarrow lev$;

istart: *$j \leftarrow perm[i]$;*

If (doms[j] & ($\tilde{d}$set)) go to idone;
perm[i] ← k;
perm[lev] ← j;
insert 1 to (j-1) digit of dset;
call this subroutine recursively
np ← np + perms(lev).
dset ← conjuction of the complement of 1 left shift j-1 digit and dset;
perm[i] ← j;
idone: If (i < K) i ← i+1; goto istart.
5. *Else*
return np; exit.

Method 2: The second method for calculating the message length of a DAG begins by describing the undirected graph, which costs $\frac{K(K-1)}{2}$ bits, and then specifies the particular direction which each arc is to assume, where this results in an acyclic graph. That is, we count the number of possible acyclic orientations; the logarithm of that number is the number of additional bits required. In order to do this count, we can subtract the number of cyclic orientations ρ from the number of total orientations, which is 2^ν, where ν is the number of undirected arcs. Hence,

$$L_2^{(s)} = \frac{K(K-1)}{2} + log(2^\nu - \rho). \tag{5}$$

Previous experimental results show that these two methods result in MML costs that are very close for a wide variety of simple graph structures[19] we tested, so we can expect that the choice of encoding method will make little difference to experimental results. In practice, thus far, our implementation of $L_1^{(s)}$ is faster. To further improve the efficiency of the discovery algorithm, we introduced the following new encoding scheme.

Method 3: (The new method) The structure of a directed acyclic graph (DAG) can be described by specifying its parents set *Parents(x)* for each one of the nodes of the DAG. This description consists of the number of parents, followed by the index of the set *Parent(x)* in some enumeration of all $\binom{K}{j}$ sets of its cardinality. So the cost for encoding the DAG can be calculated using:

$$L_3^{(s)} = \sum_i (\log K + \log \binom{K}{|Parents(X_i)|}) \tag{6}$$

To avoid intensive computational time cost in calculating $\log \binom{K}{|Parents(X_i)|}$, we use *Stirling* approximation formula $x! = x^x e^{-x}\sqrt{2\pi x}$, so we get,

$$\log \binom{K}{r_i} \approx (K - r_i) \log \frac{K}{K - r_i} + r_i \log \frac{K}{r_i} \tag{7}$$

Thus, we have,

$$L_3^{(s)} = \sum_i (\log K + (K - r_i) \log \frac{K}{K - r_i} + r \log \frac{K}{r_i} \tag{8}$$

Where, $r_i = |Parents(X_i)|$. This formula works much faster than using the formula $L_1^{(s)}$ and $L_2^{(s)}$. This can be seen from the experimental results reported in section 4.

Encoding Model Parameters To simplify the explanation, we take the case of causal models with one independent variable as an example to show how the message length is calculated. Assume that the model we are trying to find is,

$$y_n = \sum_{k=1}^{K} a_k x_{nk} + r_n \tag{9}$$

$$(n = 1, 2, \ldots, N)$$

where variable means are assumed to be zero, $r_n \sim N(0, \sigma^2)$ with $\sigma, \{a_k|_{1\le k\le K}\}$ unknown. $\hat{\sigma}^2$ and $\{\hat{a_k}\}$ $(k = 1, 2, \ldots, K)$ are the parameter estimates we need to discover. Assuming parameter variance is uniformly distributed and parameters themselves are normally distributed, Following Wallace and Freeman [17], the message length for encoding the parameters is, within a constant. We are simplifying here by ignoring a volume term[9]. We have,

$$L^{(p)} = -log(\frac{h(parameters)}{\sqrt{F}}) = \frac{1}{2}logF - log\ Prior(\sigma, \{a_k\}) \tag{10}$$

Using the formulas as provided in [19], we get,

$$L^{(p)} = \frac{1}{2}(\log 2 + \log N - 2(K+1)\log\sigma + \log|A|) + \log\sigma \tag{11}$$

$$-\sum_{k=1}^{K}\log(\frac{1}{\sqrt{2\pi}\alpha\sigma}e^{(-a_k^2/2\alpha^2\sigma^2)}) \tag{12}$$

Further we have

$$L^{(p)} = \frac{K}{2}log2\pi + Klog\alpha + \frac{1}{2\alpha^2\sigma^2}\sum_{k=1}^{K}a_k^2 + \frac{1}{2}log(2N) + \frac{1}{2}log|A| \tag{13}$$

Encoding the Data According to the formula (1), the message length for encoding the data given the model is,

$$L(D|M) = \frac{N}{2}log2\pi + Nlog\sigma + \sum_{i=1}^{N}\frac{r_i^2}{2\sigma^2} \tag{14}$$

In which we adopted the standard assumption that the data are a random sample from a normal distribution, thus the likelihood function we used in the derivation of the above formula is

$$P(y|\Theta) = P(y|\sigma, \{a_k\}) = \prod_{n=1}^{N}\frac{1}{\sqrt{2\pi\sigma^2}}e^{-\frac{1}{2\sigma^2}(y_n - \sum_k a_k x_{nk})^2} \tag{15}$$

2.2 Learning Model Parameters

Theorem 1. *Based on the Minimum Message Length Principle and the standard assumption that the data are a random sample from a normal distribution, so the parameters of the discovered causal model is given by*

$$(A+I)a = (x^T \cdot x + I)a = y \cdot x^T \tag{16}$$

where $x = (x_1, \ldots, x_K)$ *and and* $x_j = (x_{1j}, x_{2j}, \ldots, x_{Nj})$ *for* $j = 1, 2, \ldots, K$ *and* $y = (y_1, y_2, \ldots, y_N)$. *A is the* $K \times K$ *square matrix* $A = (a_{ij})_{K\times K} = (x_i \cdot x_j)_{K\times K}$ *and* I *is a* $K \times K$ *unit matrix.*

Proof. To derive the model parameters, we consider the combined message length for encoding the parameters and the data given the model,

$$\eta = L^{(p)} + L(D|M) \tag{17}$$

$$= \frac{N+K}{2} log 2\pi + K log \alpha + N log \sigma + \frac{1}{2\sigma^2}(\sum_{i=1}^{N} r_i^2 + \sum_{k=1}^{K} \frac{a_k^2}{\alpha^2}) \tag{18}$$

$$+\frac{1}{2} log(2N) + \frac{1}{2} log|A| \tag{19}$$

To minimize this value we examine its partial derivatives with respect to σ and the a_j:

$$\frac{\partial \eta}{\partial \sigma} = \frac{N}{\sigma} - \frac{1}{\sigma^3}(\sum_{i=1}^{N} r_i^2 + \sum_{k=1}^{K} \frac{a_k^2}{\alpha^2}) = 0 \tag{20}$$

Therefore,

$$\hat{\sigma^2} = \frac{1}{N}(\sum_{i=1}^{N} r_i^2 + \sum_{k=1}^{K} \frac{a_k^2}{\alpha^2}) \tag{21}$$

And for $j = 1, 2, \ldots, K$,

$$\frac{\partial \eta}{\partial a_j} = \frac{1}{\sigma^2}(-\sum_{i=1}^{N} r_i x_{ij} + \frac{a_j}{\alpha^2}) = 0 \tag{22}$$

i.e.,

$$\frac{a_j}{\alpha^2} - \sum_{i=1}^{N} r_i x_{ij} = 0 \tag{23}$$

Letting $\alpha = 1$, we have

$$(x^T \cdot x + I)a = (A+I)a = b \tag{24}$$

where

$$a = \begin{pmatrix} a_1 \\ a_2 \\ \ldots \\ a_K \end{pmatrix} \; b = \begin{pmatrix} y \cdot x_1 \\ y \cdot x_2 \\ \ldots \\ y \cdot x_K \end{pmatrix} \; x^T = \begin{pmatrix} x_1 \\ x_2 \\ \ldots \\ x_K \end{pmatrix} \tag{25}$$

and $x_j = (x_{1j}, x_{2j}, \ldots, x_{Nj})$ for $j = 1, 2, \ldots, K$ and $y = (y_1, y_2, \ldots, y_N)$. A is the $K \times K$ square matrix $A = (a_{ij})_{K\times K} = (x_i \cdot x_j)_{K\times K}$ and I is a $K \times K$ unit matrix. So $a = (A + I)^{-1}b$. Hence, the final solution is

$$\begin{cases} \hat{a} = (A + I)^{-1}b \\ \hat{\sigma^2} = \frac{1}{N}(\sum_{i=1}^{N} r_i^2 + \sum_{k=1}^{K} \frac{a_k^2}{\alpha^2}) \end{cases} \tag{26}$$

where $\alpha^2 \approx 1, r_i = y_i - \sum_{k=1}^{K} a_k x_{ik}$.

3 Search Strategies and Implementation

We know that from a given sample data set with K variables, there are $(2^\nu - \rho)$ possible candidate models which may fit the data. This is the same as the formula given in [11]. To find out the best model from the model space we have to apply a search approach. Since the model space is exponential in the number of nodes, so the search is very expensive, especially when the number of the nodes is large. A good technique to search the space of causal models is highly demanded. In this section we will examine methods for finding a causal model with minimum message length. All of our searches could begin with or without some seed model specified by the user, which may be any DAG in the measured variables from one with no links to one that is fully connected. We have tested three search strategies: (1) *Message Length Based Greedy search*, (2) *Best-first search* and (3) *Random search*. The experimental results show that the random search is not stable and the best-first does not do anything better than greedy search. So in our system, the Greedy search method is adopted.

Message Length Based Greedy search runs through each pair of nodes attempting to add an arc (in either direction) if there is none or to delete or to reverse it if there already is one. The change is effected if the MML cost goes down. The loop continues through all node pairs so long as any change was implemented.

Algorithm 31 Message Length Based Greedy Search.
We denote the model without a link from node i to node j as M_{noij}, the model with a link from node i to node j as M_{ij} and the model with a link from node j to node i as M_{ji}. In the following algorithm, $Cal(M)$ is a subroutine to calculate the message length of the model M, and $Choose(M_1, M_2, M_3)$ is the subroutine to choose the model with minimum message length among the given three models M_1, M_2, M_3.

1. *isum* $\leftarrow 0$
2. *backsearch: iconl* $\leftarrow 0$
3. *istart:* $i \leftarrow 1$
4. *jstart:* $j \leftarrow i$
5. *If* $(i \neq j)$ **then**
6. *total-length* $\leftarrow Cal(M_{noij})$
7. *If* $\neg\exists(i \rightarrow j)$ **then**

8. *le* $\leftarrow$ *total–length add*$(i \rightarrow j)$*; form* M_{ij} $l_{ij} \leftarrow Cal(M_{ij})$
9. *delete*$(i \rightarrow j)$*; add*$(j \rightarrow i)$*; form* M_{ji} $l_{ji} \leftarrow Cal(M_{ji})$
10. $Choose(M_{noij}, M_{ij}, M_{ji})$
11. **else**
12. *total-length* $\leftarrow Cal(M_{ij})$ *delete*$(i \rightarrow j)$*; form* M_{noij}
13. *le* $\leftarrow Cal(M_{noij})$ *add*$(j \rightarrow i)$*; form* M_{ji}
14. $l_{ji} \leftarrow Cal(M_{ji})$ $Choose(M_{noij}, M_{ij}, M_{ji})$
15. *If* $(j \leq K)$ *then*
16. $j \leftarrow j + 1$*; goto jstart;*
17. *If* $(i \leq K)$ **then**
18. $i \leftarrow i + 1$*; goto istart;*
19. *If (changes made) goto backsearch;*
20. *isum* $\leftarrow$ *isum* $+ 1$;
21. *If (isum* < 2*)* **then** *goto backsearch;*
22. *exit.*

Thus far the methods we examined have worked adequately for modest sized problems. A comment on our message length based greedy search algorithm is that (1) The algorithm does not search the complete model space. So it is an inexhausted search; (2) the search is guided by message length. we have an assumption that the model with minimum message length is the best model; and (3) this search algorithm itself has no mechanism to avoid the local minimum problem. In the other world, this algorithm itself can not guarantee to achieve the global minimum.

4 Experimental Results and Analysis

In this section, we report three major results: (1) the experimental results of the causal model discovery system tested on four additional causal models and compared our results with TETRAD II in addition to what was reported in [19]; (2) the comparison of the models discovered by previous version and current version of the MML-CI systems; and (3) the time complexity comparison.

Our technique is to take the model as reported, use it to stochastically generate the sample data, and use that data as input to both TETRAD II and our MML induction program. Intuitively, if a causal induction program is working perfectly, it should reproduce exactly the model used to generate the data. In practice, sampling errors will result in deviations from the original model (we have used sample sizes of 1000 joint measurements in this study), but programs which reproduce a DAG structure similar to the original, and secondarily coefficient values similar to the original, must be considered to be performing better than those which do not.

4.1 Further Examination of the Recovered Models

Figures 1 to 3 illustrate all the original causal models, MML induced models and TETRAD II derived models. In each of the figures, the case (a) is the original model, referred to as *original*, case (b) is the MML induced model, referred to as *MML Ind.* the case (c) and whatever follows are the model produced by TETRAD II (denoted as T II) when temporal constraints were provided; and the cases from (d) to (j) are the models produced by TETRAD II without temporal constraints.

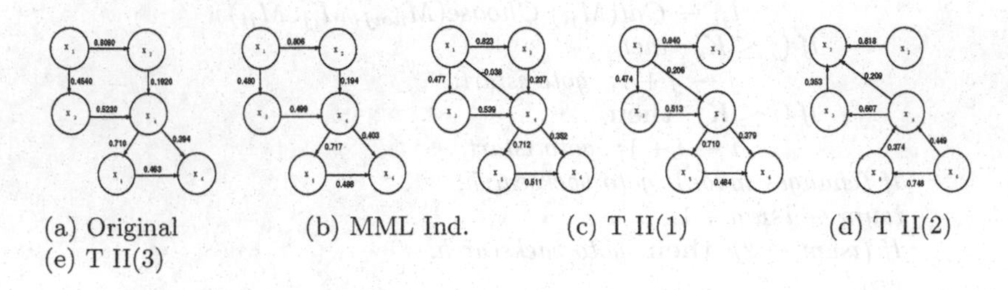

(a) Original (b) MML Ind. (c) T II(1) (d) T II(2)
(e) T II(3)

Fig. 1. Comparison of MML and TETRAD II on Goldberg's Model

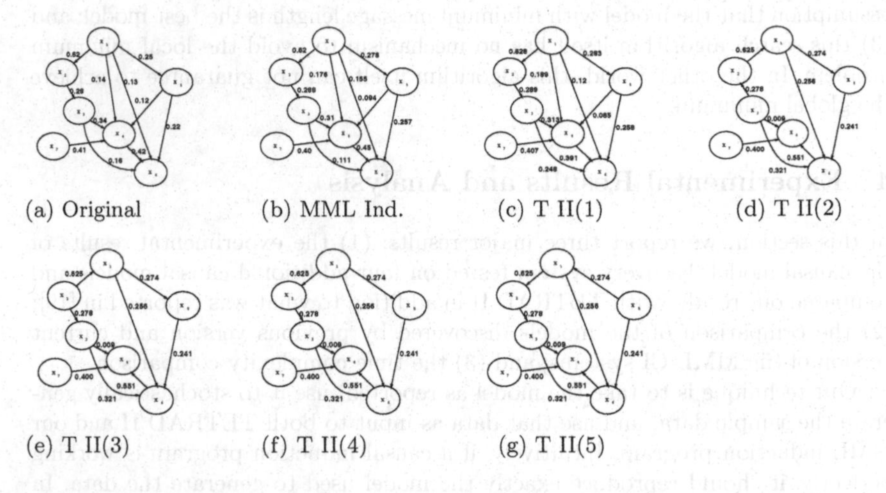

(a) Original (b) MML Ind. (c) T II(1) (d) T II(2)

(e) T II(3) (f) T II(4) (g) T II(5)

Fig. 2. Comparison of MML and TETRAD II on Rodgers and Maranto odel

4.2 Comparison of MML-CI I and MML-CI II

Figure 4 illustrates the models discovered by the MML-CI 1 and MML-CI 2 systems.

Table 1 illustrates the comparison results of the message length calculated using the encoding scheme 1 and that of scheme 2.

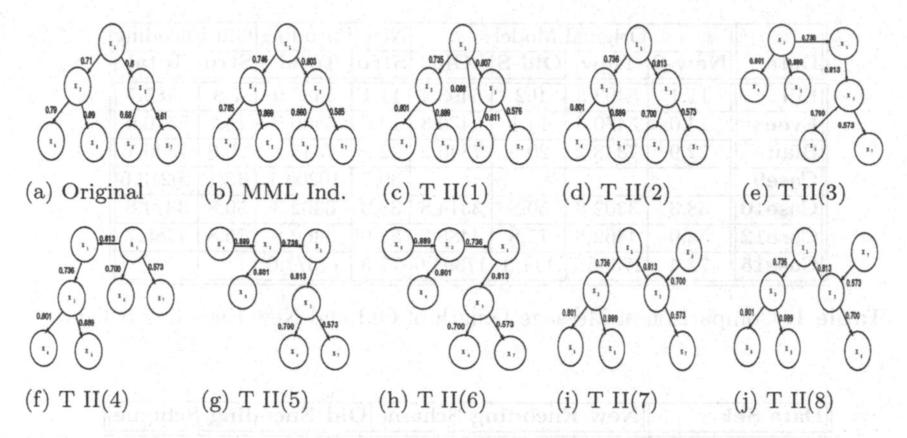

(a) Original (b) MML Ind. (c) T II(1) (d) T II(2) (e) T II(3)

(f) T II(4) (g) T II(5) (h) T II(6) (i) T II(7) (j) T II(8)

Fig. 3. Comparison of MML and TETRAD II on Verbal & Mechanical Ability odel

Table 2 compares the CPU time cost of the two MML-CI approaches in discovering causal models.

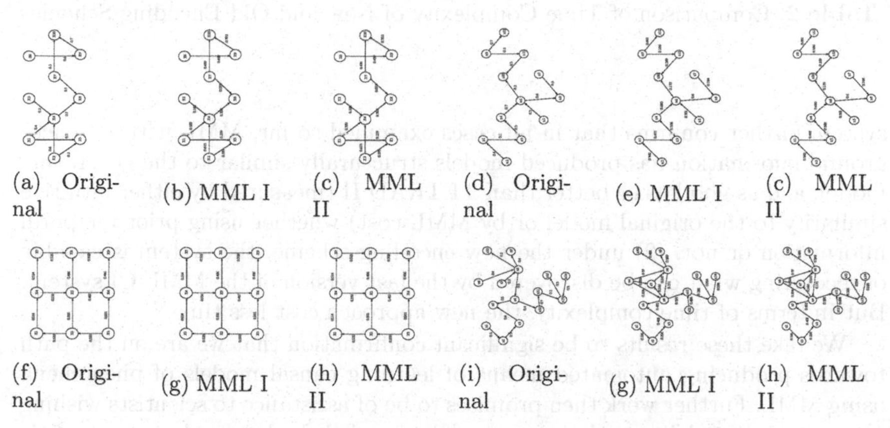

(a) Original (b) MML I (c) MML II (d) Original (e) MML I (c) MML II

(f) Original (g) MML I (h) MML II (i) Original (g) MML I (h) MML II

Fig. 4. Comparison of MML-CI Version I & II

5 Conclusion

This paper presented an improved encoding scheme for the discovery of causal models via MML principle. The experimental results reported in this paper show that (1) additional experimental results compared with MML-CI and TETRAD

Data	Original Model				New Encoding		Old Encoding	
	New-S	**New**	**Old-S**	**Old**	**Stru**	**Total**	**Stru**	**Total**
Fiji	11.1	5493.3	9.2	5493.4	11.1	5489.0	7.8	5485.7
Evens	14.0	5470.3	14.1	5470.3	14.0	5466.7	13.7	5470.0
Blau	22.9	7348.2	20.9	7346.2	22.9	7348.1	20.9	7346.1
Case9					39.7	10208.1	45.1	10213.6
Case10	38.3	3302.3	50.8	3314.8	38.3	3302.3	50.8	3314.8
Case12	53.0	4462.8	75.8	4485.7	53.0	4462.8	75.8	4485.7
Case15	72.4	17355.7	111.3	17394.6	77.3	17371.0		

Table 1. Comparison of Message Length of Old and New Encoding Scheme

Data Set	New Encoding Scheme	Old Encoding Scheme
Fiji Model	0.96 *seconds*	0.84 *seconds*
Evans Model	2.25 *seconds*	2.29 *seconds*
Blau Model	3.42 *seconds*	5.18 *seconds*
Case 9 Model	16.32 *seconds*	19.23 *seconds*
Case 10 Model	20.1 *seconds*	59.97 *seconds*
Case 12 Model	36.2 *seconds*	126.2 *seconds*
Case 15 Model	265.5 *seconds*	

Table 2. Comparison of Time Complexity of New and Old Encoding Scheme

system further confirms that in all cases examined so far, MML with no background information has produced models structurally similar to the generating model, and as good as or better than TETRAD II (measured by either intuitive similarity to the original model or by MML cost) whether using prior temporal information or not. (2) under the new encoding scheme, the system is capable of recovering what can be discovered by the last version of the MML-CI system. But in terms of time complexity, the new approach cost less time.

We take these results to be significant confirmation that we are on the path towards producing automated means of learning causal models of phenomena using MML. Further work then promises to be of assistance to scientists wishing to use causal modeling techniques to understand their data and to assess their theories, which is important particularly in the social sciences; it also promises to shed light on the nature of the enterprise within artificial intelligence to model scientific discovery.

References

1. H. L. Chin and G. F. Cooper. Stochastic simulation of Bayesian belief networks. In *Proc. of 3rd Workshop on Uncertainty in AI*, Seattle, 1987.

2. G. F. Cooper and E. Herskovits. A Bayesian method for constructing Bayesian belief networks from databases. In *Proc. of 7th Conference on Uncertainty in AI*. Morgan Kaufmann, 1991.
3. Honghua Dai, Kevin Korb, Chris Wallace, and Xindong Wu. A study of causal discovery with small samples and weak links. In *Proceedings of the 15th International Joint Conference On Artificial Intelligence* **IJCAI'97**, pages 1304–1309. Morgan Kaufmann Publishers, Inc., 1997.
4. Clark Glymour, Richard Scheines, Peter Spirtes, and Kevin Kelly. *Discovering Causal Structure: Artificial Intelligence, Philosophy of Science, and Statistical Modeling.* Academic Press, San Diego, 1987.
5. David Heckerman, Dan Geiger, and David M. Chickering. Learning bayesian networks: The combination of knowledge and statistical data. *Machine Learning*, 20(3):197–243, 1995.
6. Wai Lam and Fahiem Bacchus. Learning Bayesian belief networks: An approach based on the MDL principle. *Computational Intelligence*, 10:269–292, 1994.
7. John C. Loehlin. *Latent Variable Models: An Introduction to Factor, Path and Structural Analysis.* Lawrence Erlbaum Associates, Hillsdale, New Jersey, second edition, 1992.
8. Richard Neapolitan. *Probabilistic Reasoning in Expert Systems.* Wiley, New York, 1990.
9. J.J. Oliver and R.A. Baxter. *MML and Bayesianism: Similarities and differences.* Tech Report 206, Computer Science, Monash University, 1994.
10. Judea Pearl. *Probabilistic Reasoning in Intelligent Systems.* Morgan Kaufmann, San Mateo, California, 1988.
11. R.W. Robinson. Counting unlabelled acyclic digraphs. In C.H.C. Little, editor, *Lecture Notes in Mathematics: Combinatorial Mathematics V*, pages 28–43. Springer-Verlag, 1977.
12. Peter Spirtes, Clark Glymour, and Richard Scheines. Causality from probability. In J.E. Tiles, G.T. McKee, and G.C. Dean, editors, *Evolving Knowledge in Natural Science and Artificial Intelligence*, London, 1990. Pitman.
13. Peter Spirtes, Clark Glymour, and Richard Scheines. *Causation, Prediction, and Search.* Springer-Verlag, New York, Berlin, Heideberg, 1993.
14. Peter Spirtes, Clark Glymour, and Richard Scheines. *Causation, Prediction, and Search.* MIT Press, New York, 2000.
15. Peter Spirtes, Clark Glymour, Richard Scheines, and C. Meek. *TETRAD II: tools for causal modeling.* Lawrence Erlbaum, Hillsdale, New Jersey, 1994.
16. Chris Wallace and David Boulton. An information measure for classification. *Computer Journal*, 11:185–194, 1968.
17. Chris Wallace and P.R. Freeman. Estimation and inference by compact coding. *Journal of the Royal Statistical Society*, B, 49:240–252, 1987.
18. Chris Wallace and Michael Georgeff. A general selection criterion for inductive inference. *ECAI 84, Advances in Artificial Intelligence*, pages 1–18, 1984.
19. Chris Wallace, Kevin Korb, and Honghua Dai. Causal discovery via MML. In *Proceedings of the 13th International Conference on Machine Learning* **(ICML'96)**, pages 516–524, 1996.

SETM*-MaxK: An Efficient SET-Based Approach to Find the Largest Itemset

Ye-In Chang and Yu-Ming Hsieh

Dept. of Computer Science and Engineering
National Sun Yat-Sen University
Kaohsiung, Taiwan, Republic of China
Tel: 886-7-5252000 (ext. 4334), Fax: 886-7-5254301
changyi@cse.nsysu.edu.tw

Abstract. In this paper, we propose the SETM*-MaxK algorithm to find the largest itemset based on a high-level set-based approach, where a large itemset is a set of items appearing in a sufficient number of transactions. The advantage of the set-based approach, like the SETM algorithm, is simple and stable over the range of parameter values. In the SETM*-MaxK algorithm, we efficiently find the L_k based on L_w, where L_k denotes the set of large k-itemsets with minimum support, $L_k \neq \emptyset, L_{k+1} = \emptyset$ and $w = 2^{\lceil log_2 k \rceil - 1}$, instead of step by step. From our simulation, we show that the proposed SETM*-MaxK algorithm requires shorter time to achieve its goal than the SETM algorithm.

1 Introduction

One of the important data mining tasks, mining *association rules* in transactional or relational databases, has recently attracted a lot of attention in database communities [1,2,5]. The task is to discover the important associations among items such that the presence of some items in a transaction will imply the presence of other items in the same transaction [3]. Previous approaches to mining association rules can be classified into two approaches: low-level and high-level approaches, where a low-level approach means to retrieve one tuple from the relational database at a time, and a high-level approach means a set-based approach. For example, Apriori/AprioriTID [1] and DHP [5] are based on the low-level approach, while the SETM algorithm [4] is based on the high-level approach. A set-based approach (i.e., a high-level approach) allows a clear expression of what needs to be done as opposed to specifying exactly how the operations are carried out in a low-level approach. The declarative nature of this approach allows consideration of a variety of ways to optimize the required operations. Eventually, it should be possible to integrate rule discovery completely with the database system. This would facilitate the use of the large amounts of data that are currently stored on relational databases. The relational query optimizer can then determine the most efficient way to obtain the desired results. Finally, the set-based approach has a small number of well-defined, simple

M.-S. Chen, P.S. Yu, and B. Liu (Eds.): PAKDD 2002, LNAI 2336, pp. 316–321, 2002.

concepts and operations. This allows easy extensibility to handling additional kinds of mining, e.g. relating association rules to customer classes.

In [4], based on a high-level approach, Houtsma and Swami proposed the SETM algorithm that uses SQL for generating the frequent itemsets. Algorithm SETM is simple and stable over the range of parameter values. Moreover, it is easily parallelized. But the disadvantage of the SETM algorithm is that it generates too many invalid candidate itemsets. In this paper, we design the SETM*-MaxK algorithm to find the largest itemset based on a high-level set-oriented approach. One of the applications to find the largest itemset is that in a grocery store, we may want to know the maximum set of data items which will be bought in one transaction by the most of customers. In the SETM*-MaxK algorithm, we efficiently find the L_k based on L_w, where L_k denotes the set of large k-itemsets with minimum support, $L_k \neq \emptyset, L_{k+1} = \emptyset$ and $w = 2^{\lceil log_2 k \rceil - 1}$, instead of step by step. From our simulation, we show that the proposed SETM*-MaxK algorithm needs shorter time to achieve its goal than the SETM algorithm.

2 The SETM*-MaxK Algorithm

Sometimes, we may only want to know the maximum set of data items which will be bought in one transaction by the most of customers. That is, we only want to find out the maximum k such that $L_k \neq \emptyset$ and $L_{k+1} = \emptyset$. In this Section, we present the *SETM*-MaxK* algorithm to achieve such a goal. In the proposed algorithm, we make use of the "jump" approach and the binary search approach to efficiently find the maximum k. We use the "jump" approach to efficiently construct L_k based on L_w, until $L_k = \emptyset$, where $w = k/2$.

Table 1. Variables used in the *SETM*-MaxK* algorithm

$R_k^{'}$	A database of candidate k-itemsets(i.e., a candidate DB)
L_k	Large k-itemsets
C_k	Candidate k-itemsets
R_k	A database of large k-itemsets(i.e., a filtered DB)
half_k	Last k processed
eq_len	Record the length of the same items in the join step

Table 1 shows the variables used in the *SETM*-MaxK* algorithm. The complete algorithm are shown in Figures 2. In procedure *SETM*-MaxK*, the first step (i.e., the forward phase) is to generate $R_k^{'}$, L_k, and R_k, except $R_1^{'}$, until $L_k = \emptyset$ or $C_k = \emptyset$, where $k = 2^i$, $i \geq 0$. To generate counts for those patterns in $R_k^{'}$ that meet the minimum support constraint, we call procedure *gen-Litemset*. Before we go on to generate patterns of length $k+1$, we first have to select the tuples from $R_k^{'}$ that should be extended; that is, those tuples that meet the minimum support constraint. We also wish the resulting relation to be sorted on $R_k(trans\_id, item_1, \ldots, item_k)$. This can be done by calling procedure *filter-DB*. After L_1 and R_1 are constructed, we then apply the "jump" approach. We repeat calling procedures *gen-2k-C, gen-2k-CDB, gen-Litemset* and *filter-DB* to generate C_k, $R_k^{'}$, L_k and R_k, respectively, until $L_k = \emptyset$ or $C_k = \emptyset$, where $k \geq 2$. In procedure *gen-2k-C*, we construct C_k based on L_{half_k}, where $half\_k = k$ div 2. In procedure *gen-2k-CDB*, we could generate all lexicographically ordered patterns of length k stored on $R_k^{'}(trans\_id, item_1, \ldots, item_k)$ based on R_{half_k} only. Note that The SETM algorithm constructs $R_k^{'}$ based on R_{k-1} and the original database $SALES$. Due to this reason, the SETM algorithm generates and counts too many candidates itemsets. To reduce the size of the candidate database $R_k^{'}$, we have a new strategy to construct $R_k^{'}$ in procedure *gen-2k-CDB*. Moreover, to avoid unnecessary construction of $R_k^{'}$, L_k and R_k, we will first construct C_k before we construct L_k. That is, $R_k^{'}, L_k$ and R_k will be generated only if $C_k \neq \emptyset$.

In step 2 of procedure *SETM*-MaxK*, we will keep changing the value of $targetK$ by considering the range between k and $half\_k$ $(= k/2)$, until $MaxK \neq 0$. We apply a variant (marked with **) of the binary search, in which when $L_{targetK} = \emptyset$, in additional to updating HK, we will update $LK = LK + 1$ and compute $R_{LK}^{'}$, L_{LK} and R_{LK} if $C_{LK} \neq \emptyset$. In this way, for the next loop, $R_{targetK}^{'}$ can be generated based on a new R_{LK}. For the example as shown in Figure 1-(a), Figure 1-(b) shows the process of procedure *SETM*-MaxK* (as shown in Figure 2), where the minimum support = 3.

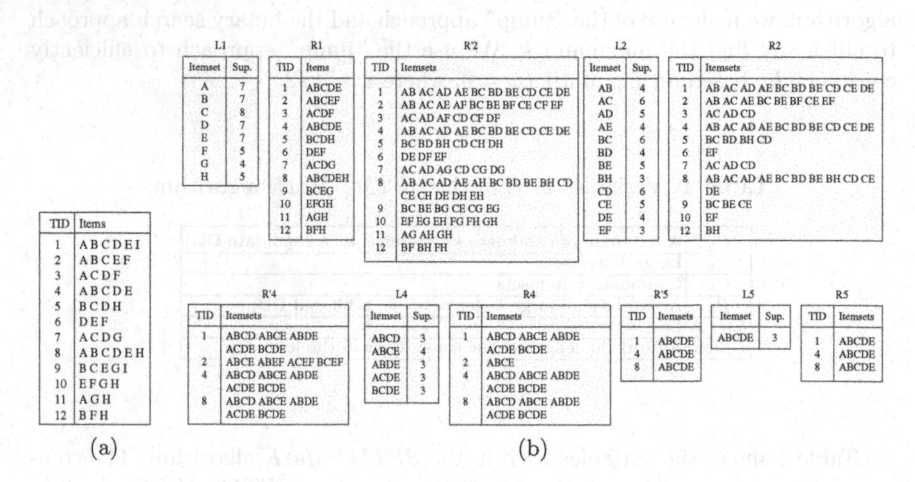

(a)

TID	Items
1	A B C D E I
2	A B C E F
3	A C D F
4	A B C D E
5	B C D H
6	D E F
7	A C D G
8	A B C D E H
9	B C E G I
10	E F G H
11	A G H
12	B F H

(b)

L1

Itemset	Sup.
A	7
B	7
C	8
D	7
E	7
F	5
G	4
H	5

R1

TID	Items
1	ABCDE
2	ABCEF
3	ACDF
4	ABCDE
5	BCDH
6	DEF
7	ACDG
8	ABCDEH
9	BCEG
10	EFGH
11	AGH
12	BFH

R'2

TID	Itemsets
1	AB AC AD AE BC BD BE CD CE DE
2	AB AC AE AF BC BE BF CE CF EF
3	AC AD AF CD CF DF
4	AB AC AD AE BC BD BE CD CE DE
5	BC BD BH CD CH DH
6	DE DF EF
7	AC AD AG CD CG DG
8	AB AC AD AE AH BC BD BE BH CD CE CH DE DH EH
9	BC BE BG CE CG EG
10	EF EG EH FG FH GH
11	AG AH GH
12	BF BH FH

L2

Itemset	Sup.
AB	4
AC	6
AD	5
AE	4
BC	6
BD	4
BE	5
BH	3
CD	6
CE	5
DE	4
EF	3

R2

TID	Itemsets
1	AB AC AD AE BC BD BE CD CE DE
2	AB AC AE BC BE BF CE EF
3	AC AD CD
4	AB AC AD AE BC BD BE CD CE DE
5	BC BD BH CD
6	EF
7	AC AD CD
8	AB AC AD AE BC BD BE BH CD CE DE
9	BC BE CE
10	EF
12	BH

R'4

TID	Itemsets
1	ABCD ABCE ABDE ACDE BCDE
2	ABCE ABEF ACEF BCEF
4	ABCD ABCE ABDE ACDE BCDE
8	ABCD ABCE ABDE ACDE BCDE

L4

Itemset	Sup.
ABCD	3
ABCE	4
ABDE	3
ACDE	3
BCDE	3

R4

TID	Itemsets
1	ABCD ABCE ABDE ACDE BCDE
2	ABCE
4	ABCD ABCE ABDE ACDE BCDE
8	ABCD ABCE ABDE ACDE BCDE

R'5

TID	Itemsets
1	ABCDE
4	ABCDE
8	ABCDE

L5

Itemset	Sup.
ABCDE	3

R5

TID	Itemsets
1	ABCDE
4	ABCDE
8	ABCDE

Fig. 1. (a) An example transaction database ; (b) The process of the SETM*-MaxK algorithm

```
procedure SETM*-MaxK;
begin
(* Step 1: Forward Phase *)
  k := 1;
  L_1 := gen-Litemset(Sales, minsup);
  R_1 := filter-DB(Sales, L_1);
  repeat
    k := 2 * k;
    half_k := k div 2;
    eq_len := 0;
    C_k := gen-2k-C(L_half_k, L_half_k);
    if C_k ≠ ∅ then
    begin
      R'_k := gen-2k-CDB(R_half_k, R_half_k);
      L_k := gen-Litemset(R'_k, minsup);
      R_k := filter-DB(R'_k, L_k);
    end;
  until (L_k = ∅ OR C_k = ∅);

(* Step 2: Find MaxK *)
  HK := k;
  LK := half_k;
  targetK := ⌈(HK + LK)/2⌉;
  MaxK := 0;
  repeat
    eq_len := 2 * LK - targetK;
    C_targetK := gen-2k-C(L_half_k, L_half_k);
    if C_targetK ≠ ∅ then
    begin
      R'_targetK := gen-2k-CDB(R_LK, R_LK);
      L_targetK := gen-Litemset(R'_targetK, minsup);
      R_targetK := filter-DB(R'_targetK, L_targetK);
    end;
    if (L_targetK = ∅) or (C_targetK = ∅) then
    begin
      HK := targetK;
      LK := LK + 1; (**)
      eq_len := LK - 2; (**)
      C_LK := gen-2k-C(L_LK-1, L_LK-1); (**)
      if C_LK ≠ ∅ then (**)
      begin (**)
        R'_LK := gen-2k-CDB(R_LK-1, R_LK-1); (**)
        L_LK := gen-Litemset(R'_LK, minsup); (**)
        R_LK := filter-DB(R'_LK, L_LK); (**)
      end; (**)
      if (L_targetK = ∅) then
        MaxK := LK - 1;
    end
    else
      LK := targetK;
    if (MaxK = 0) then
    begin
      targetK := ⌈(HK + LK)/2⌉;
      if (targetK = HK) then
      begin
        if L_targetK = ∅ then
          MaxK := targetK - 1
        else
          MaxK := targetK;
      end;
    end;
  until MaxK ≠ 0;
end;
```

```
procedure gen-Litemset(R'_k, minsup);
begin
  insert into L_k
  select p.item_1, ..., p.item_k, COUNT(*)
  from R'_k p
  group by p.item_1, ..., p.item_k
  having COUNT(*) ≥ :minsup;
end;

procedure filter-DB(R'_k, L_k);
begin
  insert into R_k
  select p.tid, p.item_1, ..., p.item_k
  from R'_k p, L_k q
  where p.item_1 = q.item_1 AND ...
      AND p.item_k = q.item_k;
end;

procedure gen-2k-C(L_half_k, L_half_k);
begin
  insert into C_k
  select p.item_1, ..., p.item_half_k,
      q.item_eq_len+1, ..., q.item_half_k
  from L_half_k p, L_half_k q
  where p.item_1 = q.item_1 AND ...
      AND p.item_eq_len = q.item_eq_len
      AND p.item_half_k < q.item_eq_len+1;
end;

procedure gen-2k-CDB(R_half_k, R_half_k);
begin
  insert into R'_k
  select p.tid, p.item_1, ..., p.item_half_k,
      q.item_eq_len+1, ..., q.item_half_k
  from R_half_k p, R_half_k q
  where p.tid=q.tid
      AND p.item_1 = q.item_1 AND ...
      AND p.item_eq_len = q.item_eq_len
      AND p.item_half_k < q.item_eq_len+1;
end;
```

Fig. 2. The *SETM*-MaxK* procedure

3 Performance

In this Section, we study the performance of the proposed SETM*-MaxK algorithm, and make a comparison with the SETM [4] algorithm by simulation. Our experiments were performed on a PentiumIII Server with one CPU clock rate of 450 MHz, 128 MB of main memory, running Windows-NT 2000, and coded in Delphi. The data resided in the Delphi relational database and was stored on a local 8G IDE 3.5" drive. Table 2 shows the parameters. The length of an

Table 2. Parameters

$\|D\|$	Number of transactions
$\|T\|$	Average size of transactions
$\|MT\|$	Maximum size of the transactions
$\|I\|$	Average size of maximal potentially large itemsets
$\|MI\|$	Maximum size of the potentially large itemsets
$\|L\|$	Number of maximal potentially large itemsets
N	Number of items

itemset in $\mathcal{F}$(potentially maximal large itemsets) is determined according to a Poisson distribution with mean μ equal to $|I|$. The size of each potentially large itemset is between 1 and $|MI|$. Items in the first itemset are chosen randomly from the set of items. To model the phenomenon that large itemsets often have common items, some fraction of items in subsequent itemsets are chosen from the previous itemset generated. We use an exponentially distributed random variable with mean equal to the *correlation level* to decide this fraction for each itemset. The remaining items are picked at random. In the datasets used in the experiments, the correlation level was set to 0.5. Each itemset in $\mathcal{F}$ has an associated weight that determines the probability that this itemset will be picked. The weight is picked from an exponential distribution with mean equal to 1. The weights are normalized such that the sum of all weights equals 1. For example, suppose the number of large itemsets is 5. According to the exponential distribution with mean equal to 1, the probabilities for those 5 itemsets with ID equal to 1, 2, 3, 4 and 5 are 0.43, 0.26, 0.16, 0.1 and 0.05, respectively, after the normalization process. These probabilities are then accumulated such that each size falls in a range. For each transaction, we generate a random real number which is between 0 and 1 to determine the ID of the potentially large itemset. To model the phenomenon that all the items in a large itemset are not always bought together, we assign each itemset in $\mathcal{F}$ a *corruption level* c. When adding an itemset to a transaction, we keep dropping an item from the itemset as long as a uniformly distributed random number (between 0 and 1) is less than c. The corruption level for an itemset is fixed and is obtained from a normal distribution with mean = 0.5 and variance = 0.1. Each transaction is stored in a file system with the form of <transaction identifier, item>. Let Case 1 denotes $|T| = 5, |MT| = 10, |I| = 4, |MI| = 8$, $|D| = $ 20K, $Size =$ 1.5MB, $N =$ 1,000 and $|L| =$ 2,000. When we choose Case 1 the synthetic dataset and with a minimum support = 0.33%, the detailed information about $|R_k^{'}|, |L_k|$,

$|R_k|$ and the execution time in these two algorithms are shown in Tables 3-(a), and 3-(b). From these tables, similarly, we observe that the execution time of SETM*-MaxK algorithm is shorter than that of the SETM algorithm.

4 Conclusion

Discovery of association rules is an important problem in the area of data mining. In order to benefit from the experience with relational databases, a set-oriented approach to mining data is needed. In this paper, to find a large itemset of a specific size in relational database, we have efficiently found the L_k based on L_w, where $L_k \neq \emptyset, L_{k+1} = \emptyset$ and $w = 2^{\lceil log_2 k \rceil - 1}$, instead of step by step. From our simulation results, we have shown that the proposed SETM*-MaxK algorithm requires shorter time to achieve their goals than the SETM algorithm.

Table 3. A comparison of storage space and execution time (Case 1): (a) SETM; (b) SETM*-MaxK.

	L_1			L_2			L_3			L_4		
	R_1'	L_1	R_1	R_2'	L_2	R_2	R_3'	L_3	R_3	R_4'	L_4	R_4
SETM	116799	209	-	323149	714	130093	198789	384	50955	49139	121	11068

	L_5			L_6			Total Time (seconds)
	R_5'	L_5	R_5	R_6'	L_6	R_6	
SETM	7938	2	147	10	0	0	62.45

(a)

SETM*-MaxK												
L_1			L_2			L_4			L_5			Total Time (seconds)
R_1'	L_1	R_1	R_2'	L_2	R_2	R_4'	L_4	R_4	R_5'	L_5	R_5	
116799	209	113491	306160	714	130093	96673	121	10068	2464	2	147	46.63

(b)

References

1. R. Agrawal and R. Srikant, "Fast Algorithms for Mining Association Rules in Large Databases," *Proc. 20th Int'l Conf. Very Large Data Bases,* pp. 490-501, Sept. 1994.
2. F. Berzal, J. Cubero, N. Marin, and J Serrano, "TBAR: An Efficient Method for Association Rule Mining in Relational Databases," *Data and Knowledge Engineering,* Vol. 37, No. 1, pp. 47-64, April 2001.
3. M.-S. Chen, J. Han, and P.S. Yu, "Data Mining: An Overview from a Database Perspective," *IEEE Trans. on Knowledge and Data Engineering,* Vol. 8, No. 5, pp. 866-882, Dec. 1996.
4. M. Houtsma and A. Swami, "Set-oriented Mining for Association Rules in Relational Databases," *Proc. 11th IEEE Int'l Conf. Data Engineering,* pp. 25-33, 1995.
5. J.-S. Park, M.-S. Chen, and P.S. Yu, "Using a Hash-Based Method with Transaction Trimming for Mining Association Rules," *IEEE Trans. on Knowledge and Data Engineering,* Vol. 9, No. 5, pp. 813-825, Sept. 1997.
6. S. Sarawagi, S. Thomas, and R. Agrawal, "Integrating Association Rule Mining with Relational Database Systems: Alternatives and Implications," *Proc. 1998 ACM SIGMOD Int'l Conf. Management of Data,* pp. 343-354, 1998.

Discovery of Ordinal Association Rules

Sylvie Guillaume[1,2]

[1] École des Mines de Nantes (*since September 2001*)
4, rue Alfred Kastler
BP 20722 – 44307 Nantes Cedex 3 - France
Sylvie.Guillaume@emn.fr
[2] IRIN – Université de Nantes
École polytechnique de l'université de Nantes
2, rue de la Houssinière – BP 92208
44322 Nantes Cedex 3 - France

Abstract. Most rule-interest measures are suitable for binary attributes and using an unsupervised usual algorithm for the discovery of association rules requires a transformation for other kinds of attributes. Given that the complexity of these algorithms increases exponentially with the number of attributes, this transformation can lead us, on the one hand to a combinatorial explosion, and on the other hand to a prohibitive number of weakly significant rules with many redundancies. To fill the gap, we propose in this study a new objective rule-interest measure called *intensity of inclination* which evaluates the implication between two ordinal attributes (*numeric or ordinal categorical attributes*). This measure allows us to extract a new kind of knowledge : ordinal association rules. An evaluation of an application to some banking data ends up the study.

Keywords. association rules, interestingness measures, numeric attributes, implicative analysis.

1 Introduction

Finding interesting measures [1] for association rules is one of the important problems in data mining. However, most objective measures [2] [3] are suitable for binary attributes and require an appropriate transformation of the initial set of attributes into binary attributes for all unsupervised usual algorithms for the discovery of association rules [4] [5]. Given that the complexity of these algorithms increases exponentially with the number of attributes, this transformation can lead us, on the one hand to a combinatorial explosion, and on the other hand to a prohibitive number of weakly significant rules with many redundancies. Moreover, the ordering of ordinal attributes (*for numeric or ordinal categorical attributes*) is lost with this transformation. To fill the gap, we have studied measures for ordinal attributes and we have proposed a new measure, ordinal intensity of implication [6], which is a generalization of intensity of propensity [7] and intensity of implication [8]. This measure has allowed us to find a new kind of knowledge : ordinal rules. However this knowledge only extracts implications between two ordinal attributes and we must extend these rules to more

M.-S. Chen, P.S. Yu, and B. Liu (Eds.): PAKDD 2002, LNAI 2336, pp. 322-327, 2002.

than an ordinal attribute in the left-hand part of the rule and/or in the right-hand part of the rule. This is why we have generalized ordinal intensity of implication for conjunctions of ordinal attributes, and we have proposed a new rule-interest measure, intensity of inclination. This paper focuses on the mining of rules discovered with this new measure.

The remainder of the paper is organized as follows. In *section 2* we present this new rule-interest measure, *intensity of inclination* and in *section 3* we give the meaning of rules extracted with this measure : ordinal association rules. This new measure is applied in *section 4* to some banking data before ending up with a set of conclusions in the final section.

2 Intensity of Inclination

Let X and Y be respectively two conjunctions of p and q ordinal attributes. We suppose that $X = X_1,..,X_p$ and $Y = Y_1,..,Y_q$, where X_1, .., X_p, Y_1, .., Y_q are ordinal attributes taking values x_{1_i}, .., x_{p_i}, y_{1_i}, .., y_{q_i} (*$i \in \{1..N\}$, N representing the number of transactions*[1] *in the database*) respectively in $[x_{1_{\min}}..x_{1_{\max}}]$, .., $[x_{p_{\min}}..x_{p_{\max}}]$, $[y_{1_{\min}}..y_{1_{\max}}]$, .., $[y_{q_{\min}}..y_{q_{\max}}]$. In order to take account of ordinal categorical attributes, an appropriate coding into numeric attributes is required.

Intensity of inclination evaluates whether the number of transactions not strongly verifying the rule $X \rightarrow Y$ (*i.e. the number of transactions verifying simultaneously a high value for each attribute X_1, .., X_p and a low value for each attribute Y_1, .., Y_q*) is significantly small compared to the expected number of transactions under the assumption that X and Y are independent. These transactions not strongly verifying the rule are called negative transactions.

Values x_i and y_i ($i \in \{1..N\}$) taken respectively by attributes X and Y are given by the following expressions : $x_i = x'_{1_i} + .. + x'_{j_i} + .. + x'_{p_i}$ and $y_i = y'_{1_i} + .. + y'_{k_i} + .. + y'_{q_i}$ where $x'_{j_i} = \dfrac{x_{j_i} - \mu_{X_j}}{\sigma_{X_j}}$ and $y'_{k_i} = \dfrac{y_{k_i} - \mu_{Y_k}}{\sigma_{Y_k}}$ with μ_{X_j}, μ_{Y_k} are respectively the arithmetic means of X_j and Y_k and σ_{X_j}, σ_{Y_k} are respectively the standard deviations of X_j and Y_k.

The proposed raw measure of non-inclination is defined by the following number of negative transactions :

$$t_o = \sum_{i=1}^{N} \left[(x'_{1_i} + .. + x'_{p_i}) - (x'_{1_{\min}} + .. + x'_{p_{\min}}) \right] \times \left[(y'_{1_{\max}} + .. + y'_{q_{\max}}) - (y'_{1_i} + .. + y'_{q_i}) \right]$$

We must compare this number of negative transactions t_0 with the expected number under an assumption of independence. Then, we have to determine the

[1] or observations or records.

cumulative distribution $F(t)=Pr(T\leq t)$ of a random variable T where t_o is an observed value.

We have demonstrated in [9] that this random variable T can be approximated asymptotically by a normal distribution $N(\mu,\sigma)$ with the mean $\mu=n\ (\mu_X-x_{min})(y_{max}-\mu_Y)$ and the variance $\sigma^2=n[v_X\, v_Y+v_Y\,(\mu_X-x_{min})^2+v_X(y_{max}-\mu_Y)^2]$.
Means and variances of attributes X and Y are given by following expressions : $\mu_X=\sum_{j=1}^{p}\mu_{X_j}$, $\mu_Y=\sum_{k=1}^{q}\mu_{Y_k}$, $v_X=\sum_{j=1}^{p}v_{X_j}+2\sum_{j=1}^{p-1}\sum_{j'=j+1}^{p}\text{cov}(X_j,X_{j'})$ and $v_Y=\sum_{k=1}^{q}v_{Y_k}+2\sum_{k=1}^{q-1}\sum_{k'=k+1}^{q}\text{cov}(Y_k,Y_{k'})$ where $\text{cov}(X_i,X_{i'})=\mu_{X_iX_{i'}}-\mu_{X_i}\mu_{X_{i'}}$ is the covariance of X_i and $X_{i'}$.

A statistical test is required for evaluating the independence between attributes X and Y.

Let H_0 be the null hypothesis of independence between X and Y and H_1 be the alternative hypothesis. Let α be the significance level (*i.e. probability of rejecting H_0 when it is true*).
The decision rule is the following :

Accept H_0 if $Pr(T\leq t_o)>\alpha$
Reject H_0 if $Pr(T\leq t_o)\leq\alpha$

If the probability $Pr(T\leq t_o)$ of having a number inferior or equal to t_o is high, we can say that t_o is not significantly small because this occurrence can arise frequently enough and then this implication $X\rightarrow Y$ is not relevant. On the other hand, if the quantity $Pr(T\leq t_o)$ is small, it means that the implication $X\rightarrow Y$ is relevant because we are unlikely to obtain so few negative transactions as compared with a random draw.

To evaluate this implication in increasing order, the measure $\varphi(X\rightarrow Y)=1-F(t_o)=Pr(T>t_o)$ has been retained. Then, the implication $X\rightarrow Y$ can be admitted at a level of confidence $(1-\alpha)$ if and only if $Pr(T\leq t_o)\leq\alpha$ or $Pr(T>t_o)\geq 1-\alpha$.

Thus, the intensity of inclination is given by :

$$\varphi(X\rightarrow Y)=\frac{1}{\sigma\sqrt{2\pi}}\int_{t_0}^{+\infty}e^{-\frac{(t-\mu)^2}{2\sigma^2}}dt$$

3 Ordinal Association Rules

In order to keep only valuable knowledge, we define the concept of surprising ordinal association rules.

An *ordinal association rule* is an implication of the form $X\rightarrow Y$ where X and Y are conjunctions of ordinal attributes and their intersection $X\cap Y$ is the empty set as defined below :

$X\rightarrow Y$ with $X=X_1,..,X_p$ and $Y=Y_1,..,Y_q$ where $X_1,..,X_p,Y_1,..,Y_q$ are ordinal attributes and $X\cap Y=\emptyset$

A *relevant ordinal association rule* is an ordinal association rule whose value of intensity of inclination $\varphi(X\to Y)$ is superior to a minimum threshold β set by the user ($\beta = 1\text{-}\alpha$).

$X\to Y$ is relevant $\Leftrightarrow \varphi(X\to Y)\geq \beta$.

A rule $X'\to Y'$ is a *general rule* of the rule $X\to Y$ if the attribute X' is a strict part of X ($X'\subset X$) and/or Y' is a strict part of Y ($Y'\subset Y$). For example, rules $X_1X_2\to Y_1$, $X_1\to Y_2$ and $X_1\to Y_1Y_2$ are general rules of the rule $X_1X_2\to Y_1Y_2$.

A *surprising ordinal association rule* is a relevant ordinal association rule whose all general rules are irrelevant.

Thus, for example the rule $X_1X_2\to Y_1$ is surprising if the rules $X_1\to Y_1$ and $X_2\to Y_1$ are not relevant. In fact, attribute X_1 (*respectively* X_2) is not close to attribute Y_1 but it is the association of these two attributes X_1X_2 which reveals an association. This rule contains valuable information. *Figure 1* shows the appearance of such a situation.

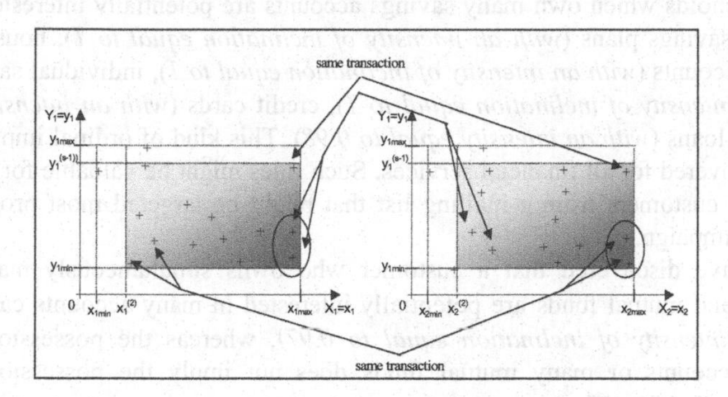

Fig. 1. Associations of transactions which reveal an implication between X_1X_2 and Y whereas associations between X_1 and Y (*respectively between* X_2 *and* Y) are not relevant.

Figure 1 shows a two-dimensional scatter diagram for attributes X_1 and Y_1 (*left part of figure 1*) and for attributes X_2 and Y_1 (*right part of figure 1*). The number of transactions having a high value for X_1 (*respectively for* X_2) and a small value for Y_1, being high, the rule $X_1\to Y_1$ (*respectively* $X_2\to Y_1$) is not relevant. However if transactions having a high value for X_1 (*respectively* X_2) and a small value for Y_1, also verify a small value for X_2 (*respectively* X_1) ; and if transactions having a high value for X_1 (*respectively* X_2) and a high value for Y_1, verify a high value for X_2 (*respectively* X_1) then the rule $X_1X_2\to Y_1$ becomes relevant. Thus, the rule $X_1X_2\to Y_1$ means that transactions verifying a high value for X_1 and X_2 also have a high value for Y_1 whereas among transactions which verify a high value for X_1 (*respectively for* X_2) most of them also have a small value for Y_1.

4 Evaluation on Banking Data

The banking database consists of *47,112* transactions described by *52* attributes, *85%* of which are numeric attributes.
Attributes can be broken down into three categories :

- information about customers (*age, number of years with bank, ...*),
- information about various accounts opened with the bank (*bonds, mortgages, savings accounts, ...*) and
- statistics about various accounts (*rate of indebtedness, total income, ...*).

We have only kept rules with one attribute in the right-hand part of the rule and at most three attributes in the left-hand part of the rule. *1,543* surprising ordinal association rules have been discovered with intensity of inclination at a *95%* level of confidence (β*=0.95*).

This experiment has allowed us to discover financial services likely to interest customers who have a given type of account. Thus for example, we have discovered that households which own many savings accounts are potentially interested in house purchase savings plans (*with an intensity of inclination equal to 1*), house purchase savings accounts (*with an intensity of inclination equal to 1*), individual savings plans (*with an intensity of inclination equal to 1*), credit cards (*with an intensity equal to 0.99*) and loans (*with an intensity equal to 0.98*). This kind of ordinal implication has been discovered for all financial services. Such rules might be valuable for identifying classes of customers from a mailing list that might be targeted most profitably in a mailing campaign.

We have discovered that a customer who owns simultaneously many savings accounts and mutual funds are potentially interested in many accounts called *"DAT" (with an intensity of inclination equal to 0.97)*, whereas the possession of many savings accounts or many mutual funds does not imply the possession of many accounts called *"DAT"*.

We have discovered that households who own simultaneously many house purchase savings accounts, *"Major"* accounts and *"DAT"* accounts are potentially interested in many *"LER-PER-PEP Insurance"* accounts ; the intensity of inclination of this rule is equal to *0.95*.

5 Conclusion and Further Work

In this paper we have proposed an alternative strategy for rule discovery : it consists in discarding the discretization step of numeric attributes and the step of complete disjunctive coding and in extracting rules on the initial set of attributes. Thus, we have defined a new interestingness measure, the *intensity of inclination*, which evaluates a "*statistical surprise*" of having so few negative transactions in a rule i.e. the small number of transactions not strongly verifying rule : the high values of the antecedent imply high values for the consequent.

Thanks to such a measure, we first obtain rules more general than the ones which are extracted on a set of binary attributes because we observe the overall behavior of an attribute, and secondly, we discard the step of transformation of attributes, which allows us to have algorithms better suited for large databases.

This study has to be extended with the exploration of specific ordinal association rules in order to refine our analysis and to extract behaviors in sub-sets. For this, we are going to continue the work already undertaken on the discovery of specific ordinal rules i.e. rules with one attribute in the antecedent and one attribute in the consequent [10].

References

1. BRIN S., MOTWANI R., ULLMAN J.D. et TSUR S.,"Dynamic Itemset Counting and Implications Rules for Market Basket Data", In *Proc. Of the 1997 ACM SIGMOD Conference*, p. 255-264, 1997, Tucson, Arizona, USA.
2. KODRATOFF Y., *Rating the Interest of Rules Induced from Data and within Texts*, proc. 12th IEEE – International Conference on Databases and Expert Systems Applications - DEXA 2001, Munich, Sept. 2001.
3. SILBERSCHATZ A. et TUZHILIN A., "What makes Patterns interesting in Knowledge Discovery Systems", *IEEE Trans. On Know. And Data Eng.* 8(6) , p. 970-974, 1996.
4. AGRAWAL R., MANNILA H., SRIKANT R., TOIVONEN H. et VERKAMO A.I.,"Fast Discovery of Association Rules", In Fayyad U.M., Piatetsky-Shapiro G., Smyth P. and Uthurusamy R. eds., *Advances in Knowledge Discovery and Data Mining*. AAAI/MIT Press. p. 307-328, 1996.
5. SRIKANT R., AGRAWAL R., „ Mining quantitative association rules in large relational tables ", *Proceedings 1996 ACM-SIGMOD International Conference Management of Data*, Montréal, Canada, Juin 1996.
6. GUILLAUME S., KHENCHAF A., BRIAND H., „ *Generalizing association rules to ordinal rules* ", Proceedings of the 2000 Conference on Information Quality IQ'2000, Boston, USA, 2000, p. 268-282.
7. LAGRANGE J.B., Analyse Implicative d'un Ensemble de Variables Numériques, Application au Traitement d'un Questionnaire à Réponses Modales Ordonnées, *rapport interne Institut de Recherche Mathématique de Rennes, Prépublication 97-32 Implication Statistique*, Décembre1997.
8. GRAS R., ALMOULOUD S.A.; BAILLEUL M.; LARHER A.; POLO M.; RATSIMBA-RAJOHN H. and TOTOHASINA A. 1996. L'implication Statistique, Ouvrage de 320 pages dans la Collection Associée à "*Recherches en Didactique des Mathématiques*", La Pensée Sauvage, Grenoble.
9. GUILLAUME S., „ *L'intensité d'inclination : une généralisation de l'intensité d'implication ordinale* ", Actes des 8ème Rencontres de la Société Francophone de Classification (SFC'2001), 17-21 décembre 2001, Pointe-à-Pitre, Guadeloupe.
10. GUILLAUME S., "*Traitement des données volumineuses – Mesures et algorithmes d'extraction de règles d'association et règles ordinales*", Thèse de Doctorat, Nantes, 15 Décembre 2000.

Value Added Association Rules

T.Y. Lin[1], Y.Y. Yao[2], and E. Louie[3]

[1] Department of Mathematics and Computer Science
San Jose State University
tylin@mathcs.sjsu.edu

[2] Department of Computer Science, University of Regina
Regina, Saskatchewan, Canada S4S 0A2
yyao@cs.uregina.ca

[3] IBM Almaden Research Center, 650 Harry Road
San Jose, CA 95120
ewlouie@almaden.ibm.com

Abstract. Value added product is an industrial term referring a minor addition to some major products. In this paper, we borrow the term to denote a minor semantic addition to the well known association rules. We consider the addition of numerical values to the attribute values, such as sale price, profit, degree of fuzziness, level of security and so on. Such additions lead to the notion of random variables (as added value to attributes) in the data model and hence the probabilistic considerations of data mining.

1 Introduction

Association rules are mined from transaction databases with the goal of improving sales and services. Two standard measures called support and confidence are used for mining association rules. However, both measures are not directly linked to the use of association rules in the context of marketing. In order to resolve this problem, many proposals have been made by adding market semantics to data model. Using first order logic, one can add semantics either by function and/or relations (functions symbols or predicates). Barber and Hamilton, and Lu *et al.* added a share measure and weight respectively to each item [1,8]. Lin and Ng *et al* considered semantic and constraint that are prescribed by binary relations (=neighborhood systems) and predicates respectively [7,6,9]. With the introduction of such semantic and constraints, the mined association rules are more suitable for marketing purpose.

In this paper, we consider a framework for value added association rules by attaching numerical values to itemsets, representing profits, importance, or benefits of itemsets. Within the proposed framework, we re-examine some fundamental issues and open upd doors for probabilistic approach to data mining.

M.-S. Chen, P.S. Yu, and B. Liu (Eds.): PAKDD 2002, LNAI 2336, pp. 328–333, 2002.

2 Semantics Added Relational Data Model

Relational database theory assumes that the universe is a classical set, namely, data is discrete and no additional structures are embedded. In practice, additional semantics often exist. For examples, there are, say monetary, values to objects, similarities among events, distance between locations, and so on. To express the additional semantics, we need to extend the expressive power of the relational model. This may be achieved by adopting the first order logic, which uses relations and functions, or predicates and function symbols, to capture such additional semantics information.

2.1 Structure Added by Relations

There are many studies on semantics modeling of relationships between objects in a database. Typically, the relationships are expressed in terms of some specific relations or predicates of logic view of databases. Details of such models can be found in [7,6,5,9,11].

2.2 Value Added by Functions

In this paper, we focus on the data model with value added by functions. For an attribute A^j, we assume that there exists a non-negative real-valued function, $f^j : Dom(A^j) \longrightarrow \Re^+$, called *value added* function, where $Dom(A^j)$ is the domain of the attribute. An attribute can be regarded as a map, $A^j : U \longrightarrow Dom(A^j)$. By composition of f^j and A^j, we have:

$$X^j = A^j \circ f^j : U \longrightarrow \Re^+,$$

which maps an object to a non-negative real number. For simplicity, we write the inverse image by $X^j_u = (X^j)^{-1}(X^j(u))$. It consists of all objects having the same value on A^j as the object u, and is called a granule, namely, the equivalence class containing u. The counting probability $P(X^j_u) = |X^j_u|/|U|$ gives:

Proposition 1. *X^j is a random variable.*

A random variable is *not* a variable varies randomly, it is merely "a function whose numerical values are determined by a chance." In other words, the chance (probability) of the function to take its individual value is known. See [2] (page 88) for connections between the mathematical notion and its intuition.

Definition 1.

1. *The system $(U, A^j, Dom(A^j), j = 1, 2, \ldots, n)$ is called a granular data model. This model allows one to generate automatically all possible attributes (features), including concept hierarchy [4].*
2. *The system $(U, A^j, X^j, j = 1, 2, \ldots, n)$ is called a value added granular data model VA-GDM.*

We will work in value added granular data model $(U, A^j, X^j, j = 1, 2, \ldots, n)$. *An itemset is a sub-tuple in a relation.* In terms of GDM, a sub-tuple corresponds to a finite intersection of elementary granules. By abuse of notations, we will use the same symbol to denote both attribute value and the corresponding elementary granule. So a sub-tuple $b = (b_1, b_2, \ldots, b_q)$ could also mean the finite intersection, $b = b_1 \cap b_2 \cap \ldots \cap b_q$, of elementary granules.

3 Value Added Association Rules

The value function f^j may be associated with intuitive interpretations such as profits. It seems intuitively natural to compute profit additively, namely, $f(A) = \sum_{i \in A} f(i)$ for an itemset in association rule mining. In general, the value may not be additive. For example, in security, the level of security of an itemset is often computed by $f(A) = Max_{i \in A} f(i)$, and integrity by $f(A) = Min_{i \in A} f(i)$. We will use the semantic neutral term and call f a value function.

Definition 2. Large value added itemsets (LVA-itemsets), *by abuse of language, we may refer to it as* **value added association rules** *(not in rule form). Let B be a subset of the attributes A, f a real-valued function that assigns a value to each itemset, and s_q be a given threshold value for q-itemset, $q = 1, 2, \ldots$.*

1. *Sum-version: A granule $b = (b_1 \cap b_2 \cap \ldots \cap b_q)$, namely, a sub-tuple $b = (b_1, b_2, \ldots, b_q)$, is a large value q-VA-itemset if $Sum(b) \geq s_q$, where*
$$Sum(b) = \sum_j x_o^j * p(x_o^j) = \sum_{j=1}^{q} f^j(b_j) * |b|/|U|, \tag{1}$$
where $x_o^j = f^j(b_j)$.
2. *Min-version: A granule $b = (b_1 \cap b_2 \cap \ldots \cap b_q)$ is a large value q-VA-itemset if $Min(b) \geq s_q$, where*
$$Min(b) = Min_j x_o^j * p(x_o^j) = Min_{j=1}^{q} f^j(b_j) * |b|/|U|. \tag{2}$$
3. *Max-version: A granule $b = (b_1 \cap b_2 \cap \ldots \cap b_q)$ is a large value q-VA-itemset if $Max(b) \geq s_q$, where*
$$Max(b) = Max_j x_o^j * p(x_o^j) = Max_{i=1}^{q} (f(b_i) * |b|). \tag{3}$$
4. *Traditional version: The Max and Min-versions become the traditional one iff the profit function is the constant = 1.*
5. *Mean version: It captures the mean trends of the data. Two attributesA^{j_1}, A^{j_2} is mean associated, if $|E(X^{j_1}) - E(X^{j_2})| \leq s_q$, where $E(\cdot)$ is the expected value, $|\cdot|$ is the absolute value.*

An LVA-itemset is an association rule without direction, since we have used only the "support". One can easily derive value added and directed association rules from LVA-itemsets.

3.1 Algorithms for Sum-Version

An immediate thought would be to mimic the classical theory. Unfortunately, "apriori" may not always be applicable. Note that counting plays a major role in classical association rules. However, in the value added case, the function values are the main concerns. Thresholds are compared against the sum, max, min, and average of the function values. Thus, the results are quite different.

Consider the case $q = 2$. Assume $s_1 = s_2$ and f is not the constant 1. Let $b = b_1 \cap b_2$ be a 2-large granule. We have,

$$Sum(b_1) = f(b_1) * |b_1|/|U|, Sum(b_2) = f(b_2) * |b_2|/|U|, \tag{4}$$

$$Sum(b) = Sum(b_1) + Sum(b_2) \geq s_2. \tag{5}$$

In classical case, $|b| \leq |b_i|, i = 1, 2$; and the apriori exploits this relationship. In the current case, such a relationship is not there; apriori criteria are not useful. Algorithm for finding value added association rules is a brutal exhaustive search: each q is computed independently.

3.2 Algorithms for Max and Min-Versions

As above, the key question is: Could we conclude any relationship among $M(b_1)$, $M(b_2)$, and $M(b)$, where M = Max and Min? Nothing for Max, but for Min, we do have:

$$Min(f(b_1), f(b_2)) \leq Min(b_i), i = 1, 2, \tag{6}$$

Hence we have apriori algorithms for Min-version.

3.3 Experiments

This section reports the experimental result of the algorithms for *LVA- itemsets.* There are three basic routines: generating candidates (potential LVA-itemsets), counting the candidates, and finally, selecting LVA-itemsets that exceed the threshold.

Finding the LVA-itemsets is an exhaustive search, the search is conducted from longest to shortest. For each q, we search all q-tuples.

The generated raw data set has 8 attributes and 500 tuples. The threshold for selecting the itemsets is 5.7. Two potential LVA-itemsets are embedded in data. Each granule is represented by (attribute, value) pair.

1. LVA-itemset (LVA-granules) of length 6 is:

$$\{(C175), (C490), (C524), (C661), (C752), (C84)\}.$$

The frequency is 2, and the sum of their weights is:

$$(0.8 + 0.7 + 0.2 + 0.3 + 0.7 + 0.2) * 2 = 5.8.$$

2. LVA-itemset (LVA-granules) of length 8 is:

$$\{(C175), (C246), (C323), (C445), (C556), (C679), (C779), (C817)\}.$$

The frequency is 1, and the sum of their weights is:

$$(0.8 + 0.5 + 0.9 + 0.6 + 0.9 + 0.3 + 0.9 + 0.8) * 1 = 5.7.$$

The result of finding LVA-itemsets based on weights is summarized as follows:

Length	Candidates	Generation time	Count time	LVA-itemsets
1	110	0.01	0.0	88
2	4491	11.20	0.561	416
3	22540	322.173	3.585	342
4	34200	601.505	6.559	42
5	27926	327.671	6.269	14
6	13997	43.562	3.606	1
7	4000	1.112	1.172	0
8	500	0.020	0.170	6
	107764			909

In this table, the first column is the length of itemsets. The second column is the number of candidates in the given data. For length 8, it is table length (500 rows). The 3rd, 4th, and 5th columns are the times needed to generate the candidates, to counting the support, to find (check the criteria) the LVA-itemsets. Generating candidates dominates most of the runtime of the algorithm. Since the dataset is converted to granules, to count the candidates is fast. The runtime is independent of the threshold. The number of candidates to be checked is the same regardless of the threshold value. LVA-itemsets are found at "random" lengths. There is six LVA-itemsets in length 8, none in length 7 and one in length 6. Unlike classical case, longer LVA-itemsets are not influenced by shorter ones. The algorithms may be improved; the performance is not our focus here.

4 Probabilistic Data Mining Theory

The VA-GDM $(U, A^j, X^j, j = 1, 2, \ldots, n)$ provides a framework for probabilistic consideration. The model naturally produces a numerical information table $(U, X^j, j = 1, 2, \ldots, n)$ so that we can immediately apply techniques in numerical databases.

Let $Y^i = X^{j_i}, i = 1, 2, \ldots, m$ and $m < n$ be the reduct [10], that is, smallest functionally independent subset. The collection:

$$V = \{(Y^1(u), Y^2(u), \ldots, Y^m(u)) \mid \forall\, u \in U\},$$

is a finite set of points in Euclidean space. Since U, and hence V, is finite, a functional dependency can take polynomial form. So the rest of X^j are polynomials over Y^i. We will regard them as random variables over V. By combining the work of [4] and [3], we can express all possible numerical attributes (features) by finitely many polynomials over Y^i. In other words, we will be able to search association rules in **all possible** attributes, **not restricted** to the given attributes, using probability theory. We will report the study in the near future.

5 Conclusions

Value added association rules extends standard association rules by taking into consideration semantics of data. Value added granular data model allows us to import probability theory into data mining. In general, there are no apriori criteria for value added cases. However, if we require the thresholds increase with the lengths, that is, $S_q \geq q * (Max(s_1, s_2, \ldots, s_q))$, there are apriori criteria: q-large implies all sub-tuples are $(q - i)$-large, where $i \geq 0$. This paper reports our preliminary findings, and more results will be presented in the near future.

References

1. Barber, B. and Hamilton, H.J. Extracting share frequent itemsets with infrequent subsets, to appear in *Data Mining and Knowledge Discovery.*
2. Halmos, P. *Measure Theory*, Van Nostrand, 1950.
3. Lin, T.Y. Attributes transformations for data mining I: theoretical explorations, to appear in *International Journal of Intelligent Systems.*
4. Lin, T.Y. The lattice structure of database and mining high level rules, presented in *International Conference on Computer Software and Applications*, Chicago, October 8-12, 2001. Printed copy is to appear as "Feature transformations and structure of attributes" in *Data Mining and Knowledge Discovery: Theory, Tools, and Technology IV*, Dasarathy, B. (Ed.), 2002; and its revision in the Workshop "Towards Foundation of Data Mining" in this conference
5. Lin, T.Y. and Louie, E. Association rules in semantically rich relations: granular computing approach, *New Frontiers in Artificial Intelligence, Lecture Notes in Computer Science 2253*, Terano, T., Nishida, T., Namatame, A., Tsumoto, S., Ohsawa, Y. and Washio, T. (Eds.), Springer, Berlin, 380-384, 2001.
6. Lin, T.Y. Data mining and machine oriented modeling: a Granular computing approach, *Journal of Applied Intelligence*, **13**, 113-124, 2000.
7. Lin, T.Y. Neighborhood systems and approximation in database and knowledge base systems, *Proceedings of the Fourth International Symposium on Methodologies of Intelligent Systems*, Poster Session, 75-86, 1989.
8. Lu, S. Hu. H. and Li, F. Mininh weighted association rules, *Intelligent data analysis*, **5**, 211-225, 2001.
9. Ng, R., Lakshmanan, L.V.S., Han, J. and Pang, A. Exploratory mining and pruning optimizations of constrained associations rules, *Proceedings of 1998 ACM-SIGMOD Conference on Management of Data*, 13-24, 1998.
10. Pawlak, Z. *Rough Sets: Theoretical Aspects of Reasoning about Data*, Kluwer Academic, Dordrecht, 1991.
11. Yao, Y.Y. and Sai, Y. On mining ordering rules, *New Frontiers in Artificial Intelligence, Lecture Notes in Computer Science 2253*, Terano, T., Nishida, T., Namatame, A., Tsumoto, S., Ohsawa, Y. and Washio, T. (Eds.), Springer, Berlin, 316-321, 2001.

Top Down FP-Growth for Association Rule Mining

Ke Wang, Liu Tang, Jiawei Han, and Junqiang Liu

School of Computing Science, Simon Fraser University
{wangk, llt, han, jliui}@cs.sfu.ca

Abstract. In this paper, we propose an efficient algorithm, called **TD-FP-Growth** (the shorthand for Top-Down FP-Growth), to mine frequent patterns. **TD-FP-Growth** searches the FP-tree in the top-down order, as opposed to the bottom-up order of previously proposed FP-Growth. The advantage of the top-down search is not generating conditional pattern bases and sub-FP-trees, thus, saving substantial amount of time and space. We extend **TD-FP-Growth** to mine association rules by applying two new pruning strategies: one is to push multiple minimum supports and the other is to push the minimum confidence. Experiments show that these algorithms and strategies are highly effective in reducing the search space.

1 Introduction

Association rule mining has many important applications in real life. An association rule represents an interesting relationship written as $A \Rightarrow B$, read as "if A occurs, then B likely occurs". The probability that both A and B occur is called the *support*, and written as *count*(AB). The probability that B occurs given that A has occurred is called the *confidence*. The association rule mining problem is to find all association rules above the user-specified minimum support and minimum confidence. This is done in two steps: step 1, *find all frequent patterns*; step 2, *generate association rules from frequent patterns*.

This two-step mining approach suffers from several drawbacks. First, only a single uniform minimum support is used, though the distribution of data in reality is not uniform. Second, the two-step process does not consider the confidence constraint at all during the first step. Pushing the confidence constraint into the first step can further reduce search space and hence improve efficiency.

In this paper, we develop a family of algorithms, called **TD-FP-Growth**, for mining frequent patterns and association rules. Instead of exploring the FP-tree in the bottom-up order as in [5], **TD-FP-Growth** explores the FP-tree in the top-down order. The advantage of the top-down search is not constructing conditional pattern bases and sub-trees as in [5]. We then extend **TD-FP-Growth** to mine association rules by applying two new pruning strategies: **TD-FP-Growth(M)** pushes multiple minimum supports and **TD-FP-Growth(C)** pushes the minimum confidence.

2 Related Work

Since its introduction [1], the problem of mining association rules has been the subject of many studies [8][9][10][11]. The most well known method is the Apriori's *anti-monotone* strategy for finding frequent patterns [3]. However, this method suffers

M.-S. Chen, P.S. Yu, and B. Liu (Eds.): PAKDD 2002, LNAI 2336, pp. 334-340, 2002.

from generating too many candidates. To avoid generating many candidates, [4] proposes to represent the database by a *frequent pattern tree* (called the FP-tree). The FP-tree is searched recursively in a bottom-up order to grow longer patterns from shorter ones. This algorithm needs to build conditional pattern bases and sub-FP-trees for each shorter pattern in order to search for longer patterns, thus, becomes very time and space consuming as the recursion goes deep and the number of patterns goes large.

As far as we know, [7][8] are the only works to explicitly deal with non-uniform minimum support. In [7], a minimum item support (MIS) is associated with each item. [8] bins items according to support and specifies the minimum support for combinations of bins. Unlike those works, our specification of minimum support is associated with the consequent of a rule, not with an arbitrary item or pattern.

3 TD-FP-Growth for Frequent Pattern Mining

As in [5], **TD-FP-Growth** first constructs the FP-tree in two scans of the database. In the first scan, we accumulate the count for each item. In the second scan, only the frequent items in each transaction are inserted as a node into the FP-tree. Two transactions share the same upper path if their first few frequent items are same. Each node in the tree is labeled by an item. An *I_node* refers to a node labeled by item *I*. For each item *I*, all *I*_nodes are linked by a *side-link*. Associated with each node *v* is a count, denoted by *count(v)*, representing the number of transactions that pass through the node. At the same time, a header table H(Item, count, side-link) is built. An entry *(I, H(I), ptr)* in the header table records the total count and the head of the side-link for item *I*, denoted by *H(I)* and *ptr* respectively. Importantly, the items in each transaction are lexicographically ordered, and so are the labels on each path in FP-tree and the entries in a header table. We use Example 3.1 to illustrate the idea of **TD-FP-Growth**.

Example 3.1 A transaction database is given as in the following Figure 3.1. Suppose that the minimum support is 2. After two scans of transaction database, the FP-tree and the header table H is built as Figure 3.1.

The *top-down mining* of FP-tree is described below. First, entry *a* at the top of H is frequent. Since *a*_node only appears on the first level of the FP-tree, we just need to output {*a*} as a frequent pattern.

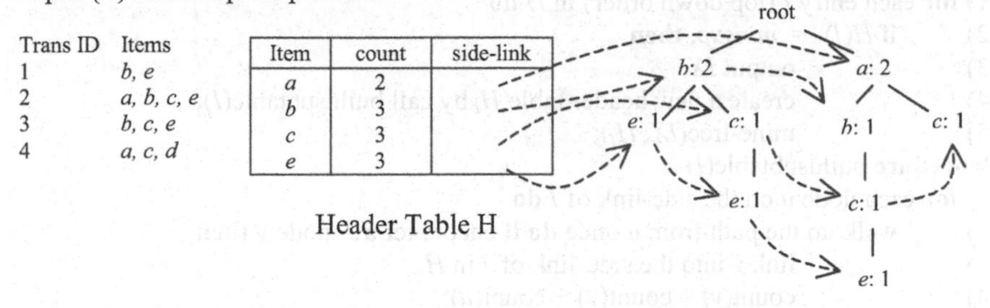

Figure 3.1 Transaction table, FP-tree and H

Then, for entry *b* in H, following the side-link of *b*, we walk up the paths starting from *b*_node in the FP-tree once to build a sub-header-table for *b*, denoted H_*b*.

These paths are in bold face in Figure 3.2. During the walk up, we link up encountered nodes of the same label by a side-link, and accumulate the count for such nodes. In Figure 3.2, there are two paths starting with *b*_node: *root-b* and *root-a-b*. By walking up these paths, we accumulate the count of the *a*_node to 1 because path *root-a-b* only occurs once in the database. We also create an entry *a* in the sub-header table H_*b*. The count of entry *a* is 1 since the count of *a* is actually the count of pattern $\{a, b\}$.

*a*_node is now linked by the side-link for entry *a* in H_*b* and the count is modified to 1. Since the minimum support is 2, pattern $\{a, b\}$ is infrequent. This finishes mining patterns with *b* as the last item. H_*b* now can be deleted from the memory. If pattern $\{a, b\}$ is frequent, we will continue to build sub-header-table H_*ab* and mine FP-tree recursively. In general, when we consider an entry *I* in H_*x*, we will mine all frequent patterns that end up with *Ix*. In this way, **TD-FP-Growth** finds out the complete set of frequent patterns.

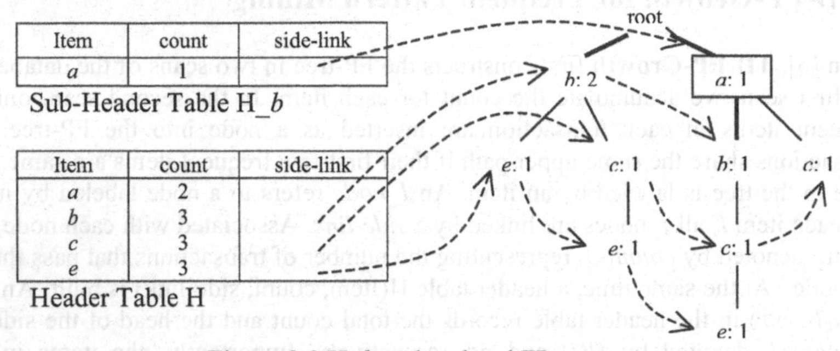

Item	count	side-link
a	1	

Sub-Header Table H_*b*

Item	count	side-link
a	2	
b	3	
c	3	
e	3	

Header Table H

Figure 3.2 H_*b* and updated FP-tree

Similarly, we find frequent patterns: $\{c\}$, $\{b, c\}$ and $\{a, c\}$ for entry *c*, and $\{e\}$, $\{b, e\}$, $\{c, e\}$ and $\{b, c, e\}$ for entry *e*.

Algorithm 1: TD-FP-Growth

Input: a transaction database, with items in each transaction sorted in the lexicographic order, a minimum support: *minsup*. **Output:** frequent patterns above the minimum support. **Method:** build the FP-tree; then call mine-tree $(\varnothing, H)$;

Procedure mine-tree(X, H)

(1) **for** each entry I (top down order) in H **do**
(2) **if** $H(I)$ >= *minsup*, **then**
(3) output IX;
(4) create a new header table H_I by call buildsubtable(I);
(5) mine-tree(IX, H_I);

Procedure buildsubtable(I)

(1) **for** each node u on the side-link of I **do**
(2) walk up the path from u once **do if** encounter a J_node v **then**
(3) link v into the side-link of J in H_I;
(4) count(v) = count(v) + count(u);
(5) $H_I(J) = H_I(J)$ + count(u);

Unlike **FP-Growth**, **TD-FP-Growth** processes nodes at upper levels before processing those at lower levels. This is important to ensure that any modification made at upper levels would not affect lower levels. Indeed, as we process a lower level node, all its ancestor nodes have already been processed. Thus, to obtain the "conditional pattern base" of a pattern (as named in [5]), we simply walk up the paths above the nodes on the current side-link and update the counts on the paths. In this way, we update the count information on these paths "in place" without creating a copy of such paths. As a result, during the whole mining process, no additional conditional pattern bases and sub-trees are built. This turns out to be a big advantage over the bottom-up **FP-Growth** that has to build a conditional pattern base and sub-tree for each pattern found. Our experiments confirm this performance gain.

4 TD-FP-Growth for Association Rule Mining

In this section, we extend **TD-FP-Growth** for frequent pattern mining to association rule mining. We consider two new strategies for association rule mining problem: push multiple minimum supports and push the confidence constraint.

In the discussion below, we assume that items are divided into *class items* and *non-class items*. Each transaction contains exactly one class item and several non-class items. We consider only rules with one class item on the right-hand side. Class items are denoted by $C_1, \ldots, C_m$.

4.1 TD-FP-Growth(M) for Multiple Minimum Supports

In a *ToyotaSale* database, suppose that *Avalon* appears in fewer transactions, say 10%, whereas *Corolla* appears in more transactions, say 50%. If a large uniform minimum support, such as 40%, is used, the information about *Avalon* will be lost. On the contrary, if a small uniform minimum support, such as 8%, is used, far too many uninteresting rules about *Corolla* will be found. A solution is to adopt different minimum supports for rules of different classes.

We adopt **TD-FP-Growth** to mine association rules using non-uniform minimum supports. The input to the algorithm is one $minsup_i$ for each class C_i, and the minimum confidence *minconf*. The output is all association rules $X \Rightarrow C_i$ satisfying $minsup_i$ and *minconf*. The algorithm is essentially **TD-FP-Growth** with the following differences. (1) We assume that the class item C_i in each transaction is the last item in the transaction. (2) Every frequent pattern XC_i must contain exactly one class item C_I, where X contains no class item. The support of X is represented by $H(I)$ and the support of XC_i is represented by $H(i, I)$. (3) The minimum support for XC_i is $minsup_i$. (4) We prune XC_i immediately if its confidence computed by by $H(i, I)/H(I)$ is below the minimum confidence.

4.2 TD-FP-Growth(C) for Confidence Pruning

A nice property of the minimum support constraint is the *anti-monotonicity:* if a pattern is not frequent, its supersets are not frequent either. All existing frequent pattern mining algorithms have used this property to prune infrequent patterns. However, the minimum confidence constraint does not have a similar property. This is the main reason that most association rule mining algorithms ignore the confidence requirement in the step of finding frequent patterns. However, if we choose a proper definition of support, we can still push the confidence requirement inside the search of patterns. Let us consider the following modified notion of support.

Definition 4.1: The *(modified) support* of a rule $A \Rightarrow B$ is *count(A)*.

We rewrite the minimum confidence constraint *C*: *count(AB)/count(A)>=minconf* to *count(AB)>=count(A)*minconf*. From the support requirement *count(A)>=minsup*, we have a new constraint *C'*: *count(AB) >= minsup * minconf*. Notice that *C'* is anti-monotone with respect to the left-hand side *A*. Also, *C'* is not satisfied, neither is *C*. This gives rise to the following pruning strategy.

Theorem 4.1: (1) If $A \Rightarrow B$ does not satisfy *C'*, $A \Rightarrow B$ does not satisfy the minimum confidence either. (2) If $A \Rightarrow B$ does not satisfy *C'*, no rule $AX \Rightarrow B$ satisfies the minimum confidence, where *X* is any set of items.

Now, we can use both the minimum support constraint and the new constraint *C'* to prune the search space: if a rule $A \Rightarrow B$ fails to satisfy both the minimum support and *C'*, we prune all rules $AX \Rightarrow B$ for any set of items *X*. In this way, the search space is tightened up by intersecting the search spaces of the two constraints.

5 Experiment Results

All experiments are performed on a 550MHz AMD PC with 512MB main memory, running on Microsoft Windows NT4.0. All programs are written in Microsoft Visual C++ 6.0. We choose several data sets from UC_Irvine Machine Learning Database Repository: http://www.ics.uci.edu/~mlearn/MLRepository.html.

dataset	# of trans	# of items per trans	class distribution	# of distinct items
Dna-train	2000	61	C_1=23.2%, C_2=24.25%, C_3= 52.55%	240
Connect-4	67557	43	C_1=65.83%, C_2= 24.62%, C_3= 9.55%	126
Forest	581012	13	C_1= 36.36%, C_2=48.76%, C_3= 6.15%, C_4: 0.47%, C_5= 1.63%, C_6= 2.99%, C_7= 3.53%	15916

Table 5.1 data sets table

Table 5.1 shows the properties for the three data sets. *Connect-4* is the densest, meaning that there are a lot of long frequent patterns. *Dna-train* comes the next in density. *Forest* is relatively sparse comparing with the other two data sets, however, it is a large data set with 581012 transactions.

5.1 Frequent Pattern Mining

In this experiment, we evaluate the performance gain of **TD-FP-Growth** on mining frequent patterns. We compare it with **Apriori** and **FP-Growth**. Figure 5.1 shows a set of curves on the scalability with respect to different minimum supports (minsup). These experiments show that **TD-FP-Growth** is the most efficient algorithm for all data sets and minimum supports tested.

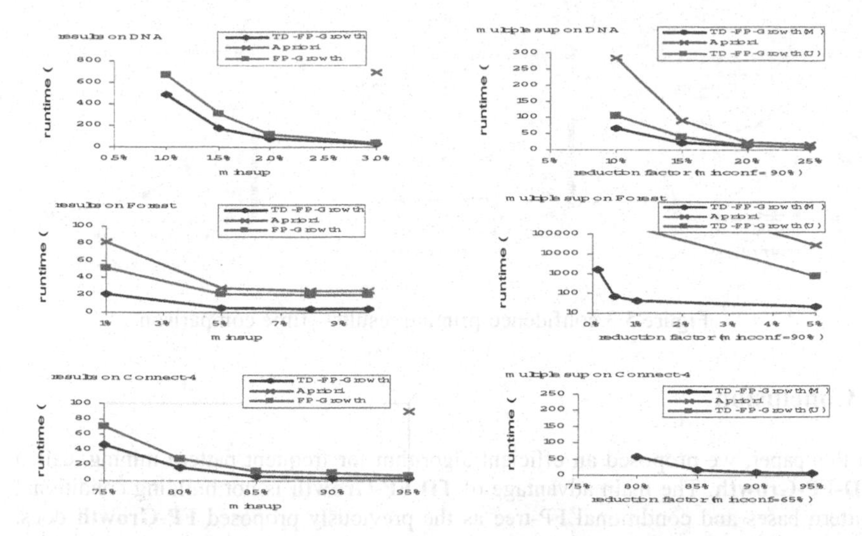

Figure 5.1 frequent pattern mining Figure 5.2 multiple support mining

5.2 TD-FP-Growth for Multiple Minimum Supports

In this experiment, we evaluate the performance gain of using multiple minimum supports. We compare three algorithms: **TD-FP-Growth(M)**, **TD-FP-Growth(U)**, and **Apriori**. **TD-FP-Growth(U)** is **TD-FP-Growth(M)** in the special case that there is only a single minimum support equal to the smallest minimum supports specified. To specify the minimum support for each class, we multiply a *reduction factor* to the percentage of each class. For example, if there are two classes in a data set and the percentage of them is 40% and 60%, with the reduction factor of 0.1, the minimum supports for the two classes are 4% and 6% respectively. And the uniform minimum support is set to 4%.

As shown in Figure 5.2, **TD-FP-Growth(M)** scales best for all data sets. Take C*onnect-4* as the example. As we try to mine rules for the rare class *draw*, which occurs in 9.55% of the transactions, the uniform minimum support at reduction factor of 0.95 is 9.55% * 95%, **TD-FP-Growth(U)** ran for more than 6 hours and generated more 115 million rules. However, with the same reduction factor, **TD-FP-Growth(M)** generated only 2051 rules and took only 5 seconds. That's why in Figure 5.2 for C*onnect-4*, the run time for the two algorithms with uniform minimum support is not included.

5.3 TD-FP-Growth for Confidence Pruning

In this experiment, we evaluate the effectiveness of confidence pruning. The results are reported in Figure 5.3. **TD-FP-Growth(NC)** is **TD-FP-Growth(C)** with confidence pruning turned off. We fix the minimum support and vary the minimum confidence. As shown in Figure 5.3, **TD-FP-Growth(C)** is more efficient than **Apriori** and **TD-FP-Growth(NC)**, for all data sets, minimum supports, and minimum confidences tested.

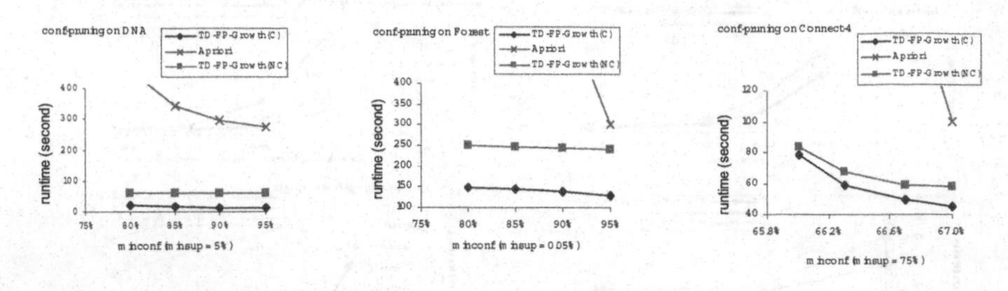

Figure 5.3 confidence-pruning results—time comparison

6 Conclusion

In this paper, we proposed an efficient algorithm for frequent pattern mining, called **TD-FP-Growth**. The main advantage of **TD-FP-Growth** is not building conditional pattern bases and conditional FP-tree as the previously proposed **FP-Growth** does. We extended this algorithm to mine association rules by applying two new pruning strategies. We studied the effectiveness of **TD-FP-Growth** and the new pruning strategies on several representative data sets. The experiments show that the proposed algorithms and strategies are highly effective and outperform the previously proposed **FP-Growth**.

Reference

[1] R. Agrawal, T. Imielinski, and A. Swami. Mining association rules between sets of items in large databases. *ACM-SIGMOD* 1993, 207-216.
[2] H. Toivonen. Sampling large databases for association rules. VLDB 1996, 134-145.
[3] R. Agrawal and S. Srikant. Mining sequential patterns. ICDE 1995, 3-14.
[4] J. Han, J. Pei and Y. Yin. Mining Frequent patterns without candidate generation. SIGMOD 2000, 1-12.
[5] R. Srikant and R. Agrawal. Mining generalized association rules. VLDB 1995, 407-419.
[6] R. Srikant, Q. Vu, and R. Agrawal. Mining association rules with item constraints. VLDB 1996, 134-145.
[7] B. Liu, W. Hsu, and Y. Ma. Mining association rules with multiple minimum supports. KDD 1999, 337-341
[8] K. Wang, Y. He and J. Han. Mining frequent patterns using support constraints. VLDB 2000, 43-52.

Discovery of Frequent Tag Tree Patterns in Semistructured Web Documents

Tetsuhiro Miyahara[1], Yusuke Suzuki[2], Takayoshi Shoudai[2], Tomoyuki Uchida[1], Kenichi Takahashi[1], and Hiroaki Ueda[1]

[1] Faculty of Information Sciences,
Hiroshima City University, Hiroshima 731-3194, Japan
{miyahara@its, uchida@cs, takahasi@its, ueda@its}.hiroshima-cu.ac.jp
[2] Department of Informatics, Kyushu University, Kasuga 816-8580, Japan
{y-suzuki,shoudai}@i.kyushu-u.ac.jp

Abstract. Many Web documents such as HTML files and XML files have no rigid structure and are called semistructured data. In general, such semistructured Web documents are represented by rooted trees with ordered children. We propose a new method for discovering frequent tree structured patterns in semistructured Web documents by using a tag tree pattern as a hypothesis. A tag tree pattern is an edge labeled tree with ordered children which has structured variables. An edge label is a tag or a keyword in such Web documents, and a variable can be substituted by an arbitrary tree. So a tag tree pattern is suited for representing tree structured patterns in such Web documents. First we show that it is hard to compute the optimum frequent tag tree pattern. So we present an algorithm for generating all maximally frequent tag tree patterns and give the correctness of it. Finally, we report some experimental results on our algorithm. Although this algorithm is not efficient, experiments show that we can extract characteristic tree structured patterns in those data.

1 Introduction

Background: Due to the rapid growth of Internet usage, Web documents have been rapidly increasing. Then, avoiding inappropriate Internet contents and searching interesting contents for users become more and more important. We need to extract the common characteristics among interesting contents for users. Then, the aim of this paper is to present a data mining technique of extracting meaningful and hidden knowledge from Web documents.

Data mining problems and main results: Web documents such as HTML files and XML files have no rigid structure. Such documents are called semistructured data. Abiteboul et al. [1] presented Object Exchange Model (OEM, for short) for representing semistructured data. Many semistructured data are represented by rooted trees with ordered children, which are called tree structured data. For example, in Fig. 1, the rooted ordered tree T represents the structure which the XML file *xml_sample* has. Then, in this paper, we use tree structured

M.-S. Chen, P.S. Yu, and B. Liu (Eds.): PAKDD 2002, LNAI 2336, pp. 341–355, 2002.

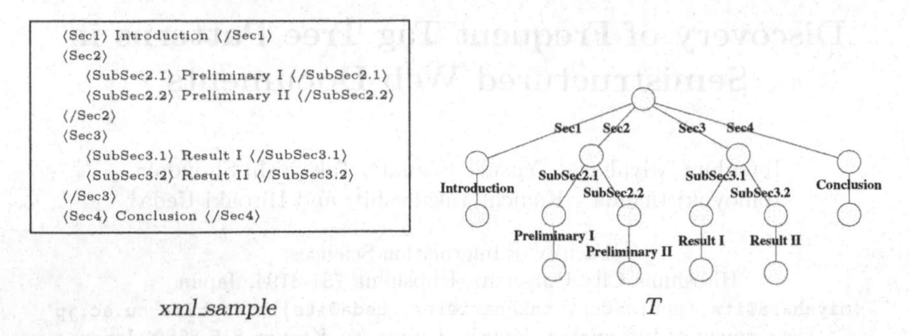

Fig. 1. An XML file *xml_sample* and a rooted ordered tree T as its OEM data.

data as OEM data. To formulate a schema on such tree structured data, we define an *ordered tag tree pattern*, or simply a *tag tree pattern*, as a rooted tree pattern with ordered children consisting of tree structures and structured variables. A tree is a rooted tree with ordered children and no variable. A variable can be substituted by an arbitrary tree.

Since a variable can be replaced by an arbitrary tree, overgeneralized patterns explaining given data are meaningless. Then, in order to extract meaningful knowledge from irregular or incomplete tree structured data such as semistructured Web documents, it is necessary to find a tag tree pattern t such that t can explain more data of given tree structured data than a user-specified threshold but any tag tree pattern obtained from t by substituting a variable of t can not. That is, we need to find one of the least generalized tag tree patterns. For example, consider to find one of the least generalized tag tree patterns explaining at least two OEM data in $\{T_1, T_2, T\}$ where T_1 and T_2 are OEM data in Fig. 2 and T in Fig. 1. The tag tree pattern t in Fig. 2 can explain all OEM data in $\{T_1, T_2, T\}$, that is OEM data T_1, T_2 and T are obtained from t by substituting the variable of t with a tree. But t is an overgeneralized pattern and is meaningless. On the other hand, the tag tree pattern t' in Fig. 2 is one of the least generalized tag tree patterns explaining two OEM data T and T_2 but not T_1. For example, T is obtained from t' by substituting the variables x_1, x_2 and x_3 with the trees g_1, g_2 and g_3 in Fig. 2, respectively.

In this paper, we consider three computational problems, **Frequent Tag Tree Pattern of Maximum Tree-size**, **Frequent Tag Tree Pattern of Minimum Variable-size**, and **All Maximally Frequent Tag Tree Patterns** over tag tree patterns. Frequent Tag Tree Pattern of Maximum Tree-size is the problem to find the maximum tag tree pattern t with respect to the number of vertices such that t can explain more data of input data than a user-specified threshold. This problem is based on the idea that the tag tree pattern, which has more vertices than any other tag tree patterns, gives more meaningful knowledge to us. In a similar motivation, we consider the second problem Frequent Tag Tree Pattern of Minimum Variable-size, which is the problem of finding the minimum tag tree pattern t with respect to the number of variables such that t

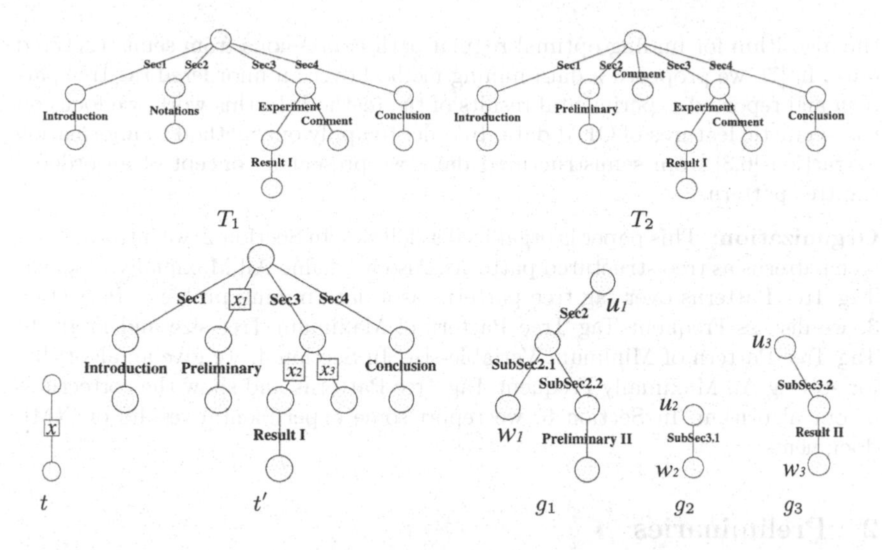

Fig. 2. A tag tree pattern t explains OEM data T_1, T_2 and T in Fig.1. A tag tree pattern t' is one of the least generalized tag tree patterns which explain OEM data T and T_2 but not T_1. A variable is represented by a box with lines to its elements. The label of a box is the variable label of the variable.

can explain more data of input data than a user-specified threshold. Firstly, we show that Frequent Tag Tree Pattern of Maximum Tree-size and Frequent Tag Tree Pattern of Minimum Variable-size are NP-complete. This indicates that it is hard to find the optimum tag tree pattern representing given data. Next, All Maximally Frequent Tag Tree Patterns is the problem to generate all maximally frequent tag tree patterns. This problem is based on the idea that meaningless tag tree patterns are excluded and all possible useful tag tree pattern are not missed. We present a data mining method from semistructured Web documents by giving an algorithm for solving All Maximally Frequent Tag Tree Patterns and show the correctness of our method.

Related works: As knowledge representations for tree structured data, a tree-expression pattern [11] and a regular path expression [4] were proposed. In our previous works [7,8], we presented the concept of a tag tree pattern with *unordered* children from the view point of the semantics of OEM data. A tag tree pattern is different from such representations in that a tag tree pattern has structured variables which can be substituted by arbitrary trees. Several Data mining methods for discovering characteristic schema from semistructured data were proposed. In [11], Wang and Liu presented the algorithm for finding maximally frequent tree-expression patterns from semistructured data. In [2], Asai et al. presented an efficient algorithm for discovering frequent substructures from a large collection of semistructured data. In [4], Fernandez and Suciu presented the algorithm for finding optimal regular path expressions from semistructured data. In [7], we proposed a data mining method over an unordered tag tree pattern and reported experimental results of the method. In this work, we focus on the syntactic features of OEM data. In order to apply our method to information extraction [6,3] from semistructured data, we present a concept of an ordered tag tree pattern.

Organization: This paper is organized as follows. In Section 2, we introduce tag tree patterns as tree structured patterns. Also we define All Maximally Frequent Tag Tree Patterns over tag tree patterns as a data mining problem. In Section 3, we discuss Frequent Tag Tree Pattern of Maximum Tree-size and Frequent Tag Tree Pattern of Minimum Variable-size. In Section 4, we give an algorithm for solving All Maximally Frequent Tag Tree Patterns and show the correctness of our algorithm. In Section 5, we report some experimental results on XML documents.

2 Preliminaries

2.1 Term Trees as Tree Structured Patterns

Let $T = (V_T, E_T)$ be a rooted tree with ordered children (or simply a *tree*) which has a set V_T of vertices and a set E_T of edges. Let E_g and H_g be a partition of E_T, i.e., $E_g \cup H_g = E_T$ and $E_g \cap H_g = \emptyset$. And let $V_g = V_T$. A triplet $g = (V_g, E_g, H_g)$ is called a *term tree*, and elements in V_g, E_g and H_g are called a *vertex*, an *edge* and a *variable*, respectively. We assume that every edges and variables of a term tree are labeled with some words from specified languages. A label of a variable is called a *variable label*. Λ and X denote a set of edge labels and a set of variable labels, respectively, where $\Lambda \cap X = \phi$. For a set S, the number of elements in S is denoted by $|S|$. For a term tree g and its vertices v_1 and v_i, a *path* from v_1 to v_i is a sequence $v_1, v_2, \ldots, v_i$ of distinct vertices of g such that for any j with $1 \leq j < i$, there exists an edge or a variable which consists of v_j and v_{j+1}. If there is an edge or a variable which consists of v and v' such that v lies on the path from the root to v', then v is said to be the *parent* of v' and v' is a *child* of v. We use a notation $[v, v']$ to represent a variable $\{v, v'\} \in H_g$ such that v is the parent of v'. Then we call v the *parent port* of $[v, v']$ and v' the *child port* of $[v, v']$. A term tree g is called *ordered* if every internal vertex u in g has a total ordering on all children of u. The ordering on the children of u is denoted by $<_u^g$. An ordered term tree g is called *regular* if all variables in H_g have mutually distinct variable labels in X.

Definition 1. In this paper, we treat only regular ordered term trees, and then we call a regular ordered term tree a **term tree** simply. In particular, an ordered term tree with no variable is called a **ground term tree** and considered to be a tree with ordered children. Let Λ be a set of edge labels. $\mathcal{OT}_\Lambda$ denotes the set of all ground term trees whose edge labels are in Λ. $\mathcal{OTT}_\Lambda$ denotes the set of all term trees whose edge labels are in Λ.

Let $f = (V_f, E_f, H_f)$ and $g = (V_g, E_g, H_g)$ be term trees. We say that f and g are *isomorphic*, denoted by $f \equiv g$, if there is a bijection φ from V_f to V_g such that (i) the root of f is mapped to the root of g by φ, (ii) $\{u, v\} \in E_f$ if and only if $\{\varphi(u), \varphi(v)\} \in E_g$ and the two edges have the same edge label, (iii) $[u, v] \in H_f$ if and only if $[\varphi(u), \varphi(v)] \in H_g$, and (iv) for any internal vertex u in f which has more than one child, and for any two children u' and u'' of u, $u' <_u^f u''$ if and only if $\varphi(u') <_{\varphi(u)}^g \varphi(u'')$.

Let f and g be term trees with at least two vertices. Let $\sigma = [u, u']$ be a list of two distinct vertices in g where u is the root of g and u' is a leaf of g. The form $x := [g, \sigma]$ is called a *binding* for x. A new term tree $f\{x := [g, \sigma]\}$ is obtained by applying the binding $x := [g, \sigma]$ to f in the following way. Let $e = [v, v']$ be a variable in f with the variable label x. Let g' be one copy of g and w, w' the vertices of g' corresponding to u, u' of g, respectively. For the variable $e = [v, v']$, we attach g' to f by removing the variable e from H_f and by identifying the vertices v, v' with the vertices w, w' of g', respectively. A *substitution* θ is a finite collection of bindings $\{x_1 := [g_1, \sigma_1], \cdots, x_n := [g_n, \sigma_n]\}$, where x_i's are mutually distinct variable labels in X.

The term tree $f\theta$, called the *instance* of f by θ, is obtained by applying the all bindings $x_i := [g_i, \sigma_i]$ on f simultaneously. Further we define a new total ordering $<_v^{f\theta}$ on every vertex v of $f\theta$ in a natural way. Suppose that v has more than one child and let u' and u'' be two children of v of $f\theta$. If v is the parent port of variables $[v, v_1], \ldots, [v, v_k]$ of f with $v_1 <_v^f \cdots <_v^f v_k$, we have the following four cases. Let g_i be a term tree which is substituted for $[v, v_i]$ for $i = 1, \ldots, k$. *Case 1*: If $u', u'' \in V_f$ and $u' <_v^f u''$, then $u' <_v^{f\theta} u''$. *Case 2*: If $u', u'' \in V_{g_i}$ and $u' <_v^{g_i} u''$ for some i, then $u' <_v^{f\theta} u''$. *Case 3*: If $u' \in V_{g_i}$, $u'' \in V_f$, and $v_i <_v^f u''$ (resp. $u'' <_v^f v_i$), then $u' <_v^{f\theta} u''$ (resp. $u'' <_v^{f\theta} u'$). *Case 4*: If $u' \in V_{g_i}$, $u'' \in V_{g_j}$ $(i \neq j)$, and $v_i <_v^f v_j$, then $u' <_v^{f\theta} u''$. If v is not a parent port of any variable, then $u', u'' \in V_f$, therefore we have $u' <_v^{f\theta} u''$ if $u' <_v^f u''$. Lastly we define the root of the resulting term tree $f\theta$ as the root of f.

Example 1. Let t and t' be two term trees described in Fig. 2. Let $\theta = \{x_1 := [g_1, \{u_1, w_1\}], x_2 := [g_2, \{u_2, w_2\}], x_3 := [g_3, \{u_3, w_3\}]\}$ be a substitution, where g_1, g_2 and g_3 are trees in Fig. 2. Then the instance $t'\theta$ of the term tree t' by θ is the tree T in Fig. 1.

Definition 2. Let Λ be a set of edge labels. The *term tree language* $L_\Lambda(t)$ of a term tree t is $\{s \in \mathcal{OT}_\Lambda \mid s \equiv t\theta \text{ for a substitution } \theta\}$. The class $\mathcal{OTTL}_\Lambda$ of all term tree languages is $\{L_\Lambda(t) \mid t \in \mathcal{OTT}_\Lambda\}$.

2.2 Tag Tree Patterns and Data Mining Problems

Definition 3. Let Λ_{Tag} and Λ_{KW} be two languages which consist of infinitely or finitely many words where $\Lambda_{Tag} \cap \Lambda_{KW} = \emptyset$. We call words in Λ_{Tag} and Λ_{KW} a **tag** and a **keyword**, respectively. A **tag tree pattern** is a term tree such that each edge label on it is any of a tag, a keyword, and a special symbol "?". A tag tree pattern with no variable is called a **ground tag tree pattern**.

For an edge $\{v, v'\}$ of a tag tree pattern and an edge $\{u, u'\}$ of a tree, we say that $\{v, v'\}$ *matches* $\{u, u'\}$ if the following conditions (1)-(3) hold: (1) If the edge label of $\{v, v'\}$ is a tag, then the edge label of $\{u, u'\}$ is the same tag or a tag which is considered to be identical under an equality relation on tags. (2) If the edge label of $\{v, v'\}$ is a keyword, then the edge label of $\{u, u'\}$ is a keyword and the label of $\{v, v'\}$ appears as a substring in the edge label of $\{u, u'\}$. (3) If the edge label of $\{v, v'\}$ is "?", then we don't care the edge label of $\{u, u'\}$.

A ground tag tree pattern $\pi = (V_\pi, E_\pi, \emptyset)$ *matches* a tree $T = (V_T, E_T)$ if there exists a bijection φ from V_π to V_T such that (i) the root of π is mapped to the root of T by φ, (ii) $\{v, v'\} \in E_\pi$ if and only if $\{\varphi(v), \varphi(v')\} \in E_T$, (iii) for all $\{v, v'\} \in E_\pi$, $\{v, v'\}$ matches $\{\varphi(v), \varphi(v')\}$, and (iv) for any two vertices $v', v'' \in V_\pi$, v' is a younger sibling of v'' if and only if $\varphi(v')$ is a younger sibling of $\varphi(v'')$. A tag tree pattern π **matches** a tree T if there exists a substitution θ such that $\pi\theta$ is a ground tag tree pattern and $\pi\theta$ matches T. Then *language* $L_\Lambda(\pi)$, which is the descriptive power of a tag tree pattern π, is defined as $L_\Lambda(\pi) = \{\text{a tree } T \text{ in } \mathcal{OT}_\Lambda \mid \pi \text{ matches } T\}$ where $\Lambda = \Lambda_{Tag} \cup \Lambda_{KW}$.

Data Mining Setting. A *set of semistructured data* $\mathcal{D} = \{T_1, T_2, \ldots, T_m\}$ is a set of trees. The *matching count* of a given tag tree pattern π w.r.t. $\mathcal{D}$, denoted by $match_{\mathcal{D}}(\pi)$, is the number of trees $T_i \in \mathcal{D}$ $(1 \le i \le m)$ such that π matches T_i. Then the *frequency* of π w.r.t. $\mathcal{D}$ is defined by $supp_{\mathcal{D}}(\pi) = match_{\mathcal{D}}(\pi)/m$. Let σ be a real number where $0 \le \sigma \le 1$. A tag tree pattern π is **σ-frequent** w.r.t. $\mathcal{D}$ if $supp_{\mathcal{D}}(\pi) \ge \sigma$. We denote by $\Pi(\Lambda')$ the set of all tag tree patterns π such that all edge labels of π are in $\Lambda' \subseteq \Lambda = \Lambda_{Tag} \cup \Lambda_{KW}$. Let Tag be a finite subset of Λ_{Tag} and KW a finite subset of Λ_{KW}. A tag tree pattern $\pi \in \Pi(Tag \cup KW \cup \{?\})$ is **maximally σ-frequent** w.r.t. $\mathcal{D}$ if (1) π is σ-frequent, and (2) if $L_\Lambda(\pi') \subsetneq L_\Lambda(\pi)$ then π' is not σ-frequent for any tag tree pattern $\pi' \in \Pi(Tag \cup KW \cup \{?\})$.

All Maximally Frequent Tag Tree Patterns
Input: A set of semistructured data $\mathcal{D}$, a threshold $0 \le \sigma \le 1$, and finite sets of edge labels Tag and KW.
Problem: Generate all maximally σ-frequent tag tree patterns w.r.t. $\mathcal{D}$ in $\Pi(Tag \cup KW \cup \{?\})$.

Example 2. As examples, we give three OEM data T_1 and T_2 in Fig. 2 and T in Fig. 1 and a maximally $\frac{2}{3}$-frequent tag tree pattern t' in $\Pi(\{\langle\text{Sec1}\rangle, \langle\text{Sec2}\rangle, \langle\text{Sec3}\rangle, \langle\text{Sec4}\rangle\}, \{\text{Introduction, Preliminary,Result I, Conclusion}\}, \text{"?"})$. The tag tree pattern t' in Fig. 2 matches T and T_2, but t' does not match T_1.

3 Hardness Results of Finding the Optimum Frequent Tag Tree Pattern

In this section, we discuss two problems of computing an expressive σ-frequent tag tree pattern. First we show that it is hard to compute the frequent tag tree pattern of maximum tree-size w.r.t. a set of semistructured data. The formal definition of the problem is as follows.

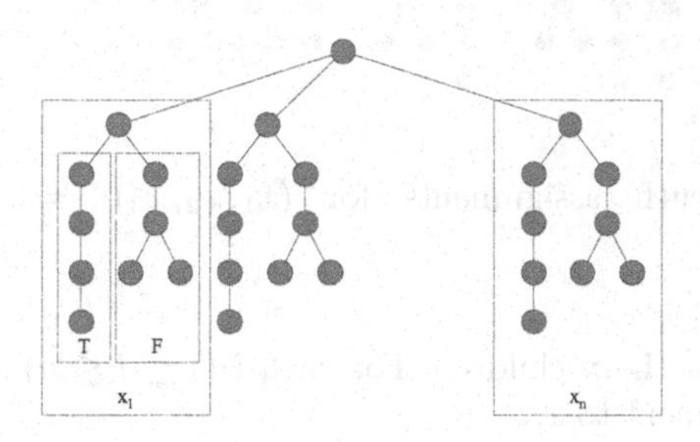

Fig. 3. Tree P_0

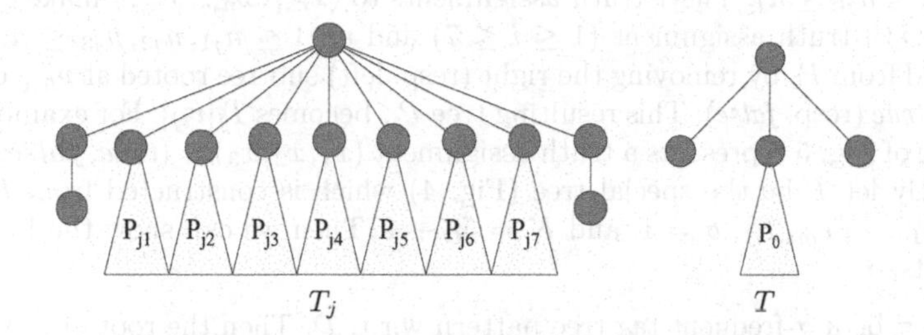

Fig. 4. Tree T_j and T

Frequent Tag Tree Pattern of Maximum Tree-size
Instance: A set of semistructured data $\mathcal{D} = \{T_1, T_2, \ldots, T_m\}$, a real number σ $(0 \leq \sigma \leq 1)$ and a positive integer K.
Question: Is there a σ-frequent tag tree pattern $\pi = (V, E, H)$ w.r.t. $\mathcal{D}$ with $|V| \geq K$?

Theorem 1. Frequent Tag Tree Pattern of Maximum Tree-size *is NP-complete.*

Proof. Membership in NP is obvious. We transform 3-SAT to this problem. Let $U = \{x_1, \ldots, x_n\}$ be a set of variables and $C = \{c_1, \ldots, c_m\}$ a collection of clauses over U with $|c_j| = 3$ for any j $(1 \leq j \leq m)$. For a tree T and a vertex u of T, we denote the subtree consisting of u and the descendants of u by $T[u]$. Let P_0 be the tree which is described in Fig. 3. The root of P_0 has n children.

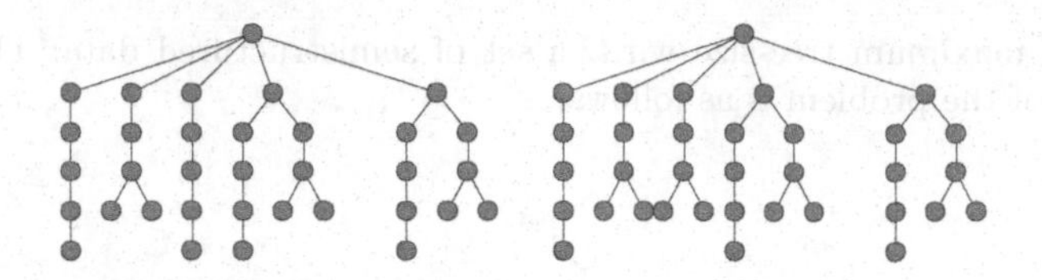

Fig. 5. Two truth assignments for $(x_1, x_2, x_3) = (true, false, true)$, $(true, false, false)$

Let $v_1, v_2, \ldots, v_n$ be the n children. For each i $(1 \leq i \leq n)$, $P_0[v_i]$ corresponds to the truth assignment to x_i.

We construct trees $T_1, \ldots, T_m$ from the tree P_0 and $c_1, \ldots, c_m$ in the following way. T_j $(1 \leq j \leq n)$ is described in Fig. 4. The root of T_j has 9 children. Let $v_{j0}, v_{j1}, \ldots, v_{j8}$ be the 9 children. The inner 7 subtrees $T_j[v_{j1}], \ldots, T_j[v_{j7}]$ correspond to the truth assignments that satisfy c_j. Each $T_j[v_{ji}]$ $(1 \leq i \leq 7)$ is constructed as follows. Let $c_j = \{\ell_{j1}, \ell_{j2}, \ell_{j3}\}$ where $\ell_{jk} = x_{n_{jk}}$ or $\bar{x_{n_{jk}}}$ $(1 \leq k \leq 3, 1 \leq n_{jk} \leq n)$. The 7 truth assignments to $(x_{n_{j1}}, x_{n_{j2}}, x_{n_{j3}})$ make c_j *true*. For the ith truth assignment $(1 \leq i \leq 7)$ and all $1 \leq n_{j1}, n_{j2}, n_{j3} \leq n$, P_{ji} is obtained from P_0 by removing the right (resp. left) subtree rooted at $v_{n_{jk}}$ of P_0 if $x_{n_{jk}}$ is *true* (resp. *false*). This resulting tree P_{ji} becomes $T_j[v_{ij}]$. For example, the left tree of Fig. 5 represents a truth assignment $(x_1, x_2, x_3) = (true, false, true)$.

Lastly let T be the special tree (Fig. 4) which is constructed from P_0. Let $S = \{T_1, \ldots, T_m, T\}$, $\sigma = 1$, and $K = 5n + 4$. Then we can show the following two facts.

1. Let π be a σ-frequent tag tree pattern w.r.t. $\mathcal{D}$. Then the root of π has just three children and the second child of the three children has just n children.
2. Let $G_1, G_2, G_3, g_1, g_2, g_3$ be trees and tag tree patterns described in Fig. 6, respectively. Then g_1 is σ-frequent w.r.t. $\{G_1, G_2, G_3\}$, g_2 is σ-frequent w.r.t. $\{G_1, G_3\}$, and g_3 is σ-frequent w.r.t. $\{G_2, G_3\}$.

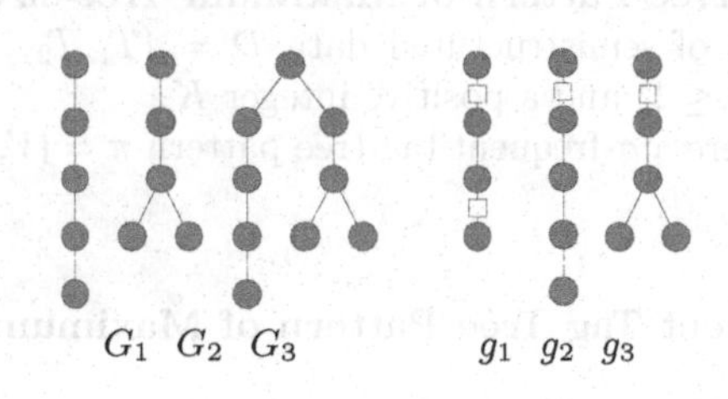

Fig. 6. Trees G_1, G_2, G_3 and tag tree patterns g_1, g_2, g_3

From these two facts, if 3-SAT has a truth assignment which satisfies all clauses in C, there is a σ-frequent tag tree pattern $\pi = (V, E, H)$ w.r.t. $\mathcal{D}$ with

$|V| = 5n + 4 = K$ (Fig. 7). Conversely, if there is a σ-frequent tag tree pattern $\pi = (V, E, H)$ w.r.t. $\mathcal{D}$ with $|V| = 5n + 4$, the numbers of the children of the vertices of depth 5 show one of the truth assignment which satisfies C. □

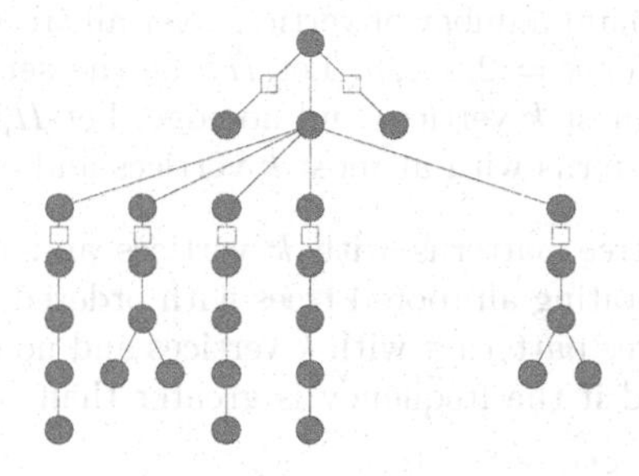

Fig. 7. A tag tree pattern π such that π is $\sigma(= 1)$-frequent w.r.t. $\mathcal{D}$

Second we show that it is hard to compute the frequent tag tree pattern of minimum variable-size w.r.t. a set of semistructured data. The formal definition of the problem is as follows.

Frequent Tag Tree Pattern of Minimum Variable-size
Instance: A set of semistructured data $\mathcal{D} = \{T_1, T_2, \cdots, T_m\}$, a real number σ $(0 \leq \sigma \leq 1)$ and a positive integer K.
Question: Is there a σ-frequent tag tree pattern $\pi = (V, E, H)$ w.r.t. $\mathcal{D}$ with $|H| \leq K$?

Theorem 2. Frequent Tag Tree Pattern of Minimum Variable-size *is NP-complete.*

Proof. (Sketch) Membership in NP is obvious. The reduction is the same as the one in Theorem 1 but $K = n + 2$. □

4 Generating All Maximally Frequent Tag Tree Patterns

4.1 Algorithm for Generating All Maximally Frequent Tag Tree Patterns

In this section, we present an algorithm for solving All Maximally Frequent Tag Tree Patterns. That is, we give an algorithm which generates all maximally σ-frequent tag tree patterns. The algorithm uses a polynomial time matching algorithm for term trees [10] to compute the frequency of a tag tree pattern and a method for generating all rooted trees with ordered children [9].

Algorithm for solving All Maximally Frequent Tag Tree Patterns
Input: A set of semistructured data $\mathcal{D}$, a threshold $0 \le \sigma \le 1$, and finite sets of edge labels Tag and KW.
Output: All maximally σ-frequent tag tree patterns w.r.t. $\mathcal{D}$ in $\Pi(Tag \cup KW \cup \{?\})$.

Let n be the maximum number of vertices over all trees in $\mathcal{D}$. We repeat the following three steps for $k = 2, \ldots, n$. Let Π_k^σ be the set of all σ-frequent tag tree patterns with at most k vertices and no edge. Let $\Pi_k^\sigma(\Lambda')$ be the set of all σ-frequent tag tree patterns with at most k vertices and edge labels in $\Lambda' \subseteq \Lambda$.

1. Generate all tag tree patterns with k vertices and no edge, by using an algorithm for generating all rooted trees with ordered children on k vertices [9]. For each tag tree pattern π with k vertices and no edge, we compute the frequency of π and if the frequency is greater than or equal to σ then we add π to Π_k^σ.
2. For each $\pi \in \Pi_k^\sigma$, we try to substitute variables of π with edges labeled with "?" as many as possible so that all σ-frequent tag tree patterns in $\Pi_k^\sigma(\{?\})$ are generated. This work can be done in a backtracking way. Then for each $\pi \in \Pi_k^\sigma(\{?\})$, we try to replace ?'s with labels in $Tag \cup KW$ as many as possible so that all σ-frequent tag tree patterns in $\Pi_k^\sigma(Tag \cup KW \cup \{?\})$ are generated. This work can be done in a backtracking way.
3. Finally we check by using the maximality test algorithm in Section 4.2 whether or not $\pi \in \Pi_k^\sigma(Tag \cup KW \cup \{?\})$ is maximally σ-frequent.

4.2 Correctness of the Generating Algorithm

In this section, we consider the following problem to complete our generating algorithm in Section 4.1. Let Λ_{Tag} and Λ_{KW} be infinite or finite languages of tags and keywords, respectively.

MAXIMALITY TEST
Instance: A set of semistructured data $\mathcal{D}$, a threshold $0 \le \sigma \le 1$, and two finite sets $Tag \subseteq \Lambda_{Tag}$ and $KW \subseteq \Lambda_{KW}$, and a tag tree pattern $\pi \in \Pi_k^\sigma(Tag \cup KW \cup \{?\})$ satisfying the following conditions:
(i) Any tag tree pattern obtained from π by replacing any variable in π with an edge which has a label in $Tag \cup KW \cup \{?\}$ is not σ-frequent w.r.t. $\mathcal{D}$.
(ii) Any tag tree pattern obtained from π by replacing any edge with a label "?" in π with an edge which has a label in $Tag \cup KW$ is not σ-frequent w.r.t. $\mathcal{D}$.
Question: Decide whether or not π is a maximally σ-frequent tag tree pattern w.r.t. $\mathcal{D}$.

We show an algorithm for solving MAXIMALITY TEST. If the target of our interest is only skeleton of given data, we ignore the edge labels of the data and find a tag tree pattern in $\Pi(\{?\})$. We note that if $|\Lambda| = |\Lambda_{Tag} \cup \Lambda_{KW}| = 1$ the label "?" is meaningless. Thus when $|\Lambda| = |\Lambda_{Tag} \cup \Lambda_{KW}| = 1$ we identify the unique label in Λ with "?". Let x_2 be a variable such that the siblings just before and after x_2 are variables and x_2 has only one child which connects to x_2 with a variable (See the right figure of Fig. 8). We call the variable like x_2 a *surrounded* variable. We omit the proof of the next lemma.

Lemma 1. *Let π' be a tag tree pattern which has a surrounded variable x_2. Let π be the tag tree pattern obtained from π' by replacing the variable x_2 with an edge which has a label "?" (Fig. 8). Then $L_\Lambda(\pi') = L_\Lambda(\pi)$.*

Our maximality test algorithm consists of the following steps.

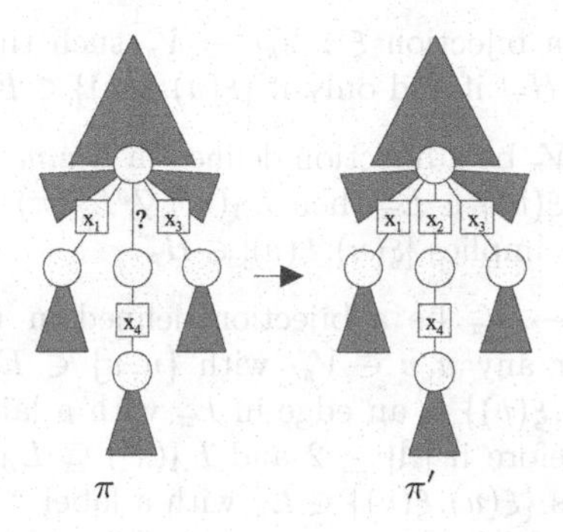

Fig. 8. This center edge of π is replaced with a variable.

1. If π has the substructure like the left figure of Fig. 8, then for all the substructures we replace the center edges with variables (See the right figure of Fig. 8). Let π' be the tag tree pattern after the replacements.

$$\theta_A(x) = \{x := [T_A, [R_A, L_A]]\}$$
$$\theta_B(x) = \{x := [T_B, [R_B, L_B]]\}$$
$$\theta_C(x) = \{x := [T_C, [R_C, L_C]]\}$$
$$\theta_D(x) = \{x := [T_D, [R_D, L_D]]\}$$

T_A T_B T_C T_D

2. If there exists a variable x in π' which is not a surrounded variable such that $\pi'\theta_X(x)$ is σ-frequent w.r.t. $\mathcal{D}$ for any $X \in \{A, B, C, D\}$, then π is not maximally σ-frequent w.r.t. $\mathcal{D}$.
3. If there exists a surrounded variable x such that $\pi'\theta_X(x)$ is σ-frequent w.r.t. $\mathcal{D}$ for any $X \in \{A, B, C\}$, then π is not maximally σ-frequent w.r.t. $\mathcal{D}$.

4. If both x_1 and x_2 are surrounded variables (See Fig. 9), we check whether or not $\pi'\theta_D(x_1)\theta_D(x_2)$ is σ-frequent w.r.t. $\mathcal{D}$. If $\pi'\theta_D(x_1)\theta_D(x_2)$ is σ-frequent w.r.t. $\mathcal{D}$, then π is not maximally σ-frequent w.r.t. $\mathcal{D}$.

If π passes all the above tests, π is maximally σ-frequent w.r.t. $\mathcal{D}$.

Lemma 2. *Let $\Lambda = \Lambda_{Tag} \cup \Lambda_{KW}$. Let $\pi = (V_\pi, E_\pi, H_\pi)$ be an input tag tree pattern which is decided to be maximally σ-frequent w.r.t. $\mathcal{D}$ by the above strategy. If there is a tag tree pattern $\pi' = (V_{\pi'}, E_{\pi'}, H_{\pi'})$ with no surrounded variable which is σ-frequent w.r.t. $\mathcal{D}$ and moreover $L_\Lambda(\pi') \subseteq L_\Lambda(\pi)$, then $\pi = \pi'$.*

Proof. (Sketch) We can show the following claims.

Claim 1. There exists a bijection $\xi : V_{\pi'} \to V_\pi$ such that for any $u, v \in V_{\pi'}$, $\{u, v\} \in E_{\pi'}$ or $[u, v] \in H_{\pi'}$ if and only if $\{\xi(u), \xi(v)\} \in E_\pi$ or $[\xi(u), \xi(v)] \in H_\pi$.

Claim 2. Let $\xi : V_{\pi'} \to V_\pi$ be a bijection defined in *Claim 1.* For any $u, v \in V_{\pi'}$, if $[u, v] \in H_{\pi'}$ and $\{\xi(u), \xi(v)\} \in E_\pi$ then $L_\Lambda(\pi') \not\subseteq L_\Lambda(\pi)$. Therefore if $L_\Lambda(\pi') \subseteq L_\Lambda(\pi)$ then $[u, v] \in H_{\pi'}$ implies $[\xi(u), \xi(v)] \in H_\pi$.

Claim 3. Let $\xi : V_{\pi'} \to V_\pi$ be a bijection defined in *Claim 1.* When $|\Lambda| = |\Lambda_{Tag} \cup \Lambda_{KW}| \geq 2$, for any $u, v \in V_{\pi'}$ with $\{u, v\} \in E_{\pi'}$, if an edge label of $\{u, v\}$ is "?" and $\{\xi(u), \xi(v)\}$ is an edge in E_π with a label in $Tag \cup KW$ then $L_\Lambda(\pi') \not\subseteq L_\Lambda(\pi)$. Therefore if $|\Lambda| \geq 2$ and $L_\Lambda(\pi') \subseteq L_\Lambda(\pi)$ then $\{u, v\} \in E_{\pi'}$ with a label "?" implies $\{\xi(u), \xi(v)\} \in E_\pi$ with a label "?" or $[\xi(u), \xi(v)] \in H_\pi$.

From the conditions of an input tag tree pattern π of MAXIMALITY TEST, for any $u, v \in V_{\pi'}$, if $[\xi(u), \xi(v)] \in H_\pi$ then $[u, v] \in H_{\pi'}$, and if $\{\xi(u), \xi(v)\} \in E_\pi$ which has a label "?" then $\{u, v\} \in E_{\pi'}$ and the label of $\{u, v\}$ is "?". Thus we conclude that $\pi = \pi'$. □

We can easily see that the above steps 1–4 run in polynomial time by using a polynomial time matching algorithm for term trees [10].

Theorem 3. **MAXIMALITY TEST** *is computable in polynomial time.*

5 Implementation and Experimental Results

We have implemented the algorithm for generating all maximally frequent tag tree patterns in Section 4 on a DELL workstation PowerEdge 6400 with Xeon 700 MHz CPU. We report some experiments on a sample file of semistructured data. The sample file is converted from a sample XML file about garment sales data such as *xml_sample* in Fig. 1. The sample file consists of 172 tree structured data. The maximum number of vertices over all trees in the file is 11, the maximum depth is 2 and the maximum number of children over all vertices is 5. In the experiments described in Fig. 10, we gave the algorithm "<Weeknumber>" and "<Designnumber>" as tags, and "Summer" and "Shirt" as keywords. The

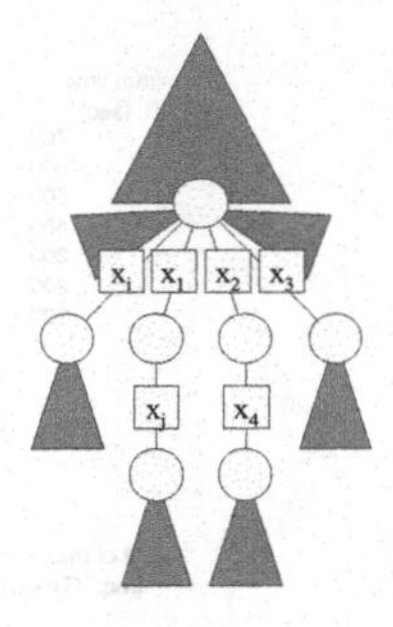

Fig. 9. Both x_1 and x_2 are surrounded variables.

algorithm generated all maximally σ-frequent tag tree patterns w.r.t. the sample file for a specified minimum frequency σ. We can set the maximum number ("max # of vertices in TTPs") of vertices of tag tree patterns in the hypothesis space.

We explain the results of Fig. 10. "TTP" means a tag tree pattern. In order to evaluate the usefulness and performance of our data mining method in this paper, we have two types of experiments. The results of "ordered TTP" are given by the algorithm for generating all maximally frequent "ordered" tag tree patterns in this work. The results of "unordered TTP" are given by the algorithm for generating all maximally frequent "unordered" tag tree patterns in our previous work [7].

Exp.1 shows the consumed run time (sec) by the two algorithms for varied minimum frequencies and the specified max # of vertices=7. Exp.2 gives the consumed run time (sec) by the two algorithms for the specified minimum frequency =0.3 and varied max numbers of vertices of TTP in the hypothesis spaces. Also, Exp.3 shows the numbers of maximally frequent TTPs obtained by the two algorithms for the specified minimum frequency =0.3 and varied max numbers of vertices of TTP in the hypothesis spaces. These experiments show that the method for generating all maximally frequent "ordered" tag tree patterns is more time-consuming than the one for generating all maximally frequent "unordered" tag tree patterns. But it is effective as compared with the size of hypothesis spaces. Also maximally frequent ordered tag tree patterns capture more precisely characteristic structures than maximally frequent unordered tag tree patterns.

6 Conclusions

In this paper, we have studied knowledge discovery from semistructured Web documents such as HTML/XML files. We have proposed a tag tree pattern which is suited for representing tree structured patterns in such semistructured data. We have shown that it is hard to compute the frequent tag tree pattern of

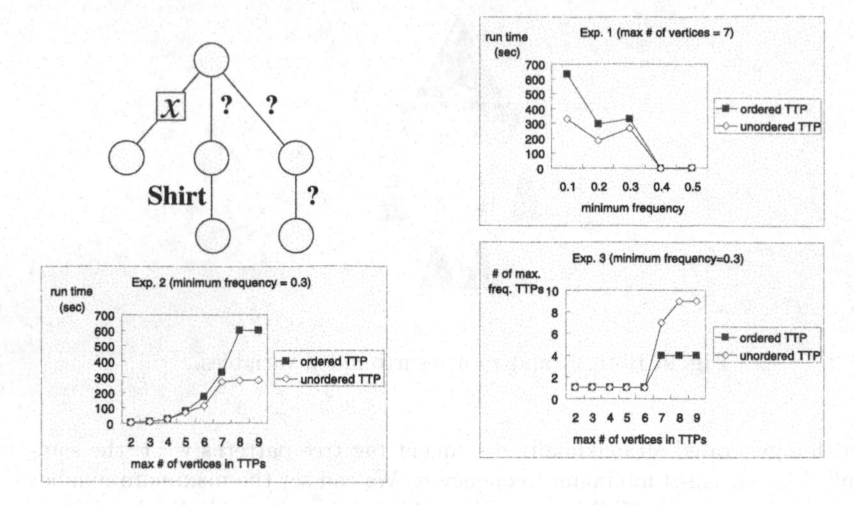

Fig. 10. Experimental results for generating all maximally frequent tag tree patterns. A maximally σ-frequent tag tree pattern obtained in the experiment.

maximum tree-size. So we have given an algorithm for generating all maximally frequent tag tree patterns. We can improve this algorithm by using the method in [5].

Acknowledgments. This work is partly supported by Grant-in-Aid for Scientific Research (C) No.13680459 from Japan Society for the Promotion of Science and Grant for Special Academic Research No.1608 from Hiroshima City University.

References

1. S. Abiteboul, P. Buneman, and D. Suciu. *Data on the Web: From Relations to Semistructured Data and XML.* Morgan Kaufmann, 2000.
2. T. Asai, K. Abe, S. Kawasoe, H. Arimura, H. Sakamoto, and S. Arikawa. Efficient substructure discovery from large semi-structured data. *Proc. 2nd SIAM Int. Conf. Data Mining (SDM-2002) (to appear)*, 2002.
3. C.-H. Chang, S.-C. Lui, and Y.-C. Wu. Applying pattern mining to web information extraction. *Proceedings of the 5th Pacific-Asia Conference on Knowledge Discovery and Data Mining (PAKDD-2001), Springer-Verlag, LNAI 2035*, pages 4–15, 2001.
4. M. Fernandez and D. Suciu. Optimizing regular path expressions using graph schemas. *Proceedings of the 14th International Conference on Data Engineering (ICDE-98), IEEE Computer Society*, pages 14–23, 1998.
5. K. Furukawa, T. Uchida, K. Yamada, T. Miyahara, T. Shoudai, and Y. Nakamura. Extracting characteristic structures among words in semistructured documents. *Proc. PAKDD-2002, Springer-Verlag, LNAI (to appear)*, 2002.

6. N. Kushmerick. Wrapper induction: efficiency and expressiveness. *Artificial Intelligence*, 118:15–68, 2000.
7. T. Miyahara, T. Shoudai, T. Uchida, K. Takahashi, and H. Ueda. Discovery of frequent tree structuted patterns in semistructured web documents. *Proc. PAKDD-2001, Springer-Verlag, LNAI 2035*, pages 47–52, 2001.
8. T. Shoudai, T. Uchida, and T. Miyahara. Polynomial time algorithms for finding unordered tree patterns with internal variables. *Proc. FCT-2001, Springer-Verlag, LNCS 2138*, pages 335–346, 2001.
9. W. Skarbek. Generating ordered trees. *Theoretical Computer Science*, 57:153–159, 1988.
10. Y. Suzuki, T. Shoudai, T. Miyahara, and T. Uchida. Polynomial time inductive inference of ordered tree patterns with internal variables from positive data. *Proc. LA Winter Symposium, Kyoto, Japan*, pages 33–1 – 33–12, 2002.
11. K. Wang and H. Liu. Discovering structural association of semistructured data. *IEEE Trans. Knowledge and Data Engineering*, 12:353–371, 2000.

Extracting Characteristic Structures among Words in Semistructured Documents

Kazuyoshi Furukawa[1], Tomoyuki Uchida[1], Kazuya Yamada[1], Tetsuhiro Miyahara[1], Takayoshi Shoudai[2], and Yasuaki Nakamura[1]

[1] Faculty of Information Sciences,
Hiroshima City University, Hiroshima 731-3194, Japan
{k_furukawa@toc.cs, uchida@cs, kazuy@toc.cs, miyahara@its, nakamura@cs}.hiroshima-cu.ac.jp
[2] Department of Informatics, Kyushu University, Kasuga 816-8580, Japan
shoudai@i.kyushu-u.ac.jp

Abstract. Electronic documents such as SGML/HTML/XML files and LaTeX files have been rapidly increasing, by the rapid progress of network and storage technologies. Many electronic documents have no rigid structure and are called semistructured documents. Since a lot of semistructured documents contain large plain texts, we focus on the structural characteristics among words in semistructured documents. The aim of this paper is to present a text mining technique for semistructured documents. We consider a problem of finding all frequent structured patterns among words in semistructured documents. Let $(W_1, W_2, \ldots, W_k)$ be a list of words which are sorted in lexicographical order and let $k \geq 2$ be an integer. Firstly, we define a tree-association pattern on $(W_1, W_2, \ldots, W_k)$. A *tree-association pattern on* $(W_1, W_2, \ldots, W_k)$ is a sequence $\langle t_1; t_2; \ldots; t_{k-1} \rangle$ of labeled rooted trees such that, for $i = 1, 2, \ldots, k-1$, (1) t_i consists of only one node having the pair of two words W_i and W_{i+1} as its label, or (2) t_i is a labeled rooted tree which has just two leaves labeled with W_i and W_{i+1}, respectively. Next, we present a text mining algorithm for finding all frequent tree-association patterns in semistructured documents. Finally, by reporting experimental results on our algorithm, we show that our algorithm is effective for extracting structural characteristics in semistructured documents.

1 Introduction

Background: In recent years, due to the rapid progress of network and storage technologies, electronic documents such as SGML/HTML/XML files and LaTeX files have been rapidly increasing. Avoiding inappropriate documents and searching interesting documents for users become more and more important. Many electronic documents have no rigid structure and are called semistructured documents. Since many semistructured documents contain large plain texts, we focus on the characteristics such as the usage of words and the structural relations among words in semistructured documents. The aim of this paper is to

M.-S. Chen, P.S. Yu, and B. Liu (Eds.): PAKDD 2002, LNAI 2336, pp. 356–367, 2002.

```
⟨REUTERS⟩
  ⟨DATE⟩ 26-FEB-1987 ⟨/DATE⟩
  ⟨TOPICS⟩ ⟨D⟩ cocoa ⟨/D⟩ ⟨/TOPICS⟩
  ⟨TITLE⟩ BAHIA COCOA REVIEW ⟨/TITLE⟩
  ⟨DATELINE⟩ SALVADOR, Feb 26 - ⟨/DATELINE⟩
  ⟨BODY⟩
    Showers continued throughout the week
    in the Behia cocoa zone, ...
  ⟨/BODY⟩
⟨/REUTERS⟩
```

xml_sample

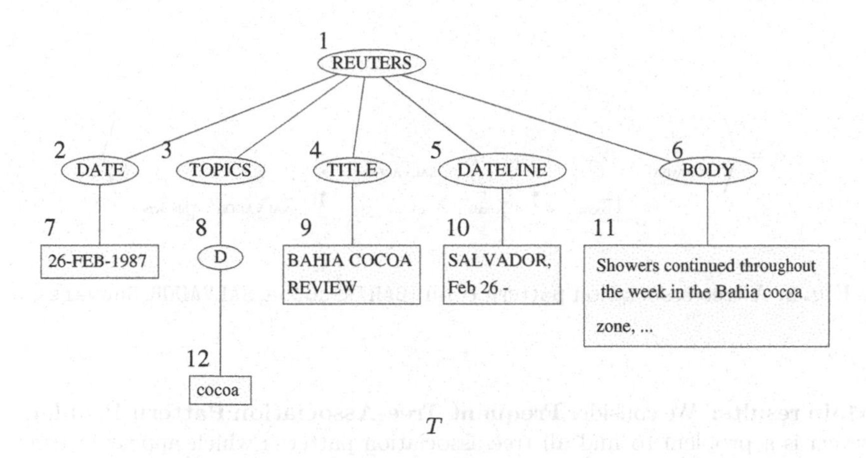

Fig. 1. An XML file *xml_sample* and a labeled rooted tree T as its OEM tree.

present an efficient text mining technique for extracting interesting structures among words in semistructured documents.

Data model and tree-association pattern: As a data model for semistructured documents, we use a variant of Object Exchange Model (OEM, for short) presented by Abiteboul et al. in [1]. As usual, a semistructured document is represented by a labeled rooted tree, which is called an *OEM tree*. For example, we give an XML file *xml_sample* and its OEM tree T in Fig. 1.

Many semistructured data have no absolute schema fixed in advance, and their structures may be irregular or incomplete. The formalization of representing knowledge is important for finding useful knowledge. In this paper, we consider a problem of discovering all frequent patterns in semistructured documents. For a semistructured document d, $\mathcal{W}(d)$ denotes the set of all words which appear in d. For a set D of semistructured documents, let $\mathcal{W}(D) = \bigcup_{d \in D} \mathcal{W}(d)$. For an integer $k \geq 2$, let $(W_1, W_2, \ldots, W_k)$ be a list of k words in $\mathcal{W}(D)$ such that words are sorted in lexicographical order. Then, a **tree-association pattern** on $(W_1, W_2, \ldots, W_k)$ (a **tree-association pattern**, for short) is a sequence $\langle t_1; t_2; \ldots; t_{k-1} \rangle$ of labeled rooted trees such that, for $i = 1, 2, \ldots, k-1$, (1) t_i consists of only one node having the pair of two distinct words W_i and W_{i+1} as its label, or (2) t_i is a labeled rooted tree which has just two leaves labeled with W_i and W_{i+1}, respectively. In Fig. 2, we give a tree-association pattern α on (`BAHIA`, `cocoa`, `SALVADOR`, `Showers`) as an example. In the concept of a tree-association pattern, we ignore the labels of the nodes except the leaves in an OEM tree and focus on the structural relations among words which appear in the leaves of the OEM tree. But, in information retrieval, we can design queries of higher quality than usual keyword queries by modifying a tree-association pattern.

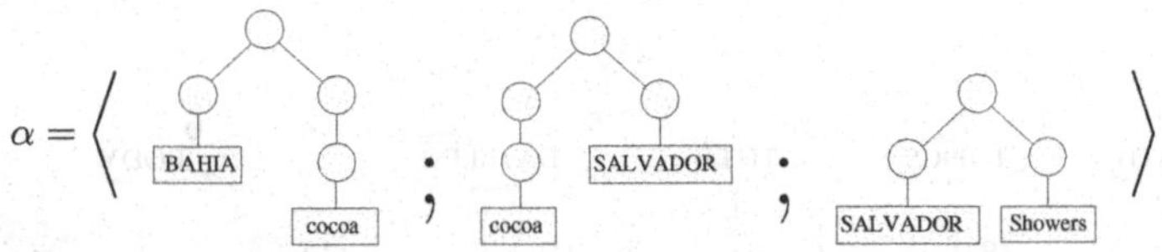

Fig. 2. A tree-association pattern α on (`BAHIA`, `cocoa`, `SALVADOR`, `Showers`).

Main results: We consider **Frequent Tree-Association Pattern Problem**, which is a problem to find all tree-association patterns which appear in given semistructured documents in a frequency of more than a user-specified threshold, which is called a minimum support. In this paper, we present a text mining algorithm for solving Frequent Tree-Association Pattern Problem. Our algorithm is an extension of the Apriori algorithm [2]. Further, the algorithm uses the technique of frequent pattern tree (FP-tree) presented by Han et al. [6,7] for keeping good performance of it in situations with a lot of frequent tree-association patterns of a low minimum support. Next, in order to show the effectiveness of our algorithm, we apply our algorithm to Reuters newswires [8] which contain 21,578 SGML documents and whose total size is 28MB. We report experimental results on our algorithm. By reporting running times of our algorithm for each user-specified threshold from 0.05 up to 0.25 and the number of documents of Reuters newswires from 1000 up to 21,578, we show that our algorithm has good performance when a user-specified threshold is low and it is applied to a set of a large number of documents.

Related works:

As a text mining method for unstructured texts, Fujino et al. [5] presented an efficient algorithm for finding all frequent phrase association patterns in a large collection of unstructured texts, where a phrase association pattern is a set of consecutive sequences of keywords which appear together in a document.

As a new knowledge representation for semistructured data, we proposed in [9,10,11] the notion of a tag tree pattern, which is a pattern consisting of structured variables and tree structures. In order to discover a frequent schema from semistructured data, several data mining methods were proposed. For example, Wang and Liu [12] presented an algorithm for finding frequent tree-expression patterns and Fernandez and Suciu [4] presented an algorithm for finding optimal regular path expressions from semistructured data. In [3], Asai et al. presented an efficient algorithm for discovering frequent substructures from a large collection of semistructured data.

In order to discover characteristic schema, many researches including above results focused on the tags and their structured relations appearing in semistructured data. But in this paper, we are more interested in words and their structural relations rather than tags in semistructured documents. The algorithm given in this work is useful for avoiding inappropriate documents and searching interesting documents for users.

Organization: This paper is organized as follows. In Section 2, we introduce a data model with which this paper deal and define a tree-association pattern. In Section 3, we formulate *Frequent Tree-Association Pattern Problem* which is a data mining problem of finding all frequent tree-association patterns for a given set of semistructured documents. Then, we give an algorithm for solving this problem. In Section 4, we show experimental results of applying our algorithm to Reuters newswires [8]. In Section 5, we conclude this paper.

2 Preliminaries

In this section, we introduce a data model of semistructured documents which is considered in this paper. Then, we define a tree-association pattern, which is a new knowledge representation for a structural characteristic in semistructured documents.

2.1 The Data Model of Semistructured Documents

As a data model for semistructured documents, we adopt a variant of Object Exchange Model (OEM, for short) presented by Abiteboul et al. in [1]. In our data model, an object o consists of an *identifier*, a *value* and a *link*, which are denoted by $\&o$, $val(\&o)$ and $link(\&o)$, respectively. The identifier $\&o$ uniquely identifies the object o. The value $val(\&o)$ is either a string such as a tag in HTML/XML files, or a text such as a text written in the field of PCDATA in XML files. The link $link(\&o)$ is a *list* $(\&o_1, \&o_2, \ldots, \&o_p)$ or a *bag* $\{\&o_1, \&o_2, \ldots, \&o_p\}$ of the identifiers of all subobjects o_i $(i = 1, 2, \ldots, p)$, where $p > 0$. As usual, a semistructured document is represented by a labeled rooted tree, which is called an *OEM tree*. That is, each node represents an object identifier $\&o$ and is labeled with the value $val(\&o)$. An edge $(\&o, \&o_i)$ represents a reference $\&o_i$ in $link(\&o)$

and has no label. If the link $link(\&o)$ of $\&o$ is a list $(\&o_1, \&o_2, \ldots, \&o_p)$ then the identifiers $\&o_1, \&o_2, \ldots, \&o_p$, which are the children of $\&o$, are ordered. On the other hand, if the link $link(\&o)$ of $\&o$ is a bag $\{\&o_1, \&o_2, \ldots, \&o_p\}$ then the identifiers $\&o_1, \&o_2, \ldots, \&o_p$, which are the children of $\&o$, are unordered. An OEM tree is said to be *ordered* and *unordered* if all inner nodes of it have ordered children and unordered children, respectively. In this paper, we represent a semistructured document by an unordered OEM tree.

Example 1. In Fig. 1, T is an unordered OEM tree for the semistructured document *xml_sample*. In T, for example, the node of identifier 1 has $val(1)$ = "REUTERS" and $link(1) = \{2, 3, 4, 5, 6\}$, and the object of identifier 9 has $val(9)$ = "BAHIA COCOA REVIEW" and $link(9) = \emptyset$.

2.2 Tree-Association Patterns

An *alphabet* is a set of finite symbols and is denoted by Σ. We assume that Σ includes the space symbol " ". A finite sequence $(a_1, a_2, \ldots, a_n)$ of symbols in Σ is called a *string* and it is denoted by $a_1 a_2 \cdots a_n$ for short. A *word* is a substring $a_2 a_3 \cdots a_{n-1}$ of $a_1 a_2 \cdots a_n$ over Σ such that both a_1 and a_n are space symbols and each a_i $(i = 2, 3, \ldots, n-1)$ is a symbol in Σ which is not the space symbol. For a set or a list S, the number of elements of S is denoted by $\#S$.

A rooted tree t is called a **2-tree** if (1) t consists of only one node or (2) the number of leaves of t is just two and the number of edges incident to the root of t is also just two. We assume that for a 2-tree t, if t consists of only one node then its node is labeled with a pair of two distinct words over Σ, otherwise each leaf of t is labeled with a word over Σ and the other nodes of t have no label. For two rooted trees s and t, s is said to be *semi-isomorphic* to t if there exists a bijection π from the nodes of s to the nodes of t such that π satisfies the following two conditions (1) and (2).

(1) For the root r_s of s and the root r_t of t, $\pi(r_s) = r_t$.
(2) s has an edge (u, v) if and only if t has an edge $(\pi(u), \pi(v))$.

For two different words W and W', if W is less than W' in lexicographical order, we denote $W < W'$. For an integer $k \geq 2$, let $(W_1, W_2, \ldots, W_k)$ be a list of k words $W_1, W_2, \ldots, W_k$ which are sorted in lexicographical order. A **tree-association pattern** on $(W_1, W_2, \ldots, W_k)$ is a sequence $\langle t_1; t_2; \ldots; t_{k-1} \rangle$ of 2-trees $t_1, t_2, \ldots, t_{k-1}$ which satisfy the following conditions. For each $i = 1, 2, \ldots, k-1$, (1) if t_i consists of only one node then the label of its node is the pair of two words W_i and W_{i+1} with $W_i < W_{i+1}$, (2) otherwise two leaves of t_i have W_i and W_{i+1} with $W_i < W_{i+1}$ as its label, respectively. It is simply called a **tree-association pattern**, if we do not need to specify the k words. We remark that, for each $i = 1, 2, \ldots, k-1$, t_i and t_{i+1} have at least one leaf having the same label.

Let T be an OEM tree. We say that a 2-tree t *appears in* T if there exists a rooted tree s satisfying the following conditions.

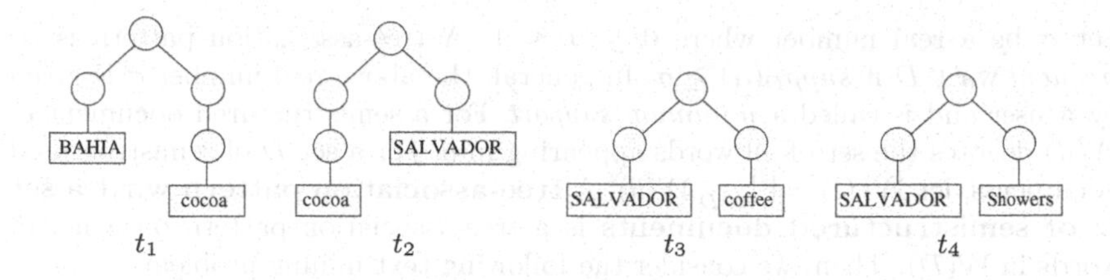

Fig. 3. 2-trees t_1, t_2, t_3 and t_4.

(1) s is a subtree of T whose leaves are those of T, and s is a 2-tree which is semi-isomorphic to t.
(2) If t consists of only one node labeled with two words, then each word occurs in the label of s as a substring. If t has two leaves l_t^1 and l_t^2, let l_s^1 and l_s^2 be leaves of s corresponding to l_t^1 and l_t^2 of t, respectively. Then, for each $i = 1, 2$, the label of l_t^i occurs in the label of l_s^i as a substring.

Let $\alpha = \langle t_1; t_2; \ldots; t_{k-1} \rangle$ be a tree-association pattern on $(W_1, W_2, \ldots, W_k)$. For each $i = 1, 2, \ldots, k-1$ and $W \in \{W_i, W_{i+1}\}$, let $l_i(W)$ be the leaf of t_i which has the word W as its label. Then, α *appears in* T if, for each $i = 1, 2, \ldots, k-1$, t_i appears in T and the leaf of T corresponding to the leaf $l_i(W_{i+1})$ of t_i is the same as one corresponding to the leaf $l_{i+1}(W_{i+1})$ of t_{i+1}.

Example 2. Three 2-trees t_1, t_2 and t_4 in Fig. 3 appear in T in Fig. 1. But the 2-tree t_3 in Fig. 3 does not appear in T. The tree-association pattern $\langle t_1; t_2; t_4 \rangle$ appears T in Fig. 1, but the tree-association pattern $\langle t_1; t_2; t_3 \rangle$ does not appear in T.

3 Text Mining Method for Extracting Tree-Association Patterns from Semistructured Text Documents

In this section, we define Frequent Tree-Association Pattern Problem, which is a text mining problem of finding all frequent tree-association patterns from semistructured documents. Next, we give a text mining algorithm for solving this problem.

3.1 Text Mining Problem

Let $D = \{d_1, d_2, \ldots, d_m\}$ be a set of semistructured documents. Let $T(D) = \{T_1, T_2, \ldots, T_m\}$ be the set of OEM trees such that each T_i $(i = 1, 2, \ldots, m)$ is the corresponding OEM tree of the document d_i in D. Then, *the occurrence count* of a given tree-association pattern α w.r.t. D, which is denoted by $Occ_D(\alpha)$, is the number of OEM trees in $T(D)$ in which the tree association pattern α appears. Then, the *frequency* of α w.r.t. D is defined by $supp_D(\alpha) = Occ_D(\alpha)/m$.

Let σ be a real number where $0 \leq \sigma \leq 1$. A tree-association pattern is σ-*frequent* w.r.t D if $supp_D(\alpha) \geq \sigma$. In general, the above real number σ is given by a user and is called a *minimum support.* For a semistructured document d, $\mathcal{W}(d)$ denotes the set of all words appearing in d. For a set D of semistructured documents, let $\mathcal{W}(D) = \bigcup_{d \in D} \mathcal{W}(d)$. A **tree-association pattern w.r.t a set D of semistructured documents** is a tree-association pattern on a list of words in $\mathcal{W}(\mathcal{D})$. Then, we consider the following text mining problem.

Frequent Tree-Association Pattern Problem
Instance: A set D of semistructured documents and a minimum support $0 \leq \sigma \leq 1$.
Problem: Find all σ-frequent tree-association patterns w.r.t. D.

3.2 Text Mining Algorithm

In Fig. 4, we present an algorithm *Find_Freq_TAPs* for solving Frequent Tree-Association Pattern Problem. Our algorithm is based on Apriori heuristic presented by Agrawal [2]. As input for our algorithm, given a set D of semistructured documents and a minimum support $0 \leq \sigma \leq 1$, our algorithm outputs the set $\mathcal{F}$ of all σ-frequent tree-association patterns w.r.t. D.

For a given set $D = \{d_1, d_2, \ldots, d_m\}$ of semistructured documents, we regard a word as an *item*, the set $\mathcal{W}(d_i)$ of all words appearing in a document d_i as a transaction, and $\mathcal{D} = \{\mathcal{W}(d_1), \mathcal{W}(d_2), \ldots, \mathcal{W}(d_m)\}$ as a set of transactions. We say that a set $f \subseteq \mathcal{W}(D)$ is a *frequent itemset* w.r.t. D and σ if the condition $\#\{i \mid f \subseteq \mathcal{W}(d_i), 1 \leq i \leq m\}/m \geq \sigma$ holds. As a preprocess, our algorithm finds all frequent itemsets w.r.t. D and σ by using a frequent pattern tree (FP-tree, for short) structure, which is a compact data structure presented by Han et al. in [6,7]. If a set f of words is not a frequent itemset w.r.t D and σ, then any tree-association pattern, which contains all words in f, is not frequent. Then, by using the resulting frequent itemsets, we can reduce the size of the search space of possible tree-association patterns. Given $\mathcal{D}$ and σ as inputs, the function *Freq_Itemsets*$(\mathcal{D}, \sigma)$ in the line 1 of Fig. 4 outputs the set of all sets S of words such that S is a frequent itemset w.r.t $\mathcal{D}$ and σ. We can design a fast algorithm for computing the function *Freq_Itemsets* by modifying the data mining algorithm presented by Han et al.[6,7]. Since the algorithm in [6,7] correctly finds the complete set of frequent itemsets in $\mathcal{D}$ and σ, we can correctly get the set Π in the line 1 of Fig. 4.

For a list L of words which are sorted in lexicographical order, $\mathcal{TAP}(L)$ denotes the set of all tree-association patterns on L. In the line 3 of Fig. 4, our algorithm constructs the set $\mathcal{F}[2]$ of all frequent tree-association patterns on a list which has just two words. For a tree-association pattern α on $(W_1, W_2, \ldots, W_k)$, $\mathcal{WL}(\alpha) = (W_1, W_2, \ldots, W_k)$. For example, for the tree-association pattern α in Fig. 2, $\mathcal{WL}(\alpha) = (\texttt{BAHIA}, \texttt{cocoa}, \texttt{SALVADOR}, \texttt{Showers})$. In while-loop from the lines 5 to 16 in Fig. 4, our algorithm can correctly generate the set $\mathcal{F}[k+1]$ of all frequent tree-association patterns α w.r.t. D and σ such that $\#\mathcal{WL}(\alpha) = k+1$ by using the levelwise search strategy of Apriori algorithm from the set $\mathcal{F}[k]$. While-loop repeats until no more frequent tree-association patterns are generated. Then the algorithm correctly computes all σ-frequent tree-association patterns w.r.t. D and σ.

Algorithm Find_Freq_TAPs

Input: A set $D = \{d_1, d_2, \ldots, d_m\}$ of semistructured documents,
and a minimum support $0 \leq \sigma \leq 1$.
Output: The set $\mathcal{F}$ of all σ-frequent tree-association patterns w.r.t. D.

1. Π =*Freq_Itemsets*$(\{\mathcal{W}(d_1), \mathcal{W}(d_2), \ldots, \mathcal{W}(d_m)\}, \sigma)$;
2. Let $\Pi[i] = \{I \mid I \in \Pi, \#I = i\}$ for $1 \leq i \leq \#\mathcal{W}(D)$;
3. $\mathcal{F}[2] = \{\alpha \mid \alpha \in \mathcal{TAP}((w_1, w_2)), \{w_1, w_2\} \in \Pi[2], w_1 < w_2, Occ_D(\alpha)/m \geq \sigma\}$;
4. $k = 2$;
5. **while** $\mathcal{F}[k] \neq \emptyset$ **do**
6.   **begin**
7. $\mathcal{F}[k+1] = \emptyset$;
8.   **for each** pair of $\alpha = \langle t_\alpha^1, t_\alpha^2, \ldots, t_\alpha^{k-1} \rangle$ and $\beta = \langle t_\beta^1, t_\beta^2, \ldots, t_\beta^{k-1} \rangle$ in $\mathcal{F}[k]$
        such that $t_\alpha^i = t_\beta^i$ for each $i = 1, 2, \ldots, k-2$, and $t_\alpha^{k-1} \neq t_\beta^{k-1}$ **do**
9.   **begin**
10.   let $\mathcal{WL}(\alpha) = (W_\alpha^1, W_\alpha^2, \ldots, W_\alpha^k)$ and $\mathcal{WL}(\beta) = (W_\beta^1, W_\beta^2, \ldots, W_\beta^k)$
          such that $W_\alpha^k < W_\beta^k$;
11.   **if** $\{W_\alpha^1, W_\alpha^2, \ldots, W_\alpha^k, W_\beta^k\} \in \Pi[k+1]$ **then**
12.   **for each** $t \in \mathcal{F}[2] \cap \mathcal{TAP}((W_\alpha^k, W_\beta^k))$ **do**
13.   add a new tree-association pattern $\gamma = \langle t_\alpha^1; t_\alpha^2; \ldots; t_\alpha^{k-1}; t \rangle$
              to $\mathcal{F}[k+1]$ if $Occ_D(\gamma)/m \geq \sigma$;
14.   **end**;
15. $k = k + 1$;
16.   **end**; /* end of while loop */
17. **return** $\mathcal{F} = \mathcal{F}[2] \cup \cdots \cup \mathcal{F}[k]$;

Fig. 4. The text mining algorithm *Find_Freq_TAPs* for solving Frequent Tree-Association Pattern Problem.

4 Experimental Results

In this section, in order to show the effectiveness of our text mining algorithm *Find_Freq_TAPs* in Fig. 4, we report some experimental results on Reuters-21578 text categorization collection in [8]. The collection contains 21,578 SGML documents and its size is about 28MB. We implemented our text mining algorithm in C and performed several experiments on a DELL workstation PowerEdge 6400 running Red Hat Linux 7J with four 700 MHz Intel Xeon processors and 2GB of main memory.

As heuristics for extracting interesting patterns in semistructured documents, a quite frequent word such as "a" or "the" is not used as a word in a tree-association pattern. Such a word is called a *stop word.* In the following experiments, as stop words, we choose symbols such as "-,+,...", numbers such as "0,1,2,...", pronouns such as "it, this, ...", articles "a, an, the", and auxiliary verbs "can, may, ..." and so on.

Experimental Setup:
We have some experiments of applying our algorithm to Reuters-21578 text categorization collection by varying the minimum support from 0.05 to 0.25 and the number of documents from 1,000 to 21,578.

We show the running times of finding all frequent tree-association patterns and computing the function *Freq_Itemsets* under the above experimental setup in Fig. 5 (a) and (b), respectively. We remark that these running times do not contain the times of parsing input documents. The running times of computing *Freq_Itemsets* and finding all frequent tree-association patterns are relatively small with the times of parsing all input documents.

Fig. 5 (c) and (d) show the numbers of all frequent tree-association patterns and all frequent itemsets under the above experimental setup, respectively. From Fig. 5 (c) and (d), we can see that both of the numbers become quickly large for lower minimum supports. This fact causes speed down of our system for lower minimum supports shown in Fig. 5, respectively. Conversely, for higher minimum supports, since these numbers become small, we can quickly get all frequent tree-association patterns.

Next, Fig. 5 (e) and (f) show the average numbers of words in tree-association patterns and frequent itemsets under the above experimental setup, respectively. From Fig. 5 (e) and (f), we can see that the average number of words in tree-association patterns is about 2, regardless of the numbers of documents and minimum supports. On the other hand, the average number of words in frequent itemsets is about 5 for lower minimum supports and less number of documents. This fact shows the significance of finding frequent tree-association patterns.

Finally, as examples of 0.05-frequent tree-association patterns found by our algorithm, we give two 0.05-frequent tree-association patterns α and β in Fig. 6, respectively. The tree-association pattern α in Fig. 6 appears in 676 documents in a set of 10,000 documents, which consists of the first document up to 10,000th document in Reuters-21578 text categorization collection. The tree-association pattern β in Fig. 6 appears in 1080 documents of the full Reuters-21578 text categorization collection.

5 Concluding Remarks

In this paper, we considered the problem of extracting structural characteristics among words from semistructured documents. We formulated a tree-association pattern on a list of words as a new knowledge representation and presented a

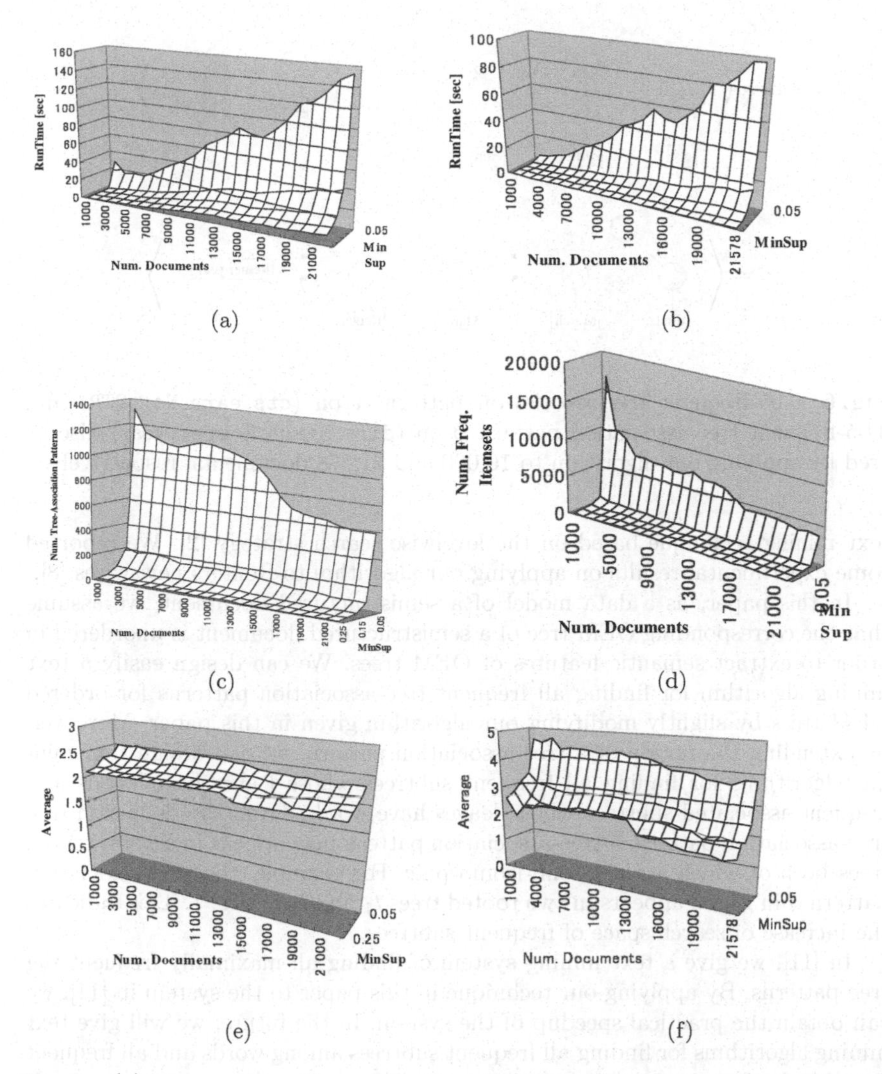

Fig. 5. (a) The running time of finding all frequent tree-association patterns, (b) the running time of computing the function *Freq_Itemsets*, (c) the number of all frequent tree-association patterns, (d) the number of all frequent itemsets, (e) the average number of words in tree-association patterns, and (f) the average number of words in frequent itemsets in applying our algorithm to Reuters-21578 text categorization collection by varying the minimum support and the number of documents. In each figure, "MinSup" denotes a minimum support, "Num. Documents" is the number of documents, "Num. Tree-Association Patterns" is the number of tree-association patterns and "Num. Freq-Itemsets" is the number of frequent itemsets.

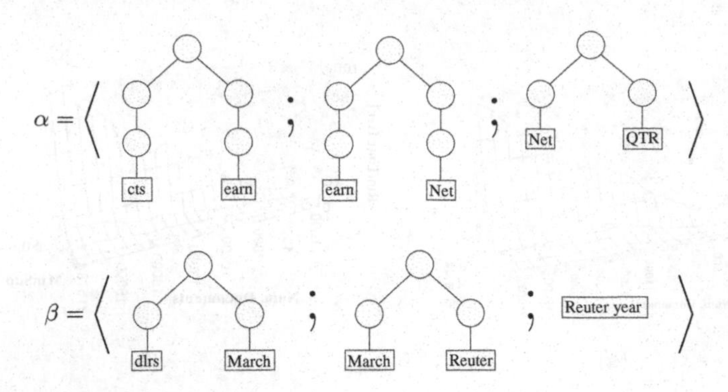

Fig. 6. 0.05-frequent tree-association pattern α on (`cts`, `earn`, `Net`, `QTR`) and 0.05-frequent tree-association pattern β on (`dlrs`, `March`, `Reuter`, year) discovered by applying our algorithm to 10,000 and 21,578 documents, respectively.

text mining technique based on the levelwise search strategy [2]. We reported some experimental results on applying our algorithm to Reuters newswires [8].

In this paper, as a data model of a semistructured document, we assume that the corresponding OEM tree of a semistructured document is unordered in order to extract semantic features of OEM trees. We can design easily a text mining algorithm for finding all frequent tree-association patterns for ordered OEM trees by slightly modifying our algorithm given in this paper. Moreover, by extending the notion of a tree-association pattern, we can design text mining algorithms for finding all frequent subtrees whose leaves have words and frequent associated subtrees whose leaves have words. From the definition of a tree-association pattern, a tree-association pattern may appear in several rooted trees both of which are not semi-isomorphic. For example, the tree-association pattern β in Fig. 6 appears in two rooted trees t_1 and t_2 in Fig. 7. This indicates the increase of search space of frequent subtrees.

In [11], we give a text mining system of finding all maximally frequent tag tree patterns. By applying our technique in this paper to the system in [11], we can obtain the practical speedup of the system. In the future, we will give text mining algorithms for finding all frequent subtrees among words and all frequent associated subtrees among words.

References

1. S. Abiteboul, P. Buneman, and D. Suciu. *Data on the Web: From Relations to Semistructured Data and XML.* Morgan Kaufmann, 2000.
2. R. Agrawal and R. Srikant. Fast algorithms for mining association rules. *Proc. of the 20th VLDB Conference*, pages 487–499, 1994.
3. T. Asai, K. Abe, S. Kawasoe, H. Arimura, H. Sakamoto, and S. Arikawa. Efficient substructure discovery from large semi-structured data. *Proc. 2nd SIAM Int. Conf. Data Mining (SDM-2002) (to appear)*, 2002.

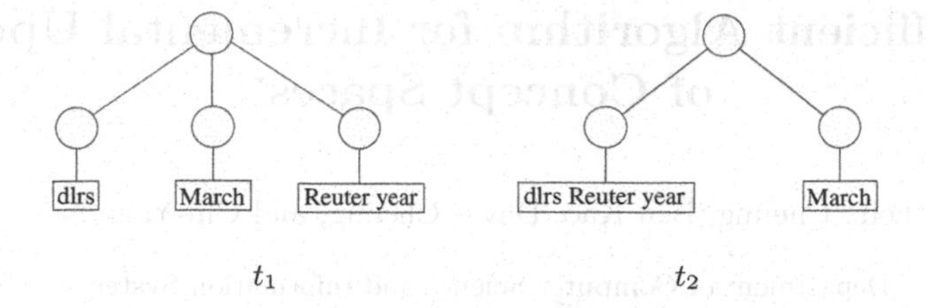

Fig. 7. Two rooted trees t_1 and t_2 in which the tree-association pattern β in Fig. 6 appears.

4. M. Fernandez and Suciu D. Optimizing regular path expressions using graph schemas. *Proc. Int. Conf. on Data Engineering (ICDE-98)*, pages 14–23, 1998.
5. R. Fujino, H. Arimura, and S. Arikawa. Discovering unordered and ordered phrase association patterns for text mining. *Proc. PAKDD-2000, Springer-Verlag, LNAI 1805*, pages 281–293, 2000.
6. J. Han and M. Kamber. *Data Mining: Concepts and Techniques.* Morgan Kaufmann Publishers, 2001.
7. J. Han, J. Pei, and Y. Yin. Mining frequent patterns without candidate generation. *Proc. ACM SIGMOD Conf.*, pages 1–12, 2000.
8. D. Lewis. Reuters-21578 text categorization test collection. *UCI KDD Archive, http://kdd.ics.uci.edu/databases/reuters21578/reuters21578.html*, 1997.
9. T. Miyahara, T. Shoudai, T. Uchida, K. Takahashi, and H. Ueda. Polynomial time matching algorithms for tree-like structured patterns in knowledge discovery. *Proc. PAKDD-2000, Springer-Verlag, LNAI 1805*, pages 5–16, 2000.
10. T. Miyahara, T. Shoudai, T. Uchida, K. Takahashi, and H. Ueda. Discovery of frequent tree structuted patterns in semistructured web documents. *Proc. PAKDD-2001, Springer-Verlag, LNAI 2035*, pages 47–52, 2001.
11. T. Miyahara, Y. Suzuki, T. Shoudai, T. Uchida, K. Takahashi, and H. Ueda. Discovery of frequent tag tree patterns in semistructured web documents. *Proc. PAKDD-2002, Springer-Verlag, LNAI (to appear)*, 2002.
12. K. Wang and H. Liu. Discovering structural association of semistructured data. *IEEE Trans. Knowledge and Data Engineering*, 12:353–371, 2000.

An Efficient Algorithm for Incremental Update of Concept Spaces*

Felix Cheung, Ben Kao, David Cheung, and Chi-Yuen Ng

Department of Computer Science and Information Systems
The University of Hong Kong
{kmcheung, kao, dcheung, cyng}@csis.hku.hk

Abstract. The vocabulary problem in information retrieval arises because authors and indexers often use different terms for the same concept. A thesaurus defines mappings between different but related terms. It is widely used in modern information retrieval systems to solve the vocabulary problem. Chen et al. proposed the *concept space* approach to automatic thesaurus construction. A concept space contains the associations between every pair of terms. Previous research studies show that concept space is a useful tool for helping information searchers in revising their queries in order to get better results from information retrieval systems. The construction of a concept space, however, is very computationally intensive. In this paper, we propose and evaluate an efficient algorithm for the incremental update of concept spaces. In our model, only *strong* associations are maintained, since they are most useful in thesauri construction. Our algorithm uses a pruning technique to avoid computing *weak* associations to achieve efficiency.
Keywords: concept space, thesaurus, information retrieval, text mining

1 Introduction

The vocabulary problem has been studied for many years [6, 3]. It refers to the failure of a system caused by the variety of terms used by its users during human-system communication. Furnes et al. studied the tendency of people using different terms in describing a similar concept. For example, they discovered that for spontaneous word choice for concepts, in certain domain, the probability that two people choose the same term is less than 20% [6]. In an information retrieval system, if the keywords that a user specifies in his query are not used by the indexer, the retrieval fails.

To solve the vocabulary problem, a thesaurus is often used. A thesaurus contains a list of terms along with the relationships between them. During searching, a user can make use of the thesaurus to design the most appropriate search strategy. For example, if a search retrieves too few documents, a user can expand his query by consulting the thesaurus for similar terms. On the other hand,

* This research is supported by Hong Kong Research Grants Council grant HKU 7035/99E

M.-S. Chen, P.S. Yu, and B. Liu (Eds.): PAKDD 2002, LNAI 2336, pp. 368–380, 2002.

if a search retrieves too many documents, a user can use a more specific term suggested by the thesaurus. Manual construction of thesauri is a very complex process and often involves human experts. Previous research works have been done on automatic thesaurus construction [5].

In [7], Chen et al. proposed the concept space approach to automatic thesaurus generation. A concept space is a network of terms and their weighted associations. The association between two terms is a quantity between 0 and 1, computed from the co-occurrence of the terms from a given document collection. Its value represents the strength of similarity between the terms. If two terms never co-exist in a document, their associations are zero. When the association from a term a to another term b is close to 1, term a is highly related to term b in the document collection. Based on the idea of concept space, Schatz et al. constructed a prototype system to provide interactive term suggestion to searchers of the University of Illinois Digital Library Initiative test-bed [8]. Given a term, the system retrieves all the terms from a concept space that has non-zero associations to the given term. The associated terms are presented to the user in a list, sorted in decreasing order of association value. The user then selects new terms from the list to refine his queries interactively. Schatz showed that users could make use of the terms suggested by the concept space to improve the recall of their queries.

The construction of a concept space involves two phases: (1) an automatic indexing phase in which a document is processed to build inverted lists, and (2) a co-occurrence analysis phase in which the associations of every term pair are computed. Since there could be hundreds of thousands of terms in a document collection, a complete concept space that lists out the association values for all the term pairs is gigantic. For the purpose of thesaurus construction, fortunately, most of the association values are not used. Chen et al. suggested that for productive user-system interaction, only highly relevant concepts should be suggested to searchers [2]. For example, their worm thesaurus originally has 1,708,551 co-occurrence pairs and each term has a few thousand associated terms. They used 100 as the maximum number of related concepts for any term. If a term has more than 100 related terms, only the 100 terms with the highest association values are retained. They successfully removed about 60% of the less relevant co-occurrence pairs. In this paper, we call a concept space that only contains highly-ranked associations a *partial concept space*.

In a dynamic environment, such as the Web, the collection of documents on which a concept space is built changes with time. To capture the dynamics, a previously built (partial) concept space needs to be updated accordingly. The simplest approach to maintaining a concept space is to reconstruct it from scratch, using the updated set of documents. For large collections, unfortunately, such a brute force approach is too time-consuming. Our goal is to study the incremental update problem of partial concept spaces. We propose and evaluate an efficient pruning algorithm that achieves a significant speedup comparing with the brute-force method.

The rest of the paper is organized as follows. In Section 2 we give a formal definition of concept spaces and the incremental update problem. Section 3 briefly discusses the brute-force method. Section 4 discusses our pruning algorithm and its implementation details. Experiment results comparing the performance of the algorithms are shown in Section 5. Finally, Section 6 concludes the paper.

2 Definitions

A concept space contains the associations, W_{jk} and W_{kj}, between any two terms j and k found in a document collection. Chen and Lynch [7] define W_{jk} by the formula:

$$W_{jk} = \sum_{i=1}^{N} d_{ijk} / \sum_{i=1}^{N} d_{ij} \times \mathit{WeightingFactor}(k). \tag{1}$$

The symbol d_{ij} represents the weight of term j in document i based on the term-frequency-inverse-document-frequency (TFIDF) measure [1]:

$$d_{ij} = tf_{ij} \times \log(N/df_j \times w_j), \text{where}$$

$$\begin{aligned} tf_{ij} &= \text{number of occurrences of term } j \text{ in document } i, \\ df_j &= \text{number of documents in which term } j \text{ occurs,} \\ w_j &= \text{number of words in term } j, \\ N &= \text{number of documents.} \end{aligned}$$

The symbol d_{ijk} represents the combined weight of both terms j and k in document i. It is defined as:

$$d_{ijk} = tf_{ijk} \times \log(N/df_{jk} \times w_j), \text{where} \tag{2}$$

$$\begin{aligned} tf_{ijk} &= \min(tf_{ij}, tf_{ik}), \\ df_{jk} &= \text{number of documents in which both terms } j \text{ and } k \text{ occur.} \end{aligned}$$

Finally, $\mathit{WeightingFactor}(k)$ is defined as:

$$\mathit{WeightingFactor}(k) = \log(N/df_k)/\log N.$$

The term $\mathit{WeightingFactor}(k)$ is used as a weighting scheme (similar to the concept of inverse document frequency) to penalize general terms (terms that appear in many documents). Terms with a high df_k value has a small weighting factor, which results in a small association value. Note that the associations are asymmetric, that is, W_{jk} and W_{kj} are not necessarily equal. Chen showed that this asymmetric similarity function (W_{jk}) gives a better association measure than the popular cosine function [7].

In the following discussion, for simplicity, we assume that $w_j = 1$ for all j (i.e., all terms are single-word ones). We thus remove the term w_j from the formula of d_{ij} and d_{ijk}.

As we have mentioned in the introduction, for the purpose of thesaurus construction, only the highly-ranked associations are needed. In particular, given a term j and a user-specified parameter n, we assume that only the n largest associations $W_{jk_1}, W_{jk_2}, \ldots, W_{jk_n}$ of j are kept. Such associations are called *strong associations*. An association W_{jk} that does not crack into the top n values of j is called a *weak association* and is ignored. Setting $n = 100$, for example, has shown to be sufficient for some chosen domains [2]. We call the set of all strong associations a *partial concept space*.

Given a document collection D, we assume that a partial concept space, CS_D is constructed. Let ΔD be a set of documents that is added to D to form a new document collection D', the problem of incremental concept space update is to compute the partial concept space $CS_{D'}$ with respect to D' given D and CS_D.

For notational convenience, the symbols (e.g., df_{ijk}) that we used in the various formulae for concept space construction refer to the quantities with respect to the original document collection D. We use the prime notation (e.g., df'_{ijk}) to denote those quantities with respect to the updated document collection D'. Also, a preceding Δ (e.g., Δdf_{ijk}) denotes a quantity with respect to ΔD.

We assume that the size of ΔD is relatively small compared with that of D. It is thus computationally inexpensive to process ΔD to obtain the various "delta" values.

3 Concept Space Construction

Concept space construction is a two-phase process. In the first phase (automatic indexing), a term-document matrix, *TF* is constructed. Given a document i and a term j, the matrix *TF* returns the term frequency, tf_{ij}. In practice, *TF* is implemented using inverted lists. That is, for each term j, a linked list of [document-id,term-frequency] tuples is maintained. Each tuple records the occurrence frequency of term j in the document with the corresponding id. Documents that don't contain the term j aren't included in the inverted list of j.

Besides the matrix, *TF*, the automatic indexing phase also calculates the quantity df_j (the number of documents containing term j) as well as $\sum_{i=1}^{N} tf_{ij}$ (the sum of the term frequency of term j over the whole document collection) for each term j. These numbers are stored in arrays for fast retrieval during the second phase (co-occurrence analysis).

In the co-occurrence analysis phase, associations of every term pair are calculated. According to Equation 1 (page 370), to compute W_{jk}, we need to compute the values of three factors, namely, $\sum_{i=1}^{N} d_{ijk}$, $\sum_{i=1}^{N} d_{ij}$, and $WeightingFactor(k)$. Note that

$$\sum_{i=1}^{N} d_{ij} = \sum_{i=1}^{N}[tf_{ij} \times \log(N/df_j)] = \log(N/df_j) \times \sum_{i=1}^{N} tf_{ij}. \tag{3}$$

Since both df_j and $\sum_{i=1}^{N} tf_{ij}$ are already computed and stored during the automatic indexing phase, $\sum_{i=1}^{N} d_{ij}$ can be computed in constant time. Similarly, $WeightingFactor(k)$ can be computed in constant time as well.

Computing $\sum_{i=1}^{N} d_{ijk}$, however, requires much more work. From Equation 2, one needs to compute df_{jk} (i.e., the number of documents containing both terms

j and k) and $\sum_{i=1}^{N} tf_{ijk}$ in order to find $\sum_{i=1}^{N} d_{ijk}$. Figure 1 shows the algorithm Weight for computing W_{jk}. The execution time of Weight is dominated by the for-loop in line 3. Basically, most of the work is spent on scanning the inverted lists of terms j and k to determine $\sum_i d_{ijk}$.

```
WEIGHT(j, k)
 1  df_jk ← 0, sum_tf_ijk ← 0
 2  for each (i, tf_ij) in the adjacency list of j
 3  do if there exists (i, tf_ik) in the adjacency list of k
 4        then df_jk ← df_jk + 1
 5             if tf_ij < tf_ik
 6                then sum_tf_ijk ← sum_tf_ijk + tf_ij
 7                else sum_tf_ijk ← sum_tf_ijk + tf_ik
 8  sum_d_ijk ← sum_tf_ijk × log(N/df_jk)
 9  sum_d_ij ← sum_tf_ij × log(N/df_j)
10  weighting_factor_k ← log(N/df_k)/ log N
11  return sum_d_ijk × weighting_factor_k/sum_d_ij
```

Fig. 1. Function Weight

To compute the partial concept space of D', a brute-force approach would be to compute the associations of all term-pairs. For any term j, the associations of j are sorted and only the n largest ones (i.e., those that are strong) are retained. The brute-force method can be made significantly more efficient by first constructing a two-dimensional triangular bit matrix C in the automatic indexing phase. Given two terms j and k, the matrix C indicates whether j and k ever co-exist in any documents. We notice that $W'_{jk} = 0$ if j and k do not co-exist in any documents of D'. The function Weight is thus only executed for those j, k pair such that the entry $C(j,k)$ is set. In typical document collections, matrix C is very sparse. For storage efficiency, the matrix is compressed using bit-vector compression techniques [4].

4 Pruning Method

The baseline brute-force algorithm isn't particularly efficient. It basically computes all possible non-zero associations before filtering out those that are weak. As an example, we ran the baseline algorithm on a collection of 191,966 documents. The execution time was 146.5 min. The main source of inefficiency lies in the Weight function, which scans two inverted lists for every non-zero association. In our collection, there are about 60 millions non-zero associations, and hence the baseline algorithm performed about 120 millions inverted lists scanning.

Our approach to a more efficient algorithm for the incremental update problem is to use an efficient method to compute an upper bound of W'_{jk} (denoted by $\widehat{W'_{jk}}$) using the information of the partial concept space constructed for the old collection D. We then decide whether the association W'_{jk} is a strong association of term j (w.r.t. D') by comparing the upper bound $\widehat{W'_{jk}}$ with a threshold σ_j. The threshold σ_j is chosen such that if $\widehat{W'_{jk}} < \sigma_j$, then W'_{jk} cannot be a strong association; the value W'_{jk} is thus not computed. As we will see later in Section 5, this pruning technique significantly reduces the execution time.

Applying the pruning method thus requires two issues be addressed: (1) how to compute an upper bound $\widehat{W'_{jk}}$, and (2) how to determine the threshold σ_j for a term j.

To determine an upper bound $\widehat{W'_{jk}}$, we assume that certain information about the old collection D is kept. In particular, we assume that for each term j, we keep two values: df_j and $\sum_{i=1}^{N} tf_{ij}$. These values allow us to compute $\sum_{i=1}^{N} d_{ij}$ and $Weightingfactor(k)$ in constant time (see Equation 3). Also, we assume that for each *weak association* W_{jk} of term j w.r.t. the old collection D, an upper bound $\widehat{W_{jk}}$ is available[1]. Finally, we assume that the values of strong associations are kept (in the partial concept space CS_D). Since $\widehat{W_{jk}}$ is an upper bound of W_{jk}, we have

$$\widehat{W_{jk}} \geq W_{jk} = \sum_{i=1}^{N} d_{ijk} / \sum_{i=1}^{N} d_{ij} \times Weightingfactor(k).$$

Define

$$K_{jk} = \widehat{W_{jk}} \times \sum_{i=1}^{N} d_{ij} / Weightfactor(k),$$

we have,

$$K_{jk} \geq \sum_{i=1}^{N} d_{ijk} = \sum_{i=1}^{N} tf_{ijk} \times \log(N/df_{jk}).$$

By definition,

$$W'_{jk} = \sum_i d'_{ijk} / \sum_i d'_{ij} \times WeightingFactor'(k).$$

Note that

$$\sum_i d'_{ij} = (\sum_i tf_{ij} + \sum_i \Delta tf_{ij}) \times \log(\frac{N + \Delta N}{df_j + \Delta df_j}),$$

$$WeightingFactor'(k) = \frac{\log((N + \Delta N)/(df_k + \Delta df_k))}{\log(N + \Delta N)},$$

$$\sum_i d'_{ijk} = (\sum_i tf_{ijk} + \sum_i \Delta tf_{ijk}) \times \log(\frac{N + \Delta N}{df_{jk} + \Delta df_{jk}}).$$

[1] If collection D is obtained by adding documents to an even older collection D^- and the partial concept space of D is obtained by applying the pruning algorithm, then the upper bound $\widehat{W_{jk}}$ is obtained as a by-product and is kept for the next incremental update.

Assuming that $\sum_i tf_{ij}$ and df_j are available and that ΔD is small enough to be processed to obtain $\sum_i \Delta tf_{ij}$, Δdf_j and Δdf_k, we can efficiently compute $\sum_i d'_{ij}$ and *WeightingFactor*$'(k)$.

Computing $\sum_i d'_{ijk}$, however, requires scanning inverted lists and thus is expensive. To compute an upper bound of W'_{jk} efficiently, we must derive an efficient method to compute an upper bound of $\sum_i d'_{ijk}$. Here, we consider 3 cases.

Case 1: terms j and k do not co-exist in any document of ΔD. In this case, $\sum_i \Delta tf_{ijk}$ and Δdf_{jk} are 0. Hence,

$$\begin{aligned}\sum_i d'_{ijk} &= \sum_i tf_{ijk} \times \log(\frac{N+\Delta N}{df_{jk}}),\\ &\leq \frac{K_{jk}}{\log(N/df_{jk})} \times \left(\log(\frac{N+\Delta N}{N}) + \log(\frac{N}{df_{jk}})\right),\\ &= K_{jk} \times \left(1 + \frac{\log((N+\Delta N)/N)}{\log(N/df_{jk})}\right).\end{aligned}$$

Notice that df_{jk} is the number of documents in D that contains both terms j and k, we have $df_{jk} \leq \min(df_j, df_k) = df_{min}$. Therefore,

$$\begin{aligned}\sum_i d'_{ijk} &\leq K_{jk} \times \left(1 + \frac{\log((N+\Delta N)/N)}{\log(N/df_{min})}\right),\\ &= K_{jk} \times \frac{\log((N+\Delta N)/df_{min})}{\log(N/df_{min})}.\end{aligned}$$

We thus define the upper bound $\widehat{W'_{jk}}$ of W'_{jk} by

$$\widehat{W'_{jk}} = \frac{K_{jk}}{\sum_i d'_{ij}} \times \frac{\log((N+\Delta N)/df_{min})}{\log(N/df_{min})} \times Weightfactor'(k).$$

Case 2: terms j and k co-exist in documents of D and ΔD. Consider the formula for $\sum_i d'_{ijk}$ again.

$$\sum_i d'_{ijk} = \underbrace{\sum_i tf_{ijk} \times \log(\frac{N+\Delta N}{df_{jk}+\Delta df_{jk}})}_{T_1} + \underbrace{\sum_i \Delta tf_{ijk} \times \log(\frac{N+\Delta N}{df_{jk}+\Delta df_{jk}})}_{T_2}.$$

With a derivation similar to that of case 1, we have

$$T_1 \leq K_{jk} \times \frac{\log(\frac{N+\Delta N}{df_{min}+\Delta df_{jk}})}{\log(N/df_{min})}.$$

For T_2, assume that all the "delta" values can be computed efficiently from ΔD, and observe that $df_{jk} \geq 1$ for case 2, we have

$$T_2 \leq \sum_i \Delta tf_{ijk} \times \log(\frac{N+\Delta N}{1+\Delta df_{jk}}).$$

We thus define the upper bound $\widehat{W'_{jk}}$ by

$$\widehat{W'_{jk}} = \left(K_{jk} \times \frac{\log(\frac{N+\Delta N}{df_{min}+\Delta df_{jk}})}{\log(N/df_{min})} + \sum_i \Delta tf_{ijk} \times \log(\frac{N+\Delta N}{1+\Delta df_{jk}}) \right) \times \frac{Weightfactor'(k)}{\sum_i d'_{ij}}.$$

Case 3: terms j and k only co-exist in documents of ΔD.

In this case, both df_{jk} and $\sum_i tf_{ijk}$ are 0. With the "delta" quantities made available by processing the small ΔD, the values of $\sum_i df'_{ijk}$ and hence W'_{jk} can be computed exactly. Thus, we set $\widehat{W'_{jk}} = W'_{jk}$.

Note that in all three cases, the computational cost of determining the upper bound $\widehat{W'_{jk}}$ is small, since we do not scan the inverted lists of the large document collection D.

Recall that, for a term j, our pruning method uses a threshold σ_j to determine whether the association W'_{jk} should be computed. In particular, if the upper bound $\widehat{W'_{jk}}$ is less than σ_j, then the association W'_{jk} must be weak and should not be computed. To determine σ_j, we compute all n associations W'_{jk_i}'s for which W_{jk_i} is strong w.r.t the old collection D. σ_j is then set to the minimum value of such W'_{jk_i}'s. Given a term k, if $\widehat{W'_{jk}} < \sigma_j$, we know that W'_{jk} must be smaller than all the n W'_{jk_i}'s. Hence, W'_{jk} must not be strong.

Our pruning algorithm for the incremental update problem then goes as follows. For a term j, and a strong association W_{jk_p} w.r.t. D, we compute the association W'_{jk_p} w.r.t. to the updated collection D' using the Weight function. Among the n such associations of term j, we determine the threshold σ_j. After that, for any term k such that W_{jk} is weak w.r.t. D, we compute the upper bound $\widehat{W'_{jk}}$ of W'_{jk}. If $\widehat{W'_{jk}}$ is larger than σ_j, W'_{jk} is computed using the Weight function. Finally, only the n largest associations of j are kept in the partial concept space of D'. The upper bounds calculated in the algorithm are also retained for the next incremental update.

4.1 Quantization

With our pruning method, if the association W_{jk} is weak w.r.t. to D, we have to determine an upper bound $\widehat{W'_{jk}}$ of W'_{jk}. To compute $\widehat{W'_{jk}}$, we require that an upper bound $\widehat{W_{jk}}$ (w.r.t. D) be available[2]. Since the number of such bounds is quadratic with respect to the total number of keywords, the amount of storage required for storing all the $\widehat{W_{jk}}$'s is very big. Fortunately, we do not need to represent the bounds in high precision. Quantization techniques can be applied so that a bound is represented by a small number of bits.

Let W_{jk_n} be the n-th largest association of term j. We quantize the *differences* between bounds by a 4-bit codeword. Given a term j, we first compute all the bounds of the weak associations of j. These bounds are then sorted in decreasing value forming a sequence $\widehat{W_{jk_{n+1}}}, \widehat{W_{jk_{n+2}}}, \ldots$ We put the n-th largest

[2] $\widehat{W'_{jk}}$ is expressed in terms of K_{jk}, which is in turn expressed in terms of $\widehat{W_{jk}}$.

association (W_{jk_n}) of j in front of the sequence and compute the differences between successive values in the sequence as shown below.

$$W_{jk_n} \underbrace{-}_{\rho_{n+1}} \widehat{W_{jk_{n+1}}} \underbrace{-}_{\rho_{n+2}} \widehat{W_{jk_{n+2}}} \cdots$$

The largest difference ρ_{max} is determined and we divide the interval $[0,\rho_{max}]$ into 16 levels. Each difference ρ_i can then be represented by a 4-bit codeword. Knowing the value of W_{jk_n} and the differences, all the bounds can be re-computed. We call this quantization scheme the *differential quantization algorithm*, DQA.

5 Performance Evaluation

In this section we evaluate the performance of our pruning algorithm and its variants DQA. We applied the algorithm on "The Ohsumed Test Collection" [9], which is a medical document collection. The document collection consists of 348,566 abstracts with 240,247 terms. The document database is 169 MB large (after stop-word removal and stemming). We ran the algorithms on a 700 MHz Pentium III Xeon machine.

In our experiment, half of the documents are randomly picked as the original collection D. Collection D contains 174,566 documents. We compute the partial concept space of D. Also, for any weak association W_{jk} of D, we compute its upper bound, $\widehat{W_{jk}}$. For the experiment evaluating DQA, the upper bounds are quantized according to the quantization schemes described in the last section. We partition the other half (174,000 documents) into 10 equal parts, with 17,400 documents apiece. These parts are added to D successively and cumulatively. The first update thus increases the collection size by 10%, while the 10th update increases the collection size by about 5%.

Figure 2 shows the runtime of the pruning algorithm (without quantization) over the 10 updates under different values of n. (Recall that n is the number of strong associations per term that are kept in a partial concept space.) From the figure, we see that the execution time is larger when the update number increases. This is because the collection size is made bigger by the updates. For example, the collection before the 10th update contains 331,166 documents, which is about 90% larger than the collection before the 1st update. We see that the execution time of the pruning algorithm is linearly proportional to the size of the collection.

Figure 2 also shows that a larger n increases the execution time of the pruning algorithm. Recall that in the pruning algorithm, for a given term j, we use the n-th largest association value W_{jk_n} of j as the pruning threshold. A larger n means a smaller pruning threshold and thus more associations have to be computed. This fact is illustrated by Figure 3, which shows the number of associations computed by the pruning algorithm under different values of n.

Figure 4 compares the performance of the 4 algorithms when n is set to 100. We observe that the pruning algorithm and its quantization variants DQA

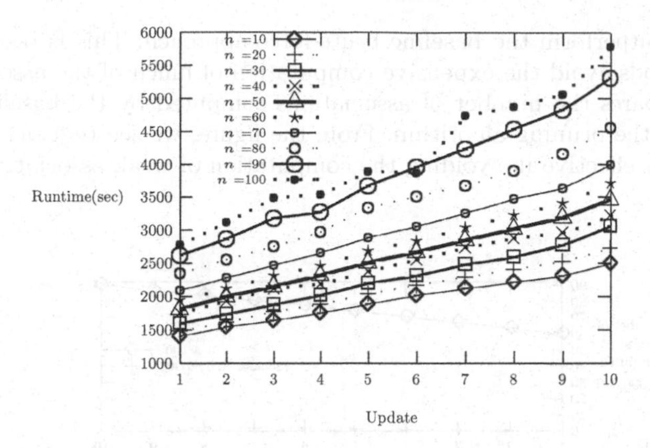

Fig. 2. Runtime of the pruning algorithm over the 10 updates under different values of n

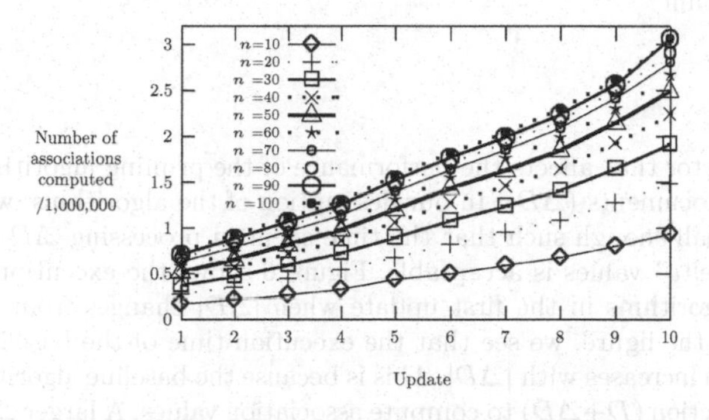

Fig. 3. The number of associations computed by the pruning algorithm under different values of n

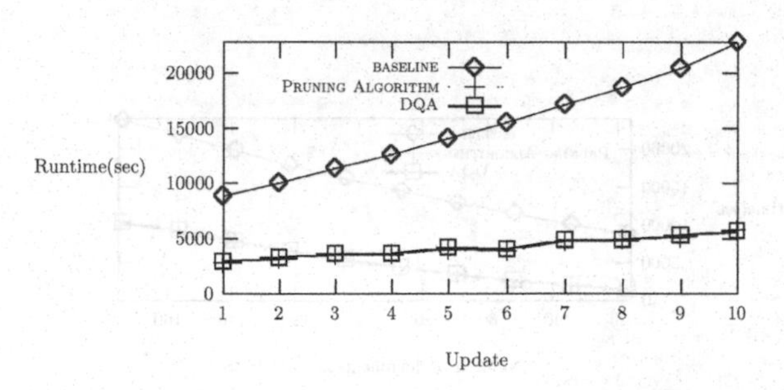

Fig. 4. Algorithms' runtimes over the 10 updates, $n = 100$

significantly outperform the baseline brute-force approach. This is because the pruning methods avoid the expensive computation of much of the associations. Figure 5 compares the number of associations computed by the baseline algorithm and by the pruning algorithm. From the figure, we see that our pruning method is very effective in avoiding the computation of weak associations.

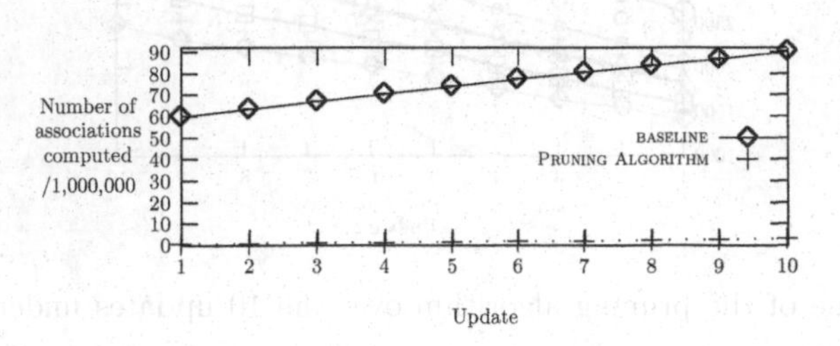

Fig. 5. The number of associations computed by the pruning algorithm and the baseline algorithm

Another factor that affects the performance of the pruning algorithms is the size of added documents (ΔD). In our description of the algorithms, we assume that ΔD is small enough such that the time spent in processing ΔD to obtain the various "delta" values is acceptable. Figure 6 shows the execution times of the various algorithms in the first update when $|\Delta D|$ changes from 17,400 to 174,000. From the figure, we see that the execution time of the baseline brute-force algorithm increases with $|\Delta D|$. This is because the baseline algorithm scans the whole collection ($D+\Delta D$) to compute association values. A larger ΔD means a larger collection and hence more work for the baseline algorithm. Also, since most of the work done by the pruning algorithms is on processing ΔD, we observe a similar increase in execution times for the pruning algorithms as ΔD increases.

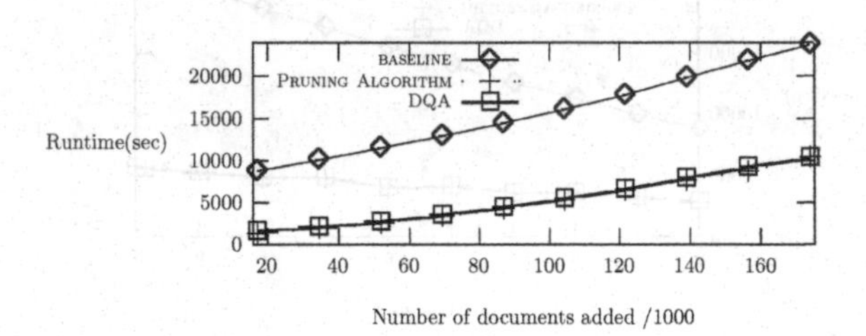

Fig. 6. Algorithms' performance vs. $|\Delta D|$, $n = 10$ and $|D| = 174,566$

Finally, we remark that the baseline algorithm is more storage efficient than the pruning algorithms. Basically, the baseline algorithm only maintains a partial concept space that stores only the strong associations. The basic pruning method, on the other hand, stores the upper bounds of the weak associations as well. This storage overhead could be significant. To reduce the storage cost, DQA quantize the bounds. In our experiment, the storage requirement of DQA ranges from 300MB to 400MB. This is about 1/4 to 1/3 of the storage required to maintain a full concept space.

6 Conclusion

This paper studied the problem of incremental update of concept spaces. Previous studies have shown that the concept space approach to automatic thesaurus construction is a useful tool for information retrieval. The construction and incremental update of concept spaces, however, are very time consuming. In many applications, a full concept space is not needed, in particular, only a few strong associations per keyword are used. We proposed a pruning algorithm for incremental update of concept spaces that contain only strong associations. To reduce the storage requirement of the pruning algorithm, we propose a quantization variant, namely, DQA. Our experiment shows that the pruning algorithms are very effective in avoiding the computation of weak associations. As an example, if the number of associations to be maintained for each keyword (i.e., n) is 10, and that the size of the added documents is about 10% of the original collection, our experiment registered a 9-time speedup of the pruning algorithms compared with the brute-force approach.

References

[1] R. Baeza-Yates and Berthier Ribeiro-Neto. *Modern Information Retrieval.* Addison Wesley, 1999.

[2] H. Chen, T. Yim, D. Fye, and B. Schatz. Automatic thesaurus generation for an electronic community system. *Journal of American Society for Information Science*, 46(3):175–193, 1995.

[3] Hsinchun Chen, Joanne Martinez, Tobun D. Ng, and Bruce R. Schatz. A concept space approach to addressing the vocabulary problem in scientific information retrieval: an experiment on the worm community system. *Journal of American Society for Information Science*, 48(1):17–31, 1997.

[4] Y. Choueka, A.S. Fraenkel, S.T. Klein, and E. Segal. Improved hierarchical bit-vector compression in document retrieval systems. In *In Proc. 9th ACM-SIGIR Conference on Information Retrieval*, pages 88–97, Pisa, Italy, September 1986.

[5] W.B. Frakes and R. Baeza-Yates. *Information Retreival: Data Structures and Algorithms.* Prentice Hall, 1992.

[6] G.W. Furnas et al. The vocabulary problem in human-system communicaiton. *Comm. ACM*, 30(11):964–971, 1987.

[7] H.Chen and Kevin J. Lynch. Automatic construction of networks of concepts characterizing document databases. *IEEE Transaction of Systems, Man, and Cybernetics*, 22(5):885–902, Sep/Oct 1992.

[8] Bruce R. Schatz, Eric H. Johnson, Pauline A. Cochrane, and Hsinchun Chen. Interactive term suggestion for users of digital libraries: using subject thesauri and co-occurrence lists for information retrieval. In *Digital Library 96*, Bethesda MD, 1996.
[9] Hersh WR, Buckley C, Leone TJ, and Hickam DH. Ohsumed: An interactive retrieval evaluation and new large test collection for research. In *Proceedings of the 17th Annual ACM SIGIR Conference*, pages 192–201, 1994. http://www1.ics.uci.edu/pub/machine-learning-databases/ohsumed/.

Efficient Constraint-Based Exploratory Mining on Large Data Cubes

Cuiping Li[1], Shengen Li[1], Shan Wang[2], and Xiaoyong Du[2]

[1]Institute of Computing Technology, Chinese Academy of Sciences, Beijing, China
{cuiping_li, sheng_en_li}@263.net
[2]Information School, Renmin University of China, Beijing, China
{ushan, duyong }@public.ruc.edu.cn

Abstract. Analysts often explore data cubes to identify anomalous regions that may represent problem areas or new opportunities. Discovery-driven exploration (proposed by S.Sarawagi et al [5]) automatically detects and marks the exceptions for the user and reduces the reliance on manual discovery. However, when the data is large, it is hard to materialize the whole cube due to the limitations of both space and time. So, exploratory mining on complete cube cells needs to construct the data cube dynamically. That will take a very long time. In this paper, we investigate optimization methods by pushing several constraints into the mining process. By enforcing several user-defined constraints, we first restrict the multidimensional space to a small constrained-cube and then mine exceptions on it. Two efficient constrained-cube construction algorithms, the NAIVE algorithm and the AGOA algorithm, were proposed. Experimental results indicate that this kind of constraint-based exploratory mining method is efficient and scalable.

1 Introduction

With recent developments of data warehouse and OLAP technology, it is expected that a tremendous amount of data will be integrated, preprocessed, and stored in large data warehouses. Currently, most of the data warehouses are being used for multidimensional On-Line Analytical Processing. However, with the rapid progress in data mining research, it is expected that powerful data mining tools will be introduced for sophisticated data analysis in data warehouses [1].

In the field of data mining, substantial research has been performed for integrating existing mining algorithms within OLAP products: decision tree classifiers to find the factors affecting profitability of products is used by Cognos [2], clustering customers based on buying patterns to create new hierarchies is used in Pilot Software [3], and association rules at multiple levels of aggregation to find progressively detailed correlation between members of a dimension is suggested by Han et al [4].

In the same direction, S.Sarawagi et al proposed a different approach in that they first investigate how and why analysts currently explore the data cube and next automate them using several new advanced mining primitives [5]. They proposed discovery-driven exploration mining approach in which precomputed measures indicating

M.-S. Chen, P.S. Yu, and B. Liu (Eds.): PAKDD 2002, LNAI 2336, pp. 381-392, 2002.

data exceptions are used to guide the user in the data analysis process. Unlike the batch process of existing mining algorithms, they wish to enable interactive invocation so that an analyst can use them seamlessly with the existing simple operations.

This form of exploratory mining method significantly reduces the reliance on manual discovery and takes the OLAP tools to the next stage of intelligent interactive analysis. It is interesting and has broad applications. However, it also poses serious challenges on both computational efficiency and scalability for the following three reasons.

First, for *d* dimensions in a cube, not only are there 2^d cuboids, but also there are numerous cells to be computed. In addition, dimensions usually are associated with hierarchies that specify aggregation levels. Mining on such a huge amount of data may take a very long time. Since a data mine task is usually relevant to only a portion of the data cube, selecting the relevant set of data not only makes mining more efficient, but also derives more meaningful results than mining on the entire data cube. Thus, it is expected to introduce certain *data constraints* to select a subset of cells from the cube. By doing so, a manager in charge of sales in the United States can specify that only the data relating to customer purchases in United States need to be mined.

Second, due to the limitations of both space and time, the large data cube in a practical data warehouse is hard to be completely materialized even when considering only task-relevant set of data. When mining on a data cube, it is desirable to compute all cells on-the-fly. Thus the response time will increase. On the other hand, in practice, a user may prefer some dimensions to remain at rather high abstraction levels while others are specialized to lower levels. The control of how low a dimension should be drill-down is typically quite subjective. So *level constraints* can be introduced to limit the degree to which the given data set should be specialized in order to find further exceptions. Enforcing the level constraint can solve the previous problem and lead to efficient methods for exploratory mining because we can utilize the materialized cuboids to construct a "constrained-cube" with a reasonable storage and processing cost. Although further drill-down beyond these levels will still require another computation, this "divide and conquer" method can help the system to reach a real on-line performance.

Third, there are separate criteria for defining exceptions. For example, top executives may be interested in only those cells whose actual average profits increase by more than 30% compared to that of the expected values, while to a local store manager, slight increase may be very important. Such difference can be used as a threshold in the form of ratio/difference between certain measure values of the cells under comparison.

Based on above discussions, one can find that to explore interesting patterns in a data cube, it is often necessary to have the following three kinds of constraints: a) *data constraint* ensures that we examine only the cells in which we really interested; b) *level constraint* confines the set of cells that our exploratory mining will focus on; c) *exception constraint* specifies the user's interest threshold for a exception. Enforcing these constraints may lead to efficient methods for exploratory analysis in a multidimensional space.

The remaining of the paper is organized as follows. Section 2 gives the background information of our study and defines the constrained exploratory mining problem. Section 3 introduces the constraint-based exploratory mining approach and presents two algorithms for construct a constrained-cube efficiently using materialized cuboids at the minimal cost. A performance analysis of our method is presented in section 4. We compare our work with other related works in section 5 and conclude in section 6.

2 Preliminaries and Problem Definition

Data warehouse is a data repository in which data is organized along a set of dimensions $D=\{d_1, d_2, \dots, d_n\}$. Each dimension d_i is organized as a hierarchy H_i, which is an ordered set of dimension levels. Each dimension level is used as a criterion by which raw data are grouped for aggregation. The Cartesian product of all dimension hierarchies $\Gamma=H_1 \times H_2 \times \dots \times H_n$, can be represented as a cube lattice [6], which is a directed graph whose nodes represent cuboids that aggregate data over the attributes present in those nodes and edges express the interdependencies among nodes. In practice, in order to reduce on-line query execution time, some of the nodes in Γ have been materialized.

Usually, a user or analyst can search for interesting patterns in a data cube by specifying a number of OLAP operations, such as drill-down, roll-up, slice, and dice. While these tools are available to help the user explore the data, the discovery process is not automated. It is the user who, following her own intuition or hypotheses, tries to recognize exceptions or anomalies in the data. This form of manual exploration of data can get tedious and error-prone for large datasets that is commonly appear in real-life [5].

Discovery-driven exploration is an alternative approach that automatically detects and marks the exceptions for the user with visual cues. Intuitively, an exception is a data cube cell that is significantly different from the value anticipated, based on a statistical model. The model considers variations and patterns in the measure value across all of the dimensions to which a cell belongs. The exception mining process consists of three phases. The first step involves the computation of the aggregate values defining the cube, such as sum or count, over which exceptions will be found. The second phase consists of model fitting, in which the coefficients of the statistical model are determined and used to compute the standardized residuals. The Third phase computes the exception indicators and uses them to identify data anomalies. The last two phases can be overlapped with the first phase, therefore the computation of data cubes for discovery-driven exploration can be done efficiently.

However, since the data is too large and it is hard to materialize the whole data cube with high dimensionality due to the limitations of both space and time, exploratory mining on complete cube cells is still too expensive. In this study, we will investigate the optimization methods by pushing several constraints into the mining process.

Let M be the set of all materialized nodes in Γ. Assume that the base-cuboid (the fact table) is always materialized and included in M. As mentioned in section 1, the specification of a constraint-based exploratory mining requires three constraints: a data constraint C_{data}, a level constraint C_{lev}, and an exception constraint C_{exc}. Data constraints can be specified by condition-based data filtering, slicing, or dicing of the data cube. Level constraints specify the degree to which the data set can be drilled-down. These two kinds of constraints determine the multidimensional space of a "*con-*

strained-cube" which is composed of those cells that satisfy the condition $C_{data} \cap C_{lev}$=true. Such a cell is called a *constrained cell.* Exception constraints C_{exc} give the user the flexibility to choose the criteria to define the exception to be mined from the "constrained-cube".

Given the set of all materialized cuboids M, a data constraint C_{data}, a level constraint C_{lev}, and an exception constraint C_{exc}, the *constraint-based exploratory mining problem* is to find all anomalous cells such that $C_{data} \cap C_{lev} \cap C_{exc}$=true, and at the same time the total response time for mining these cells must be minimized.

Example 1 Consider a data warehouse schema that consists of three dimensions: *Time, Store, and Product.* Dimension *Time* has a hierarchy of three levels: *day→month→year*. Dimension *Store* has a hierarchy of three levels: *state→country→continent*. The schema includes a single measure: *Sales,* which describes the sales value of particular product sold in a particular store on a particular date.

Suppose the data constraint is $C_{data} \equiv$ (country = "United States"). This means the mining task is focused on the sales of United States. All cells with the store attribute not belong to United States are regarded as irrelevant to the mine task. Suppose the level constraint is $C_{lev} \equiv$ (time = 2, store = 1, product = 1). The number denotes the lowest level to which a dimension should be drilled down. With these two constraints, the multidimensional space is restricted to a small constrained-cube that consists of three dimensions: time, store, and product. Time is a hierarchy that has the levels year, month, and store is a hierarchy that has the levels state, country (it is meaningless to aggregate to continent level due to the data constraint). Thus, the number of cuboids degrease from 32 to 18. Let exception constraint $C_{exc} \equiv (r \geq 2.5)$. r is the standardized residual defined as $r = |y-y@|/\sigma$, y is the actual cell value, y@ is the anticipated cell value, and σ is the standard deviation [5]. The constraint-based exploratory mining problem is to find all cells in the constrained-cube where the standardized residual is greater or equal to 2.5.

If the data cube is completely materialized, to construct the above constrained-cube is relatively simple since it only needs to retrieve those cells that satisfy the constraints. Unfortunately, as mentioned in section 1, the number of cells is too huge to be materialized and stored. Thus we assume that some selected cuboids are materialized, and it is our work to use them to construct the constrained-cube efficiently and then detect all exceptions on it.

3 Constraint-Based Exploratory Mining

The constraint-based exploratory mining approach consists of two logical phases:

1. The first phase is to specify the data on which mining is to be performed. The set of task-relevant data can be collected via applying the data constraint and the level constraint on a data cube. The data collection results in a constrained-cube. The construction process of the constrained-cube can be seen as queries on the source cube and all cells on which exception will be found are computed.

2. The second phase is to identify exceptions (defined by the exception constraint C_{exc}) on the constrained-cube through a statistical method.

The second phase can be overlapped with the first phase [5]. So it is very important to develop good algorithms to construct the constrained-cube efficiently. In the following, we will first introduce our constrained-cube construction algorithms and then explain how to identify exceptions on the constrained-cube.

3.1 Constrained-Cube Construction Algorithms

After applying the data and level constraint, we get the definition of a constrained-cube. Now, our task is to construct this constrained-cube as quickly as possible. We use the set of all cuboids CID to represent the constrained-cube C and assume the materialized cuboid set is M. We will develop algorithms to find a materialized cuboid $m \in M$ (we call it a input) for every cuboid $cid \in CID$ (we call it a target) from which that cid can be answered and the total response time for answering all cids must be minimized.

Computing a cuboid in CID consists of two phases: the first phase scans the input materialized cuboid. The second phase uses the values from the input and the dimensional information to aggregate the result. So the response time for computing a cuboid consists of two parts: the time of loading necessary data and the time of computing result cells.

The load time for each materialized cuboid $m \in M$ is:

$$t_{load}(m) = \frac{size(m)}{pagesize} \times t_{I/O} \tag{1}$$

where size(m) is the number of cells of cuboid m, pagesize is the number of cells of a disk page, $t_{I/O}$ is the transfer time of a single page between the memory and the disk.

The computation time for each cuboid $cid \in CID$ is:

$$t_{compute}(cid, m) = t_{map}(m) + size(m) \times t_{cpu}(cid, m) \tag{2}$$

in which the cells of cid can be computed using the cells in m (we note it as $cid \prec m$), $m \in M$, $t_{map}(m)$ is the sum of time for mapping all dimension information from one dimension level to another , $t_{cpu}(cid, m)$ is the time to aggregate the target using the values from the input and the dimensional information.

Hence, the cost of computing a cuboid $cid \in CID$ using a materialized cuboid $m \in M$ is:

$$\cos t(cid, m) = t_{load}(m) + t_{compute}(cid, m) \tag{3}$$

Therefore, given a cuboid set CID and a materialized cuboid set M, the total time for computing all cuboids in CID is:

$$T_{total}(CID, M) = \sum_{cid \in CID, m \in M} \cos t(cid, m) \tag{4}$$

Now, the problem is to minimize the above expression. It is akin to the minimal Steiner tree problem and is NP-hard [16]. To obtain an efficient solution within an acceptable time, we develop two practical approximate algorithms to construct the constraint-cube efficiently.

3.1.1 The NAIVE Algorithm

The most natural and straightforward way to compute the constrained-cube from materialized cuboids is to pick the best input independently for each target. The limitation of this method is that it cannot create the sharing of input if each individual target uses a different input. For example, suppose we need to compute cuboids $A^2B^3C^3$ and $A^3B^2C^2$ (A^2 means the second level of A dimension, A^3 means the third level of A dimension, and so on), the materialized cuboids include $A^2B^2C^3$ and $A^2B^2C^2$. Naturally, We select $A^2B^2C^3$ to compute $A^2B^3C^3$ and $A^2B^2C^2$ to compute $A^3B^2C^2$. But if both $A^2B^3C^3$ and $A^3B^2C^2$ select $A^2B^2C^2$ as their input, though $A^2B^3C^3$ uses a sub-optimal input, the total response time may be reduced since they can share scan in the computation. This motivates us to find a global optimal method to compute all cuboids in CID simultaneously.

3.1.2 The AGOA Algorithm

The basic idea of the Approximate Global Optimal Algorithm (AGOA) is to create more opportunities for global sharing. It explores two optimization techniques: 1) share-sort to reduce CPU time and I/O time, 2) share-scan to reduce I/O time. In order to share sort, cuboids, which have the same prefixes, are organized as a tree. All nodes at a tree can be computed simultaneously. In order to share scan, AGOA adds the target to the global plan one by one according to the number of its usable inputs. The heuristic used here is to first meet the requirements of those targets that have fewer chances. The larger the number of a target's usable inputs, the more likely it can share the input with other targets.

First, all cuboids in CID are organized as trees. Construct an auxiliary graph $G_1=(V_1, E_1)$. V_1 is the set of all cuboids in CID, i.e., $V_1=$ CID. There is a directed edge $<i, j> \in E_1$ from $i \in V_1$ to $j \in V_1$ if j can be computed from i without sorting i. In this step multiple cuboids are organized into several trees by using the share-sort optimization and each tree is a set of cuboids all of which can be computed in a single scan of the input. We use X to represent those vertices whose input degrees equal to zero in G_1. Obviously, the problem now is to select a proper input for each cuboid in X.

Second, construct another auxiliary graph $G_2=(V_2, E_2)$. At first, V_2 is equal to X. For each $m \in M$, if at least one vertex $x \in X$ can be answered from it, add m to G_2. For each vertex $x \in X$, if x can be answered by m then add a directed edge $<m, x>$ (from m to x) to G_2. We represent the set of m, $m \in M$, in G_2 as Y. The object now is to find a subset $Y< \subseteq Y$ in G_2 such that every vertex $x \in X$ chooses a vertex $y \in$ $Y<$ which can be used to answer x and $\sum_{y \in Y'} t_{load}(y) + \sum_{x \in X, y \in Y', x \prec y} t_{compute}(x, y)$ is minimized.

Third, all targets in G_2 are sorted by their input degree. At each iteration, the target with the smallest input degree is picked and the best usable input is chosen based on a cost metric. The program continues until all targets are processed.

During the optimization of multiple cuboids, we also maintain two sets like [17]. The shared input set contains inputs shared by the previously processed targets. The unused input set includes all inputs that are not used by any chosen target. When the algorithm picks a new target cid and adds it to the global plan, it chooses the best input m from the shared input set or the unused input set. It calculates the cost of the best query plan in each set and picks the most efficient plan from each set.

The cost metric is defined as follows:

$$\cos t(cid,m) = \begin{cases} t_{compute}(cid,m) & if \quad m \in SharedInputSet \\ t_{load}(m) + t_{compute}(cid,m) & if \quad m \in UnusedInputSet \end{cases}$$

Thus we have the best query plan P_s which uses a shared input s and P_u which uses a unused input u. if $P_s > P_u$, the algorithm adds the target cid to the global plan and update the two sets by adding the input m to the shared input set and deleting it from the unused input set. Otherwise, it only needs to add the target cid to the global plan.

Now, we present the AGOA algorithm in greater detail.

Algorithm 1. (Outline of the AGOA algorithm)

Input: A cuboid set CID and a materialized cuboid set M

Output: A constraint-cube construction plan PLAN

Begin

1. Organize all cuboids in CID as trees, generate the first auxiliary graph G_1
2. Construct the second auxiliary graph G_2
3. $P = \emptyset$; SharedInputSet = $\emptyset$; UnusedInputSet = M;
4. **for** (i =1; i $\le$ MaxInputDegreeInG_2; i++)
5.   **while** there exists a target cid in G_2 whose input degree equals to i **do**
6.   Let m be the usable input for cid which incurred the most minimum cost
7.   **if** m$\in$ SharedInputSet
8.   **then** $P = P \cup cid \cup$ {the edge from m to cid}
9.   **else**
10. $P = P \cup cid \cup m \cup$ {the edge from m to cid}
11.   SharedInputSet = SharedInputSet $\cup$ m
12.   UnusedInputSet = UnusedInputSet – m
13.   **end if**
14.   **end while**
15. **end for**
16. Integrate P and G_1, generate the constraint-cube construction plan PLAN
17. Output PLAN and stop

End

The complexity of the AGOA algorithm is polynomial. In the worst case, the selection of the most beneficial input is executed in O(|M|) time, the while-loop is executed O(|CID|) times, and the for-loop is executed O(|M|) times. Hence, the complexity of the algorithm is $O(|M|^2|CID|)$.

Example 2 Let's examine how to construct the constrained-cube specified in Example 1, using AGOA algorithm. We use A, B, C to represent dimension *Time, Store, Product* respectively. The cuboids set CID is $\{A^2B^1C^1, A^3B^1C^1, A^2B^2C^1, A^3B^2C^1, A^2B^1, A^3B^1, A^2B^2, A^3B^2, A^2C^1, A^3C^1, B^1C^1, B^2C^1, A^2, A^3, B^1, B^2, C^1, all\}$. Assume the materialized cuboids set M is $\{A^1B^1C^1, A^1B^2C^1, A^1C^1\}$. After the first step we get the auxiliary graph G_1 shown in Fig. 1.

According to G_1, we can get X that is the set of all tree roots, i.e. $\{A^2B^1C^1, A^2B^2C^1, A^2C^1, B^1C^1, C^1\}$. After the second and the third steps, we get the second auxiliary graph G_2 shown in Fig. 2 in which solid lines indicate the computing paths selected by the algorithm. According to G_2, we get the best input Y<={$A^1B^1C^1$, A^1C^1} for targets in X. Finally, integrating these two graphs we obtain a global optimal plan to construct the constrained-cube specified in Example 1.

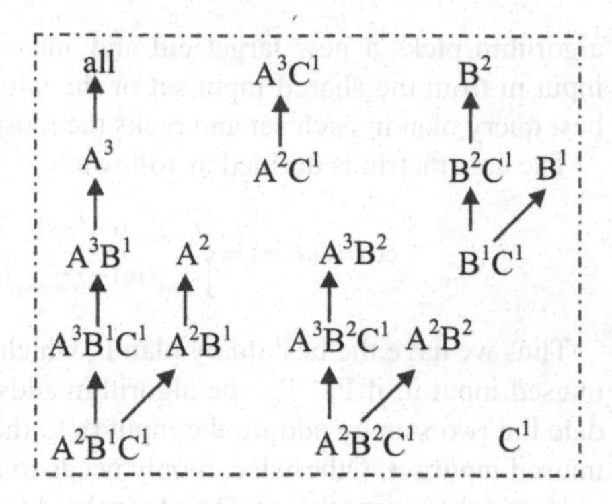

Fig. 1. The first auxiliary graph

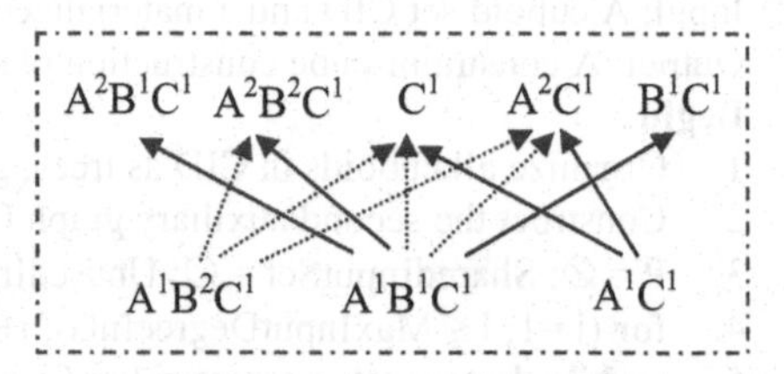

Fig. 2. The second auxiliary graph

3.2 Identify Exceptions on the Constrained-Cube

The same exception definition and computation techniques are used in this paper as in [5]. The difference is that we extend the statistical model to handle hierarchies on dimensions. At the same time, we give users the flexibility to choose the criteria of an exception by setting the exception constraint.

Assume a constrained-cube C has n dimensions, and each dimension d_i has a hierarchy that has l_i levels. For a value $y_{i_1i_2...i_n}$ in a cuboid at position i_{rj} of the j_{th} level of the r_{th} dimension (1≤r≤n, 1≤j≤l_r), we define the anticipated value $\hat{y}_{i_1i_2...i_n}$ as a function f of contributions from various higher-level cuboids as:

$$\hat{y}_{i_1i_2...i_n} = f(\gamma^G_{i_{rj}|d_{rj}\in G} \mid G \subset \{d_{rj} \mid 1 \le r \le n, 1 \le j \le l_r\}) \tag{5}$$

We will refer to the γ terms as the coefficients of the model equation. These coefficients reflect how different the values at more detailed levels are, based on generalized impression formed by looking at higher-level aggregations. The way of these coefficients are derived and the form of function f is explained in [5]

For example, consider a data cube that has two dimensions A and B. A has two levels of hierarchies: $A^1 \to A^2$ and B also has two levels $B^1 \to B^2$. To calculate an anticipated value $\hat{y}_{ij}$ for the i_{th} member of level 1 of dimension A, j_{th} member of level 1 of dimension B, the equation is:

$$\hat{y}_{ij} = f(\gamma + \gamma_i^{A^1} + \gamma_j^{B^1} + \gamma_{ij}^{A^1B^1} + \gamma_i^{A^2} + \gamma_j^{B^2} + \gamma_{ij}^{A^2B^1} + \gamma_{ij}^{A^1B^2} + \gamma_{ij}^{A^2B^2}) \tag{6}$$

The absolute difference between the actual value, $y_{i_1i_2...i_n}$ and the anticipated value $\hat{y}_{i_1i_2...i_n}$ is termed as the residual $r_{i_1i_2...i_n}$ of the model. Thus,

$$r_{i_1i_2...i_n} = | y_{i_1i_2...i_n} - \hat{y}_{i_1i_2...i_n} | \tag{7}$$

Intuitively, the larger the residual, the more the given cell value is an exception. The comparison of residual requires us to scale the values based on the expected standard deviation associated with the residuals. Thus, we call a value an exception if the standardized residual, $s_{i_1 i_2 \ldots i_n}$, defined as

$$s_{i_1 i_2 \ldots i_n} = \frac{| y_{i_1 i_2 \ldots i_n} - \hat{y}_{i_1 i_2 \ldots i_n} |}{\sigma_{i_1 i_2 \ldots i_n}} \quad (8)$$

is higher that the exception constraint C_{exc}. Then using computation techniques developed in [5], we can identify all exceptions on the constrained-cube.

4 Performance Analysis

In this section, we report our performance analysis on exploratory mining based on a constrained cube.

All experiments were conducted on a PC with an Intel Pentium III 933MHz CPU and 128M main memory, running Microsoft Window/NT. All programs were coded in Microsoft Visual C++ 6.0. The experiments were conducted on synthetic data sets generated by the APB-1 benchmark data generator. We report here results on some typical data sets, with 5 attributes (4 dimensional attributes and 1 measure attribute) and between 10,000-20,000 tuples.

Four dimensions are: Product (P), Time (T), Custom (C), and cHannel (H). Dimension P has a hierarchy $P^1 \rightarrow P^2 \rightarrow P^3 \rightarrow P^4 \rightarrow P^5 \rightarrow P^6$; Dimension T has the hierarchy $T^1 \rightarrow T^2 \rightarrow T^3$; Dimension C has a hierarchy $C^1 \rightarrow C^2$; Dimension H has the hierarchy H^1. The single measure is *Sales*. We assume the materialized cuboids set is $\{P^1T^1C^1H^1, P^2T^2C^2H^1, P^3T^2H^1, T^2C^2H^1, T^1C^2\}$ and the disk transfer rate is 1.5 MB/sec as in [10]. A disk–page is 8-kbytes long and each tuple will take 256 bytes.

Fig. 3 shows the scalability of two algorithms, NAIVE and AGOA, with respect to the number of constrained cells. We set various data constraint and level constraint on the data cube, the number of constrained cells varies from 1,000 to 10,000 approximately, and the exception constraint is 2.5. When the number of constrained cells is small, both algorithms have similar performance. However, as the number of constrained cells increases, the optimizing power of the AGOA algorithm takes effect. It takes all tasks into account simultaneously and keeps the total runtime low.

Fig. 4 shows the scalability of both algorithms with respect to various numbers of tuples varying from 10,000 to 20,000. We fix all level constraints to 1 and do not set any data constraint. The exception constraint is still 2.5. One can see that in this case the constrained-cube is exactly the source cube. The AGOA algorithm achieves good scalability and global efficiency by picking the best input for all targets whereas the NAIVE algorithm considers each target independently thus just gets local profit and requires more time. This experiment simultaneously shows that even without any data and level constraints setting on the data cube our algorithms can perform efficiently since it reduces the total execution time by sharing scan and computation.

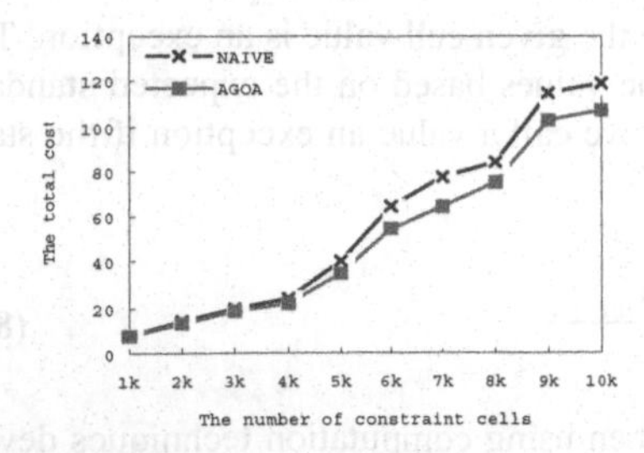

Fig.3. Scalability over number of constrained cells

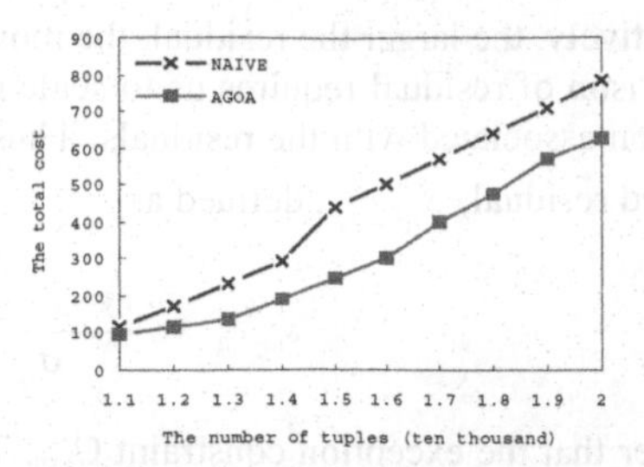

Fig.4. Scalability over number of tuples

5 Related Works

The closest work related to our study is the discovery-driven exploration of OLAP data cubes by S.Sarawagi et al [5]. It computes anticipated value for a cell using the neighborhood values, and a cell is considered an exception if its value is significantly different from its anticipated value. It automates the tedious discovery work for the analysts and takes the OLAP tools to a new stage of intelligent interactive analysis. Exploratory mining can also have constraints that restrict the multidimensional space to a small subset of the source cube. Thus adding user-defined constraints will actually lead to more efficient processing.

There are also a few other studies on efficient exploration of interesting cells in data cubes or interesting rules in multi-dimensional space. [7] proposes the cube-grade problem that studies how changes in a set of measures of interest are associated with the changes in the underlying characteristics of sectors. [8, 9, 10] presented several new automatic operators in order to enhance multidimensional database systems with advanced mining primitives.

Our study is also closely related to constraint-based data mining methods, such as [11, 12, 13, 14, 15]. This study can be considered as an integration of both intelligent exploratory mining and constraint-based mining mechanisms towards efficient multi-dimensional data analysis.

6 Conclusion

In this paper, we have studied issues and methods on effectively exploring large OLAP data cubes. The key disadvantage of mining exceptions on the whole data cube is that it needs to construct the data cube dynamically. Since the data is too large to be completely pre-computed, mining on the entire data cube will take a very long time. To make mining more efficient, we show that it is necessary to introduce three kinds of constraints: data constraints, level constraints, and exception constraints. These constraints restrict the multidimensional space to a small constrained-cube. As the user typically navigates the data cube top-down, this "divide and conquer" strategy helps the system reach a real on-line performance.

Two algorithms have been developed to construct the constrained-cube efficiently. The AGOA algorithm performs better since it selects the best set of materialized cuboids as inputs and achieves "maximal" sharing among different tasks. Our performance study shows that this kind of constraint-based exploratory mining is efficient and scalable. However, there are many interesting issues that call for further studies, such as the method to help the user set constraints, the development of more efficient mining algorithms, and so on.

References

1 J. Han, S. Chee, and J. Chiang.: Issues for On-Line Analytical Mining of Data Warehouses. In the Proc. of 1998 SIGMOD'96 Workshop on Research Issues on Data Mining and Knowledge Discovery (DMKD), 1998

2 Cognos Software Corporation. Power Play 5, special edition. http://www.cognos.com/powercubes/index.html, 1997

3 Pilot Software. Decision support suite. http://www.pilotsw.com/.

4 J. Han and Y. Fu.: Discovery of multiple-level association rules from large databases. In Proc. of the 21st Int'l Conference on Very Large Databases(VLDB), 1995.

5 S. Sarawagi, R. Agrawal, and N. Megiddo.: Discovery-driven exploration of OLAP data cubes. In proc. of the 6th Int'l Conference on Extending Database Technology (EDBT), 1998

6 V. Harinarayan, A. Rajaraman and J.: Ullman. Implementing data cubes efficiently, In *Proc. Of ACM-SIGMOD Int'l Conference on Management of Data,* 1996

7 T. Imielinski, L.Khachiyan, and A.Abdulghani.: Cubegrades: generalizing association rules. Technique Report, Dept. Computer Science, Rutgers Univ., Aug. 2000

8 S. Sarawagi.: Explaining differences in multidimensional aggregates. In Proc. of the 25st Int'l Conference on Very Large Databases (VLDB), 1999.

9 S. Sarawagi.: User-adaptive exploration of multidimensional data. In Proc. of the 26st Int'l Conference on Very Large Databases (VLDB), 2000.

10 G. Sathe, S. Sarawagi.: Intelligent Rollups in Multidimensional OLAP data. In Proc. of the 27st Int'l Conference on Very Large Databases (VLDB), 2001

11 G. Dong, J. Han, J. lam, and K. wang.: Mining Multi-Dimensional Constrained Gradients in Data Cubes, In Proc. of the 27st Int'l Conference on Very Large Databases (VLDB), 2001

12 R. Bayardo, R. Agrawal, and D. Gunopulos.: Constraint-based rule mining on large, dense data sets. In Proc. of 1999 Int'l Conf. on Data Engineering (ICDE), 1999.

13 R. Ng, L.Lakshmanan, J. han, and A. Pang.: Exploratory mining and pruning optimizations of constrained association rules. In Proc. of ACM-SIGMOD Int'l Conference on Management of Data, 1998.

14 J. Pei, J. Han, and L. Lakshmanan.: Mining frequent itemsets with convertible constraints. In Proc. of 2001 Int'l Conf. on Data Engineering (ICDE), 2001.

15 R. Srikant, Q. Vu, and R. Agrawal.: Mining association rules with item constraints. In Proc. 1997 Int'l Conf. on Data Mining and Knowledge Discovery (KDD), 1997

16 W. Liang, M. E. Orlowska, and J.X.Yu.: Optimizing multiple dimensional queries simultaneously in multidimensional databases, VLDB Journal, 8(3-4), 2000

17 Y. Zhao, P. Deshpande, J. Naughton, and A. Shukla.: Simultaneous optimization and evaluation of multiple dimensional queries, In Proc. of ACM-SIGMOD Int'l Conference on Management of Data, 1998

Efficient Utilization of Materialized Views in a Data Warehouse

Don-Lin Yang, Man-Lin Huang, and Ming-Chuan Hung

Department of Information Engineering
Feng Chia University, Taichung, Taiwan 407
dlyang@fcu.edu.tw, mlhuang@fcu.edu.tw, mchong@fcu.edu.tw

Abstract. View Materialization is an effective method to increase query efficiency in a data warehouse. However, one encounters the problem of space insufficiency if all possible views are materialized in advance. Reducing query time by means of selecting a proper set of materialized views with a lower cost is crucial for efficient data warehousing. In addition, the costs of data warehouse creation, query, and maintenance have to be taken into account while views are materialized. The purpose of this research is to select a proper set of materialized views under the storage and cost constraints and to help speedup the entire data warehousing process. We propose a cost model for data warehouse query and maintenance along with an efficient view selection algorithm, which uses the gain and loss indices. The main contribution of our paper is to speedup the selection process of materialized views. The second one is to reduce the total cost of data warehouse query and maintenance.

Keywords: Data warehouse, OLAP, materialized view, gain index, and loss index.

1 Introduction

A data warehousing system is not resource-oriented but subject-oriented, which is arranged upon users' view and can be conveniently visualized via on-line analytical processing (OLAP) tools. It is opposite to an on-line transaction processing (OLTP) system that focuses on transaction throughput and maintaining consistency. A data warehouse (DW) serves as a large warehouse containing data from local, remote, and heterogeneous sources. In addition to better utilization of storage space, researchers are concerned about the problem of view materialization [1]. Materialization of views is strongly related to query effectiveness in a DW system. However, storage limitation prohibits the materialization of all possible views. In this research, we look for optimal query effectiveness on the premise that source data might change over time and the total cost of DW query and maintenance is minimized.

As indicated before, if available space is sufficient, we would like to generate all possible views in advance to speed up DW query process. For practical information services, one must also allow periodical updates (i.e., additions) to the DW for data validity and correctness. This will increase the maintenance cost of the materialized views. We should consider the following constraints when dealing with optimal

M.-S. Chen, P.S. Yu, and B. Liu (Eds.): PAKDD 2002, LNAI 2336, pp. 393-404, 2002.

utilization of materialized views: (1) available storage space, (2) query processing efficiency, and (3) costs of DW query and maintenance.

In this paper, we use the vector data structure to determine the relationship of lattice and develop an algorithm to materialize an optimal pool of shared views with the lowest cost. There are three characteristics in this algorithm. First, "indices of gain and loss" and "candidate view" are introduced in the optimization of adding and deleting materialized views. Second, in addition to query cost, maintenance cost is also taken into account. Third, we improve the query process by adopting a vector data configuration to speedup the lattice operation in searching dependent views.

The outline of our paper is as follows. Section 2 discusses materialized views and reviews some related work. In Section 3, we present the architectural design of our algorithm. Section 4 describes the cost model used in our algorithm. In section 5, we detail our algorithm for materialized view selection. Section 6 discusses the experiment results. We conclude the paper in Section 7 and sketch some ideas for future work.

2 Literature Review

Selection of optimal materialized views is an NP-complete problem. Since 1995, researchers have been interested in materialized views. However, they were in an all or nothing mode for materializing views. In 1996, Harinarayan proposed the "greedy algorithm" [11], in which one materializes the top view that other views depend on. Then new views are added according to the benefit from the query efficiency of materialization in steps. In the same year, Ross [8] considered the use of additional views to reduce maintenance cost.

Unfortunately, if storage space is short of containing all the views that are dependent by other views, then greedy algorithm has the worst result. In 1997, Chen [2] used the two-phase algorithm to select materialized views. The purpose of the first phase is to reduce the space need, and the second phase is to reduce the query cost under the remaining space. Yang [7] also proposed an analysis framework for materialized views by using Multiple View Processing Plan (MVPP). The MVPP method considers the cost as the factor of view materialization. It does not take required space into account.

In 1999, Lin [9] proposed an improved solution on performance using genetic algorithms. Gupta [5] thinks storage devices are cheap. Under given maintenance cost, he developed a method to select materialized views effectively without worrying about space. Theodoratos [4] proposed the simple views and auxiliary views. The simple views are for users' queries. The auxiliary views are used in order to maintain the simple views. They used the exhaustive incremental algorithm to analyze materialized views. Since the algorithm takes too much time, authors use the r-greedy algorithm to restrict the search space and later use heuristics algorithms to delete the unneeded search space. Liang [12] devised an algorithm for finding such an auxiliary view set by exploiting information shared among the auxiliary views and materialized views themselves. This can reduce the total size of auxiliary views.

Yang and Huang [13] figured out available space for materialization, which includes the minimal space and views as well as the maximal space and views. The minimal space is for the materialized views to meet all required queries. The maximal space indicates the space in which all views are materialized. They define k as the value of (available space – minimal space) / (maximal space – minimal space). When k is not less than 0.5, all the query views will be materialized. Then less productive views are removed step by step until space limit is met. When k is less than 0.5, minimal views are materialized. Higher productive views are added until no more space is available. Although this is a good way to select materialized views, there is no cost model associated with it to minimize the total cost.

In this research, we develop a cost model and use the cutting-point position of given space between candidate space and minimal space to make better decisions. Based on the cutting-point position we can determine whether to delete some materialized views from the candidate space or to add more materialized views to the minimal space.

3 Architectural Design

3.1 Lattice Frameworks

We use the lattice framework [11] as a tool in the analysis of materialized view selection. It is because lattice frameworks are well suited for views with dependency. Here we adopt some of their symbols to define dependent views:

For queries Q_1 and Q_2, $Q_1 \subseteq Q_2$ is defined as "if only if the query Q_1 can be answered by using the result of querying Q_2." For example, in Example 1, the query of PS can also be answered by using the PSC view. That is, PS $\subseteq$ PS and PS $\subseteq$ PSC.

In a Data Cube (DC), a dependency expressed in a lattice framework should satisfy the following conditions: First, $\subseteq$ is in partial order. Second, there should have a top element or view as a base of other dependent views in a lattice.

Example 1: Assume we have sales transactions in a DW system. A DC consists of three major dimensions, parts (P), suppliers (S), and customers (C), and represents the sale of parts from suppliers to customers. Each cell of a DC contains the total sale of a part supplied by a supplier to a customer. The relationship of parts, suppliers, and customers is illustrated in Figure 1.

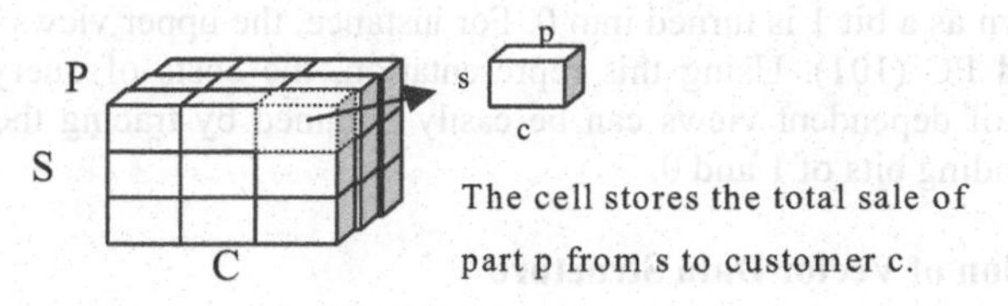

Fig. 1. The three-dimensional relationship of parts, suppliers, and customers

The lattice framework of the DC in Example 1 is illustrated in Figure 2. Dependencies of views are expressed with connection lines. For instance, the connection line from PSC to PS means the dependency PS $\subseteq$ PSC. PSC is a top view, on which other views are dependent. The numbers next to a view indicate the data amount and query frequency. For example, it shows an amount of 6M resultant data and a query frequency of 0.05 (out of 1.00) in the query of the PSC dimension.

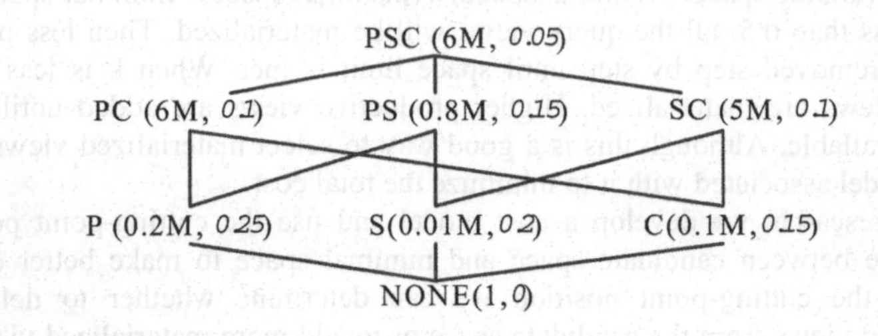

Fig. 2. Lattice framework of the data cube in Example 1

3.2 Vector Data Structure

We use a lattice to express a DC. The data representation of the lattice is expressed in vectors. For example, a lattice consisting of three dimensions has eight possible views in binary bits: 111 for PSC, 101 for PC, and so on. This is shown in Figure 3.

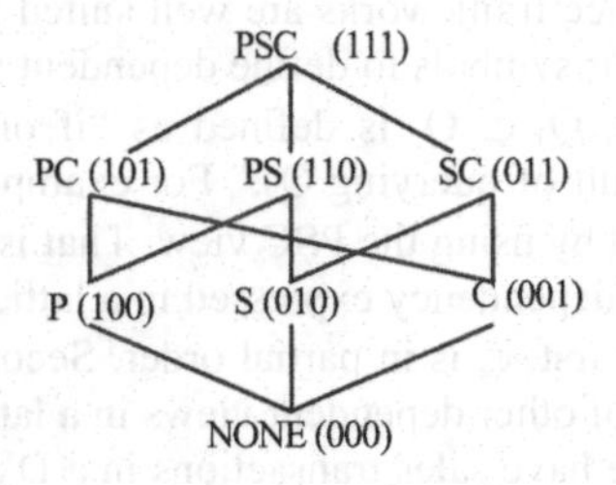

Fig. 3. Lattice representation for a data cube

Here we can reach upper levels of dependent views as a bit 0 is turned into 1. We can also drill down as a bit 1 is turned into 0. For instance, the upper views of P (100) are PS (110) and PC (101). Using this representation, the costs of query process and maintenance of dependent views can be easily obtained by tracing their hierarchies and corresponding bits of 1 and 0.

3.3 Application of Vector Data Structure

Since view materialization must be able to "answer every query," we will use vector data structure described in the last section to find dependent materialized views. For example, the materialized views we obtain from Figure 3 are PS (110), PC (101), and C (001). Assume that a query view is S (010). Its dependent views are PS (110), SC (011), and PSC (111). In this case, we can use the materialized view PS (110) to answer S (010). To improve the efficiency of query, we select views of smaller capacity for processing if the number of dependent materialized views is more that one.

4 Our Cost Model

The costs of query process and maintenance are different when various views are used in queries. The selection of a view may influence various query result in a DC. Therefore, we need to select the optimal views for materialization. In general, we generate views with better contribution to speedup query time, and remove views of less contribution to save storage space. However, the total cost and space saving must be further analyzed when two or more views have the same contributions in the view selection.

4.1 Cost Formulas

Referring to [7], we compute the total cost by using a weight with the ratio of maintenance frequency to query frequency (w = maintenance frequency / query frequency) in a given period of time. The weight is over one if maintenance is more often than query while the weight is under one if maintenance frequency is less than query, and the weight is one if both frequencies are the same. In general, w is under one because query frequency is often larger than maintenance frequency.

Assume that CS, C, and P are user's query views. There are a variety of materialization options including CPS, CS & PS, CPS & PS, or all views. Which view materialization is better if C is changed very often? The answer depends on the available storage space and total costs of query and maintenance. In this research, we intend to minimize the total cost under the limited storage availability. That is, we try to find out the desirable materialized views in which the sum of query processing and maintenance costs is minimal. Referring to the definitions in [7], our algorithm is illustrated with the following symbols in Table 1.

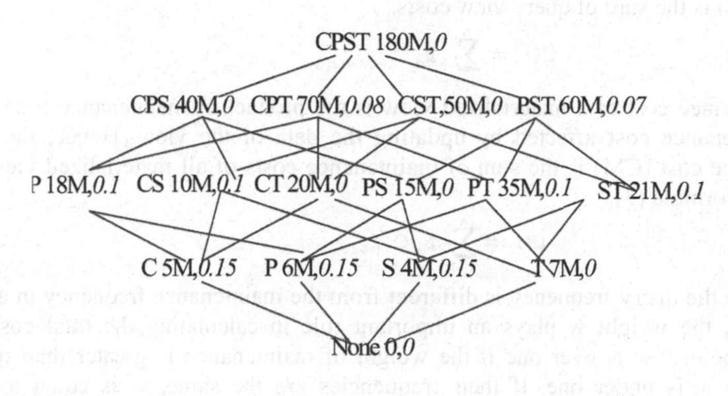

Fig. 4. A more complex lattice framework

Table 1. Definition of symbols in our cost model

Symbol	Definition
l	Subset of data source
r	Subset of view
S_i	Space of the "i"th view
fu_i	Frequency of updating the "i"th view
fq_i	Frequency of querying the "i"th view
CQ	Query cost
CM	Maintenance cost
w	Weight of maintenance and query
mv	Set of materialized views
$mv \rightarrow r_i$	The "i"th view affected by materialized view
$l \rightarrow mv_j$	The "j"th materialized view affected by data updating
S_s	Actual DW space
S_{mv}	Space of materialized views
n	Number of query views
m	Number of maintenance views
$C_{(mv \rightarrow r_i)}$	Cost of querying the "i"th query view affected by materialized view
$C_{(l \rightarrow mv_j)}$	Cost of the "j"th materialized view affected by data updating

The sum of all subsets of materialized views (S_i) equals the total materialized views (S_{mv}) as shown in Formula (1).

$$S_{mv} = \sum_{i \in mv} S_i \tag{1}$$

The query cost of each view is the product of query frequency and materialization cost of the view or its dependent view. Therefore, the total query cost (CQ) of Formula (2) is the sum of query view costs.

$$CQ = \sum_{i=1}^{n} f_{q_i} C_{(mv \rightarrow r_i)} \tag{2}$$

Maintenance cost of a materialized view is the product of maintenance frequency and maintenance cost affected by updating the data of the view. Hence, the total maintenance cost (CM) is the sum of maintenance costs of all materialized views as shown in Formula (3).

$$CM = \sum_{j=1}^{m} f_{u_j} C_{(l \rightarrow mv_j)} \tag{3}$$

Because the query frequency is different from the maintenance frequency in a data warehouse, the weight w plays an important role in calculating the total cost. As indicated before, w is over one if the weight of maintenance is greater than query. Otherwise, w is under one. If their frequencies are the same, w is equal to one. Therefore, the total cost is shown with the weight in Formula (4).

$$Cost(mv) = CQ + w * CM = \sum_{i=1}^{n} f_{q_i} C_{(mv \rightarrow r_i)} + w * \sum_{j=1}^{m} f_{u_j} C_{(l \rightarrow mv_j)} \tag{4}$$

We can figure out the total space of materialized views by means of Formula (1) to decide if further addition or deletion of views is possible. That is, we can add views if more space is available for view materialization. On the other hand, we will delete some views to fit actual space if materialized views exceed available space.

Formula (4) is taken as an indicator to determine if materialization of a view is necessary or not. The lower the indicator is, the smaller the total cost represents. Our focus is to lower the total cost by selecting an optimal set of materialized views.

We use either the method of addition or deletion to select materialized views. For addition, the total cost of materializing a new view is calculated accordingly to become a candidate. Then we select the lowest cost from the candidates to become a materialized view. For deletion, the total cost of materializing the remaining views is calculated for each candidate view to be deleted. We also select the candidate that leads to the lowest total cost.

In many cases, we may merely select the view with a large space in which the total cost is the lowest. However, no additional views can be materialized to get a better solution. Since an optimal solution is an NP-complete problem, we need more help to improve our result. Next, we propose the "gain index" and "loss index" in the next two sections that take view space into account. Through measurement of unit space, we have a more cost-effective process of view materialization.

4.2 Gain Index

The gain index is an indicator allowing the addition of a view for materialization. In addition method for view selection, we estimate the total cost of materialization for un-materialized nodes (views) in each lattice. Using "gain index", we assess the cost-effectiveness of each node and select views with a higher gain index as candidates for materialization. The definition of a gain index is shown as follows:

Under the limitation of available space ($S_{mv+\{v\}} \leq S_s$), if the total cost is reduced in response to the increased number of materialized views (v), we call the level of gain as a "gain index". Here is its mathematic definition:

$$\{\, Cost(mv) - Cost(mv + \{v\}) \,\} / S_{\{v\}}$$

That is, to compute the difference between the initial total cost and the total cost after view addition and divide the result by the space required for additional views. In our proposed algorithm described later in Section 5, we prefer selecting the views of higher gain index for materialization to have better performance of cost reduction. Nevertheless, the views of higher gain index are not selected if required space exceeds actual space.

4.3 Loss Index

The loss index is an indicator for removing materialized view candidates in the deletion method for view selection. In our proposed algorithm in Section 5, we will estimate the total cost of materialization after deleting materialized nodes (views) from a lattice. Using "loss index", we assess the cost-effectiveness of each node and select views with a lower loss index as candidates for deletion. The definition of a loss index is shown as follows:

Under the limitation of available space ($S_{mv-\{v\}} \leq S_s$), if the total cost is increased in response to the decreased number of materialized views (v), we call the level of loss as a "loss index". Here is its mathematic definition:

$$\{\, Cost(mv - \{v\}) - Cost(mv) \,\} / S_{\{v\}}$$

That is, to calculate the difference of the total cost after deletion and initial total cost, and divide the result by the space of final views. In the proposed algorithm, we prefer views of a lower loss index for deletion because of their lower total cost.

However, the views of lower loss index are not selected if the required space after deletion exceeds actual space.

5 Our Proposed Selection Algorithm

In [13] we presented an early version of this algorithm based on the cutting-point position of any given space lying between the maximum space and minimal space. After that, according to the cutting-point position one can decide whether to move from the maximum space to delete materialized views or from the minimal space to add materialized views. Additionally, a view exceeding available space is considered in this improved algorithm. We also eliminate the problem of erroneous judgment of a mid-point.

The purpose of our proposed algorithm is to minimize the total cost by selecting an optimal set of materialized views under a limited space (S_s). Cost calculations are critical indices in our cost model, in which the total cost includes query processing cost and maintenance cost. Figure 5 depicts a simplified visual presentation of our algorithm. First, we find out the minimal space (S_{ss}) and its views (M_{ss}) for materialization. The minimal space [2] means the least materialization views to answer all possible queries. The maximal space (S_{ls}) implies the space in which all views are materialized (M_{ls}). All views in the set M_{ss} must be materialized and need not be considered in the optimization procedure. Then, we define the available space (S_{as}), which equals to S_s - S_{ss} . In other words, the available space is the difference between the given space and the minimal space.

Our goal of optimization is to fully utilize the available space. The views that do not belong to the set M_{ss} may be materialized in the optimization procedure. To speedup the optimization procedure, a view larger than available space does not need to be considered. Based on the above discussion, we define a view that should be considered in the optimization procedure as a candidate view. The candidate space (S_{cs}) is the total size of all candidate views. We defined a ratio r as the value of (candidate space) / 2. When r is less than S_{as}, all the candidate views are materialized. Then views of the lowest loss index are deleted step by step until space limit is reached. This will increase query cost and reduce maintenance cost. When r is not less than S_{as}, materialized views are minimized. Then views of the highest gain index are added progressively until no space is available. It reduces query cost and increases maintenance cost. This algorithm includes ten steps as described below.

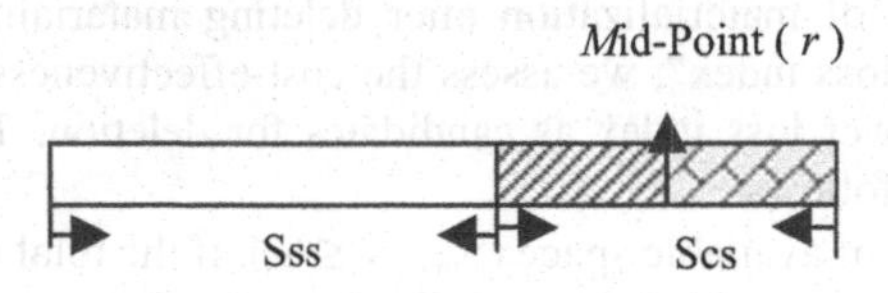

Fig. 5. Visual presentation of the proposed algorithm

Our Proposed Algorithm:

Step1: Figure out minimal materialization views (M_{ss}) , and minimal space (S_{ss}).
Step2: Figure out all materialization views (M_{ls}), and maximal space (S_{ls}).
Step3: According to the given space (S_S), we compute the available space (S_{as}= S_S- S_{ss}).

```
    If S_S ≥ S_ls then
        mv=M_ls /* Materialize all views*/
        go to Step10
    End if
    IF S_S < S_ss then
        mv=∅ /* Minimal Ss should be > S_ss*/
        go to Step10
    End if
    IF S_S = S_ss then
        mv= M_ss /*Minimal views*/
        go to Step10
    End if
```

Step4: Figure out all candidate materialization views (M_{cs}) , candidate space (S_{cs}) , and r (=S_{cs} / 2).
Step5: If r >= S_{as} then

```
        mv=M_ss
```

$$S_{mv} = \sum_{i \in mv} S_i$$

```
    else
        mv=M_cs
```

$$S_{mv} = \sum_{i \in mv} S_i$$

```
        go to Step8
    End if
```

Step6: Figure out remaining space, and candidate views. Using vector data structure to figure out cost(mv) of lattice resulted from each candidate view, and find out views of highest gain index (v).

```
    If candidate view = ∅ then
        go to Step10
    End if
```

Step7: mv = mv + {v}

$$S_{mv} = \sum_{i \in mv} S_i$$

```
    go to Step6
```

Step8: Using vector data structure to figure out cost(mv) of lattice resulted from each candidate view, and find out views of lowest loss index (v).
Step9: mv = mv – {v}

$$S_{mv} = \sum_{i \in mv} S_i$$

```
    If S_S < S_mv then
        go to Step8
    End if
```

Step10: Result in materialized view set (mv)

6 Experiment Results and Discussion

Because the two-phase algorithms [2] have better results than the greedy algorithm [11], we compare our approach with the two-phase algorithms.

Our experiments include two parts. First, without considering maintenance cost, we compare our approach with [2] and [13]. Secondly, by giving different weights of maintenance cost, we compare our approach with [13].

6.1 $W = 0$ Eexperiments

The test case is from [14]. Without considering maintenance cost, that is w = 0, we measured looping times to compare results. Assuming that views of minimal space are known, we get remaining views space (100M, 5M, 5M, 5M, 5M, 5M). The available space is the difference between given space and minimal space. We increase the available space gradually. The result of comparison with [2] and [13] is shown in Figure 6.

Selection of materialized views has a very high cost in calculating nodes of a lattice because the cyclic calculation of cost (mv) is very time-consuming. It keeps calculating until optimal materialized views are extracted. That is to say, Step 6 (or Step 8) needs to sum up cost (mv) of the entire lattice. By taking the advantage of the candidate space, we can reduce looping times in our algorithm. In the experiments, we have an average improvement rate of 57.1% if compared with [2] and [13].

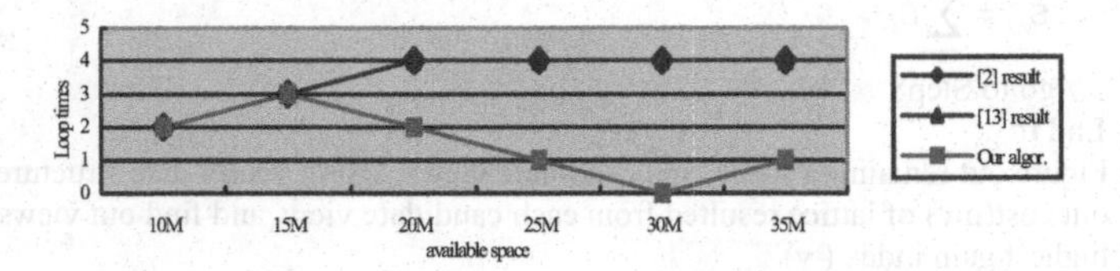

Fig. 6. Result of 6.1 experiments

6.2 $W \neq 0$ Experiments

The test case is from [13]. The details of maintenance frequency and cost data are also described in [13]. Assume that available space is 190MB. With the consideration of maintenance cost, that is w ≠ 0, we obtained the results in comparison with [2] and [13] as shown in Figure 7 and Figure 8 for two sets of experiments. The X-axis in Figure 7 and Figure 8 is the weight of maintenance cost over query.

The result reveals that our algorithm requires less cost (mv) and looping times than [2] and [13]. An average improvement of 9% (with the best improvement of 33%) better than [13] in looping times is found for the case of considering maintenance cost. And, there is an average improvement of 47.4% better than [2] in looping times. In the total cost of data warehousing, there is an average improvement rate of 25.5% better than [2].

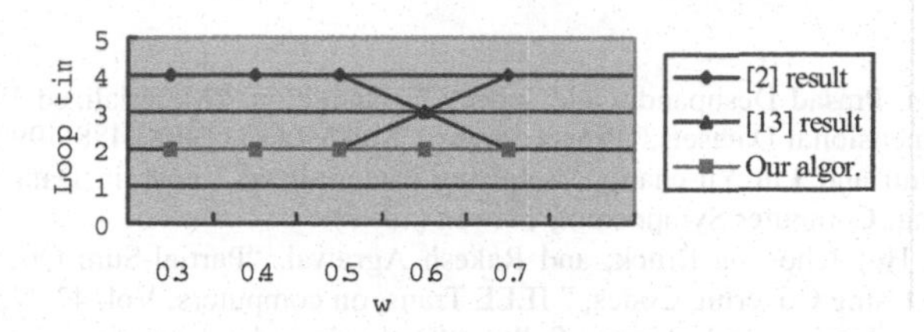

Fig. 7. Result of 6.2 experiment 1

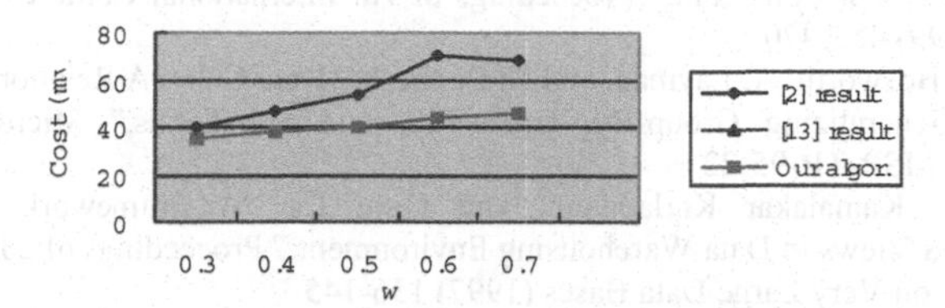

Fig. 8. Result of 6.2 experiment 2

7 Conclusions and Future Work

In this research, we introduced a cost model of query process and maintenance in efficient use of materialized views in a data warehouse. The gain index and loss index are used in selecting materialized views, and the vector data structure is employed in searching for dependency of lattice. Efficiency of view selection is greatly improved as a result.

The cost of calculating lattice nodes, cost (mv), in our proposed algorithm is the most time-consuming step. We delete views in order to decrease cost (mv) while available space is close to the required space of candidate views. In contrast, we add minimal materialized views one step at a time to accelerate selection process while available space is far less than the space required for candidate views. A significant performance improvement has been obtained in our experiments.

We are subsequently interested in adding and deleting materialized views in responding to on-line queries. The question is how the materialized views evolve to meet the new requirements with minimal impact on existing DW processing. Moreover, we need further investigations on utilization of residual space in materialized views.

In recent e-commerce episodes, huge amounts of data are generated everyday. To be able to digest and analyze discovered information, e-commerce applications need to employ efficient data warehousing and OLAP systems [3]. In the future, data warehousing and decision support systems will become indispensable application for any enterprise. To prompt the quality of effective decision support, a data warehouse supporting query optimization is an inevitable trend. Therefore, view materialization plays an important role and deserves further performance study.

References

1. Amit Shukla, Prasad Deshpande, and Jeffrey F. Naughton, "Materialized View Selection for Multidimensional Datasets," Proceedings of 24th VLDB (1998) 488-499
2. Chen Yao-hui and Liu Yu-chang, "Selecting Materialized Views in Data Warehouses," 1997 National Computer Symposium, Taiwan (1997)
3. Ching-Tien Ho, Jehoshua Bruck, and Rakesh Agrawal, "Partial-Sum Queries in OLAP Data Cubes Using Covering Codes, " IEEE Trans. on computers, Vol. 47, No. 12 (1998)
4. Dimitri Theodoratos and Timos Sellis, "Designing data warehouses, " Data and Knowledge Engineering, 31 (1999) 279-301
5. Himanshu Gupta and Inderpal Singh Mumick, "Selection of Views to Materialize Under a Maintenance Cost Constraint," Proceedings of 7th International Conference on Database Theory (1999) 453-470
6. J. Gray, A. Bosworth, A. Layman, and H. Pirahesh, "Data Cube: A Relational Aggregation Operator Generalizing Group-By, Cross-Tab, and Sub-Totals," Microsoft Technical Report No. MSR-TR-95-22
7. Jian Yang, Kamalakar Karlapalem, and Qing Li, "A Framework for Designing Materialized Views in Data Warehousing Environment," Proceedings of 23rd International Conference on Very Large Data Bases (1997) 136-145
8. Kenneth A. Ross, Divesh Srivastava, and S. Sudarshan, "Materialized View Maintenance and Integrity Constraint Checking: Trading Space for Time," Proceedings of the 1996 ACM SIGMOD International Conference on Management of Data (1996) 447-458
9. Wen-yang Lin and Yi-zhong Guo, "Configuring Data Cubes in Data Warehousing Environment with Genetic Algorithms,"1999 National Computer Symposium, Taiwan (1999)
10. Cheng-fan Qiu and Zhi-lin Lin, "A study of Selecting Materialized Views in Data Warehouses," Proceedings of the 11th Workshop of National Information Management Research (2000)
11. Venky Harinarayan, Anand Rajaraman, and Jeffrey D. Ullman, "Implementing Data Cubes Efficiently," Proceedings of the 1996 ACM SIGMOD International Conference on Management of Data (1996) 205-216
12. Weifa Liang, Hui Li, and Maria E. Orlowska, "Making multiple views self-maintainable in a data warehouse," Data and Knowledge Engineering, 30 (1999) 121-134.
13. Don-Lin Yang and Mon-Lin Huang, "Considering Maintenance Costs in the Selection of materializing Data Warehouse Views," Proceedings of The 6th Workshop of Information Management Research and Practice, Taiwan (2000)
14. http://www.iecs.fcu.edu.tw/material_view/data.htm
15. Stephen Morse and David Isaac, "Parallel Systems in the Data Warehouse," Prentice Hall, ISBN 0-13-680604-X (1998)
16. Jiawei Han and Micheline Kamber, "Data Mining: Concepts and Techniques," Morgan Kaufmann Publishers, ISBN 1-55860-489-8 (2001)
17. Jorng-Tzong Horng, Yu-Jan Chang, Baw-Jhiune Liu, and Cheng-Yan Kao, "Materialized View Selection Using Genetic Algorithms in a Data Warehouse System," Proceedings of World Congress on Evolutionary Computation(1999) 2221-2227
18. Carlos A. Hurtado, Alberto O. Mendelzon, and Alejandro A. Vaisman, "Maintaining Data Cubes under Dimension Updates," IEEE ICDE(1999) 346-355
19. Dimitri Theodoratos and Timos Sellis, "Answering Multidimensional Queries in Cubes Using Other Cubes," IEEE SSDBM'00(2000) 109-121
20. Sanjay Agrawal, Surajit Chaudhuri, and Vivek Narasayya, "Materialized view and Index Selection Tool for Microsoft SQL Server 2000," ACM SIGMOD (2001)

Mining Interesting Rules in Meningitis Data by Cooperatively Using GDT-RS and RSBR

Ning Zhong and Juzhen Dong

Department of Information Engineering
Maebashi Institute of Technology
460-1, Kamisadori-Cho, Maebashi-City, 371, Japan
zhong@maebashi-it.ac.jp

Abstract. This paper describes an application of two rough sets based systems, namely GDT-RS and RSBR respectively, for mining *if-then* rules in a meningitis dataset. GDT-RS (Generalized Distribution Table and Rough Set) is a soft hybrid induction system, and RSBR (Rough Sets with Boolean Reasoning) is used for discretization of real valued attributes as a preprocessing step realized before the GDT-RS starts. We argue that discretization of continuous valued attributes is an important pre-processing step in the rule discovery process. We illustrate the quality of rules discovered by GDT-RS is strongly affected by the result of discretization.

Keywords: Rough Sets, Meningitis Data Mining, Class Selection, Hybrid Systems.

1 Introduction

Rough set theory constitutes a sound basis for Knowledge Discovery and Data Mining. It offers useful tools to discover patterns hidden in data in many aspects [11,12,7]. It can be used in different phases of knowledge discovery process such as attribute selection, attribute extraction, data reduction, decision rule generation, and pattern extraction (templates, association rules).

This paper describes an application of two rough sets based systems, namely GDT-RS and RSBR respectively, for mining *if-then* rules in a meningitis dataset. The core of the rule discovery process is GDT-RS that is a soft hybrid induction system for discovering classification rules from databases with uncertainty and incompleteness [14,3,4]. The system is based on the combination of *Generalization Distribution Table (GDT)* and the rough set methodology. A GDT is a table in which the probabilistic relationships between concepts and instances over discrete domains are represented. The GDT provides a probabilistic basis for evaluating the strength of a rule. Furthermore, the rough set methodology is used to find minimal relative reducts from the set of rules with larger strengths.

Furthermore, in the pre-processing before using GDT-RS, a system called RSBR is used for discretization of real valued attributes. The system is based on the combination of the rough set method and Boolean reasoning proposed

M.-S. Chen, P.S. Yu, and B. Liu (Eds.): PAKDD 2002, LNAI 2336, pp. 405–416, 2002.

by Nguyen and Skowron [8,9,15,16]. A variant of the rule selection criteria in GDT-RS is used in RSBR. Thus, the process of the discretization of real valued attributes does not only mean to find the minimal relative reduct, but also considers the effect of the discretized attribute values on the performance of our induction system GDT-RS.

We argue that discretization of continuous valued attributes is an important pre-processing step in the rule discovery process. Rules induced without discretization are of low quality because they will not recognize many new objects. We illustrate the quality of rules discovered by GDT-RS is strongly affected by the result of discretization.

2 Rule Discovery by GDT-RS

GDT-RS is a soft hybrid induction system for discovering classification rules from databases with uncertain and incomplete data [14,3]. The system is based on a hybridization of *Generalization Distribution Table (GDT)* and the *Rough Set* methodology. The main features of GDT-RS are the following:

- Biases for search control can be selected in a flexible way. Background knowledge can be used as a bias to control the initiation of GDT and in the rule discovery process.
- The rule discovery process is oriented toward inducing rules with high quality of classification of unseen instances. The rule uncertainty, including the ability to predict unseen instances, can be explicitly represented by the rule strength.
- A minimal set of rules with the minimal (semi-minimal) description length, having large strength, and covering of all instances can be generated.
- Interesting rules can be induced by selecting a discovery target and class transformation.

In [14,4], we illustrated the first two features. This paper discusses the last two features of the GDT-RS.

2.1 GDT and Rule Strength

Any GDT consists of three components: *possible instances*, *possible generalizations* of instances, and *probabilistic relationships* between possible instances and possible generalizations. Here the *possible instances* are all possible combinations of attribute values in a database; the *possible generalizations* for instances are all possible cases of generalization for all possible instances; the *probabilistic relationships* between possible instances and possible generalizations, represented by entries G_{ij} of a given GDT, are defined by means of a probabilistic distribution describing the strength of the relationship between any possible instance and any possible generalization. The prior distribution is assumed to be uniform, if background knowledge is not available. Thus, it is defined by Eq. (1)

$$G_{ij} = p(PI_j|PG_i) = \begin{cases} \dfrac{1}{N_{PG_i}} & \text{if } PG_i \text{ is a generalization of } PI_j \\ 0 & \text{otherwise} \end{cases} \quad (1)$$

where PI_j is the j-th possible instance, PG_i is the i-th possible generalization, and N_{PG_i} is the number of the possible instances satisfying the i-th possible generalization, i.e.,

$$N_{PG_i} = \prod_{k \in \{l|\ PG_i[l]=*\}} n_k \quad (2)$$

where $PG_i[l]$ is the value of the l-th attribute in the possible generalization PG_i. n_k is the number of different attribute values in attribute k. "*", which specifies a wild card, denotes the generalization for instances[1]. Certainly we have $\sum_j G_{ij} = 1$ for any i.

Assuming $E = \prod_{k=1}^{m} n_k$, Eq. (1) can be rewritten in the following form:

$$G_{ij} = p(PI_j|PG_i) = \begin{cases} \dfrac{\prod_{k \in \{l|\ PG_i[l]\neq *\}} n_k}{E} & \text{if } PI_j \in PG_i \\ 0 & \text{otherwise.} \end{cases} \quad (3)$$

Let us recall some basic notions for rule discovery from databases represented by decision tables. A decision table is a tuple $T = (U, A, C, D)$, where U is a nonempty finite set of objects called the universe, A is a nonempty finite set of primitive attributes, and $C, D \subseteq A$ are two subsets of attributes that are called condition and decision attributes, respectively [11,12]. By $IND(B)$ we denote the indiscernibility relation defined by $B \subseteq A$, $[x]_{IND(B)}$ denotes the indiscernibility (equivalence) class defined by x, and U/B the set of all indiscernibility classes of $IND(B)$.

In our approach, the rules are expressed in the following form:

$$P \rightarrow Q \text{ with } S$$

i.e., "**if** P **then** Q with the strength S" where P denotes a conjunction of conditions (i.e. $P \subseteq C$), Q denotes a concept that the rule describes (i.e. $Q \subseteq D$), S is a "measure of strength" of the rule. Furthermore, S consists of three parts: $s(P)$, *accuracy* and *coverage*, where $s(P)$ is the strength of the generalization P (i.e., the condition of the rule), the *accuracy* of the rule is measured by a noise rate function: $r(P \rightarrow Q)$, *coverage* denotes how many instances are covered by the rule. If some instances covered by the rule also belong to another

[1] For simplicity, we would like to omit the wild card in some places in this paper.

class, the *coverage* is a set: {number of instances belonging to the class, number of instances belonging to another class}.

The strength of a given rule reflects the incompleteness and uncertainty in the process of rule inducing influenced both by unseen instances and noise. The strength of the generalization $P = PG$ is given by Eq. (4) under that assumption that the prior distribution is uniform

$$s(P) = \sum_l p(PI_l|P) = card([x]_{IND(P)}) \times \frac{1}{N_P} \tag{4}$$

where $card([x]_{IND(P)})$ is the number of observed instances satisfying the generalization P. The strength of the generalization P represents explicitly the prediction for unseen instances since possible instances are considered. On the other hand, the noise rate is given by Eq. (5)

$$r(P \to Q) = \frac{card([x]_{IND(P)} \cap [x]_{IND(Q)})}{card([x]_{IND(P)})} \tag{5}$$

where $card([x]_{IND(Q)})$ is the number of all instances from the class Q within the instances satisfying the generalization P. It shows the quality of classification measured by the number of instances satisfying the generalization P which cannot be classified into class Q. The user can specify an allowed noise level as a threshold value. Thus, the rule candidates with the larger noise level than a given threshold value will be deleted.

One can observe that the rule strength we are proposing is equal to its confidence [1] modified by the strength of the generalization appearing on the left hand side of the rule, and using the notation on the rough membership function [11,12].

2.2 A Searching Algorithm for Optimal Set of Rules

We now describe an idea of a searching algorithm for a set of rules developed in [3]. We use a sample database shown in Table 1 to illustrate the idea. Let T_{noise} be a threshold value.

Step 1. Create the GDT.
If prior background knowledge is not available, the prior distribution of a generalization is calculated using Eq. (1) and Eq. (2).

Step 2. Consider the indiscernibility classes with respect to the condition attribute set C (such as u_1, u_3, and u_5 in the sample database of Table 1) as one instance, called a *compound instance* (such as $u_1' = [u_1]_{IND(a,b,c)}$ in the following table). Then the probabilities of generalizations can be calculated correctly.

Table 1. A sample database

U \ A	a	b	c	d
u1	a_0	b_0	c_1	y
u2	a_0	b_1	c_1	y
u3	a_0	b_0	c_1	y
u4	a_1	b_1	c_0	n
u5	a_0	b_0	c_1	n
u6	a_0	b_2	c_1	n
u7	a_1	b_1	c_1	y

U \ A	a	b	c	d
$u_1'(u_1, u_3, u_5)$	a_0	b_0	c_1	y,y,n
u_2	a_0	b_1	c_1	y
u_4	a_1	b_1	c_0	n
u_6	a_0	b_2	c_1	n
u_7	a_1	b_1	c_1	y

Step 3. For any compound instance u' (such as the instance u_1' in the above table), let $d(u')$ be the set of the decision classes to which the instances in u' belong. Furthermore, let $X_v = \{x \in U : d(x) = v\}$ be the decision class corresponding to the decision value v. The rate r_v can be calculated by Eq. (5). If there exists a $v \in d(u')$ such that $r_v(u') = min\{r_{v'}(u')|v' \in d(u')\} < T_{noise}$ then we let the compound instance u' to point to the decision class corresponding to v. If does not any $v \in d(u')$ exist such that $r_v(u') < T_{noise}$, we treat the compound instance u' as a contradictory one, and set the decision class of u' to $\perp(uncertain)$. For example,

U \ A	a	b	c	d
$u_1(u_1, u_3, u_5)$	a_0	b_0	c_1	$\perp$

Let U' be the set of all the instances except the contradictory ones.

Step 4. Select one instance u from U'. Using the idea of discernibility matrix, create a discernibility vector (i.e., the row or the column with respect to u in the discernibility matrix) for u. For example, the discernibility vector for instance $u_2 : a_0b_1c_1$ is as follows:

U \ U	$u_1'(\perp)$	$u_2(y)$	$u_4(n)$	$u_6(n)$	$u_7(y)$
$u_2(y)$	b	$\emptyset$	a, c	b	$\emptyset$

Step 5. Compute all the so called local relative reducts for the instance u by using the discernibility function. For example, from instance u_2:$a_0b_1c_1$, we obtain two reducts $\{a, b\}$ and $\{b, c\}$:

$$f_T(\mathrm{u2}) = (b) \wedge \top \wedge (a \vee c) \wedge (b) \wedge \top$$
$$= (a \wedge b) \vee (b \wedge c).$$

Step 6. Construct rules from the local reducts for the instance u, and revise the strength of each rule using Eqs. (4) and (5). For example, the following rules are acquired

$$\{a_0b_1\} \rightarrow y \text{ with } S = 1 \times \frac{1}{2} = 0.5, \text{ and}$$

$$\{b_1c_1\} \rightarrow y \text{ with } S = 2 \times \frac{1}{2} = 1$$

for the instance u_2:$a_0b_1c_1$.

Step 7. Select the best rules from the rules (for u) obtained in *Step* 6 according to its priority [14]. For example, the rule "$\{b_1c_1\} \rightarrow y$" is selected for the instance u_2:$a_0b_1c_1$ because it matches more instances than the rule "$\{a_0b_1\} \rightarrow y$".

Step 8. $U' = U' - \{u\}$. If $U' \neq \emptyset$, then go back to *Step* 4. Otherwise, go to *Step* 9.

Step 9. If any rule selected in *Step* 7 is covering exactly one instance then STOP, otherwise, select a minimal set of rules covering all instances in the decision table.

The following table shows the result for the sample database shown in Table 1.

U	rules	s(P)	accuracy	coverage
u_1, u_3, u_5	$b_0 \rightarrow y$	0.25	0.67	{2, 1}
u_2, u_7	$b_1 \wedge c_1 \rightarrow y$	1	1	2
u_4	$c_0 \rightarrow n$	0.17	1	1
u_6	$b_2 \rightarrow n$	0.25	1	1

One can see that the discovered rule set is a minimal one having large strength and covering all instances. Furthermore, the searching algorithm can be conveniently used to discover a rule set with respect to an interesting class (or a subset of classes) selected by the user as a discovery target. Thus, by using class selection/transformation, and combining with the some preprocessing steps such as discretization, we can obtain more interesting results.

3 Discretization Based on RSBR

Discretization of continuous valued attributes is an important pre-processing step in the process for rule discovery in the databases with mixed type of data including continuous valued attributes. In order to solve the discretization issues, we have developed a discretization system called RSBR that is based on hybridization of rough sets and Boolean reasoning proposed in [8,9].

A great effort has been made (see e.g. [6,2,5,10]) to find effective methods for discretization of continuous valued attributes. We may obtain different results by using different discretization methods. The results of discretization affect directly the quality of the discovered rules. Some of discretization methods totally ignore the effect of the discretized attribute values on the performance of the induction algorithm. RSBR combines discretization of continuous valued attributes and classification together. In the process of the discretization of continuous valued attributes, we should also take into account the effect of the discretization on the performance of our induction system GDT-RS by using a variant of the rule selection criteria in GDT-RS.

Roughly speaking, the basic concepts of the discretization based on RSBR can be summarized as follows:

- Discretization of a decision table, where $V_c = [v_c, w_c)$ is an interval of real values taken by attribute c, is a searching process for a partition P_c of V_c for any $c \in C$ satisfying some optimization criteria (like minimal partition) preserving some discernibility constraints [8,9].
- Any partition of V_c is defined by a sequence of the so-called *cuts* $v_1 < v_2 < \ldots < v_k$ from V_c.
- Any family of partitions $\{P_c\}_{c \in C}$ can be identified with a set of cuts.

Table 2 shows an example of discretization. The discretization process returns a partition of the value sets of condition attributes into intervals:

$$P = \{(a, 0.9), (a, 1.5), (b, 0.75), (b, 1.5)\}.$$

Table 2. An example of discretization

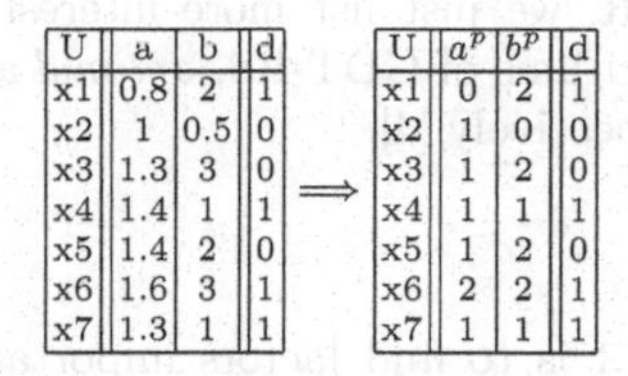

U	a	b	d
x1	0.8	2	1
x2	1	0.5	0
x3	1.3	3	0
x4	1.4	1	1
x5	1.4	2	0
x6	1.6	3	1
x7	1.3	1	1

$\Longrightarrow$

U	a^P	b^P	d
x1	0	2	1
x2	1	0	0
x3	1	2	0
x4	1	1	1
x5	1	2	0
x6	2	2	1
x7	1	1	1

4 Application in Meningitis Data Mining

This section shows the results of mining in a meningitis dataset by using cooperatively GDT-RS and RSBR. There are three main analyses [13]. These are

- diagnosis (DIAG2 and DIAG)
- detection of bacteria or virus (CULT_FIND and CULTURE) and
- predicting prognosis (COURSE and C_COURSE).

Furthermore, in each analysis, there are two attributes (e.g. DIAG2 and DIAG for Analysis 1) that can be used as the decision attribute, respectively. One of them is a multi-class attribute (e.g. DIAG = {ABSCESS, BACTERIA, BACTERIA(E), TB(E), VIRUS, VIRUS(E)}), the other one is a binary-class attribute that is obtained by grouping the corresponding multi-class attribute (e.g. DIAG2 = {BACTERIA, VIRUS}).

In the meningitis dataset, 19 of 38 attributes are continuous valued attributes that must be discretized by RSBR before rule induction by GDT-RS. Since the quality of rules discovered by GDT-RS is strongly affected by the result of discretization of continuous valued attributes, we need to do the discretization of continuous valued attributes carefully.

In this experiment, for each decision attribute with multi-class, we used two different modes of cooperatively using GDT-RS and RSBR:

1. All classes in a decision attribute are considered simultaneously when using RSBR for discretization.
2. Focus on an interesting class selected by a user as positive class (+) and other classes are considered as negative class (-). The GDT-RS and RSBR are cooperatively used to find the rules with respect to the focused positive class. After that, a class with respect to negative class is selected as a new interesting positive class, and then the RSBR and GDT-RS are cooperatively used again. Repeat this process until all interesting classes are selected as positive class.

In the following subsections, we show the results of three analyses respectively. Here, "Used" means the coverage, i.e., the number of instances covered by a rule. We just consider the rule candidates with respect to "Used $\geq$ 5". Although many reasonable rules were generated from the meningitis dataset by using GDT-RS and RSBR, we just list more interesting ones evaluated by a medical doctor. Two algorithms of GDT-RS, *optimal* and *sub-optimal* ones, are used for each analysis respectively [4].

4.1 Analysis 1

The objective of analysis 1 is to find factors important for diagnosis (DIAG2 and DIAG).

DIAG2: BACTERIA, VIRUS The following 4 of 7 reasonable rules are interesting ones in the result of using the GDT-RS optimal algorithm.

$r_1 : SEX(M) \wedge CRP(\geq 4) \rightarrow Diag2(BACTERIA)$
with coverage = 16.

$r_{2.1} : CRP(< 4) \wedge Cell\_Poly(< 221) \wedge Cell\_Mono(\geq 15) \rightarrow Diag2(VIRUS)$
with coverage = 91.

$r_{2.2} : ESR(< 16) \wedge Cell\_Poly(< 221) \wedge Cell\_Mono(\geq 15) \rightarrow Diag2(VIRUS)$
with coverage = 89.

$r_{2.3} : CRP(< 4) \wedge CT\_FIND(normal) \wedge Cell\_Poly(< 221) \rightarrow Diag2(VIRUS)$
with coverage = 80.

DIAG (Considering All Classes When Discretization) The following 7 of 36 reasonable rules are interesting ones in the result of using the GDT-RS optimal algorithm.

$r_{3.1}: AGE[35-48) \wedge Cell\_Poly[221-11000) \rightarrow Diag(BACTERIA)$
with coverage = 8.

$r_{3.2}: AGE[48-58) \wedge NAUSEA(\geq 1) \wedge Cell\_Poly[221-11000)$
$\rightarrow Diag(BACTERIA)$
with coverage = 6.

$r_{3.3}: SEX(M) \wedge STIFF(2) \wedge CT\_FIND(abnormal)$
$\wedge\ CSF\_PRO(<445) \wedge CULT\_FIND(T)$
$\rightarrow Diag(BACTERIA)$
with coverage = 5.

$r_{3.4}: COLD(<2) \wedge FOCAL(-) \wedge CRP(\geq 0.3) \wedge EEG\_WAVE(abnormal)$
$\wedge\ CSF\_GLU(<44) \rightarrow Diag(BACTERIA)$
with coverage = 5.

$r_{4.1}: SEX(F) \wedge CRP(<0.3) \wedge EEG\_FOCUS(+) \rightarrow Diag(VIRUS\_E)$
with coverage = 10.

$r_{4.2}: SEX(F) \wedge FEVER(\geq 5) \wedge KERNIG(0) \wedge FOCAL(+) \rightarrow Diag(VIRUS\_E)$
with coverage = 10.

$r_{4.3}: SEX(F) \wedge STIFF(2) \wedge EEG\_FOCUS(+) \rightarrow Diag(VIRUS\_E)$
with coverage = 10.

DIAG (Focusing on an Interesting Class) The following 3 of 28 reasonable rules are interesting ones in the result of using the GDT-RS optimal algorithm.

$r_{5.1}: SEX(F) \wedge WBC(<6300) \wedge EEG\_FOCUS(+) \rightarrow Diag(VIRUS\_E)$
with coverage = 10.

$r_{5.2}: SEX(F) \wedge CRP(<0.4) \wedge EEG\_FOCUS(+) \rightarrow Diag(VIRUS\_E)$
with coverage = 10.

$r_{5.3}: SEX(F) \wedge STIFF(2) \wedge EEG\_FOCUS(+) \rightarrow Diag(VIRUS\_E)$
with coverage = 10.

4.2 Analysis 2

The objective of analysis 2 is to find factors important for detection of bacteria or virus (CULT_FIND and CULTURE).

CULT_FIND: F,T The following 6 of 17 reasonable rules are interesting ones in the result of using the GDT-RS sub-optimal algorithm.

$r_{6.1}: FEVER(\geq 7) \wedge Cell\_Mono(\geq 429) \rightarrow CULT\_FIND(F)$
with coverage = {16, 0}, accuracy = 88%.

$r_{6.2}: FEVER(\geq 7) \wedge FOCAL(+) \rightarrow CULT\_FIND(F)$
with coverage = {15, 2}, accuracy = 88%.

$r_{7.1}: COLD(\geq 9) \wedge CSF\_PRO(\geq 93) \rightarrow CULT\_FIND(T)$
with coverage = {5, 0}, accuracy = 100%.

$r_{7.2}: LOC(\geq 1) \wedge Cell\_Mono(\geq 429) \rightarrow CULT\_FIND(T)$
with coverage = {5, 2}, accuracy = 71%.

$r_{7.3}: AGE(\geq 41) \wedge FEVER[3-7) \wedge LOC\_DAT(+) \rightarrow CULT\_FIND(T)$
with coverage = {5, 2}, accuracy = 71%.

$r_{7.4}: FEVER[3-7) \wedge EEG\_FOCUS(+) \wedge CSF\_PRO(\geq 93) \rightarrow CULT\_FIND(T)$
with coverage = {4, 1}, accuracy = 80%.

CULTURE (Considering All Classes When Discretization) The following 2 of 11 reasonable rules are interesting ones in the result of using the GDT-RS sub-optimal algorithm.

$r_{8.1}: FEVER(\geq 8) \wedge BT(< 37.1) \rightarrow CULTURE(-)$
with coverage = {13, 2}, accuracy = 86%.

$r_{8.2}: FOCAL(+) \wedge CT\_FIND(normal) \rightarrow CULTURE(-)$
with coverage = {12, 2}, accuracy = 85%.

CULTURE (Focusing on an Interesting Class) The following 2 of 26 reasonable rules are interesting ones in the result of using the GDT-RS optimal algorithm.

$r_{9.1}: COLD(< 9) \wedge BT(\geq 37.1) \wedge LOC\_DAT(-)$
$\wedge\ Cell\_Mono(< 429) \rightarrow CULTURE(-)$
with coverage = 31.

$r_{9.2}: COLD(< 9) \wedge LOC\_DAT(-) \wedge Cell\_Poly(\geq 32)$
$\wedge\ CSF\_PRO(< 93) \rightarrow CULTURE(-)$
with coverage = 30.

According to a medical doctor opinion, both the rule set, $r_{8.1}, r_{8.2}$, shown in Section 4.2.2, and the rule set, $r_{9.1}, r_{9.2}$, are reasonable, but $r_{9.1}, r_{9.2}$ are much better than $r_{8.1}, r_{8.2}$.

This example shows that more interesting rules can be generated by selecting an interesting discovery target.

4.3 Analysis 3

The objective of analysis 3 is to find factors important for predicting prognosis (C_COURSE and COURSE).

COURSE: n, p The following 2 of 31 reasonable rules are interesting ones in the result of using the GDT-RS optimal algorithm.

$r_{10.1}$: $Diag2(VIRUS) \wedge HEADACHE(\geq 5) \wedge STIFF(< 3) \wedge FOCAL(+)$
$\wedge\ CSF\_GLU(< 60) \rightarrow COURSE(p)$
with coverage = 8.

$r_{10.2}$: $LOC\_DAT(+) \wedge FOCAL(+) \wedge CSF\_GLU(< 60) \wedge RISK(n) \rightarrow COURSE(p)$
with coverage = 7.

5 Conclusion

We have presented an application of two rough sets based systems, GDT-RS and RSBR, for mining *if-then* rules from a meningitis dataset. The experimental results illustrate that the quality of rules discovered by GDT-RS is strongly affected by the results of discretization of continuous valued attributes. We need to do the discretization of continuous valued attributes carefully. It is a good way that using cooperatively RSBR and GDT-RS for rule discovery in the datasets with mixed type of attributes and multi-class.

Acknowledgements

The authors would like to thank Prof. S. Tsumoto for providing the meningitis dataset and evaluating the experimental results.

References

1. Agrawal, R., Mannila, H., Srikant, R., Toivonen, H., Verkano, A. "Fast Discovery of Association Rules", Fayyad U.M., et al. (eds.) *Advances in Knowledge Discovery and Data Mining*, The MIT Press (1996) 307-328.
2. Chmielewski, M.R. and Crzymala-Busse, J.W. "Global Discretization of Attributes as Preprocessing for Machine Learning", *Proc. Thrid Inter. Workshop on Rough Sets and Soft Computing*, (1994) 294-301.
3. Dong, J.Z., Zhong, N., and Ohsuga, S. "Probabilistic Rough Induction: The GDT-RS Methodology and Algorithms", Z.W. Ras and A. Skowron (eds.) *Foundations of Intelligent Systems.* LNAI 1609, Springer (1999) 621-629.
4. Dong, J.Z., Zhong, N., and Ohsuga, S. "Rule Discovery by Probabilistic Rough Induction", *Journal of Japanese Society for Artificial Intelligence*, Vol.15, No.2 (2000) 276-286.
5. Dougherty, J, Kohavi, R., and Sahami, M. "Supervised and Unsupervised Discretization of Continuous Features", *Proc. 12th Inter. Conf. on Machine Learning* (1995) 194-202.
6. Fayyad, U.M. and Irani, K.B. "On the Handling of Continuous-Valued Attributes in Decison Tree Generation", *Machine Learning*, Vol.8 (1996) 87-102.
7. Lin, T.Y. and Cercone, N. (ed.) *Rough Sets and Data Mining: Analysis of Imprecise Data*, Kluwer (1997).

8. Nguyen, H. Son, Skowron, A. "Quantization of Real Value Attributes", P.P. Wang (ed.) *Proc Inter. Workshop on Rough Sets and Soft Computing* at Second Joint Conference on Information Sciences (JCIS'95) (1995) 34-37.
9. Nguyen, H. Son, Skowron, A. "Boolean Reasoning for Feature Extraction Problems", Z.W. Ras, A. Skowron (eds.), *Foundations of Intelligent Systems*, LNAI 1325, Springer (1997) 117-126.
10. Nguyen H. Son and Nguyen S. Hoa "Discretization Methods in Data Mining", L. Polkowski, A. Skowron (eds.) *Rough Sets in Knowledge Discovery*, Physica-Verlag (1998) 451–482.
11. Pawlak, Z. *Rough Sets, Theoretical Aspects of Reasoning about Data*, Kluwer (1991).
12. Skowron, A. and Rauszer, C. "The Discernibility Matrixes and Functions in Information Systems", R. Slowinski (ed.) *Intelligent Decision Support*, Kluwer (1992) 331-362.
13. Tsumoto, S. "The Common Medical Data Sets to Compare and Evaluate KDD Methods", *Journal of Japanese Society for Artificial Intelligence*, Vol.15, No.5 (2000) 751-758.
14. Zhong, N., Dong, J.Z., and Ohsuga, S. "Data Mining: A Probabilistic Rough Set Approach", L. Polkowski and A. Skowron (eds.) *Rough Sets in Knowledge Discovery*, Vol.2, Physica-Verlag (1998) 127-146.
15. Zhong, N., Dong, J.Z., and Ohsuga, S. "A Rough Sets Based Knowledge Discovery Process", *Proc. Fourth Asian Fuzzy Systems Symposium* (2000) 415-420.
16. Zhong, N. and Skowron, A. "Rough Sets in KDD: Tutorial Notes", *Bulletin of International Rough Set Society*, Vol. 4, No. 1/2 (2000) 7-42.

Evaluation of Techniques for Classifying Biological Sequences*

Mukund Deshpande and George Karypis

University of Minnesota, Computer Science Department, Minneapolis, MN 55455
{deshpand,karypis}@cs.umn.edu

Abstract. In recent years we have witnessed an exponential increase in the amount of biological information, either DNA or protein sequences, that has become available in public databases. This has been followed by an increased interest in developing computational techniques to automatically classify these large volumes of sequence data into various categories corresponding to either their role in the chromosomes, their structure, and/or their function. In this paper we evaluate some of the widely-used sequence classification algorithms and develop a framework for modeling sequences in a fashion so that traditional machine learning algorithms, such as support vector machines, can be applied easily. Our detailed experimental evaluation shows that the SVM-based approaches are able to achieve higher classification accuracy compared to the more traditional sequence classification algorithms such as Markov model based techniques and K-nearest neighbor based approaches.

1 Introduction

The emergence of automated high-throughput sequencing technologies has resulted in an exponential increase in the amount of DNA and protein sequences that is available in public databases. At the time of writing, GenBank [BBL+99] had over 16 billion DNA base-pairs, SWISS-PROT [BA99] contained around 100,000 protein sequences, and PIR [BGH+99] contained over 230,000 protein sequences. As the amount of sequence information continues to increase genome researchers are shifting their biological investigation to understand the function of the different genes and proteins, determine their regulatory mechanisms, and discover how the different proteins interact to form the genetic network [Dha00].

The sheer volume of existing and anticipated data makes data mining techniques the only viable approach to analyze this data within the genome of a single specie and across genomes of different species. Classification algorithms applied on sequence data can be used to gain valuable insights on the function of genes and proteins and their relations, since throughout evolution nature has reused similar building blocks, and it is well-known that strong sequence similarity often translates to functional and structural relations [Gus97]. In particular,

* This work was supported by NSF CCR-9972519, EIA-9986042, ACI-9982274, NASA NCC 21231, and Army High Performance Computing Research Center contract number DAAD19-01-2-0014.

M.-S. Chen, P.S. Yu, and B. Liu (Eds.): PAKDD 2002, LNAI 2336, pp. 417–431, 2002.

sequence classification algorithms can be used to determine whether a DNA sequence is part of a gene-coding or a non-coding region, identify the introns or exons of a eukaryotic gene sequence, predict the secondary structure of a protein, or assign a protein sequence to a specific protein family.

Over the years, algorithms based on K-nearest neighbor, Markov models, and Hidden Markov models have been used extensively for solving various sequence-based classification problems in computational biology [Gus97, DHK+98, Mou01]. The popularity of these approaches is primarily due to the fact that they can directly (or easily modified to) take into account the sequential nature present in these data sets, which is critical to the overall classification accuracy. However, there have been few attempts in trying to use some of the more traditional machine learning classification algorithms (*e.g.*, decision trees, rule-based systems, support vector machines) [KH98, LZO99, ZLM00, WZH00, KDK+01], primarily because these algorithms were thought of not being able to model the sequential nature of the data sets.

The focus of this paper is to evaluate some of the widely used sequence classification algorithms in computational biology and develop a framework to model the sequences in such a fashion so that traditional machine learning algorithms can be easily used. To this end, we focus on three classes of classifiers. The first class is based on the K-nearest neighbor algorithm that computes the similarity between sequences using optimal pairwise sequence alignments. The second class is based on Markov models. In particular, we evaluate the performance of simple Markov chains of various orders, as well as the performance of sequence classifiers that were derived from Interpolated Markov Models (IMM) [SDKW98] and Selective Markov Models (SMM) [DK01a]. The third class is based on representing each sequence as a vector in a derived feature space and then using support vector machines (SVM) [Vap98] to build a sequence classifier. The key property of the derived feature space is that it is able to capture exactly the same sequential relations as those exploited by the different Markov model based schemes. In fact, in this derived space, the various Markov model based schemes are nothing more than linear classifiers, whose decision hyperplane was derived from the maximum likelihood estimates.

We experimentally evaluate the various classification algorithms on five different datasets that cover a wide-range of biologically relevant classification problems. Our results show that the SVM-based approaches are able to achieve higher classification accuracy than that achieved by the different Markov model based techniques as well as the much more expensive K-nearest based approaches. For most of the datasets, the improvements in classification accuracy is significant ranging from 5% to 10%.

2 Sequence Classification Techniques

In this section we discuss various techniques used for classifying sequence datasets. We start off by defining some terminology that will be used throughout this paper followed by a detailed description of three classes of algorithms.

Definitions A sequence S_r is an ordered collection of symbols and is represented as $S_r = \{x_1, x_2, x_3, \ldots x_l\}$. The alphabet $\sum$ for symbols (x) is known in advance and of fixed size N. Each sequence S_r has a class label C_r associated with it. The set of class labels (C) is also fixed and known in advance. To keep the discussion of this paper simple we will assume C contains only two class labels (C_+, C_-); however, the classification techniques discussed here are not restricted to two class problems but can handle multi-class classification problems as well.

2.1 *K* Nearest Neighbor Classifier

K-nearest neighbor (KNN) classification techniques are very popular in the biological domain because of their simplicity and their ability to capture sequential constraints present in the sequences. To classify a test sequence KNN first locates K training sequences which are most similar to the test sequence. It then assigns the class label that occurs the most in those K sequences (majority function) to the test sequence. The key part of the KNN classifier is the method used for computing the similarity between two sequences.

In context of biological applications the similarity is computed by either ***global alignment score*** or ***local alignment score***. In global alignment, the sequences are aligned across their entire length, whereas in local alignment only portions of the two sequences are aligned. Since global alignment aligns complete sequences, it is able to capture position specific patterns, *i.e*, the symbols which occur at the same relative position in the two sequences. However, this score is biased by the length of the sequence and hence needs to be normalized. Local alignment score can effectively capture small subsequences of symbols which are present in two sequences but may not necessarily be at the same position. The optimal alignment between two sequences (both global and local) is computed using a dynamic programming algorithm.

2.2 Markov Chain Based Classifiers

Markov chain based techniques are extremely popular for modeling sequences because of their inherent ability to capture sequential constraints present in the data. In this section we discuss different classification techniques based on Markov chains and their variants.

Simple Markov Chain Classifier To build a Markov chain based classification model, we first partition the training sequences according to the class label associated with each sequence and then a simple Markov chain (M) is built for each of these smaller datasets. A test sequence S_r is classified by first computing the likelihood/probability of that sequence being generated by each of those Markov chains *i.e*, computing the conditional probability $P(S_r|M)$. The Markov chain which gives the highest likelihood is used and its associated class label is assigned to the test sequence. For a two class problem this is usually done by computing the log-likelihood ratio given by

$$L(S_r) = \log \frac{P(S_r|M_+)}{P(S_r|M_-)}, \tag{1}$$

where M_+ and M_- are the Markov chains corresponding to the positive and negative class, respectively. Using Equation 1 the value of C_r is C^+ for $L(S_r) \geq 0$ and C^- if $L(S_r) < 0$.

The conditional probability, $P(S_r|M)$, that the sequence is generated by the Markov chain M is given by

$$\begin{aligned} P(S_r|M) &= P(x_l, x_{l-1}, x_{l-2}, \ldots, x_1|M) \\ &= P(x_l|x_{l-1}, x_{l-2}, \ldots, x_1, M)P(x_{l-1}|x_{l-2}, \ldots, x_1, M) \cdots P(x_1|M). \end{aligned}$$

The different conditional probabilities in the above equation are calculated using the *Markov principle* that states that the symbol in a sequence depends only on its preceding symbol *i.e.*, $P(x_l|x_{l-1}, x_{l-2}, \ldots, x_1) = P(x_l|x_{l-1})$. In this case, $P(S_r|M)$ is given by

$$P(S_r|M) = P(x_l|x_{l-1}, M) \cdots P(x_1, M) = P(x_1|M) \prod_{i=2}^{l} P(x_i|x_{i-1}, M). \tag{2}$$

The Markov chain M estimates the probability of $P(x_i|x_{i-1})$ by taking the ratio of the frequency of the sequence (x_{i-1}, x_i) in the training dataset to the frequency of x_{i-1}. This conditional probability is usually referred as the *transition probability* (α_{x_{i-1},x_i}); that is, the probability of making the transition from symbol x_{i-1} to symbol x_i. Furthermore, in Markov chain terminology every symbol in the alphabet is associated with a state and the transition probability (α_{x_{i-1},x_i}) gives the probability of making a transition from state with symbol x_{i-1} to state with symbol x_i. These transition probabilities for all pairs of symbols are usually stored in a matrix referred to as the Transition Probability Matrix (TPM).

Substituting Equation 2 in Equation 1 and using the new notation for transition probabilities we get,

$$L(S_r) = \log \frac{\prod_{i=1}^{l} \alpha^+_{x_{i-1},x_i}}{\prod_{i=1}^{l} \alpha^-_{x_{i-1},x_i}} = \sum_{i=1}^{l} \log \frac{\alpha^+_{x_{i-1},x_i}}{\alpha^-_{x_{i-1},x_i}}, \tag{3}$$

where $a^+_{x_{i-1},x_i}$ and $a^-_{x_{i-1},x_i}$ are the transition probabilities of the positive and negative class, respectively. Note that to avoid the inhomogeneity in the Equation 2 due of the probability $P(x_1|M)$, it is common to add an extra symbol, *begin*, to the alphabet, which is assumed to be present at the beginning of each sequences [DEKM98]. Equation 3 contains these extra states.

Higher Order Markov Chains The Markov chain (M) we have discussed so far is referred to as the first order Markov chain, since it looks one symbol in the past to compute the transition probability. A generalization of the first order Markov chain is the kth order Markov chain in which the transition probability for a symbol x_l is computed by looking at the k preceding symbols *i.e.*, $P(x_l|x_{l-1}, x_{l-2}, \ldots, x_1) = P(x_l|x_{l-1}, x_{l-2}, \ldots, x_{l-k})$. The kth order Markov chain will have N^k states each associated with a sequence of k symbols. For example, for a second order Markov chain the transition probability of a symbol x_i, $(\alpha_{x_{i-2}x_{i-1},x_i})$, will depend on two preceding symbols x_{i-2} and x_{i-1}. Moreover, the second order Markov chain will have N^2 states and a TPM of size $N^2 \times N$.

In general, these higher order models have better classification accuracy as compared to lower order models because they are able to capture longer ordering constraints present in the dataset. However, the number of states in higher order Markov chains grows exponentially with the order of the chain. Consequently, it is harder to accurately estimate all the transition probabilities from the (often limited) training set. To overcome these problems many variants of the Markov chains have been developed. In the next section we discuss two such variants: the Interpolated Markov Models [SDKW98] and the Selective Markov Models [DK01a]. The basic idea of these approaches is to combine portions of different order Markov chains so that to improve the overall classification accuracy.

Interpolated Markov Models Interpolated Markov models (IMM) [SDKW98] address the problem of accuracy in the higher order models due to the presence of a large number of states with low accuracies. The solution that was proposed by IMMs is to build a series of Markov chains starting for the 0th order Markov chain up to the kth order, where k is user defined and fixed before hand. IMM computes the transition probability for a symbol as a linear combination of the transition probabilities of the different order Markov chains starting from 0th order to k^{th} order Markov chain. That is, $P(x_i|x_{i-1}, x_{i-2}, \ldots, x_1, IMM_k)$ is given by

$$P(x_i|x_{i-1}, x_{i-2}, \ldots, x_1, IMM_k) = \sum_{j=0}^{k} \gamma_{j,i} P(x_i|x_{i-1}, x_{i-2}, \ldots, x_1, M_j),$$

where $P(x_i|x_{i-1}, x_{i-2}, \ldots, x_1, M_j)$ is the transition probability for the jth-order Markov chain and $\gamma_{j,i}$ is a constant indicating how the particular transition probability of the jth-order model will be weighted in the overall calculation. The motivation behind this approach is that even though higher order Markov states capture longer context they have lower support. Hence, using probabilities from lower order states which have higher support helps in correcting some of the errors in estimating the transition probabilities for higher order models. Furthermore, by using the various γ parameters, IMMs can control how the various states are used while computing the transition probabilities. A number of methods for computing the various γ factors where presented in [SDKW98] that are based on the distribution of the different states in the various order models. However, the right method appears to be dataset dependent.

The IMM techniques were originally developed for the problem of predicting the next symbol in the sequence [SDKW98] and not for sequence classification. However, IMMs can be easily used for sequence classification, as well. In particular, we can build an IMM model for the positive and negative class and then classify a sequence by taking the log-likelihood ratio of their conditional probabilities. That is,

$$L(S_r) = \sum_{i=1}^{l} \log \frac{P(x_i|x_{i-1}, x_{i-2}, \ldots, x_1, IMM_k^+)}{P(x_i|x_{i-1}, x_{i-2}, \ldots, x_1, IMM_k^-)},$$

where IMM_k^+ and IMM_k^- are the interpolated Markov models for the positive and negative class, respectively.

Selective Markov Models Selective Markov models (SMM) [DK01a] is another set of techniques which address the problems associated with higher order Markov chains and like IMMs they were developed to predict the next symbol in the sequence. These techniques first build varying order Markov chains and then prune the non-discriminatory states from the higher order Markov chains. The key task in building a SMM is to decide which states are non-discriminatory so they can be pruned. A number of techniques have been proposed for deciding which states to prune [DK01a]. The simplest method uses a *frequency threshold* and prunes all the states which occur less than that frequency threshold. The most advanced method uses a validation set to estimate the error rates of the states of the different order Markov chains and uses these estimates to prune the higher-error states for each type of prediction. Once the non-discriminatory states have been pruned, the condition probability $P(x_i|x_{i-1}, x_{i-2}, \ldots, x_1, SMM_k)$ is equal to the conditional probability that corresponds to the highest order Markov chain present amongst the remaining states. Note that unlike IMMs that combine different order Markov chains, SMMs select the appropriate order Markov chain for each prediction. SMM-based techniques can be easily extended to solve the sequence classification problem, as was done for the IMM in Section 2.2. In our experiments we used the simple frequency-threshold based method to prune the infrequent occurring states of the various Markov chains. This pruning was controled by a user specified parameter δ that indicates that a particular state-transition pair is kept only if it occurs δ times more frequently than its expected frequency, when uniform distribution is assumed.

2.3 Feature Based Sequence Classification

The various algorithms that we described so far were specifically designed to take advantage of the sequential nature of the data sets. However, traditional machine learning classification algorithms such as decision trees, rule-based classifiers, and support vector machines can also be used to classify sequence data sets, provided that the sequences are first modeled into a form that is suitable to these algorithms, that take into account their sequential nature. In the rest of this section we present one such class of transformations that was motivated by analyzing the classification model used by the various Markov chain-based techniques described in the previous section.

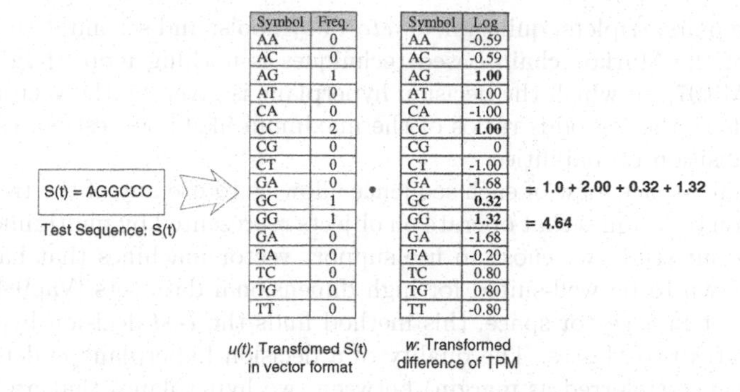

Fig. 1. Sequence classification as a dot product of training sequence vector $S(t)$ and the weight vector w obtained from the Transition probability matrices.

Recall from Equation 3, that the simple first order Markov chain will classify a particular sequence S_i by looking at the sign of

$$L(S_i) = \sum_{i=1}^{l} \log \frac{a^{+}_{x_{i-1},x_i}}{a^{-}_{x_{i-1},x_i}}.$$

If the number of distinct symbols in the different sequences is N (*e.g.*, four for nucleotides and 20 for amino-acids), then the above equation can be re-written as the dot-product of two vectors, *i.e.*,

$$L(S_i) = u^t w, \tag{4}$$

where u and w are of length N^2. Each one of the dimensions of these vectors corresponds to a unique pair of symbols. If $\alpha\beta$ is the symbol-pair corresponding to the jth dimension of these vectors, then $u(j)$ is equal to the number of times $\alpha\beta$ appears in S_i, and $w(j)$ is equal to $\log(a^{+}_{\alpha,\beta}/a^{-}_{\alpha,\beta})$. This re-write of Equation 3 is illustrated in Figure 1.

A similar transformation can be performed for all different variants of the Markov chains that were described in Section 2.2. In the case of higher order Markov chains, the dimensionality of the new space will become equal to N^{k+1}, where k is the order of the model. In the case of interpolated Markov models, the new space will have a dimensionality that is equal to $N + N^2 + \cdots N^{k+1}$, whereas in the case of selective Markov models, the dimensionality of the space will be equal to the number of states that were retained from the different order Markov chains after pruning. For higher order Markov chains and interpolated Markov models, each sequence will be represented as a frequency vector, whereas in the case of selective Markov models, the frequency along the various dimensions will be computed following the rules described in Section 2.2 for selecting the various states and transitions of the different order Markov chains.

There are two important aspects of these transformations. First, it allows us to view each sequence as a vector in a new space whose dimensions are made of all possible pairs, triplets, quintuplets, *etc*, of symbols, and second it shows that each one of the Markov chain based techniques is nothing more than a linear classifier [Mit97], in which the decision hyperplane is given by the vector w that corresponds to the log-odds ratios of the maximum likelihood estimates for the various transition probabilities.

The vector-space view of each sequence allow us to use any of the traditional classification techniques that operate on objects represented by multidimensional vectors. In our study we chose to use support vector machines that have been recently shown to be well-suited for high dimensional data sets [Vap98]. Given a training set in a vector space, this method finds the *best* decision hyperplane that separates two classes. The quality of a decision hyperplane is determined by the distance (referred as margin) between two hyperplanes that are parallel to the decision hyperplane and touch the closest data points of each class. The *best* decision hyperplane is the one with the maximum margin. The maximum margin hyperplane can be found using quadratic programming techniques. SVM extends its applicability on the linearly non-separable data sets by either using soft margin hyperplanes, or by mapping the original data vectors into a higher dimensional space in which the data points are linearly separable. The mapping to higher dimensional spaces is done using appropriate kernel functions, resulting in efficient algorithms.

3 Experimental Evaluation

We experimentally evaluate the different sequence classification schemes presented in Sections 2.1, 2.2 and 2.3 on different sequence datasets. In this section we describe the various datasets, our experimental methodology, and the performance of the different classification algorithms.

3.1 Dataset Description

The performance of the various sequence classification algorithms was evaluated on five different datasets. Each of the datasets addresses a well identified sequence classification problem in computational biology. The sequences can be divided in two broad types: nucleotide sequences which have an alphabet of size four, and amino-acid sequences which have an alphabet of size twenty. Table 1 shows some general statistics on the various datasets. As can be seen, the datasets vary a lot in both the number of sequences as well as the length of the sequences present in the dataset. The detailed description about the datasets and the classification problems can be obtained from [DK01b].

3.2 Experimental Methodology

The performance of the various classification algorithms was measured using the classification accuracy. The accuracy for each scheme is estimated using ten-way cross validation. To make the evaluation of different schemes easier we restricted

Table 1. Preliminary dataset statistics.

Nucleotide Sequences			*Amino Acid Sequences*		
Dataset	*# Sessions*	*Avg. Length*	*Dataset*	*# Sessions*	*Avg. Length*
Splice (S-EI)	1, 527	60.0	**Peptidias** (P-)	1,584	511.3
Exon	762	60.0	cysteine	416	854.3
Intron	765	60.0	metallo	580	512.6
Mouse Gen.(MG-GN)	10,918	361.6	serine	775	500.5
exon	6,593	215.6	**Protein Struct.**(PS-)	16,154	5.2
intron	4,325	584.4	Helix	4,636	9.0
Ecoli Gen.(EC-CN)	3,370	74.9	Turn	6,079	2.3
coding region	1,700	75	Strand	5,439	5.2
non-coding region	1,670	74.9			

our experiments to two class problems; however, the proposed schemes can be easily extended to handle multiple classes. Since the datasets which are used for sequence classification have different characteristics the performance of each scheme is also not uniform across the datasets. To ease comparison of different schemes we have displayed the maximum accuracy attained for every dataset (across all the classification schemes) in bold font.

3.3 Performance of the KNN Techniques

Our first set of experiments was focused on evaluating the performance of the K-nearest-neighbor algorithm. Table 2 shows the average classification accuracy achieved by the KNN algorithm for $K = 1$, $K = 5$, and $K = 20$. For each value of K, the table shows three sets of experiments each corresponding to a different similarity function. The columns labeled "Global" determine the similarity between sequences by using the score of the global sequence alignment, whereas the columns labeled "Local" use the local alignment score instead. Also, for comparison purposes, the results shown in the columns labeled "Cosine" correspond to the accuracy achieved by the KNN algorithm when each sequence is represented as a frequency vector of the different symbols that it contains, and the similarity between sequences was computed as the cosine of the two vectors. Note that this representation is similar to the feature-space of the 0th order Markov chain and does not take into account any sequential constraints. Also, in our experiments we only used the simple binary scoring matrices between the different nucleotides or amino-acids. This was primarily done to make it easier to compare the various KNN and Markov chains based methods, as both methods can take advantage of relations between different amino-acids.

Looking at these results we can see that for most datasets the scheme that uses global alignment scores as the similarity measure outperforms the other two schemes on almost all the datasets and for all the three values of K. Cosine similarity based techniques tend to perform significantly worse than the global alignment method for most datasets. The only exception are the PS-HT and PS-TS datasets for which the cosine-based similarity approaches achieve comparable classification accuracy to that achieved by global alignment. This can be attributed to the fact that the length of sequences for these datasets is

Table 2. Results of K Nearest Neighbor scheme for different similarity schemes at different values of k.

Dataset	*K = 1*			*K = 5*			*K = 20*		
	Cosine	*Global*	*Local*	*Cosine*	*Global*	*Local*	*Cosine*	*Global*	*Local*
S-EI	0.72	0.92	0.91	0.73	**0.93**	0.92	0.74	0.91	0.88
PS-HT	0.86	0.88	0.45	0.80	0.88	0.68	0.68	0.85	0.43
PS-TS	0.81	0.82	0.60	0.71	0.81	0.71	0.76	0.61	0.82
PS-HS	0.73	0.81	0.56	0.73	0.80	0.51	0.73	0.75	0.48
EC-CN	0.61	0.67	0.69	0.62	0.67	0.67	0.66	0.52	0.62
MG-GN	0.67	0.84	0.36	0.69	0.86	0.48	0.70	0.86	0.50
P-CM	0.89	0.96	0.60	0.87	0.94	0.47	0.80	0.90	0.41
P-CS	0.92	0.97	0.76	0.88	0.95	0.66	0.85	0.88	0.44
P-MS	0.87	**0.97**	0.66	0.85	0.95	0.58	0.83	0.90	0.57

extremely short (shown in Table 1). As a result, these datasets have limited sequential information which can be exploited by alignment based schemes. Also, the local alignment scheme performs very poorly especially on the protein sequences, indicating that basing the classification only on a single subsequence is not a good strategy for these datasets. Finally, note that unlike the behavior usually observed in other datasets, the classification accuracy does not improve as K increases but actually decreases. In fact, when $K = 1$, the KNN approach tends to achieve the best results in seven out of the nine experiments.

3.4 Performance of the Simple Markov Chains and Their Feature Spaces

Our second set of experiments was focused on evaluating the performance achieved by simple Markov chains as well as by SVM when operating on the same feature space used by the simple Markov chains. Table 3 shows the classification accuracy achieved by the zeroth-, first-, second-, and third-order Markov chains, and by the SVM algorithm in the corresponding feature spaces. Note that all the SVM results were obtained using linear kernel functions. The columns labeled "Markov" correspond to the results obtained by the Markov chains, and the columns labeled "SVM" correspond to the SVM results.

Table 3. Results of SVM & Markov chain based classifier at different order values.

Dataset	*Order = 0*		*Order = 1*		*Order = 2*		*Order = 3*	
	SVM	*Markov*	*SVM*	*Markov*	*SVM*	*Markov*	*SVM*	*Markov*
S-EI	0.74	0.73	0.78	0.77	0.83	0.80	0.87	0.84
PS-HT	0.87	0.74	0.87	0.74	0.95	0.92	0.85	0.81
PS-TS	0.84	0.80	0.85	0.78	0.83	0.80	0.78	0.73
PS-HS	0.75	0.72	0.80	0.74	0.84	0.80	0.68	0.65
EC-CN	0.66	0.66	0.71	0.69	0.73	0.70	0.73	0.73
MG-GN	0.69	0.64	0.81	0.72	0.89	0.79	0.91	0.81
P-CM	0.73	0.72	0.89	0.83	0.94	0.92	0.94	0.94
P-CS	0.73	0.67	0.91	0.79	0.96	0.94	0.96	0.96
P-MS	0.72	0.67	0.88	0.77	0.94	0.90	0.96	0.95

Two key observations can be made by looking at the results in this table. First, for most datasets, the classification accuracy obtained by either the Markov chains or SVM improves with the order of the model. The only exception are the experiments performed with the datasets corresponding to secondary structure prediction (PS-HT, PS-TS, and PS-HS), in which the classification accuracy for Markov chains peaks at the first-order model, and for SVM peaks at the second-order model. The reason for that is that these datasets are made of very short sequences, that depending on the particular class, range on the average from 2.3 to 9.0 nucleotides. Consequently, higher order models and their associated feature spaces contain very few examples for the accurate calculation of the transition probabilities. Second, comparing the performance of the Markov chains against the performance achieved by SVM, we can see that almost across-the-board SVM is able to achieve higher classification accuracies. In fact, for many experiments, this improvement is in the range of 5% to 10%. This should not be surprising because as discussed in Section 2.3, both Markov chains and SVM with a linear kernel function learn a linear model. In the case of Markov chains this model is learned by using a simple maximum likelihood approach, whereas in the case of SVM the model is learned so that it best separates the two classes, which is inherently more powerful.

3.5 Performance of the Interpolated Markov Models and Their Feature Spaces

The third set of experiments is focused on evaluating the performance of the interpolated Markov models and the performance of SVM when the various sequences are represented in the feature space used by these models. The classification accuracy achieved by these techniques is shown in Table 4. This table contain three sets of experiments. The experiments labeled *"Order = 1"* correspond to the IMM that combines the zeroth- and the first-order Markov chains, the ones labeled *"Order = 2"* correspond to the IMM that combines the zeroth-, first-, and second-order model, and so on. In our IMM experiments, we weighted the different order Markov chains equally. Better results can potentially be obtained, by weighting the conditional probabilities of the various order Markov chains differently. We did not pursue this any further, since as our discussion in Section 2.3 shows, this is done automatically by SVM.

From the results shown in Table 4 we can see that the SVM classifier outperforms the IMM based techniques for the majority of the datasets. The only class of datasets that IMM does better than SVM are the ones derived from peptidias (P-CM, P-CS, P-MS), in which the higher order IMM models do considerably better than the corresponding SVM models. We are correctly investigating the reason for this reduction in the accuracy of the SVM models. Second, we can see that the traditional Markov chain based classifiers presented in the previous section tend to outperform IMM based techniques on most of the datasets. The only exception to that are the protein structure datasets for which IMM derived classifiers achieve the highest accuracy amongst all the classifiers. This can be attributed to the fact that these datasets have sequences which are comparatively

Table 4. Results of IMM based classifiers on different orders of Markov chain.

Dataset	*Order = 1*		*Order = 2*		*Order = 3*	
	SVM	*Markov*	*SVM*	*Markov*	*SVM*	*Markov*
S-EI	0.77	0.75	0.80	0.78	0.82	0.81
PS-HT	0.88	0.75	**0.95**	0.93	0.86	0.84
PS-TS	**0.86**	0.81	0.84	0.83	0.78	0.77
PS-HS	0.80	0.74	**0.86**	0.84	0.72	0.72
EC-CN	0.70	0.68	0.71	0.70	0.73	0.72
MG-GN	0.81	0.69	0.88	0.73	0.89	0.76
P-CM	0.76	0.80	0.78	0.92	0.79	0.95
P-CS	0.81	0.75	0.84	0.91	0.86	**0.97**
P-MS	0.78	0.73	0.81	0.85	0.82	0.93

short hence there is a greater benefit in using different order Markov states for classification.

3.6 Performance of the Selective Markov Models and Their Features Spaces

Last set of results evaluate the performance of selective Markov models and SVM using the pruned feature space of SMM. The classification accuracies of these two schemes is presented in Table 5. The experiments were conducted at three different order values. The result in column *"Order =1"* correspond to SMM built using zeroth and the first order Markov chain which are then pruned using frequency threshold. Similarly the column *"Order = 2"* contains the result for SMM which was built using zeroth, first and second order Markov chain and so on. For each SMM we let δ take the values of 0.0, 1.0, and 2.0. The value of $\delta = 0.0$ indicates no pruning and leads to models that are identical to the simple Markov chains presented in Table 3, $\delta = 1.0$ and $\delta = 2.0$ indicate that progressively more number of states in the Markov chain are pruned.

Table 5. Results of SMM based classifiers on orders of Markov chain and varying values of δ.

Dataset	*Order = 1*						*Order = 2*						*Order=3*					
	SVM, $\delta =$			*Markov*, $\delta =$			*SVM*, $\delta =$			*Markov*, $\delta =$			*SVM*, $\delta =$			*Markov*, $\delta =$		
	0.0	1.0	2.0	0.0	1.0	2.0	0.0	1.0	2.0	0.0	1.0	2.0	0.0	1.0	2.0	0.0	1.0	2.0
S-EI	0.78	0.78	0.78	0.77	0.77	0.72	0.83	0.83	0.80	0.80	0.81	0.76	0.87	0.87	0.87	0.84	0.84	0.85
PS-HT	0.87	0.87	0.87	0.74	0.74	0.73	0.95	0.95	0.95	0.93	0.94	0.94	0.85	0.85	0.85	0.84	0.85	0.85
PS-TS	0.85	0.85	0.85	0.78	0.78	0.77	0.83	0.83	0.83	0.81	0.81	0.81	0.78	0.78	0.78	0.76	0.76	0.76
PS-HS	0.80	0.79	0.78	0.74	0.73	0.72	0.84	0.84	0.84	0.80	0.81	0.81	0.68	0.68	0.68	0.68	0.68	0.68
EC-CN	0.71	0.71	0.69	0.69	0.69	0.65	0.73	0.73	0.71	0.70	0.71	0.70	0.73	0.73	0.72	0.73	**0.73**	0.72
MG-GN	0.81	0.81	0.80	0.72	0.66	0.68	0.89	0.89	0.88	0.79	0.78	0.73	0.91	**0.91**	0.90	0.81	0.80	0.80
P-CM	0.89	0.88	0.86	0.83	0.81	0.80	0.94	0.94	0.94	0.92	0.92	0.91	0.94	0.94	0.72	0.95	**0.95**	0.94
P-CS	0.91	0.91	0.90	0.79	0.77	0.76	0.96	0.96	0.95	0.94	0.94	0.92	0.96	0.96	0.97	0.96	0.96	0.96
P-MS	0.88	0.88	0.86	0.77	0.77	0.74	0.94	0.94	0.93	0.89	0.89	0.87	**0.96**	0.96	0.74	0.95	0.95	0.94

Looking at the results of Table 5 we can see that as it was the case with both the simple Markov chains and IMMs, using SVM on SMM's feature space leads to higher classification accuracy than that obtained by SMM. Also, even though for many problems the accuracy achieved by SMM does improve as δ increases, the gains in classification accuracy are rather small. One of the reasons for this behavior may be the relatively simple strategy that we followed for pruning the various states of the different order Markov chains.

3.7 Discussion of the Results

Studying the results across the different classification schemes we can make three key observations. First, for most of the datasets, the SVM classifier used on the feature spaces of the different Markov chains and its variants achieves substantially better accuracies than the corresponding Markov chain classifier. This suggests that the linear classification modeled learned by SVM is better than the linear models learned by the Markov chain-based approaches. The classification accuracy of SVM can be improved even further if higher order kernels (*e.g.*, polynomial and Gaussian) are used. Our preliminary results with a second-order polynomial kernel suggest a considerable improvement in accuracy.

Second, the performance of the SVM classifier is influenced by the nature of the feature space on which the SVM classifier is built. This is apparent by comparing the results achieved on the IMM's feature space with the results achieved on the simple Markov chain's feature space. The maximum accuracy attained by the SVM classifier on simple Markov chains is always higher than that obtained by SVM on the IMM's feature space (the protein structure dataset is an exception because of its peculiar sequence length characteristics). This shows that increasing the feature space and in the process increasing the information available for the classifier does not necessarily improve the accuracy. At the same time, we see that even if we do even a simple frequency based feature selection, as it is done in SMM, we can achieve some improvement in the overall accuracy. These results suggest that proper feature selection can lead to improved performance, even when a powerful classifier, such as SVM, is used.

Third, we observe that the KNN scheme for the splice dataset outperforms all other classifiers by a significant margin—the nearest classifier lagging behind by approximately 6%. Our analysis of that dataset showed that the reason for this performance difference is that the KNN algorithm by computing global alignments is able to take advantage of the relative position of the aligned sequences. In fact, we performed a simple experiment in which we represented each object as a binary vector with 60×4 dimensions, in which the ith location of the sequence was mapped to the $(4i \ldots 4i + 3)$ dimensions of that vector, and based on the particular nucleotide, one of the four entries was one and the remaining zero. Given this representation, we used SVM to learn a linear model that was able to achieve an accuracy of over 97%, which is the highest ever reported for this dataset. This illustrates the benefit of incorporating information about the position of the symbols in the sequences, something that cannot currently be done with the Markov chain-based techniques described in this paper.

4 Conclusion

In this paper we presented different sequence classification techniques and evaluated their performance on variety of sequence datasets. We show that the traditional Markov chain based sequence classifier can be thought as a linear classifier operating on a feature space generated by the various state-transition pairs of the various Markov chains. This formulation of sequence classification allows use to use sophisticated linear classifiers in place of the Markov classifier leading to substantial improvements in accuracy.

References

[BA99] A. Bairoch and R. Apweiler. The SWISS-PROT protein sequence data bank and its supplement TrEMBL in 1999. *Nucleic Acids Res.*, 27(1):49–54, 1999.

[BBL+99] Dennis .A. Benson, Mark . S. Boguski, David J. Lipman, James Ostell, B. F. Francis Ouellette, BArabra A. Rapp, and David L. Wheeler. GenBank. *Nucleic Acids Research*, 27(1):12–17, 1999.

[BGH+99] W. C. Barker, J. S. Garavelli, D. H. Haft, L. T. Hunt, C. R. Marzec, B. C. Orcutt, G. Y. Srinivasarao, L.S. L. Yeh, R. S. Ledley, H.W. Mewes, F. Pfeiffer, and A. Tsugita. The PIR-International protein sequence database. *Nucleic Acids Res.*, 27(1):27–32, 1999.

[DEKM98] Richard Durbin, Sean Eddy, Anders Krogh, and Graeme Mitchinson. *Biological sequence analysis.* Cambridge University Press, 1998.

[Dha00] Ritu Dhand. *Nature Insight: Functional Genomics*, volume 405. 2000.

[DHK+98] A. L. Delcher, D. Harmon, S. Kasif, O. White, and S. L. Salzberg. Improved microbial gene identification with glimmer. *Nucleic Acid Research*, 27(23):4436–4641, 1998.

[DK01a] M. Deshpande and G. Karypis. Selective markov models for predicting web-page accesses. In *First International SIAM Conference on Data Mining*, 2001.

[DK01b] Mukund Deshpande and George Karypis. Evaluation of techniques for classifying biological sequence. Technical Report TR-01-033, University of Minnesota, 2001.

[Gus97] Dan Gusfield. *Algorithms on Strings, Trees, and Sequences.* Cambridge University Press, 1997.

[KDK+01] Michihiro Kuramochi, Mukund Deshpand, George Karypis, Qing Zhang, and Vivek Kapur. Promoter prediction for prokaryotes. In *Passific Symposium on Bioinformatics (submitted)*, 2001. Also available as a UMN-CS technical report, TR# 01-030.

[KH98] Daniel Kudenko and Haym Hirsh. Feature generation for sequence categorization. In *In proceedings of AAAI-98*, 1998.

[LZO99] Neal Lesh, Mohammed J. Zaki, and Mitsunari Ogihara. Mining features for sequence classification. In *5th ACM SIGKDD International Conference on Knowledge Discovery and Data Mining (KDD)*, 1999.

[Mit97] T.M. Mitchell. *Machine Learning.* WCB/McGraw-Hill, 1997.

[Mou01] David W. Mount. *Bioinformatics: Sequence and Genome Analysis.* CSHL Press, 2001.

[SDKW98] Steven L Salzberg, Arthur L. Delcher, Simon Kasif, and Owen White. Microbial gene identification using interpolated markov models. *Nucleic Acids Research*, 1998.

[Vap98] V. Vapnik. *Statistical Learning Theory.* John Wiley, New York, 1998.

[WZH00] K Wang, S. Zhou, and Y. He. Growing decision trees on support-less assoication rules. In *Proceedings of SIGKDD 2000*, 2000.

[ZLM00] Mohamed J. Zaki, Neal Lesh, and Ogihara Mitsunari. Planmine: Predicting plan failures using sequence mining. *Intelligence Review, special issue on the Application of Data Mining*, 2000.

Efficiently Mining Gene Expression Data via Integrated Clustering and Validation Techniques*

Vincent S. M. Tseng Ching-Pin Kao

Department of Computer Science and Information Engineering
National Cheng Kung University, Tainan, Taiwan, R.O.C.
tsengsm@mail.ncku.edu.tw

Abstract. In recent years, the *microarray* techniques have received extensive attentions due to its wide applications in biomedical industry. The main advantage of microarray technique is it allows simultaneous studies of the expressions of thousands of genes in a single experiment. Analyzing the microarray data is a challenge that arises the applications of various clustering methods used for data mining. Although a number of clustering methods have been proposed, they can not meet the requirements of automation, high quality and high efficiency at the same time in analyzing gene expression data. In this paper, we propose an automatic and efficient clustering approach for mining gene expression data produced via microarray techniques. Through performance experiments on real data sets, the proposed method is shown to achieve higher efficiency, clustering quality and automation than other clustering methods.

1 Introduction

DNA microarray techniques [3, 10] enable the possibility of examining the expressions of thousands of genes in a single experiment. An important research issue underlying microarray applications is the capability of analyzing and interpreting the obtained gene expression data [2]. Currently most researchers adopt clustering-based methods to analyze the gene expression data, with the aim to identify genes with similar expression patterns for further biological studies.

Although a number of clustering methods have been proposed [1, 4-9, 11], they incur problems in the following aspects: 1) Automation, 2) Quality, and 3) Efficiency. In the aspect of automation, most clustering algorithms request the users to set up some parameters for conducting the clustering task. However, in real applications, it is difficult for a biologist to determine the right parameters for clustering manually. Hence, an automated clustering method is required. In the aspect of quality, the existing clustering algorithms aim to produce the best clustering result based on the input parameters. Consequently, the quality of the clustering is considered only in the local view instead of a universal view. In the aspect of efficiency, the existing clustering algorithms may not perform well when the optimal or near-optimal clustering result is required from the universal view.

* This research was partially supported by National Science Council, R. O. C., under grant NSC 89-2218-E006-112.

M.-S. Chen, P.S. Yu, and B. Liu (Eds.): PAKDD 2002, LNAI 2336, pp. 432-437, 2002.

In this paper, we propose an automatic and efficient approach for mining gene expressions under multi-conditions microarray experiments. This approach integrates the density-based clustering method with the validation techniques to provide automation and accuracy for the clustering. Furthermore, an iterative computing process is utilized to reduce the computation in clustering such as to meet the efficiency requirement. Through performance experiments on real data sets, the proposed method is shown to outperform other methods in terms of efficiency and clustering quality.

The rest of the paper is organized as follows: In section 2, the proposed clustering method is introduced; Experimental results for evaluating performance of the proposed method is described in Section 3; a conclusion is made in Section 4.

2 Proposed Method

In this section, we describe first the definition for the problem of gene expression clustering, then we present the basic principles and the computation reduction technique in our approach.

2.1 Problem Definition

The problem of gene expression clustering can be described briefly as follows. Given a set of genes with unique identifiers, a vector $E_i = \{E_{i1}, E_{i2}, \ldots, E_{in}\}$ is associated with each gene i, where E_{ij} is a numerical data that represents the response of gene i under experiments j. The goal of gene expression clustering is to group together genes with similar expressions over the corresponding vectors, i.e., to classify genes such that those with similar expressions over the all experimental conditions will be put in the same cluster.

2.2 Basic Principles

The main ideas of the proposed clustering method are as follows. First, we use CAST [1] algorithm as the basic clustering method. Basically, CAST generates a clustering result based on the value of an input parameter named *affinity threshold t*, where $0 < t < 1$. An unique feature of CAST is the utilization of the *similarity matrix*, which records the similarity of genes between each other. Since the similarity matrix can be computed in advance, the overall execution of CAST is very efficient.

Secondly, we use a quality validation method to find the best clustering result. To validate the quality of the clustering result, we adopt *Hubert's Γ statistic* [6] to measure it.

Let $X=[X(i,j)]$ and $Y=[Y(i,j)]$ be two $n \times n$ matrix, $X(i,j)$ indicates the similarity of genes i and j, $Y(i,j)$ is defined as follows:

$$Y(i,j) = \begin{cases} 1 & \text{if genes } i \text{ and } j \text{ are in same cluster,} \\ 0 & \text{otherwise} \end{cases} \quad (1)$$

Hubert's Γ statistic represents the point serial correlation between the matrix X and Y, and is defined as follows:

$$\Gamma = \frac{1}{M}\sum_{i=1}^{n-1}\sum_{j=i+1}^{n}\left(\frac{X(i,j)-\overline{X}}{\sigma_X}\right)\left(\frac{Y(i,j)-\overline{Y}}{\sigma_Y}\right) \quad (2)$$

where $M = n(n-1)/2$ and Γ is between [-1, 1]. A higher value of Γ represents the better clustering quality

Therefore, it is clear that the best clustering result can be obtained by applying a number of values for the *affinity threshold t* to be input as parameters to the CAST algorithm, and choose the one with the clustering result showing the highest value of *Hubert's Γ statistic*. For example, as shown in Figure 1, the X axis represents the values of *affinity threshold t* input to CAST and the Y axis shows the obtained *Hubert's Γ statistic* for each of the clustering result. The highest peak in the curve corresponds to the best clustering result. To determine the values for the *affinity threshold t*, the easiest way is to fix the increment of the value of *affinity threshold t*. For example, we may vary the values of t from 0.05 to 0.95 in steps of 0.05. We call this approach CAST-FI (Fixed Increment) in the following.

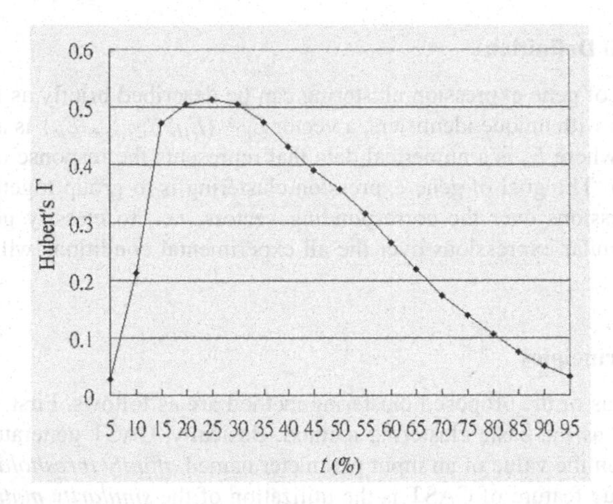

Figure 1. *Hubert's Γ statistic* vs. values of t

This approach is feasible in that CAST executes very fast once the similarity matrix of the gene expressions was obtained in advance. However, it will take a number of iterations to find the best clustering result still. In the next section, we propose an approach to reduce the computations.

2.3 Computation Reduction

The idea behind our approach is to reduce the computations by eliminating unnecessary executions of clustering such as to obtain a "nearly optimal" result instead of the optimal one. That is, we try to make the executions of CAST as less as possible.

Therefore, we need to narrow down the range of the parameter *affinity threshold t* effectively. Our approach works as follows:

1. Initially, the testing range R is set as [0, 1]. Divide R equally into m parts by points $P_1, P_2, \ldots, P_{m-1}$, where $P_1 < P_2 < \ldots < P_{m-1}$, m•3. Then, the value of each of P_i is taken as the *affinity threshold t* for executing CAST and the *Γ statistic* of the clustering result for each of P_i is calculated. We call this process a "run".
2. When a run of executing the clustering is completed, the clustering at point P_b that produces the highest *Γ statistic* is considered as the best clustering. The testing range R is replaced by the range $[P_{b-1}, P_{b+1}]$ that contains the point P_b.
3. The above process is repeated until the testing range R is smaller than a threshold δ or the difference between the maximal value and minimal values of the quality is smaller than another threshold σ.
4. The clustering result with the best quality during the tested process is output as the answer.

In this way, we can obtain the clustering result that has a "nearly optimal" clustering quality with much less computation.

3 Experimental Evaluation

To evaluate the performance of the proposed approach, we use the microarray expression data of yeast saccharomyces cerevisiae obtained from Lawrence Berkeley National Lab (LBNL) (http://rana.lbl.gov/EisenData.htm). The dataset contains the expressions of 6221 genes under 80 experimental conditions, and we call this the base dataset. Based on the base dataset, we generate two more datasets for testing. First, we choose 2000 genes from the dataset randomly as the test dataset I. The average similarity of this dataset is 0.137 by using Pearson's correlation coefficient [6] as similarity measurement. Thus we call this a low similarity dataset. Secondly, we generate dataset II with higher similarity based on dataset I. We select a large cluster from dataset I and mixed it with 100 outliers to generate the final of 2000 genes in dataset II. The obtained average similarity of dataset II is 0.696.

We compare the proposed method with the well-known clustering method, namely k-means [6]. For the proposed method, the parameters m, δ and σ are default as 4, 0.01 and 0.01, respectively. For k-means, the value of k is varied from 3 to 21 in step of 2, and from 3 to 39 in step of 2, respectively. The quality of clustering results was measured by using Hubert's Γ statistic.

The experimental results on the base dataset show that in the aspect of the execution time, our approach takes about 530 seconds which is 8 times faster than CAST-FI and 4~11 times faster than k-means; in the aspect of clustering quality, the Hubert's Γ of our approach is 0.508 which is similar to that of CAST-FI and much better than k-means (about 0.446). The above results show that our approach outperforms other methods substantially in both of clustering quality and the execution time. The experimental results on dataset I and II are described in the following sections.

3.1 Dataset I: Low Similarity Dataset

The total execution time and the best clustering quality of the tested methods on dataset I are listed in Table 1. The notation "CAST-FI" indicates the approach running CAST iteratively by setting *affinity threshold t* from 0.05 to 0.95 in fixed increment of 0.05, while the notation "Our Approach" indicates the one described in Section 2.3 using the computation reduction method.

Table 1. Experimental results (dataset I)

methods	time(sec)	#cluster	Γ statistic
Our Approach	27	57	0.514
CAST-FI	246	57	0.514
k-means (k=3~21)	404	5	0.447
k-means (k=3~39)	1092	5	0.447

Table 2. Distribution of clusters (dataset I)

cluster size / methods	1~10	11~100	101~400	401~600
Our Approach	38	15	2	2
CAST-FI	38	15	2	2
k-means (k=3~21)	0	0	5	0
k-means (k=3~39)	0	0	5	0

It is obvious that our approach and CAST-FI outperform k-means substantially in both of execution time and clustering quality. In particular, our approach performs 15 times to 40 times faster than k-means with k ranged as [3, 21] and [3, 39], respectively. The results also show that the best Γ statistic for our method is very close to that of CAST-FI, meaning that the clustering quality of our approach is as good as CAST-FI even though the computation time of our approach is reduced substantially.

Table 2 shows the distribution of the best clustering results generated by each method. It is shown that k-means produced 5 clusters sized between 101 and 400. In contrast, our approach produced 57 clusters as the best clustering result, in which 4 main clusters are generated (two sized between 101 to 400 and two sized between 401 to 600), and a number of clusters are produced with small size (1~10 and 11~100), which are mostly outliers (or noise). This means that our method is superior to k-means in filtering out the outliers from the main clusters. This can provide more accurate clustering result and insight for gene expression analysis.

3.2 Dataset II: High Similarity Dataset

The experimental results show again that our approach outperforms other methods greatly in both of clustering quality and the execution time. Specifically, in the aspect of the execution time, our approach is 3 times faster than CAST-FI and 6~20 times faster than k-means; in the aspect of clustering quality, our approach is similar to CAST-FI (about 0.833) and much better than k-means (about 0.309). Compared to the results of dataset I, the clustering quality of our method is much better than that of k-means. This means that our method is much more effective than k-means under high similarity dataset. In fact, by observing the distribution of generated clusters, we found that both of our method and CAST-FI produce a main cluster with large size and many other small clusters with outliers. This matches the real distribution of dataset II. In contrast, k-means partitions the large cluster in the original dataset into several clusters distributed more uniformly on the cluster size. Consequently, the clustering result distracts with the original data distribution. This indicates that k-means can not perform well under high similarity dataset.

4 Conclusions

In this paper, we propose an automatic and efficient clustering method for mining gene expressions in forms of microarray data under multi-conditions. The proposed method was shown to achieve higher efficiency and clustering quality than other methods under datasets with different degree of similarities. Therefore, the proposed approach is shown to provide high degree of automation, efficiency and clustering quality, which are lacked in other gene expression clustering methods.

In the future, we will try to reduce the initial range of *affinity threshold t* for executing CAST. This will significantly reduce the computation further once the correct range can be estimated initially.

References

1. Amir Ben-Dor and Zohar Yakhini, "Clustering gene expression patterns." *Proc. of the 3rd Annual Int'l Conf. on Computational Molecular Biology (RECOMB '99)*, 1999.
2. Ming-Syan Chen, Jiawei Han, and Philip S. Yu, "Data mining : An Overview from a Database Perspective." *IEEE Transactions on Knowledge and Data Engineering*,Vol. 8, No.6, December 1996.
3. J. DeRisi, et al, "Use of a cDNA microarray to analyze gene expression patterns in human cancer." *Nature Genetics* 14: 457-460, 1996.
4. Sudipto Guha, Rajeev Rastogi, and Kyuseok Shim, "CURE: An efficient clustering algorithm for large databases." *Proc. of ACM Int'l Conf. on Management of Data*, pp. 73-84, New York, 1998.
5. Sudipto Guha, Rajeev Rastogi, and Kyuseok Shim, "ROCK: a robust clustering algorithm for categorical attributes." *Proc. of the 15th Int'l Conf. on Data Eng.*, 1999.
6. Anil K. Jain and Richard C. Dubes, "Algorithms for Clustering Data." Prentice Hall, 1988.
7. Teuvo Kohonen, "The self-organizing map." *Proc. of the IEEE*, Vol. 78, No 9, pp. 1464-1480, September 1990.
8. Mark S. Aldenderfer and Roger K. Blashfield, "Cluster Analysis." Sage Publications, Inc., 1984.
9. J. B. McQueen, "Some Methods of Classification and Analysis of Multivariate Observations." *Proc. of the 5th Berkeley Symposium on Mathematical Statistics and Probability*, pp. 281-297, 1967.
10. M. Schena, D. Shalon, R. W. Davis and P. O. Brown, "Quantitative monitoring of gene expression patterns with a complementary DNA microarray." *Science* 270: 467-470, 1995.
11. Tian Zhang, Raghu Ramakrishnan, and Miron Livny, "BIRCH: An Efficient Data Clustering Method for Very Large Databases," *Proc. of the 1996 ACM SIGMOD Int'l Conf. on Management of Data*, pp. 103-114, Montreal, Canada, 1996.

Adaptive Generalized Estimation Equation with Bayes Classifier for the Job Assignment Problem

Yulan Liang, King-Ip Lin, Arpad Kelemen

Department of Mathematical Sciences
The University of Memphis
Memphis TN 38152, USA
Yuliang@memphis.edu,
linki@msci.memphis.edu,
kelemena@msci.memphis.edu

Abstract. We propose combining advanced statistical approaches with data mining techniques to build classifiers to enhance decision-making models for the job assignment problem. Adaptive Generalized Estimation Equation (AGEE) approaches with Gibbs sampling under Bayesian framework and adaptive Bayes classifiers based on the estimations of AGEE models which uses modified Naive Bayes algorithm are proposed. The proposed classifiers have several important features. Firstly, it accounts for the correlation among the outputs and the indeterministic subjective noise into the estimation of parameters. Secondly, it reduces the number of attributes used to predict the class. Moreover, it drops the assumption of independence made by the Naive Bayes classifier. We apply our techniques to the problem of assigning jobs to Navy officers, with the goal of enhancing happiness for both the Navy and the officers. The classification results were compared with nearest neighbor, Multi-Layer Perceptron and Support Vector Machine approaches.

Keywords: Job assignment, Adaptive Generalized Estimation Equation, Gibbs Sampling with Bayesian, Naive Bayes algorithm, Adaptive Bayes Classifier

1 Introduction

One crucial problem many organizations, especially the military forces, face is the job assignment problem. Typically, a service personnel is posted to an assignment for a fixed period of time (like 2-3 years) and then move on to some other job. It is crucial that the right job is assigned to the right person to achieve optimal performance. However, many, often conflicting, criteria complicate the assignment process: on one hand, jobs needed to be assigned according to the ability of the personnel; on the other hand, the needs of the assignee, like the reluctance to be posted to jobs far away from home for too long, have to be addressed also. This makes job assignment a challenging, yet crucial task. For instance, in the United States Navy, groups of experts, called detailers, are responsible for the job assignment task for sailors.

M.-S. Chen, P.S. Yu, and B. Liu (Eds.): PAKDD 2002, LNAI 2336, pp. 438-449, 2002.

Currently the job is done manually. Any technique that can automate this process can prove invaluable for the Navy's personal management unit.

The task of job assignment can be viewed as a classification problem. Thus, to automate the classification process we need to build a model. However, this task is complicated by many factors. Firstly, the various criteria, or constraints, can be soft, hard and semi-hard. Secondly, different detailers may have different views on the importance of constraints, as their views are based on the particular sailor community they handle and may change from time to time as the environment changes. Thirdly, the data is highly correlated. On one hand many attributes in the databases contain overlapping data or data that can be directly derived from others. For example, "reverse paygrade" can be derived from "paygrade"; other attributes maybe naturally correlated such as "paygrade" and "Navy Enlisted Classification codes" (trained skills). On the other hand detailers typically offer exactly one job for sailors, therefore the decisions themselves correlate too. All these included indeterminate subjective components, making the optimization of classification and prediction of the job assignment a very sophisticated task.

Given the above description, one approach is to model the problem as a constraint satisfaction problem. However, this is not preferable for this problem. Firstly, typical constraint satisfaction models such as the Gale-Shapley model or linear programming do not work well for data with high level of noise. Moreover, our model aims at being adaptive to the ever-changing environment, as well as the ever-evolving demands towards the decision maker, which should be learned and adapted to also.

In this paper, we propose Adaptive Generalized Estimation Equation (AGEE) with additive noise to model the job assignment problem [2], [4]. The AGEE models can be directly used to estimate the effects of the changes over time in covariates on the response (decision) with the correlated measurement and give us more insight in what criteria are important to decision-makers and the weight of each. To handle the noise and the outlier data, we use Gibbs sampling under Bayesian framework with some prior information [5], [12], [13]. Based on the estimations of AGEE with additive noise model, Adaptive Bayes (AB) classifiers are proposed which uses modified Naive Bayes algorithm [3]. We believe that combining advanced statistical approaches with data mining techniques through using data from human detailers to build new classifiers can enhance the performance of the job-assignment process.

The rest of the paper is divided as follows. Section 2 gives the problem definition, as well as describes the detail of our model. Section 3 presents the results of our techniques. We survey related work in section 4 and provide some concluding remarks and future direction in section 5.

2 Building the Classifier for the Job Description Problem

2.1 Problem Description

The United States Navy faces the task of assigning sailors to different assignment. Each assignment lasted for 2-5 years and then the sailor will be reassigned to another job. The Navy's Assignment Policy Management System maintains data on jobs and sailors. Experts in the Navy, called detailers, use this data to offer jobs to sailors. In assigning jobs, certain hard constraints have to be satisfied. These constraints ensure that jobs are offered to sailors with the required skill set, as well as to maintain various other Navy policies. The four constraints are:

- Sea/shore rotation: If a sailor's previous job was on shore then he/she is only eligible for jobs at sea and vice versa
- Dependents match: If a sailor has more than 3 dependents then he is not eligible for overseas jobs
- Navy Enlisted Classification (NEC) Match: The sailor must have an NEC (trained skill) that is required by the job
- Hard Paygrade Match: The sailor's paygrade can't be off by more than one from the job's required paygrade

Apart from the hard constraints, there are also soft constraints that need to be satisfied as closely as possible. Soft constraints are designed to increase sailor happiness and to satisfy Navy policies. Each constraint is represented by a function with range [0,1], with 1 being the optimal value. The functions were set up and normalized through knowledge given by Navy detailers. The four soft constraints are:

- Job Priority Match (f_1): The higher the job priority, the more important to fill the job
- Sailor Location Preference Match (f_2): It is better to send a sailor to a place he wants to go
- Soft paygrade match (f_3): The sailor's paygrade should exactly match the job's required paygrade
- Geographic Location Match (f_4): Certain moves are preferable than others

Notice that some details about the constraints are omitted. The detailed constraints are used in the experiment to construct the data set and to build the classifiers.

2.2 Overall Procedure

We obtained data from the Navy's Assignment Policy Management System's job and sailor databases and from surveys of Navy experts. The databases contain data about 467 sailors and 167 possible jobs for them. Each database has more than 100 attributes, such as sailor name, sailor paygrade, sailor location preferences, job paygrade, job location code and so on. We first pre-process and clean the data, eliminating unnecessary and superfluous attributes. We end up with 18 attributes from the sailor database and 6 from the job database. Then we employ the hard constraints to filter away the inappropriate "matches" between sailors and jobs. After that, the four functions representing the soft constraints are applied. Then the data is sent through the classification algorithm to build the model for classification. We set aside one-third of the data for testing purpose, and based on the test set we obtain the accuracy of the classifier.

2.3 Classification Model

2.3.1 Overview

We propose AGEE with Gibbs sampling under Bayesian framework with AB classifier for the job assignment problem. The similarity between AGEE with AB classifiers and SVM with RBF as model and Adatron as learning algorithm is that both methods find out the confusion region, which we defined as the misclassified region [9], [10], [11]. The training processes focus on this region using mixture Gaussian distribution to model it. The advantage of AGEE with AB classifiers is that it can identify attributes most useful for classification through GEE model selection according the Akaike Information Criteria (AIC) [8]. It also counts the correlation structure among the outputs into the estimation of the mean and standard deviation of each class. The Gibbs sampling with Bayesian framework can deal with noise and outliers (which both come from the virtually non-deterministic, subjective nature of human decision making) more efficiently and robustly through treating some output as missing value and adding prior distribution to it. This can overcome drawbacks of the survey and time delay effects on the output 'decision'. Moreover it estimates the noise through assessments of uncertainty in the posterior probabilities of belonging to classes. All these important pieces of information are partially ignored in other classification methods, which may make biased estimation and lower the classification accuracy. One assumption for our proposed approach is that the distribution for each class of data follows Gaussian distribution. From the Central Limit Theorem it can easily be shown that it holds in most cases, especially for non-deterministic data. We will show that the Gaussian distribution holds in our case using Skewness and Kurtosis statistics.

2.3.2 GEE Model for Correlated Data

We model binary outcome with correlated data with Generalized Estimation Equation model, which was introduced by Liang and Zeger [2]. The marginal expectation (average response for observations sharing the same covariates) is modeled as a logistic function of exploratory variables. Considering stationary process, the model:

$$\hat{y} = P(decision = 1 \mid w) = g(w^T f) \tag{2.3.2.1}$$

where g uses a logistic function, w is a column vector of weights to be estimated, and f is a column vector of inputs "$f_1,\ldots,f_5$", where f_5 is the sailor group ID.

The marginal mean:

$$P(decision = 1 \mid w) = \frac{\exp(w' f)}{1+\exp(w' f)} \tag{2.3.2.2}$$

Since the outcomes are binary the estimated variance from (2.3.2.1) is

$$\mathrm{var}(P(decision = 1 \mid w)) = \frac{\exp(w' f)}{(1+\exp(w' f))^2} \tag{2.3.2.3}$$

The covariance correlation of the correlated outcome on a given subject suggested by Horton [4]:

$$V_i = \phi A_i^{1/2} R(\alpha) A_i^{1/2} \quad (2.3.2.4)$$

where A_i is a diagonal matrix of the variance function and $R(\alpha)$ is the working correlation matrix of the outcome index by vector parameters α.

Because of the lack of logical ordering of the observations within groups and the unbalanced size of groups in our data set, an exchangeable structure maybe most appropriate. Different working correlation structures like independent, M-dependent, unstructured, etc. were also employed through the comparisons of the estimates and standard error to see the sensitivity of the misspecification of the variance structure. We also employed the weight function with and without interaction terms to find out whether there are confounded patterns among the exploratory variables.

To avoid a sequence of hypothesis testing, we use the AIC for the model selection.

$$AIC(K_a) = -2\log(L_{ml}) + 2K_a \quad (2.3.2.5)$$

where L_{ml} is the maximum likelihood of the model and K_a is the number of adjustable parameters.

The advantage of using AIC as proved by Stone [15] is that there is an asymptotic equivalence of the choice by cross-validation and the AIC. Therefore we can avoid the computational drawback of cross-validation and do not need to separate the data into cross-validation set when selecting the model.

Using conditional probability computed according to (2.3.2.2), we get the estimated mean and standard deviation of each class and proportion of each class such that the estimation from the GEE model can also be written as mixture Gaussian model form:

$$y = \sum_{i=1}^{2} \mathrm{p_i}\, \phi(\mu_i, \sigma_i^2) \quad (2.3.2.6)$$

where μ_i, σ_i are means and standard deviations of classes, p_i are proportions of each class.

2.3.3 Modeling the Noise and Outliers Using Gibbs Sampling Under Bayesian Framework

As proposed by Fraley and Raftery in [1], the data with noise and outliers can be handled with an additive mixture Gaussian distribution in which each component represents the noise of class.

$$\varepsilon = \sum_{i=1}^{n} \mathrm{p_i}\, \varphi(\lambda_i, \sigma_i^2) \quad (2.3.3.1)$$

For the estimation of model parameters of the Gaussian distributions we employed Gibbs sampling with Bayesian framework. The merit of this approach is that the non-deterministic subjective data (human decision) can be more efficiently and robustly estimated through treating some output as missing value and adding prior distribution to it. This can overcome drawbacks of survey and time delay effects on the output "decision". Also, using Gibbs sampling with Bayesian framework to estimate the noise enables assessing the uncertainty in the posterior probabilities of belonging to classes. Starting from the simple case of modeling our noised data, we fit a mixture of two normal distributions with common variance so that each estimated conditional probability is y_j is assumed drawn from one of the two classes and is misclassified. T_j=1,2 is the true class of the i'th observation, where class T_j has a normal distribution with mean λ_{Tj} and standard deviation σ. We assume unknown mixture coefficients P of observations are in group 2 and 1-P in group 1. The model is thus:

$$\begin{aligned} & y_j \sim \text{Normal}(\lambda_{Tj}, \sigma^2), \\ & T_j \sim \text{Categorical}(P), \\ & \lambda_2 = \lambda_1 + \theta, \theta > 0 \end{aligned} \quad (2.3.3.2)$$

λ_1, θ, σ, P are given independent "noninformative" priors, including a uniform prior for P on (0,1). A re-parameterization procedure was employed to avoid the data go to the same component through shift λ_2 with θ.

The equation (2.3.2.6) provides estimation of the parameters of two classes served as initial values for Gibbs sampling under Bayesian framework, which can speed up the convergence of Gibbs sampling. The prior distribution applied Dirichlet function.

2.3.4 Adaptive Bayes Classifiers and Algorithm for Correlated Data Combining GEE with Additive Noise Model

The test class of the sample can be separate using likelihood metric and modified Naive Bayes (NB) algorithm. The computed class of the model is the model the sample has greatest likelihood

$$class(x) = \arg_i \max(\log p(x \mid M_i)) \quad (2.3.4.1)$$

The original NB algorithm uses the assumption of independence, which assumes given the class model the value for each attributes are independent of one another, then the class of the test sample can be given according to the following form:

$$class(x) = \arg_i \max\{\sum_k \log p(x_k \mid M^k{}_i)\} \quad (2.3.4.2)$$

where k's are attributes.

The modified NB classifier is directly based on the conditional probability estimated from AGEE with addictive model and (2.3.2.6). When substituting each class for a Gaussian distribution with estimated mean μ_i and standard deviation σ_i. The test sample of the class can be decided by

$$class(x) = \arg_i \max\{-\log(\sigma_i) - 0.5((x_i - \mu_i)/\sigma_i)^2\} \quad (2.3.4.3)$$

Furthermore, if we assume equal prior probabilities for all models, the relative log probabilities between class a and class b with respect to sample x can be expressed simply as the difference between their log likelihoods. The difference between log likelihoods can be used as the confidence measure for one class over another.

$$\log p(M_a \mid x) - \log p(M_b \mid x) =$$
$$\{-\log(\sigma_a) - 0.5((x-\mu_a)/\sigma_a)^2 + \log(\sigma_b) + 0.5((x-\mu_b)/\sigma_b)^2\} \qquad (2.3.4.4)$$

where μ_a, σ_a, μ_b, σ_b are means and standard deviation of class a and class b estimated from (2.3.2.6).

In addition, the relative entropy can also be used for confidence measure of class a over class b if given different prior probabilities.

2.3.5 Algorithm Summary

The full classification algorithm is iterative. At each step, we use the current classifier to predict the class of each observation. For those observations that are misclassified, we moved them to the D_{noise} data set. Then we build a new classifier by modeling the current denoised data ($D_{denoised}$) and noise data (D_{noise}) separately.

Algorithm: AGEE with additive noise models with Adaptive Bayer classifier for classification for correlated data from the training data.
Input: Training set D from the job database
Output: Classifier C for the job assignment process.
Method:

1. **$D_{denoised} \leftarrow D$, $D_{noise} \leftarrow \varnothing$**
2. **Apply the GEE model on $D_{denoised}$ to obtain initial weight estimation. Use the AIC criteria to decide which models are most appropriate i.e. add interaction term or use different correlation structure.**
3. **Compute the mean, variance and conditional probability of each class using (2.3.2.2) and (2.3.2.3). Proportion the mean and variance such that the estimation from the AGEE model can be written as mixtural Gaussian model form:**
$$y = \sum_{i=1}^{n} p_i\, \phi(\mu_i, \sigma_i^{\,2}) \qquad (2.3.5.1)$$
where μ_i, σ_i and p_i are mean standard deviation and proportions of each class.
4. **Model D_{noise} as a mixture Gaussian distribution. Estimate the mixture coefficients and parameters by Gibbs sampling with Bayesian framework. The prior distribution uses Dirichlet function.**
5. **Use the modified Naive Bays (NB) classifier to predict the class of each observation x in $D_{denoised}$. If an observation is misclassified, move it to D_{noise}. The classifier can be written as**
$$\hat{y} = P(decision = 1 \mid w) = y + \varepsilon \qquad (2.3.5.2)$$
where ε is the model of the noise data.
6. **Use AIC criterion and lowest misclassification rate to determine whether the current classifier is satisfactory. If not, repeat step (2) – (5). The resulting model is updated AGEE with an additive noise**

This model is used for further classifications and predictions.

3 Experiment Results and Discussion

The weight estimations using AGEE with additive model are shown in table 1. The four coefficients: Job Priority Match, Sailor Location Preference Match, Paygrade Match, Geographic Location Match is 5.0002, 0.5133, 3.9618, and 4.2431 respectively. It shows that job priority, paygrade and geographic location are far more important than sailor location preference match. The best model for our case is without interaction and using exchangeable correlation structure according to the AIC criterion. Table 2 gives the estimated exchangeable correlation matrix.

Table 1. Parameter Estimates (std: Standard Deviation, CI: 95% confidence interval) with exchangeable working correlation matrix using AGEE with additive model

Parameter	Estimate	std	95% CI	Z	Pr >|Z|
Intercept	-12.1456	1.0975	-14.2967, -9.9945	-11.07	<.0001
Job priority	5.0002	0.9668	3.1053, 6.8951	5.17	<.0001
Location preference	0.5133	0.2466	0.0299, 0.9968	2.08	<0.0374
Paygrade	3.9618	0.3256	3.3236, 4.6000	12.17	<.0001
Geographic location	4.2431	0.5696	3.1266, 5.3595	7.45	<.0001

Table 2. Exchangeable correlation matrix

Parameters	Intercept	Job priority	Location preference	Paygrade	Geographic location
Intercept	1.0000	-0.7287	-0.0958	-0.5107	-0.5940
Job priority	-0.7287	1.0000	0.0861	0.0568	0.0816
Location preference	-0.0958	0.0861	1.0000	0.0870	-0.1805
Paygrade	-0.5107	0.0568	0.0870	1.0000	0.1333
Geographic location	-0.5940	0.0816	-0.1805	0.1333	1.0000

Table 3 shows the mixture mean and standard deviation estimation of two classes for denoised data from AGEE model. As we can see the statistics of Skewness and Kurtosis of two classes are smaller than 3, so the Gaussian assumption for AB classifier holds. Table 4 provides mean, standard deviation, median and Monte-Carlo simulation error with 1000 iterations for noised data using Gibbs sampling with Bayesian framework.

Table 3. Mixture Gaussian Distribution Estimation from denoised data using GEE with Additive Model. (D: Decision, obs: observation)

D	obs	mean	std	T Value	Pr > |t|	Skewness	Kurtosis
0	1032	0.108368	0.154699	22.50	<.0001	1.764932	2.467487
1	161	0.513594	0.186938	34.86	<.0001	-0.100617	1.185754

Table 4. The noise data estimation using GIBBS sampling under Bayesian framework

node	mean	std	MC err.	2.5%	median	97.5%	start	sample
P[1]	0.484	0.097	0.003	0.299	0.483	0.671	1	1000
P[2]	0.515	0.097	0.003	0.329	0.516	0.701	1	1000
λ[1]	0.754	0.038	0.001	0.675	0.756	0.821	1	1000
λ[2]	0.815	0.037	9.6E-4	0.742	0.814	0.897	1	1000
σ	0.141	0.023	8.9E-4	0.104	0.138	0.191	1	1000

The following figures show some simulation results of noise estimation using BUGS.

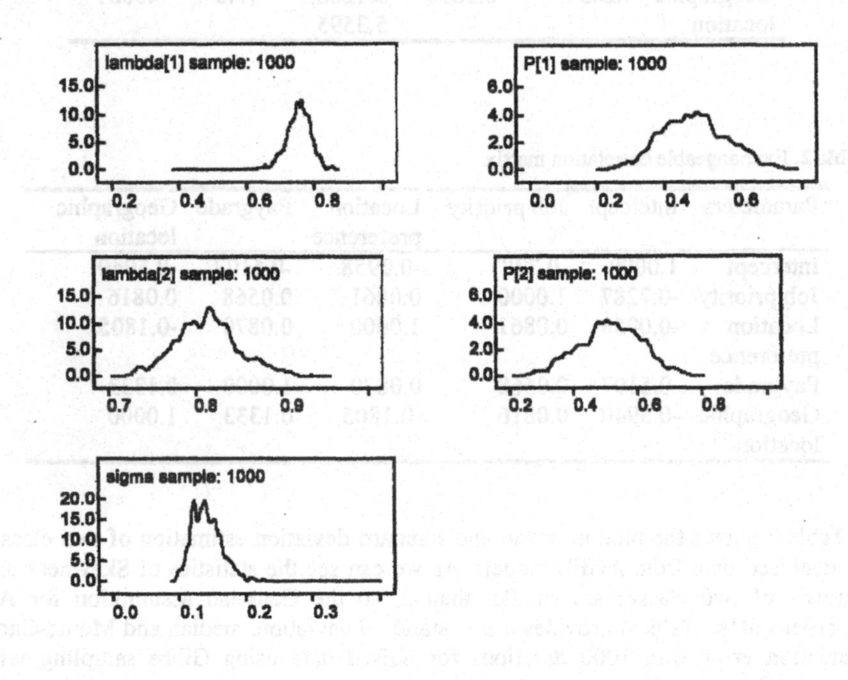

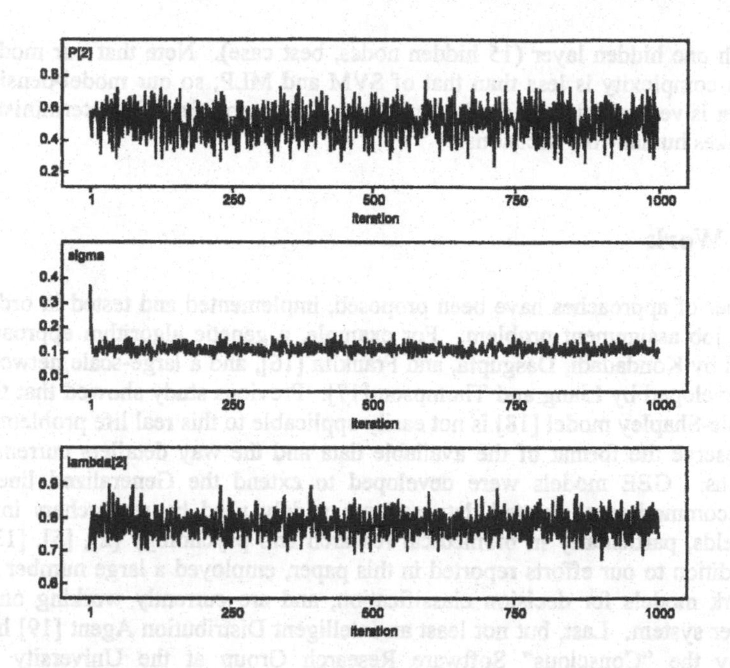

Table 5. Correct classification rates for different approaches

Method	Correct Classification Rate
NN using normal kernal density	76.05%
NN using Mahalanobis distances	77.52%
NN using biweight kernal density estimation	81.65%
NB without GEE	84.81%
NB with GEE	85.75%
AB with AGEE	93.42%
MLP (15nodes, best case)	93.14%
SVM	93.50%

Table 5 compares different classification approaches with correct classification rates. Nearest neighbor using normal density estimation, using Biweight density estimation and Mahalanobis distances, data mining approaches like Multi-layer Percepton (MLP) with back-propagation algorithms, Support Vector Machine (SVM) with Adatron algorithms and Naive Bayes classifier are used. Comparison of NB with and without (directly computing mean and standard deviation from data) GEE and Adaptive Bayes classifiers combining with AGEE with additive models are also given in the same table. The table shows that the Adaptive Bayes classifiers combining with AGEE with additive models give much higher accuracy than three Nearest Neighbor (NN), and also higher than the original Naive Bayes algorithms with or without using GEE with additive model. The performance is very close to the best SVM with Adatron algorithm and MLP with back-propagation and momentum algorithm with one hidden layer (15 hidden nodes, best case). Note that our model and algorithm complexity is less than that of SVM and MLP, so our model-density based classifier is very efficient and robust to capture the properties of indeterministic data and it makes human-like decisions.

4 Related Work

A large number of approaches have been proposed, implemented and tested in order to model our job assignment problem. For example, a genetic algorithm approach was discussed by Kondadadi, Dasgupta, and Franklin [16], and a large-scale network model was developed by Liang and Thompson [17]. Previous study showed that the traditional Gale-Shapley model [18] is not easily applicable to this real life problem if we are to preserve the format of the available data and the way detailers currently make decisions. GEE models were developed to extend the Generalized linear models to accommodate correlated data and are widely used by researchers in a number of fields, particularly in biomedical research and psychology [2], [5], [13]. We, as an addition to our efforts reported in this paper, employed a large number of neural network models for decision classification, and are currently working on a fuzzy classifier system. Last, but not least an Intelligent Distribution Agent [19] has been built by the "Conscious" Software Research Group at the University of Memphis, which uses standard operations research techniques and cognitive theories of mind for decision-making.

5 Conclusion and Further Work

We can extend our method to multi-class probability estimation and classification problems. The approach can be used for any kind of decision-making problem, like financial prediction problems, medical diagnoses, medical prediction and so on. As we mentioned earlier, the classifier had several additional properties compared with other classification methods. These include fewer assumptions, less complexity of the model and algorithms, less likely to be trapped at local minimum, can deal with data with high noise level and data with large number of attributes. The computed density estimation could give a very good representation of the true distribution due to the correlation structure of the output accounted for in the model and also we model the noised data using Gibbs sampling with Bayesian framework, so it is more efficient and it can posses good generalization performance. Further study will consider the stochasitic process as nonstationary case and build Bayesian algorithms combining Bayesian theory with regularization theory into dynamic GEE model.

References

1. Fraley, C., Raftery, A. E.: Model-Based Clustering, Discriminant Analysis, and Density Estimation, Technical Report NO.380, 2000.

2. Liang, K. Y., Zeger S. L.: Longitudinal Data Analysis Using Generalized Linear Models, Biometrika, 73,13-22, 1986.
3. Keller, A. D., Schummer, M., Hood L., Ruzzo W. L.: Bayesian Classification of DNA Array Expression Data, Technical Report UW-CSE-2000-08-01, 2000.
4. Horton, N. J., Lipsitz, S. R.: Review of Software to Fit Generalized Estimation Equation Regression Models, The American Statistician, Vol.53, May 1999.
5. Albert I., Jais, J. P.: Gibbs Sampler for the Logistic Model In the Analysis of Longitudinal Binary Data, Statist. Med.17, 2905-2921, 1998
6. Friedman, J., Hastie T., Tibshirani, R.: Additive Logistic Regression: a Statistical View of Boosting. Technical Report, 2000.
7. Dietterich, T. G.: Ensemble methods in machine learning, In Multiple Classier Systems. First International Workshop, MCS 2000, Cagliari, Italy, pages 1-15. Springer-Verlag, 2000.
8. Akaike, H.: A new look at the statistical model identification, IEEE Trans. Automatic Control, Vol. 19, No. 6, pp. 716-723, 1974.
9. Cortes C., Vapnik, V.: Support vector machines, Machine Learning, 20, pp. 273-297, 1995.
10. Friess, T. T., Cristianini N., Campbell, C.: The kernel adatron algorithm: a fast and simple learning procedure for support vector machine, In Proc. 15th International Conference on Machine Learning, Morgan Kaufman Publishers, 1998.
11. Boser, B., Guyon, I., Vapnik, V.: A Training Algorithm for Optimal Margin Classifiers, In Proceedings of the Fifth Workshop on Computational Learning Theory, pp. 144-152, 1992.
12. Gelfand, A. E., Hills, S. E., Racine-Poon, A., Smith, A. F. M.: Illustration of Bayesian inference in normal data models using Gibbs sampling. Journal of the American Statistical Association, 85(412):pp. 972–985,1990.
13. Spiegelman, D., Rosner B., Logan, R.: Estimation and Inference for logistic Regression with Covariate Misclassification and Measurement Error in main Study/Validation Study Designs, Journal of the American Statistical Association, 95(449): pp. 51–61, 2000.
14. Han, J., Kamber, M.: Data Mining Concepts and Techniques, 2000.
15. Stone, M.: Cross-validatory choice and assessment of statistical predictions (with discussion), Journal of the Royal Statistical Society, Series B, 36, pp. 111-147, 1974.
16. Kondadadi, R., Dasgupta, D., Franklin, S.: An Evolutionary Approach For Job Assignment, Proceedings of International Conference on Intelligent Systems-2000, Louisville, Kentucky, 2000.
17. Liang T. T., Thompson, T. J.: Applications and Implementation – A large-scale personnel assignment model for the Navy, The Journal For The Decisions Sciences Institute, Volume 18, No. 2 Spring, 1987.
18. Gale D., Shapley, L. S.: College Admissions and stability of marriage, The American Mathematical monthly, Vol 60, No 1, pp. 9-15, 1962.
19. Franklin, S., Kelemen, A., McCauley, L.: IDA: A cognitive agent architecture, In the proceedings of IEEE International Conference on Systems, Man, and Cybernetics '98, IEEE Press, pp. 2646, 1998.

GEC: An Evolutionary Approach for Evolving Classifiers

William W. Hsu[1] and Ching-Chi Hsu[1,2]

[1]Department of Computer Science and Information Engineering,
National Taiwan University,
Taipei 106, Taiwan
{r7526001, cchsu}@csie.ntu.edu.tw

[2]Kai Nan University
Taoyuan 338, Taiwan

Abstract. Using an evolutionary approach for evolving classifiers can simplify the classification task. It requires no domain knowledge of the data to be classified nor the requirement to decide which attribute to select for partitioning. Our method, called the Genetic Evolved Classifier (GEC), uses a simple structured genetic algorithm to evolve classifiers. Besides being able to evolve rules to classify data in to multi-classes, it also provides a simple way to partition continuous data into discrete intervals, i.e., transform all types of attribute values into enumerable types. Experiment results shows that our approach produces promising results and is comparable to methods like C4.5, Fuzzy-ID3 (F-ID3), and probabilistic models such as modified Naïve-Bayesian classifiers.

1 Introduction

Many data mining tasks require classification of data in to classes, i.e., bank loaning applications can be classified into either 'approve' or 'disapprove' classes. A *classifier* provides functions that map/classifies a data item/instance into one of the several predefined classes [4]. The automatic induction of classifiers from data not only provides a classifier that can be used to map new instances into their classes, but may also provide a human-comprehensible characterization of the classes.

Genetic algorithms [5] have been used successfully in a variety of search and optimization problems. Two general approaches of genetic algorithm-based learning have been used. The *Pittsburg* approach [10] uses a traditional genetic algorithm in which each entity in the population is a set of rules representing a complete solution to the learning problem. On the other hand, the *Michigan* approach [6] uses a distinctly different evolutionary mechanism in which the population consists of individual rules, each of which represents a partial solution to the overall learning task.

In this work, we focus on using genetic algorithms to evolve classification rules, which in a micro view, may be a set of rules. We shall call our method *Genetic Evolved Classifiers* (GEC). Our technique here is to evolve rules for each class

M.-S. Chen, P.S. Yu, and B. Liu (Eds.): PAKDD 2002, LNAI 2336, pp. 450-455, 2002.

separately, and combine these separately evolved rules into a complete classifier. We throw data into this classifier and it will tell us which class this data belongs to.

2 Encoding the Problem

As for categorical data, our approach is enumerating each possible value one by one as in GABIL [3]. For numerical attributes that are continuous over a range, directly enumerating each one of the possible values is impossible and impractical. Our approach here is to provide a method for partitioning these continuous values in to a limited number of discrete partitions. We do the following procedure fore each continuous attribute to convert it into discrete partitions. First, gather the numerical data from the data set. From the gathered data, we obtain the following values:

(N_{max}, N_{min}): The maximum and minimum value of the numerical attribute.
R: The range of the numerical attribute, i.e., N_{max} - N_{min}
σ: The standard deviation of the gathered numerical values.
μ: The mean of the gathered numerical values

$$\ldots,\left(\mu-\frac{3}{2}\rho,\mu-\frac{1}{2}\rho\right)\left(\mu-\frac{1}{2}\rho,\mu+\frac{1}{2}\rho\right)\left(\mu+\frac{1}{2}\rho,\mu+\frac{3}{2}\rho\right)\ldots \quad (1)$$

$$\left\lceil\frac{R}{\rho}\right\rceil+1 \quad (2)$$

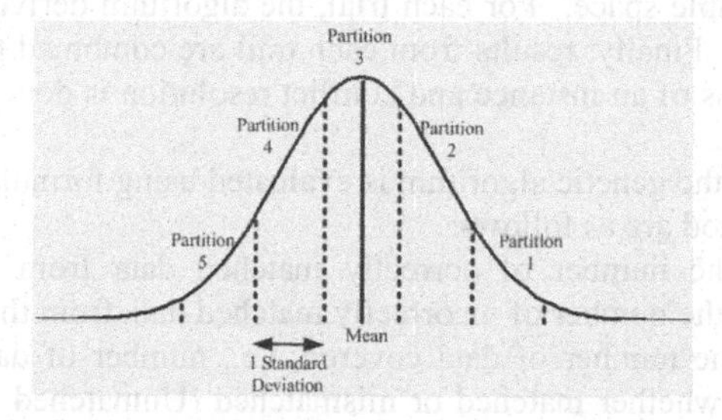

Figure 1. Illustrated partitioning of a normally distributed continuous attribute value

Partition is done with the mean μ as the center and the standard deviation σ as an interval. The new discrete intervals generated will look like (1), and the total number of partition generated will be estimated to (2). N_{max} will lie in the last interval and N_{min} will lie in the first interval. This partitioning method is done under the assumption that many natural phenomena carries the normal distribution property. Visualization of our partitioning method is shown in Figure 1.

3 The Genetic Algorithm for Evolving the Classifier

The outline of our GEC is shown in Figure 2. Rules are evolved for each class separately; each executing multiple runs of genetic algorithms (GAs) in hope to discover rules to cover the whole domain. Although this framework is like the one proposed by Liu [9], our method is better due to that it can handle continuous attributes and that independent runs can be combined to find maximum coverage of the problem domain. The outline of the inner GA evolution is shown in the right side of Figure 2. Our approach is basically the genetic algorithm framework proposed by Holland [5] with some modifications of our own.

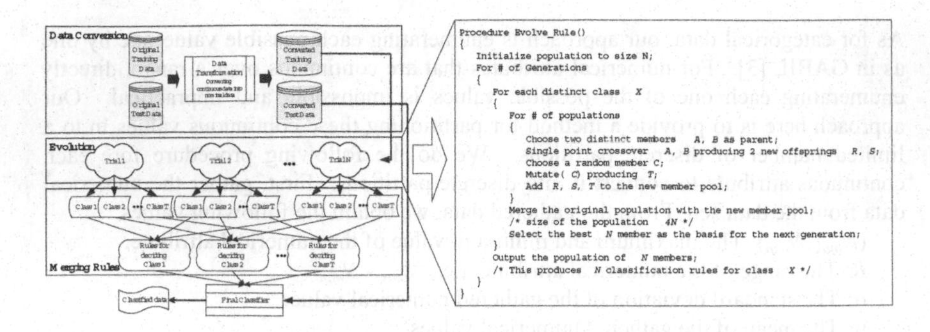

Figure 2. Outline of the GEC

We run independent trials to produce rules in hope to obtain the maximum coverage of the sample space. For each trial, the algorithm derives equal number of rules for each class. Finally, results from each trail are combined to form a complete classifier. Final class of an instance and conflict resolution is done using the majority voting scheme.

Fitness value for the genetic algorithm is evaluated using formula (3). The ideas of this evaluation method are as follows:

1. Maximize the number of correctly matched data from the test set while minimizing the number of incorrectly matched data from the test set.
2. Maximize the number of data covered, i.e., number of data decided by this certain rule whether matched or mismatched (Unmatched = Total number of data minus both the number of data matched correctly and incorrectly).

$$\text{Fitness} = \frac{\#\,Matched - \#\,Mismatched}{\#\,Unmatched} \tag{3}$$

We can see that the more data matched, the less data mismatched, and the more data covered, the larger the fitness value will be. Besides, rule that matches the same number of times as it mismatches are considered useless. What we want is to let the rule match more often than mismatch. This is also reflected in formula (3).

4 Experiment Results

We have used the following parameters in our experiment:

1. Population size is set equivalent to the number of attributes, i.e., for the adult census database, there are 15 attributes (including the 'class') and so the population size is set to 15.

2. Maximum number of generations per trial is set to 50. The sampling of the result is done every 10 generations.
3. Uniform single point crossover is used and the mutation rate is set to 0.5 for each locus position.
4. Elitism is collected, i.e., the best rule of each independent trial is gathered.

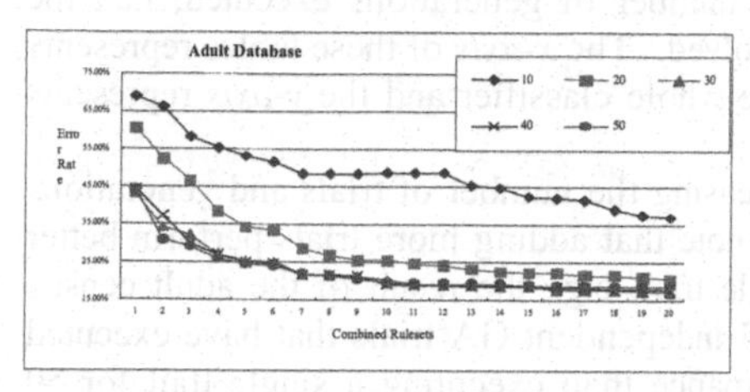

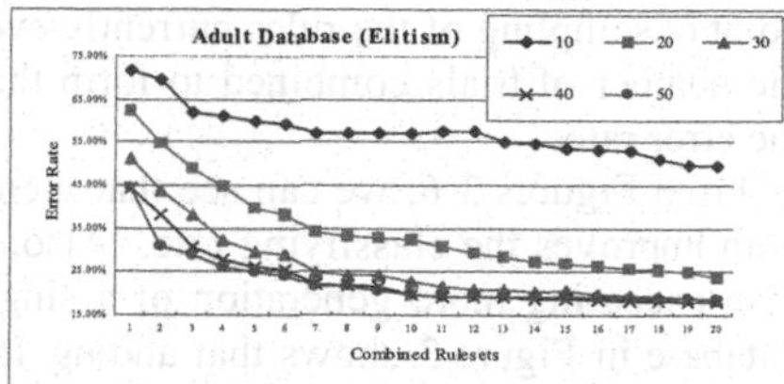

Figure 3. Result of the adult census database

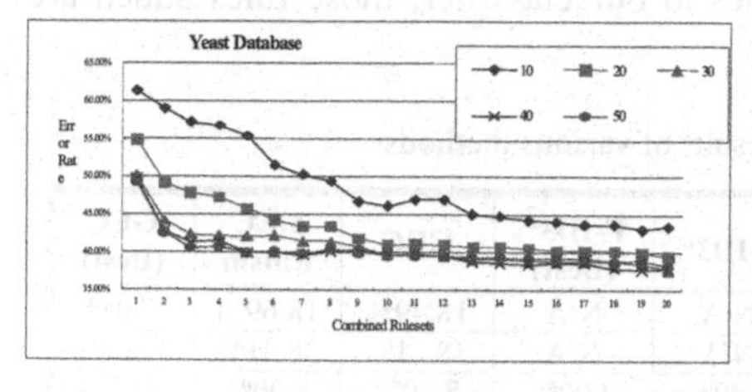

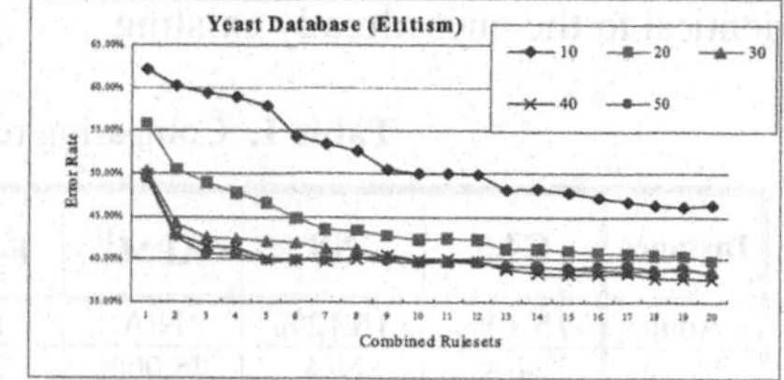

Figure 4. Result of the yeast database

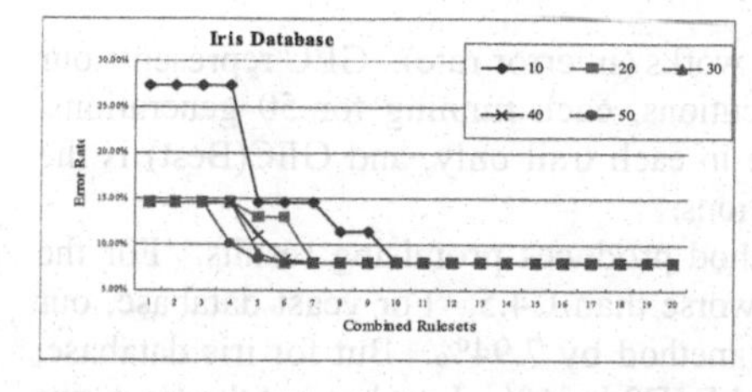

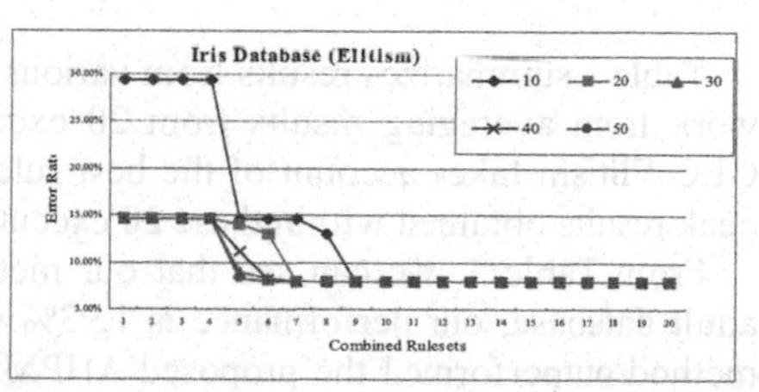

Figure 5. Result of the iris database

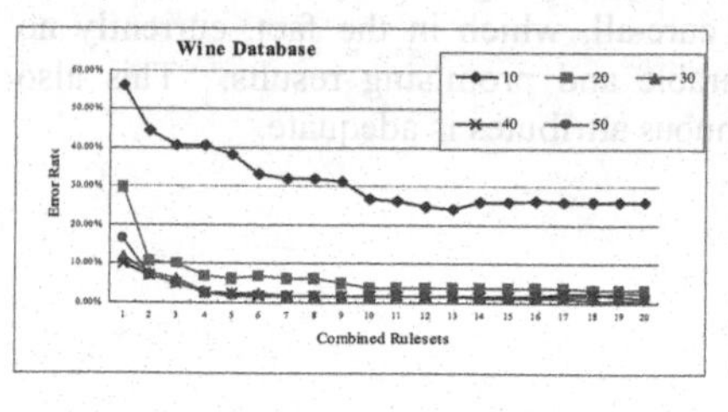

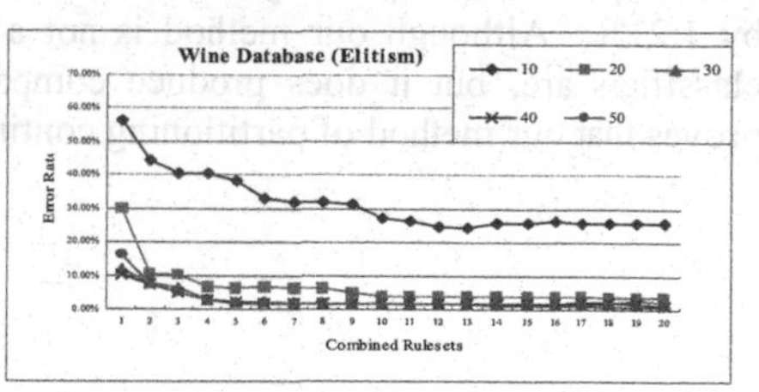

Figure 6. Result of the wine database

Experiments are carried out on 4 databases and are to be compared with past results using 3 fold cross validation (except for the adult census database, which training set and test set are already provided): they are the adult census database from [8], yeast classification database from [7], iris and wine database from [2]. Results are averaged from 20 independent executions (All of the instances can be found in [1]). Elitism uses only the best rule evolved in each independent trial. The legend of the following Figures 3-6 represents the number of generations executed, i.e., the point of sampling of the rules currently evolved. The *x-axis* of these tables represents the number of trials combined to form the whole classifier and the *y-axis* represents the error rate.

From Figures 3-6, we can see that increasing the number of trials and generations both improves the classifying rate. Also, note that adding more trials perform better than executing more generation of a single trial, e.g., the result of the adult census database in Figure 3 shows that adding 15 independent GA trials that have executed 10 generations together has better performance than executing a single trail for 50 generations. We can also see that in both axis, convergence can be reached. This represents that although we add more rules to our classifier, those rules added are identical to the ones already existing.

Table 1. Comparing results of various methods

Instance	C4.5	NB	APM[1]	F-ID3[2]	F-ID3[2] (Best)	GEC	GEC Elitsm	GEC (Best)
Adult	15.14%	16.12%	N/A	N/A	N/A	18.49%	18.69%	17.07%
Yeast	N/A	N/A	45.00%	N/A	N/A	38.21%	38.41%	37.06%
Iris	5.00%	N/A	N/A	4.00%	2.00%	8.00%	8.00%	8.00%
Wine	5.50%	N/A	N/A	7.70%	3.50%	5.31%	2.29%	2.29%

Table 1 summarizes results from various works (in error rate). GEC represents our work from averaging results from 20 executions, each running for 50 generations. GEC Elitism takes account of the best rule in each trail only, and GEC(Best) is the peak results obtained within these 20 executions.

From Table 1, we can see that our method produces promising results. For the adult database, our performance is 1.93% worse than C4.5. For yeast database, our method outperformed the proposed AHPM method by 7.94%. But for iris database, our method underperformed C4.5 by 3% and F-ID3 by 6%. Last but not the least, our method performed superiorly in the wine database, winning C4.5 by 3.21% and F-ID3 by 1.21%. Although our method is not a cure-all, which in the fact, currently no classifiers are, but it does produce comparable and promising results. This also proves that our method of partitioning continuous attributes is adequate.

[1] The Ad Hoc Structured Probability Model. Results directly obtained from Horton [7].

[2] Fuzzy ID3, decision tree method. Results directly obtained from Chen [2].

5. Conclusions

The GEC is capable of discovering rules for classifying data into multi-classes. It is simple, elegant and easy to use. Compared to C4.5 which requires the decision of which attribute to use for the next branch, our method is a much more direct approach. After we encode the problem, the rules are directly evolved without the need of finding which attribute to use for decision. Our approach is much more like F-ID3, but the difference is that F-ID3 uses GA to calibrate fuzzy rules first and then use these fuzzy rules to do the classification job. GEC evolves rules directly for the classification task. Although the final classifier created by GEC may contain many identical and redundant rules, many are found to be redundant. By purging these rules, we can compact our classifier.

Besides, from the experiment results, we can see that GEC produces comparable results. With simple transformation of continuous data into discrete intervals, GEC can work on any data given to it.

Finally, if it is required to list out the rules found for classification, our encoding method is reversible. We can obtain the classification rules by mapping the genetic encoding back to the attribute values without difficulty.

References

1. C. L. Blake and C. J. Merz. UCI Repository of machine learning databases [http://www.ics.uci.edu/~mlearn/MLRepository.html]. Irvine, CA: University of California, Department of Information and Computer Science, 1998.
2. H. M. Chen and S. Y. Ho, "Designing an Optimal Evolutionary Fuzzy Decision Tree for Data Mining", *Proceedings of the Genetic and Evolutionary Computation Conference*, pp. 943-950, 2001.
3. K. A. De Jong, W. M. Spears, D. F. Gordon, "Using Genetic Algorithms for Concept Learning", *Machine Learning*, vol. 13, no. 2, pp. 161-188, 1993.
4. U. M. Fayyad, G. Piatetsky-Shapiro, P. Smyth, "From data mining to knowledge discovery: An overview", *Advances in Knowledge Discovery and Data Mining*, chap. 1, pp. 1-34, AAAI Press and MIT Press, 1996.
5. J. H. Holland, *Adaptation in Natural and Artificial Systems*, Univ. of Michigan Press (Ann Arbor), 1975.
6. J. H. Holland, "Escaping brittleness: the possibilities of general-purpose learning algorithms applied to parallel rule-based systems", *Machine Learning, an artificial intelligence approach*, *2*, 1986.
7. "P. Horton and K. Nakai, "A Probablistic Classification System for Predicting the Cellular Localization Sites of Proteins", *Intelligent Systems in Molecular Biology*, pp. 109-115, 1996.
8. R. Kohavi, "Scaling Up the Accuracy of Naïve-Bayes Classifiers: a Decision-Tree Hybrid", *Proceedings of the Second International Conference on Knowledge Discovery and Data Mining*, pp. 202-207, 1996.
9. C. H. Liu, C. C. Lu and W. P. Lee, "Document Categorization by Genetic Algorithms", *IEEE International Conference on Systems, Man and Cybernetics (SMC)*, pp. 3868-3872, 2000.
10. S. F. Smith, *A Learning System Based on Genetic Adaptive Algorithms*, PhD Thesis, Univ. of Pittsburgh, 1980.

An Efficient Single-Scan Algorithm for Mining Essential Jumping Emerging Patterns for Classification

Hongjian Fan and Ramamohanarao Kotagiri

Department of CSSE, The University of Melbourne
Parkville, Vic 3052, Australia
{hfan,rao}@cs.mu.oz.au

Abstract. Emerging patterns (EPs), namely itemsets whose supports change significantly from one class to another, were recently proposed to capture multi-attribute contrasts between data classes, or trends over time. Previous studies show that EP/JEP(jumping emerging patterns)-based classifiers such as CAEP[2] and JEP-classifier[6] have good overall predictive accuracy. But they suffer from the huge number of mined EPs/JEPs, which makes the classifiers complex.
In this study, we propose a special type of EP, essential jumping emerging patterns (eJEPs), which are believed to be high quality patterns with the most differentiating power and thus are sufficient for building accurate classifiers. Existing algorithms such as border-based algorithms and consEPMiner[7] can not directly mine such eJEPs. We present a new single-scan algorithm to effectively mine eJEPs of both data classes (both directions). Experimental results show that the classifier based exclusively on eJEPs, which uses much fewer JEPs than JEP-classifier, achieves the same or higher testing accuracy and is often also superior to other state-of-the-art classification systems such as C4.5 and CBA.
Key words: emerging patterns, classification, data mining, decision trees, complexity

1 Introduction

Data classification is the process which finds the common properties among a set of objects in a database and classifies them according to a classification model. A new type of knowledge pattern, emerging patterns (EPs) [1] can serve as such a classification model. By aggregating the most expressive jumping emerging patterns (JEPs) [1], JEP-Classifier(JEP-C) [6] achieves surprisingly higher accuracy than other state-of-the-art classifiers such as C4.5 and CBA. However, it suffers from a huge number of JEPs.

We are only interested in those JEPs which represent the essential knowledge to discriminate between different classes, so we propose essential jumping emerging patterns (eJEPs).

[1] A JEP is a special type of EP (also a special type of discriminant rule), defined as an itemset whose support increases abruptly from zero in one dataset, to non-zero in another dataset — the ratio of support-increase being infinite.

M.-S. Chen, P.S. Yu, and B. Liu (Eds.): PAKDD 2002, LNAI 2336, pp. 456–462, 2002.

Definition 1. *Given two classes of datasets D' and D'', and $\mu > 0$ as a minimum support threshold, an essential jumping emerging pattern from D' to D'', denoted $eJEP(D', D'')$, or simply $eJEP(D'')$, is an itemset X satisfying the following conditions:*
1. $supp_{D'}(X) = 0$ and $supp_{D''}(X) \geq \mu$, and
2. Any proper subset of X does not satisfy condition 1.

An eJEP covers at least a predefined number of training instances. eJEPs are also shortest: any proper subset of an eJEP is no longer an eJEP. If all the itemsets satisfying supports in one class are 0 but in another above μ are to be represented by border description [1], eJEPs are those minimal in the left bounds. eJEPs are more useful in classification because:

- The set of eJEPs is the subset of the set of JEPs, after removing JEPs containing noise and redundant information. JEPs have been proved to have sharp discriminating power. eJEPs maintain such power by having infinite growth rates, and improve it by having a minimum coverage on training data.
- eJEPs are minimal itemsets. If less attributes can distinguish two data classes, using more may not help and may even add noise.

To get useful new insights and guidance from training data, μ should be set very low (e.g. 1%). Since useful Apriori property no longer holds for eJEPs and there are usually too many candidates, naive algorithms are too costly.

Inspired by FP-tree [5], a successful structure to mine frequent patterns without candidate generation, we propose a pattern tree (P-tree). A P-tree is an extended prefix-tree structure storing the quantitative information about eJEPs. The counts of an item for both data classes are registered. Items with larger support ratios are closer to the root. So eJEPs, itemsets with infinite growth rate made up of more likely items with large support ratio, will appear near the root. From the root to search eJEPs, we will always find the shortest ones first.

We develop a P-tree based pattern fragment growth mining method. It searches a P-tree depth-first to discover eJEPs from the root, which is completely different from FP-Growth [5]. While searching, nodes are merged, which ensures the complete set of eJEPs are generated. The pattern growth is achieved via concatenation of the prefix pattern with the new ones at deeper level. Since we are interested in the shortest eJEPs, the depth of the search is not very deep (normally 5-10). Another advantage is that it is a single-scan algorithm, which can mine eJEPs from D' to D'' and those from D'' to D' at the same time. Previous approaches such as border-based algorithms [1] and consEPMiner [7] will call the corresponding algorithm twice using D' and D'' as target databases separately.

We build classifiers based on eJEPs to measure their quality in classification. Experimental results show that our classifier uses much fewer eJEPs than the JEPs used in JEP-Classifier, while maintaining the accuracy close to JEP-Classifier. Thus, eJEPs are sufficient for building accurate classifiers.

2 Using the Pattern Tree Structure to Mine eJEPs

2.1 Pattern Tree

A pattern tree (P-tree) is an *ordered multiway* tree structure. If a node N has k items $N.items_i$, $(i = 1, \cdots, k)$, N has k positive counts $N.positive\text{-}counts_i$, k negative counts $N.negative\text{-}counts_i$, and at most k branches (child nodes), $N.childnodes_i$. For $N.items_i$, $N.positive\text{-}counts_i$ and $N.negative\text{-}counts_i$ register the number of positive(D') and negative (D'') instances represented by the portion of the path reaching $N.items_i$, respectively; $N.childnodes_i$ refers to the subtree with the root of $N.items_i$ ($N.items_i$'s subtree). P-tree is *ordered*: the items inside each node are *ordered*; and the items between parent and child nodes are *ordered.* The order is items' supports-ratio-descending order.

Consider data class 1={ {a,c,d,e}, {a}, {b,e}, {b,c,d,e} } and data class 2={ {a,b}, {c,e}, {a,b,c,d}, {d,e} }. The supports-ratio-descending order is $e > a > b > c > d$. The initially constructed P-tree is shown in Figure 1 (a). R denotes its root and N denotes $R.e$'s subtree. After calling $merge(N, R)$, P-tree is as Figure 1 (b). Let the minimum count to be an eJEP be 1 for either datasets, we will get a set of eJEPs: {e,a}(1:0), {e,b}(2:0), {e,c,d}(2:0), {a,b}(0:2).

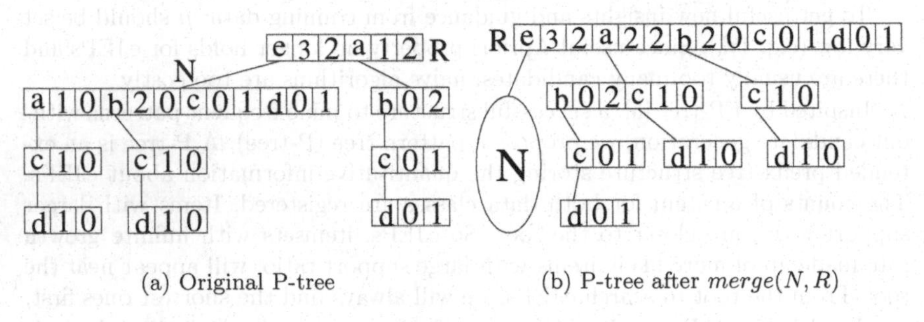

(a) Original P-tree (b) P-tree after $merge(N, R)$

Fig. 1. P-tree and nodes merge.

2.2 P-tree Construction

Based on P-tree definition, we have algorithm 1 to construct P-tree.
Algorithm 1: (P-tree construction)
Input:A training dataset DBTRAINING containing two data classes
Output: Its P-tree
Procedure:

1. Scan DBTRAINING once. Collect the set of items whose support ratios are more than zero, denoted as J, and their support ratios. Sort J in supports-ratio-descending order as L.

2. Create the root of a P-tree, R. For each instance Ins in DBTRAINING do the following: Select and sort the J items in Ins according to the order of L; Let the sorted J item list in Ins be $[p|P]$, where p is the first element and P is the remaining list, call $insert\_tree([p|P], R)$.

Function `insert_tree`$([p|P], T)$

---

search T for $T.items_i = p$;
if $T.items_i$ *not found* **then**
  insert p at the right place in T, $T.items_i$, with both counts zero;
  ;;This ensures all items in T are always in supports-ratio-descending order
end
if $Ins([p|P]) \in$ *positive class* **then** increment $T.count\text{-}positive_i$ by 1;
else increment $T.count\text{-}negative_i$ by 1;
if P *is nonempty* **then**
  if $T.items_i$*'s subtree is empty* **then** create a new node N ;
  Let N be $T.items_i$'s subtree, call $insert\_tree([P], N)$
end

2.3 Mining eJEPs from P-tree

While the P-tree structure is somewhat similar to FP-tree [5], the mining process is completely different. Because every training instance is sorted by support ratio between both classes when inserting into P-tree, items with high ratio, which are more likely to appear in an eJEP, are close to the root. We start from the root to search P-tree depth-first for eJEPs. We have algorithm 2 for mining eJEPs using P-tree.

Algorithm 2: (Mining eJEPs with P-tree)
Input:P-tree R, a minimum support threshold μ
Output:eJEPs
Procedure: call $mine\_tree(R, null)$ to produce a set of JEPs;
then select those minimal.

Assume we arrive at node M. When the path from the root to $M.items_k$, which represents an itemset, forms a JEP P_1, we no longer need to go deeper into $M.childnodes_k$, because those paths are supersets of P_1. All the JEPs that P-tree contains can be represented by a border description $< L, R >$. Although the algorithm can not exclusively generate JEPs in the left-hand bound L, it will not generate all either. Therefore, we usually get a relatively small set of JEPs, we choose eJEPs from them by selecting minimal ones.

3 Experimental Results

All the experiments were performed on a 500Mhz Pentium III PC with 512Mb of memory. The accuracy was obtained by using the methodology of ten-fold cross-

```
Function mine_tree(T, α)

  ;;T is a subtree of P-tree and α is an accumulating itemset
  ;;min_positive= μ * |D'|; min_negative= μ * |D''|
  foreach item of T, T.items_i do
  |  if T.items_i'subtree M is not empty then merge(M,T);
  |  β = α ∪ T.items_i;
  |  switch conditions do
  |  |  case T.count-positive_i = 0∧ T.count-negative_i ≥ min_negative
  |  |  |  generate an eJEP β of D'' with supp(β) = T.count-negative_i
  |  |
  |  |  case T.count-negative_i = 0∧ T.count-positive_i ≥ min_positive
  |  |  |  generate an eJEP β of D' with supp(β) = T.count-positive_i
  |  |
  |  |  case (T.items_i's subtree N is not empty) ∧
  |  |  (T.count-positive_i ≥ min_positive ∨
  |  |  T.count-negative_i ≥ min_negative) call mine_tree(N, β)
  |  end
  end
```

```
Function merge_tree(T1, T2)

  ;;Given two subtrees of P-tree, T1 and T2, merge T1's nodes into T2.
  ;;T2 is updated while T1 remains unchanged.
  foreach item of T1, T1.items_i do
  |  search T2 for T2.items_j = T1.items_i;
  |  if T2.items_j found then
  |  |  add T1.items_i's both counts to T2.items_j's counts, respectively
  |  else
  |  |  copy and insert T1.items_i with its both counts and childnode at the
  |  |  right place in T2, denoted as T2.items_j;
  |  end
  |  if T1.items_i's subtree, M is not null then
  |  |  if T2.items_j's subtree is nonempty then create a new node N ;
  |  |  else N ← T2.items_j's subtree;
  |  |  call merge_tree(M, N);
  |  end
  end
```

validation. For datasets containing continuous attributes, they are discretized using the Entropy method in (Fayyad and Irani 1993).
From Table 1, we highlight some interesting points as follows: eJEP-C always uses much fewer eJEPs than JEP-C uses JEPs; eJEPs-C maintains accuracy close to JEP-Classifier, in most datasets, and sometimes achieves higher accuracy, such as Cleve, Crx, Heart; both of them are more accurate than C4.5 and CBA in general; the low standard deviation shows eJEP-C is very consistent; the running time(including training and testing) of eJEP-C is much shorter than JEP-C. We conclude that eJEP-C can be used effectively for large classification problems.

Dataset	#JEPs	#eJEPs	eJEP-C accuracy	eJEP-C S.D.	accuracy ratio	JEP-C runtime	eJEP-C runtime	runtime ratio
Australian	9,806	819	85.8	0.031	0.994	7.213s	3.140s	3.3
Cleve	8,633	1,478	84.7	0.062	1.006	0.823s	1.564s	0.5
Crx	9,880	2,131	86.1	0.048	1.002	9.023s	1.983s	4.6
Diabete	4,581	62	75.4	0.053	0.983	0.587s	0.031s	18.9
German	32,510	1,137	74.1	0.037	0.985	35.287s	2.670s	13.2
Heart	7,596	299	83.0	0.058	1.005	0.357s	0.045s	7.9
Horse	22,425	1,605	83.6	0.029	1.005	22.807s	2.306s	9.9
Iono	8,170	1,391	91.7	0.056	0.985	386.290s	24.507s	15.8
Mushroom	2,985	71	99.9	0.001	0.999	296.973s	1.535s	193.0
Pima	54	54	74.2	0.056	0.932	0.563s	0.030s	18.8
Sonar	13,050	3,434	79.0	0.010	0.920	25.840s	9.677s	2.7
Tic-tac	2,926	1,568	98.8	0.014	0.990	3.843s	3.420s	1.1
Waveform	4,096,477	7937	83.3	0.017	-	14932.920s	1380.270s	10.8

Table 1. Number of JEPs and eJEPs used in classifiers, eJEP-C's average accuracy in 10 folds and the standard deviation, the accuracy ratio(eJEP-C vs JEP-C), JEP-C runtime, eJEP-C runtime and the runtime ratio(JEP-C vs eJEP-C)

4 Conclusions

This paper proposes a special emerging patterns, called essential jumping emerging patterns (eJEPs), which have strong discriminating power. An efficient single-scan algorithm is presented to generate all eJEPs of both data classes. Our experiments on databases in the UCI machine learning database repository show that the classifier based on eJEPs is consistent, highly effective at classifying of various kinds of databases and has better average classification accuracy in comparison with CBA and C4.5, and is more efficient and scalable than other EP-based classifiers.

References

1. G. Dong and J. Li. Efficient mining of emerging patterns: Discovering trends and differences. In *Proc. of KDD'99*, pages 15–18, 1999.
2. G. Dong, X. Zhang, L. Wong, and J. Li. CAEP: Classification by aggregating emerging patterns. In *Proc. of the 2nd Int'l Conf. on Discovery Science (DS'99)*, Tokyo, Japan, Dec. 1999.
3. E. Keogh C. Blake and C.J. Merz. UCI repository of machine learning databases. http://www.ics.uci.edu/~mlearn/MLRepository.html, 1998.
4. R. Kohavi, G. John, R. Long, D. Manley, and K. Pfleger. MLC++: a machine learning library in C++. In *Tools with artificial intelligence*, pages 740–743, 1994.
5. J. Han, J. Pei, and Y. Yin. Mining frequent patterns without candidate generation. In *SIGMOD'00*, Dallas, TX, May 2000.
6. J. Li, G. Dong, and K. Ramamohanarao. JEP-Classifier: Classification by Aggregating Jumping Emerging Patterns. In *Knowledge and Information Systems*, Volume 3 Issue 2, 131-145, 2001.
7. X. Zhang, G. Dong, K. Ramamohanarao. Exploring Constraints to Efficiently Mine Emerging Patterns from Large High-dimensional Datasets. In *Proc. 2000 ACM SIGKDD Conf.*, pages 310-314, Boston, USA, Aug. 2000.

A Method to Boost Support Vector Machines

Lili Diao, Keyun Hu, Yuchang Lu, and Chunyi Shi

The State Key Laboratory of Intelligent Technology and System, Dept. of Computer Science and Technology, Tsinghua University, Beijing 100084, China
diaolili@mails.tsinghua.edu.cn
http://www.cs.tsinghua.edu.cn

Abstract. Combining boosting and Support Vector Machine (SVM) is proved to be beneficial, but it is too complex to be feasible. This paper introduces an efficient way to boost SVM. It embraces the idea of active learning to dynamically select "important" samples into training sample set for constructing base classifiers. This method maintains a small training sample set with settled size in order to control the complexity of each base classifier. Other than construct each base SVM classifier directly, it uses the training samples only for finding support vectors. This way to combine boosting and SVM is proved to be accurate and efficient by experimental results.

1 Introduction

Boosting is a novel, powerful machine learning method for classification [1]. It is an iterative procedure that successively classifies a weighted version of the sample by its base classifiers generated one for each time of iteration, and then re-weights the sample dependent on how successful the classification was. Support Vector Machine (SVM) [2] first maps the original sample space to a high-dimensional linear space via a nonlinear mapping defined by a dot product function, and then finds a separating hyperplane, which is optimal and is decided by support vectors, in the linear space. Finding the hyperplane can be casted into a quadratic optimization problem subject to an in-equation. We conducted series of experiments to examine the results of two different ideas for combining SVM and boosting. The first idea is to directly boost weighted SVMs, which we later will refer to as BSVM. The second is to integrate the constituent classifiers of boosting by SVM, which we will latter refer to as SBoost. Both BSVM and SBoost can have better performance than any single SVM classifier, and could be better than many other learning methods. Whereas according to our experiments for text categorization, these two algorithms are very inefficient. The reason is that, the computational complexity of SVM is substantial, and so is boosting iteration. In the sense, though these methods for combining boosting and SVM have the ability to achieve very high accuracy, we may still prefer other easier and quicker

---

Supported by the National Grand Fundamental Research 973 Program of China under Grant No.G1998030414 and the National Natural Science Foundation of China under Grant No.79990580.

M.-S. Chen, P.S. Yu, and B. Liu (Eds.): PAKDD 2002, LNAI 2336, pp. 463-468, 2002.

ways for solving classification problems though their accuracy may be worse. In this paper we will introduce an efficient way to boost SVMs, which is not only much quicker than BSVM and SBoost, but also has similar average error rates with them. In this method, we use a settled size (usually very small related to the number of all samples) of training set to construct the base classifiers (SVMs), and then adjust the training set by choosing the currently most important examples from outside into the training set and at the same time delete the most unimportant examples from it to maintain the size of training set. Once the training set is updated, a new SVM as a base classifier of boosting will be generated. Finally these SVMs will be integrated by boosting. Small number of training sample set is designed for constructing SVM quickly, while adjusting of the training sample set is for making different base classifiers.

2 SVM

Let $(x_1, y_1), \ldots, (x_N, y_N) \in \aleph \times \{-1, +1\}$ denote the training samples of size N, where x come from some set $\aleph \subseteq R^m$ (m is the dimension) and y is the class label.

SVM first maps the original sample space to a high-dimensional linear space via a nonlinear mapping defined by a dot product function, and then finds a separating hyperplane, which is optimal and is decided by support vectors, in the linear space. Consider an m-dimensional feature space $\hbar$, which is a subset of R^m and is spanned by a mapping ϕ. In a support vector setting, any ϕ corresponds to a Merceer kernel $k(x, x') = (\phi(x) \cdot s\phi(x'))$ implicitly computing the dot product in $\hbar$. The goal of SVMs is to find some separating hyperplane described by a vector ω in feature space $\hbar$:

Finding the hyperplane can be casted into a quadratic optimization problem:

$$\min_{\omega \in \hbar} \quad \frac{1}{2}\|\omega\|_2^2$$
$$subject \quad to \quad y_n(\omega \cdot \phi(x_n) + b) \geq 1 \quad n = 1, \ldots, N$$

One selects the hyperplane with minimal VC capacity [2], which is in this case can be achieved by maximizing the margin. Here, the margin $\rho = \frac{2}{\|\omega\|_2}$ is defined as the minimum l_2-distance of a training point to the separating hyperplane. Here support vector (SV) are the points (x_i, y_i) that satisfy $y_n(\omega \cdot \phi(x_n) + b) = 1$. The decision rule of SVM can be re-defined as $f(x) = \operatorname{sgn}\left(\sum_{(x_i, x_j) \in SV} y_i \alpha_i^* \cdot K(x_i \cdot x_j) - b_0\right)$. Here SV denotes the set of support vectors. $\alpha^* = \left(\alpha_1^*, \alpha_2^*, \ldots, \alpha_{|SV|}^*\right)$ can be computed by maximizing: $W(\alpha) = \sum_{i=1}^{N} \alpha_i - \frac{1}{2}\sum_{i,j=1}^{N} \alpha_i \alpha_j y_i y_j K(x_i, x_j)$, and $b_0 = \frac{1}{2}\left[\omega_0 \cdot x^*(1) + \omega_0 \cdot x^*(-1)\right]$ where

$(x^*(1), y^*(1)) \in SV$ is any positive example and $(x^*(-1), y^*(-1)) \in SV$ is any negative example.

3 How to Boost SVM Efficiently?

There are two different ideas for combining SVM and boosting. The first idea is to directly boost weighted SVMs, which we call BSVM. The second is to integrate the constituent classifiers of boosting by SVM, which we call SBoost.

- BSVM: for each boosting round, a weight distribution over the samples is incorporated into SVM as base classifier. Along with the adjusting of the weight distribution, the performance of SVMs becomes very different and thus boosting can improve their performance as a whole.
- SBoost: this is more complex a way to realize the goal of combining boosting and SVM. SBoost has three steps. The first step is using boosting to construct a certain number of base classifiers $\{H_s(x)\}$; the second step uses all the base classifiers generated in the first step to classify all the samples. Therefore, each sample will correspond to a set of classification results, one for each base classifier. So the original sample space $(x_1, y_1), \dots, (x_N, y_N) \in \aleph \times \{-1, +1\}$ is mapped to a new space $(z(x_1), y_1), \dots, (z(x_N), y_N) \in R^T \times \{-1, +1\}$ where $z_i(x) = (H_1(x_i), \dots, H_T(x_i))$, if we run the boosting algorithm for T rounds. The third step is to construct a SVM in the new space. The performance of this method is slightly better than BSVM.

Both BSVM and SBoost can have better performance than any single SVM classifier, and could be better than many other learning methods. Whereas according to our experiments for text categorization, these two algorithms are very inefficient: only for training 100-300 sample documents, it will take several hours to achieve reasonable performance. The reason is that, the computational complexity of SVM is substantial, and so is boosting iterations. To combine these two methods is hence very expensive. In the field of classification, especially text categorization, the problem space is always represented by high dimensional vectors, and usually a great number of examples are needed to establish the training set prepared for learning process. $10^3 - 10^4$ is only a middling size of training sets. Therefore, SBoost and BSVM are not pragmatic in practice. In the sense, though these methods for boosting SVMs have the ability to achieve very high accuracy, we may still prefer other easier and quicker ways for solving classification problems though their accuracy may be worse. However, in many cases the requirements for accuracy are as critical as those for efficiency. The most possible solution for reducing the computational complexity of combining boosting and SVM is to reduce the samples used for constructing the constituent classifiers. As we know, in many real world problems, the number of Support Vectors (SVs) is much smaller than that of training samples [3]. So, it would be a good idea to only use support vectors for simplifying the construction of classifiers. Although support vectors can represent many samples, and hence facilitate the construction of classifiers, it is still very costive to search for them because we have to use all the training samples to solve a quadratic optimization problem. In the sense, we need to further consider how to reduce the training samples needed for finding support vectors.

Active learning is an efficient method of supervised learning in reducing the number of samples [4]. Instead of learning from the original training sample set passively, it can actively select "useful" samples to form better training sample set as learning procedure progresses, which will save many computations since it does not need to process all the training samples. Active learning is proved to be very pragmatic in text classification [5]. Here "useful" samples mean the samples with large uncertainty to current classifier, which are measured by their information entropy. This idea might be also beneficial for finding the support vectors with less training samples. We can maintain a small training set for finding support vectors where the member samples are adaptively selected or deleted according to different conditions. The method to evaluate the "usefulness" or "importance" of training sample for SVM should be to find the location of sample related to the separating hyperplane. The sample that is not support vectors but is closest to the separating hyperplane would be important or useful and should be included in the training set to find support vectors. Specifically, the sample lies in the classification margin is the most important since its inclusion in the training set would definitely change the separating hyperplane. Besides, even there are no samples fall in the classification margin, some samples may be still important since the hyperplane may also be changed, especially for the separating hyperplane of weighted SVM. In this way, we can dynamically change the training set via always searching for important samples to join it and thus generate different support vectors. Finally, we combine these classifiers in a boosting style. We call this algorithm as IBS (Interactively boosted SVM). The convergence of active learning is proved by Park J.M. [6].

4 New Algorithm

Given original training sample set:

$\{(x_1, y_1), (x_2, y_2), \cdots, (x_N, y_N)\}$

Here x_i is any training sample, $y \in \{-1,+1\}$ denotes the classes x_i belongs to. N is the number of training samples. $i = 1,2,\cdots,N$.

This algorithm maintains a set of weights as a distribution W over samples, i.e., for each $x_i \in x$, and at each boosting round s, there is an associated real value $w_s[i]$.

Given a settled number of training samples used for finding support vectors $n = \frac{1}{10}N$.

Given a settled number of boosting rounds T.

Choose a kernel function $K(x, x')$ to compute the dot product of two vectors.

Step #1: Initialize $w_1[i] = 1/N$ for all $i = 1,2,\cdots,N$; set $s = 0$;

Step #2: Select randomly n samples from original sample set according to current weight distribution W (if we regard the normalized weight of each sample as its apriori probability). These selected samples establish a training set of size n for finding support vectors. We call this small training set as *STS*.

Step #3: Use each sample $(x, y) \in STS$ to find the support vectors. We first find the coefficients $\alpha^* = \{\alpha_1^*, \ldots, \alpha_n^*\}$ that maximize: $W(\alpha) = \sum_{i=1}^{n} \alpha_i - \frac{1}{2} \sum_{i,j=1}^{n} \alpha_i \alpha_j y_i y_j K(x_i, x_j)$. If $\alpha_i^* \neq 0$, then (x_i, y_i) is the support vector we are looking for. Set $s = s + 1$ to compute the number of base classifiers already generated. If $s \geq T$ then go to step #8.

Step #4: Combine these support vectors to two representational points: $x^+ = \frac{1}{C} \sum_{(x_i, y_i) \in SV^+} \alpha_i^* x_i$ for positive samples, and $x^- = \frac{1}{C} \sum_{(x_i, y_i) \in SV^-} \alpha_i^* x_i$ for negative samples. Here SV^+ and SV^- denote all the positive and negative support vectors respectively. $C = \sum_{y_i = -1} \alpha_i^* = \sum_{y_i = +1} \alpha_i^*$. Based on these two points, we establish a nearest neighbor classifier, whose decision rule is, if an unknown point is nearer to x^+ (or x^-) than to x^- (or x^+), it will be classified as the positive (or negative) class.

Step #5: Classify all the samples of original sample set. Compute the weighted error $\varepsilon = \sum_{i=1}^{N} w[i] \cdot [(x_i, y_i) \;\; is \;\; misclassified]$. $[\![\bullet]\!]$ is a function that maps its content to 1 if it is true, otherwise, maps to 0. Increase the samples' weights if they are misclassified. Otherwise, decrease:

$$w[i] = w[i] \div \begin{cases} 2\varepsilon & if \;\; (x_i, y_i) \;\; is \;\; misclassified \\ 2(1-\varepsilon) & otherwise \end{cases}$$

Normalize these weights because we hope they also can be regarded as probability distribution.

Step #6: If, between any two support vectors of negative positive pair, there are samples not in the *STS*, then we choose the sample nearest to the separating hyperplane to join in the *STS*. Before that, the sample in the *STS* with smallest weight value will be deleted from *STS*. Go to step #3. if there are no sample in the margin, go to step #7.

Step #7: Select randomly from the original sample set according to current weight distribution W. If the sample is not in *STS* and is misclassified by current base classifier, then the sample with smallest weight value in the *STS* will be deleted, and the selected sample will be added to *STS*. Go to Step #3; otherwise, go to step #7.

Step #8: Combine the T base classifiers by weighted votes:

$$H_B(x) = \operatorname{sgn}\left(\sum_{s=1}^{T} \ln(\varepsilon_s / (1 - \varepsilon_s)) \cdot H_s(x) \right).$$

It is important to note here that the distance between any two points is decided by the kernel function, say, all the dot products in the original space are replaced by kernel functions to compute them. i.e. $\|x - x'\| = K(x, x) - 2K(x, x') + K(x', x')$ denotes the distance between x and x' in the new space.

5 Experiments

We realized four methods to compare their performances: SVM, BSVM, SBoost and IBS. For testing their capabilities we conducted series of experiments. Though generally speaking IBS is not the most accurate one in these datasets, its performance is only slightly worse than the best one, and has already been good enough to be employed in solving text categorization problems. From our further study, as the boosting algorithm goes on, IBS still have great potential to achieve better precision and recall, whereas BSVM and SBoost cannot improve themselves by more iterations.

Though IBS is capable of obtaining high accuracy, its most valuable merit for us is its efficiency. From our experiments we can see that a reasonable accuracy can be easily achieved within 40 minutes, while BSVM and SBoost are nearly impractical to achieve such a high accuracy. The most possible reason is that the size of training samples for constructing each base classifier is strictly limited to a small scale, thus it is relatively easy for IBS to generate its constituent classifiers.

6 Conclusion

Combining boosting and SVM is proved to be beneficial, but it is too complex to be feasible. In this paper we introduce an efficient way to boost SVM. We embrace the idea of active learning to dynamically select "important" or "useful" samples into training sample set for constructing base classifiers. Our method maintains a small training sample set with settled size in order to control the complexity of each base classifier. Other than construct each base SVM classifiers directly, we only use the training samples to find support vectors. From the experiments we can conclude that, IBS is an efficient way to combine boosting and SVM, and is proved to be able to have high accuracy with high speed. IBS has great potency in text categorization and filtering.

References

1. Freund, Y., Schapire, R.: A Decision-theoretic Generalization of On-line Learning and an Application to Boosting. Journal of Computer and System Sciences (1997), 55(1), 119-139
2. Vapnik, V.: The Nature of Statistical Learning Theory. Springer Verlag, New York (1995)
3. Schobkopf B., Burges C., Vapnik V.: Extracting Support Data for a Given Task. Proceedings of the First International Conference on Knowledge Discovery and Data-mining, AAAI Press, German (1996)
4. Hu, Y.: From Pattern Classification to Active Learning. IEEE Signal Processing Magazine (1997), 11, 39-43
5. Park J.M.: Intelligent Information Query and Browsing System Using Active Learning. PhD Dissertation. University of Wisconsin, Dept. of Electronic and Computer Engineering (1997)
6. Park J.M., Hu Y.: Online Learning for Active Pattern Recognition. IEEE Signal Processing Letters (1996), 3(11), 301-303

Distribution Discovery: Local Analysis of Temporal Rules[1]

Xiaoming Jin, Yuchang Lu, Chunyi Shi

The State Key Laboratory of Intelligent Technology and System
Computer Science and Technology Dept., Tsinghua University, Beijing, China
xmjin00@mails.tsinghua.edu.cn, lyc@tsinghua.edu.cn,
scy@est4.cs.tsinghua.edu.cn

Abstract. In recent years, there has been increased interest in using data mining techniques to extract temporal rules from temporal sequences. Local temporal rules, which only a subsequence exhibits, are actually very common in practice. Efficient discovery of the time duration in which temporal rules are valid could benefit KDD of many real applications. In this paper, we present a novel problem class that is the discovery of the distribution of temporal rules. We simplify the mining problem and depict a model that could represent this knowledge clearly, uniquely and efficiently. Our methods include four online dividing strategies for different mining interest, an incremental algorithm for measuring rule-sets, and an algorithm for mining this knowledge. We have analyzed the behavior of the problem and our algorithms with both synthetic data and real data. The results correspond with the definition of our problem and reveal a kind of novel knowledge.

1 Introduction

Temporal sequences that are lists of transaction records ordered by transaction time constitute a large part of data stored in many information systems. Well-known examples could be derived from sales records, stock prices, weather data, medical data, etc. In recent years, there has been increased interest in using data mining techniques to discover temporal rules from temporal sequences or from the discretized versions of time series data [14]. Simple examples include, "The price of stock A goes up and falls the next day, then it goes up the third day" and "A customer bought book A and B in one transaction, followed by book C in a later transaction".

Previous work on temporal rule discovery has mainly considered finding global rules, where every record in the temporal sequence contributes to the rules. However, in some applications, interesting rules are valid only in some time durations, i.e. some subsequences. This kind of rules that are time varying are actually very common in practice. Easily understood examples include, "In the sale records of a supermarket, a customer always buys biscuits followed by soda in the summer, but biscuits followed by milk in winter"; "During the period when film A is shown on TV, if a custom buys

[1] The research has been supported in part of Chinese national key fundamental research program (no, G1998030414) and Chinese national fund of natural science (no. 79990580)

M.-S. Chen, P.S. Yu, and B. Liu (Eds.): PAKDD 2002, LNAI 2336, pp. 469-480, 2002.

video CD A, then he buys video CD B in one month, where A and B have the same theme". Knowing in which time period a rule is frequent, i.e. the rule distribution, could be equally if not more useful than simply knowing whether it is frequent.

Another promotion of our research is that there have been a large amount of prevalent rules or broad regularities that are already known to domain experts. It's a broad consensus that the success of data mining will depend critically on the ability to go beyond obvious patterns and find novel and useful ones [1][12]. Mining known rules to discover the distribution of these rules is of crucial important for many applications. For example, "In summer, if the stock price of a game producer goes up and stays about level for two days, then it will go up the third day". We observe that there may be some correlation of price behavior from July to September so we can plan buy-sell strategy appropriately in that season while we will not be confused by this rule when we make a decision for the rest time.

To our knowledge, this problem has not been well considered in the KDD field. Nor has it been formally defined. In this paper, we present a novel problem class that is the discovery of the distribution of temporal rules in temporal sequences. The problem has a two-dimensional solution space consisting of rules and temporal features. From a practical point of view, it is impractical that use traditional methods to find all rules with their distributional information from large databases in terms of time complexity and analyzing difficulty. In this paper, we simplify the mining problem into restricted mining tasks, and use a model to represent this knowledge clearly, uniquely and efficiently. Another aspect of temporal sequences is it's appended frequently in the end, which makes it important that design an online mining method. For this reason, we propose four online dividing strategies for different mining interest. In addition, we propose an incremental algorithm for measuring rules and a mining algorithm to support efficient discovery of this knowledge. We have analyzed the behavior of the problem and our algorithms with both synthetic data and real data. The results correspond with the definition of our problem and reveal a kind of novel knowledge.

This paper is organized as follows: Section 2 discusses some related work. Section 3 formally defines the problem. Section 4 describes our dividing strategies. Section 5 presents the mining algorithms. Section 6 presents our experimental results. Finally, Section 7 offers some concluding remarks.

2 Related Work

In [10], the discretized version of the time series was used to discover time series rules with the format "if event *A* occurs, then event *B* occurs within time *T*". We use the same rule format in this paper, but to mine local behaviors instead of global ones. Discovery of temporal association rules [5][9][16] is where temporal features are considered as components of association rules. The problem seems alike to our approach, but the formats of both the database and the knowledge are essentially different. The problem of mining sequential patterns [2][3][4] has been well researched. [8] introduced the problem of discovering temporal patterns in multiple granularities. [6][13] proposed methods for discovering frequent episodes in sequences, where an episode consists of a set of events and their order. In [7], interval-based events were considered, where the duration of events is regarded as a temporal constraint in the discovery process. The problems considered in the works described above are different, and the algorithms were designed and optimized for the different purpose to ours. Applying these methods to preprocessed data can derive an applicable algorithm. However, it often leads to extremely poor time complexity. For example, we may retrieve all the possible subsequences, and use one of the above algorithms to mine rules in each subsequence. However, the number of possible subsequences of an N length sequence is $(1+2+\ldots+N)=(N+1)N/2$, which means we need to run the algorithm of mining global rules for $(N+1)N/2$ times.

3 Problem Descriptions

A temporal sequence $S=S_1, S_2, \ldots, S_N$ is a list of records ordered by position number. The subsequence $s=S_m, \ldots, S_n$ is denoted by $S[m,n]$. $|s|=n-m+1$ denote the length of s. In an actual application, a record is a description of interesting events at a time point. Without losing generality, we represent S by a sequence of event identifiers from an alphabet $\sum=\{a_1,\ldots,a_k\}$, where each symbol a_i uniquely represents a record.

In this paper, we consider a simple temporal rule format: if A occurs, then B occurs within time T where A and B are two event identifiers. We denote it as $R{:}A_{(T)}\rightarrow B$. Note that, any other temporal rule format could be used in our distributional representation mechanism with little modification needed on either problem definition or mining algorithm. This ensures our approach adaptive to a wild range of applications.

Definition 1: The *local frequency* of event A in subsequence s is the number of occurrences of A in s. The *local frequency* of rule $A_{(T)}\rightarrow B$ in subsequence s is the number of occurrences of the rule in s. We denote it as:

$$\mathrm{Lf}(A, s) = |\{i \mid S_i = A \wedge S_i \in s\}|$$

$$\mathrm{Lf}(A_{(T)}\rightarrow B, s) = |\{<i,j> \mid S_i=A \wedge S_j = B \wedge S_i \in S[j-T+1, j-1] \wedge S[i,j] \subseteq s\}|$$

Definition 2: The *local support* of A in subsequence s is the ratio of the *local frequency* of A in s over the length of s. The *local support* of rule $A_{(T)}\rightarrow B$ in subsequence s is the ratio of the *local frequency* of the rule in s over the length of s. The *local confidence* of rule $A_{(T)}\rightarrow B$ in subsequence s is the ratio of the *local support* of the rule in s over the *local support* of A in s. We denote it as:

$$\mathrm{Lsupp}(A,s) = \mathrm{Lf}(A,s)/|s|$$

$$\mathrm{Lsupp}(A_{(T)}\rightarrow B,s) = \mathrm{Lf}(A_{(T)}\rightarrow B, s)/|s|$$

$$\mathrm{Lconf}(A_{(T)}\rightarrow B,s) = \mathrm{Lsupp}(A_{(T)}\rightarrow B,s)/\mathrm{Lsupp}(A,s)$$

Definition 3: Given a rule-set $RS=\{R_k\}$, the *local support* of RS in subsequence s is the minimum value of all $\mathrm{Lsupp}(R_k,s)$. The *local confidence* of RS in subsequence s is the minimum value of all $\mathrm{Lsupp}(R_k,s)$.

Definition 4: Given minimum support MS_R, minimum confidence MC_R, rule $R{:}A_{(T)}\rightarrow B$ and subsequence s, if $\mathrm{Lsupp}(R,s) > MS_R \wedge \mathrm{Lconf}(R,s) > MC_R$, we consider $<R, s>$ a *local frequent rule* (LFR). Given a rule-set $RS=\{R_k\}$ and a subsequence s, if any $<R_k, s>$ is LFR, we consider $<RS, s>$ a *local frequent rule-set* (LFRS).

Example 1. Consider the sequence $S=$ "*ududududdddsusussu*" and MS=0.4, MC=0.5, $<u_{(2)}\rightarrow d, S[1,8]>$ is a LFR, because $\mathrm{Lsupp}(u_{(2)}\rightarrow d, S[1,8])=0.5\geq MS$ and $\mathrm{Lconf}(u_{(2)}\rightarrow d, S[1,8])=1\geq MC$. But $u_{(2)}\rightarrow d$ is not a frequent pattern for the whole sequence, because Lsupp($u_{(2)}$→d, S) = 4/18<0.4 .We could find some other LFRs such as <$s_{(2)}$→u, S[12,18]>, <$d_{(3)}$→d, S[1,11]>, <$d_{(2)}$→d, S[8,11]>, etc.

The *local support* and *local confidence* of a rule-set is different in different subsequence. The goal of distribution discovery is to find in which subsequences s, and for which rule-sets RS, <RS,s> are LFRSs. This problem has a two-dimensional solution space consisting of rules and temporal features. It is impractical that use traditional methods to find all rules with their distributional information from large databases in terms of time complexity and analyzing difficulty. Our strategy is simplifying the mining problem into restricted mining tasks, in which only distributional information of interest is considered. In order to describe this knowledge clearly, uniquely and efficiently, we propose using a model, which we term representation of rule distribution. The formal definition of the model is as follows:

Let ARS denote the set that include all interesting rules. Rule-set RS_m is a subsets of ARS, i.e. $RS_m=\{R_{mj}\}$ ($0<m\leq M$), $RS_m \subseteq ARS$.

Definition 5: A dividing candidate of sequence S is ordered continuous subsequences ($S[D_1,D_2$-1], $S[D_2, D_3$-1], … , $S[D_{n-1}, D_n$-1]) where for any k ($1<k\leq n$), we can find a m ($0<m\leq M$) such that <RS_m, $S[D_{k-1}, D_k$-1]> is LFRS.

Definition 6: Given a sequence S and a set of rule-sets RS_m, representation of rule distribution (RRD) of S is a dividing candidate that is chosen by a given strategy. It has the format: RRD (S) = (<RS_m, $S[D_{k-1}, D_k$-1]>)

Form the analyst's viewpoint, (D_n) is the change points that divide the sequence into segments with different valid rule-sets. Given a sequence S and a set of rule-sets $\{RS_m\}$, there might be several diving candidates. Among them, some are useless and others reveal different aspects about the sequence. The problem of discovery of rule distribution is: examine the sequence to find instantiations of rule-sets in $\{RS_m\}$ that appear frequently and generate a unique representation RRD (S) for analyst.

Example 2. Consider the sequence in example 1: S= "*ududududdddsusussu*" and rule-sets {{$u_{(2)}$→d}{$d_{(3)}$→d}{$s_{(2)}$→u}}. Given MS=0.4, MC=0.5, we could find dividing candidates such as (S[1,8] S[9,11] S[12,18]) and (S[1,11] S[12,18]). Then we can use a strategy to choose one from the two candidates as RRD(S). For example, RRD(S) = (<$u_{(2)}$→d,S[1,8]> <$d_{(3)}$→d,S[9,11]> <$s_{(2)}$→u,S[12,18]>).

To acquire the knowledge that we defined, there are two main problems that must be solved. First, we need an efficient algorithm for measuring the rules in a certain subsequence. Second, we need a method that can generate a unique distributional representation, i.e. a dividing strategy, for candidates choosing.

4 Dividing Strategies

A common approach for generating RRD is first get all the dividing candidates, then one of them is chosen manually based on the domain expert's analysis and explanation. This method is simple to understand or to implement. However, it has some fatal drawbacks. First, the database is always with extremely huge size. From a practical point of view, it is too expensive to find and explore all candidates. In addition, this method need consider sequence context during the process of generating all the candidates, so it can only perform offline.

Our approach is: generate RRD automatically along with the process of scanning the sequence by using a dividing strategy. In this section, four dividing strategies are proposed. We denote the considered sequence by *S*; denote the number of all rules by *M*; and denote the number of rules in a valid rule-set by *m*.

4.1 Piecewise Validation

Piecewise validation (PV) is the simplest dividing strategy. The method is: divide the sequence into time windows with the same width; then find valid rule-sets in every window; and sew up adjoining time windows with the same valid rule-sets.

This strategy can approximately identify valid time period of rule-sets. The quality of this strategy depends mainly on the window width. If the window is too large, we may miss the rules of which the valid period is a small portion in a window.

Measuring of a rule *R* in subsequence *s* can be completed in time $O(|s|)$ by a simple pass through s. So the time complexity of piecewise validation strategy is $O(WM|S|/W)=O(M|S|)$ where *W* is the window width.

4.2 Length First Strategy

The length first strategy (LF) discovers the longest valid subsequence, while the rules that are valid only in a small portion are ignored. The strategy is: expand the tentative subsequence while phase-out rule-sets that are not valid until at last find the expected subsequence. The detail process is as follows:

1. Let *RSS* be a set includes all the rule-sets, *s* be the next minimal subsequence that has not been considered.
2. Go through the sequence, and expand *s* in order.
3. If *s* does not support a rule-set *RS*, delete *RS* from *RSS*.
4. Back to step 2 until *RSS* become an empty set after deleting or the length of *s* exceed a predefined threshold.
5. Output the sequence before the last expanding as a resulting subsequence. Back to step 1 to find the next subsequence until the whole sequence has been considered.

The time complexity of LF strategy varies with the distribution of rule-sets. The best case is when one rule-set is valid during the whole time duration of *S*, and none of other rule-sets is valid. In this occasion, all rule-sets except one are deleted from the beginning, thus only one rule-set need to be considered. As we will discuss in section 5.1, rule-set can be measured incrementally in $O(1)$ time during the expanding process. So the time complexity of the best case is $O(m|S|)$. The worst case is when all the rule-sets are valid in the whole time duration. Then all rule-sets need measured at each expanding step. The time complexity is $O(M|S|)$.

4.3 Frequency First Strategy

In some applications, user might be interested only in the most remarkable rules. Frequency first strategy (FF) discovers subsequence with the strongest rules by expanding the tentative subsequence and deleting the rule-sets of which the *local support* or *local confidence* is minor. The detail process is:

1. Let *RSS* be a set includes all the rule-sets, *s* be the next minimal subsequence that has not been considered.

2. Go through the sequence, and expand *s* in order.

3. Delete every rule-set of which the *local support* is less than $\varepsilon \cdot maxsupp$ or the *local confidence* is less than $\varepsilon \cdot maxconf$, where *maxsupp* is the maximum local support and *maxconf* is the maximum local confidence of all the rule-sets.

5. Back to step 3 until $maxsupp<ms$ or $maxconf<mc$ or the length of *s* exceed a predefined threshold.

6. Output the sequence before the last expanding as a resulting subsequence. Back to step 1 until the whole sequence has been considered.

The time complexity of FF strategy is similar to that of the LF strategy. The best case is when one rule-sets is valid during the whole time duration, and the support and confidence of other rule-sets are small enough in the initial subsequence. Then only one rule-set need considered. The time complexity is $O(m|S|)$. The worst case is when all the rule-sets are with similar support and similar confidence in the whole time duration. The time complexity is $O(M|S|)$.

4.4 Quantity First Strategy

Quantity first strategy (QF) focuses on discovering subsequence with most rule-sets. The strategy finds subsequence as follows: given a subsequence with minimal length, get all the rule-sets that are valid in it; then expand the subsequence until one of the rule-sets is not valid. The detail process is:

1. Let *s* be the next minimal subsequence that has not been considered. Let *RSS* be a set includes all the rule-sets that are valid in *s*, i.e. $RSS=\{RS|\langle RS, s\rangle \text{ is LFRS}\}$.

2. Go through the sequence and expand *s* in order until at least one rule-set in *RSS* is not valid in *s* or the length of *s* exceed a predefined threshold.

6. Output the sequence before the last expanding as a resulting subsequence. Back to step 1 until the whole sequence has been considered.

The time complexity of this strategy is same to that of the LF strategy. The QF strategy gives similar results to that of the PV strategy. But the dividing positions using QF strategy, which are not constrained to be with the same interval, is more flexible.

4.5 Discussion

The four dividing strategy differ in the ways of time complexity, meaning of the results and the effect of the initial subsequence. Which strategies are used is based on the mining aims, the application aspects and the characters of the datasets.

For datasets with common rule distribution, the time complexity of PV strategy is poorer than that of the other strategies. But it can give approximately all the valid rule-sets in each time period by a careful choosing window width.

The aim of the LF FF and QF strategies is to find subsequence with valid rule-sets of different user interest. The three strategies are "expanding and deleting" methods.

We first get a group of rule-sets that are valid in the first few records, and then use some methods to select some of them as interesting rule-sets at the beginning or while expanding the subsequence. So the initial subsequence determines which rule-sets are considered. This methodology degreases the searching space and gives concise results. But some interesting rule-sets may be ignored. To cope with this problem, we use the threshold *maximum length* to force the tentative subsequence reinitialized. This helps more potential valid rule-sets involved. In addition, the representing sequence accomplishes a significant space reduction. This makes it possible that domain experts use different strategies to generate multi distributional representations with different aspects. This could fill up the gaps of our "expanding and deleting" strategies.

The four dividing strategies don't need future data involved. Furthermore, additional post-procession is not needed. So our methods could online mine the data.

5 Discovery Method

5.1 Incremental Frequency Measurement

The *local support* and *local confidence* of rules need repeatedly calculated in the mining process. Using the "expanding and deleting" strategies, Lf($R{:}A_{(T)}\rightarrow B$, $S[m, n]$) and Lf(A, $S[m, n]$) are available when Lsupp($R{:}A_{(T)}\rightarrow B$, $S[m, n+1]$) and Lconf($R{:}A_{(T)}\rightarrow B$, $S[m, n+1]$) are calculated during the expanding process. This enables the implementation of an incremental algorithm for rule measurement. Because *local support* and *local confidence* can be derived from *local frequency* by one-step calculation, we only give the Lf algorithm in Fig.1.

```
Input: R:A(T)→B, S[m, n], Lf (R, S[m, n-1]), Lf (A, S[m, n-1])
Output: Lf (R, S[m, n]), Lf (A, S[m, n])

If Sn = A then
Lf (R, S[m, n]) = Lf (R, S[m, n-1]);
Lf (A, S[m, n]) = Lf (A, S[m, n-1]) + 1
Else if Sn = B
Lf (R, S[m, n]) = Lf (R, S[m, n-1]) + |{p|n-T<p<n ∧ Sp = A}|
Lf (A, S[m, n]) = Lf (A, S[m, n-1])
Else
Lf (R, S[m, n]) = Lf (R, S[m, n-1])
Lf (A, S[m, n]) = Lf (A, S[m, n-1])
End if
```

Fig. 1. Lf Algorithm

5.2 Mining Algorithm

Our method is: scan the temporal sequence and expand a time window step by step if needed; after each expanding, we calculate the *local support* and *local confidence* for each possible rule-set incrementally; delete chosen rule-sets if the dividing strategy need to; the expanding process is repeated until the terminating condition of the dividing strategy is satisfied or the width of the time window exceed *maximum length*; after one expanding process have terminated, we reset the window to the next position and start another expanding process. After the whole sequence has been scanned, we get unique RRD sequence. The detail of our algorithm is shown in Fig.2.

Let *s* be the window before expanding, *RSS* be the set of current remained rule-sets. When a subsequence is outputted, there exists $RS \in RSS \Leftrightarrow \langle RS, s \rangle$ is LFRS. So the results accord with definition 4 in section 2. In addition, each time only one subsequence need outputted. So we get a unique result, which accords with definition 5.

```
Input: temporal sequence S, dividing strategy D, minimal length of a
subsequence (also serve as the window width when use PV as D), maxi-
mum length of a subsequence.
Output: RRD (S)

m =first position in S
Repeat
   W=S[m, m+minimal length-1]
   RSS = initial rule sets which are generated based on D
   Repeat
     If D∈{LF FF QF} then
       For every RS_k∈{RSS}
           If the deleting condition of D is satisfied
               RSS=RSS-RS_k
           else
               lastRS = RS_k
           End if
        Expand W to right
      Else
        LastRS = RSS
     End if
   Until the terminating condition of D or |W|>maximum length
   If |W| = minimal length and D∈{LF FF QF} then
     Output <Ø, S[m,m+minimal length-1]>;
     m = m + minimal length
   Else
     Output < lastRS, W>;
     m = m + |W|
   End if
Until m > |S|
```

Fig. 2. RRD Algorithm

6 Experimental Results

We implemented a mining system for the purpose of discovering rule distributions in stock price data. The inputting database stores a set of time series, which is a sequence of real numbers. First, we discretize it into symbol pattern sequences in which each symbol represents the series behavior at that time point. The method used is discretizing by clustering windows [10]: Given time series $x = x_1, \ldots, x_N$, window $W_i=(x_{ik}, x_{ik+1},\ldots, x_{ik+w-1})$ of width w and step k is a contiguous sub-series. Then use recursive k-means cluster method and a distance matrix to cluster all the subsequences into sets $C_1, \ldots, C_M$. Then the discretized version D(x) of x is obtained by looking for each widows W_i the cluster ID $j(i)$ such that $W_i \in C_{j(i)}$, i.e. D(x) = $j(1), \ldots, j(\lfloor N/k \rfloor)$.

Rule-sets, which are input parameters, are derived by three ways: 1) the results of mining global rules with minor support and minor confidence, 2) the large amount of prevalent rules or broad regularities that are already known to domain experts, and 3) all the possible rules. The number of possible rules with the format $A_{(T)} \rightarrow B$ is mp^2, where p is the number of letters in the alphabet and m is the number of different possibilities for T.

After the discretized sequences and rule-sets have been generated, the distributional representation is generated using RRD algorithm. We evaluated the behavior of our algorithm with both synthetic data and real data from stock market, of which the goal is to find whether the method is able to correctly identify the distribution of rule-sets. In this section, we present the results of these evaluations.

6.1 Experiments with Synthetic Data

The synthetic dataset was generated using the following function:

$$x(n) = \begin{cases} L + A\sin(2\pi n/T) + Rg & 0 < n \le N/6 \\ kn - Nk/6 + Rg & N/6 < n \le N/3 \\ H + Rg & N/3 < n \le N/2 \\ H + A\sin(2\pi n/T) + Rg & N/2 < n \le 2N/3 \\ -kn + 2Nk/3 + Rg & 2N/3 < n \le 5N/6 \\ L + Rg & 5N/6 < n \le N \end{cases}$$

where $k=6(H-L)/N$. g is Gaussian noise with zero mean and unit variance. H, L, A, T control the shape of the function; N controls the length; R controls the noise-to-signal ratio. An example (without noise), with the coefficients L=1, H=2, A=1/15, N=600, T=25 is depicted in Fig 3. During the discretizing process, the window width was set to be 16; the window step was set to be 1. Corresponding discretized sequences D(x) with R=0.5 and R=1 are depicted in Fig.4 and Fig.5.

We used the setting *minimal length*=50, *maximum length*=∞ and applied all the four dividing strategies on each D(x). Rule-sets, together with corresponding MF_R, MC_R and the meanings of the rule-sets are shown in Table 1.

The results are shown in Table 2. From the visual analysis, the answer should be (<RS_4,S[1,100]> <RS_2,S[101,200]> <RS_1,S[201,300]> <RS_4, S[301,400]> <RS_3,S[401, 500]> <RS_1,S[501, 584]>). The experiments using PV strategy presented the exact results. When LF, FF and QF were used, since the initial rule-sets were same, the resulting RRDs were same. And the results are similar to the right answer. The slight difference comes from the randomness of the method of generating experimental data.

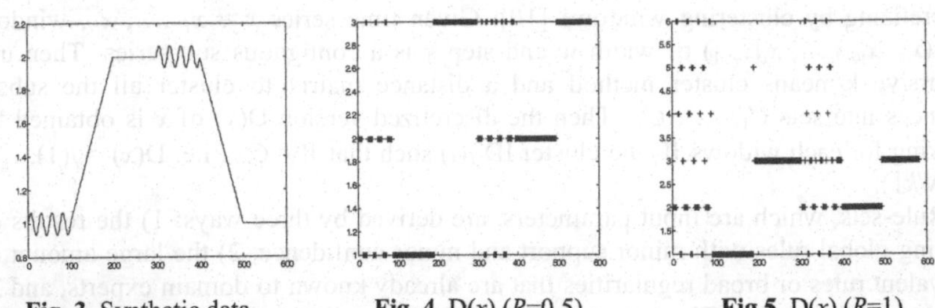

Fig. 3. Synthetic data Fig. 4. D(x) (R=0.5) Fig.5. D(x) (R=1)

m	RS_m	Meaning of rule-set	MF_R	MC_R
1	$\{3_{(3)}\rightarrow 3\}$	stand ⇒ stand	0.7	0.8
2	$\{1_{(3)}\rightarrow 1\}$	up ⇒ up	0.7	0.8
3	$\{2_{(3)}\rightarrow 2\}$	down ⇒ down	0.7	0.8
4	$\{1_{(15)}\rightarrow 2, 2_{(15)}\rightarrow 1\}$	up ⇒ down, down ⇒ up	0.1,0.1	0.4,0.4

Table 1. Rule-sets and corresponding meanings

R	Dividing Strategy	Results
0.5	PV	(<RS_4, S[1,100]> <RS_2, S[101,200]> <RS_1, S[201,300]> <RS_4, S[301,400]> <RS_3,S[401,500]> <RS_1,S[501,584]>)
1	PV	(<RS_4, S[1,100]> <RS_2, S[101,200]> <RS_1, S[201,300]> <RS_4, S[301,400]> <RS_3,S[401,500]> <RS_1,S[501,584]>)
0.5	LF,FF,QF	(<RS_4, S[1,106]> <RS_2, S[107,189]> <RS_1, S[190,325]> <RS_4, S[326,396]> <RS_3,S[397,494]> <RS_1,S[495,584]>)
1	LF,FF,QF	(<RS_4, S[1,109]> <RS_2, S[110,187]> <RS_1, S[188,300]> <RS_4, S[301,396]> <RS_3,S[397,493]> <RS_1,S[494,584]>)

Table 2. Results on synthetic data

6.2 Experiments with Real Data

The real data was time series of daily prices that extracted from Shenzhen stock market. During the discretizing process, the window width was set to be 30; the window step was set to be 10. One of the dataset is shown in Fig. 6; corresponding D(x) is shown in Fig. 7; the center of each cluster, which represents the price movement, is shown in Fig. 8. We used the setting *minimal length*=10 and *maximum length*=∞. Rule-sets, together with corresponding MF_R, MC_R are shown in Table 3. The results are shown in Table. 4.

The results give us a concise and meaningful representation of novel knowledge about the price movement. For example, from the results using LF, we see during the most time, a period of fluctuating was always followed by a period of fall in two days. From the results using PV, we could get more information such as: During the time [31,60], there is another frequent rule that is: a period of "wave hollow" was always followed by a period of fluctuating in two days. By analyzing the results using FF, we realize that the latter one happened with more possibility during that time. After the RRD have been discovered, it is possible to obtain other novel knowledge through the comparisons with the rule distribution derived from the market environment or from price movement of other stocks.

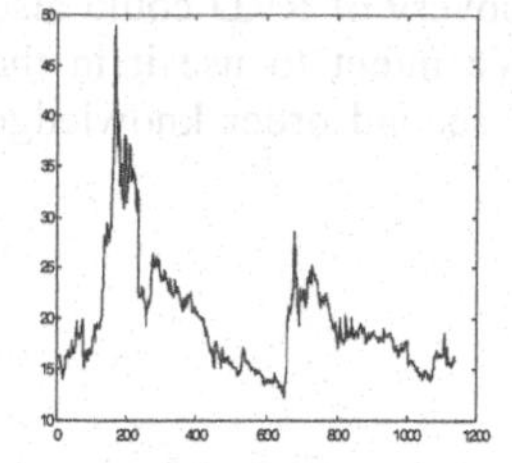

Fig. 6. Real data

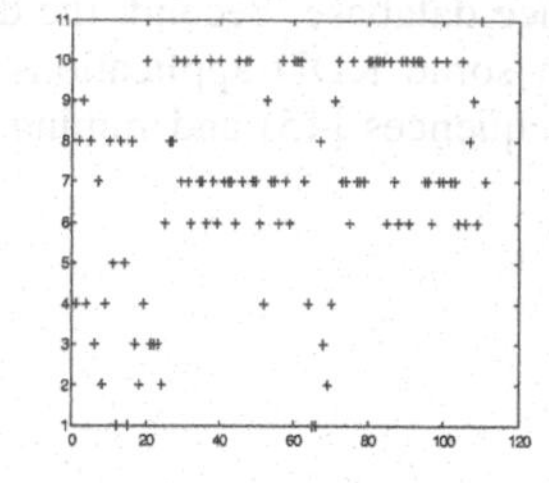

Fig. 7. Discretized sequence D(x)

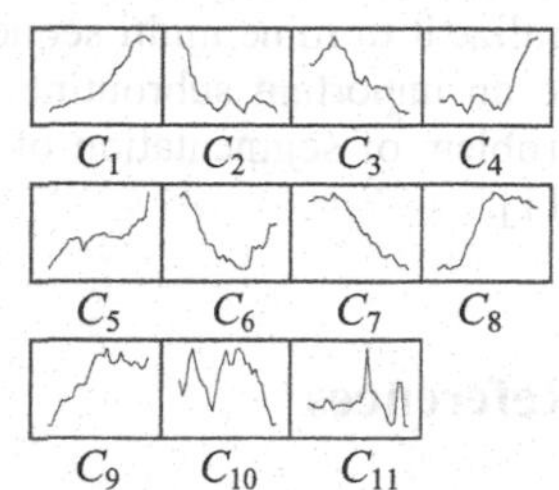

Fig. 8. Center of each cluster

m	RSm	MF_R	MC_R
1	$\{10_{(2)} \rightarrow 10\}$	0.2	0.6
2	$\{6_{(2)} \rightarrow 10\}$	0.2	0.6
3	$\{10_{(2)} \rightarrow 7\}$	0.2	0.6
4	$\{7_{(2)} \rightarrow 6\}$	0.2	0.6

Table 3. Rule-sets

Dividing Strategy	Results
PV	$(<\varnothing, S[1,30]> <RS_2\ RS_3, S[31,50]> < RS_2\ RS_3\ RS_4, S[51,60]>$ $<RS_3, S[61,80]> < RS_1\ RS_2\ RS_3, S[81,100]> <\varnothing, S[101,110]>)$
LF	$(<\varnothing, S[1,30]> <RS_3, S[31,102]> <\varnothing, S[103,110]>)$
FF	$(<\varnothing, S[1,30]> <RS_2, S[31,60]> <RS_3, S[61,110]>$
QF	$(<\varnothing, S[1,30]> <RS_2\ RS_3, S[31,60]> <RS_3\ S[61,110]>)$

Table 4. Results on stock price data

7 Conclusion

In this paper, we present a novel problem class that is the discovery of the distribution of temporal rules. We depict a model that could represent this knowledge clearly, uniquely and efficiently. And we address our methods for mining of this knowledge.

The time complexity of our mining algorithm varies depending on the datasets and the rule-sets. Compared to standard mining algorithms of which the goal is the discovery of global temporal rules, our method involves additional computations. However, our method could ignore mundane knowledge [1][12] and discover novel one. This merits the added expense. Furthermore, The distributional representation is generated along with the process of scanning the sequence. This enables us to develop an online mining system. We analyzed the behavior of the problem and our algorithm with both synthetic data and real data. The results correspond with the definition of our problem and reveal a kind of novel knowledge.

In future, we intend to extend the problem in the following two directions. First, in this paper, we confined the database consists of a single sequence. We intend to generalize it to mine multi sequence database. Second, the discovery of RRD could also be an important subroutine in some KDD applications. We intent to use it in the problem of segmentation of sequences [15] and mining of second order knowledge [11].

References

1. M. Klemettinen, H. Mannila, P. Ronkainen, H. Toivonen, and A. I. Verkamo. Finding interesting rules from large sets of discovered association rules. The 3rd International Conference on Information and Knowledge Management, pages 401-407, 1994.
2. T. Dietterich, R. Michalski. Discovering patterns in sequences of events. Artificial Intelligence, Vol.25, 1985.
3. R. Agrawal, R. Srikant. Mining sequential patterns. International Conference On Data Engineering. Taipei, 1995.
4. R. Srikant, R. Agrawal. Mining sequential patterns: generalizations and performance improvements. The Fifth International Conference on Extending Database Technology, 1996.
5. S. Ramaswamy, S. Mahajan, A. Silberschatz. On the discovery of interesting patterns in association rules, The VLDB Journal, pages 368-379, 1998
6. H. Mannila, H. Toivonen, and A.I. Verkamo. Discovering frequent episodes in sequences. The 1st International Conference on Knowledge Discovery and Data Mining, Canada. 1995.
7. P. Kam, A. Fu. Discovering temporal patterns for interval-based events. The 2nd International Conference on Data Warehousing and Knowledge Discovery (DaWaK 2000). 2000.
8. Y. Li, X. Wang and S. Jajodia. Discovering temporal patterns in multiple granularities. International Workshop on Temporal, Spatial and Spatio-Temporal Data Mining. Lyon, France. 2000.
9. X. Chen, I. Petrounias. An Integrated query and mining system for temporal association rules. The 2nd International Conference on Data Warehousing and Knowledge Discovery (DaWaK 2000), London, UK. 327-336. 2000.
10. G. Das, K. Lin, H. Mannila, G. Renganathan, P. Smyth. Rule discovery from time series. the 4th International Conference on KDD. 1998.
11. M. Spiliopoulou, J.F.Roddick. Higher order mining: modelling and mining the results of knowledge discovery. Data Mining II - Second International Conference on Data Mining Methods and Databases. 2000.
12. U. Fayyad, G. Piatetsky-Shapiro, P. Smyth. From data mining to knowledge discover: and overview. Advances in Knowledge Discovery and Data Mining. 1996.
13. H. Mannila, H. Toivonen. Discovering generalised episodes using minimal occurences. The 2nd International Conference on Knowledge Discovery and Data Mining (KDD-96). 1996.
14. J. Roddick, M. Spiliopoulou. A bibliography of temporal, spatial and spatio-temporal data mining research. SIGKDD Explorations, Vol 1, No. 1. 1999.
15. V. Guralnik, J. Srivastava. Event detection from time series data. The 5th International Conference on Knowledge Discovery and Data Mining, USA. 1999.
16. A. Tansel, N. Ayan. Discovery of association rules in temporal databases. 4th International Conference on Knowledge Discovery and Data Mining (KDD'98) Distributed Data Mining Workshop, NewYork, USA, August 1998.

News Sensitive Stock Trend Prediction

Gabriel Pui Cheong Fung, Jeffrey Xu Yu, and Wai Lam

Department of Systems Engineering & Engineering Management
The Chinese University of Hong Kong
Shatin, N.T., Hong Kong
{pcfung,yu,wlam}@se.cuhk.edu.hk

Abstract. Stock market prediction with data mining techniques is one of the most important issues to be investigated. In this paper, we present a system that predicts the changes of stock trend by analyzing the influence of non-quantifiable information (news articles). In particular, we investigate the immediate impact of news articles on the time series based on the Efficient Markets Hypothesis. Several data mining and text mining techniques are used in a novel way. A new statistical based piecewise segmentation algorithm is proposed to identify trends on the time series. The segmented trends are clustered into two categories, `Rise` and `Drop`, according to the slope of trends and the coefficient of determination. We propose an algorithm, which is called guided clustering, to filter news articles with the help of the clusters that we have obtained from trends. We also propose a new differentiated weighting scheme that assigns higher weights to the features if they occur in the `Rise` (`Drop`) news-article cluster but do not occur in its opposite `Drop` (`Rise`).

1 Introduction

Autoregressive and moving average are some of the famous stock trend prediction techniques which have dominated the time series prediction for several decays [1,6,17]. With the help of data mining, several approaches using inductive learning for prediction have also been developed, such as k-nearest neighbor and neural network. However, their major weakness is that they rely heavily on structural data, in which they neglect the influence of non-quantifiable information. Fawcett and Provost [5] formulated the stock forecasting problem as an activity-monitoring problem by monitoring the relationship between news articles and stock prices. However, a detailed procedure is absent in their paper. Lavrenko et al. [13] proposed a language modeling approach for the stock trend forecasting problem. In their approach, all news articles that are announced five hours before the happening of any given trend are aligned to that trend, which is then used as the basis for generating the prediction model. However, this approach will not only lead to many conflicts [1], but also lead to contradiction against the Efficient Markets Hypothesis (EMH).

[1] For example, the same article may align to more than one type of trend.

M.-S. Chen, P.S. Yu, and B. Liu (Eds.): PAKDD 2002, LNAI 2336, pp. 481–493, 2002.

According to EMH, it states that the current market is an efficient information processor which immediately reflects the assimilation of all of the information available [2,3,17]. Thus, a long time lag is normally impossible.

In this paper, we present a system, which is based on EMH, to predict the future behavior of the stock market using non-quantifiable information (news articles). No fixed periods are needed in our system. Predictions are made according to the contents of the news articles. Thus, our system is event-driven. The unique features of our system are summarized as follows: first, a new t-test based piecewise segmentation algorithm for trend discovery is proposed. Second, interesting trends are clustered using an agglomerative hierarchical clustering algorithm based on slopes and coefficient of determination. Third, news articles are clustered and aligned to the interesting trends according to a newly proposed algorithm based on incremental K-mean. It is used to filter the news articles which are announced under the trend but do not support it. Fourth, a new differentiated weighting scheme is proposed for assigning higher weights to those features that are present in the set of articles which are aligned to only one trend type.

The reminding of this paper is organized as follows. Section 2 presents the detailed system architecture. Section 3 evaluates the performance of our system. A conclusion of this paper is given in Section 4.

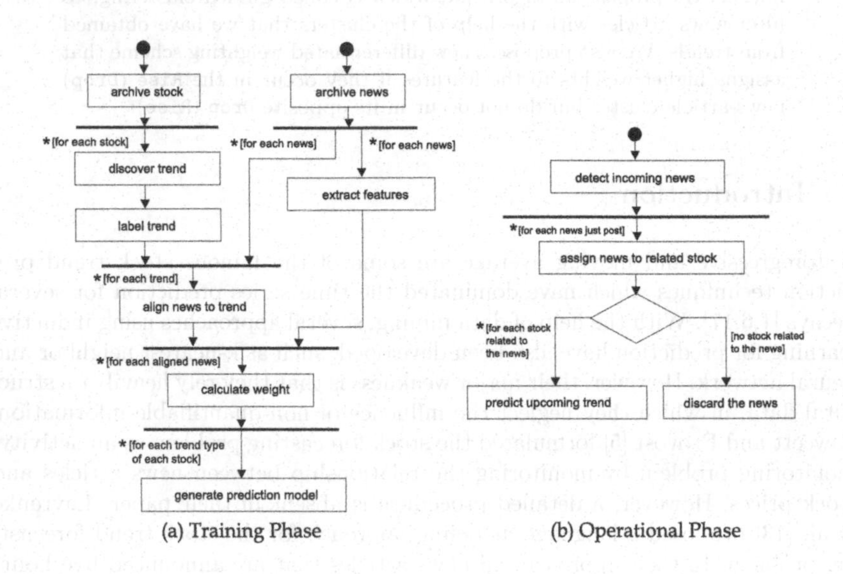

Fig. 1. The System Overview

2 A News Sensitive Stock Prediction System

The overview of our system is shown in Figure 1(a) and Figure 1(b) using the Unified Model Language. It consists of two phases: system training phase and operational phase. The system training phase includes six main procedures: 1) trend discovery, which identifies trends within different periods; 2) trend labeling, which clusters similar trends together; 3) feature extraction, which extracts the main features in the news articles; 4) articles-trend alignment, which associates related news articles with trends; 5) feature weighting, which assigns different weights to different features according to their importance; and 6) model generation, which generates the desire prediction model. The operational phase is used to predict the future trends according to the contents of the newly broadcasted news articles. The unique features and details of these procedures are discussed in the following subsections in details.

2.1 Stock Trend Discovery

The general trend on the time series is always more interesting to study than the exact fluctuation, as stock traders are always adhere to the trends only. Some well-known time series segmentation techniques include Fourier coefficients [4] and parametric spectral models [19]. However, if a time series consists of transient behavior or unstationary with respect to time, such as stock prices, it will possess very weak spectral signatures even locally [10]. Thus, instead of using a spectral representation, we propose a new piecewise segmentation algorithm. This algorithm adopts the similar ideas given in [16] for finding the minimum k segments with the error norm below a threshold. Our piecewise segmentation algorithm consists of two phases: splitting phase and merging phase.

The splitting phase is aimed at discovering all of the interesting trends on the time series regardless of the duration of the trend, while the merging phase is aimed at avoiding over-segmentation. The details of the algorithm are as follows.

Initially, the time series is regarded as a single segment, and a regression line is formulated to represent it. For every point (x_i, y_i) on the time segment, its error norm with respect to the regression line is:

$$E_i = |\sin\theta \cdot x_i + \cos\theta \cdot y_i - d| \tag{1}$$

where θ is the angle between the regression line and the x-axis; d is the perpendicular distance from the origin to the regression line. The error norm calculated is the Euclidean distance between (x_i, y_i) and the regression line. Note that E_i has a normally independent distribution. In order to determine whether the time series should be split, an one tailed t-test is formulated:

$$\begin{aligned} H_0 &: \mu = 0 \\ H_1 &: \mu > 0 \end{aligned} \tag{2}$$

where μ is the mean error norm of interest. The required t-statistics is:

$$t = \frac{\overline{E} - \mu}{\frac{s}{\sqrt{n}}} \tag{3}$$

where, n is the number of points within the segment, s is the standard deviation of the error norm, and $\overline{E}$ is the mean error norm. The t-statistics is compared with t-distribution with $n - 1$ degree of freedom using $\alpha = 0.05$. If the null hypothesis is rejected, then the time series will be split at the point where the error norm is maximum, i.e. $\max\{E_i\}$, and the whole process will be executed recursively on each segment.

Since the splitting phase is carried out on each segment independently, the resulting segments may not be globally optimal. The merging phase is aimed at avoiding over-segmentation by merging adjacent segments, provided that the error norm after merging will not exceed the threshold which is determined by t-test. The whole process of the merging phase is exactly the same as the splitting phase, except that the adjacent segments are merged. An example is given in Figure 2. If the hypothesis is rejected, it implies that the error norm will exceed the threshold after merging, therefore merging would not be performed.

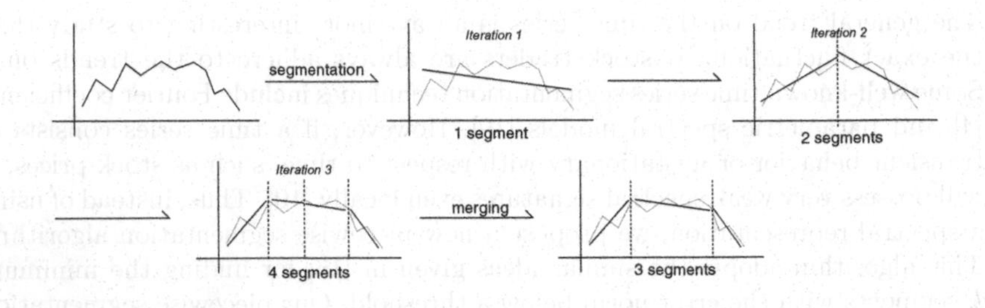

Fig. 2. t-test based piecewise segmentation

2.2 Stock Trend Labeling

In reality, a stock trader will never be interested in the trends that are relatively steady, as they provide neither opportunities nor threats. In order to cluster the trends into different interesting categories, a two dimensional agglomerative hierarchical clustering algorithm is formulated, which is based on: 1)the slope of the segment (m) and 2) the coefficient of determination (R^2). Slope is chosen because it is one of the most concerning issues for the stock traders. Coefficient of determination is chosen because it is a measurement for the goodness of fit of the regression model. R^2 is defined as:

$$R^2 = \frac{\sum_{i=1}^{n}(\widehat{y}_i - \overline{y})^2}{\sum_{i=1}^{n}(y_i - \overline{y})^2} \tag{4}$$

where y_i is the original data; $\widehat{y}_i$ is the segmented data corresponding to y_i; and $\overline{y}$ is the mean of the original data in that segment. Thus, the molecule is the regression sum of squares and the denominator is the corrected sum of squares. All slopes are normalized within 1 and -1 in order to have consistent clustering results across different stocks. Note that R^2 is always between 0 and 1. Each segment is thus represented by (m, R^2), and is regarded as an individual cluster object. The segments are merged according to the minimum group average distance (GAD):

$$GAD(C_i, C_j) = \frac{\sum_{i \in C_i} \sum_{j \in C_j} d_{ij}(i,j)}{|C_i||C_j|} \tag{5}$$

where $|C_i|$ and $|C_j|$ are the magnitudes of the cluster C_i and C_j respectively; $d(i,j)$ is the Euclidean distance between the objects inside C_i and C_j:

$$d_{ij}(i,j) = \sqrt{(m_i - m_j)^2 + (R_i^2 - R_j^2)^2} \tag{6}$$

The clustering procedure terminates when the number of clusters are equal to three. This is because we are interested in clustering the trends into three classes (`Rise`/`Drop`/`Steady`). Based on the average slope in the three clusters, those segments in the cluster having the maximum average slope are labeled as `Rise`. Similarly, those segments in the cluster having the minimum average slope are labeled as `Drop`. Segments in the remained cluster are labeled as `Steady`. Note that simple threshold could also be used for classifying the trends. However, we claim that it is not scientific enough and the results may vary from stock to stock.

2.3 Article and Trend Alignment

After trends are grouped into clusters, relevant news articles are then aligned to them. By alignment, we mean that the contents of the news articles would support and account for the happening of the trends. Obviously, not every piece of news article announced under the time series would support the happening of the trends. In this section, we propose a new algorithm, named guided clustering, which can filter out news articles that do not support the trends. Our algorithm is an extension of the incremental K-Means [11], together with the help of the `Rise`/`Drop` trends, in which they are regarded as "guides". In other words, we use trends to govern the clustering of the news articles. Incremental K-Means is chosen because recent research findings showed that K-Means outperforms the hierarchical approach for textual documents clustering [20], and is more efficient and robust [12].

In the following discussions, let T-clusters be the clusters of trends and let N-clusters be the clusters of news articles. T-cluster `Rise` (`Drop`) is regarded as guide-`Rise` (guide-`Drop`). A guide-`Rise` (guide-`Drop`) includes all of the trends in the T-cluster `Rise` (`Drop`). The guided clustering algorithm works as follows. First, all of the news articles that are broadcasted within T-cluster `Rise` (`Drop`) are grouped together. Each article is represented by a normalized vector-space model:

$$d_i = (w_1, w_2, ..., w_n) \tag{7}$$

where the element w_t corresponds to the score of the term t in the article d_i, and it is calculated by the standard $tf \cdot idf$ scheme:

$$w_t = tf_{d,t} \times \log \frac{N}{df_t} \tag{8}$$

where $tf_{d,t}$ is the frequency of the term t in the article d; df_t is the number of article(s) containing the term t; N is the total number of articles contained in the particular T-cluster (Rise/Drop). Incremental K-Means is then used for splitting the weighted articles into two clusters [20]. The centroid of the cluster C_i is defined as:

$$C_i = \frac{1}{|S_i|} \sum_{d \in S_i} d \tag{9}$$

where S_i is the set of articles within the cluster C_i and $|S_i|$ is the number of articles in this set. The similarity between the article d_i and the centroid C_j is determined by the cosine measure which is recommended by most researchers in document categorization:

$$\cos(d_i, C_j) = \frac{d_i \cdot C_j}{|d_i||C_j|} \tag{10}$$

where $|d_i|$ and $|C_j|$ is the magnitude of the article d_i and the cluster C_j respectively.

The above procedure is taken out by both guide-Rise and guide-Drop. Thus, on completion, four N-clusters exist: two N-clusters for guide-Rise, C_{r_1} and C_{r_2}, and two N-clusters for guide-Drop, C_{d_1} and C_{d_2}. In order to align the correct N-cluster to the corresponding T-cluster, the following formula is used for comparison:

$$\cos(C_i, C_j) = \frac{C_i \cdot C_j}{|C_i||C_j|} \tag{11}$$

The Rise N-cluster, C_{r_i} (for $i = 1, 2$), which has the highest average similarity with the two Drop N-clusters, C_{d_k} (for $k = 1, 2$), is regarded as insufficient to differentiate the trend Rise. Thus, it will be removed. Suppose we removed C_{r_1}, then all of the news articles within C_{r_2} will be aligned to T-cluster Rise. Similarity, the Drop N-cluster, C_{d_k} (for $k = 1, 2$), which has the highest average similarity with the two Rise N-clusters, will also be removed. Filtering is thus achieved.

2.4 Differentiated News Article Weighting

After the alignment process, a set of news articles are aligned to the trends. In order to distinguish the importance of each feature in each N-cluster, the weight of each feature has to be re-calculated. For most of the existing weighting algorithms, the basic idea is that features which rarely occur over a collection of articles are valuable. However, we are interested in finding out the features which frequently occur in one of the N-cluster but rarely occur in the other one. Thus, none of the existing weighting algorithms fully fit into our system.

A new weighting scheme is proposed here. Note that word independence is assumed under this scheme, which is a common practice in text classification research. Furthermore, preserving word dependence may not necessarily help to improve the accuracy of the model, or to some extents may even make it worse [18].

In order to differentiate the features appearing in one of the cluster but not the other, two coefficients are introduced: inter-cluster discrimination coefficient (*CDC*) and intra-cluster similarity coefficient (*CSC*):

$$CDC = (\frac{n_{i,t}}{N_t})^2 \tag{12}$$

$$CSC = \sqrt{\frac{n_{i,t}}{n_i}} \tag{13}$$

where $n_{i,t}$ is the number of articles in the N-cluster i containing the term t; N_t is the total number of articles containing the term t; n_i is the total number of different terms appearing within the N-cluster i.

The intuition of *CDC* and *CSC* is, in fact, according to the special distribution of words across news articles [7]. As noted in [7], the occurrence of any keyword across news articles is extremely rare. Thus, instead of assuming a linear relationship, a quadratic relationship should be expected. Note that both *CDC* and *CSC* are always between 0 and 1. For *CDC*, the higher the value of it, the more powerful for it to discriminate the feature, t, across clusters. For *CSC*, the higher the value of it, the more of the articles in the cluster contains the specific feature. The weight of each feature in each document is finally calculated as follows:

$$w(t,d) = tf_{t,d} \times CDC \times CSC \tag{14}$$

where $tf_{t,d}$ is the frequency of the term t in the article d. Term frequency is used along with the weighting algorithm in order to improve the recall of the model [21]. Finally, each news article is represented by a vector-space model in which it is normalized to unit length, so as to account for documents with different lengths.

2.5 Learning and Prediction

The associations between different features and different trend types are generated based on Support Vector Machine (SVM)[8]. SVM is a new learning algorithm proposed by Vapnik to solve the two-class pattern recognition problem using the structural risk minimization principle [22]. It obtains very accurate result in text classification, and outperforms many other techniques such as neural network and Naive Bayes [9,23]. Since SVM is a binary classifier, therefore we need to have a pair of classifiers. One classifier is responsible for classifying whether a news article will trigger the rise event, the other classifier is responsible for the event of drop.

For the prediction, we simply pass the newly collected news article to the pair of classifiers and decide to which class the article should belong. For example, an article is believed to signal a rise (drop) event if the output value of the rise (drop) classifier is positive. If the output values of both classifiers are negative, we will classify that article as no recommendation, as it belongs to neither trend types. If the output values of the two classifiers are positive, then it certainly leds to ambiguous, and therefore it is classified as no recommendation as well.

3 Evaluation

We have developed a prototype system to evaluate our system. Stock data and news articles are archived through Reuters Market 3000 Extra[2], and are stored into IBM DB2. Features of the articles are extracted using IBM Intelligent Miner^{TM} for Text. The training of the classifiers, as well as the prediction task, is carried out using the package of SVM^{light}. The whole system is implemented using Java under Unix platform.

The data and news articles that we used for testing are 614 stocks in Hong Kong exchange market during seven consecutive months. The total number of news articles archived is about 350,000. The number of stock data varies from stock to stock, and is around 2,000 ticks for each stock. The total number of stock data archived is about 1,228,000. The data/news from the first six months are used for training, while those from the last month are used for testing.

3.1 Trends Discovery and Labeling

A typical result after the time series segmentation is shown in Figure 3. It is easy to see that the shape of the time series after segmentation is preserved, while the number of data points would be reduced up to one-tenth of the original one.

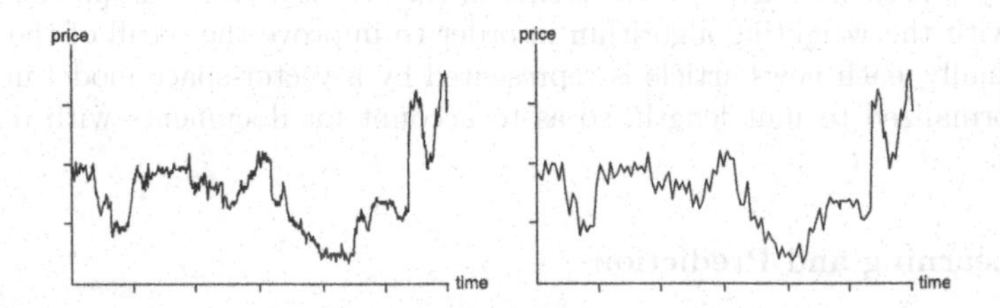

Fig. 3. A time series before and after segmentation. On the left: original time series. On the right: t-test based piecewise segmented time series

One of the biggest concerns about trend discovery process is its quality and reliability. It is a great danger of using a model that is a poor approximation

[2] Reuters has maintained a list of stock symbols that are relevant to each news article.

of the true functional relationship. In our trend discovery algorithm, each trend is the regression line of the original data points in that segment. A common practice in statistics for measuring the adequacy of a regression line is the use of coefficient of determination (R^2). Having a higher value of R^2 implies that the regression line formulated is more fit to the original data points. Figure 4 shows a typical result of the plots of m versus R^2.

Each symbol in the graph corresponds to a distinct segment. Note the special "T" shape of the graph, in which those segments with steep slopes will have high values of R^2. The "T" shape shows that the quality of our stock trend labeling is high. Misclassification is unlikely to happen, since we are interested in those steep slopes only. It is very important because it could maintain sufficient high quality training examples.

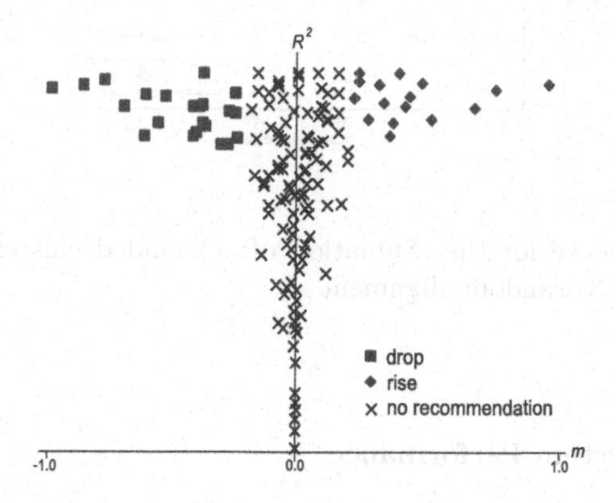

Fig. 4. A plot of slope (m) Vs coefficient of determination (R^2) for the same time series in Figure 3

3.2 Trends Alignment and Weighting Scheme

To evaluate the robustness of the guided clustering algorithm, receiver operating characteristic (ROC) curve is chosen. ROC is a common evaluation method for the classification problem. In order to have an unbiased estimator of the mean performance, a ten-fold cross-validation experiment is conducted. For all of the news aligned to a stock, 90% of them are randomly grouped into a training set and the remaining 10% of them are grouped into a testing set. The prediction model is built according to the training set, which is then used to evaluate the testing set. This procedure is repeated ten times with different training sets and testing sets. The average result obtained is plotted against the ROC curve.

Another alignment approach without using guided clustering is also conducted to examine the necessity of our guided clustering. Here, all of the news articles announced within a given trend are aligned to that trend. A typical result is shown in Figure 5, in which the x-axis denotes the false positive rate and y-axis denotes the true positive rate. The true positive rate is the percentage of news articles that were classified correctly, while the false positive rate is the percentage of news articles that were misclassified. The performance of guided clustering outperforms either random or without a guide.

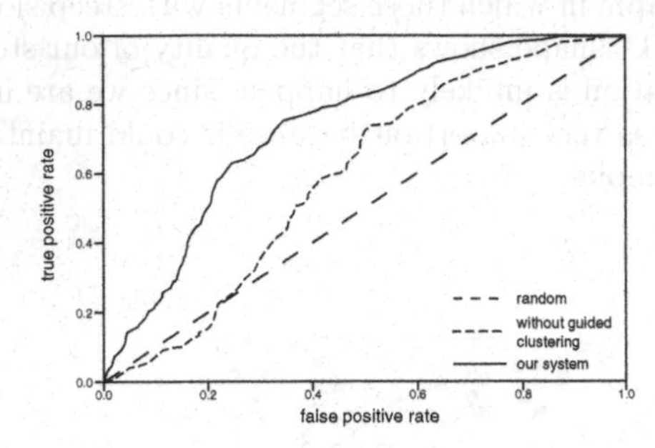

Fig. 5. A ROC curve for the evaluation of: 1) guided clustering; 2) without guided clustering; 3) random alignment

3.3 Overall System Performance

The best way to evaluate the reliability of a prediction model is certainly to conduct a market simulation which could mimic the behaviors of a stock trader using real life data. A Buy-and-Hold test[3] is conducted like [6]. The rate of return r of the test is:

$$r = \sum_{i=0}^{t} \frac{y_{i+1} - y_i}{y_i} \tag{15}$$

In this test, profit (loss) is made when shares are sold (short are bought). The assumption of zero transaction cost is taken out. Two strategies that are used to govern the decisions of buy, sell and hold are:

- if the prediction of the upcoming trend is positive, then shares of that stock are purchased. If a profit of 1% or more could be made within an hour, then all shares are sold immediately; otherwise they are sold at the end of that hour.
- if the prediction of the upcoming trend is negative, then shares of that stock are sold for short. If the trading price is dropped 1% or more than the shorted

[3] Some relaxation about the stock exchange restriction is assumed under the buy-and-hold test, such as the acceptance of shorting the stock.

price within an hour, then shares of that stock are purchased immediately; otherwise they are purchased at the end of that hour.

Since we are concerned with the rate of return, how much shares are bought in each transaction can be ignored.

We compare our system with the fixed period alignment approach[13]. As noted before, the underlying assumption between the fixed period approach and ours is different. Our model is based on the EMH which states that the market reacts immediately according to the new information arises. However, the fixed period approach assumes that every related piece of information has an impact on the market after a fixed time interval.

Since all of the predictions made are based on the news articles, the frequency of the news articles broadcast must be a critical factor of affecting the prediction performance. Figure 6 shows a comparison between our system and the fixed period approach with the frequency of news announced versus resulting profit. All of the stocks are ranked based on the total number of news articles that are associated with in the training period. They are further divided into different categories in which every category contains the same number of stocks.

In the figure, the smaller the number of the x-axis is, the fewer number of news articles are aligned to that stock. In general, our approach is highly superior to the fixed period approach. The reason is that we use all news articles, but fixed period approach only uses the news articles within a fixed interval preceding the happening of a trend. However, when a stock received too many news articles, fixed period approach outperforms us due to the fact that the probability of having noise would be higher.

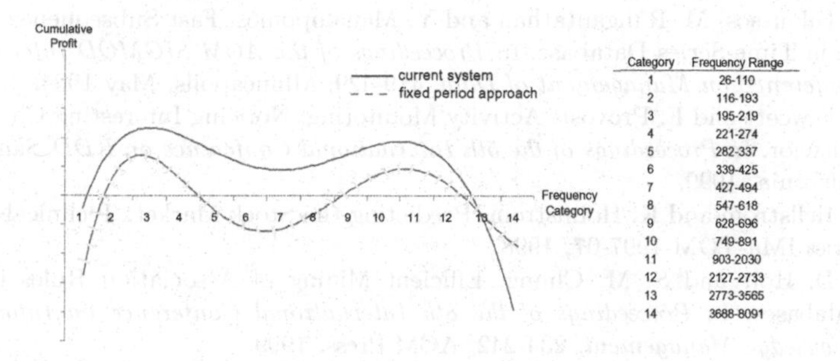

Category	Frequency Range
1	26-110
2	116-193
3	195-219
4	221-274
5	282-337
6	339-425
7	427-494
8	547-618
9	628-696
10	749-891
11	903-2030
12	2127-2733
13	2773-3565
14	3688-8091

Fig. 6. The relationship between the frequency of news announced and the resulting profit

4 Conclusions

In this paper, we demonstrated a sophisticated system which monitors the stock market and predicts its future behaviors. The major difference between our system and the existing forecasting techniques is that we take immeasurable factors or non-quantifiable information into account during prediction. News articles are served as the major source of the non-quantifiable information that we rely on. Our approach does not need any assumptions that require a fixed period for aligning news articles to a trend. Several data mining and text mining techniques are incorporated in the system architecture. A t-test based piecewise segmentation algorithm and an agglomerative hierarchical clustering algorithm are used for discovering and labeling trends respectively. Two new algorithms are introduced: guided clustering and new weighting scheme. The guided clustering is developed based on incremental K-means. It is used for article filtering and article-trend alignment. The new weighting scheme is formulated to identify the important features within the collection of articles. Finally, a market simulation using a very simple trading strategy based on the Buy-and-Hold test is carried out, and the results indicated that our approach is profitable.

References

1. S. B. Achelis. *Technical Analysis from A to Z.* Irwin Professional Publishing, Chicago, 2nd edition, 1995.
2. P. A. Adler and P. Adler. *The Social Dynamics of Financial Markets.* Jai Press Inc., 1984.
3. W. J. Eiteman, C. A. Dice and D. K. Eiteman. *The Stock Market.* McDGraw-Hill Book Company, 4th edition, 1966.
4. C. Faloutsos, M. Rangantathan and Y. Manalopoulos. Fast Subsequence Matching in Time-Series Database. In *Proceedings of the ACM SIGMOD International Conference on Management of Data*, 419-429, Minneapolis, May 1994.
5. T. Fawcett and F. Provost. Activity Monitoring: Noticing Interesting Changes in Behavior. In *Proceedings of the 5th International Conference on KDD*, San Diego, California, 1999.
6. T. Hellstrom and K. Holmstrom. Predicting the Stock Market. Technical Report Series IMa-TOM-1997-07, 1998.
7. J. D. Holt and S. M. Chung. Efficient Mining of Association Rules in Text Databases. In *Proceedings of the 8th International Conference on Information Knowledge Management*, 234-242, ACM Press, 1999.
8. T. Joachims. *Making large-Scale SVM Learning Practical.* Advances in Kernel Methods - Support Vector Learning. B. Sholkopf and C. Burges and A. Smola, MIT-Press, 1999.
9. T. Joachims. Text Categorization with Support Vector Machines: Learning with many relevant features. In *Proceedings of the European Conference on Machine Learning*, Springer, 1998.
10. E. Keogh and P. Smyth. A Probabilistic Approach to Fast Pattern Matching in Time Series Databases. In *Proceedings of the 3rd International Conference of KDD*, 24-40, AAAl Press, 1997.

11. L. Kaufman and P. J. Rousseeuw. *Finding Groups in Data - An Introduction to Cluster Analysis.* John Wiley & Sons, Inc., 1990.
12. B. Larsen and C. Aone. Fast and Effective Text Mining Using Linear-time Document Clustering. In *Proceedings of the 5th International Conference on KDD*, San Diego, California, 1999.
13. V. Lavrenko, M. Schmill, D. Lawire, P. Ogilvie, D. Jensen and J. Allan. Mining of Concurrent Text and Time Series, In *Proceedings of the 6th International Conference on KDD*, Boston, MA, 2000.
14. W. Mendenhall and T. Sincich. *A Second Course in Business Statistics: Regression Analysis.* Dellen Publishing Company, 1989.
15. D. C. Montgomery and G. C. Runger. *Applied Statistics and Probability for Engineers.* John Wiley & Sons, Inc., 2nd edition, 1999.
16. T. Pavlidis and S. L. Horowitz. Segmentation of Plan Curves. *IEEE Transactions on Computers*, Vol. c-23, No. 8, August 1974.
17. C. Pratten. *The Stock Market.* Cambridge University Press, 1993.
18. C. J. van Rijsbergen. A Theoretical Basis for the use of Co-occurance Data in Information Retrieval. *Journal of Documentation*, 33:106-119, 1977.
19. P. Smyth. Hidden Markov Models for Fault Detection in Dynamic Systems. *Pattern Recognition*, 27(1), 149-164, 1994.
20. M. Steinbach, G. Karypis and V. Kumar. A Comparison of Document Clustering Techniques. Technical Report, 2000.
21. T. Takenobu and I. Makoto. Text Categorization Based on Weighted Inverse Document Frequency. Technical Report, ISSN 0918-2802, 1994.
22. V. N. Vapnik, *The Nature of Statistical Learning Theory.* Springer, 1995.
23. Y. Yang and X. Liu. A Re-examination of Text Categorization Methods. In *Proceedings of the 22nd Annual International ACM SIGIR Conference on Research and Development in Information Retrieval*, 42-49, 1999.

User Profiling for Intrusion Detection Using Dynamic and Static Behavioral Models*

Dit-Yan Yeung and Yuxin Ding

Department of Computer Science, Hong Kong University of Science and Technology
dyyeung@cs.ust.hk

Abstract. Intrusion detection has emerged as an important approach to network security. In this paper, we adopt an anomaly detection approach by detecting possible intrusions based on user profiles built from normal usage data. In particular, user profiles based on Unix shell commands are modeled using two different types of behavioral models. The dynamic modeling approach is based on *hidden Markov models* (HMM) and the principle of *maximum likelihood*, while the static modeling approach is based on event occurrence frequency distributions and the principle of *minimum cross entropy*. The *novelty detection* approach is adopted to estimate the model parameters using normal training data only. To determine whether a certain behavior is similar enough to the normal model and hence should be classified as normal, we use a scheme that can be justified from the perspective of hypothesis testing. Our experimental results show that static modeling outperforms dynamic modeling for this application. Moreover, the static modeling approach based on cross entropy is similar in performance to instance-based learning reported previously by others for the same dataset but with much higher computational and storage requirements than our method.

1 Introduction

Intrusion detection, which refers to a certain class of system attack detection problems, is a relatively new area in computer and information security. Many intrusion detection systems built thus far are based on the general model proposed by Denning in a seminal paper [6]. From a high-level view, the goal is to find out whether or not a system is operating normally. Abnormality or anomaly in the system behavior may indicate the occurrence of system intrusions caused by successful exploitation of system vulnerabilities.

Host-based intrusion detection systems detect possible attacks into individual host computers. Such systems typically utilize information specific to the operating systems of the target computers. On the other hand, *network-based* intrusion detection systems monitor network behavior by examining the content as well as the format of network data packets, which typically are not specific

* This research was supported by the Hong Kong Innovation and Technology Commission (ITC) under project AF/223/98 and the Hong Kong University Grants Committee (UGC) under Areas of Excellence research grant AoE98/99.EG01.

M.-S. Chen, P.S. Yu, and B. Liu (Eds.): PAKDD 2002, LNAI 2336, pp. 494–505, 2002.

to the exact operating systems used by individual computers as long as these computers can communicate among themselves using the same network protocol.

Two general approaches are commonly used for building intrusion detection systems. *Misuse detection* systems detect evidence of attacks based on prior knowledge about known attacks. *Anomaly detection* systems, on the other hand, model normal system behavior to provide a reference against which deviations are detected. In other words, the major difference between the two approaches is on whether normal or abnormal (i.e., intrusive) behavior is modeled explicitly. In this paper, data mining methods based on the anomaly detection approach are proposed for host-based intrusion detection. We consider normal user profiling based on shell command sequences from audit logs [20,9,14,16,18,21].

Typical *classification* problems studied in pattern recognition can be formulated as classifying each pattern into one of c (≥ 2) classes with as low classification error as possible. To build such a discriminative classifier, training examples from all c classes are needed. While this formulation is commonly used in pattern recognition, there also exists another formulation called *novelty detection* [4,2,11]. In a probabilistic sense, it corresponds to deciding whether an unknown test pattern is produced by the underlying data distribution that corresponds to the training set of normal patterns. While novelty detection problems appear to be similar to 2-class classification problems, with the two classes corresponding to normal and abnormal patterns respectively, a major difference is that novelty detection methods typically use only training examples from the class corresponding to normal patterns to build a generative model of normal behavior. The novelty detection approach is particularly attractive under situations where novel or abnormal patterns are expensive or difficult to obtain for model construction. In this paper, the novelty detection approach is adopted.

2 Dynamic versus Static Behavioral Models

Normal user behaviors are profiled by building behavioral models using data collected from normal operations. There are generally two categories of behavioral models. *Dynamic models* explicitly model temporal variations essential for discriminating abnormal behavior from normal behavior. *Static models*, on the other hand, do not explicitly model temporal variations. They could be used for anomaly detection problems if the normal system behavior does not exhibit significant temporal variations, or if the temporal sequences are first converted into some non-temporal feature representation.

Different anomaly detection methods have been used, e.g., instance-based learning [15,16], multi-layer perceptrons [7,20], decision trees [9], hidden Markov models (HMM) [14,23], frequent episodes [18], correlation analysis [8], statistical likelihood analysis [7], rule learning [10,17,23], and uniqueness method [21]. Of these methods, only HMMs are intrinsically dynamic in nature. Also, not all of them are based on the preferred novelty detection approach and hence they need both normal and intrusion data for model construction.

In this paper, intrusion detection systems based on profiling shell command sequences are first studied with the dynamic modeling approach using HMMs. Afterwards, we will propose an information-theoretic static modeling approach based on the usage frequencies of shell commands. Comparative studies will then be performed.

3 Dynamic Modeling Approach

3.1 Hidden Markov Models

HMMs are stochastic models of sequential data. Each HMM contains a finite number of unobservable states. State transitions are governed by a stochastic process to form a Markov chain. At each state, some state-dependent events can be observed. The emission probabilities of these observable events are determined by a probability distribution, one for each state. Of interest here are discrete HMMs in which the observed events are discrete symbols, as opposed to other models not studied here, e.g., continuous-density HMMs.

Fully-connected or ergodic HMMs allow state transitions between all state pairs. On the other hand, left-to-right HMMs do not allow state transition back to any state to the left of the current state. In fact, most left-to-right HMMs used in practice only allow state transition from a state to itself (called self-transition), to the immediate neighbor to the right, and to the neighbor two steps to the right. In this paper, our left-to-right HMMs are further restricted to only the first two types of state transition.

To estimate the parameters of an HMM for modeling normal behavior, sequences of normal events (shell commands in our case) collected from normal system usage are used as training examples. An *expectation-maximization* (EM) algorithm [5] known as the *Baum-Welch re-estimation algorithm* [1] for mixture density estimation is used to find the maximum-likelihood (ML) parameter estimate. More details of the algorithm can be found in [19].

3.2 Sample Likelihood with Respect to Model

Given a trained HMM M, the sample likelihood of an observation sequence S with respect to M can be computed using either the *forward algorithm* or the *backward algorithm* [19]. From a generative point of view, this can be seen as computing the probability that a given observation sequence is generated by the model. Alternatively, we can also consider it as providing a quantitative measure for assessing how well the model matches the sequence.

Ideally, a well-trained HMM can give sufficiently high likelihood values only for sequences that correspond to normal behavior. Sequences that correspond to intrusive behavior should give significantly lower likelihood values. By comparing the sample likelihood of S with respect to M against a certain threshold, one can decide whether S deviates significantly from M and hence should be considered a possible intrusion. We will describe how to determine the threshold in Section 6.3.

4 Static Modeling Approach

4.1 Occurrence Frequency Distributions

Suppose the occurrence frequencies of different events (shell commands) are measured over a period of time. A probability distribution can represent the overall occurrence pattern. Since the event order is not considered, we refer to this as a static modeling method. Using this scheme, a normal user behavioral model is simply represented as an occurrence frequency distribution, with which possible system intrusions can be detected.

Let $P(M)$ denote the probability distribution characterizing a normal model M and let $P_i(M)$ denote the occurrence probability of event i among N possible events. Similarly, $Q(S)$ and $Q_i(S), i = 1, 2, \ldots, N$ denote the probability distribution and individual event probabilities, respectively, for some behavior S being monitored. For simplicity, the dependencies on M and S will not be shown.

4.2 Cross Entropy between Distributions

We need a dissimilarity measure between distributions. An information-theoretic measure that can serve this purpose is *cross entropy* [22,12], which is also related to *Kullback-Leibler information measure* [13].

We use this definition of cross entropy: $C(P,Q) = \sum_{i=1}^{N}(Q_i - P_i)\log(Q_i/P_i)$. Note that these properties hold: (a) $C(P,Q) = C(Q,P)$; (b) $C(P,Q) \geq 0$; (c) $C(P,Q) = 0 \Leftrightarrow P = Q$. Thus, by checking whether the cross entropy between P and Q is larger than a certain threshold, one can decide if S should be considered a possible intrusion with respect to the model M. We will describe how to determine the threshold later in Section 6.3.

5 Data Preprocessing and Partitioning

5.1 Preprocessing of Shell Command Data

The shell command datasets are available at the public-domain KDD archive.[1] Since it is difficult to obtain real intrusion data, only normal data were collected via the history file mechanism from eight different Unix users. For each user login session, each word typed by the user was recorded as a token. Since many Unix commands are followed by parameters (e.g., `ls -laF Paper Notes letter.tex`), the set of all distinct tokens would become too large. To reduce the token set size, only a count of the files or directories is represented as a token (e.g., `ls -laF <3>`). All tokens in a login session form a *trace*.

Note that the datasets contain no real intrusion data because it is difficult to collect such data for this kind of intrusion detection applications. In our experiments, (normal) data from other users were used as if they were "intrusive"

[1] `http://kdd.ics.uci.edu/databases/UNIX_user_data/UNIX_user_data.html`

data for a given user. Thus, by its very nature, this problem is more like a classification problem than a novelty detection problem, although we still use a novelty detection approach as it is more desirable in practice.

5.2 Partitioning of Datasets

Each set of data is partitioned into three subsets: *training set* (normal data only), *threshold determination set* (normal data only), and *test set* (both normal and intrusion data). The training set of data is for estimating model parameters. Only normal data are needed for the novelty detection approach. As the model is built using normal data only, we need a criterion to decide when an observed behavior should be considered normal or intrusive. In particular, it corresponds to finding a threshold for some similarity measure (e.g., likelihood) or dissimilarity measure (e.g., cross entropy). The threshold determination set of normal data is used for determining this threshold. After the model parameters and the threshold have been estimated using the training and threshold determination sets, respectively, the test set can be used for evaluating model performance. More details about the performance measures used will be discussed in Section 6.1.

Table 1 summarizes the dataset sizes in our experiments. For each user, the (normal) data of all other users were treated as if they were "intrusive" data for that user. Since the available datasets are quite large, we used disjoint sets of data for training, threshold determination, and testing. Table 1 also shows the number of distinct tokens found in the data for each user. When the datasets for all eight users are combined together, we have a total of 2356 distinct tokens.

Table 1. Shell command datasets

	Training set		Threshold determination set		Test set		No. of distinct tokens in training set	Total no. of distinct tokens
User	No. of traces	No. of tokens	No. of traces	No. of tokens	No. of traces	No. of tokens		
0	171	5733	170	6802	147	6316	151	286
1	196	5702	194	5327	365	5571	152	308
2	134	6776	134	3844	216	5120	174	484
3	313	13626	312	10309	286	11970	291	476
4	213	10826	212	11850	121	10981	375	561
5	832	20285	831	18009	762	19020	360	607
6	419	4485	418	5062	502	4738	228	447
7	562	16784	562	16586	466	16706	406	704

6 Model Construction and Performance Evaluation

6.1 Performance Criteria

We use two performance measures, namely, *true detection rate* (TDR) and *false detection rate* (FDR):

$$\text{TDR} = \Pr(\text{intrusive} \mid \text{intrusive}) = \frac{\#\ \text{intrusive testing traces detected as intrusive}}{\#\ \text{intrusive traces in test set}}$$

$$\text{FDR} = \Pr(\text{intrusive} \mid \text{normal}) = \frac{\#\ \text{normal testing traces detected as intrusive}}{\#\ \text{normal traces in test set}}$$

We prefer these two measures because both relate reporting the occurrence of an intrusive event to the ground truth (i.e., normal or intrusive nature) of that event. This is in line with the convention used in [23] although they refer to the two measures as *true positives* and *false positives*, respectively. We use the term 'detection' to make the meaning of detecting intrusions more explicit. *Hit rate* and *false alarm rate* may also be used in place of TDR and FDR.

6.2 Model Training

To train an HMM, fixed-length sequences of events are extracted from each trace of the training set by moving a window of the specified width (i.e., sequence length) through the entire trace with a step size of 1. Identical sequences extracted are represented by only a single copy in the training set. In our experiments, both fully-connected and left-to-right HMMs (denoted as FC-HMM and LR-HMM, respectively) were used. For the static modeling approach, all traces from the training set are used to create a distribution-based behavioral model.

6.3 Threshold Determination

After model training, the threshold determination set is used to determine an appropriate threshold for use later as a criterion for detecting possible intrusions.

For HMM-based dynamic modeling, fixed-length sequences are extracted from each trace of the threshold determination set in the same way as before for the training data. The sample likelihood of each sequence with respect to the model can then be computed. For the static modeling approach, each trace of the threshold determination set is used to compute a distribution as well as the cross entropy between this distribution and the reference distribution computed from the training data.

For each chosen FDR for the threshold determination set, a corresponding threshold value can be obtained. In our experiments, different threshold values were tried by choosing different FDR values.

6.4 Model Testing

To test whether a trace in the test set is intrusive, fixed-length sequences extracted from the trace are presented to a trained HMM to compute the likelihood values. If at least one sequence has a likelihood value that is lower than the threshold, the trace is said to be intrusive. In other words, we can conclude that a trace under investigation is intrusive as soon as the first intrusive sequence is found inside the trace.

For the static modeling approach, in order to perform timely detection of possible intrusions, it would be desirable if a decision could be made as soon as sufficient data have been collected to compute a reasonably reliable distribution. Since a trace may be very long (if a user login session is long), we do not want to make a decision only after the trace ends. Instead, a distribution is computed for each sub-trace sequence. The cross entropy between this distribution and the reference distribution of the model will be compared with the threshold to determine whether it is an intrusive sequence. The extraction of variable-length sequences from a trace is illustrated in Figure 1 below. The detection of possible intrusions in a trace can be performed immediately after the first K events have arrived. We refer to K as the *minimum sequence length.*

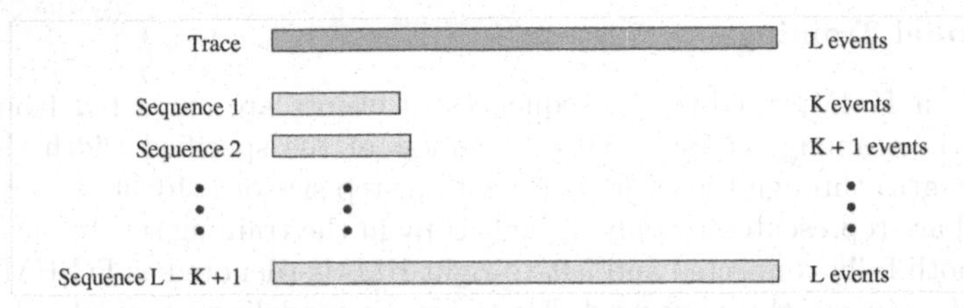

Fig. 1. Extraction of variable-length sequences from a trace

6.5 Hypothesis Testing Perspective

In this section, we will justify the scheme above from a hypothesis testing perspective. Although our explanation is based on HMMs, it also holds for the static modeling method.

Let M denote an HMM learned from the training data. Given a sequence S from the test set, we want to decide whether it is likely to be generated by M and hence is a normal sequence. This problem can be formulated as applying a statistical test [3]. Let us generate a sufficiently large sample $\mathcal{R}$ of (normal) sequences from M. For an arbitrary sequence $R \in \mathcal{R}$, the log-likelihood of R with respect to M is denoted as $L(R) = \log \Pr(R|M)$. Similarly, the log-likelihood of S is denoted as $L(S) = \log \Pr(S|M)$. Based on the empirical probability distribution of $L(R)$ over $\mathcal{R}$, we then test the hypothesis that $L(S)$ is drawn from the probability distribution of the log-likelihood of the sequences in $\mathcal{R}$, i.e., $\Pr(L(R) \leq L(S)) > \psi$, for some threshold $0 < \psi < 1$. We reject the null hypothesis if the probability is not greater than ψ, implying that S is not a normal sequence with respect to model M.

In our case, the threshold determination set of normal data plays the role of $\mathcal{R}$ although $\mathcal{R}$ is not actually generated by M. If M is a well-trained model representing the training set, the underlying distributions of the training and threshold determination sets are close enough to each other, and the threshold determination set is sufficiently large, then it is reasonable to use the threshold determination set as $\mathcal{R}$. Apparently, we can see that the threshold ψ is just the FDR chosen for the threshold determination set.

7 Experiments

7.1 Results

Table 2 shows the results. For each method, only two choices of the sequence length or minimum sequence length are included to show the effect of varying the parameter, although more choices were actually tried. The threshold was chosen such that the FDR of the threshold determination set was equal to 5%, 10%, 15% or 20%. For each FDR value, the TDR shown is the average over the individual TDR values with the data from other users treated as intrusion data. The number of states shown is the minimum value that maximizes the TDR.

Increasing the sequence length always increased the discrimination power of both FC-HMM and LR-HMM in detecting intrusions. Since traces shorter than the sequence length chosen were eliminated and there exist many short shell command traces in the datasets, increasing the sequence length would eliminate the shorter traces which could be partially responsible for the performance improvement because these traces could not model the behavior well. This is also a possible reason for the observed performance improvement of the static modeling method as the minimum sequence length increases.

7.2 Discussions

In our experiments, the static modeling method performed significantly better than both FC-HMM and LR-HMM, typically 10%-20% higher in the TDR. A possible reason is that the temporal dependencies between shell commands are weak and hence are not very useful for intrusion detection. The static shell command distribution seems to be sufficient for many users.

Although FC-HMM is usually slightly better than LR-HMM, increasing the number of states in an LR-HMM can approach the performance of an FC-HMM with fewer states. For example, the FC-HMM with sequence length 30 is similar in performance to the LR-HMM with sequence length 50. Note that the time complexity of each training iteration of an FC-HMM is $O(W^2T)$ for W states and sequence length T. As a comparison, the time complexity of each training iteration of an LR-HMM is only $O(WT)$.

We also measured the CPU execution time for different methods. All the tasks were run on a Sun UltraSPARC 30 workstation with 256MB memory. Table 3 shows the CPU time required for the training and testing stages in the experiments reported in Table 2. Due to page limit, only the average statistics over all users are reported. It can be seen that LR-HMM is faster than FC-HMM for both the training and testing stages. Our static modeling method is impressive in that its training time is always negligible because it simply requires

Table 2. Results for shell command data

User	TDR (%) of FC-HMM (sequence length = 10)					TDR (%) of FC-HMM (sequence length = 30)				
	No. of states	FDR =5%	FDR =10%	FDR =15%	FDR =20%	No. of states	FDR =5%	FDR =10%	FDR =15%	FDR =20%
0	10	31.9	50.3	62.3	67.2	10	45.0	52.8	60.9	66.6
1	50	57.9	80.9	84.0	90.1	5	73.1	79.4	89.5	93.7
2	30	46.1	62.6	70.1	79.1	20	49.8	63.1	83.3	89.4
3	10	34.2	45.7	54.8	64.4	10	40.7	64.4	73.9	75.8
4	20	11.1	18.9	36.8	44.5	20	24.8	27.5	45.3	52.0
5	20	49.2	71.4	73.8	78.0	20	57.1	70.8	79.1	82.3
6	20	13.0	26.2	42.4	53.6	20	14.3	43.0	58.1	60.6
7	20	28.7	44.5	61.2	75.0	20	39.9	55.8	65.7	81.0
Average		34.0	50.1	60.7	69.0		43.1	57.1	69.5	75.2

User	TDR (%) of LR-HMM (sequence length = 30)					TDR (%) of LR-HMM (sequence length = 50)				
	No. of states	FDR =5%	FDR =10%	FDR =15%	FDR =20%	No. of states	FDR =5%	FDR =10%	FDR =15%	FDR =20%
0	5	40.8	48.2	64.0	68.0	5	45.1	60.3	64.7	68.3
1	10	71.3	79.1	87.2	89.7	30	73.5	81.4	82.4	83.8
2	5	43.2	62.1	79.6	88.0	10	61.8	69.4	79.4	82.9
3	10	23.9	59.9	71.7	76.4	10	35.5	58.5	73.4	78.4
4	20	21.1	22.4	37.4	49.4	20	14.8	36.0	38.8	52.3
5	10	59.0	67.7	77.5	83.5	5	49.6	63.3	69.7	76.3
6	10	12.0	39.8	50.3	55.8	10	15.5	19.7	21.1	44.6
7	5	32.7	47.2	56.8	74.7	10	52.9	63.6	72.4	75.0
Average		38.0	53.3	65.6	73.2		43.6	56.5	62.7	70.2

User	TDR (%) of cross entropy (min. sequence length = 30)				TDR (%) of cross entropy (min. sequence length = 50)			
	FDR =5%	FDR =10%	FDR =15%	FDR =20%	FDR =5%	FDR =10%	FDR =15%	FDR =20%
0	52.9	62.7	79.3	82.2	46.2	78.5	80.5	81.1
1	56.0	81.2	93.2	95.4	54.4	71.4	89.8	92.7
2	43.3	49.4	73.4	95.9	43.3	49.4	73.4	96.0
3	46.8	76.5	90.7	95.1	46.8	76.5	88.5	95.1
4	48.4	76.1	85.7	88.0	55.0	73.9	81.0	82.2
5	56.7	74.3	78.7	82.8	50.3	70.2	81.2	89.4
6	25.6	44.0	57.2	58.8	36.1	40.1	56.0	60.0
7	60.8	85.7	93.9	97.1	83.4	95.4	97.6	99.5
Average	48.8	68.7	81.5	86.9	51.9	69.4	81.0	87.0

Table 3. Execution time statistics for shell command data

CPU time (sec.) of FC-HMM (sequence length = 10)		CPU time (sec.) of FC-HMM (sequence length = 30)	
Training	Testing	Training	Testing
14537	14.3	19932	14.4

CPU time (sec.) of LR-HMM (sequence length = 30)		CPU time (sec.) of LR-HMM (sequence length = 50)	
Training	Testing	Training	Testing
12518	5.8	15591	8.9

CPU time (sec.) of cross entropy (min. sequence length = 30)		CPU time (sec.) of cross entropy (min. sequence length = 50)	
Training	Testing	Training	Testing
0	12.9	0	11.8

the computation of a distribution based on the training data. The testing time is also comparable to that for HMMs. Our method would be particularly attractive if new models have to be built regularly due to changes in the system behavior.

7.3 Comparison with Previous Work

We also compared our results with those from previous work. To facilitate comparison, we performed another experiment using the same setup as in [15,16]. The datasets were partitioned into training, parameter selection, and test sets as shown in Table 4. Moreover, in their work, the TDR and FDR were computed based on sequences. We think it makes more sense to measure TDR and FDR according to traces as in our earlier experiments. However, to facilitate comparison here, we used the same scheme as theirs for this experiment.

Table 5 shows the classification results obtained by [15] using *instance-based learning* (IBL), giving an average TDR of 34.2% with an average FDR of 5.3%, as well as our results using the static modeling method. The average TDR is 35.6% at an average FDR of 5.5%. Thus, it can be concluded that the two methods can achieve very similar performance in terms of the TDR and FDR measures.

However, there are major differences between the two methods in terms of computational and storage requirements. Although data reduction techniques can alleviate the problems to a certain extent, the high computational and storage requirements are still the major limitations of IBL methods. Our method is clearly superior in this aspect because the training examples are summarized as a distribution, the storage requirement of which does not depend on the training set size. Similarly, during testing, the computational requirement is very low.

Table 4. Data partitioning for shell command datasets in comparative study

	Training set		Parameter selection set		Test set	
User	No. of tokens	No. of traces	No. of tokens	No. of traces	No. of tokens	No. of traces
0	1557	49	1487	37	11992	356
1	1502	64	1714	63	11833	442
2	1995	76	1137	39	11877	330
3	1551	42	1474	40	12696	314
4	1500	45	1739	7	12255	311
5	1555	35	1507	55	11980	558
6	1500	90	1508	111	11277	1138
7	1590	52	1423	52	12250	456

8 Concluding Remarks and Future Research

In this paper, we have presented two different anomaly detection approaches for a host-based intrusion detection problem based on shell command sequences. It

Table 5. Results for shell command data in comparative study

Tested user	User model (instance-based learning)								User model (cross entropy)							
	0	1	2	3	4	5	6	7	0	1	2	3	4	5	6	7
0	99.3	57.0	31.7	61.0	75.1	0.6	38.5	10.1	99.0	71.1	26.6	27.8	20.9	2.0	11.4	44.8
1	14.9	92.9	12.4	64.2	16.3	0.9	4.0	6.0	25.2	92.8	50.6	56.5	43.1	12.0	27.0	75.8
2	41.3	58.7	94.7	43.6	71.1	0.3	47.9	8.3	8.7	54.3	94.7	48.5	17.6	2.3	12.6	43.6
3	64.8	91.7	46.7	90.0	86.4	0.6	69.0	15.1	21.3	87.2	56.8	90.0	29.5	12.0	16.7	19.9
4	34.4	21.2	18.6	72.1	92.7	1.3	8.6	3.0	24.2	75.4	66.6	30.2	92.5	17.9	16.6	16.1
5	50.4	68.3	39.7	70.3	78.0	99.9	57.2	29.4	9.7	68.0	15.9	25.1	15.2	99.0	8.6	56.9
6	41.8	15.4	17.7	82.3	48.7	0.6	91.7	4.7	22.7	77.4	44.6	54.4	27.5	8.2	91.7	76.2
7	24.7	11.0	8.7	40.7	22.1	0.6	5.8	96.2	32.4	99.8	73.2	20.3	16.1	12.0	53.6	96.1
FDR	0.7	7.1	5.3	10.0	7.3	0.1	8.3	3.8	1.0	7.2	5.3	10.0	7.5	1.0	8.3	3.9
Average TDR	38.9	46.2	25.1	62.0	56.8	0.7	33.0	10.9	20.6	76.2	47.8	37.5	24.3	9.5	20.9	47.6

was found that static behavioral models can give better results. It can be speculated that temporal dependencies are not very useful or may even be harmful for this problem. Our information-theoretic static modeling method based on cross entropy is simple and computationally cheap, yet it can outperform the more sophisticated dynamic modeling method based on HMMs. A lesson to learn is that one should be careful in finding the best match between problems and methods.

A closer look at Table 5 reveals the fact that IBL is better for some users (0, 3, 4, 6) while the cross-entropy method is better for other users (1, 2, 5, 7). This shows that the two methods are complementary to each other. A potential future research direction is to combine these two methods and possibly also some other methods to further improve the discrimination power. Besides host-based intrusion detection problems, we are also conducting research on network-based intrusion detection. Some of the ideas learned from this research may also be relevant to network-based intrusion detection.

References

1. L.E. Baum, T. Petrie, G. Soules, and N. Weiss. A maximization technique occurring in the statistical analysis of probabilistic functions of Markov chains. *Annals of Mathematical Statistics*, 41(1):164–171, 1970.
2. C.M. Bishop. Novelty detection and neural network validation. *IEE Proceedings: Vision, Image and Signal Processing*, 141(4):217–222, 1994.
3. P.R. Cohen. *Empirical Methods for Artificial Intelligence.* MIT Press, Cambridge, MA, USA, 1995.
4. W.J. Daunicht. Autoassociation and novelty detection by neuromechanics. *Science*, 253(5025):1289–1291, 1991.
5. A.P. Dempster, N.M. Laird, and D.B. Rubin. Maximum likelihood from incomplete data via the EM algorithm (with discussion). *Journal of the Royal Statistical Society, Series B*, 39:1–38, 1977.
6. D.E. Denning. An intrusion-detection model. *IEEE Transactions on Software Engineering*, 13(2):222–232, 1987.
7. D. Endler. Intrusion detection: applying machine learning to Solaris audit data. In *Proceedings of the Fourteenth Annual Computer Security Applications Conference*, pages 268–279, Phoenix, AZ, USA, 7–11 December 1998.

8. S. Forrest, S.A. Hofmeyr, A. Somayaji, and T.A. Longstaff. A sense of self for Unix processes. In *Proceedings of the IEEE Symposium on Security and Privacy*, pages 120–128, Oakland, CA, USA, 6–8 May 1996.
9. D. Gunetti and G. Ruffo. Intrusion detection through behavioral data. In *Proceedings of the Third International Symposium on Intelligent Data Analysis*, pages 383–394, Amsterdam, Netherlands, 9-11 August 1999.
10. G.G. Helmer, J.S.K. Wong, V. Honavar, and L. Miller. Intelligent agents for intrusion detection. In *Proceedings of the 1998 IEEE Information Technology Conference - Information Environment for the Future*, pages 121–124, Syracuse, NY, USA, 1-3 September 1998.
11. N. Japkowicz, C. Myers, and M. Gluck. A novelty detection approach to classification. In *Proceedings of the Fourteenth International Joint Conference on Artificial Intelligence*, volume 1, pages 518–523, Montréal, Quebec, Canada, 20–25 August 1995.
12. R.W. Johnson and J.E. Shore. Comments on and correction to 'axiomatic derivation of the principle of maximum entropy and the principle of minimum cross-entropy' (Jan 80 26–37). *IEEE Transactions on Information Theory*, 29(6):942–943, 1983.
13. S. Kullback and R.A. Leibler. On information and sufficiency. *Annals of Mathematical Statistics*, 22:79–86, 1951.
14. T. Lane. Hidden Markov models for human/computer interface modeling. In *Proceedings of the IJCAI-99 Workshop on Learning about Users*, pages 35–44, Stockholm, Sweden, 31 July 1999.
15. T. Lane and C.E. Brodley. Temporal sequence learning and data reduction for anomaly detection. In *Proceedings of the Fifth ACM Conference on Computer and Communications Security*, pages 150–158, San Francisco, CA, USA, 2–5 November 1998.
16. T. Lane and C.E. Brodley. Temporal sequence learning and data reduction for anomaly detection. *ACM Transactions on Information and System Security*, 2(3):295–331, 1999.
17. W. Lee and S.J. Stolfo. Data mining approaches for intrusion detection. In *Proceedings of the Seventh USENIX Security Symposium*, pages 79–93, San Antonio, TX, USA, 26-29 January 1998.
18. W. Lee, S.J. Stolfo, and K.W. Mok. A data mining framework for building intrusion detection models. In *Proceedings of the IEEE Symposium on Security and Privacy*, pages 120–132, Oakland, CA, USA, 9-12 May 1999.
19. L.R. Rabiner. A tutorial on hidden Markov models and selected applications in speech recognition. *Proceedings of the IEEE*, 77(2):257–286, 1989.
20. J. Ryan, M.J. Lin, and R. Miikkulainen. Intrusion detection with neural networks. In M.I. Jordan, M.J. Kearns, and S.A. Solla, editors, *Advances in Neural Information Processing Systems 10*, pages 943–949. MIT Press, 1998.
21. M. Schonlau and M. Theus. Detecting masquerades in intrusion detection based on unpopular commands. *Information Processing Letters*, 76(1/2):33–38, 2000.
22. J.E. Shore and R.W. Johnson. Axiomatic derivation of the principle of maximum entropy and the principle of minimum cross-entropy. *IEEE Transactions on Information Theory*, 26(1):26–37, 1980.
23. C. Warrender, S. Forrest, and B. Pearlmutter. Detecting intrusions using system calls: alternative data models. In *Proceedings of the IEEE Symposium on Security and Privacy*, pages 133–145, Oakland, CA, USA, 9-12 May 1999.

Incremental Extraction of Keyterms for Classifying Multilingual Documents in the Web

Lee-Feng Chien[1], Chien-Kang Huang[2], Hsin-Chen Chiao[1], and Shih-Jui Lin[1]

[1] Institute of Information Science, Academic Sinica
{lfchien, hcchiao, sjlin}@iis.sinica.edu.tw
[2] Department of Computer Science and Information Engineering,
National Taiwan University, Taiwan
ckhuang@mars.csie.ntu.edu.tw

Abstract. With the rapid growth of the Web, there is a need of high-performance techniques for document collection and classification. The goal of our research is to develop a platform to discover English, traditional and simplified Chinese documents from the Web in the Greater China area and classify them into a large number of subject classes. Three major challenges are encountered. First, the collection (i.e., the Web) is dynamic: new documents are added in and the features of subject classes change constantly. Second, the documents should be classified in a large-scale taxonomy. Third, the collection contains documents written in different languages. A PAT-tree-based approach is developed to deal with document classification in dynamic collections. It uses PAT tree as a working structure to extract keyterms from documents in each subject class and then update the features of the class accordingly. The feedback will contribute to the classification of the incoming documents immediately. In addition, we make use of a manually-constructed keyterms to serve as the base of document classification in a large-scale taxonomy. Two sets of experiments were done to evaluate the classification performance in a dynamic collection and in a large-scale taxonomy respectively. Both of the experiments yielded encouraging results. We further suggest an approach extended from the PAT-tree-based working structure to deal with classification in multilingual documents.

1 Introduction

With the rapid growth of the Web, more and more research documents are available online. Although there are successful Web document discovery systems, such as automatic citation system CiteSeer [3] and special-purpose search engine Cora [6], organizing online research documents still faces several challenges. First, Web is dynamic: new documents are added into the Web constantly. Second, Web documents contain various topics and lack of subject organization. Classifying these documents into appropriate subject domains is thus highly demanded. The third challenge is that the documents are written in various kinds of languages. New frameworks must be developed in order to deal with multilingual access issues.

M.-S. Chen, P.S. Yu, and B. Liu (Eds.): PAKDD 2002, LNAI 2336, pp. 506–516, 2002.

The goal of our research is to develop a platform to discover English, traditional and simplified Chinese documents from the Web in the Greater China area and classify them into a large number of subject classes. This task of document classification encounters all the three challenges mentioned above: it is performed within a dynamic collection (i.e., documents are added incrementally, and features of classes changes from time to time), it consists of a large-scale taxonomy, and it should be able to deal with multilingual documents. Moreover, classifying Chinese documents has to face the additional challenge of word segmentation. Also, there are certain differences between the terminologies used in traditional Chinese documents (e.g., the documents in Taiwan and Hong Kong) and simplified Chinese (e.g., the documents in mainland China). The classification of Chinese documents in the Web needs new approaches and additional techniques.

This paper proposes a PAT-tree-based working structure to overcome these challenges in document classification. The proposed approach has three important features[1], [2]. First, it overcomes the problem of word segmentation in Chinese documents. It can successfully extract keyterms in arbitrary lengths. Second, it is suitable for document classification in dynamic collections. The PAT-tree technique extract the longest string that appears frequently in the training data without checking the dictionary. Hence, it is not limited to the vocabulary provided in the dictionary. This is especially suitable for extracting keyterms in a dynamic collection, which often contains new keyterms that are not included in dictionaries. The third feature of this approach is that it is language independent. As all keyterms are regarded as strings without looking up the dictionary, the approach is able to extract keyterms written in any languages and perform classification of multilingual documents.

The paper also proposes an approach to classifying documents in a large-scale taxonomy. Current Web document discovery systems seldom classify documents with respect to a large number of subject classes. One of the reasons is the lack of sufficient training data (i.e., manually classified documents) to learn the features of each class. Instead of training features of each class from the beginning, we perform the classification by using of a set of manually-constructed keyterms, together with the keyterms extracted by PAT trees.

Two sets of experiments were done to evaluate the classification performance in a dynamic collection and in a large-scale taxonomy respectively. In the first set of experiments, we collected news documents from the Web and classified them into 21 subject classes. The experimental results show that extracting keyterms incrementally by the PAT-based structure can improve the classification performance. In the second set of experiments, we classified scientific documents into a manually-constructed subject taxonomy with 1,011 subject classes and a set of 410, 557 keyterms in both traditional Chinese and English.

As the experimental results obtained were encouraging, we plan to use the taxonomy and the keyterm files to organize technical documents from the Web in the Greater China area. As the PAT-tree-based approach is language independent, we are able to extract keyterms from English, traditional Chinese, and simplified Chinese documents. We further suggest an approach extended from the PAT-tree-based working structure to acquire the keyterm files for simplified Chinese documents by taking English as the medium. This extended approach would be able to deal with classification in multilingual documents.

In the following sections, Section 2 is a review of existing Web document discovery systems and the PAT-tree-based technique. Section 3 proposes the PAT-based working structure for document classification. The experiments on classification in a dynamic collection are presented in Section 4, and the experiments on classification in a large-scale taxonomy are presented in Section 5. An extended approach dealing with classification of multilingual documents is suggested in Section 6, followed by the proposal of Web document discovery in the Greater China area.

2 Related Work

2.1 Web Document Discovery

There have been document discovery systems that collect and organize documents in the Web. Generally speaking, they all use focused spiders to collect new documents from the Web and are able to update the subject classes dynamically. However, as far as we know, none of the current systems provides document classification in large-scale taxonomies (e.g., a taxonomy containing more than 1000 classes). Neither do these systems deal with documents written in different languages.

Cora [6] provides a taxonomy, which consists of 3 levels of subject classes. This hierarchical structure is also used in OCLC [4]. However, their collections are limited to a specific domain, and the number of subject classes is not large. One reason is that classifying documents into a large number of subjects needs a great deal of training data.

Citeseer [3] collects scientific documents in the Web and especially focuses on the document citations. It analyzes the bibliography entries in each document and establishes forward and backward reference links among documents. Citeseer also provides classification for computer science; however, it hasn't contained classification for other subject domains.

2.2 PAT-tree-Based Working Structure and Keyterm Extraction

PAT tree is an efficient indexing structure successfully used in the area of information retrieval [5]. Using this structure for indexing full-text content of a document collection, all possible data strings including their frequency counts in the collection can be retrieved and updated in a very efficient way, whereas only different suffix strings with arbitrary lengths need to be stored. Considering the inherent difficulty of word segmentation and unknown proper noun extraction in Chinese, a *character* rather than a *word* as in English is a more proper indexing unit for Chinese document storage and classification with the PAT tree indexing mechanism. In this way each distinct character string in the collection that occurs within a sentence fragment can be conceptually recorded. We usually use punctuation marks such as "." and "," as delimiters to determine boundaries of strings. For example, the sentence "abcd,cd" (each alphabet stands for a Chinese character) will be broken into two strings "abcd" and "cd". The PAT tree only needs to record the distinct *suffix strings* of them, namely, "abcd", "bcd","cd", and "d". For each string, all its sub-strings can be easily detected if the sub-string is exactly a prefix of a suffix string recorded in the PAT tree. For example, if we check whether "bc" is a sub-string of the string "abcd" in the PAT tree, we find that the answer is yes because "bc" can be found as a prefix of the suffix string "bcd".

In the implementation, each distinct suffix string is represented as a node in the PAT tree and has only one pointer pointing to its position in the document. Each node consists of three parameters: the number of external nodes, the frequency count, and a comparison bit. The number of external nodes indicates the number of different suffix strings in the sub-trees, the frequency count indicates the frequency of occurrence of the corresponding string in the data stream, and the comparison bit indicates which bit will be compared when branching into sub-trees during the tree traversal. The frequencies of every string can be derived from the frequency count and the number of external nodes.

The PAT-tree-based keyterm extraction technique is a statistics-based approach used to find the representative keyterms of a document [1]. It consists of three stages. In the first stage, an efficient working structure extended from the PAT-tree is created for the given document. Within the working structure, all possible character strings with frequency counts in the document can be retrieved and updated very efficiently. In this way, all possible sub-strings can be considered as keyterm candidates, which is dictionary-free and language-independent. In the second stage, a mutual-information-based filtering algorithm is used to estimate the association among the component sub-strings for the character strings stored in the extended PAT-tree. In this way, all impossible character segments will be deleted. In the third stage, a keyterm selection policy based on a general-domain corpus and a commonly-used-word lexicon is applied, so over-general words and phrases which cannot represent the document will be deleted.

3 PAT-tree-Based Document Classification

Traditional document classification usually deals with static collections, in which documents used for training the feature values of each subject class remain the same during the classification process. However, documents gathered from the Web make a dynamic collection: new documents and keyterms are added into the collection incrementally, and the features of subject classes change as time goes on. Therefore, conventional classification approaches are hardly suitable. This section presents a PAT-tree-based approach for document classification in dynamic collections.

The proposed PAT-tree-based approach is not only an algorithm for classification but also a flexible working structure for feature selection. The working structure consists of three major components: PAT-tree-based classification, PAT-tree updating and PAT-tree-based keyterm extraction as shown in Fig. 1. Initially, documents are classified by an arbitrary classifier. The features of each subject class is represented by a PAT tree. These features can be variable length n-grams or extracted terms [7]. The PAT-tree is derived or updated according to the documents belonging to the class. Terms with high frequency in the PAT-tree are extracted as the keyterms that represent the subject class. These new keyterms can feed back the classifier and contribute to the classification of incoming documents immediately. With this working structure, the classification system is able to extract the new features of subject classes incrementally and hence performs up-to-date classification for dynamic collections.

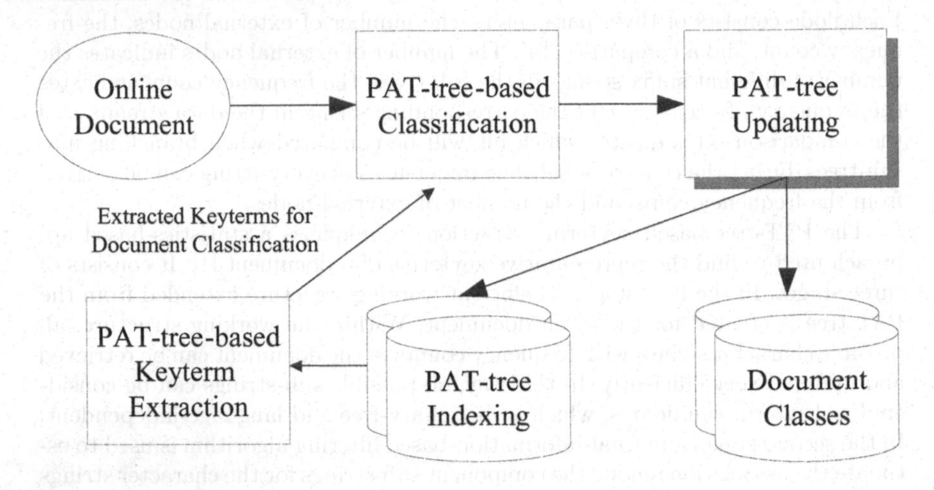

Fig. 1. The PAT-tree-based working structure for online document classification

4 Experiments on Classification in a Dynamic Collection

Based on the PAT-tree working structure, we implemented three well-known classifiers: TF-IDF classifier, Naive Bayes classifier, and Naive Bayes method with Kullback-Leibler Divergence classifier. Section 4.1 includes a short review of these classifiers, and Section 4.2 presents two experiments on document classification in a dynamic collection. The first experiment in Section 4.2 examines the classification performance of the PAT-tree-based keyterm extraction technique, while the second experiment examines the effect of incremental keyterm extraction used in the PAT-tree-based working structure. These experimental results show that the proposed PAT-tree-based approach performs well in document classification, and the performance can be further improved by incremental keyterm extraction.

4.1 Document Classifiers

TF-IDF Classifier The basic idea of TF-IDF classifier is to calculate how frequent a keyterm (a sequence of words or sequence of Chinese characters) in the testing examples appeared in the training ones. A simplified scoring formula as (1) is squaring the length of the examined keyterms as the term weighting.

For each online document d to be classified into class C, the features in d is represented by a set $W = (w_1, w_2, \ldots, w_n)$. Any variable length keyterms in W is examined to access its occurrence in the PAT tree. The classifier can be defined as the following form:

$$\arg\max_C \left[\sum_{i=1}^{n} \frac{L^2(w_i) * tf^2(w_i)}{N(w_i|C) * h^2} \right] \quad (1)$$

where w_i is in W, h is the number of classes where w_i ever occurs, $L(w_i)$ is the length of pattern w_i, $tf(w_i)$ is the term frequency of w_i in document d, and $N(w_i|C)$ is the frequency of w_i in class C.

Naive Bayes Classifier Naive Bayes Classifier is a well-known classifier. To classify a document with n extracted keyterms $(w_1, w_2, \ldots, w_n)$ into one of a set of classes C, the calculation function of the Naive Bayes classifier is

$$\arg\max_C \left[Pr(C) \prod_{i=1}^{n} Pr(w_i|C) \right] \quad (2)$$

where $Pr(w_i|C) = \frac{N(w_i|C)+1}{N(C)+T}$, is the probability of drawing w_i given a document from C, $N(w_i|C)$ is the number of times term w_i appearing in the training set of class C, $N(C)$ is the total number of terms in the training set of class C and T is the total number of unique terms in the document collection.

Naive Bayes Classifier with Kullback-Leibler Divergence The classifier uses the Naive Bayes method but with minor modifications based on Kullback-Leibler Divergence. This formula implies the minimum difference between the distribution of features in the class and the distribution of features in the document.

$$\arg\max_C \left[\frac{\log Pr(C)}{N} + \sum_{i=1}^{T} Pr(w_i|d) * \log \frac{Pr(w_i|C)}{Pr(w_i|d)} \right] \quad (3)$$

where N is the number of terms in d and $Pr(w_i|d)$ represents the frequency of occurrence of w_i in document d.

4.2 Experiments in a Dynamic Collection

This subsection presents two experiments on document classification in a dynamic collection based on the PAT-tree-based approach. The first experiment evaluates the performance of the PAT-tree-based keyterm extraction technique, and the second experiment evaluates the performance of incremental keyterm extraction used in the PAT-tree-based working structure.

Environment The document collection consists of Chinese full-story news gathered from Central News Agency (CNA) during Jan. 1996 to Dec. 1996. The data set is a good benchmark for classification of Chinese documents because it reflects real world situations in document classification. The statistics of training and testing corpus are listed in Table 1. The full-story news were classified into 21 subject classes according to the classification rules provided by CNA.

Table 1. Statistics of the training and testing document sets

Class Name	Sep-96	Oct-96	Nov-96	Dec-96
Congress	416	458	382	360
Culture (Domestic)	530	591	487	438
Culture (Foreign)	141	128	138	114
Culture (Mainland)	66	68	55	74
Economics (Domestic)	988	988	1000	1009
Economics (Foreign)	449	421	386	411
Economics (Mainland)	236	248	295	298
Local Government	1640	1774	1726	1734
Society (Domestic)	1508	1682	1686	1622
Society (Foreign)	468	475	464	477
Society (Mainland)	110	124	126	114
Sports (Domestic)	324	468	285	301
Sports (Foreign)	110	87	66	44
Sports (Mainland)	6	13	42	13
Politics (Domestic)	1154	1353	1142	1311
Politics (Foreign)	1742	1796	1658	1361
Politics (Mainland)	510	504	460	501
Transportation (Domestic)	282	262	262	264
Transportation (Foreign)	52	70	78	70
Transportation (Mainland)	31	22	49	37
Weather	0	18	27	87
Total	10763	11550	10814	10640

Document Classification In this experiment, the training set consists of the news documents from Sep-96 to Nov-96, and the testing set consists of the news documents in Dec-96. It is worthy of noticing that the keyterms were extracted in a batch mode from the training set only. The feature set of each subject class is generated in the training stage and remains the same during in testing stage. After classifying each document in the testing set, the frequencies of existing

Table 2. Classification results of the three different classifiers, where P stands for precision rate and R stands for recall rate

Class Name	Test Size	TF-IDF		Naive Bayes		Kullback	
		R	P	R	P	R	P
Congress	360	0.372	0.854	0.656	0.641	0.786	0.496
Culture (Domestic)	438	0.377	0.782	0.731	0.582	0.852	0.519
Culture (Foreign)	114	0.044	0.833	0.149	0.630	0.474	0.524
Culture (Mainland)	74	0.068	0.714	0.041	1.000	0.297	0.786
Economics (Domestic)	1009	0.758	0.871	0.833	0.844	0.825	0.843
Economics (Foreign)	411	0.324	0.816	0.664	0.592	0.815	0.556
Economics (Mainland)	298	0.742	0.837	0.839	0.694	0.906	0.692
Local Government	1734	0.865	0.627	0.787	0.786	0.712	0.831
Society (Domestic)	1622	0.850	0.722	0.773	0.851	0.750	0.878
Society (Foreign)	477	0.199	0.736	0.633	0.524	0.744	0.502
Society (Mainland)	114	0.079	0.900	0.026	1.000	0.360	0.651
Sports (Domestic)	301	0.721	0.908	0.877	0.857	0.944	0.821
Sports (Foreign)	44	0.227	1.000	0.114	1.000	0.386	0.895
Sports (Mainland)	13	0.000	NA	0.000	NA	0.000	NA
Politics (Domestic)	1311	0.762	0.686	0.829	0.672	0.726	0.747
Politics (Foreign)	1361	0.931	0.518	0.744	0.690	0.567	0.910
Politics (Mainland)	501	0.261	0.775	0.529	0.690	0.842	0.655
Transportation (Domestic)	264	0.258	0.944	0.598	0.840	0.826	0.673
Transportation (Foreign)	70	0.443	1.000	0.443	1.000	0.529	0.804
Transportation (Mainland)	37	0.081	1.000	0.027	1.000	0.270	0.909
Weather	87	0.989	1.000	1.000	1.000	1.000	1.000
Micro-average		0.679	0.679	0.731	0.731	0.734	0.734

features will be updated accordingly; however, no new features will be extracted. Table 2 shows the obtained classification results. In this experiment, Naive Bayes with Kullback-Leibler Divergence classifier performed better.

We compared this result with a manual classification test. Totally 9 journalists were invited to perform the manual classification test. Among them, 4 were informed; that is, they completed the testing by following the classification rules provided by CNA. The performance of the informed journalists was 0.655 (measured by micro-average precision/recall) while that of the uninformed ones was 0.530. The result shows that machine classification can even perform better than professional human agents.

Incremental Keyterm Extraction In this experiment, we want to evaluate the effect of incremental keyterm extraction for document classification. We used the news from Sep-96 to Nov-96 as the training set and the news on Dec-96 as the testing set. The Naive Bayes with Kullback-Leibler Divergence classifier was used in the classification. After each document in the test set was classified, it was added into the corresponding PAT tree. New keyterms were extracted,

Table 3. Classification results with keyterm extraction, where P stands for precision rate and R stands for recall rate

Class Name	Test Size	Incremental		Upper-bound	
		R	P	R	P
Congress	360	0.678	0.683	0.764	0.674
Culture (Domestic)	438	0.760	0.737	0.760	0.727
Culture (Foreign)	114	0.254	0.460	0.281	0.525
Culture (Mainland)	74	0.581	0.754	0.541	0.741
Economics (Domestic)	1009	0.810	0.877	0.818	0.868
Economics (Foreign)	411	0.703	0.685	0.681	0.685
Economics (Mainland)	298	0.856	0.823	0.856	0.812
Local Government	1734	0.796	0.812	0.830	0.804
Society (Domestic)	1622	0.861	0.757	0.858	0.795
Society (Foreign)	477	0.363	0.689	0.392	0.719
Society (Mainland)	114	0.368	0.656	0.325	0.649
Sports (Domestic)	301	0.917	0.841	0.937	0.849
Sports (Foreign)	44	0.455	0.714	0.500	0.759
Sports (Mainland)	13	0.077	1.000	0.077	1.000
Politics (Domestic)	1311	0.769	0.721	0.748	0.770
Politics (Foreign)	1361	0.825	0.738	0.847	0.723
Politics (Mainland)	501	0.778	0.763	0.747	0.773
Transportation (Domestic)	264	0.716	0.788	0.742	0.781
Transportation (Foreign)	70	0.557	0.736	0.557	0.709
Transportation (Mainland)	37	0.378	0.778	0.405	0.750
Weather	87	1.000	1.000	1.000	1.000
Micro-average		0.766	0.766	0.775	0.775

and frequencies of existing keyterms were updated accordingly. The feedback contributes to the classification of the incoming documents immediately.

The results are shown in Table 3, labeled as "Incremental". In comparison to the results in Table 2, it shows that extracting new keyterms by PAT-tree-based working structure enhances the performance of document classification.

The results labeled as "Upper-bound" were conducted as follows. The test documents were added into the document classes according to their *correct classes* rather than automatic classification results. Therefore, the PAT trees were updated according to the correctly-classified documents. Therefore, these values indicate the best performance this working structure can obtain. The current performance was encouraging: it slightly decreases from the upper-bound value 0.775 to 0.766.

5 Experiments on Classification in a Large-Scale Taxonomy

This section presents the experiments on scientific document classification into a large-scale taxonomy. We utilized a manually-constructed subject taxonomy and a modest number of keyterms to perform initial classification. The Naive Bayes with Kullback-Leibler Divergence classifier was used in this experiment.

The taxonomy was developed by STIC (Science and Technology Information Center) of National Science Council in Taiwan. It provides hierarchical classification of technical research documents. The hierarchy consists of three levels. The highest level contains 19 classes; the second level contains 188 levels; the lowest level contains 1,011 classes. For each class, there is a set of traditional Chinese keyterms and a set of English keyterms, which describe the feature of that class. The total keyterms were 410,557 for traditional Chinese and English respectively.

Initially, we used a set of 25,717 manually-classified Chinese documents as the training set to create domain-specific PAT trees. For each class, the feature set includes manually-classified keyterms, as well as new keyterms extracted from the corresponding PAT tree. 1,000 technical documents were randomly selected from the STIC collection to test the performance of the classification into the 1,011 classes. The obtained precision rate is 0.713, which is similar to the performance obtained in the experiments shown in Section 4.2. We believe the good performance on classifying documents into such a large-scale taxonomy is mainly due to both the high quality of the keyterm files and the PAT-tree-based working structure.

6 Web Document Discovery and Future Work

Our goal is to discover the technical documents from the Web in the Greater China area, index the documents written in English, traditional and simplified Chinese, and classify them into the STIC taxonomy. The experimental results in Section 4 and Section 5 show that the PAT-tree-based working structure has great potential to overcome the classification challenges in dynamic collections and in large-scale taxonomies. In the following, we extend the working structure to deal with classification of multilingual documents. Our future work is described at the end of this section.

6.1 Classification of Multilingual Documents

In the current stage, we are able to classify technical documents written in traditional Chinese and English based on the STIC taxonomy and keyterm files. As the PAT-tree technique is language independent, we can use them to extract keyterms in simplified Chinese documents.

The basic idea of our approach is to take English as the medium. Nowadays, a large number of simplified Chinese documents in the Web contain English abstracts. We can classify these documents into the STIC taxonomy according to the English abstracts. For each class in the taxonomy, we apply the PAT-tree-based approach and extract simplified Chinese keyterms from these documents. By the mediation of their English abstracts, we can obtain the simplified Chinese keyterms corresponding to the STIC taxonomy. Once we have lots of simplified Chinese documents classified, we may obtain sufficient keyterms and create the simplified Chinese version of feature sets in the STIC taxonomy. In this stage, we would be able to classify documents according to the simplified Chinese keyterms even if the documents don't contain English abstracts.

This approach could be applied to other languages. We could classify documents written in all kinds of languages as long as they contain English abstracts. By the PAT-tree-based extraction, we could add multilingual keyterms into the taxonomy. We could obtain a universal taxonomy, which is able to classify all documents written in different languages.

6.2 Future Work

The prototype of this system is still under development. The approach proposed to deal with multilingual documents should be evaluated, and larger-scale experiments should be done on classification in the dynamic collection and the large-scale taxonomy. Future efforts will be committed to collecting technical documents from the Web and improving the performance of document classification. We need to collect sufficient documents and extract new keyterms in order to evaluate the proposed approach and make the system more practicable.

References

1. L.-F. Chien. PAT-tree-based keyword extraction for chinese information retrieval. In *Proceedings of ACM SIGIR'97 Conference*, 1997.
2. L.-F. Chien. PAT-tree-based adaptive keyphrase extraction for intelligent chinese information retrieval. *Informatin Processing and Management*, 35:501–521, 1999.
3. C. L. Giles, K. Bollacker, and S. Lawrence. Citeseer: An automatic citation indexing systm. In *Proceedings of 1998 ACM Conference on Digital Library*, 1998.
4. C. J. Godby and R. Reighart. Using machine-readable text as a source of novel vocabulary to update the dewey decimal classification. In *Proceedings of the 1998 ASIS Classification Workshop*, 1998.
5. G. H. Gonnet, R. Baeza-Yates, and T. Snider. *Information Retrieval Data Structures and Algorithms*, pages 66 – 82. London : Prentice Hall International, 1992.
6. A. McCallum and K. Nigam et. al. A machine learning approach to building domain-specific search engines. In *Proceedings of the Sixteenth International Joint Conference on Artificial Intelligence (IJCAI'99)*, 1999.
7. O. Zamir and O. Etzioni. Web document clustering: A feasibility demonstration. In *Proceedings of SIGIR'98*, 1998.

k-nearest Neighbor Classification on Spatial Data Streams Using P-trees[1, 2]

Maleq Khan, Qin Ding, and William Perrizo

Computer Science Department, North Dakota State University
Fargo, ND 58105, USA
{Md.Khan, Qin.Ding, William.Perrizo}@ndsu.nodak.edu

Abstract. Classification of spatial data streams is crucial, since the training dataset changes often. Building a new classifier each time can be very costly with most techniques. In this situation, *k*-nearest neighbor (KNN) classification is a very good choice, since no residual classifier needs to be built ahead of time. KNN is extremely simple to implement and lends itself to a wide variety of variations. We propose a new method of KNN classification for spatial data using a new, rich, data-mining-ready structure, the *Peano-count-tree* (*P-tree*). We merely perform some AND/OR operations on P-trees to find the nearest neighbors of a new sample and assign the class label. We have fast and efficient algorithms for the AND/OR operations, which reduce the classification time significantly. Instead of taking exactly the *k* nearest neighbors we form a closed-KNN set. Our experimental results show closed-KNN yields higher classification accuracy as well as significantly higher speed.

1. Introduction

There are various techniques for classification such as Decision Tree Induction, Bayesian Classification, and Neural Networks [7, 8]. Unlike other common classifiers, a *k-nearest neighbor* (KNN) classifier does not build a classifier in advance. That is what makes it suitable for data streams. When a new sample arrives, KNN finds the k neighbors nearest to the new sample from the training space based on some suitable similarity or distance metric. The plurality class among the nearest neighbors is the class label of the new sample [3, 4, 5, 7]. A common similarity function is based on the Euclidian distance between two data tuples [7]. For two tuples, $X = \langle x_1, x_2, x_3, \ldots, x_{n-1}\rangle$ and $Y = \langle y_1, y_2, y_3, \ldots, y_{n-1}\rangle$ (excluding the class labels), the *Euclidian* similarity function is $d_2(X,Y) = \sqrt{\sum_{i=1}^{n-1}(x_i - y_i)^2}$. A generalization of the Euclidean function is the *Minkowski* similarity function is

[1] Patents are pending on the bSQ and Ptree technology.
[2] This work is partially supported by NSF Grant OSR-9553368, DARPA Grant DAAH04-96-1-0329 and GSA Grant ACT#: K96130308.

M.-S. Chen, P.S. Yu, and B. Liu (Eds.): PAKDD 2002, LNAI 2336, pp. 517-528, 2002.

$d_q(X,Y) = \sqrt[q]{\sum_{i=1}^{n-1} w_i |x_i - y_i|^q}$. The Euclidean function results by setting q to 2 and each weight, w_i, to 1. The *Manhattan* distance, $d_1(X,Y) = \sum_{i=1}^{n-1} |x_i - y_i|$ result by setting q to 1. Setting q to ∞, results in the *max* function $d_\infty(X,Y) = \max_{i=1}^{n-1} |x_i - y_i|$.

In this paper, we introduced a new metric called *Higher Order Bit* (*HOB*) similarity metric and evaluated the effect of all of the above distance metrics in classification time and accuracy. HOB distance provides an efficient way of computing neighborhoods while keeping the classification accuracy very high.

KNN is a good choice when simplicity and accuracy are the predominant issues. KNN can be superior when a residual, trained and tested classifiers, such as ID3, has a short useful lifespan, such as in the case with data streams, where new data arrives rapidly and the training set is ever changing [1, 2]. For example, in spatial data, AVHRR images are generated in every one hour and can be viewed as spatial data streams. The purpose of this paper is to introduce a new KNN-like model, which is not only simple and accurate but is also fast – fast enough for use in spatial data stream classification.

In this paper we propose a simple and fast KNN-like classification algorithm for spatial data using P-trees. P-trees are new, compact, data-mining-ready data structures, which provide a lossless representation of the original spatial data [6, 9, 10]. In the section 2, we review the structure of P-trees and various P-tree operations.

We consider a space to be represented by a 2-dimensional array of locations. Associated with each location are various attributes, called ***bands***, such as visible reflectance intensities (blue, green and red), infrared reflectance intensities (e.g., NIR, MIR1, MIR2 and TIR) and possibly crop yield quantities, soil attributes and radar reflectance intensities. We refer to a location as a pixel in this paper.

Using P-trees, we presented two algorithms, one based on the max distance metric and the other based on our new HOB distance metric. HOB is the similarity of the most significant bit positions in each band. It differs from pure Euclidean similarity in that it can be an asymmetric function depending upon the bit arrangement of the values involved. However, it is very fast, very simple and quite accurate. Instead of using exactly k nearest neighbor (a KNN set), our algorithms build a *closed-KNN* set and perform voting on this closed-KNN set to find the predicting class. Closed-KNN, a superset of KNN, is formed by including the pixels, which have the same distance from the target pixel as some of the pixels in KNN set. Based on this similarity measure, finding nearest neighbors of new samples (pixel to be classified) can be done easily and very efficiently using P-trees and we found higher classification accuracy than traditional methods on considered datasets. Detailed definitions of the similarity and the algorithms to find nearest neighbors are given in the section 3. We provided experimental results and analyses in section 4. The conclusion is given in Section 5.

2. P-tree Data Structures

Most spatial data comes in a format called *BSQ* for *Band Sequential* (or can be easily converted to BSQ). BSQ data has a separate file for each band. The ordering of the data values within a band is raster ordering with respect to the spatial area represented in the dataset. We divided each BSQ band into several files, one for each bit position of the data values. We call this format *bit Sequential* or *bSQ* [6, 9, 10]. A Landsat Thematic Mapper satellite image, for example, is in BSQ format with 7 bands, $B_1,\ldots,B_7$, (Landsat-7 has 8) and ~40,000,000 8-bit data values. A typical *TIFF* image aerial digital photograph is one file containing ~24,000,000 bits ordered by it-position, then band and then raster-ordered-pixel-location.

We organize each bSQ bit file, B_{ij} (the file constructed from the jth bits of ith band), into a tree structure, called a *Peano Count Tree* (*P-tree*). A P-tree is a quadrant-based tree. The root of a P-tree contains the 1-bit count, called *root count*, of the entire bit-band. The next level of the tree contains the 1-bit counts of the four quadrants in raster order. At the next level, each quadrant is partitioned into sub-quadrants and their 1-bit counts in raster order constitute the children of the quadrant node. This construction is continued recursively down each tree path until the sub-quadrant is *pure* (entirely 1-bits or entirely 0-bits). Recursive raster ordering is called the Peano or Z-ordering in the literature – therefore, the name Peano Count trees. The P-tree for a 8-row-8-column bit-band is shown in Fig. 1.

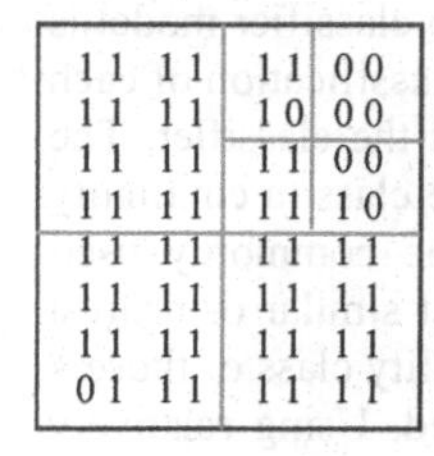

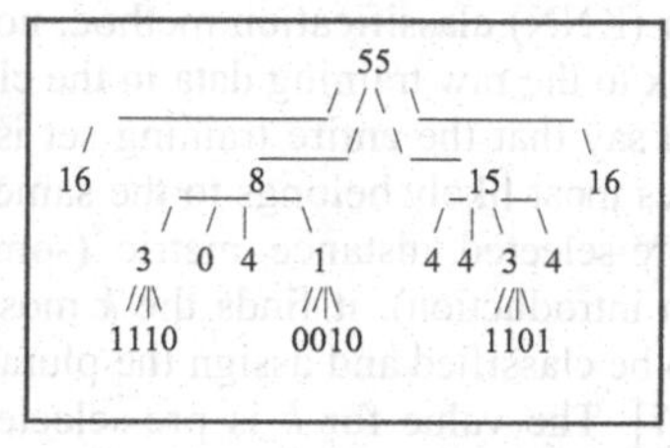

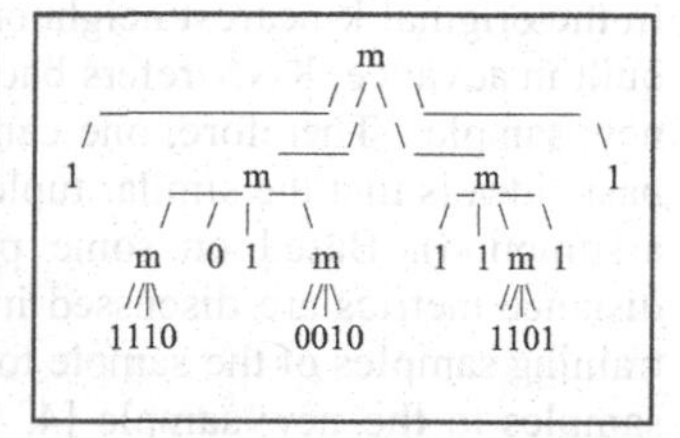

Fig. 1. 8-by-8 image and its P-tree (P-tree and PM-tree)

In this example, root count is 55, and the counts at the next level, 16, 8, 15 and 16, are the 1-bit counts for the four major quadrants. Since the first and last quadrant is made up of entirely 1-bits, we do not need sub-trees for these two quadrants.

For each band (assuming 8-bit data values), we get 8 *basic P-trees*. $P_{i,j}$ is the P-tree for the jth bits of the values from the ith band. For efficient implementation, we use variation of basic P-trees, called *Pure Mask tree* (*PM-tree*). In the PM-tree, we use a 3-value logic, in which 11 represents a quadrant of pure 1-bits, *pure1* quadrant, 00 represents a quadrant of pure 0-bits, *pure0* quadrant, and 01 represents a mixed quadrant. To simplify the exposition, we use 1 instead of 11 for pure1, 0 for pure0, and *m* for mixed. The PM-tree for the previous example is also given in Fig. 1.

P-tree algebra contains operators, AND, OR, NOT and XOR, which are the pixel-by-pixel logical operations on P-trees. The AND/OR operations on PM-trees are shown in Fig. 2. The AND operation between two PM-trees is performed by ANDing the corresponding nodes in the two operand PM-trees level-by-level starting from the root node. A pure0 node ANDed with any node produces a pure0 node. A pure1 node ANDed with any node, *n*, produces the node *n*; we just need to copy the node, *n*, to the resultant PM-tree in the corresponding position. ANDing of two mixed nodes produces a mixed node; the children of the resultant mixed node are obtained by ANDing children of the operands recursively. The details of P-tree operations can be found in [10, 11].

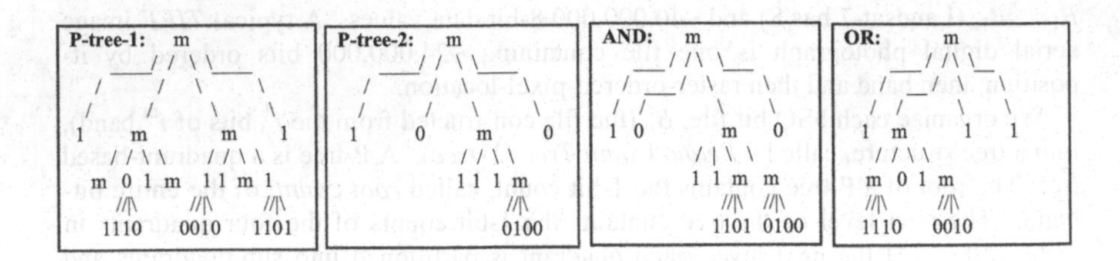

Fig. 2. P-tree Algebra

3. The Classification Algorithms

In the original k-nearest neighbor (KNN) classification method, no classifier model is built in advance. KNN refers back to the raw training data in the classification of each new sample. Therefore, one can say that the entire training set is the classifier. The basic idea is that the similar tuples most likely belongs to the same class (a continuity assumption). Based on some pre-selected distance metric (some commonly used distance metrics are discussed in introduction), it finds the k most similar or nearest training samples of the sample to be classified and assign the plurality class of those k samples to the new sample [4, 7]. The value for k is pre-selected. Using relatively larger k may include some not so similar pixels and on the other hand, using very smaller k may exclude some potential candidate pixels. In both cases the classification accuracy will decrease. The optimal value of k depends on the size and nature of the data. The typical value for k is 3, 5 or 7. The steps of the classification process are:

1) Determine a suitable distance metric.
2) Find the k nearest neighbors using the selected distance metric.
3) Find the plurality class of the k-nearest neighbors (voting on the class labels of the NNs).
4) Assign that class to the sample to be classified.

We provided two different algorithms using P-trees, based two different distance metrics *max* (Minkowski distance with $q = \infty$) and our newly defined *HOB* distance. Instead of examining individual pixels to find the nearest neighbors, we start our initial *neighborhood* (neighborhood is a set of neighbors of the target pixel within a specified distance based on some distance metric, not the spatial neighbors, neighbors with respect to values) with the target sample and then successively expand the neighborhood area until there are k pixels in the neighborhood set. The expansion is done in such a way that the neighborhood always contains the closest or most similar pixels of the target sample. The different expansion mechanisms implement different distance functions. In the next section (section 3.1) we described the distance metrics and expansion mechanisms.

Of course, there may be more boundary neighbors equidistant from the sample than are necessary to complete the k nearest neighbor set, in which case, one can either use the larger set or arbitrarily ignore some of them. To find the exact k nearest neighbors one has to arbitrarily ignore some of them.

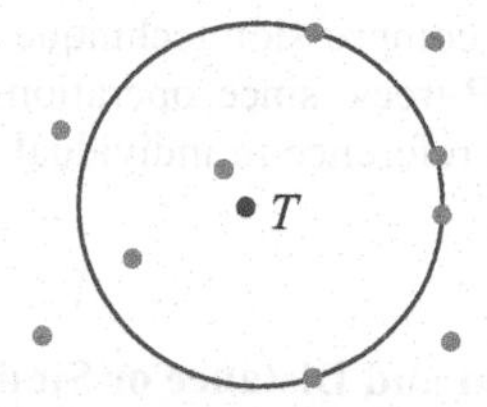

Fig. 3. *T*, the pixel in the center is the target pixels. With $k = 3$, to find the third nearest neighbor, we have four pixels (on the boundary line of the neighborhood) which are equidistant from the target.

Instead we propose a new approach of building nearest neighbor (NN) set, where we take the closure of the *k*-NN set, that is, we include all of the boundary neighbors and we call it the *closed-KNN* set. Obviously closed-KNN is a superset of KNN set. In the above example, with $k = 3$, KNN includes the two points inside the circle and any one point on the boundary. The closed-KNN includes the two points in side the circle and all of the four boundary points. The inductive definition of the closed-KNN set is given below.

Definition 1. a) If $x \in$ KNN, then $x \in$ closed-KNN

b) If $x \in$ closed-KNN and $d(T,y) \leq d(T,x)$, then $y \in$ closed-KNN where, $d(T,x)$ is the distance of x from target T.

c) Closed-KNN does not contain any pixel, which cannot be produced by steps a and b.

Our experimental results show closed-KNN yields higher classification accuracy than KNN does. The reason is if for some target there are many pixels on the boundary, they have more influence on the target pixel. While all of them are in the nearest neighborhood area, inclusion of one or two of them does not provide the necessary weight in the voting mechanism. One may argue that then why don't we use a higher *k*? For example using $k = 5$ instead of $k = 3$. The answer is if there are too few points (for example only one or two points) on the boundary to make k neighbors in the neighborhood, we have to expand neighborhood and include some not so similar points which will decrease the classification accuracy. We construct closed-KNN only by including those pixels, which are in as same distance as some other pixels in the neighborhood without further expanding the neighborhood. To perform our experiments, we find the optimal k (by trial and error method) for that particular dataset and then using the optimal k, we performed both KNN and closed-KNN and found higher accuracy for P-tree-based closed-KNN method. The experimental results are given in section 4. In our P-tree implementation, no extra computation is required to find the closed-KNN. Our expansion mechanism of nearest neighborhood automatically includes the points on the boundary of the neighborhood.

Also, there may be more than one class in plurality (if there is a tie in voting), in which case one can arbitrarily chose one of the plurality classes. Without storing the raw data we create the basic P-trees and store them for future classification purpose. Avoiding the examination of individual data points and being ready for data mining these P-trees not only saves classification time but also saves storage space, since data is stored in compressed form. This compression technique also increases the speed of ANDing and other operations on P-trees, since operations can be performed on the pure0 and pure1 quadrants without reference to individual bits, since all of the bits in those quadrants are the same.

3.1 Expansion of Neighborhood and Distance or Similarity Metrics

We begin searching for nearest neighbors by finding the exact matches. If the number of exact matches is less than k, we expand the neighborhood. The expansion of the neighborhood in each dimension are done simultaneously, and continued until the number pixels in the neighborhood is greater than or equal to k. We develop the following two different mechanisms, corresponding to max distance and our newly defined HOB distance, for expanding the neighborhood. They have trade offs between execution time and classification accuracy.

Higher Order Bit Similarity (HOBS): We propose a new similarity metric where we consider similarity in the most significant consecutive bit positions starting from the left most bit, the highest order bit. Consider the following two values, x_1 and y_1, represented in binary. The 1st bit is the most significant bit and 8th bit is the least significant bit.

Bit position:	1 2 3 4 5 6 7 8		1 2 3 4 5 6 7 8
x_1:	0 1 1 0 1 0 0 1	x_1:	0 1 1 0 1 0 0 1
y_1:	0 1 1 1 1 1 0 1	y_2:	0 1 1 0 0 1 0 0

These two values are similar in the three most significant bit positions, 1st, 2nd and 3rd bits (011). After they differ (4th bit), we don't consider anymore lower order bit positions though x_1 and y_1 have identical bits in the 5th, 7th and 8th positions. Since we are looking for closeness in values, after differing in some higher order bit positions, similarity in some lower order bit is meaningless with respect to our purpose. Similarly, x_1 and y_2 are identical in the 4 most significant bits (0110). Therefore, according to our definition, x_1 is closer or similar to y_2 than to y_1.

Definition 2. The similarity between two integers A and B is defined by

$$\text{HOBS}(A, B) = \max\{s \mid 0 \le i \le s \Rightarrow a_i = b_i\}$$

in other words, HOBS(A, B) = s, *where for all* $i \le s$ *and* $0 \le i$, $a_i = b_i$ *and* $a_{s+1} \ne b_{s+1}$. a_i and b_i are the ith bits of A and B respectively.

Definition 3. The distance between the values *A* and *B* is defined by

$$d_v(A, B) = m - \text{HOBS}(A, B)$$

where *m* is the number of bits in binary representations of the values. All values must be represented using the same number of bits.

Definition 4. The distance between two pixels *X* and *Y* is defined by

$$d_p(X,Y) = \max_{i=1}^{n-1}\{d_v(x_i,y_i)\} = \max_{i=1}^{n-1}\{m - \text{HOBS}(x_i,y_i)\}$$

where n is the total number of bands; one of them (the last band) is the class attribute that we don't use for measuring similarity.

To find the closed –KNN set, first we look for the pixels, which are identical to the target pixel in all 8 bits of all bands i.e. the pixels, *X*, having distance from the target *T*, $d_p(X,T) = 0$. If, for instance, x_1=105 ($01101001_b = 105_d$) is the target pixel, the initial neighborhood is [105, 105] ([01101001, 01101001]). If the number of matches is less than *k*, we look for the pixels, which are identical in the 7 most significant bits, not caring about the 8th bit, i.e. pixels having $d_p(X,T) \leq 1$. Therefore our expanded neighborhood is [104,105] ([01101000, 01101001] or [0110100-, 0110100-] - don't care about the 8th bit). Removing one more bit from the right, the neighborhood is [104, 107] ([011010--, 011010--] - don't care about the 7th or the 8th bit). Continuing to remove bits from the right we get intervals, [104, 111], then [96, 111] and so on.

Computationally this method is very cheap. However, the expansion does not occur evenly on both sides of the target value (note: the center of the neighborhood [104, 111] is (104 + 111) /2 = 107.5 but the target value is 105). Another observation is that the size of the neighborhood is expanded by powers of 2. These uneven and jump expansions include some not so similar pixels in the neighborhood keeping the classification accuracy lower. But P-tree-based closed-KNN method using this HOBS metric still outperforms KNN methods using any distance metric as well as becomes the fastest among all of these methods.

Perfect Centering: In this method we expand the neighborhood by 1 on both the left and right side of the range keeping the target value always precisely in the center of the neighborhood range. We begin with finding the exact matches as we did in HOBS method. The initial neighborhood is [*a*, *a*], where *a* is the target band value. If the number of matches is less than *k* we expand it to [*a*-1, *a*+1], next expansion to [*a*-2, *a*+2], then to [*a*-3, *a*+3] and so on.

Perfect centering expands neighborhood based on max distance metric or L_∞ metric (discussed in introduction). In the initial neighborhood $d_\infty(X,T)$ is 0. In the first expanded neighborhood [*a*-1, *a*+1], $d_\infty(X,T) \leq 1$. In each expansion $d_\infty(X,T)$ increases by 1. As distance is the direct difference of the values, increasing distance by one also increases the difference of values by 1 evenly in both side of the range.

This method is computationally a little more costly because we need to find matches for each value in the neighborhood range and then accumulate those matches but it results better nearest neighbor sets and yields better classification accuracy. We compare these two techniques later in section 4.

3.2 Computing the Nearest Neighbors

For HOBS: $P_{i,j}$ is the basic P-tree for bit j of band i and $P'_{i,j}$ is the complement of $P_{i,j}$. Let, $b_{i,j} = j^{th}$ bit of the ith band of the target pixel, and define

$$Pt_{i,j} = P_{i,j}, \quad \text{if } b_{i,j} = 1,$$
$$= P'_{i,j}, \quad \text{otherwise.}$$

We can say that the root count of $Pt_{i,j}$ is the number of pixels in the training dataset having as same value as the jth bit of the ith band of the target pixel. Let,

$$Pv_{i,1\text{-}j} = Pt_{i,1} \ \& \ Pt_{i,2} \ \& \ Pt_{i,3} \ \& \ \ldots \ \& \ Pt_{i,j}, \text{ and}$$

$$Pd(j) = Pv_{1,1\text{-}j} \ \& \ Pv_{2,1\text{-}j} \ \& \ Pv_{3,1\text{-}j} \ \& \ \ldots \ \& \ Pv_{n\text{-}1,1\text{-}j}$$

where & is the P-tree AND operator and n is the number of bands. $Pv_{i,1\text{-}j}$ counts the pixels having as same bit values as the target pixel in the higher order j bits of ith band. We calculate the initial neighborhood P-tree, $Pnn = Pd(8)$, the exact matching, considering 8-bit values. Then we calculate $Pnn = Pd(7)$, matching in 7 higher order bits; then Then $Pnn = Pd(6)$ and so on. We continue as long as root count of Pnn is less than k. Pnn represents closed-KNN set and the root count of Pnn is the number of the nearest pixels. A 1 bit in Pnn for a pixel means that pixel is in closed-KNN set. The algorithm for finding nearest neighbors is given in Fig. 4.

Input: $P_{i,j}$ for all bit i and band j, the basic P-trees
and $b_{i,j}$ for all i and j, the bits for the target pixels
Output: Pnn, the P-tree representing closed-KNN
// n - # of bands, m - # of bits in each band
FOR i = 1 TO n-1 D
    FOR j = 1 TO m DO
        IF $b_{i,j} = 1$ $Pt_{ij} \leftarrow P_{i,j}$
        ELSE $Pt_{i,j} \leftarrow P'_{i,j}$
FOR i = 1 TO n-1 DO
	$Pv_{i,1} \leftarrow Pt_{i,1}$
	FOR j = 2 TO m DO
	    $Pv_{i,j} \leftarrow Pv_{i,j\text{-}1} \ \& \ Pt_{i,j}$
	$s \leftarrow m$
	REPEAT
	    $Pnn \leftarrow Pv_{1,s}$
	    FOR r = 2 TO n-1 DO
	        $Pnn \leftarrow Pnn \ \& \ Pv_{r,s}$
	    $s \leftarrow s - 1$
	UNTIL RootCount(Pnn) $\geq k$

Fig. 4. Algorithm to find closed-KNN set based on HOB metric

For Perfect Centering: Let v_i is the value of the target pixels for band i. The *value P-tree*, $P_i(v_i)$, represents the pixels having value v_i in band i. The algorithm for computing the value P-trees is given in Fig. 5(b). For finding the initial nearest neighbors (the exact matches), we calculate

$$Pnn = P_1(v_1) \ \& \ P_2(v_2) \ \& \ P_3(v_3) \ \& \ \ldots \ \& \ P_{n\text{-}1}(v_{n\text{-}1})$$

that represents the pixels having the same values in each band as that of the target pixel. If the root count of $Pnn \leq k$, we expand neighborhood along each dimension. For each band i, we calculate *range P-tree* $Pr_i = P_i(v_i\text{-}1) \mid P_i(v_i) \mid P_i(v_i\text{+}1)$. '|' is the P-tree OR operator. Pr_i represents the pixels having any value in the range $[v_i\text{-}1, v_i\text{+}1]$ of band i. The *AND*ed result of these range P-trees, Pr_i, for all i, produce the expanded neighborhood. The algorithm is given in Fig. 5(a).

```
Input: P_{i,j} for all i and j, basic P-trees and
       v_i for all i, band values of target pixel
Output: Pnn, closed-KNN P-tree
// n - # of bands, m- #of bits in each band
FOR i = 1 TO n-1 DO
    Pr_i ← P_i(v_i)
Pnn ← Pr_1
FOR i = 2 TO n-1 DO
    Pnn ← Pnn & Pr_i //initial neighborhood
d ← 1    // distance for the first expansion
WHILE RootCount(Pnn) < k DO
    FOR i = 1 to n-1 DO      // expansion
        Pr_i ← Pr_i | P_i(v_i-d) | P_i(v_i+d)
    Pnn ← Pr_1          // '|' - OR operator
    FOR i = 2 TO n-1 DO
        Pnn ← Pnn AND Pr_i
    d ← d + 1
```

Fig. 5(a). Algorithm to find closed-KNN set based on Max metric (Perfect Centering).

```
Input: P_{i,j} for all j, basic P-trees of all
the bits of band i and the value v_i for
band i.
Output: P_i(v_i), the value p-tree for the
value v_i
// m is the number of bits in each band
// b_{i,j} is the j^th bit of value v_i
FOR j = 1 TO m DO
    IF b_{i,j} = 1 Pt_{ij} ← P_{i,j}
    ELSE Pt_{i,j} ← P'_{i,j}

P_i(v) ← Pt_{i,1}
FOR j = 2 TO m DO
    P_i(v) ← P_i(v) & Pt_{i,j}
```

5(b). Algorithm to compute value P-trees

3.3 Finding the Plurality Class among the Nearest Neighbors

For the classification purpose, we don't need to consider all bits in the class band. If the class band is 8 bits long, there are 256 possible classes. Instead, we partition the class band values into fewer, say 8, groups by truncating the 5 least significant bits. The 8 classes are 0, 1, 2, …, 7. Using the leftmost 3 bits we construct the value P-trees $P_n(0)$, $P_n(1)$, …, $P_n(7)$. The P-tree Pnn & $P_n(i)$ represents the pixels having a class value i and are in the closed-KNN set, *Pnn*. An i which yields the maximum root count of Pnn & $P_n(i)$ is the plurality class; that is

$$\text{predicted class} = \arg\max_i \{RootCount(Pnn \,\&\, P_n(i))\}.$$

4. Performance Analysis

We performed experiments on two sets of Arial photographs of the Best Management Plot (BMP) of Oakes Irrigation Test Area (OITA) near Oaks, North Dakota, United States. The latitude and longitude are 45°49'15"N and 97°42'18"W respectively. The two images "29NW083097.tiff" and "29NW082598.tiff" have been taken in 1997 and 1998 respectively. Each image contains 3 bands, red, green and blue reflectance values. Three other separate files contain synchronized soil moisture, nitrate and yield values. Soil moisture and nitrate are measured using shallow and deep well lysimeters. Yield values were collected by using a GPS yield monitor on the harvesting equipments. The datasets are available at http://datasurg.ndsu.edu/.

Yield is the class attribute. Each band is 8 bits long. So we have 8 basic P-trees for each band and 40 (for the other 5 bands except yield) in total. For the class band, we considered only the most significant 3 bits. Therefore we have 8 different class labels. We built 8 value P-trees from the yield values – one for each class label.

The original image size is 1320×1320. For experimental purpose we form 16×16, 32×32, 64×64, 128×128, 256×256 and 512×512 image by choosing pixels uniformly distributed in the original image. In each case, we formed one test set and one training set of equal size and tested KNN with Manhattan, Euclidian, Max and HOBS distance metrics and our two P-tree methods, Perfect Centering and HOBS. The accuracies of these different implementations are given in the Fig 6.

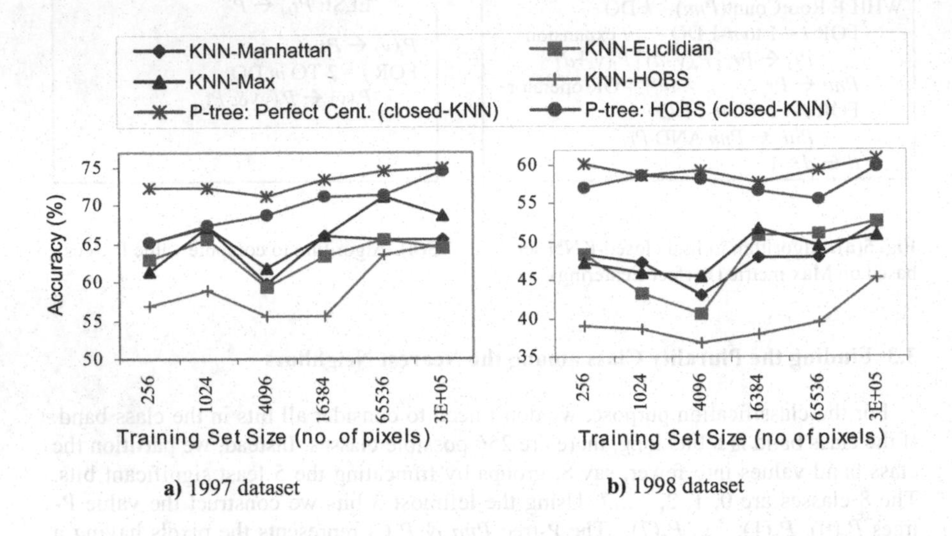

Fig. 6. Accuracy of different implementations for the 1997 and 1998 datasets

We see that both of our P-tree based closed-KNN methods outperform the KNN methods for both of the datasets. The reasons are discussed in section 3. The perfect centering methods performs better than HOBS as expected. The HOBS metric is not suitable for a KNN approach since HOBS does not provide a neighborhood with the target pixel in the exact center. Increased accuracy of HOBS in P-tree implementation is the effect of closed-KNN. In a P-tree implementation, the ease of computability for closed-KNN using HOBS makes it a superior method. The P-tree based HOBS is the fastest method where as the KNN-HOBS is still the poorest (Fig. 8).

Another observation is that for 1997 data (Fig. 6), in KNN implementations, the max metric performs better than other three metrics. For the 1998 dataset, max is competitive with other three metrics. In many cases, as well as for image data, max metrics can be the best choice. In our P-tree implementations, we also get very high accuracy with the max distance (perfect centering method). We can understand this by examining the shape of the neighborhood for different metrics (Fig. 7).

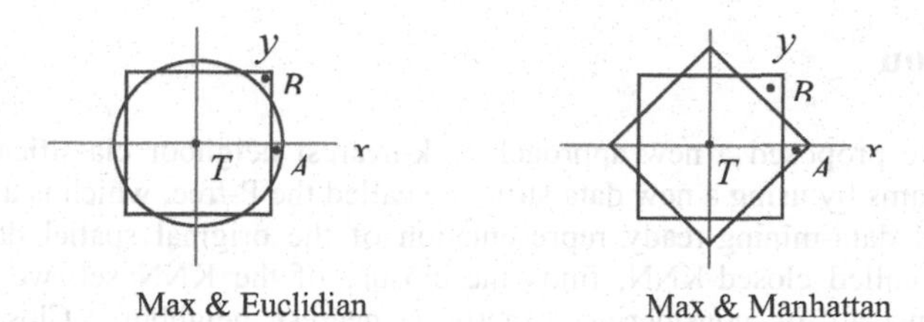

Fig. 7. Considering two dimensions the shape of the neighborhood for Euclidian distance is the circle, for max it is the square and for Manhattan it is the diamond. T is the target pixel.

Let, A be a point included in the circle, but not in the square and B be a point, which is included in the square but not in the circle. The point A is very similar to target T in the x-dimension but very dissimilar in the y-dimension. On the other hand, the point B is not so dissimilar in any dimension. Relying on high similarity only on one band while keeping high dissimilarity in the other bands may decrease the accuracy. Therefore in many cases, inclusion of B in the neighborhood instead of A, is a better choice. That is what we have found for our image data. For all of the methods classification accuracy increases with the size of the dataset since inclusion of more training data, the chance of getting better nearest neighbors increases.

On the average, the perfect centering method is five times faster than the KNN, and HOBS is 10 times faster (Fig. 8). P-tree implementations are more scalable. Both perfect centering and HOBS increases the classification time with data size at a lower rate than the other methods. As dataset size increases, there are more and larger pure-0 and pure-1 quadrants in the P-trees; that makes the P-tree operations faster.

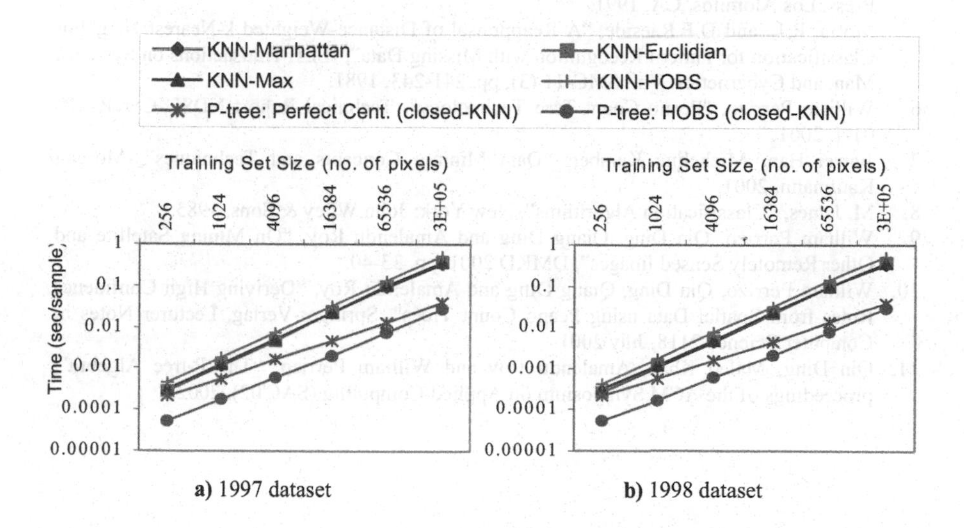

Fig. 8. Classification time per sample of the different implementations for the 1997 and 1998 datasets; both of the size and classification time are plotted in logarithmic scale.

5. Conclusion

In this paper we proposed a new approach to k-nearest neighbor classification for spatial data streams by using a new data structure called the P-tree, which is a lossless compressed and data-mining-ready representation of the original spatial data. Our new approach, called closed-KNN, finds the closure of the KNN set, we call the closed-KNN, instead of considering exactly *k* nearest neighbors. Closed-KNN includes all of the points on the boundary even if the size of the nearest neighbor set becomes larger than *k*. Instead of examining individual data points to find nearest neighbors, we rely on the expansion of the neighborhood. The P-tree structure facilitates efficient computation of the nearest neighbors. Our methods outperform the traditional implementations of KNN both in terms of accuracy and speed.

We proposed a new distance metric called Higher Order Bit (HOB) distance that provides an easy and efficient way of computing closed-KNN using P-trees while preserving the classification accuracy at a high level.

References

1. Domingos, P. and Hulten, G., "Mining high-speed data streams", Proceedings of ACM SIGKDD 2000.
2. Domingos, P., & Hulten, G., "Catching Up with the Data: Research Issues in Mining Data Streams", DMKD 2001.
3. T. Cover and P. Hart, "Nearest Neighbor pattern classification", IEEE Trans. Information Theory, 13:21-27, 1967.
4. Dasarathy, B.V., "Nearest-Neighbor Classification Techniques". IEEE Computer Society Press, Los Alomitos, CA, 1991.
5. Morin, R.L. and D.E.Raeside, "A Reappraisal of Distance-Weighted k-Nearest Neighbor Classification for Pattern Recognition with Missing Data", IEEE Transactions on Systems, Man, and Cybernetics, Vol. SMC-11 (3), pp. 241-243, 1981.
6. William Perrizo, "Peano Count Tree Technology", Technical Report NDSU-CSOR-TR-01-1, 2001.
7. Jiawei Han, Micheline Kamber, "Data Mining: Concepts and Techniques", Morgan Kaufmann, 2001.
8. M. James, "Classification Algorithms", New York: John Wiley & Sons, 1985.
9. William Perrizo, Qin Ding, Qiang Ding and Amalendu Roy, "On Mining Satellite and Other Remotely Sensed Images", DMKD 2001, pp. 33-40.
10. William Perrizo, Qin Ding, Qiang Ding and Amalendu Roy, "Deriving High Confidence Rules from Spatial Data using Peano Count Trees", Springer-Verlag, Lecturer Notes in Computer Science 2118, July 2001.
11. Qin Ding, Maleq Khan, Amalendu Roy and William Perrizo, "The P-tree Algebra", proceedings of the ACM Symposium on Applied Computing (SAC'02), 2002.

Interactive Construction of Classification Rules

Jianchao Han and Nick Cercone

Department of Computer Science, University of Waterloo
Waterloo, Ontario, N2L 3G1, Canada
{j2han, ncercone}@math.uwaterloo.ca

Abstract. We introduce an interactive classifier construction system, CVizT, in which the entire process is visualized based on a multidimensional visualization technique, *Table Lens*. The CVizT system is a fully interactive approach. The appropriate visualization-based interaction capabilities are provided for the user to include human perception into the construction process. Our experiments with data sets from the UCI repository demonstrates that the CVizT system is straightforward and easily learned. The user's preference and domain knowledge can also be integrated into the construction process.
Keywords: Interactive rule construction, visual data mining.

1 Introduction

Most existing classification approaches developed in the machine learning and data mining communities are based on sophisticated algorithms and require using probability theory and statistical methods, fuzzy set and rough set theories, etc. to find a way to assign a new object to one of a number of predetermined classes[5]. These approaches are recognized as *algorithm-based*. Alternatively, visualization-based approaches have also been developed recently[1,2]. These approaches address data and pattern visualization as well as the human-machine interaction. Visual representations of the data set portray the data items in a two- or three-dimensional space, and help the user gain better insight into multidimensional data[6,7], while pattern visualization helps the user understand and interpret the constructed classifiers. Moreover, by means of human-machine interaction, the system can include a role for the user's perception and domain knowledge to assist the construction of classifiers.

We present a novel approach for constructing classifiers by visualizing the construction process and introduce our implemented system CVizT. The CVizT system is fully interactive and allows us to integrate the domain knowledge of experts and user perception on demand during classifier construction. Our performance evaluation with data sets from the UCI repository demonstrates that this interactive approach is useful to easily build understandable classifiers with no required *a prior* knowledge about the datasets but high prediction accuracy.

M.-S. Chen, P.S. Yu, and B. Liu (Eds.): PAKDD 2002, LNAI 2336, pp. 529–534, 2002.

2 The CVizT System

According to the RuleViz model[2], the CVizT system is divided into five components, shown in Figure 1a. The *data visualization* component visualizes the

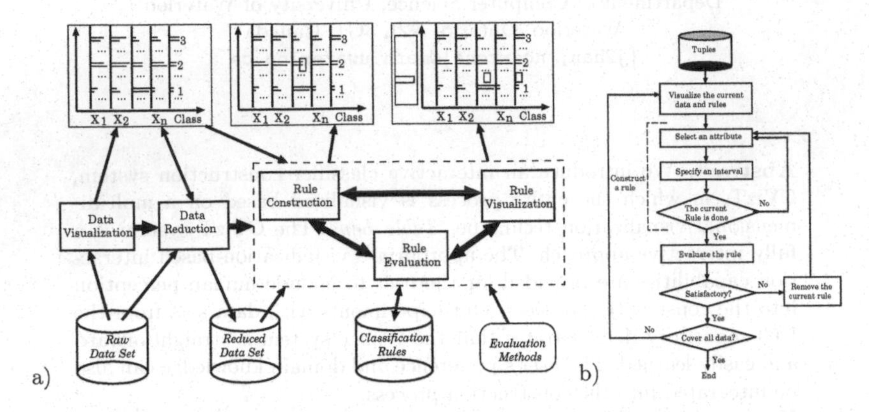

Fig. 1. a) The CVizT System b)The process of buliding Rules in CTViz

original data using the Table Lens [6,7] where the representation space is divided into columns with each column corresponding to an attribute of the data set. Each tuple is represented in the visual form as a row consisting of a set of "fields". Each field is filled with an attribute value and drawn as a color line with the length proportional to the value with respect to the column width and the value range of the attribute. The color is obtained in terms of the class label of the tuple. Users can operate on the tuples according to the specified attributes and use the features of the Table Lens to view the data distribution and the relationships between attributes from different angles so as to perceive the characteristics of the data tuples.

The original data can be cleaned through the *data reduction* component. Based on the operation and user perception on the dataset, the user can reduce the data set if necessary by removing some attributes that do not provide useful or helpful information to the user, aggregating attributes strongly correlated, and/or interactively drawing tuples closely related to the class labels.

The last three components form the classification rule generation loop, in which the user can iteratively construct classification rules until all original data are covered by a set of rules. To construct a classification rule, the condition attributes are chosen one by one. For each of chosen attributes, the remaining data tuples are sorted with respect to the chosen attributes, and either a value or an interval of the chosen attribute is specified depending on whether it is a continuous or categorical attribute. Each constructed classification rule is evaluated according to the evaluation criteria. The *bad* rules with low evaluation can be removed or reconstructed.

Finally, the classification rules with their evaluation are visualized on a variant of the Table Lens, where a rule is visually represented as a set of rectangles, each rectangle corresponding to a condition attribute. Rules are located in the corresponding column with height representing the rules' *coverage* and length representing its value or interval. The rule evaluation (prediction accuracy and quality) is visually displayed on the right column of the Table Lens.

3 Visualization Based on the Table Lens

In order to support interactive construction of classifiers, the original dataset and the intermediate result are visualized with the **Table Lens** for the user to observe and control the construction process. Basically, the Table Lens integrates the relational data table or spreadsheet with graphical representations to support browsing of the values for hundreds of tuples and tens of columns on a typical workstation display[6,7]. Each tuple is represented as a row in the graph, and each attribute (variable) is represented as a column. Assume attribute A has maximum and minimum values m and M, respectively. The width of the column corresponding to A is L, and a tuple t has value a of A. Then the length of the line corresponding to t in column A is determined by $l(t) = \frac{a-m}{M-m} \times L$ pixels.

Each row is rendered according to the HSI color model and the class label of the corresponding tuple, and the color scale is derived from the PBC system[1]. Assume N class labels, for the i-th class label, $i = 1, 2, \ldots, N$, hue$=0.5+2\times\frac{i-1}{N}$, intensity=saturation=0. Particular rows and columns can be assigned variable widths according to the number of points in attribute domains.

The Table Lens also provides a small set of direct manipulation operators for discovering correlations among the observed variables, such as finding the attribute mean, standard deviation, variables (columns) permutation, etc. Most operations have been defined for multivariate data analysis[7]. Sorting columns is a major operation on the Table Lens. Once a column is sorted, properties of the batch of values can be estimated by graphical perception and some amount of display manipulation. In addition, the correlations among the columns can be observed if other columns are also sorted. In our method, tuples are aggregated in terms of their class labels, and the number of tuples in each class is displayed on the right side. The strongly correlated features can be aggregated together to reduce the number of features.

The rules obtained can also be visualized in the Table Lens, with conditions corresponding to attribute intervals and decision to the class label. The rule accuracy is displayed as the area whose height is proportional to the rule coverage and whose width is proportional to the rule accuracy. In addition, the color of the rules can be used to represent the rule quality[2].

Rules are rendered in terms of rule evaluation and the HSI color model. Assume that the prediction accuracy and coverage of a rule are a and s, respectively. Then the rule color is calculated as follows: hue $= 0.5 + 2 \times a$, intensity $= 0.5 + 2 \times \frac{s}{n}$, saturation $= 1.0$, where n is the size of the test or training data set, depending on which evaluation method the user chooses.

4 Interactive Rule Construction

To construct a classification rule, one can interactively specify the intervals for the attributes strongly correlated to the class labels to form the condition part of the rule. Building classifiers is to interactively and iteratively construct classification rules. This process is illustrated in Figure 1b and described below.

Visualize the current data and rules: The classifier construction process operates on the cleaned data set. The current available data and the rules constructed already are visualized on the Table Lens. By looking into the distribution of the available dataset, one can perceive correlations between attributes.

Construct a rule: If some tuples have not been covered by the current classifier, one can construct a classification rule iteratively as follows: Select an attribute as the current attribute. Tuples within each class are sorted by the selected attribute in ascending order. Use a "rubber band" to draw an interesting interval for the selected attribute. The current attribute with the interval encompassed by the rubber band is displayed at the left bottom corner. If the selected attribute or the current interval is not good enough, it can be canceled.

Evaluate the rule: The current rule constructed in the previous step must be evaluated before it is put in the classifier. Evaluating a rule consists of computing its accuracy, coverage, and quality[2]. Two methods of rule evaluation are provided, *resubstitution estimate* and *test sample estimate*. However, CVizT does not automatically divide a data set into training set and test set. If *test sample estimate* is used, the user must specify the test data set.

Remove the rule: If the rule accuracy is equal to or greater than a prespecified threshold, then it is acceptable, otherwise it is discarded automatically. The tuples covered by the current rule are marked and are not be displayed. The display space saved by the marked tuples is used to visualize the rule. The rule condition is displayed as a set of attribute-interval pairs. If the current rule is not satisfactory, one can delete it from the classifier, and the marked tuples are then restored to be available.

Update the data set: If the current rule is accepted, it is appended to the classifier, the available data set is updated to cover all unmarked tuples, and the marked tuples are ruled out from the current data set.

Repeat above steps until all rules produced have high accuracy and quality and cover all tuples. The final classifier is composed of these classification rules. The main idea of learning classification rules behind the CVizT system is based on the sequential covering learning strategy, proposed by Michalski, et al.[3].

5 CVizT Implementation and Experiment

We experimented the CVizT system using some UCI datasets[4], including Glass, Diabetes, Iris, Monks, Parity5+5, etc. In this section, we illustrate our experiments with the the Glass Identification Database which contains 214 instances, and 10 attributes (including an Id number) plus the class attribute. Except the ID and the class label, all attributes are continuously valued.

The class attribute (Type) has 7 possible values which are labeled as 1 through 7. Each class contains $70, 76, 17, 0, 13, 9$, and 29 instances, respectively.

Initially, 11 attributes including ID number and class label are visualized, shown in Figure 2a, which illustrates some attributes are strongly correlated with the class label, while the others are not. Actually, the ID number is absolutely not correlated with the class label. The attributes uncorrelated to the class label can be removed to save display space for the correlated attributes. For example, attributes Na, Mg, Al, and Ba have high correlation strength with the class label, and the other attributes have low correlation strength (positive strength means positive correlation while negative strength means negative correlation). Thus, select attributes Na, Mg, Al, and Ba, as well as the class label, illustrated in Figure 2b, which also shows that the instances are sorted by attribute Ba.

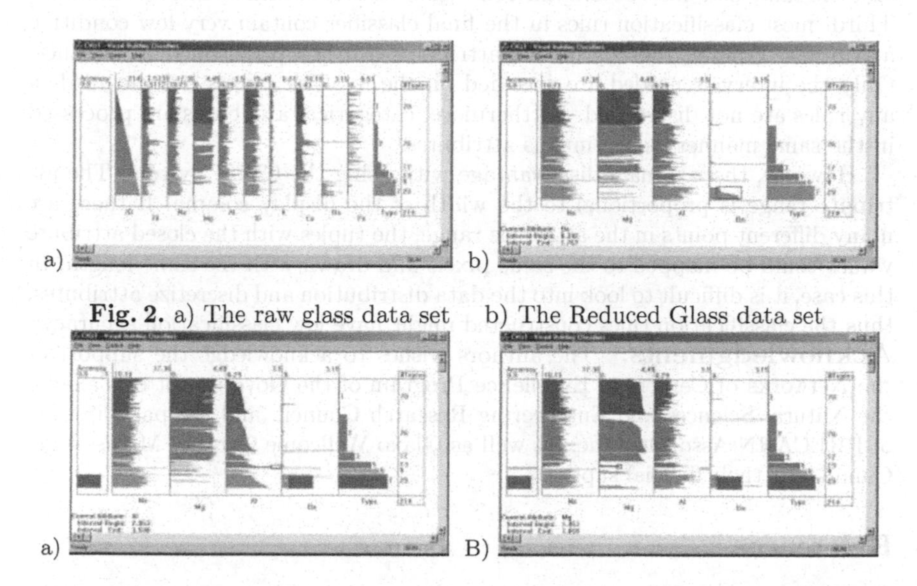

Fig. 2. a) The raw glass data set b) The Reduced Glass data set

Fig. 3. The Reduced Glass data set: a) one rule b) two rules

Clearly, most of instances belonging to Type 7 have middle values of Ba. To construct a classification rule to reflect such a fact, the attribute Ba is selected and the middle interval is specified. The rule consisting of the condition part as *Ba in the middle* and the decision part *Type 7* is formed and illustrated in Figure 3a. The accuracy of this rule is about 96%. Furthermore, Figure 3a also shows that only part of Type 5 instances have attribute values of Al close to the maximum value 3.5. Select attribute Al and draw the interval close to the maximum end. A new rule is formed and illustrated in Figure 3b. This rule has a pretty high accuracy, 100%.

This process can be repeated until a group of rules are achieved with high evaluation. No specific stopping criterion is defined.

6 Conclusion

We presented the interactive classifier construction system with Table Lens based visualization, CVizT, which consists of five components, including original data visualization, interactive data reduction, rule construction, rule evaluation, and rule visualization.

Compared to other approaches for building classifiers, The CVizT system has the following characteristics: First, the CVizT system is ease to learn. We trained a group of non computer science students for about ten minutes with the IRIS flower data set. They built classifiers with an average accuracy higher than 90%. Second, for the same dataset, the generated classifier is uncertain and user-dependent. The accuracy of generated classifiers also depends on the user, and the same user may obtain different accuracies of classifiers in different runs. Third, most classification rules in the final classifier contain very few condition attributes. Finally, the continuous attributes are not partitioned in advance. Only the intervals needed are specified on the fly. The ranges not included in any rules are not discretized. Furthermore, categorical attributes are processed in the same manner as continuous attributes.

However, there a main disadvantage with using the CVizT system. The attribute range is proportional to the width of the display column. If there are many different points in the attribute range, the tuples with the closed attribute values could be mapped to the same pixels and drawn with the same length. In this case, it is difficult to look into the data distribution and discretize attributes, thus the classification rules constructed might have low classification accuracy.

Acknowledgments: The authors wishes to acknowledge the support of the Networks of Centers of Excellence Program of the Government of Canada, the Natural Sciences and Engineering Research Council, and the participation of PRECARN Associates Inc., as well as Glaxo Wellcome Corp. of Mississauga, Canada for their finacial support.

References

1. M. Ankerst, M. Ester, and H. P. Kriegel, Towards an Effective Cooperation of the User and the Computer for Classification, *Proc. of KDD-2000*, pp.179-188, 2000.
2. J. Han and N. Cercone, RuleViz: A Model for Visualizing Knowledge Discovery Process, *Proc. of KDD-2000*, pp.223-242, Boston, USA, August 2000.
3. R. S. Michalski, I. Mozetic, J. Hong, and N. Lavrac, The Multi-Purpose Incremental Learning System AQ15 and Its Testing Application to Three Medical Domains, *Proc. of AAAI-86*, pp. 1041-1045, 1986.
4. P. M. Murphy and D. W. Aho, UCI Repository of Machine Learning Databases, URL: *http://www.ics.uci.edu/m̃learn/MLRepository.html*, 1996.
5. G. Nakhaeizadeh and C. C. Taylor, Machine Learning and Statistics: The Interface, *John Wiley & Sons, Inc.*, 1997.
6. P. Pirolli and R. Rao, Table Lens as a Tool for Making Sense of Data, *Proc. of AVI, Workshop on Advanced Visual Interfaces*, pp. 67-80, 1996.
7. R. Rao and S. K. Card, The Table Lens, *Proc. of ACM Conference on Human Factors in Computing Systems*, 318-322, New York, NY, 1994.

Enhancing Effectiveness of Outlier Detections for Low Density Patterns

Jian Tang[1]*, Zhixiang Chen[2], Ada Wai-chee Fu[1], and David W. Cheung[3]

[1] Department of Computer Science and Engineering
Chinese University of Hong Kong
Shatin, Hong Kong
[2] Department of Computer Science
University of Texas at Pan-America
Texas, U.S.A
[3] Department of Computer Science and Information Systems
University of Hong Kong
Pokfulam, Hong Kong

Abstract. Outlier detection is concerned with discovering exceptional behaviors of objects in data sets. It is becoming a growingly useful tool in applications such as credit card fraud detection, discovering criminal behaviors in e-commerce, identifying computer intrusion, detecting health problems, etc. In this paper, we introduce a connectivity-based outlier factor (COF) scheme that improves the effectiveness of an existing local outlier factor (LOF) scheme when a pattern itself has similar neighbourhood density as an outlier. We give theoretical and empirical analysis to demonstrate the improvement in effectiveness and the capability of the COF scheme in comparison with the LOF scheme.

1 Introduction

Outlier detection is an important branch in the area of data mining. It is concerned with discovering the exceptional behaviors of certain objects. Revealing these behaviors is important since it signifies that something out of ordinary has happened and shall deserve people's attention. In many cases, such exceptional behaviors will cause damage to users and must be stopped. In other cases, there can be "good" outliers which can help users to make profits. Therefore, in some sense detecting outliers is at least as significant as discovering general patterns. Outlier detection is becoming a growingly useful tool in applications to which people have already paid attention, such as credit card fraud detection, calling card fraud detection, discovering criminal behaviors in e-commerce, discovering computer intrusion, and etc. [4, 6].

Hawkins [7] characterizes an outlier in a quite intuitive way as follows:

An outlier is an observation that deviates so much from other observations as to arouse suspicion that it was generated by a different mechanism.

* On leave from Memorial University of Newfoundland, Canada.

M.-S. Chen, P.S. Yu, and B. Liu (Eds.): PAKDD 2002, LNAI 2336, pp. 535–548, 2002.

Following the spirit of this definition, researchers have proposed various schemes for outlier detection. A large amount of the work was under the general topic of clustering [5, 10, 13, 14, 16]. These algorithms can also generate outliers as by-products. However, the outliers discovered this way are highly dependent on the clustering algorithms used and hence subject to the clusters generated. Most methods in the early work that detects outliers independently have been developed in the field of statistics [2]. These methods normally assume that the distribution of a data set is known in advance and try to detect outliers by examining the deviations of individual data objects based on such a distribution. In reality, however, a priori knowledge about the distribution of a data set is not always obtainable. Besides, these methods do not scale well for even modest number of dimensions as the size of a data set increases.

More recently, researchers proposed distance based schemes, which distinguish objects that are likely to be outliers from those that are not based on the number of objects in the neighborhood of an object [8, 9, 11]. These schemes do not make any assumptions about the distribution of a data set. Furthermore, since the counting process is restricted only to the neighborhood of an object, the scalability of these methods is better than that of their predecessors. As a result, distance based schemes are more appropriate for detecting outliers in large data sets without assuming a priori knowledge about their distributions.

Knorr and Ng [8] propose a distance based scheme, called $DB(n, q)$-outlier. In this scheme, if the neighborhood with the radius of q (called "q-neighborhood") of an object contains less than n objects, then it is called an outlier with respect to n and q, otherwise it is not. The advantage of this scheme is its simplicity while capturing the basic intuition given in Hawkins' definition. Its weakness is that it cannot deal with data sets that contain patterns with diverse characteristics. The scheme proposed by Ramaswamy, et al. [11], called (t, k)-nearest neighbor scheme, considers for each point its k-distance, i.e., the distance to its kth nearest neighbor(s). It ranks the top t objects with the maximum k-distances as the outliers. If there are multiple objects with the same k-distance ranked as the top k, they are all considered as outliers. Therefore, the number of outliers returned may be greater than t. This scheme is actually a special case of $DB(n, q)$-outlier. Thus it shares the same weakness as $DB(n, q)$-outlier has.

Recently, Breuning, et al, [3] proposed a density based formulation scheme as follows.

Let $p, o \in \mathcal{D}$ and k be a positive integer. Let *k-distance*(o) be the distance from o to its k-th nearest neighbor, where if two neighbors are at same distance from o, the ordering of "nearest" for them is arbitrary. The k-distance neighbourhood of an object p is denoted by $N_{k\text{-}distance(p)}(p)$ and is the set of objects whose distance from p is not greater than k-distance.

The reachability distance of p with respect to o for k is defined as:

$$reach\text{-}disk_k(p, o) = max\{k\text{-}distance(o), dist(p, o)\}.$$

The reachability distance smoothes the fluctuation of the distances between p and its *"close"* neighbors. The local reachability density of p for k is defined as:

$$lrd_k(p) = \left(\frac{\sum_{o \in N_{k\text{-}distance(p)}(p)} reach\text{-}dist_k(p,o)}{|N_{k\text{-}distance(p)}(p)|} \right)^{-1}.$$

That is, $lrd_k(p)$ is the inverse of the average reachability distance from p to the objects in its k-distance neighborhood. For simplicity, we shall refer to the local reachability density of a point p as the *density* of p. The local outlier factor of p is defined as

$$LOF_k(p) = \frac{\sum_{o \in N_{k\text{-}distance(p)}(p)} \frac{lrd_k(o)}{lrd_k(p)}}{|N_{k\text{-}distance(p)}(p)|}.$$

The value on the right side is the average fraction of the reachability densities of p's *k-distance* neighbors and that of p. Thus, as pointed out in [3], the lower the density of p, or the higher the densities of p's neighbors, the larger the value of $LOF_k(p)$, which indicates p has a higher degree of being an outlier.

Note that the density based scheme does not explicitly categorize the objects into either outliers or non-outliers. (If desired, a user can do so by choosing a threshold value to separate the LOF values of the two classes.) It uses the LOF to measure how strong an object can be an outlier. Since the LOF value of an object is obtained by comparing its density with those in its neighborhood, it has stronger modeling capability than a distance based scheme, which is based only on the density of the object itself.

In [3] the authors give an example, which we have duplicated in Figure 1. The data set contains an outlier o, and $C1$ and $C2$ are two clusters with very different densities. The authors show that the DB(n,q)-outlier method cannot distinguish o from the rest of the data set no matter what values the parameters take. However, LOF method can handle it successfully.

The weakness of the density based scheme is that it considers solely the difference between the density of an object and those of its neighbors (we shall show such an example in the next section). Thus its effectiveness will diminish if the density of an outlier is close to those of its neighbors. In this paper, we introduce a connectivity-based outlier factor (COF) scheme for outlier formulation, We use empirical analysis to demonstrate the improvement in effectiveness and the capability of the COF scheme over the LOF scheme.

The rest of this paper is organized as follows. In Section 2 we propose a definition of ON-compatibility for the goodness of an outlier detection method. In Section 3, we revisit the density based schemes. In Section 4, we introduce the connectivity-based scheme. In Section 5, we compare the connectivity-based and density-based schemes using the experimental data. In Section 6, we discuss the complexity involved in calculating the COF. Finally in Section 7 we conclude the paper by summarizing the main results.

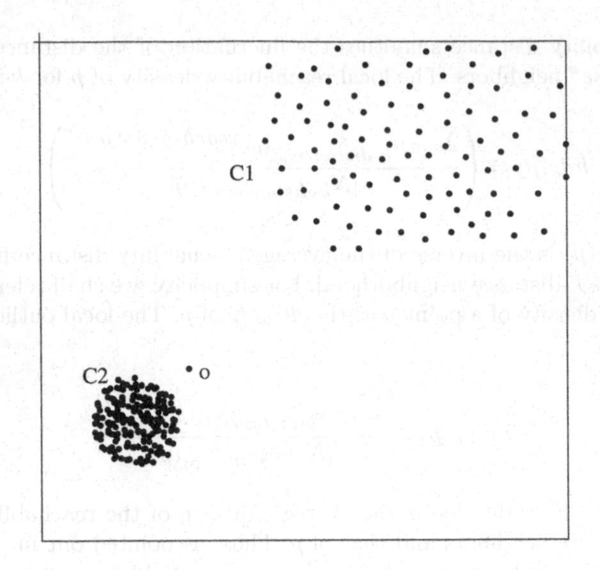

Fig. 1. A Data Set showing the strength of LOF

2 ON-Compatibility

All the previously introduced methods utilize an outlier measure function and a number of parameter settings. For the DB(n,q)-outlier method, the measurement for an object is the number of objects within a radius of q, and the outlier decision is based on whether the number is less than n. The parameters are q and n. For the (t, k)-nearest neighbors, the measurement for an object is the distance to the k-th nearest neighbor, and decision is based on whether the distance is among the top t such values. The LOF measurement is a little more complex and also utilize parameters of k in k-distance. In all the above, the measurement is typically for a data point p and its value depends on some set of parameters S, hence it can be denoted by $f(p, S)$. We would like $f(p, S)$ to be large when p is an outlier, and small if it is not. Therefore for the DB(n,q)-outlier method, $f(p, S)$ can be set as n divided by the number of objects within a radius of q.

In [15] we have developed a stack of measurements to evaluate the capabilities of outlier measure functions for different formulation schemes, with the increasingly relaxed requirements down the stack. Due to the space limitation we introduce only the measurement on the top of the stack, termed *ON-Compatibility* (ON stands for Outliers and Non-outliers). We will use ON-compatibility to evaluate the effectiveness of the density-based scheme and the connectivity based scheme.

Definition 1 *The outlier measure function $f(p, S)$ is ON-compatible with a given set of data with outliers and non-outliers (we call this an interpretation I), if there exists a parameter setting S, and a value u, such that*

(1). *for each outlier o, the measure $f(o, S)$ has a value above u.*
(2). *for each of the non-outliers n, the measure $f(n, S)$ has a value below u.*

The value u is called cut-off value.

ON-compatibility indicates the capability of an outlier measure function to use a single parameter setting to detect all outliers for a given interpretation. It is most desirable, but not often attainable. An interpretation is given by a set of data $\mathcal{D}$ together with its partitioning into the set of outliers $\mathcal{D}_o$ and the set of non-outliers $\mathcal{D}_n$. With the following theorem, we introduce a method to determine when a function is not compatible with some given interpretation.

Theorem 1 *Let $f(p, S)$ be an outlier measure function and I be an interpretation: $\mathcal{D} = \mathcal{D}_o \cup \mathcal{D}_n$, then the following holds for $f(p, S)$: It is not ON-compatible with I if for any setting S, there exist an object $a \in \mathcal{D}_n$, and an object $o \in \mathcal{D}_o$, such that $f(o, S) \leq f(a, S)$.*

3 Density Based Schemes Revisited

We have seen from the previous sections, the density based scheme, such as LOF, is more powerful than the previous methods. However, one weakness of the density based scheme is that it may rule out outliers close to some non-outliers pattern that has low density. To understand the problem, let us first take a closer look at the concept of pattern. According to the **Concise Oxford Dictionary**, a pattern is

"a regular or logical form, order or arrangement of parts ...".

We observe that although a high density can reflect such a logical form, order or arrangement, it nonetheless is not a necessary condition, at least in the form defined in the current literature. As a result, an outlier does not always have to be of a lower density than a pattern it deviates from. A typical example is shown in Figure 2.

In this figure, the pattern, C_1, is a straight line, which is of low density in a two dimensional space. Point o1 and the points in C_2 are outliers. Since o1 shifts away from a low density pattern, the density based outlier measure function will not be effective to identify it, unless we use a small k. On the other hand, using too small a k will rule out the outliers in C_2, which must be identified using a value for k larger than its cardinality. In the following, we assume some specific values for the variables of the data set.

EXAMPLE 1 *C_1 contains 91 points, with distance one between adjacent points. o1 is a point closest to the middle of C_1. The circle C_2 with radius of one contains eight points evenly positioned on its circumference. The center of C_2, o1 and the middle point in C_1 are on a line. (Note that the circle for C_2 has been much enlarged in Figure 2.) The distance from o1 to C_2 is 1000. The distance from o1 to the middle point of C_1 is two. Let I be the interpretation: $\mathcal{D} = \mathcal{D}_o \cup \mathcal{D}_n$ where $\mathcal{D}_o = \{o1\} \cup C_2$ and $\mathcal{D}_n = C_1$.*

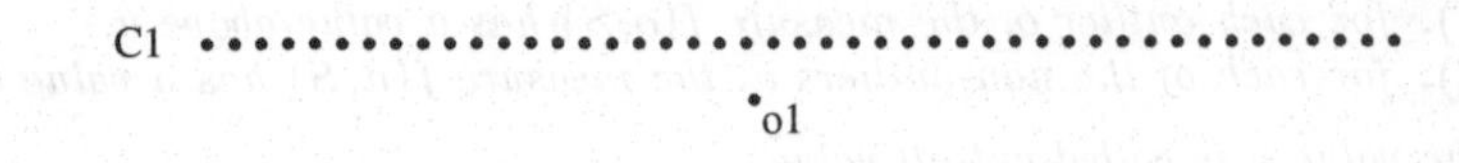

Fig. 2. Failure of Outliers detection for LOF

For the values given in the above example, we can formally prove

ASSERTION 2 *The LOF outlier measure is not ON-compatible for I in the data set of Example 1.*

The proof of the above assertion is based on Theorem 1, and is omitted here. In a later section, we will show the ineffectiveness of LOF in handling a similar case. In the next section, we will introduce a scheme that can handle low density patterns such as the line of points in Figure 2, while at the same time does not compromise detecting a group of stay-together outliers like those in Figure 2.

4 Connectivity-Based Outliers

Our solution is based on the idea of differentiating *"low density"* from *"isolativity"*. While low density normally refers to the fact that the number of objects in the *"close"* neighborhood of an object is (relatively) small, isolativity refers to the degree that an object is *"connected"* to other objects. As a result, isolation can imply low density, but the other direction is not always true. For example, in Figure 2 point $o1$ is isolated, while any point p in C_1 is not. But both of them are of roughly equal low density. In the general case a low density outlier results from deviating from a high density pattern, and an isolated outlier results from deviating from a connected pattern. An outlier indicator should take into consideration of both cases.

We observe that patterns that possess low densities usually exhibit low dimensional structures. For example, a pattern shown in Figure 2 is a line in the two dimensional space. The isolativity of an object, on the other hand, can be described by the distance to its nearest neighbor. In the general case we can also talk about the isolativity of a group of objects, which is the distance from the group to its nearest neighbor.

We first introduce some notations and then formulate our connectivity-based outlier scheme.

Definition 2 *Let* $P, Q \subseteq \mathcal{D}$, $P \cap Q = \emptyset$ *and* $P, Q \neq \emptyset$. *We define* $dist(P,Q) = min\{dist(x,y) : x \in P \ \& \ y \in Q\}$, *and call* $dist(P,Q)$ *the distance between* P *and* Q. *For any given* $q \in Q$, *we say that* q *is the nearest neighbor of* P *in* Q *if there is a* $p \in P$ *such that* $dist(p,q) = dist(P,Q)$.

In the following definitions, let $G = \{p_1, p_2, \ldots, p_r\}$ be a subset of $\mathcal{D}$.

Definition 3 *A set based nearest path, or SBN-path, from* p_1 *on* G *is a sequence* $\langle p_1, p_2, \ldots, p_r \rangle$ *such that for all* $1 \leq i \leq r-1, p_{i+1}$ *is the nearest neighbor of set* $\{p_1, \ldots, p_i\}$ *in* $\{p_{i+1}, \ldots, p_r\}$.

Imagine that a set initially contains object p_1 only. Then it goes into an iterative expansion process. In each iteration, it picks up its nearest neighbor among the remaining objects. If its nearest neighbor is not unique, we can impose a pre-defined order among its neighbors to break tie. Thus an *SBN*-path is uniquely determined. An *SBN*-path indicates the order in which the nearest objects are presented.

Definition 4 *Let* $s = \langle p_1, p_2, \ldots, p_r \rangle$ *be an SBN-path. A set based nearest trail, or SBN-trail, with respect to* s *is a sequence* $\langle e_1, \ldots, e_{r-1} \rangle$ *such that for all* $1 \leq i \leq r-1$, $e_i = (o_i, p_{i+1})$ *where* $o_i \in \{p_1, \ldots, p_i\}$, *and* $dist(e_i) = dist(o_i, p_{i+1}) = dist(\{p_1, \ldots, p_i\}, \{p_{i+1}, \ldots, p_r\})$. *We call each* e_i *an edge and the sequence* $\langle dist(e_1), \ldots, dist(e_{r-1}) \rangle$ *the cost description of* $\langle e_1, \ldots, e_{r-1} \rangle$.

Again, if o_i is not uniquely determined, we should break tie by a pre-defined order. Thus the *SBN*-trail is unique for any *SBN*-path.

Definition 5 *Let* $s = \langle p_1, p_2, \ldots, p_r \rangle$ *be an SBN-path from* p_1 *and* $e = \langle e_1, \ldots, e_{r-1} \rangle$ *be the SBN-trail with respect to* s. *The average chaining distance from* p_1 *to* $G - \{p_1\}$, *denoted by* $ac\text{-}dist_G(p_1)$, *is defined as*

$$ac\text{-}dist_G(p_1) = \sum_{i=1}^{r-1} \frac{2(r-i)}{r(r-1)} \cdot dist(e_i).$$

The average chaining distance from p_1 to $G - \{p_1\}$ is the weighted sum of the cost description of the *SBN*-trail for some *SBN*-path from p_1. Since this cost description is unique for p_1, our definition is well defined. Rewriting

$$ac\text{-}dist_G(p_1) = \frac{1}{r-1} \cdot \sum_{i=1}^{r-1} \frac{2(r-i)}{r} \cdot dist(e_i)$$

and viewing the fraction following the summation sign as the weight, the average chaining distance can then be viewed as the average of the weighted distances in the cost description of the *SBN*-trail. Note that larger weights are assigned to the earlier terms. Thus if the edges close to p_1 are substantially larger than those away from p_1, then they contribute more in the $ac\text{-}dist_G(p_1)$. This is consistent with our motivation. In the special case where $dist(e_i)$ is the same for all e_i, we have $ac\text{-}dist_G(p_1) = dist(e_i)$.

Definition 6 *Let $p \in \mathcal{D}$ and k be a positive integer. The connectivity-based outlier factor (COF) at p with respect to its k-neighborhood is defined as*

$$COF_k(p) = \frac{|N_k(p)| \cdot ac\text{-}dist_{N_k(p)}(p)}{\sum_{o \in N_k(p)} ac\text{-}dist_{N_k(o)}(o)}.$$

The connectivity-based outlier factor at p is the ratio of the average chaining distance from p to $N_k(p)$ and the average of the average chaining distances from p's k-distance neighbors to their own k-distance neighbors. It indicates how far away a point shifts from a pattern. We now use an example to highlight the motivation behind it.

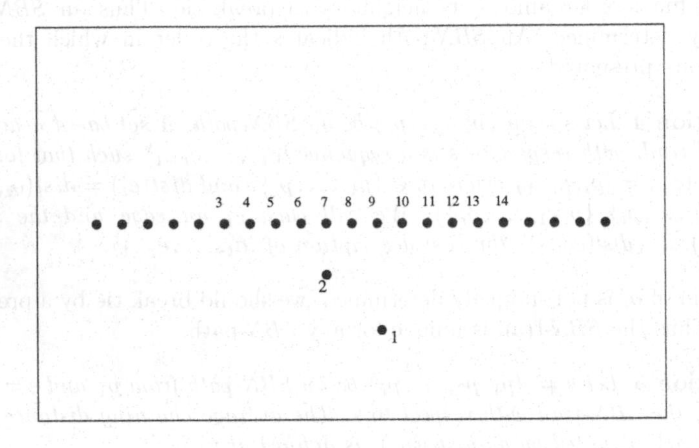

Fig. 3. Calculating COF

Consider the data set in Figure 3. The pattern is a single line and two points shift away from it. Suppose $dist(1,2) = 5, dist(2,7) = 3$, and the distance between any two adjacent points in the line is 1. Let $k = 10$. We now calculate the average chaining distances for three sample points to show how the COF values of those sample points reflect *"shifting from pattern"* in an appropriate way.

For point 1: $N_k(1) = \{2, 9, 10, 8, 11, 7, 12, 6, 13, 5\}$. The *SBN*-path from 1 on $N_k(1) \cup \{1\}$ is

$$s_1 = \langle 1, 2, 7, 6, 5, 8, 9, 10, 11, 12, 13 \rangle.$$

The *SBN*-trail for s_1 is

$$tr_1 = \langle (1,2), (2,7), (7,6), (6,5), (7,8), (8,9), (9,10), (10,11), (11,12), (12,13) \rangle.$$

The cost description of tr_1 is $c_1 = \langle 5, 3, 1, 1, 1, 1, 1, 1, 1, 1 \rangle$, and $ac\text{-}dist_{N_k(1) \cup \{1\}}(1) = 2.05$.

For point 2: $N_k(2) = \{7, 6, 8, 5, 9, 4, 10, 3, 11, 1\}$. The *SBN*-path from 2 on $N_k(2) \cup \{2\}$ is

$$s_2 = \langle 2, 7, 6, 5, 4, 3, 8, 9, 10, 11, 1 \rangle.$$

The *SBN*-trail for s_2 is

$$tr_2 = \langle (2,7), (7,6), (6,5), (5,4), (4,3), (7,8), (8,9), (9,10), (10,11), (2,1) \rangle.$$

The cost description of tr_2 is $c_2 = \langle 3, 1, 1, 1, 1, 1, 1, 1, 1, 5 \rangle$, and $ac\text{-}dist_{N_k(2) \cup \{2\}}(2) = 1.46$.

For point 7: $N_k(7) = \{6, 8, 5, 9, 4, 10, 2, 3, 11, 12\}$. The *SBN*-path from 7 on $N_k(7) \cup \{7\}$ is

$$s_3 = \langle 7, 6, 5, 4, 3, 8, 9, 10, 11, 12, 2 \rangle.$$

The *SBN*-trail for s_3 is

$$tr_3 = \langle (7,6), (6,5), (5,4), (4,3), (7,8), (8,9), (9,10), (10,11), (11,12), (7,2) \rangle.$$

The cost description of tr_3 is $c_3 = \langle 1, 1, 1, 1, 1, 1, 1, 1, 1, 3 \rangle$, and $ac\text{-}dist_{N_k(7) \cup \{7\}}(7) = 0.98$.

The average chaining distances for the other points on the line can be calculated similarly. The above results show that for points that shift more from the pattern, such as points 1 and 2, the first few items in their cost description lists tend to be larger values than those for points that shift less, such as point 7. Since earlier items in a cost description list are assigned larger weights, they contribute more to the corresponding average chaining distance, which is the weighted sums of the values in the cost description. Thus, strongly shifted points will have larger average chaining distances than weakly shifted ones. In the general case, most points in the k-distance neighborhood of a strongly shifted point should have small average chaining distances. This results in a larger connectivity-based outlier factor for such a strongly shifted point. On the other hand, for a weakly shifted point, most points in its k-distance neighborhood should have comparable average chaining distance values, resulting in a smaller connectivity-based outlier factor for such a point. The weakest shifted points are those that belong to the pattern itself. Their connectivity-based outlier factors should be close to 1. For the three sample points in the above example, we have the following:

$$COF_k(1) = 2.1, \;\; COF_k(2) = 1.35 \;\; and \;\; COF_k(7) = 0.96.$$

5 Comparison of COF and LOF

The connectivity-based scheme has some important properties like the density-based scheme, for example the COF value for an object deep inside a cluster being close to 1. For instance, for an object p and a cluster C such that $N_k(p) \subseteq C$, we can prove that $\frac{1}{1+\epsilon} \leq \text{COF}_k(p) \leq 1 + \epsilon$ where ϵ is a small value. We follow the approaches in [3] to show those similar bounds for *COF*. But first we give the following definition.

Definition 7 *Given any object $p \in \mathcal{D}$, let $s = \{e_1, \ldots, e_{r-1}\}$ be the SBN-trail with respect to the SBN-path from p on $N_k(p)$. We define*

$$path\text{-}min(p) = min\{dist(e_1), \ldots, dist(e_{r-1})\},$$
$$path\text{-}max(p) = max\{dist(e_1), \ldots, dist(e_{r-1})\}.$$

Theorem 3 *Given any set $C \subseteq \mathcal{D}$, let $path\text{-}min = min\{\ path\text{-}min(p) : p \in C\}$ and $path\text{-}max = max\{\ path\text{-}max(p) : p \in C\}$. Let $\epsilon = \frac{path\text{-}max}{path\text{-}min} - 1$, then for every object $p \in C$ such that*

(i) $N_k(p) \subseteq C$, and

(ii) for every $q \in N_k(p)$, $N_k(q) \subseteq C$, we have

$$\frac{1}{1+\epsilon} \leq COF_k(p) \leq 1 + \epsilon.$$

The above theorem, together with the illustration in the previous section, indicate that the connectivity based scheme has the similar power to that of the density based scheme in detecting outliers which deviate from high density patterns. On the other hand, recall that the motivation for introducing the connectivity based scheme is to handle outliers deviating from low density patterns. We showed previously in Figure 3 an example of a low density pattern. We now present a similar example in Figure 4. (We have assumed special geometric shapes and distances for the data set. These are used only for convenience of plotting the results.)

Example 1. In Figure 4, C_2 contains 8 points lying on the circle with its center at $(1, 0)$ and a radius of 1. Distances between any two adjacent points on the circle are the same. C_1 contains 91 points lying on two straight lines l_1 and l_2. The two lines meet at the point $p = (20, 0)$. Line l_1 and the x-axis form an angle of $\frac{\pi}{2}$, and so do line l_2 and the x-axis. C_1 contains p and 45 points on each of the lines l_1 and l_2. Moreover, the distance between any two adjacent points on each line is $\sqrt{2}$. Finally, $o = (23, 0)$. According to Hawkins' definition, it is easy to understand that point o and the points in C_2 can be considered as outliers while others are non-outliers. Thus, we have an interpretation $I : \mathcal{D} = \mathcal{D}_o \cup \mathcal{D}_n$, where $\mathcal{D}_o = \{o\} \cup C_2$ and $\mathcal{D}_n = C_1$.

Our result is contained in the following assertion.

ASSERTION 4 *The LOF outlier measure is not ON-compatible for I in the above example.*

We support the above assertion by the experimental data. We choose two non-outlier points $p = (20, 0)$ and $q = (65, 45)$ from $\mathcal{D}_n$ and two outlier points $w = (0, 0)$ and $o = (23, 0)$ from $\mathcal{D}_o$. The four points are illustrated in Figure 4. Note that q is the end point of C_1 on line l_1. Note also that the total number of points in the data set is 100. We have calculated the LOF values for all

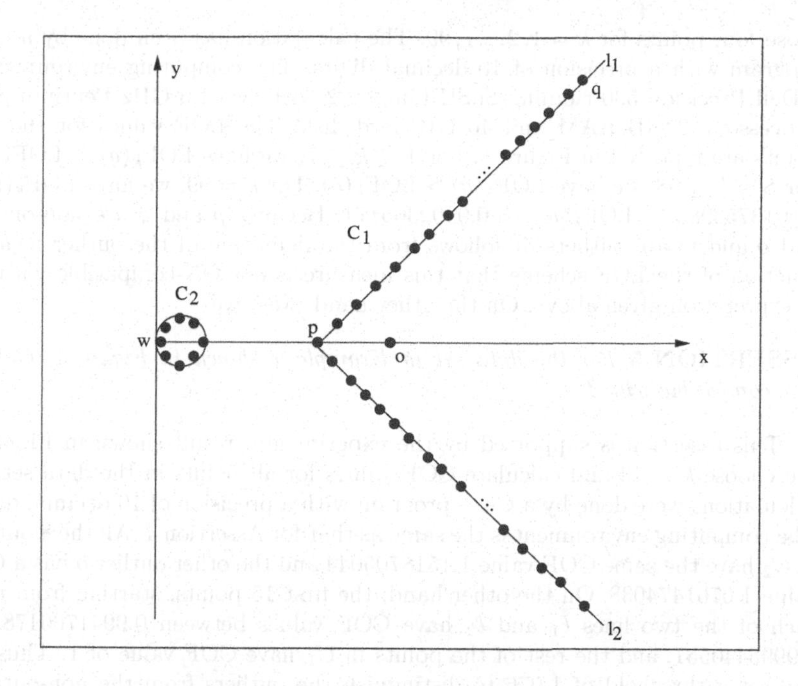

Fig. 4. Data Set for Comparison

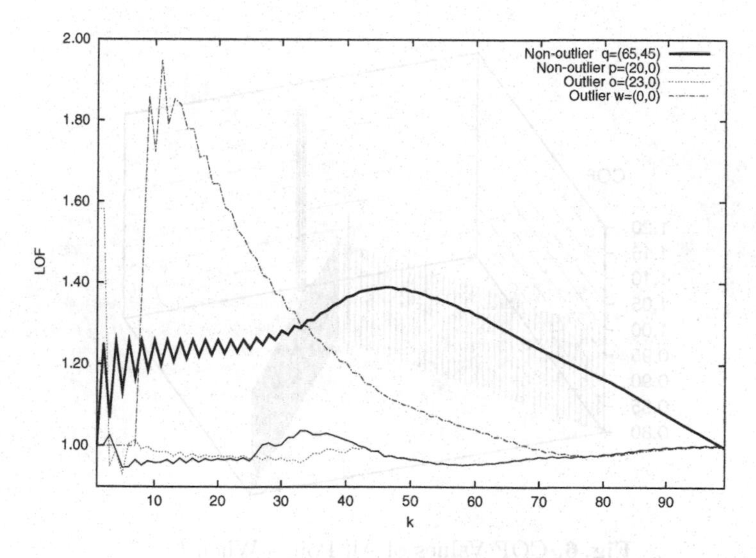

Fig. 5. LOF Values for Four Points on Different Settings

those four points for $k = 1, 2, \ldots, 99$. The calculation has been done by a C++ program with a precision of 10 decimal digits. The computing environment is a Dell Precision 530 running SuSE Linux 7.2 with two 1.5 GHz Pentium Xeon Processors, 2 GB RAM and 40 GB hard disk. The LOF values for the four points are reported in Figure 5. For $1 \le k \le 7$, we have $\mathrm{LOF}_k(q) > \mathrm{LOF}_k(w)$. For $8 \le k \le 98$, we have $\mathrm{LOF}_k(q) > \mathrm{LOF}_k(o)$. For $k = 99$, we have $\mathrm{LOF}_k(p) = 1.0013753983 > \mathrm{LOF}_k(w) = 0.9992365171$. Because p and q are non-outliers and o and w are outliers, it follows from the definition of the outlier measure function of the LOF scheme that this measure is not ON-compatible with the interpretation given above. On the other hand, we have

ASSERTION 5 *For the data set in Example 1 shown in Figure 4, COF is ON-compatible with* $\mathcal{I}$

This assertion is supported by the experimental result shown in Figure 6. We choose $k = 13$ and calculate COF values for all points in the data set. All calculations were done by a C++ program with a precision of 10 decimal digits. The computing environment is the same as that for Assertion 4. All the 8 outliers in C_2 have the same COF value 1.1518705044 and the other outlier o has a COF value 1.0761474038. On the other hand, the first 15 points, starting from p, on each of the two lines ℓ_1 and ℓ_2 have COF values between 0.9941766178 and 0.9995440551, and the rest of the points in C_1 have COF value of 1. Thus, we can set a threshold of 1.076 to distinguish the outliers from the non-outliers. Hence, by Definition 2, COF is ON-compatible with $\mathcal{I}$ as defined in Assertion 4.

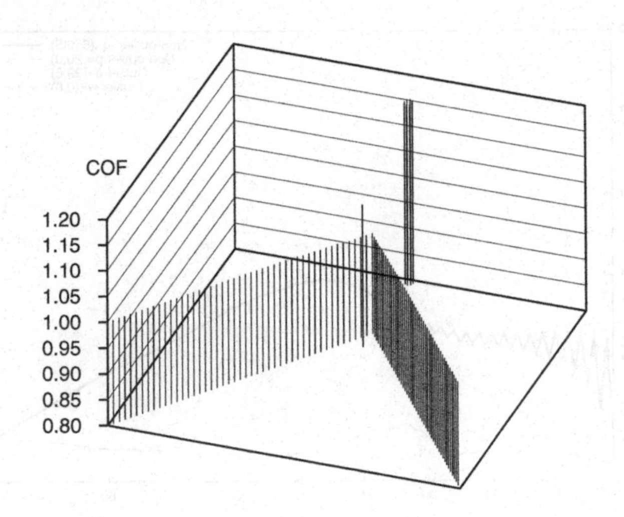

Fig. 6. COF Values of All Points When k=13

6 Time Complexity

Suppose that the database $\mathcal{D}$ has n objects. As the LOF algorithm in [3], the COF algorithm can be divided into two steps. In the first step, the COF method finds the k-nearest neighborhoods and the SBN-trails. Precisely, the COF algorithm finds, for any object $p \in \mathcal{D}$, the k-nearest neighborhoods together with their distances to p, and the SBN-trail together with the costs of the objects in the trail. The result of this step is materialization database $\mathcal{M}$ of size $2 \times n \times k$. Similar to the LOF algorithm, the size of this intermediate database is independent of the dimensionality of the original data. The running time of this step is $O(n \times \text{(time for a } k\text{-nn query))}$. Depending on the particular implementations of the *k-nn* query, its time complexity can vary from constant time for low-dimensional data, to $\log n$ for medium-dimensional data, and to n for extremely high-dimensional data. Hence, the time complexity of the COF algorithm in the first step can vary from $O(n)$ for low-dimensional data, to $O(n \log n)$ for medium dimensional data, and to $O(n^2)$ for extremely high-dimensional data.

In the second step, the COF method computes the COF values with the help of the materialization database $\mathcal{M}$. The original database is not needed in this step. The COF algorithm scans the database $\mathcal{M}$ twice. In the first scan, the algorithm finds the average chaining distance for every objects. In the second scan, the COF values of every objects are computed and written to a file. The time complexity of this step is $O(n)$. Notice that the time complexity for computing COF is similar to that for LOF.

7 Conclusions

While the field of data mining has been studied extensively, most work has concentrated on the discovery of patterns. Outlier detection as a branch of data mining has many important applications, and deserves more attention from data mining community. The existing work on outlier detection is either distance based or density based. In essence, these schemes all assume patterns have high (relative) densities. Therefore they do not work adequately where the patterns are of low densities. We propose a scheme that overcomes this weakness. Our scheme separates the notion of density from that of isolation. It can therefore detect outliers independently of the densities of the patterns from which they deviate. To measure the capabilities of outlier detection schemes, we introduce a notion of ON-compatibility. We show that while our scheme preserves the same nice properties as those of the density based method, it can achieve better results for data sets with connectivity characteristics in the data patterns.

ACKNOWLEDGMENT. The authors thank Ms. Yin Ling Cheung for preparing several *gnuplot* templates for plotting. This research is supported by the RGC (the Hong Kong Research Grants Council) grant UGC REF.CUHK 4179/01E.

References

[1] A. Arning, R. Agrawal, P. Raghavan: "A Linear Method for Deviation detection in Large Databases", Proc. of 2nd Intl. Conf. On Knowledge Discovery and Data Mining, 1996, pp 164 - 169.
[2] V. Barnett, T. Lewis: "Outliers in Statistical Data", John Wiley, 1994.
[3] M. Breuning, Hans-Peter Kriegel, R. Ng, J. Sander: "LOF: Identifying density based Local Outliers", Proc. of the ACM SIGMOD Conf. On Management of Data, 2000.
[4] W. DuMouchel, M. Schonlau: "A Fast Computer Intrusion Detection Algorithm based on Hypothesis Testing of Command Transition Probabilities", Proc.of 4th Intl. Conf. On Knowledge Discovery and Data Mining, 1998, pp. 189 - 193.
[5] M. Ester, H. Kriegel, J. Sander, X. Xu: "A Density-Based Algorithm for Discovering Clusters in Large Spatial Databases with Noise", Proc. of 2nd Intl. Conf. On Knowledge Discovery and Data Mining, 1996, pp 226 - 231.
[6] T. Fawcett, F. Provost: "Adaptive Fraud Detection", Data Mining and Knowledge Discovery Journal, Kluwer Academic Publishers, Vol. 1, No. 3, 1997, pp 291 - 316.
[7] D. Hawkins: "Identification of Outliers", Chapman and Hall, London, 1980.
[8] E. Knorr, R. Ng: "Algorithms for Mining Distance based Outliers in Large Datasets", Proc. of 24th Intl. Conf. On Very Large Data Bases, 1998, pp 392 - 403.
[9] E. Knorr, R. Ng: "Finding Intensional Knowledge of Distance-based Outliers", Proc. of 25th Intl. Conf. On Very Large Data Bases, 1999, pp 211 - 222.
[10] R. Ng, J. Han: "Efficient and Effective Clustering Methods for Spatial Data Mining", Proc. of 20th Intl. Conf. On Very Large Data Bases, 1994, pp 144 - 155.
[11] S. Ramaswamy, R. Rastogi, S. Kyuseok: "Efficient Algorithms for Mining Outliers from Large Data Sets", Proc. of ACM SIGMOD Intl. Conf. On Management of Data, 2000, pp 427 - 438.
[12] N. Roussopoulos, S. Kelley, F. Vincent, "Nearest Neighbor Queries", Proc. of ACM SIGMOD Intl. Conf. On Management of Data, 1995, pp 71 - 79.
[13] G. Sheikholeslami, S. Chatterjee, A. Zhang: "WaveCluster: A multi-Resolution Clustering Approach for Very Large Spatial Databases", Proc. of 24th Intl. Conf. On Very Large Data Bases, 1998, pp 428 - 439.
[14] S. Guha, R. Rastogi, K. Shim: "Cure: An Efficient Clustering Algorithm for Large Databases", In Proc. of the ACM SIGMOD Conf. On Management of Data, 1998, pp 73-84.
[15] J. Tang, Z. Chen, A. Fu and D. Cheung: "A General Framework for Outlier Formulations: Density versus Connectivity", Manuscript.
[16] T. Zhang, R. Ramakrishnan, M. Linvy: "BIRCH: An Efficient Data Clustering Method for Very Large Databases", Proc. of ACM SIGMOD Intl. Conf. On Management of Data, , 1996, pp 103 - 114.

Cluster-Based Algorithms for Dealing with Missing Values

Yoshikazu Fujikawa and TuBao Ho

Japan Advanced Institute of Science and Technology
Tatsunokuchi, Ishikawa 923-1292 Japan

Abstract. We first survey existing methods to deal with missing values and report the results of an experimental comparative evaluation in terms of their processing cost and quality of imputing missing values. We then propose three cluster-based mean-and-mode algorithms to impute missing values. Experimental results show that these algorithms with linear complexity can achieve comparative quality as sophisticated algorithms and therefore are applicable to large datasets.

1 Introduction

The objective of this research is twofold. One is to evaluate several well-known missing value methods in order to get a better understanding in their usage. The other is to develop algorithms to deal with missing values in large datasets. The key idea of these algorithms is to divide a dataset with missing values into clusters beforehand and replace missing values on each attribute by mean or mode value of the corresponding cluster depending on the attribute is numeric or categorical, respectively.

2 Evaluation of Existing Algorithms for Missing Values

2.1 Classification of Missing Values Cases

Generally, missing values can occur in data sets in different forms. We roughly classify missing values in datasets into three cases: Case 1: Missing values occur in several attributes (columns); Case 2: Missing values occur in a number of instances (rows); Case 3: Missing values occur randomly in attributes and instances. The occurrence cases of missing values can affect the result of missing value methods, so the selection of suitable missing value methods in each case is significant. For example, a method that ignores instances having missing values cannot be used when most of instances have one or more missing values.

2.2 Existing Methods for Dealing with Missing Values

We classify methods to deal with missing values into two groups: (i) *pre-replacing* methods that replace missing values before the data mining process, and (ii) *embedded* methods that deal with missing values during the data mining process.

M.-S. Chen, P.S. Yu, and B. Liu (Eds.): PAKDD 2002, LNAI 2336, pp. 549–554, 2002.

Table 1. Comparative evaluation of methods to deal with missing values

Method	Group	Cost	Attributes	Case
Mean-and-mode method	Pre-replacing	Low	Num & Cat	Case 2
Linear regression	Pre-replacing	Low	Num	Case 2
Standard deviation method	Pre-replacing	Low	Num	Case 2
Nearest neighbor estimator	Pre-replacing	High	Num & Cat	Case 1
Decision tree imputation	Pre-replacing	Middle	Cat	Case 1
Autoassociative neural network	Pre-replacing	High	Num & Cat	Case 1
Casewise deletion	Embedded	Low	Num & Cat	Case 2
Lazy decision tree	Embedded	High	Num & Cat	Case 1
Dynamic path generation	Embedded	High	Num & Cat	Case 1
C4.5	Embedded	Middle	Num & Cat	Case 1
Surrogate split	Embedded	Middle	Num & Cat	Case 1

Among missing value methods from the literature that we consider in this work, statistics-based methods include linear regression, replacement under same standard deviation [6] and mean-mode method [2]; and machine learning-based methods include nearest neighbor estimator, autoassociative neural network [6], decision tree imputation [7]. All of these are pre-replacing methods. Embedded methods include case-wise deletion [4], lazy decision tree [1], dynamic path generation [8] and some popular methods such as C4.5 and CART. Table 1 summarize our evaluation of these methods in terms of their group, computation cost, attributes types and missing value cases applicable.

3 Cluster-Based Algorithms to Deal with Missing Values

3.1 Basic Idea of the Algorithms

For large datasets with missing values, complicated methods are not suitable because of their high computation cost. We tend to find simple methods that can reach performance as good as complicated ones.

The results and experience obtained in the previous session suggested us that mean-and-mode method can be efficient and effective for large datasets with necessary improvements. The basic idea of our method is the cluster-based filling up of missing values. Instead of using mean-and-mode on the whole dataset we use mean-and-mode in its subsets obtained by clustering. The method consists of three variants of the mean-and-mode algorithm:

1. Natural Cluster Based Mean-and-Mode algorithm (NCBMM),
2. attribute Rank Cluster Based Mean-and-Mode algorithm (RCBMM),
3. K-Means Clustering based Mean-and-Mode algorithm (KMCMM).

NCBMM uses the class attribute to divide instances into natural clusters and uses the mean or mode of each cluster to fill up missing values of instance

Table 2. Algorithm RCBMM

1. For each missing attribute a_i
2. Make a ranking of all n categorical attributes $a_{j_1}, a_{j_2}, \ldots, a_{j_n}$ in decreasing order of distance between a_i and each attribute a_{j_k}
3. Divide all instances into clusters based on the values of a_h, where a_h is the attribute, which has the highest rank among $a_{j_1}, a_{j_2}, \ldots, a_{j_n}$.
4. Replace missing value on attribute a_i of an instance by the mode of each cluster to which it belongs to.
5. Repeat steps from 2 to 4 until all missing values on attribute a_i are replaced where h changes to the next number of ranking.

belongs to the cluster depending on the attribute id numeric or categorical, respectively. This algorithm is the simplest improvement of the mean-and-mode algorithm in case of supervised data but as shown in next sections it is very efficient if applicable. RCBMM and KMCMM divide a dataset into subsets by using the relationship between attributes. The starting point of these algorithms came from the question whether the class attribute is always the key attribute for clustering an arbitrary descriptive attribute? For some attributes, some of descriptive attributes may be better for clustering than the class attribute. The remark that to cluster instances concerning one missing attribute, the key attribute selected from all attributes may give better results than NCBMM.

3.2 The Proposed Algorithms

Natural Cluster Based Mean-and-Mode Algorithm (NCBMM). NCBMM algorithm can be applied to supervised data where missing value attributes can be either categorical or numeric. It produces a number of clusters equal to the number of values of the class attribute. At first, the whole instances are divided into clusters, where instances of each cluster have the same value of the class attribute. Then, in each cluster, the mean value is used to fill up missing values for each numeric attribute, and the mode value is used to fill up missing values for each categorical attribute.

Attribute Rank Cluster Based Mean-and-Mode Algorithm (RCBMM). RCBMM (Table 2) can be applied to filling up missing values for categorical attributes independently with the class attribute. It can be applied to both supervised and unsupervised data. Firstly, for one missing attribute, this method ranks attributes by their distance to the missing value attribute. The attribute that has smallest distance is used for clustering. Secondly, all instances are divided into clusters each contains instances having the same value of the selected

Table 3. Algorithm KMCMM

1. For each missing attribute a_i,
2. Make a ranking of all n numeric attributes $a_{j_1}, a_{j_2}, \ldots, a_{j_n}$ in increasing order of absolute correlation coefficients between attribute a_i and each attribute a_{j_k}.
3. Divide all instances by k-means algorithm based on the values of a_h that is the attribute of highest rank among $a_{j_1}, a_{j_2}, \ldots, a_{j_n}$.
4. Replace missing value on attribute a_i by the mean of each cluster.
5. Repeat steps from 2 to 4 till all missing values on attribute a_i are replaced where h changes to the next number of ranking.

attribute. Thirdly, the mode of each cluster is used to fill up missing values. This process is applied to each missing attribute.

We can have several ways to calculate distances between attributes. Our idea is to measure how two attributes have similar distributions of values. For this purpose, we used the distance proposed in [5] for two partitions P_A and P_B of n and m subsets of values of attributes A and B.

K-means Clustering Based Mean-and-Mode Algorithm (KMCMM). KMCMM can be applied to filling up missing values for numeric attributes independently with the class attribute. Therefore, it can be applied to both supervised and unsupervised data. We describe the algorithm for KMCMM in Table 3. The *correlation coefficient* r used in KMCMM is calculated from p pairs of observations (x, y). The k-means clustering algorithm first randomly selects k of the objects, each of which initially represents a cluster mean or center. Each remaining object is assigned to the cluster who mean is most similar with it. It then computes the new mean for each cluster. This process iterates until the criterion satisfied.

3.3 Evaluation

Methodology. It consists of two phases: (1) filling up missing values on dataset by pre-replacing methods and measuring the executing time, (2) evaluating the quality of the replaced datasets in terms of error rate in classification. Each missing value dataset is filled up by suitable pre-replacing methods to yield datasets without missing values. Then the same data mining methods (See5) are applied to the non-missing value datasets as well replaced datasets and their results are compared in order to evaluate the quality of replacing methods.

Six replacing methods, NCBMM, RCBMM, KMCMM, nearest neighbor estimator, autoassociative neural network and decision tree imputation and one embedded method, C4.5, are used for this comparative evaluation. For evaluating the quality of replaced datasets, three classification methods, C4.5, Naive

Table 4. Properties of datasets used in experiments

dataset	inst.	miss inst.	atts	miss atts	class	type of miss att	miss values case
adt	22,747	1,335	13	2	2	cat	case 1&2
att	10,000	2,430	9	8	2	cat	case 1&2
ban	5,400	2,610	30	25	2	mixed	case 2&3
bcw	6,990	160	9	1	2	num	case 1
bio	2,090	150	5	2	2	num	case 3
bld	3,450	560	6	3	2	num	case 1
bos	5,060	860	13	3	3	num	case 1
bpr	3,600	75	16	3	3	cat	case 1
census	299,285	156,764	41	8	2	cat	case 1&2
cmc	14,730	2,002	9	3	3	mixed	case 1
crx	6,900	337	15	7	2	mixed	case 1&2
der	3,660	74	34	1	6	num	case 1
dna	2,372	354	60	3	3	cat	case 1
ech	1,310	230	6	5	2	num	case 3
edu	10,000	9,990	12	10	4	mixed	case 1&2
hab	3,060	562	3	3	2	num	case 3
hco	3,680	3,290	19	18	2	mixed	case 2
hea	3,030	60	13	2	2	cat	case 1
hep	1,550	750	19	15	2	mixed	case 2
hin	10,000	4,050	6	5	3	cat	case 1
hur	2,090	220	6	2	2	num	case 1
hyp	3,772	3,772	29	8	5	mixed	case 1
imp	2,050	80	22	5	5	mixed	case 1
inf	2,380	250	18	2	6	cat	case 1
lbw	1,890	240	8	3	2	mixed	case 1
led	6,000	1,770	7	7	10	cat	case 3
pima	7,680	3,750	8	4	2	num	case 1
sat	6,435	1,195	36	2	6	num	case 1
seg	23,100	3,240	11	3	7	num	case 1&3
smo	28,550	5,340	8	2	3	mixed	case 1
tae	1,510	120	5	5	3	cat	case 3
usn	11,830	10,402	27	26	3	num	case 1&2
veh	8,460	1,550	18	3	4	num	case 1
vot	4,350	420	16	1	2	cat	case 1
wav	3,600	674	21	3	3	num	case 1

Bayesian classifier, k-nearest neighbor classifiers are used. Fifteen UCI datasets were replaced and classified.

The properties of each dataset are summarized in Table 4, and the error rates after treating missing values are summarized in Table 5. The experimenting result shows that NCBMM, RCBMM and KMCMM are much faster than other methods and higher accuracy than others (number in bold).

We also compare the methods on the large census dataset containing 299,285 instances with 156,764 instances having missing values. This set has eight numeric and thirty-three categorical attributes except the class attribute and missing values occur on only categorical attributes. We compared the error rate of See5 obtained directly, and that of See5 obtained after replacing missing values by NCBMM and RCBMM. And we measured the execution time for preparing at each replacing method. Experiments were done on a ultra spark machine and Sun-UNIX operating system.

In our experiments, See5, NCBMM and RCBMM have the same error of 4.6%, and time to replace missing values for each of our two methods is around 3 and 11 minutes respectively. These show that the proposed cluster-based algorithms could deal with missing values as good as other complicated methods and with low cost, and they can be applied to large datasets.

Table 5. Error rates of datasets after treating missing values

dataset	See5	NCBMM NCBMM	NCBMM RCBMM	KMCMM NCBMM	KMCMM RCBMM	NN NCBMM	NN RCBMM	decision tree	ANN
adt	**15.1**	*15.2*	**15.1**					*15.2*	
att	*2.2*	1.2	1.2					*2.3*	**1.1**
ban	*3.0*	2.5	2.5	2.6	2.5			2.5	
bcw	*0.9*	**0.7**		0.8		0.8		0.8	0.8
bio	*2.2*	1.8		1.8		1.8		1.8	1.8
bld	*4.7*	**3.3**		3.5		3.7		3.7	3.7
bos	*3.8*	2.9		2.9		2.9		2.9	2.9
bpr	3.8	**3.7**	3.8					3.8	*3.9*
cmc	6.7	5.2	5.5	5.2	5.5			*11.2*	**4.7**
crx	*2.5*	2.4	2.3	2.4	2.3	2.3	2.3	2.3	2.3
der	1.4	1.4						1.4	
dna	4.7	4.7	4.7					**4.6**	
ech	*5.4*	**3.5**		**3.5**				4.0	3.9
edu	5.2	**0.0**	**0.0**	**0.0**	**0.0**			*7.7*	1.8
hab	4.3	**2.9**		**2.9**				*7.6*	3.8
hco	0.9	**0.0**	**0.0**	**0.0**	**0.0**			*2.8*	*2.7*
hea	3.0	3.0	3.0					3.0	3.0
hep	*3.8*	**1.3**	**1.3**	**1.3**	**1.3**	2.5	2.6	2.2	2.2
hin	*12.9*	**9.0**	10.6					10.4	11.3
hur	*2.8*	**2.0**		**2.0**		2.2		2.2	2.1
hyp	0.4	0.4	0.4	0.4	0.4			0.4	0.4
imp	3.0	3.0	3.0	3.0	3.0	3.0	3.0	3.0	
inf	2.4	**2.3**	**2.3**					2.4	2.4
lbw	5.1	**3.9**	**3.9**	**3.9**	**3.9**	5.2	5.2	*8.6*	4.7
led	29.1	**25.1**	**25.1**					25.2	*32.0*
pima	*3.7*	**1.5**		**1.5**		3.0		2.9	2.9
sat	13.1	13.6		*14.1*		13.5		**9.0**	13.1
seg	*1.2*	**0.6**		**0.6**		0.7		0.7	0.7
smo	2.2	**1.3**	**1.3**	**1.3**	**1.3**	1.9	1.9	*7.3*	1.6
tae	**2.8**	**2.8**	4.0					*11.2*	
usn	*4.2*	**1.3**		2.4				2.8	2.9
veh	*3.0*	**2.7**		**2.7**		**2.7**		2.8	2.8
vot	*1.9*	**1.0**	**1.0**					1.1	1.4
wav	*24.5*	24.0		23.5		*24.6*		**13.1**	24.4

References

1. Friedman, J. H., Khavi, R., Yun, Y.: Lazy Decision Trees. Proceedings of the 13th National Conference on Artificial Intelligence, 717-724, AAAI Pres/MIT Press, 1996.
2. Han, J., Kamber, M.: Data Mining: Concepts and Techniques, Morgan Kaufmann Publishers, 2001.
3. Kononenko, I., Bratko, I., Roskar, E.: Experiments in automatic learning of medical diagnostic rules. Technical Report. Jozef Stefan Institute, Ljubjana, Yogoslavia, 1984.
4. Liu, W.Z., White, A.P., and Thompson S.G., Bramer M.A.: Techniques for Dealing with Missing Values in Classification. In IDAf97, Vol.1280 of Lecture notes, 527-536, 1997.
5. Mantaras, R. L.: A Distance-Based Attribute Selection Measure for Decision Tree Induction. Machine Learning, 6, 81-92, 1991.
6. Pyle, D.: Data Preparation for Data Mining. Morgan Kaufmann Publishers, Inc, 1999.
7. Quinlan, J.R.: Induction of decision trees. Machine Learning, 1, 81-106, 1986.
8. White, A.P.: Probabilistic induction by dynamic path generation in virtual trees. In Research and Development in Expert Systems III, edited by M.A. Bramer, pp. 35-46. Cambridge: Cambridge University Press, 1987.

Extracting Causation Knowledge from Natural Language Texts

Ki Chan, Boon-Toh Low, Wai Lam, and Kai-Pui Lam

Department of Systems Engineering and Engineering Management
The Chinese University of Hong Kong
Shatin, N.T., Hong Kong
{kchan,btlow,wlam,kplam}@se.cuhk.edu.hk

Abstract. SEKE2 is a semantic expectation-based knowledge extraction system for extracting causation relations from natural language texts. It is inspired by capitalizing the human behavior of analyzing information with semantic expectations. The framework of SEKE2 consists of different kinds of generic templates organized in a hierarchical fashion. All kinds of templates are domain independent. They are robust and enable flexible changes for different domains and expected semantics. By associating a causation semantic template with a set of sentence templates, SEKE2 can extract causation knowledge from complex sentences without full-fledged syntactic parsing. To demonstrate the flexibility of SEKE2 for different domains, we study the application of causation semantic templates on two domain areas of news stories, namely, Hong Kong stock market movement and global warming.

1 Introduction

With vast amount of information readily available to us, extracting useful information becomes a primary issue in handling textual information. Human can extract information rather easily. One can obtain needed information from a lengthy text effortlessly. However, it is not such an easy task for machines. A way to develop an effective information extraction system is to investigate how human extract information and learn from corresponding human behavior. There are different relations expressed in natural language, such as, whole-part, conditional, causation. Causation relations, which greatly influence people in decision making, play an important position in human cognition.

Information extraction refers to the identification of instances of a particular class of events or relationships in natural language texts and the extraction of the relevant arguments of the events or relationships [2]. Many previous researches focused on extraction of a particular event, or domain area. Few have looked into general causation knowledge from texts [1,5,7,6].

Our research focuses on using a semantic expectation-based approach to extract causation relations. We believe the same semantics are preserved for a certain type of relation in different domains. With the use of a causation semantic template, causation knowledge of different domains can be obtained. In our

M.-S. Chen, P.S. Yu, and B. Liu (Eds.): PAKDD 2002, LNAI 2336, pp. 555–560, 2002.

previous papers [3,4], we investigated the Semantic Expectation-based Causation Information Extraction (SEKE) approach on the Hong Kong Stock market movement domain. In this paper, we further generalize the concepts of causation knowledge into different kinds of generic templates organized in a hierarchical fashion and the new approach is named SEKE2. The sentence templates are also re-designed to improve a wider coverage of sentence styles and complex sentence structures. Moreover, we investigate the application of SEKE2 on two domain areas of news articles, namely, Hong Kong stock market movement and global warming, to demonstrate its flexibility on different domains.

2 Semantic Expectation-Based Knowledge Causation Extraction

Human make decisions from expectations, which is important for understanding of knowledge. When human read from texts, they could extract needed information readily and easily because there is always an expected semantic in mind and it is used to guide the search and understanding. Therefore, we can learn from relevant human behavior to design an effective knowledge extraction system.

There are many relations expressed in natural language, while causation relation is an important one. For long, human preserve their knowledge in texts and much of the knowledge are related to causation knowledge which helps us understand our world. Therefore, our goal is to find a way to extract the causation knowledge for effective understanding and reasoning.

Even though there are many different ways to present the information, the same semantics are preserved for a particular type of relations. The encapsulated knowledge based on the expected semantics can be extracted by the use of semantic templates which specifies the linking of actions. A causation semantic template states the linkage between reasons and consequences. It represents the highest level of templates. Besides, it is domain and language independent. In the next level of the hierarchy, it is associated with some sentence templates. The sentence templates act as a middle level to reconcile the semantic template, consequence and reason templates to a particular language. With some expected concepts of consequences and reasons, in form of consequence and reason templates, we can further obtain detailed information.

2.1 Semantic and Sentence Templates

A causation relation typically has two entities, reason(s) and consequence, linked by a directional causation indication. For causation knowledge, the semantics expected are reasons and consequences. Hence, its semantic template can be illustrated in Fig. 1, which captures the fact that one or more reasons cause the occurrence of the consequence.

The semantic template is language independent. However, we need to deal with natural texts written in a particular language. In particular, a variety of

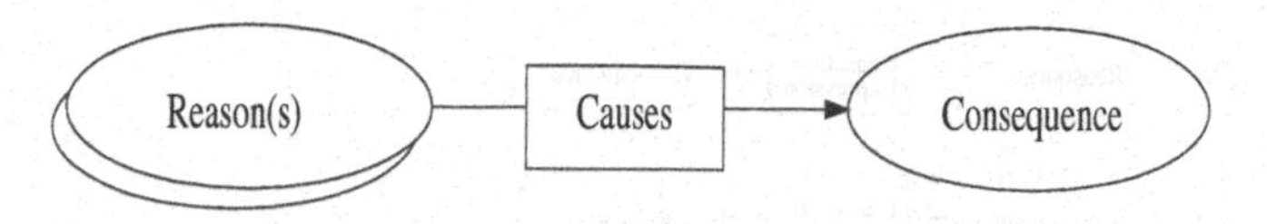

Fig. 1. Causation Semantic Template

sentence styles can express causality relationship. Sentence templates, associated with the semantic templates, are introduced to handle different styles of sentences. They represent the characteristics of expressing a relation in texts.

The causation knowledge, expressed in English sentences in texts, can be categorized into simple and complex sentences according to the organization of the reasons and consequences.

1. Simple Sentences consists of single or multiple reasons, for example:
   "The benchmark Hang Seng Index extended losses on Monday morning, sinking 654.77 points or 4.05 percent to 15,531.17 on Wall Street weakness and interest rate jitters."
2. Complex sentences, with complicated structure, consists of single or multiple reasons like simple sentences. But the reason itself is more complex as it contains a causation relation within,
   "Hong Kong stocks fell on Monday, taking a cue from Friday's dive in U.S. stocks as investors heeded Federal Reserve Chairman Alan Greenspan's warning that interest rates will probably rise."

To represent the sentence structure of simple sentences, the first two of the following sentence templates can be used. However, for complex sentence, it is observed that causation semantic templates have to be used also for extracting causation relation among reasons and consequences, and so are the sentence templates. Hence, recursive existence of causation relation among reasons is also a characteristic in our sentence templates.

In Fig. 2, sentence template 1 conveys that some reasons cause a consequence, while sentence template 2 conveys that a consequence is caused by some reasons. One can see that sentence template 1 is in the same order as the causation semantic template in Fig. 1, whereas sentence template 2 is in the reverse order. Sentence template 3 states some reasons cause a consequence, and those reasons can come before and after the consequence. For the remaining sentence templates, "Complex Consequence" means the consequence may consist one of the first three sentence templates. "Complex Reason(s)" refers to the existence of one of the first three sentence templates in the reason(s). "Causation Expression" refers to phrases for linking consequences to reasons. Some examples are {*as, due to, because, cause*}. Reasons are joined among themselves with the conjunction terms such as {*and*}.

Here are two simple sentence templates in the Hong Kong Stock market domain, which the orders of reason and consequence are different. It shows that there are different sentence templates associated with the same causation semantic template.

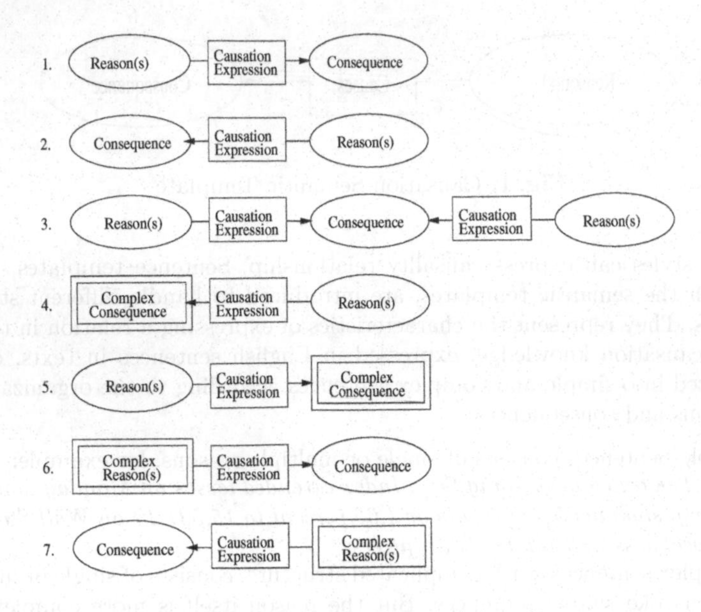

Fig. 2. Sentence Templates

1. **Factor(s) with its movement causes the movement of stock**
   "The increase of interest rate caused Hang Seng Index surged."
2. **The movement of stock is caused by factor(s) with its movements**
   "Hang Seng Index rose as Wall Street gained."

Example of sentence template for complex sentence in global warming:
A factor with action caused by global warming causes a factor with movement.
"An increase in temperatures as a result of global warming may lead to significantly higher levels of carbon dioxide being released into atmosphere."

2.2 Consequence and Reason Templates

Similarly, reasons and consequences usually contain expected concepts. We design reason and consequence templates to represent their semantics. A variety of concepts or information can exist among consequences or reasons. For example, they may have information including factors, movements, modifier of movements, time, etc. The elements for a consequence or a reason depends on what are the expected concepts, and the focus of the causation relation.

Using the Hong Kong stock market as an example, we focus on finding the set of possible reasons on the stock movement. The consequence template includes "Hang Seng Index" or equivalent as the main element, and other elements such as what the market movement is, at when the movement takes place and how it moves. The reason template includes the factor(s) as the main element and how it moves as the secondary element.

3 Investigations on Hong Kong Stock Market Movement and Global Warming Domains

Two studies using the SEKE2 framework are carried out on English news articles. With the expected semantics collected, raw news articles are taken in and irrelevant information are filtered out. Reasons and consequences are then extracted by parsing and matching semantically. News articles are divided into two sets. One is used as training data, which prerequisite information is obtained. The other set is used for testing of the system.

The first study is to extract causation knowledge affecting the movement of Hang Seng Index (HSI) in Hong Kong stock market. A total of 313 relevant news articles, collected from Reuters news, and two local newspapers, South China Morning Post and Hong Kong I-Mail between February to May 2000, were used as training data. News articles from January to July 2000 were used as the testing data. They were classified and analyzed manually and served as the correct answer for evaluation, A total of 178 sentences were classified to have expected causation knowledge of reasons. The resulted recall and precision of extracted knowledge by SEKE2 are 70.74 % and 78.06 %.

The second study is to extract the causation knowledge concerning global warming. The causation semantics behind are the influences and the causes for global warming. News articles are collected from CNN and Reuters, which 26 of them are used as training data articles and another 191 of them were used as testing data. 84 sentences are identified manually as correct answers. The resulted recall and precision of knowledge extracted are 63 % and 81 % respectively.

Some examples of discovered causation relations of the two studies are given in Table 1. The main reasons for causing the Hong Kong stock market to go upwards are the upward movements of Wall Street market, internal factors, such as property and technical stocks, and the downward movement of interest rates. Moreover, the downward movement of Wall Street market, internal factors, and the upward movement of interest rates cause Hang Seng Index to go downwards. For causing global warming, the emission of greenhouse gases and the burning of fossil fuels are two most important reasons. For the consequences of global warming, they can be droughts, decrease of crops, etc.

4 Conclusions

We have developed SEKE2, a framework for extracting causation information from natural language texts, It composed of different kinds of generic templates organized in a hierarchical fashion. We have applied our framework on two different domains, Hong Kong stock market movement domain and global warming domain. By using the expected semantic knowledge approach, we have shown that it is feasible to mainly use semantic knowledge, and simple linguistic clues in extracting causation knowledge from texts. It also illustrated the domain independence characteristic of our approach, as the semantics for a certain relation are preserved across different domains.

Table 1. Causation Knowledge Discovered

Reasons for HSI up	Wall Street	16.9%	1.5%
	Internal factors	56.9%	4.6%
	Interest rates	1.5%	4.6%
	other factors	12.3%	1.5%
Reasons for HSI down	Wall Street	0%	4.5%
	Interest rate	15.9%	2.3%
	Internal factors	11.4%	50%
	other factors	4.5%	11.4%
Reasons for Global warming	Greenhouse gases	emmission	47.6%
		decrease/increase	11.9%/11.9%
	Fossil fuels	burning	7.1%
	Others	–	21.5%
Consequences for Global Warming	Droughts		26.0%
	Crops	decrease	10.5%
	Greenhouse gases	increase/decrease	10.5%/15.8%
	Others(e.g.Ozone)	–	37.1%

References

1. Garcia, D. (1997). COATIS, an NLP System to Locate Expressions of Actions Connected by Causality Links. Proceedings of the 10th European Workshop in Knowledge Acquisition, Modeling and Management, EKAW '97, 347-352
2. Grishman, R. (1997). Information Extraction: Techniques & Challenges in "Information Extraction: A Multidisciplinary Approach to an Emerging Information Technology", International Summer School on Information Extraction, SCIE-97, 10-27
3. Low, B.T., Chan, K., Chin, M.Y., Choi, L.L. & Lay, S.L.(2001). A Semantic-based Acquisition Study on Hong Kong Stock Market Movement. Proceedings of 5th World Multiconference on Systemics, Cybernetics and Informatics, SCI 2001
4. Low, B.T., Chan, K., Chin, M.Y., Choi, L.L. & Lay, S.L.(2001). Semantic Expectation-based Causation Knowledge Extraction: A Study on Hong Kong Stock Market Movement Analysis. Proceedings of 5th Pacific-Asia Conference on Knowledge Discovery and Data Mining, PAKDD 2001, 114-123
5. Kaplan, R.M. & Berry-Bogghe, G.(1991). Knowledge-based Acquisition of Causal Relationships in Text. Knowledge Acquisition, 3(3), 317-337
6. Khoo, C.S.G., Chan, S. & Niu, Y.(2000). Extracting Causal Knowledge from a Medical Database Using Graphical Pattern. Proceedings of 38th Annual Meeting of the Association for Computational Linguistics
7. Khoo, C.S.G., Kornfit, J., Oddy, R.N. & Myaeng, S.H. (1998). Automatic Extraction of Cause-effect Information from Newspaper Text Without Knowledge-based Inferencing. Literary and Linguistic Computing, 13(4), 177-186
8. WordNet, an on-line lexical reference system, developed by the Cognitive Science Laboratory at Princeton University `http://www.cogsci.princeton.edu/~wn/`

Mining Relationship Graphs for Effective Business Objectives*

Kok-Leong Ong, Wee-Keong Ng, and Ee-Peng Lim

Centre for Advanced Information Systems, Nanyang Technological University,
Nanyang Avenue, N4-B3C-14, Singapore 639798, SINGAPORE
ongkl@pmail.ntu.edu.sg

Abstract. Modern organization has two types of customer profiles: active and passive. Active customers contribute to the business goals of an organization, while passive customers are potential candidates that can be converted to active ones. Existing KDD techniques focused mainly on past data generated by active customers. The insights discovered apply well to active ones but may scale poorly with passive customers. This is because there is no attempt to generate know-how to convert passive customers into active ones. We propose an algorithm to discover relationship graphs using both types of profile. Using relationship graphs, an organization can be more effective in realizing its goals.

1 Introduction

The modern organization has two types of customer profiles: active and passive. We define an *active customer* as one that contributes to the business objectives of the organization. For example, an active customer may engage in a purchase [4], surfed an organization's Website [2,3], or signed up for trial/free products and services. A passive customer is thus one that has no contributions other than its profile. In the early era of KDD, passive customers do not really exist. However, the ubiquity of e-Commerce and data acquisition technologies has made such information readily available. Essentially, passive customers are acquired by means of third party information providers, partnerships with other organizations, or registration of trial products and services.

This presents an important issue. We have, in fact, been looking at data generated by active customers, and the insights obtained are better catered towards them. For example, the knowledge used in the bundling of items in a supermarket [1] is likely to work well for active customers. However, the same know-how may not reach a passive customer since they do not contribute any data for KDD. More importantly, the knowledge obtained does not provide any indication about how passive customers may be converted into active ones. With the belief that there are insights in data that can help convert passive customers, we propose the mining of relationship graphs (or simply graphs in this paper) from both types of customer profiles. In a graph, nodes represent "customers"

* This work was supported by SingAREN under project M48020004.

M.-S. Chen, P.S. Yu, and B. Liu (Eds.): PAKDD 2002, LNAI 2336, pp. 561–566, 2002.

and edges represent "relationships". By finding relationships between active and passive customers, we can capitalize on the active customer's ability to influence the passive party to make a contribution to the organization's goal.

Example: The insurance industry in Singapore is a very competitive market. Ad-hoc approaches such as selling on the street or calling a party in the house are attempts that is difficult and ineffective. Hence, agents depend largely on their immediate family members, friends and relatives, and in the later stage, the relationships from the current pool of people. To help their agents reach beyond relatives and friends, an insurance company teamed with an information provider to allow its brokers to query a particular existing customer (i.e., active) and have the graph of the active customer shown. From the graph, a broker may select a related passive customer via the active customer under its portfolio, and thus, path an opportunity to a potential customer.

Certainly, there are other possibilities. The focus is the use of relationships between two people to create a business opportunity. In Asia, the use of relationships is sometimes more effective in achieving business objectives then plain marketing campaigns. This is how relationship graphs differ from other techniques – they find insights in both types of customer profiles that help organizations target potential customers effectively. Formally, the mining of relationship graphs involves finding the set of passive customers in a database that have the *strongest* relationship with a given active customer. In the next section, we give a formal definition of the problem and its parameters. Section 3 presents the algorithm for mining relationship graphs. We then conclude our discussion in Section 4.

2 Problem Formulation

Let $C = \{c_1, c_2, \ldots, c_i\}$ be the set of all passive and active customers in the organization and $R = \{r_1, r_2, \ldots, r_j\}$ be the set of all possible relationships that exist between two customers. A tuple t in the database D is of the form $\langle c_\alpha, c_\beta, r \rangle$ where $c_\alpha, c_\beta \in C$ and $r \in R$, that defines the existence of a relationship r between c_α and c_β. A **relationship graph (G)** is a set of tuples $\{t_1, t_2, \ldots, t_k\} \subset D$ that represents a set of relationship paths from the active customer c_a to a set of passive customers $\{c_{p_1}, c_{p_2}, \ldots, c_{p_j}\}$. A **relationship path** is a set of tuples $P = \{t_x, t_y, \ldots, t_z\} \subset G$ where $t_1.\alpha = c_a \vee t_1.\beta = c_a$ and $t_k.\alpha = c_{p_i} \vee t_k.\beta = c_{p_i}$ such that $\forall t_x \in P - \{t_1, t_k\}, \exists t_y \in G,\ t_x.\beta = t_y.\alpha$.

For simplicity and compactness of the database, all relationships are undirected. Even with this simplification, the potential relationships possible from an organization warehouse is huge[1]. To limit the number of relationship paths in G, only those strongest in the goal of the organization's objectives are considered

[1] A domain expert may choose to relate two entities by their address, interest, working organization, profession, geographic region, expertise etc., and this generates huge number of tuples.

for a given c_a. The **minimum strength** (δ) is the score that a relationship path must achieve in order to be selected into the graph. This is a quantitative value that act as a measure on the likely effectiveness of a relationship path in the view point of the domain expert. This is computed by a user defined function $\mathcal{S}(\{t_0, t_1, \ldots, t_k\})$ where the strength is determined by the evaluation of the relationships $t_0.r_0$, $t_1.r_1$, ..., $t_k.r_k$. The motivation behind $\mathcal{S}$ is to model complex interactions between different relationships that cannot be represented by the direct assignment of weights used in typical graph exploration algorithms (e.g., shortest path [5]).

We also define the distance to determine the "radius" of the graph G with the c_a as the "centriod". Intuitively, as the distance gets larger, the effectiveness of the relationship weakens. Therefore, it may not be cost effective to mine beyond a certain distance. Formally, the **distance** (π) is the maximum allowable distance for a relationship path from the active customer to a passive customer. Together with the notion of strength, we obtain the **size** (φ) of a relationship graph. The size is thus the maximum number of allowable relationship paths that satisfy δ and are within π. Therefore, φ represents the top φ strongest relationship paths.

3 Mining Relationship Graphs

Figure 1 shows the algorithm for mining a single relationship graph given an active customer. This is done in two steps. First, tuples that are involved in the current computation are identified and selected into the respective ordered list. Once the lists are constructed, the next step is to scan the lists to construct the top φ strongest path for a given active customer.

Upon entry to the algorithm, π ordered list are created (line 3) and represented as $b_0, b_1, \ldots, b_{\pi-1}$ where each holds tuples selected from D. Starting with b_0, we select all tuples in D that contain the active customer c_a. Instead of the lexicographical ordering in the database, we order the tuples such that the first customer (i.e., $t.\alpha$) is c_a using the function "ArrangeTuple" in line 4. We then compute the strength using $\mathcal{S}$ and then sort, in descending order, the strength of their relationship (line 5). This generates the relationship graph with a distance of 1. The loop from lines 6-11 constructs the remaining list up to the distance specified by π. Notice the construction of the kth list is dependent on the results of the $(k-1)^{th}$ list. This is equivalent to a breadth-first search of tuples in D to construct the "super-graph" with c_a as the "centroid" up to the distance of π. In line 7, we identify the selection criteria for the next list and stores it in B_c. We then select tuples (line 8) containing a customer in B_c. Line 11 prune tuples containing all active customers. Such tuples are unnecessary by the definition and should be removed to minimize memory consumption and faster computation.

The second half of the algorithm finds the subgraph of c_a that satisfies φ and δ. Since all tuples in b_0 contains c_a, all tuples in this group must be considered regardless of the relationship strength. For each tuple in b_0 (line 12), we construct the path up to the maximum distance possible by finding matching tuples to form a relationship path. In the algorithm, T holds the set of tuples

```
01 procedure MineGraph
02 begin
03     let B = { b_0, b_1, ..., b_{π-1} } and G = Ø and B_c = { c_a }
04     let b_0 = ArrangeTuple({ t ∈ D | t.α = c_a ∨ t.β = c_a }, B_c)
05     foreach t ∈ b_0 do t.δ = S({t}); Sort b_0 using δ in descending order
06     for (k = 1; k < π; k++) begin
07         let B_c = B_c ∪ { t.β | t ∈ b_{k-1} }
08         let b_k = ArrangeTuple({ t ∈ D − {b_0, b_1, ..., b_{k-1}} | t.α ∈ B_c ∨ t.β ∈ B_c }, B_c)
09         foreach t ∈ b_k do t.δ = S({t}); Sort b_k using δ in descending order
10     endfor
11     b_k = b_k − { t ∈ b_k | t.α is active customer ∧ t.β is active customer }
12     foreach tuple t ∈ b_0 do
13         let T = { t } and t_c = t
14         for (k = 1; k < π; k++) begin
15             if ( FindTuple(t_c, b_k) = Ø) then exit for-loop else T = T ∪ { t_c }
16         endfor
17         if (S(T) ≥ δ ∧ |T| ≤ π) then
18             G = G ∪ { T }; φ = φ − 1
19             if (φ = 0) then end by outputting solution G
20         endif
21     endfor
22 end

23 procedure ArrangeTuple(list of tuples T, B_c)
24 begin
25     foreach tuple t ∈ T do
26         if (t.α ∉ B_c) then Swap contents of t.α and t.β
27     endfor
28     return T
39 end

30 procedure FindTuple(tuple t_c, ordered list b_k)
31 begin
32     foreach tuple t ∈ b_k do
33         if (t.α = t_c.β) then return t' | t'.α = t.β ∧ t'.β = t.α
34         if (t.α = t_c.α) then return t
35     endfor
36     return Ø
37 end
```

Fig. 1. Algorithm for mining relationship graph of a given active customer.

that forms a relationship path (line 13) and t_c is the tuple under consideration. The function "FindTuple" attempts to find a corresponding tuple that joins to the current tuple t_c and thus, extends the distance of the current path by 1 in the next bin b_k (line 14-16). Once the path is obtained, we measure its strength (line 17). If it satisfies δ, it is added to the solution G and φ is decremented (line 18-19). This repeats for all tuples or until the top φ solution is discovered. The algorithm then terminates by printing the solution G in line 19.

3.1 Design Heuristics

In most cases, approximately φ tuples of b_0 need to be scanned if φ is smaller than the number of strong relationships. For the remaining lists, an average of

$\varphi \times log_2(|b_k|)$ tuples are scanned if a binary search is assumed. This reduction in the scan is based on a simple heuristic that allows the relationship graph to be discovered quickly. Specifically, tuples in the list are sorted by their relationship strength. For a given path P, we observe that if $\mathcal{S}(P) \geq \delta$, then adding the next tuple t_c, regardless of its strength, is likely to be stronger than both T and t_c being weak. As such, sorting each list translates to a higher probability of finding strong paths without traversing every tuples in the lists.

In the best case (assuming a binary search), only $\frac{1}{h} \times (\varphi + log_2(|b| \times (\pi - 1))$ tuples need to be scanned, where h is the probability of finding a matching tuple such that the length of the current relationship path is extended by 1 and b is the average number of tuples in each ordered list. Compare this to the trivial approach in which $b_0 \times b_1 \times \ldots \times b_{\pi-1}$ (i.e., $|b|^{\pi}$) tuples are scanned and its relationship strength computed to find the top φ list. Clearly, the use of such heuristics is desirable when the relationship database and the number of active customers to discover in each run is large.

3.2 Scoring Relationships

Earlier, we mentioned that $\mathcal{S}$ is user-defined. To illustrate how $\mathcal{S}$ may be defined, let R be the set of all relationships in the relationship database. We partition R into $R_1, R_2, \ldots, R_j$ such that $R_1 \cup R_2 \cup \ldots \cup R_j = R$ and $R_1 \cap R_2 \cap \ldots \cap R_j = \emptyset$. For each $R_k \subset R$, $0 \leqslant k \leqslant j$, we define a function $f_{R_k}(T)$ as

$$f_{R_k}(T) = \begin{cases} \exists t \in T, t.r \in R_k & v_k \\ otherwise & 0 \end{cases} \tag{1}$$

where v_k is a real value that is assigned to each group of relationships that are considered to be of the same level of importance. Then, a possible score for any relationship may be defined as

$$\mathcal{S}(T) = C + f_{R_1}(T) + f_{R_2}(T) - f_{R_3}(T) \times \frac{1}{f_{R_4}(T)} + \ldots \times f_{R_j}(T) \tag{2}$$

In the above, the default score is given by C. Note that one possible interpretation of the above model is that normal relationships add score to $\mathcal{S}$ and relationships that are weak in the goal of the organization objective subtract scores from $\mathcal{S}$. In addition, high influence relationships contribute by multiplying the score while very negative relationships can be modelled by reducing the score through division. Most important of all, as T grows, the value of $\mathcal{S}$ changes. This allows a relationship discovered later to override a preceding relationship and hence modify the overall strength of the path.

4 Summary

We have introduced the problem of mining relationship graphs using active and passive customers profiles. For a given active customer, we are interested in finding a subgraph containing strong relationship paths that reach passive customers.

Since a relationship graph is essentially an undirected graph [6], our intuition is to look at existing graph exploration algorithms. However, algorithms such as the "shortest path" [5] returns a single solution between two vertices and designates that as the optimum. In the case of mining relationship graphs, there can be more than one solution between two vertices as long as the solution satisfies δ. In addition, there can be no input from the domain analyst and there is no provision for freedom in the solution that is necessary to model relationships. Furthermore, the costing function assumes that the cost of an edge cannot overwrite any preceding edges traversed. In the case of relationship graphs, certain relationships may have a high influence over another preceding relationship that it is worth the effort (i.e., by computation of $\mathcal{S}$) to take an alternative solution.

The work reported in this paper is the result of a discussion with a company looking into visualizing its customer database for effective channelling of their business efforts. The results of visualization is overly complex and motivated the use of algorithmic methods to discover appropriate relationships in the customer database. The algorithm is currently implemented as part of a trial to assess its effectiveness over plain visualization.

Although the paper is set in the context of an active and passive customer, we would like to point out that the graph is equally applicable in other situations. For example, two active customers may purchase the same software. Using the graph, the software company may use the influence ability of one who upgraded to a newer version to create an upgrade on the other. This achieves the business objective of the organization as well. It is thus important to note that the application of the relationship graph depends largely on the goals and nature of the organization.

References

1. Agrawal, R. and Srikant, R.: Fast Algorithm for Mining Association Rules. Proc. 20th Int. Conf. on Very Large Databases, Santiago, Chile, Sep. 1994.
2. Cooley, R., Mobasher, B. and Srivastava, J.: Web Mining – Information and Pattern Discovery on the World Wide Web. Proc. 9th IEEE Int. Conf. on Tools with Artificial Intelligence, Nov. 1997.
3. Dua, S., Cho, E. and Iyengar, S. S.: Discovery of Web Frequent Patterns and User Characteristics from Web Access Logs – A Framework for Dynamic Web Personalization. Proc. Int. Conf. on Application-Specific Systems and Software Engineering Technology, pp. 3-8, 2000.
4. R. D. Lawrence, G. S. Almasi, V. Kotlyar, M. S. Viveros and S. S. Duri.:Personalization of Supermarket Product Recommendations. Applications of Data Mining to e-Commerce – a Special Issue of the Int. J. on Data Mining and Knowledge Discovery, 2001.
5. Swamy, M. N. S. and Thulasiraman, K.: Graphs, Networks, and Algorithms. John Wiley & Sons, Inc., 1981.
6. Wilson, R. J. and Wakins, J. J.: Graphs – An Introductory Approach. John Wiley & Sons, Inc., 1990.

Author Index

Lecture Notes in Artificial Intelligence (LNAI)

Lecture Notes in Computer Science

Vol. 2298: I. Wachsmuth, T. Sowa (Eds.), Gesture and Language in Human-Computer Interaction. Proceedings, 2001. XI, 323 pages. (Subseries LNAI).

Vol. 2299: H. Schmeck, T. Ungerer, L. Wolf (Eds.), Trends in Network and Pervasive Computing – ARCS 2002. Proceedings, 2002. XIV, 287 pages. 2002.

Vol. 2300: W. Brauer, H. Ehrig, J. Karhumäki, A. Salomaa (Eds.), Formal and Natural Computing. XXXVI, 431 pages. 2002.

Vol. 2301: A. Braquelaire, J.-O. Lachaud, A. Vialard (Eds.), Discrete Geometry for Computer Imagery. Proceedings, 2002. XI, 439 pages. 2002.

Vol. 2302: C. Schulte, Programming Constraint Services. XII, 176 pages. 2002. (Subseries LNAI).

Vol. 2303: M. Nielsen, U. Engberg (Eds.), Foundations of Software Science and Computation Structures. Proceedings, 2002. XIII, 435 pages. 2002.

Vol. 2304: R.N. Horspool (Ed.), Compiler Construction. Proceedings, 2002. XI, 343 pages. 2002.

Vol. 2305: D. Le Métayer (Ed.), Programming Languages and Systems. Proceedings, 2002. XII, 331 pages. 2002.

Vol. 2306: R.-D. Kutsche, H. Weber (Eds.), Fundamental Approaches to Software Engineering. Proceedings, 2002. XIII, 341 pages. 2002.

Vol. 2307: C. Zhang, S. Zhang, Association Rule Mining. XII, 238 pages. 2002. (Subseries LNAI).

Vol. 2308: I.P. Vlahavas, C.D. Spyropoulos (Eds.), Methods and Applications of Artificial Intelligence. Proceedings, 2002. XIV, 514 pages. 2002. (Subseries LNAI).

Vol. 2309: A. Armando (Ed.), Frontiers of Combining Systems. Proceedings, 2002. VIII, 255 pages. 2002. (Subseries LNAI).

Vol. 2310: P. Collet, C. Fonlupt, J.-K. Hao, E. Lutton, M. Schoenauer (Eds.), Artificial Evolution. Proceedings, 2001. XI, 375 pages. 2002.

Vol. 2311: D. Bustard, W. Liu, R. Sterritt (Eds.), Soft-Ware 2002: Computing in an Imperfect World. Proceedings, 2002. XI, 359 pages. 2002.

Vol. 2312: T. Arts, M. Mohnen (Eds.), Implementation of Functional Languages. Proceedings, 2001. VII, 187 pages. 2002.

Vol. 2313: C.A. Coello Coello, A. de Albornoz, L.E. Sucar, O.Cairó Battistutti (Eds.), MICAI 2002: Advances in Artificial Intelligence. Proceedings, 2002. XIII, 548 pages. 2002. (Subseries LNAI).

Vol. 2314: S.-K. Chang, Z. Chen, S.-Y. Lee (Eds.), Recent Advances in Visual Information Systems. Proceedings, 2002. XI, 323 pages. 2002.

Vol. 2315: F. Arhab, C. Talcott (Eds.), Coordination Models and Languages. Proceedings, 2002. XI, 406 pages. 2002.

Vol. 2316: J. Domingo-Ferrer (Ed.), Inference Control in Statistical Databases. VIII, 231 pages. 2002.

Vol. 2317: M. Hegarty, B. Meyer, N. Hari Narayanan (Eds.), Diagrammatic Representation and Inference. Proceedings, 2002. XIV, 362 pages. 2002. (Subseries LNAI).

Vol. 2318: D. Bošnački, S. Leue (Eds.), Model Checking Software. Proceedings, 2002. X, 259 pages. 2002.

Vol. 2319: C. Gacek (Ed.), Software Reuse: Methods, Techniques, and Tools. Proceedings, 2002. XI, 353 pages. 2002.

Vol.2320: T. Sander (Ed.), Security and Privacy in Digital Rights Management. Proceedings, 2001. X, 245 pages. 2002.

Vol. 2322: V. Mařík, O. Stěpánková, H. Krautwurmová, M. Luck (Eds.), Multi-Agent Systems and Applications II. Proceedings, 2001. XII, 377 pages. 2002. (Subseries LNAI).

Vol. 2323: À. Frohner (Ed.), Object-Oriented Technology. Proceedings, 2001. IX, 225 pages. 2002.

Vol. 2324: T. Field, P.G. Harrison, J. Bradley, U. Harder (Eds.), Computer Performance Evaluation. Proceedings, 2002. XI, 349 pages. 2002.

Vol 2326: D. Grigoras, A. Nicolau, B. Toursel, B. Folliot (Eds.), Advanced Environments, Tools, and Applications for Cluster Computing. Proceedings, 2001. XIII, 321 pages. 2002.

Vol. 2327: H.P. Zima, K. Joe, M. Sato, Y. Seo, M. Shimasaki (Eds.), High Performance Computing. Proceedings, 2002. XV, 564 pages. 2002.

Vol. 2329: P.M.A. Sloot, C.J.K. Tan, J.J. Dongarra, A.G. Hoekstra (Eds.), Computational Science – ICCS 2002. Proceedings, Part I. XLI, 1095 pages. 2002.

Vol. 2330: P.M.A. Sloot, C.J.K. Tan, J.J. Dongarra, A.G. Hoekstra (Eds.), Computational Science – ICCS 2002. Proceedings, Part II. XLI, 1115 pages. 2002.

Vol. 2331: P.M.A. Sloot, C.J.K. Tan, J.J. Dongarra, A.G. Hoekstra (Eds.), Computational Science – ICCS 2002. Proceedings, Part III. XLI, 1227 pages. 2002.

Vol. 2332: L. Knudsen (Ed.), Advances in Cryptology – EUROCRYPT 2002. Proceedings, 2002. XII, 547 pages. 2002.

Vol. 2334: G. Carle, M. Zitterbart (Eds.), Protocols for High Speed Networks. Proceedings, 2002. X, 267 pages. 2002.

Vol. 2335: M. Butler, L. Petre, K. Sere (Eds.), Integrated Formal Methods. Proceedings, 2002. X, 401 pages. 2002.

Vol. 2336: M.-S. Chen, P.S. Yu, B. Liu (Eds.), Advances in Knowledge Discovery and Data Mining. Proceedings, 2002. XIII, 568 pages. 2002. (Subseries LNAI).